CALCIUM BINDING PROTEINS

WILEY SERIES IN PROTEIN AND PEPTIDE SCIENCE

VLADIMIR N. UVERSKY, Series Editor

Metalloproteomics • Eugene A. Permyakov

Protein Misfolding Diseases: Current and Emerging Principles and Therapies • Edited by Marina Ramirez-Alvarado, Jeffery W. Kelly, and Christopher M. Dobson

Instrumental Analysis of Intrinsically Disordered Proteins: Assessing Structure and Conformation • Vladimir Uversky and Sonia Longhi

Calcium Binding Proteins • Eugene A. Permyakov and Robert H. Kretsinger

CALCIUM BINDING PROTEINS

EUGENE A. PERMYAKOV
Institute for Biological Instrumentation
Russian Academy of Sciences
Moscow, Russia

ROBERT H. KRETSINGER
Department of Biology
University of Virginia
Charlottesville, Virginia

WILEY

A JOHN WILEY & SONS, INC., PUBLICATION

Published by John Wiley & Sons, Inc., Hoboken, New Jersey.
Published simultaneously in Canada.

Library of Congress Cataloging-in-Publication Data:

Permyakov, E. A. (Evgeniĭ Anatol'evich)

Calcium binding proteins / Eugene Permyakov and Robert H. Kretsinger.
p. cm. – (Wiley series in protein and peptide science ; 5)
Includes index.
Summary: "Calcium Binding Proteins explains the unique and highly diverse functions of calcium in biology, which are realized by calcium binding proteins. The structures and physical characteristics of these calcium binding proteins are described, as well as their functions and general patterns of their evolution. Techniques that underlie the description of proteins are diccussed, including NMR, circular dichroism, optical rotatory dispersion spectroscopy, calorimetry, and crystallography. The book discusses the patterns of biochemical phenomena such as calcium homeostasis, mineralization, and cell signaling and involve specific proteins. It summarizes ongoing research and presents general hypotheses that help to focus future research, and also provides a conceptual framework and a description of the underlying techniques that permits someone entering the field to become conversant."– Provided by publisher.
Summary: "This book explains the unique and highly diverse functions of calcium in biology, which are realized by calcium binding proteins "– Provided by publisher.
ISBN 978-0-470-52584-5 (hardback)
1. Calcium-binding proteins. 2. Calcium in the body. I. Kretsinger, Robert H. II. Title.
QP552.C24P457 2010
572′.69–dc22

2010026572

10 9 8 7 6 5 4 3 2 1

CONTENTS

PREFACE

Calcium plays a unique role in eukaryotes as an intracellular second messenger. Information regarding a pulse of Ca^{2+} ions is transduced to a change in conformation and function of a calcium modulated protein. However, calcium binds specifically and essentially to many other proteins intra- and extracellularly, as well as to proteins of eubacteria, archae, and viruses. Further, calcium is the essential constituent of many biominerals. All of these interrelated functions involve calcium binding proteins, the subject of the book.

To put these descriptions of calcium binding proteins in context, we have summarized the history of these studies, the physiological processes involved, the techniques used to study the proteins, the chemistry of calcium and related metals, as well as the proteins that bind these metals. These descriptions should incorporate the recent literature and be valuable to the research community. There should also be enough background information in biology to make the discussions meaningful to advanced undergraduates in the physical sciences and, at the same time, enough description of the fundamentals of inorganic and physical chemistry to permit a critical reading by students in biology.

Acknowledgment

This work was supported by grants from the Molecular and Cellular Biology program in the Russian Academy of Sciences.

EUGENE A. PERMYAKOV
ROBERT H. KRETSINGER

1

HISTORICAL PERSPECTIVE

This historical perspective of the functions of calcium in biology lays the foundation for understanding the central role of calcium binding proteins. Ten interrelated themes bring us to about 1970. The successes and frustrations of those concepts help us understand recent decades. For clarity, these themes are each treated chronologically; however, there are myriad, complex interactions among them. Much of that integration is left for subsequent chapters and future research. The themes include biomineralization, coagulation, secondary messengers, colloids, cross-linking, pumps, calcium binding proteins, mitochondria, hormones, and measurements.

1.1. BIOMINERALIZATION

The study of calcium in biology began with bone. "Examination of the red stained human bones, excavated in the cemetery of the Community of the Dead Sea Scrolls at Qumran, showed that the pigmentation was due to alizarin. Its characteristic anatomical location on the extremities and in the medullary cavities was consistent with the location of intravital staining due to a diet containing madder root" (Steckoll et al., 1971). Although we cannot say who first observed this staining by madder root, it provides a fascinating introduction to biomineralization, one of the many functions that involve calcium binding proteins.

Calcium Binding Proteins, By Eugene A. Permyakov and Robert H. Kretsinger

In 1581, Lemnius wrote in *De Miraculis Occultis Naturae*: "So some penetrate into the remotest and farthest parts, and are carried to the Nerves, as Hermodactyls, sweet Mullens, commonly called Palsey-hearb, Madder that colours the bones of the Cattle red, if they chance to eat it green, though they touch not the root that is red, which may be seen in the boyled flesh of that cattle, and in sheep that are painted red with decoction of it, as Wood dyes them blew, wild Saffron, and yellow Ocre."

Stephan Hales (1727) performed a remarkably broad range of experiments; growth and development were explored. "I took a half-grown chick, whose leg-bone was then two inches long; with a sharp-pointed iron, at half an inch distance, I pierced two small holes thro' the middle of the scaly covering of the leg and shin bone; two months after I killed the chick and upon laying the bone bare, I found on it obscure remains of the two marks I had made, at the same distance of half an inch: so that that part of the bone had not at all distended lengthwise, since that time that I marked it; notwithstanding the bone was in that time grown an inch more in length, which growth was mostly at the upper end of the bone, where a wonderful provision is made for its growth at the joining of its head to the shank, called by anatomists symphysis."

In 1736, Mr. Belchier, an English surgeon, was dining at the home of a dyer. He noted that the pork bones were red. After investigating, he wrote: "The diet with which the Hogs are fed is Bran, after it have been boiled in a Copper with printed Callicoes, in order to clean them from a dirty red Colour occasioned by an infusion of Madder Root, which was made use of to fix the Colours printed on the Cloth...." "The Brass having absorbed the red Colour discharged from the Cloth is mixed with the common food of the Hogs, and produces this effect on the Bones." "...the Bones of several Hogs of a different Breed, changed to a deep red Colour merely Aliment. And what makes this still more surprising is that neither the fleshy nor cartilaginous parts suffer the least alteration in Colour or in Taste."

Mr. Sloane, president of the Royal Society, wrote of Belchier's observation to M. Geoffry, who communicated a letter to the Académie Royal des Sciences in Paris. This letter motivated the du Hamel's research. None of these men appear to have heard of Hales' earlier results. From 1739 through 1743, du Hamel (1739) presented seven "Memoirs on Bone," in which he pursued the analogy between "...the formation and the structure of the two types of living bodies, plants and animals...." In the fourth (1743, p. 87): "...one proposes to report new evidence that establishes that bones grow in size by the addition of bony layers which originate in the periosteum as the woody parts of trees increase in size by the addition of woody layers, which are formed in the bark." "One obtains by use of madder red layers and white layers distinct from one another ..." (1743, p. 11). In the fifth (1743, p. 111): "...one proposes to clarify by new experiments how the growth of bones follows their length and show that this growth operates by a mechanism very similar to that which one observes in nature for the elongation of the woody parts in the buds of trees." "...one takes a six week old chick, the bone of its leg is two inches long; one pierces it with a drill a half inch from the ankle, one makes another hole a half inch higher, finally one makes a third hole yet half a inch higher & this last hole being a half inch from the knee, so that the length of the bone was divided by half inches." "At the end of the experiment the tibia of this animal was 3 inches long instead of the 2 inches at the beginning of the experiment.. . ." "The first (hole), which at the beginning of the experiment was 6 twelfths from the lower end, was at the end of the experiment at 9 twelfths; thus it was elongated 3 twelfths in this region. The second hole was at the end of the experiment as at the beginning 6 twelfths from the first and from the third hole; there had been no elongation between the first and the third hole. But the third hole which at the beginning of the experiment was located at the termination 15 twelfths away. . . ." "When the bones have not been well hardened, they elongate in all their parts. . . ." For all his ingenuity and eloquence, he missed the significance of the epiphysis.

"Mémoires sur les Os": ". . . la génération & la structure des deux espèces de corps vivants, les plantes & les animaux. . . ." In the fourth (1743, p. 87): ". . . on se propose de rapproter de nouvelles preuves qui établissent que les Os croissent en grosseur par l'addition de couches osseuses qui tirent leur origine du périoste, comme le corps ligneux des Arbes augmente en grosseur par l'addition de couches ligneuses qui se forment dans l'ecorce." "On obtient par le moyen de la garence des couche rouges & des couches blanches fait distinctes les unes des autres. . ." (1743, p. 11). In the fifth (1743, p. 111): ". . . on se propose d'eclaircir par de nouvelles experiences comment se fait la crûe des Os suivant leur longeur, & de prouver que cet accroissement s'opère par un mèchanisme très approchant de celui qu'observe la Nature pour l'alongement du corps ligneux dans les bourgeons des Arbes." ". . . on choisit un poulet âgé d'environ six semaines, l'os de sa jambe avait 2 pouces de longueur; on le perca avec un fout à un demi-pouce de l'articulation du pied, on fit un autre trou un demi-pouce plus haut, enfin on fit un troisième trou encore un demi-pouce plus haut, & ce dernier trou étoit éloingé de l'articulation du genou d'un demi-pouce, de sorte que toute la longeur de l'os étoit divisée par demi-pouces." "A la fin de l'expérience" (treize semaines) "l'os tibia de cet animal avoit 3 pouces de longeur au lieu de 2 pouces qu'il avoit au commencement de l'expérience. . . ." "Le premier, qui au commencement de l'éxperience étoit à 6 lignes de l'extrémité infèrieure, étoit à la fin de l'expérience à 9 lignes; ainsi il s'étoit alongé de 3 lignes en cet endroit. Le deuxième trou étoit à la fin de l'expérience comme au commencement, à 6 lignes du premier & du troisième trou; il n'y avoit donce eu aucun alongement entre le premier & le troisième trou. Mais ce troisième trou qui au commencement de l'expérience étroit éloigné de 6 lignes de l'extrèmité supérieure du tibia en étoit éloigné à la fin de 15 lignes. . . ." "Que quand les os ne sont pas bien endurcis, ils s'alongent dans toutes leurs parties. . . ."

The "Experiments and Observations on the Growth of Bones" of the famous surgeon John Hunter were read some years after his death in 1798 by Mr. Horne (J. Hunter's Works, 1837). "The colouring principle of the Rubia tincorum [madder] has strong affinity to phosphate of lime, which earth, if artificially precipitated from a solution of madder, carries down with it the colouring matter in a state of combination, which water does not disturb." Two pigs were fed madder for a fortnight. One was killed; "...the exterior part was found to be principally coloured, and the interior was much less tinged. The other pig was allowed to live a fortnight longer, but had no madder in its food; it was then killed, and the exterior part of the bone was found of the natural colour, but the interior was red." "Accordingly, although an inference may be safely drawn with respect to the part of a growing bone which receives the accessions of osseous substance, by observing the part which is coloured with madder, yet we cannot too certainly conclude that a superficial colourless layer, in an animal killed after remission of the madder, is a new deposit, since it may be the old, from which the madder has been removed, after having been redissolved in the serum." "To ascertain that the cylindrical bones are not elongated, by new matter being interposed in the interstices of the old he made the following experiment: he bored two holes in the tibia of a pig, one near the upper end, and the other near the lower: the space between the holes was exactly two inches: a small leaden shot was inserted into each hole. When the bone had been increased in its length by the growth of the animal, the pig was killed, and the space within the two shot was also exactly two inches." "Bones, according to Mr. Hunter's doctrine grow by two processes going on at the same time, and assisting each other; the arteries bring the supplies to the bone for its increase; the absorbents at the same time are employed in removing portions of the old bone, so as to give the new proper form." He contrasted Hunter's explanation with that of du Hamel, who "...placed a ring of silver wire round the middle of the shaft of the thigh-bone of a young pigeon; and found at a subsequent period the ring in the medullary cavity of the bone. ..." Hunter's interpretation would be "...that the arteries of the periosteum had deposited new bone on the external surface of the ring, while the absorbents had removed the old bone in contact with the internal surface of the ring. ..." "Du Hamel explains the facts on mechanical principles; assuming that the bony layers of the shaft of the thigh bone were expanded by the interposition of additional osseous matter, and that the layers were cut through in this process of expansion by the unyielding wire which had been placed around them."

In these days of nearly miraculous genetic engineering, it gives one pause to contemplate that we still do not understand the basic mechanisms of biomineralization. Other aspects of calcium physiology have fared better.

1.2. COAGULATION

During the latter decades of the nineteenth century, the effects of calcium on various organisms, organs, and extracts were tested. Analogies were sought among these systems; not all were apt.

In 1873, Hammarsten examined the effects of various heat and salt treatments on milk coagulation and subsequent cheese formation. "We have thus found that calcium salts were the necessary component for cheese formation in solution no. 1; the proteins previously existing in the same solution were without any significance and thus it is generally proven, that the chemical course of casein coagulation by ferment does not consist of the fact, that there are found two proteins in milk; a caseinogen and caseinoplastic substance, which by the action of a ferment can join themselves to a new protein body, cheese."

"Vi hafva sålunde funnit att kalksalterna voro une för ostbildningen nödvändiga beständsdelen af lösningen no 1; den i samma lösning förhandenvarande ägghvitan var utan någon betydelse och härmedär dat sålunda bevisadt, att det chemiska förloppet vid caseinets coagulation med löpe ej bestor deri, att I mjölken finnas tvenna ägghvitekropper, en caseogen och en caseoplastisk substans, som genom löpets inverkan förena sig till en ny ägghvitekropp, osten."

In 1887, Green noted that "there is considerable similarity between the coagulation of milk and that of blood." He also presented preliminary experiments showing that the addition of $CaSO_4$ solution to diluted plasma increases its coagulation rate markedly. Apparently unaware of Green's work, but following the earlier suggestion of Hammarsten, Arthus and Pagès in 1890 described the digestion of milk by "labferment," gastic mucosa of calf.

First they distinguished the "coagulation," as it was then called, caused by acid, from that caused by the enzyme rennin, or chymosin. By complexing the calcium with oxalate, they showed that lab ferment contains a calcium-dependent enzyme, one that cleaves caseine. Most perceptively, they distinguished two effects: first, the transformation caused by lab ferment from the subsequent "precipitation" of the casein fragments by calcium. In the second (1890), Arthus and Pagès, working from the physiology laboratory in the Sorbonne, discussed the analogy between the "...caseinification of milk and the spontaneous coagulation of blood ..." and summarized this result: "Oxalates, fluorides, and alkaline soaps make the blood spontaneously coagulatable."

"...caséification du lait et la coagulation spontenée du sang. ..."

"Les oxalates, fluoruses et savons alcalin rendent le sang non spontanément coagulable."

Again they distinguished among calcium activation of the enzyme(s), fibrin ferment, and the involvement of calcium in fibrin precipitaion or clot.

As discussed in Chapter 13, many extracellular enzymes bind calcium and can be inactivated by removal of free Ca^{2+} ions usually found in the extracellular environment at about 10^{-3} M. Further, this extracellular calcium is ideally suited to cross-linking protein, carbohydrates, and phospholipids, either as molecules free in solution or as attached to cell surfaces. These early studies of Arthus and Pagès addressed two themes that reappear as functions of extracellular calcium: enzyme activation and cross-linking of molecules.

1.3. SECONDARY MESSENGERS (ANTICIPATED)

During the same period, even more important observations were being made by Ringer and by Locke. In 1881, Ringer described the optimal concentrations of sodium, potassium, and ammonia to maintain eel and frog heart contractility. The next year he observed: "After the publication (just mentioned) I discovered that the saline solution which I had used had not been prepared with distilled water, but with pipe water supplied by the New River Water Company of London. Gratis they furnished 38 mg calcium per liter of water." "I conclude, therefore, that a lime salt is necessary for the maintenance of muscular contractility of both eel heart and frog skeletal muscle."

Locke (1894) found that he could cause the dissected frog sartorius muscle to contract by both direct electrical stimulation and via stimulation of the attached motor nerve. The latter indirect response depended critically on the presence of calcium in the bathing medium. Only today are the mechanisms underlying Ringer's and Locke's observations becoming understood. They had the good fortune to use systems in which contraction and secretion could be controlled by altering the concentrations of extracellular calcium. We now know that these cytosolic processes are regulated or modulated by the concentrations of the free Ca^{2+} ion within the cytosol.

Ringer fully appreciated the significance of his observations. Yet, like most of his contemporaries, he attempted to explain the effect of calcium on muscle by a very reasonable but false analogy. Earlier, M. Foster had noted the similarities between muscle contraction and coagulation. In the fourth edition of *Text-book of Physiology* (1883, p. 66) he stated a theme to be explored for 60 years: "We may in fact speak of rigor mortis as characterized by a coagulation of blood plasma but differing from it in as much as the product is not fibrin but myosin." Subsequent results of Ringer (Ringer and Buxton, 1887) and of Arthus and Pagés (1890) seemed to confirm Foster's suggestion. Ringer himself turned to the study of calcium in blood coagulation.

1.4. COLLOIDS

Heilbrunn championed the view held by many physiologists of his day: "Living substance is colloidal—it is universally so. The activity of living things is almost certainly bound up with some sort of colloidal change" (1927). In one of his first papers (1915), he asked: "Is the egg [of the sea urchin, *Arbacia*] essentially fluid or is it a more or less rigid jelly?" He argued that the cell membrane is not lipoid, but later came to accept its lipoid nature. He made no mention of calcium but concluded: "It is also possible to believe that the primary effect of all the parthenogenic reagents is a coagulation effect." Pursuing this idea, he wrote in 1920: "In my work on artificial parthenogenesis, I showed that all substances which incite the sea-urchin egg to divide mitotically produce a marked increase in the viscosity of the cytoplasm." "Accordingly I held the view that some sort of 'solidification' was the essential factor which initiated mitosis." By 1923 he was led to consider the effects of ions. "Perhaps the most important characteristic of a colloid is its electrical charge." He discussed experiments with sea urchin eggs and with the paramecium, *Stentor*, that "... show clearly that the bivalent cations, calcium and magnesium, do not exert as powerful a coagulative action as sodium, potassium and ammonium." First at the University of Michigan and subsequently at Penn, he and his students found system after system in which calcium plays an important role. For the next three decades he was the voice of calcium.

In his 1928 *The Colloid Chemistry of Protoplasm*, he argued that the particles or molecules within protoplasm carry a net positive charge while the surface membrane has a net negative charge. "It is thus apparent that the greater the amount of cation absorbed, the more fluid is the protoplasm." "The effect in decreasing the viscosity of the protoplasmic suspension..." ($Ca > Mg > Na > K > NH_4$) is equal to "... the order of decreasing absorbability."

He also argued that "the experimental study of the surface precipitation reaction has shown that in various cells, the presence of calcium is a necessary prerequisite." The fact that the viscosity of protoplasm decreases briefly following fertilization and his observation on vacuole formation led him to speculate: "Perhaps the calcium within the egg is not in an available form." He went on to suggest: "If we make the assumption that the calcium within the egg is for the most part bound with some fatty or lipoid substance, then we can offer an explanation which will fit practically all of the known facts." He suggested that free calcium might be increased by (1) cell rupture, (2) decrease in cell volume, (3) long exposure to isotonic calcium solutions, or (4) dissolution of protoplasmic lipoids by heat or fat solvents.

In his proposed mechanism, "(1) Calcium set free in cell interior (2) Calcium reacts with the pigment granule of some constituent of it to produce a substance which we have called ovothrombin (3) Ovothrombin reacts with a substance in the protoplasm, presumably a protein, to cause vesicle formation."

Heilbrunn and his colleagues studied the action of ultraviolet rays on *Arbacia* eggs (1930) and on amoeba (1933). They argued a general model in which ultraviolet "... stimulation of amoeba protoplasm consists first of the breakdown of a calcium gel in the cortex, and second in the entry of this calcium into the interior, where it produces a preliminary liquefaction followed by a gelation."

Mazia, a student of Heilbrunn, investigated "... free calcium in the action of stimulating agents on *Elodea* cells ..." (1936) and "... the release of calcium in *Arbacia* eggs on fertilization" (1937). The Penn group focused on the release of calcium: "The results are shown to be consistent with the idea that when stimulants act on these cells, they cause a release of Ca from combinations located in the periphery of the cells" (1936). "It is intended that a primary effect of a stimulating agent on a cell is to cause a release of calcium from organic combinations in the cortex of the cell into the main body of protoplasm" (1937).

In Heilbrunn's *An Outline of General Physiology* (third edition, 1952), calcium played a crucial role. "There is a large amount of corrobative evidence to show that when cells are stimulated, calcium is set free" (1952, p. 606). The cell "... cortex contains, as an essential part, calcium proteinate...." "... the clotting action of the calcium ion is important both for the cell exterior and for the interior." These ideas were expressed in his 1940 paper "The Action of Calcium on Muscle Protoplasm." "On stimulation calcium is released from the cortex and enters the cell interior. There it produces gelation as clotting. This clotting reaction is essentially the same type of reaction as that which occurs when naked protoplasm is exposed by teasing or cutting a cell (surface precipitation reaction)."

At Woods Hole, Heilbrunn was well known for his comment: "Yes, calcium is everything!" His 1940 paper contains another characteristic passage: "The muscle physiologist, adept as he is at chemical analysis and various intricate types of physical recording, has hardly ever considered the fact that the protoplasm of the muscle cell is not markedly different from the protoplasm of less differentiated cells...." Heilbrunn's reluctance to accept modern biochemistry and the remarkable variety and specificity of protein reactions prevented him from anticipating the mechanisms whereby calcium affects the protoplasm. Cell division, growth, and adhesion have now been examined in many systems; many of the calcium-modulated proteins have been identified.

1.5. CROSS-LINKING AND CELL SURFACES

Starting with men like Loeb and continuing today, scientists have searched for general laws or formulas relating cell response to the ionic composition of the bathing medium. There were many attempts to explain calcium effects in terms of general mass law characteristics. Concepts of cross-linking, like those of colloids, held sway in the first half of the twentieth century. Only gradually have two important generalizations emerged. The $[Ca^{2+}]_{cyt}$ within the cytosol under resting conditions is almost 100,000 times lower than it is in the bathing plasma, $\sim 10^{-7.5}$ vs. $\sim 10^{-2.7}$ M. Elaborate mechanisms have evolved to establish and control such concentration gradients. No one anticipated the incredible specificity of protein–addend interactions.

Hamburger, in 1910, assayed "The Influence of Small Amounts of Calcium on the Motion of Phagocytes" by placing a small capillary-filled *Bacillus coli* under the skin of a rabbit and after a fixed time measuring the length of leucocytes in the column. He concluded that a slight amount of calcium caused an "... acceleration of the amoeboid motion." His following discussion should have been heeded by many of his colleagues over the next 30 years. "We might be inclined to think of a modification in the agglomeration of the colloid protoplasm particles as a consequence of the electric charge, caused by the entering of a number of bi-valent calcium ions. This explanation however can hardly be the correct one. For the experiment teaches that other bi-valent cations namely barium, strontium, magnesium do not cause an acceleration of the amoeboid motion. It must be assumed then, that the action of calcium in this case, is based upon a specific, hitherto unknown, biochemical property."

In the same year (1910), Osterhout pointed out that the Overton theory of a lipoid outer layer (membrane) demanded that the inorganic salts not penetrate the membrane. He grew seeds of *Dianthus barbatus* on the surface of distilled water. These cells normally produce a great deal of oxalate. Following addition of 5.0 mM calcium to the external medium, crystals of calcium oxalate appeared in the cytosol. At least in these cells some calcium can cross the cell membrane. Until the discovery of the photoprotein, aequorin, some 50 years later, there existed no method of measuring directly the concentration of the free Ca^{2+} ion within the cytosol.

Galstoff (1925) studied the adhesion of sponge cells following dissociation by squeezing them through bolting cloth. "According to their effect on amoeboid movement and coalescence, the cations can be put in the following order...: $Ca > NH_3 > Li > K > Mg > Na$."

In addition, calcium might alter the properties of membranes by interacting with either proteins or phopholipids. Wilbrant (1940) studied the increase in osmotic resistance of the erythrocyte to various salts and found that "$CaCl_2$ actually acts more strongly and between NaCl and $CaCl_2$ concentrations of the same effectiveness the relationship $Na/Ca^{1/2}$ = const. also holds approximately." In 1948, he and Koller extended these comparative studies to the frog heart.

"$CaCl_2$ wirkt tatsächlich viel stärker and zwischen Konzentration gleicherchen Wirksamkeit von NaCl and $CaCl_2$ besteht auch hier annähernd die Beziehung $Na/Ca^{1/2}$ = const."

In 1944, Carruthers and Suntzeff reported that "estimations of the calcium content of the mouse epidermis during the process of experimental carcinogenesis reveal two distinct phases: an immediate reduction in the calcium content which persists at a fairly constant level for many weeks and a further reduction when the epithelial cells have been transformed into cancer cells."

Coman (1944) found that "malignant neoplastic squamous cells from carcinomas of the lip and from carcinomas of the cervix showed mean values of adhesiveness far below that of the normal cells." Although he had not yet done any calcium measurements, from the literature he suggested "...that decreased mutual adhesiveness in cells of squamous cell carcinoma may be related to a lowered calcium content of these cells." De Long et al. (1950) studied rapidly dividing, nonneoplastic cells. "As compared with normal rat liver, the calcium and sodium contents [of regenerating rat liver] were not significantly altered, but potassium was increased by an average of 11 per cent. It is concluded that the increased potassium content of cancers is at least in part an expression of cellular multiplication, but that the diminution in calcium is peculiar to cancer and is partially responsible for decreased mutual adhesiveness of cancer cells. ..." Coman (1954) observed tissues in the electron microscope. "In liver perfused with versenate (EDTA) cells were no longer closely opposed but were separated to great or lesser degree." "These findings suggest a molecular bond of the calcium to the carboxyl groups of the proteins and to the phosphate groups of lipoids, as the basis of cellular adhesiveness." These interpretations are still controversial.

Steinberg (1962) extended these ideas to embryology. In the amphibian gastrulae, "calcium is shown not to be functioning as a desolvating agent by experiments in which other desolvating agents, substituted for Ca^{++}, fail to duplicate its effects." "The supposition that Ca^{++} functions by countering the negative charge on cell surfaces, thereby reducing mutual electrostatic repulsion, is demonstrated to be incorrect by the observation that suppression of these negative charges by a reduction in pH prevents, rather than encourages, cell adhesion, in consonance with the prevention of Ca^{++} binding. These observations are shown to be compatible with theories of Ca^{++} bridging between cells and with those invoking 'extracellular' cementing material. . . ."

In 1962, Whitfield and Dixon described postirradiation mitotic delay in cultures of L cells: "Not only does calcium prevent radiation-inducted mitotic delay, its chromatin condensing effect is also in opposition to a radiation-induced elongation of prophase chromosomes." Balk in 1971 described another system that was to become very significant: "Whereas multiplication of normal chicken fibroblasts is limited and controlled by the concentration and access of ionic calcium, cells transformed by the Rous sarcoma virus appear to have gained independence from these environmentally imposed controls."

Alford (1970) reported on the metal cation requirements for phytohemagglutinin-induced transformation of human peripheral blood lymphocytes: "Citrate inhibition of PHA-induced lymphocyte transformation as induced by diminished ^{3}H-thymidine uptake was due to chelation of ionized calcium and could be reversed by added calcium ion." At 20 mM, magnesium is 75% as effective as calcium.

1.6. SECONDARY MESSENGERS (UPDATED)

Steinhardt and Epel (1974) confirmed and extended Mazia's observations. "Micromolar amounts of the divalent ionophore A23187 can activate echinoderm eggs." ". . . eggs preloaded with ^{45}Ca show a 20-fold increase in ^{45}Ca efflux when activated by inonophore A23187 or fertilization." "We propose that both normal fertilization and ionophore activations affect the metabolism of the egg by releasing Ca sequestered in intercellular stores."

Brachet et al. (1975) studied the induction of maturation (meiosis) in *Xenopus laevis* oocytes: "Organomercurial- and progesterone induced maturation have many features in common: they do not occur when the inducer is injected into the oocyte, they require the presence of Ca^{++} in the medium, they are inhibited by cycloheximide but not by actinomycin D."

Since the studies of Ringer, contractile systems have provided many insights into the biological functions of calcium. The gel nature of protoplasm is far more complex than Heilbrunn anticipated. In 1939, Engelhardt and Ljubimowa in Moscow described an ATPase function for myosin: "Thus the mineralization of adenosintriphosphate, often regarded as the primary exothermic reaction in muscle contraction, proceeds under the influence and with the direct participation of the proteins considered to form the main basis of the contractile mechanism of the muscle fiber." Bailey (1942) explored the ATPase activity of myosin activated by calcium: "We suggest that the essential feature of excitation and contraction—we cannot at present dissociate the two phases—is the liberation of the Ca ion in the neighbourhood of the ATPase grouping, which can thus by the almost instantaneous catalysis of ATP breakdown make available a large amount of energy." Bailey was right, but because of the complexity of the muscle system it was some years before the role of calcium was generally accepted.

In 1949, Hill found that frog muscle at 0°C reaches maximum tension within 40 ms of stimulus to the cell surface. "... it is impossible, therefore to assume that only a fraction of the cross-section of each fibre is involved in it" [the twitch]. He calculated that nearly a half second would be required for a small molecule to diffuse from the surface to the center of a 100-μm-diameter muscle cell. "We must look for some physical or physico-chemical process which is released by excitation at the surface then propagated inwards."

One of the keys to understanding the function of calcium in muscle was to separate the calcium-sequestering system from the calcium-activated system. In 1948, Kielly and Meyerhof described "a new magnesium-activated adenosine triphosphatase in a particulate fraction free of actin and of myosin." Marsh (1951, 1952) found "a factor modifying muscle fiber synaeresis." "Thus there appears to be a substance or substances in muscle, easily removable by dilute potassium chloride solution, the presence or absence of which determines the effect adensoine triphosphate will produce." If ATP is added to muscle brei in the presence of the "Marsh factor" or "relaxing factor" there is a low level of ATPase activity and the volume of the actomyosin gel increases slowly and reversibly by 40%. With the relaxing factor removed, ATP causes a rapid, irreversible contraction of 30%. Bendell (1952), a colleague of Marsh at Cambridge, found stabilizing conditions for the relaxing factor and noted that calcium causes contraction and synaeresis; similar supporting results were published by Hasselbach and Weber (1953) in Tübingen. Bozler (1954) used Albert Szent-Györgi's glycerinated muscle preparation to demonstrate that physiological levels of Mg-ATP can make such muscle fibers "extensible and plastic"; they then "... give strong contraction on addition of small amounts of $CaCl_2$...." "The results indicate that relaxation is caused by the inactivation of bound calcium... by either the relaxing factor or by EDTA." "Previous evidence that the relaxed state is due to the formation of an enzymatically inactive ATP–protein complex was confirmed."

In 1952, Sandow reviewed the evidence that the action potential, per se, does not initiate contraction. He focused on the link between excitation and contraction and suggested "... that, in the living muscle, activation of the contractile material may be attributed to the enzymatic activation of the myosin-ATPase system by Ca^{++}." Two years later, A. F. Huxley and Niedergerke and H. E. Huxley and Hanson presented their sliding filament model.

1.7. PUMPS, CHANNELS, AND IONOPHORES

During this period, morphologists were using new OsO_4 fixation procedures and methacrylate embedding to obtain higher-contrast electron micrographs. Porter and Palade (1957) summarized their and others' work on the sarcoplasmic reticulum, its proximity to the Z band, and the "triad" appearing to connect it to the sarcolemma: "... it is proposed that the membrane limiting the system [the sarcoplasmic reticulum] is polarized like the sarcolemma and that the corresponding potential difference is utilized in the intercellular distribution of the excitatory impulse."

Anne Maria Weber (1959), the daughter of H. H. Weber, suggested that "... it might be worthwhile to investigate whether the Marsh–Bendall factor, which is particulate and presumably does not interact directly with the myofibrils, acts by binding Ca^{++}." In retelling such a story one easily imparts the false impression of a straightforward, linear progression. In the muscle field especially, many distinguished scientists have pursued the logic of reasonable experiments to incorrect conclusions. For example, Nagai et al. (1960) found that the inhibition of myosin ATPase ceased when relaxing granules were removed. "Thus the granules do not withdraw any functionally essential substances from the contractile proteins, nor do they release any stable contraction inhibiting substance to the actomyosin or to the solution."

In 1961, Ebashi, still working at the Rockefeller Institute (now Rockefeller University), before returning to Tokyo "... demonstrated that a purified preparation of the relaxing factor of skeletal muscle, shown by electron microscopy to be a vesicular fraction, probably the endoplasmic reticulum, is able to strongly bind calcium and furthermore that this binding of calcium by the fraction depends on the presence of ATP. We have supposed that the calcium binding represents the physiological action, or the mechanism of the relaxing factor. The results demonstrated in the present paper support this concept, and suggest that the calcium ion is the main controlling factor in muscle contraction."

By 1964, Franzini-Armstrong and Porter, now using glutaraldehyde as a fixative, could better reconstruct the sarcoplasmic reticulum (SR) as well as the connecting T-tubule system. "... the T system is a sarcolemmal derivative that retains its continuity with the sarcolemma and limits a space that is in direct communication with the extracellular environment. These structural features favour the conclusion that the T system plays a prominent role in the fast intracellular conduction of the excitatory pulse." The membranes of the SR enclose a volume topologically distinct from the sarcosol, which bathes the thick and thin filaments. When the SR is homogenized, it reseals to form closed vesicles. Better preparations of such vesicles were becoming available. Hasselbach and Makinose (1963) showed that two Ca^{2+} ions were pumped per ATP and that isolated SR vesicles could lower $[Ca^{2+}]$ to 10^{-8} M or pCa 8 (Chapter 12).

1.8. CALCIUM BINDING PROTEINS

If calcium does, in fact, couple excitation to contraction, what does it activate? What is its target? We now know that calcium relieves the effects of an inhibition; under appropriate conditions, myosin ATPase is active with no calcium present. Ebashi, by 1963 in Tokyo, described a "third component participating in the superprecipitation of 'natural Actomyosin.'" He noted that "synthetic" actomyosin as prepared by Perry and Grey (1956) had ATPase activity without the addition of calcium, in contrast to "natural" actomyosin, which requires calcium for superprecipitation. The factor imparting "calcium sensitivity" is in the tropomyosin fraction described by Bailly, but tropomyosin would not restore calcium sensitivity. Finally, in 1967, Ebashi et al. characterized "troponin as the Ca^{++} receptive protein in the contractile system." "It is conceivable that binding and detaching of Ca^{++} to and from troponin might be of primary importance in regulation of muscle contraction, i.e. some conformational change of the troponin molecule induced by the removal of Ca^{++} might inhibit interaction of adjoining actin molecule with myosin and this inhibition might be cancelled by Ca^{++}." They also anticipated an as yet unanswered problem: "However, before accepting this explanation, we must answer the question as to how the troponin molecules, which are distributed along the thin filaments at a 400 Å periodicity, can exert their influence on those actin molecules which are located at some distance from adjacent two troponin molecules."

So, finally, in one system we approach a molecular interpretation for one intracellular function of calcium. Its concentration is very low in the resting muscle cell. Stimulation causes calcium inflow from a storage vesicle or from the extracellular space. This calcium interacts with a target, a calcium-modulated protein, which, in turn, "activates" an enzyme. The cycle is complete when the calcium is pumped out and the enzyme is no longer active.

While biochemical studies on troponin progressed, Kretsinger et al. determined the crystal structure of parvalbumin found in the cytosols of many vertebrate cells. In 1972, Kretsinger published "Gene Triplication Deduced from the Tertiary Structure of a Muscle Calcium Binding Protein" and suggested that "both troponin and MCBP are acidic, pI about 4.5; both have high phenylalanine contents and high calcium affinities. The molecular weight of TNC, however, is 19,000 while that of MCBP is 11,500. I consider it possible that MCBP exists in mammals in some reduplicated form, considering its tendency to duplication." The next year Collins et al. published the article "The Amino Acid Sequence of Rabbit Skeletal Muscle Troponin C: Gene Duplication and Homology with Calcium-Binding Protein from Carp and Hake Muscle." "The correspondence of α-helices and hydrophobic residues suggests that each of the four regions of TNC has a three-dimensional structure very similar to the CD and EF regions of MCBP." The basic structural and evolutionary domain, about 30 amino acids long, consists of two turns of α-helix, a calcium binding loop, and two more turns of α-helix: the EF-hand. In 1975, Kretsinger published "Hypothesis: Calcium Modulated

Proteins Contain EF Hands." Chapter 11 treats this homolog family, which now includes calmodulin and over 70 other distinct subfamilies.

1.9. SECONDARY MESSENGERS (YET AGAIN)

Calcium controls motility in systems other than those based on actomyosin. The ciliate *Paramecium caudatum* normally swims forward. By 1900 it had been established that a variety of chemical, mechanical, and electrical stimuli cause it to reverse its direction of ciliary beat and swim backwards. In 1926, Mast and Nadler measured the duration of reversal as a function of the cations in the medium. "The amount of $CaCl_2$ required to neutralize KCl is not proportional to the concentration of the KCl. Weber's law does not hold. The results seem to indicate that ciliary reversal is associated with differential absorption of the cations and subsequent changes in electrical potential, but that there are also other factors involved." Since the early 1930s, Kamada, in Tokyo, had studied the effects of electrical currents. In 1938 he published "Intracellular Calcium and Ciliary Reversal in Paramecium." "If a paramecium is immersed in a medium relatively low in [Ca]/[K] ratio, the direction of ciliary beat can be reversed. However, since the cilium itself is not provided with the mechanism to reverse its effective stroke in a direct response to stimulation, some intercellular change must be involved in this reaction." His intracellular injections were very damaging. "It may however be concluded that the diminution of the intercellular calcium ions causes ciliary reversal."

Naitoh (1968) extended the observations on the effects of extracellular cations and hypothesized: "Externally applied cations bind to the anionic sites on *Paramecium* in exchange for bound calcium in a manner consistent with the law of mass action. The calcium ions which are thus liberated are effective in activating, directly or indirectly, a contractile system which is energized by ATP." "It is not known why only those calcium ions liberated from the cellular binding sites activate the reversal system. The reason may be a difference in the effective diameter of liberated and free calcium ions resulting from a difference in the degree of hydration." A reasonable speculation, but wrong; the mechanism of calcium release and its mode of action are discussed in Chapter 2.

The study of secretion developed in parallel to that of contraction. Douglas and Rubins' "stimulus–secretion coupling" is analogous to the "excitation–contraction coupling" of Sandow. Calcium is the common coupler. As we shall see, many of the mechanisms are homologous. The act of secretion, or more specifically, exocytosis, is not observed as easily as is muscle contraction. In 1893, Locke published "Die Wirkung der Physiologischen Kochsalzlö sung auf quergestreifte Muskel." His nerve, muscle (frog sartorius) preparation in NaCl responded to "direct" stimulation over 15 hours; however, it lost its response to "indirect" stimulation (via the nerve) within an hour. If calcium was then added to 5 mM, the indirect response was regained. Calcium is not required for neurotransmission along the axon but is required for the nerve–muscle coupling and/or the actual contractile event. Over the next 40 years this observation was repeated and the calcium contents of nervous tissues were determined. However, it was only after acetylcholine (ACh) was established as (one of) the chemicals involved in nerve transmission that the function of calcium could be established. Harvey and MacIntosh in 1940 declared that "calcium ions are necessary for the transmission of the excitatory state at the neuromuscular junction." They continued: "Our experiments show that, when calcium is absent, there is no release of ACh from the preganglionic nerve endings, either during stimulation of the sympathetic trunk or following the injection of potassium salts."

Recalling Ringer's observations of the frog heart, one might ask whether these effects are exerted on the cell surface and/or in the cell interior. The studies of muscle that led to the characterization of troponin showed that calcium has a direct effect on the contractile proteins. However, it also has significant and less well understood interactions with cell membranes. First, as discussed for the sarcoplasmic reticulum and the mitochondrion, membranes can pump calcium either directly using ATP as an energy source, as in the sarcoplasmic reticulum or via preformed gradients of Na^+, as in the axon, or of H^+ ions, in mitochondria and bacteria.

An understanding of calcium's role must await other discoveries. In 1952, Fatt and Katz wrote that "... the end-plates of resting muscle fibres are the seat of spontaneous electrical discharges which have the character of miniature end-plate potentials." "The results point to the conclusion that some terminal spots of the motor nerve endings are spontaneously active and release ACh in the same impulsive manner as they do after the arrival of a normal motor nerve impulse." In their 1957 review, Birks and MacIntosh noted that some ACh is "free" in the protoplasm and some is "packaged" for release, after Katz's idea of quantal release. Calcium does not affect spontaneous release but does determine the amount of ACh released under stimulation. "The problem of what determines an effective collision [of a package with the surface membrane] is unsolved: the calcium concentration of the presynaptic axoplasm may be an important factor."

Douglas made the generalization; he coined the phrase. In 1961, he and Rubin wrote: "Our experiments show that the excitant action of Ach on the adrenal medulla is dependent on the presence of calcium, and suggest that Ach evokes secretion by causing calcium ions to penetrate the adrenal medullary cortex." This was later confirmed. They argued the analogy with muscle. It is interesting that both the release of ACh and the subsequent secretion of catechol amines by the adrenal medulla involve exocytosis. In "A Possible Mechanism of neurosecretion: Release of Vasopressin by Depolarization and Its Dependence on Calcium," Douglas (1963) argued that in "... neurones and endocrine glands of nervous origin ... calcium acts as a crucial link in the process of stimulation–secretion coupling ..." and that "... the appearance of free calcium ions somewhere in the endings then causes the release of the stored hormone." This generalization would prove valid for exocytosis by all cells.

In hindsight, so much is obvious. By 1970 the role of calcium in muscle excitation and contraction was generally accepted. That calcium is the link in stimulus–secretion coupling had been inferred in ten systems. Calcium was suspected of playing a control function in metabolic processes such as glycogenolysis. Yet much of the creative thinking in cellular control processes was focused on cyclic adensine monophosphate (cAMP). In 1958, Rall and Sutherland identified cAMP as the heat-stable factor synthesized in particulate fractions of liver, heart, skeletal muscle, and brain following addition of ATP and stimulation by epinephrine and glucagon. They cautioned: "However, the only biochemical event which is at present known to be influenced by cyclic 3,5-AMP is the phosphorylation of dephosphorylase [now called phosphorylase] and, as yet, it is difficult to understand the multitude of physiological effects of epinephrine in the light of this one reaction." By 1965, Sutherland and his colleagues had extended his studies of cAMP to other hormone-sensitive systems. "In brief summary, the hormone (the first messenger) interacts with a component of the cell membrane to initiate increased accumulation of a mediator (the second messenger), which then acts upon components of the effector cell." They anticipated the extension of this concept. "To date cyclic $3', 5'$-AMP is the only second messenger which has been identified. It is proposed other such messengers may exist, for example, to mediate the action of insulin, and that these may, or may not be other cyclic $3', 5'$-nucleotides."

In 1970, Rasmussen published "Cell Communication, Calcium Ion, and Cyclic Adenosine Monophosphate," one of the most frequently cited papers in the literature of biology. He summarized his extensive review and analysis as follows: "The basic elements of this system are two interrelated intercellular messengers, $3'$-$5'$-AMP and Ca^{++}. Activation or excitation of the cell leads to an increase in both."

1.10. MITOCHONDRIA

In 1970, Lehninger addressed the problem of mitochondria and calcium ion transport and suggested two possible functions, which although they remain unproven, have not yet been replaced by anything better. He emphasized the high affinity and specificity of calcium transport and "its primacy over oxidative phosphorylation in most tissues." He first considered a possible role in muscle activation and/or relaxation but conceded that "Very little is known about the mechanism of release of Ca^{2+} from mitochondria, which must be a very fast process if it is to serve a primary physiological role in the excitation process." "One possible mechanism for the release of Ca^{2+} from the mitochondria in red muscle is a sudden depolarization or collapse of the electrochemical gradient across the membrane, triggered by depolarization of the T-system and the sarcoplasmic reticulum, with which mitochondria appear to make contact in some red muscles." Such functional contact has not been confirmed and remains speculative.

Then Lehninger considered a more promising alternative: applicable, however, only to mineralizing tissue. "We have adopted the working hypothesis that what living cells 'do' to Ca^{2+} and phosphate is to bring about their accumulation in the mitochondria to such concentrations as to exceed the solubility product of tricalcium phosphate ...." "The end products of this stage are suggested to be 'micro-packets' of insoluble amorphous tricalcium phosphate in the mitochondrial matrix, which we regard as the essential precursors of extracellular hydroxyapatite." As with his first suggestion of muscle activation, we still face the problem of communication with the plasma membrane. "We postulate that micro-packets of amorphous calcium phosphate, which are colloidal in dimensions, may dissociate from the larger aggregates in the matrix and then depart in essentially intact form from the mitochondria to the cytoplasm, each micro-packet being stabilized by one or more molecules of inhibitor." The micropacket is extruded by reverese phagocytosis and "...converted to crystalline hydroxyapatite...." The mitochondrion probably plays a role in intracellular calcium buffering and possibly is involved in mineralization. The mechanisms remain unknown.

Forty years ago no one anticipated that the mitochondrion plays a key role in calcium metabolism; today, no one seems to understand what that role is. Research on cytochrome and respiratory enzymes began with Keilin (1929). Twelve years later, Claude (1941) could prepare by differential centrifugation "...formed elements [that] might represent well known cellular constituents, i.e. mitochondria or fragments of mitochondria." Another decade passed before functionally active mitochondria were prepared and their enzymatic characteristics defined. Lehninger and Kennedy (1948) wrote: "We have found that all of the demonstrable fatty acid oxidase activity of rat liver can be recovered in that fraction of 30 per cent sucrose homogenates of rat liver stated by Hogeboom et al. (1948) to consist of morphologically intact mitochondria, free of extraneous intercellular elements." The initial observations on calcium uptake by mitochondria indicated that calcium might inhibit normal respiration by displaying a metal ion prosthetic group (Slater and Cleland, 1953). "All this calcium in the heart is found in the sarcosomal (mitochondrial) preparations." "The destruction of DPN, accelerated by calcium prevents the formation of oxaloacetate." "In the presence of EDTA and the reaction mixture, sacrcosomes oxidized α-ketoglutarate at practically a uniform rate for an hour or more at 25°." "...sarcosomes took up large amounts of calcium from solutions. The localization of the calcium in the isolated sarcosomes does not; therefore, necessarily reflect the position in the intact heart." This admonition is still valid. The report of DeLuca and Engstrom (1961) suggested a more physiological process: "This communication indicates that kidney mitochondria take up large quantities of calcium by a process which requires adenosine triphosphate (ATP), and oxidizable substrate, and magnesium ions, but which is not directly dependent upon oxidative phosphorylation." Vasington and Murphy (1962) wrote: "The evidence presented in this paper shows that rather large quantities of Ca^{++} may be actively bound by rat kidney mitochondria in a process that is directly dependent on respiration and that requires adenosine triphosphate, Mg^{++}, and inorganic phosphate in the medium." "Ca^{++} itself uncouples oxidative phosphorylation in concentrations in which it is actively bound." Rossi and Lehninger (1964) extended these findings: "From the observations it appears that activation of the energy-conserving site by Ca^{++} is the first stage in ion accumulation. This does not require phosphate." If Mg-ATP is added to the reaction suspension, 1.0 phosphate and nearly 2.0 calcium ions are accumulated for each pair of electrons at each of three energy-converting sites.

In parallel with these biochemical studies, histological work indicated the involvement of mitochondria in bone formation and/or resorption. B. L. Scott and Pease (1956) examined epiphyses and concluded: "There is a fibrous pre-osseous zone between osteoblasts and calcified matrix. This is thought to indicate a lag between the deposition of the organic matrix and subsequent calcification." "Calcification of both the cartilagenous and preosseous matrices is a progressive accretion and aggregation of inorganic crystals." The "...abundance of mitochondria in the osteoclast, and local changes in the bone in contact with the ruffled border suggest that a 'collagen dissolving substance' is probably secreted through the ruffled border." Gonzalez and Karnovsky (1961) described osteoclasts in healing fractures. "Mitochondria containing clusters of fine granules, were abundant." "...osteoclasts phagocytize crystals of bone salts and these crystals are very probably free of collagen."

Recalling Ringer's observations on the frog heart, one might ask whether the effects are exerted on the cell surface and/or in the cell interior. The studies of muscle that led to the characterization of troponin showed that calcium has a direct effect on the contractile proteins. However, it also has "significant and less well understood interactions with cell membranes." First, as discussed for the sarcoplasmic reticulum and the mitochodrion, membranes can pump calcium either directly using ATP as an energy source, as in the endoplasmic reticulum or via preformed gradients of Na^+, as in the axon, or of H^+, as in mitochondria and bacteria. In addition, calcium can alter the properties of membranes by interacting with either proteins or phospholipids.

1.11. PUMPS, CHANNELS, AND IONOPHORES

Brink (1954) cast his review "The Role of Calcium Ions in Neural Processes" in terms of more specific interactions instead of mass colloidal effects (Wilbrant, 1940). He noted that sodium and potassium ions function primarily; "in contrast, calcium ions seem to affect primarily the constraints imposed upon these ionic movements." From 1955 through 1957, Keynes and his co-workers published a series of important papers on the content of calcium and its function in squid giant axon. Hodgkin and Keynes (1957) injected ^{45}Ca into axons and observed little diffusion with or without stimulus, "...suggesting that most of the calcium is in a relatively immobile form." "...it is attractive to suppose that the calcium entry during the spike may not be an accidental accompaniment of activity, but is somehow connected with the development of the state of increased sodium permeability." Niedergerke (1963) extended these studies and reported that "both Ca uptake and release were increased during activity" or "...on application of the contracture fluids." Quite reasonably, but incorrectly, he argued: "According to this hypothesis a compound CaR is present which in some way activates tension whereas the compound Na_2R is inactive." He anticipated separate pools of calcium: "The results have been interpreted by assuming (1) that entry of Ca occurs after combination of this ion at the cell surface with the carrier molecule whose presence is suggested by the Ca-Na antagonism of the heart, (2) that intracellular exchangeable Ca is present in two interacting forms, the activator-Ca which induces contraction and a larger store of inactive Ca." The problem was well posed. Does the Ca^{2+} ion that binds to troponin come directly from the bathing medium immediately before that contraction event? Does the Ca^{2+} ion have other effects within the cytosol or on membranes?

These questions have not yet been fully answered, but additional insights have come from other systems. Reuter and Seitz (1968) recalled that "Ca influx in cardiac muscle is greatly increased during excitation." Also, "quantitatively, Ca efflux from auricles has been shown to depend to a large extent on the ratio $[Ca^{2+}]^{1/2}/[Na^{+}]$. The affinity for Na of the activation site for Ca efflux (carrier) is much less than for Ca." The distinction between calcium pores and calcium pumps and between calcium effects on sodium and/or potassium pumps and pores is difficult. In 1961, Dunham and Glynn wrote: "The ATP-ase activity of human red cell ghosts has been shown to consist of two components. The first component requires the presence of magnesium ions but occurs in the absence of alkali metal ions and is not inhibited by cardiac glycosides. In the presence of magnesium, activity is greatly increased by small amounts of calcium but inhibited by larger amounts. The second component requires the presence of magnesium and also of both sodium and potassium ions. It is completely inhibited by cardiac glycosides in concentrations sufficient to inhibit ion transport in intact red cells. Ca^{2+} ions inhibit at both low and high concentrations." Schatzman (1966) confirmed these findings and suggested that "...red cells are able to maintain low intracellular Ca^{++} concentrations ...." Whitham in 1968 found that extracellular "calcium markedly raised passive potassium efflux but did not affect potassium influx." "Another view is that an increase in concentration of ionic calcium in cells raised permeability of the potassium."

The squid axon has served neurobiology well. Baker et al. (1969) found that "the rise in Na efflux resulting from partial replacement of NaCl by dextrose or choline chloride consisted of two components one of which was oubain-insensitive and calcium-dependent and the other was inhibited by oubain but calcium insensitive." Blaustein and Hodgkin (1969) continued: "After injecting ^{45}Ca along the axis, the efflux of calcium reached its maximum much more rapidly in a cyanide treated axon than in an unpoisoned axon." "A possible explanation of the cyanide effect is that, after poisoning, calcium ions are released from a store and can then exchange at a higher rate with external sodium or calcium." "The experiments suggest that part of the calcium efflux may be coupled to sodium entry." In 1972, Meech published "Intracellular Calcium Injection Causes Increased Potassium Conductance in *Aplysia* Nerve Cells," in which he stated: "Furthermore, when a muscle fibre is stimulated electrically the threshold of contraction is close to the membrane potential at which the increase in potassium conductance called 'delayed rectification' begins." Krnjević and Lisiewicz

(1972) injected calcium into spinal motor neurons and "... it was concluded that the fall in membrane resistance caused by intracellular Ca^{2+} is mainly due to an increase in gK" (potassium conductance).

The dark current of the retina is a well-studied system in which the sodium channel appears to be under calcium control. In 1926, Feenstra studied the retinal current of frogs: "By lowering the calcium in the Ringer solution and thereby removing as much as possible this element from the tissues, one can cause the resting current as well as the action current to disappear."

"En suppriment le sel de calcium dans le solution de Ringer et en enlevant ainsi autant que possible cet element aux tissus, on put faire disparâitre aussi bien le courant de repos que le courant d'action."

Both Fulpius and Baumann (1969), working with honeybees, and Millecchia and Mauro (1969), working with the horseshoe crab, *Limulus*, found that low external calcium increases the dark current, while low bathing sodium decreases it. H. M. Brown et al. (1970) studied another arthropod, the barnacle: "The results indicate that illumination increases the membrane permeability mainly to Na^+ ions and that the primary effect of Ca^{2+} ions is suppression of the permeability increase; Ca^{2+} permeability may increase slightly during illumination." Lisman and Brown (1972) gave a different interpretation of their results from *Limulus*: "We propose that an increase in $[Na^+]_{in}$ leads to an increase in $[Ca^{++}]_{in}$ and that an increase in $[Ca^{++}]_{in}$ by any means leads to a reduction in responsiveness to light." Hagins (1972) presented a different hypothesis for the vertebrate rods and cones. "(I) The activity of Ca^{++} ion is maintained by pumps at a much lower level in the cytoplasm of the outer segments than in the intradisk spaces as in the external fluid. (II) The Na^+ conductance of the rod envelope membrane decreases as the cytoplasmic Ca^{++} activity increases, possibly because internal Ca^{++} reacts reversibly with sites at which the dark current enters, and blocks their Na^+ permeability. (III) Light transiently increases the permeability of the disk membranes and the envelope membrane to Ca^{++}, allowing many more than one calcium ion to enter the cytoplasmic space per photon absorbed."

In 1966, Lowenstein reported his initial studies on the junctional membranes of the epithelial cells of the *Chironomus* salivary gland. "A primary factor in controlling junctional membrane permeability is Ca^{++}." "When its concentration is raised above 10^{-4} M, the junctional membranes become as impermeable as the non-junctional ones; each cell seals itself off as a unit." The mechanism(s) whereby calcium controls these various gates, pores, or pumps remain the topic of many grant requests. It is, though, a safe guess that the basic concept will differ little from that involved in excitation–contraction or stimulus–secretion coupling.

It became ever more apparent that one had to determine the concentration of the free Ca^{2+} ion within the cytosol. As early as 1934, Tipton reported the total content of frog sciatic nerve as 7.3 milliequivalents per kilogram net weight, but he lacked techniques to determine its cellular distribution. Hodgkin and Katz

(1949) wrote that "...extruded axoplasm was dispersed by millimolar calcium." "...but it apparently contrasts with the coagulation which calcium is known to produce in other colloidal materials." In 1957, Gilbert and Fenn reported a thorough study of calcium equilibrium in frog muscle. "The percentages of total calcium in the whole muscle immersed in Ringer's solution [for five hours] was as follows: 10 percent in the surface phase; 12 percent in the extracellular water space; 17 percent in the dry connective tissue; 24 percent in the extracellular water space; and 37 percent as nonexchangeable calcium." "It appears that there is a calcium pump pushing calcium out of the cell against an electrochemical gradient of about 4 cal/mM of calcium." Portzehl et al. (1964) wrote: "When buffers containing calcium and EGTA were injected [into crab muscle fibers] it was found that contraction was not obtained if the ionized calcium in the buffer was below a certain level" (pCa $\sim$ 6.2). Similar experiments in barnacle by Hagiwara and Nakajima (1966) yielded similar results. "The threshold concentration for contraction was about 8×10^{-7} M." At rest the concentration is lower. A more direct determination of $[Ca^{2+}]$ would have to await the availability of aequorin.

Most bacteria, when they form spores, synthesize dipicolinic acid (DPA) to about 10% of their total dry weight. The molar ratio of calcium to DPA is usually in the range 1.5 to 0.7. Although compositional analyses of bacteria had been performed at the beginning of the twentieth century, Curran et al. (1943) were the first to apply the newly available spectrograph. "Spores were materially higher in Ca and lower in K than the vegetative cells from which they were derived." "While there was apparently no direct correlation between elemental inorganic composition and degree of resistance to heat, in general, Slepecky and Foster (1959) found that high concentrations of Ca were associated with enhanced heat tolerance and resistance." "The results showed that the content of individual metals in spores is flexible within a very wide range and is dependent on the relative concentration of the particular metal in the growth medium." The thermosensitivity studies were somewhat inconclusive but tended to suggest "...that calcium is essential for highest thermal resistance of spores."

In 1971, Bronner et al. established that there is an active uptake system for calcium in sporulating *Bacillus megaterium*. Its cellular localization remains unknown. Subsequently two groups, in 1975, Tsuchiya and Rosen and Silver et al., demonstrated that under normal growth conditions, bacteria actively extrude calcium by a mechanism resembling that used in mitochondria. Although there are yet only a few examples from bacteria and lower plants, it appears that the maintenance of a low cytosolic Ca^{2+} ion concentration is a characteristic of all cells, prokaryote or eukaryotic.

Ordal (1977) suggested that the direction of flagellar rotation is controlled by cytosolic calcium concentration, with the switch occurring about pCa 7.5. A report on actin- and myosinlike proteins in the bacteria (Nakamura and Watanabe, 1978) promised more messenger functions for calcium in prokaryotes.

Quite obviously, an experiment in which one adds or removes an ion to the extracellular medium and then observes a cellular response gives, at best, only an indirect suggestion as to the effect of the test ion in the cytosol. Nature has provided another approach to the problem: antibiotics that function as ionophores. An ionophore facilitates the diffusion of an ion, usually a metal monovalent or divalent cation (Me^+ or Me^{2+}) from an aqueous phase into a low-dielectric environment or through a membrane. Most ionophores (I or I^-) are highly lipid soluble. At the lipid–water interface, they complex the cation, displacing all or most of its water of hydration. The ionophore–cation complex then diffuses across the membrane as the cation complex (I^*M^+) or as the neutral complex (I^-M^+ or $I^{2-}M^{2+}$). On the other side of the membrane, the metal cation is hydrated and the ionophore diffuses back to the original side, either with an alternative cation (electroneutral exchange, diffusion, usually involving I^-) or without (electrogenic exchange usually involving I). Other natural antibiotics, and related synthetics, form channels or pores in membranes. These pores permit the selective diffusion of specific cations, wholly or partially dehydrated.

Valinomycin was the first antibiotic to be recognized to function as an ionophore (Moore and Pressman, 1964). This neutral, cyclic decapeptide has a K^+/Na^+ preference of 10^4 : 1. In 1951, Berger et al. isolated three antimicrobially active molecules, one each from three previously unidentified strains of *Steptomyces*: X-206, X-464 (product, later called nigericin), and X-537A. Johnson et al. (1970) determined the crystal structure of X-537A as $Ba(C_{34}H_{53}O_8)_2 \cdot H_2O$ and suggested that it is a general ionophore for divalent cations. In 1972, Reed and Hardy characterized A23187, from *Streptomyces chartreusensis*, as a rather specific Ca^{2+} or Mg^{2+} ionophore.

Experimentalists can use such ionophores to equilibrate the ion content of the intracellular medium with the concentration that they set in the external medium. Initially, it was not appreciated that these ionophores will also "transport" ions to or from any intracellular stores, such as mitochondria and endoplasmic reticulum. The challenge for chemists is to understand the nature of the ion selectivity and to synthesize more specific ionophores and ion channels to be used in ion-selective electrodes as well as in biological experiments.

It is certainly true that the longer one works with any biological system, the more one is impressed with the significance of membranes and their abilities to establish and maintain selective compartmentalization. Calcium has specific interactions with membrane-embedded channels, ionophores, enzymes, and pumps. There also exists the possibility that calcium may affect the general properties of membranes by interacting directly with the phospholipid component. The 1966 study of Bangham and Papahadjopoulos was one of the first to explore this possibility: "...phosphatidylserine interacts with Ca^{2+} in the presence of physiological concentrations of univalent ions ...." "...a structural modification of the Ca^{2+} phosphatidylserine complex occurs at approx. 1 mM Ca^{2+} when approx. 1 equiv. Ca^{2+} is bound on the surface for each phosphotidylserine molecule." Such calcium synthetic phospholipid bilayer interactions have now been explored in great detail. It is still not known what the physiological implications of these effects are, either on cytosolic surfaces, where $[Ca^{2+}]$ seldom exceeds 10^{-5} M, or on external surfaces, where the bathing $[Ca^{2+}]$ exceeds 10^{-3} M.

1.12. HORMONES

Given these diverse functions of calcium, one is hardly surprised to find that the body has developed elaborate mechanisms for controlling serum levels of calcium. Although anatomists and embryologists such as Remak and Owen had previously referred to "accessory thyroids," Sandström (1880) was the first to describe their anatomy and histology in humans and other mammals. "...I suggest therefore employing for these the designation glandulae parathyreodiea; a name in which the relationship of the accessory gland to the thyroid is expressed, in the same way as by analogy parovium, paradidymis leads the thoughts to images of an embryonic nature." In fact, both the development and function of the parathyroid are quite different from those of the thyroid.

"...föreslår jag derföre att för dessa använda benämningen Glandualæ parathyreoideæ; ett namn, hvari deras egenskap af bikörtlar till sköldörteln fått sitt uttryck, på samma gång som det genom analogierna parovarium, paradidymis leder tanken på bildningar af en embryonal natur."

Gley (1891) rediscovered the small glands and argued that it was their unnoticed removal that was responsible for the "incessant muscular twitches and clonic convulsions" that usually brought death within a day of total thyroidectomy. Incorrectly, he, too, speculated that "the structure of this organ is analogous to the body of the thyroide in an embryonic state."

"secousses musculaires incessantes et convulsions cloniques"

La structure de cet organe est analogue à celle du corps thyroide à l'état embryónnair.''

Moussu (1898) first claimed an active glycerine extract to relieve postoperative tetany in dogs: "I have indicated the facts and experiments that have forced me to acknowledge the existence of distinct functions of the thyroid and of the parathyroid." A real demonstration of the parathyroid hormone would await Collip's work in 1925. Vassole and Generali (1900) described the effects of parathyroidectomy in much greater detail. "In the dog and in the cat, the abolition of the function of the parathyroid (complete parathyroidectomy) gives rise to a mortal tetany ...." The characteristics included fibrillary contractions and muscular spasms, psychial depression, rigid and uncertain gait, anorexia, tachycardia, rapid emaciation, and fall in body temperature. About this time, Loeb (1901) reported "on an apparently new form of muscular irritability by anions liable to form insoluble calcium compounds." This and other bits of evidence inspired the experiments of MacCallum and Voegtlin (1909) "on the relation of tetany to the parathyroid glands and to calcium metabolism." Parathyroidectomized animals have lowered calcium contents in their tissues and elevated calcium in their urine; their symptoms can be relieved by injection of calcium. Collip in 1925 succeeded in "the extraction of a parathyroid hormone which will prevent or control parathyroid tetany and which regulates the level of blood calcium." He summarized his findings as follows: "1. An extract has been made from the parathyroid glands of oxen by the use of which parathyroid tetany in dogs can be prevented or controlled. 2. The active principle in this extract produces its effect by causing the calcium content of the blood serum to be restored within normal limits." "A rise in the level of blood calcium in the normal dog has been observed following the injection of parathyroid extract." What controls the synthesis and release of this protein, and subsequently how it mobilizes calcium and phosphate from bone, are still under investigation. Additional homeostatic systems are involved.

Many urban children in temperate zones suffered from rickets through the first quarter of the twentieth century. McCollum et al. at John Hopkins wrote in 1920: "During the past 15 years studies made on the metabolism of children suffering from rickets has made it reasonably certain that the administration of cod liver oil alters the calcium balance in such a manner that calcium will be retained by the body. He suggested that "... some substance or substances in the oil cause calcium to be deposited in the same fashion in which deposition occurs in spontaneous healing of rachitis in man."

In London, Mellanby (1921) was reaching similar conclusions: "... it will be generally admitted that experimental results on animals and clinical experience are opposed to calcium deficiency as being the main cause of rickets." "The following conditions tend to prevent rickets in puppies: ... (2) Something associated with certain fats probably identical with the fat soluble vitamine." Fresh air and sunlight seemed to help. Hume and Smith (1923) examined "the effect of air, which has been exposed to the radiations of the mercury-vapour quartz lamp, in promoting the growth of rats, fed on a diet deficient in fat-soluble vitamins." They concluded that "... it is the air and not some property impressed on the glass jars which is active." Their general idea was right, but the controls and conclusions were wrong. By 1924, two groups, apparently unaware of one another's work, had a better grasp of the effect. Steenbock and Black wrote: "By irradiation with the quartz mercury vapour lamp, rat rations can be activated, making them growth-promoting and bone-calcifiying, to the same degree as when the rats are irradiated directly." "It suggests itself that, in ultimate analysis, both light and the antirachitic vitamin may represent the same antirachitic agent—possibly a form of radiant energy." Hess and Weinstock came a bit closer: "It was found that cottonseed oil and linseed oil [and green vegetables] could be rendered specifically active by this means. ..." "... An antirachitic factor therefore had been produced in vitro and outside the living organism." Work then proceeded on the characterization of and conversion of vitamin D as well as its general site of action. In 1937, Nicolaysen separated the phosphorus from the calcium effects: "... the action of vitamin D, in the gut is confined to a direct action on the absorption of Ca. The well-known reduced absorption of P in vitamin D deficiency is due to a precipitation by the increased amount of Ca in the bowel." The effects of 1,25-dihydroxycholecalciferol on protein synthesis and calcium transport are discussed in Chapter 12.

For some years, PTH and vitamin D seemed to account for calcium metabolism. Quite reasonably, McLean (1957) suggested a negative feedback control: "The setting of the regulator of the parathyroid glands is fixed under physiological conditions; it is raised in hyperparathyroidism and lowered in hypoparathyroidism." P. H. Sanderson et al. (1960) examined "calcium and phosphorus homeostasis in the parathyroidectomized dog" and concluded that "...the surviving dogs became adapted to existence without any parathyroid tissue whatsoever." Their results, which remain unexplained, implied "...that the parathyroid glands play an important part in dealing with acute disturbances of the serum calcium level. However, a secondary mechanism, capable of more sluggish regulation, seems to be revealed when the parathyroid glands are removed." Rasmussen (1961) agreed with McLean that "...on the basis of the present evidence, it appears that the primary factor regulating the secretory activity of the parathyroid glands is the calcium ion activity of the plasma." However, "if, as McLean postulates, the bone were the only means of regulating the Ca^{++} of plasma in conjuction with the parathyroids, the resulting feedback system would lead to wide oscillations in the level of Ca^{++} in the plasma." He then reasonably, but mistakedly, suggested that "the kidney is an organ admirably suited, both anatomically and physiologically, to the task of responding rapidly to minor fluctuations of parathyroid concentrations." Copp (1964) reviewed his "...evidence that hypercalcemia stimulates release from the glands of a fast-acting hormone that lowers plasma calcium. This hormone, would account for the lack of oscillation in the feedback mechanism and for the control of hypercalcemia by Sanderson et al." It remained for G. B. Foster et al. (1964) to establish the "thyroid origin of calcitonin": "Perfusion of the parathyroid glands in isolation from thyroid tissue is not possible in the dog, but this can be accomplished in the goat." "The results establish the thyroid as the source of this new hormone." Now, finally, when we thought that we had all of the components of calcium homeostasis, the most intriguing of all is just being explored. DiSaia (1966) and Kerber et al. (1968) reported fetal anomalies possibly correlated with warfarin administration during pregnancy. Postmenopausal osteoporosis is a pathological condition of bone characterized by excessive loss of calcium due to resorptive processes; vitamin K treatment has significantly reduced the negative calcium balance (Tomita, 1971). In 1975, Pettifor and Benson reported "...three additional cases of nasal hypoplasia, associated in two with skeletal abnormalities not previously described; it may be that these anomalies are related to anticoagulant [of the coumarin group] therapy during pregnancy." Independently, Warkany (1975) reviewed various clinical reports and answered yes to his title "A Warfarin Embryopathy?"

In 1974, three groups reported γ-carboxyglutamic acid (Gla) as a component of prothrombin (Chapter 13). It was soon found that all the vitamin K–dependent clotting factors contain Gla as a component essential to their calcium binding and subsequent function. Hauschka et al. (1975) and Price et al. (1976) characterized a bone protein, osteoclastin, that contains Gla. It is the seventh most abundant protein in vertebrates. At a calcium/osteoclastin molar ratio of

1 000:1, osteoclastin strongly inhibits the transition of brushite ($CaHPO_4 \cdot 2H_2O$) to hydroxyapatite [$Ca_5(PO_4)_3OH$]. Vitamin K is somehow involved in the control of mineralization in vertebrates and thereby affects calcium balance.

1.13. MEASUREMENT

As is so often the case in science, advanced understanding depends on improved physical chemical concepts and techniques—and, of course, the biological questions drive the physical research. The addition of two protons and two neutrons to argon ($^21s^22s^62p^23s^63p$) does not make its chemistry all that much more interesting. There is no 3*d* character to any of its orbitals. It has no visible or ultraviolet absorption spectrum. The interactions of the cation are purely ionic. The two positive charges are within a volume defined by the crystal radius of 0.99 Å. Strangely enough, those very characteristics that make the Ca^{2+} ion so dull for the chemist probably account for its unique role in biology. Most of our understanding of calcium chemistry, which really means calcium coordination (Chapter 3), has come from general studies of inorganic chemistry. Historically, most of the chemistry devoted specifically to calcium has been concerned with calcium analysis or calcium buffering. Many calcium salts are insoluble. The quantity of calcium present as CaO can be determined in such precipitates by ashing and weighing if the precipitating anion is volatile and if no other nonvolatile cations are present. Přibram (1871) first exploited the insolubility of calcium oxalate in an article translated as "A New Method for Determination of Calcium and Phosphoric Acid in Blood Serum." After centifuging out the cells and adding ammonium: "Since no precipitation of calcium phosphate occurred, which as such cannot easily exist in alkaline serum, ammonium oxalate was added whereupon clouding follows immediately." For over half a century, precipation by oxalate was the first step in calcium analysis. Richards et al. (1901) removed coprecipitating magnesium oxalate by washing the oxalate precipitate with ammonium oxalate. The method of Kramer and Tisdall (1921), as optimized by Clark and Collip (1925), was to be the standard for 30 years. The washed and acid-dissolved calcium oxalate precipitate was titrated with potassium permangenate; oxidation of the oxalate was followed by clearing of the blue permangenate.

"Eine neue Methode zur Bestimmung des Kalkes and der Phosphorsäure in Blutserum"

"Da sich keine Fallung von Calcium phosphat ergab, das als solches wohl Überhaupt in alkalischen Serum nicht leicht vorhanden sein kann, so wurde Ammoniumoxalat zugesetzt, worauf sogleich Trübung erfolgt."

There were numerous, less satisfactory variations on this basic theme of oxalate precipitation and permanganate titration:

Ligand	Reference
sulfate (EtOH ppt).	Aron, 1907
ferric thiocyanate	Marriot and Howland, 1916
alizarinate	Laidlaw and Payne, 1922
picrolonate (indicator: methylene blue)	Bolliger, 1935

Calcium-specific dyes were also developed:

stearate (turbidity)	Lyman, 1917
alizarinate	Laidlaw and Payne, 1922
sulforicinicum (turbidity)	Rona and Kleinmann, 1923
8-hydroxypinoline	Yoshimatsu, 1929
murexide	Schwarzenbach and Gysling, 1949
eriochrome black T	Gilbert and Fenn, 1957
glyoxal bis(*t*-hydroxyanil)	Goldstein and Stark-Mayer, 1958
"calcichrome"	Close and West, 1960
phthalein complexan	Herrero-Lancina and West, 1963
kalcion urea	Bezdekova and Budesinsky, 1965
antipyrylazo III	Budesinsky, 1974

These methods all exhibit disadvantages. Vines (1921) tried to use the blood coagulation reaction. "The principle of the method is the recalcification of oxalate blood by the addition of the material to be tested, and comparison of the action of the latter with that of calcium chloride solutions of known strength." Their results were not reproducible. Two valuable bioassays were developed to measure the free, as opposed to total, calcium ion. McLean and Hastings (1934) turned to the problem of a direct method, or "a biological method for the estimation of calcium ion concentration." They measured the amplitude of contraction of frog heart ventricle as a function of calcium ion, or test solution, in the bathing solution. One of the early successes was recorded in 1939 by Joseph, who described a $PbHg/PbCaO_4-CaC_2O_4$ electrode sensitive to free Ca^{2+} ions. "It has been found to come to equilibrium in 1 hour or less, and to yield stable, reproducible potentials."

By far the most significant bioassay was introduced by Shimomura et al. (1962) in the paper "Extraction, Purification and Properties of Aequorin, a Bioluminescent Protein from the Luminous Hydromedusan, *Aequorea*." This coelenterase, 28,000 Da, emits light when exposed to $[Ca^{2+}]$ over 10^{-7} M. It is quite stable, calcium specific, and nontoxic. The physical techniques now available for measuring total and free calcium and for evaluating the effect of calcium on ligands are discussed in Chapter 6.

Even though the oxalate–permanganate technique is tedious, it is accurate and it certainly yields the total calcium content of a fluid such as serum. The initial ammonium precipitation releases all or most of the protein-bound calcium. The distinction between free and total calcium was appreciated early. Přibram (1871) anticipated protein involvement: "Physiological consequences of the deposition of calcium in the bones, the passage of calcium and phosphoric acid in the urine, etc. can be deduced only if it is established that the calcium is bound to protein." Sabbatani (1904) argued the interdependence of the various body stores of calcium and suggested that citrate prevents coagulation by forming a soluble complex with calcium. Rona and Takahashi (1911) first determined the nondiffusable, or protein-bound, calcium: "From the general result of this investigation it undoubtedly appears that a significant amount of calcium (about 25 to 35%) exists in serum in a nondiffusable form."

"Physiologishe Folgerungen über die Ablagerung des Kalkes in den Knocken, den Übergang des Kalkes and des Phosphorsäure in den Harn etc. werden sich erst ableiten lassen, wenn es feststeht, dass der Kalk an Eiweiss gebunden ist"

"Aus der Gesamtheit dieser Versuche geht unzweifelhaft hervor dass eine nennenswerte Menge des Calciums (etwa 25 bis 35%) in Serum in nicht diffusible Form vorhanden ist."

In 1915, Van Slyke and Bosworth used a Pasteur–Chamberland filtering tube to demonstrate that "calcium caseinate (casein Ca_4) and dicalcium phosphate ($CaHPO_4$) are not in combination." In 1921, von Meysenburg reported a careful study of dialysis of serum through colloidin membranes. He concluded that there is "...no alteration in the percentage of diffusible calcium of the serum..." between pH 7.0 and 7.6 or between rachitic or tetanic dogs and normal. "The diffusible calcium of the serum of normal men and dogs was found to comprise from 60 to 70 per cent of the total serum calcium."

Of course, it was appreciated that the concept of "nondiffusable" calcium was valid only for very slowly exchanging calcium. Greenberg and Gunther (1930) found that half of the serum calcium was retained by a colloiden membrane. They admitted that their results "...were not completely theoretically interpretable" but nonetheless, were consistent and reasonable.

In 1926, Greenwald offered an incorrect but stimulating and testable idea. "It is suggested that the calcium content of the plasma is, normally, maintained at a constant level by equilibration between inorganic calcium and an organic compound of calcium. Resemblances between this organic compound and calcium citrate are indicated, but the substances are not identical. It is suggested that the parathyroid hormone is necessary to the preparation of this organic constituent."

McLean and Hastings (1935) applied their frog heart assay to a study of calcium metabolism. They demonstrated by graded additions that "The frog's heart is insensitive to CaCit and Cit_3." They refuted Greenwald's suggestion that "agreement of the values for the mass law constant obtained from observations upon purified serum proteins and protein containing fluids indicates that it is unnecessary to postulate any appreciable amount of other calcium-binding substances in human fluids." Total serum calcium corresponded to values published by other workers: 9.0 to 11.5 mg per 100 mL of serum in normal and rachitic adults, 12.0 to 16.0 in hypoparathyroidism and multimyeloma, and 4.0 to 8.5 in hypoparathyroidism and infantile tetany. Most important, the level of free calcium ion could be expressed in terms of a nomograph varying total serum calcium and total serum protein. The average protein dissociation constant $K_{dis} = [Pr] \times [Ca]/[PrCa]$. They were justified in referring to the serum content of $[Ca^{2+}]$ as "... one of nature's physiological constants."

To distinguish free from total calcium within the cell was a greater challenge. Pollack (1928) injected alizarin sulfonate into amoebae. Unfortunately, alizarin sulfonate is moderately toxic; the resulting red crystals form slowly, are difficult to see, and do not represent a quantitative response. His failure defined the three criteria of a good cytosolic indicator. The microinjection of arsenazo III, antipyrylazo III, and aequorin have proven reasonably satisfactory. The development of calcium-specific microelectrodes is progressing well (Chapter 6).

It is also valuable to determine the cellular or organelle distribution of total calcium. G. H. Scott's microincineration method (1932) gave an accurate spatial distribution of the total metal distribution within the cell. His principle (should have) guided many subsequent refinements. "It is necessary to use a fixative which neither removes nor increases the inorganic elements of the tissue." The limits of spatial resolution of electroprobe analysis now approach 100 μm.

1.14. BIOMINERALIZATION: REDUX

Having left bone a century ago, we might now ask what progress has been made. Much of this research can be considered in terms of physiological and histological studies on the dynamics of bone deposition and resorption and the compositions and solubilities of the bone salts. Papillon (1870) examined the dynamics and "... the limits and the variations of physiological determinism." These "... demonstrate that one can substitute a certain quantity of strontium, of magnesium, of aluminium for the calcium normally contained in bone."

"... les limites et les variations du déterminitisme physiologique." His experiments "... démontrent que l'on peut substituer une certaine quantité de stroniane, de magnésie, d'alumine á la chaux normalement contemu dans les os."

"En résumé, l'ostéoclaste, cellule géante á noyau unique et géante à noyaux multiple, ressortissant à la classe des cellules de nature connective, a un protoplasma commun rempli d'un nombre incroyable de mitochondries, sans préjudice des vacuoles à lipoïdes et des vacuoles colorables par le rouge neuter."

Dubreuil (1910) was one of the first workers to examine bone histology and to implicate the mitochondrion in bone metabolism. "In summary, the osteoclast, giant cell with a single giant nucleus or multiple nuclei, arose from the class of cells of connective nature, and has a protoplasm completely filled with an incredible number of mitochondria, without damage to the lipoidic vacuoles and the vacuole that can be coloured by neutral red."

In his 1911 Harvey Lecture, Wells emphasized an important distinction. "In calcification we have deposited in dead tissues, or in tissues of low vitality, a considerable quantity of inorganic calcium salts, which appear at first in granular form, although later there may be more or less fusion and resulting areas of homogeneity." "In normal ossification, however, the homogeneous calcium deposits are closely related to living cells, which not only determine the form of the deposits, but which also are able to dissolve the insoluble salts or to cause their deposition as may be needed, thus rendering the inorganic salts of bone the reserve supply of a tissue of active metabolism...."

Of course, for centuries it had been realized that calcium and phosphate were the primary inorganic constituents. What is the nature of this mineral? In 1931, Bogert and Hastings suggested that "the chief inorganic consituent of bone is probably a crystalline salt, $CaCO_3 \cdot nCa_3(PO_4)_2$, where n approximates the value 2 in untreated bone." In 1937, Huggins determined the Ca/P ratio as 1 : 0.6, with variation for diet, bone type, and stage of development. He argued that $CaCO_3$ is of little significance. McConnell (1938) surveyed the wide range of compositions and unit cell dimensions available for apatite-type minerals and concluded that they were all approximately isomorphous with the fluorapatite ($Ca_{10}P_6O_{24}F_2$) structure proposed in 1930 independently by Náray-Szabó and by Mehmel. The cations Mg^{2+}, Mn^{2+}, Sr^{2+} K^+, and Na^+ could replace Ca^{2+} to varying extents; $AlO_4{}^{3-}$, $YO_4{}^{3-}$, $SO_4{}^{2-}$, and $CO_3{}^{2-}$ could partially replace $PO_4{}^{3-}$; and Cl^- and OH^- could replace F^-. Neuman and Neuman (1953) summarized the evidence supporting an apatite structure, $Ca_5(PO_4)_3$(OH or F), in bone. They explored the seemingly innocent, yet still not understood subject of calcium phosphate solubility. "It has been shown that the K_{sp} of $CaHPO_4$ must be exceeded for precipitation to occur, yet calcification occurs in individuals whose blood levels of calcium and phosphate are well below this critical product."

Levinskas and Neuman (1955) continued their solubility studies of synthetic hydroxyapatite. "This crystalline solid exhibited incongruent solubility, i.e. the solutions gave a higher Ca : P ratio than that present in the solid phase in most instances." "The most significant finding was that solubility varied with the amount of solid phase added to a given volume of solution. This often observed phenomenon, in the present instance, is real evidence that the hydroxylapatite lattice exhibits incongruent solubility; it does not conform to a solubility product. This is not surprising since the solid phase is not of constant composition. The common ion effect supports this conclusion. There was no simple relation between calcium and phosphate concentration after equilibrium. In contradiction to the conclusions of Neuman and Neuman, McLean and Budy (1959) concluded that "the body fluids are supersaturated with respect to the final product hydroxyapatite. It is certain that the crystal domains are so small, the surface to volume ratio is so enormous, the crystalline disorder so great, and the contact with organic components so extensive that our traditional concepts of crystallinity and solubility are inadequate." Various polypeptides, in particular the vitamin K–dependent osteocalcin (Hauschka et al., 1975; Price et al., 1976), which stabilize supersaturated solutions of calcium phosphate, might also alter the apparent equilibrium.

Taves (1963) determined the crystal structure $Ca_4H(PO_4)_3$ and noted its similarity to that of $Ca_5(PO_4)_3OH$. "This view of the two crystals suggests an explanation for the observation that octacalcium phosphate grows more rapidly than hydroxyapatite, even though the latter is more stable" (Chapters 3 and 4). Apatite is well named, α'πατα'ω, "I deceive."

Even when we do understand the solubility of calcium phosphate, we will still be ignorant of the cell's contribution. The abundance of phosphatase in ossifying cartilage led Robison in 1923 to suggest that the hydrolysis of "hexosephosphoric esters" resulted in increased local concentrations of phosphate and resulting calcium phosphate precipitation. There is little supporting evidence for the many variations of this postulate. The preformed precipitate and/or packet idea of Lehninger (1970) still motivates much research; yet confirmation is still lacking. A popular idea is that some organic component, collagen being the leading candidate, provides a nucleation site. Why all collagen in contact with serum does not nucleate remains a mystery. Yet no matter whether one argues selective solubilization or selective nucleation, the problem remains: How does the cell cheat equilibrium thermodynamics in the extracellular environment?

Ossification, after Well's usage, is certainly not limited to the vertebrate skeleton. For example, filtrates of cultures of *Leptothrix buccalis*, an organism always present in dental tartar, does not induce precipitation of saturated solutions of calcium phosphate, whereas filtrates from macerated cells do. In 1925, Bulleid concluded that "it is therefore probable that the precipitating action of the *Leptothrix* is due to some property in the actual organism, not in any way to the production of an extotoxin." In 1967, Ennever and Creamer summarized their own (Ennever, 1960) and others' work supporting the interpretation that calcum phosphate crystals are formed intracellularly in *Bacterionema matruchottii*. This implies that if, in fact, bacteria do extrude calcium from the cytosol, some bacteria have special vacuoles for calcium acccumulation.

Algae appear to have such organelles. Watabe (1967) described his electron microscopic observations. "Coccoliths are simple or complex aggregates of microcrystals of calcite [$CaCO_3$] covering the cells of algae belonging to the Coccolithineae." "...the coccoliths of *Hymenomonas* (*Cricosphacra*) are found in vacuoles [within the cytoplasm] the outlines of which do not differ greatly from the coccolith-forming region of *Coccolithus* (*huxleyi*)." Pienaar (1971) continued: "It was found that the Golgi body was directly involved in the production of the baseplate scale, the organic matrix membrane and the deposition of calcium carbonate to produce the coccolith." Such ossification by primitive organisms had already provided new insights to students of evolution (Margulis, 1975). Biochemists will find them equally rewarding.

It is natural that early investigators should seek reasons for the unique distributions and functions of the metal ions of biological fluids. Quinton (1900) first noted a correspondence that has since intrigued physiologists: "Osmosis establishes, from the point of view of minerals, a remarkable communication between the internal milieu of the marine invertebrate and the external milieu." "...in the natural states a great number of animals have as their internal milieus for minerals that same milieu as the ocean itself."

"L'osmose établit donc, au point da vue minéral, une communication remarquable entre le milieu intérieur de l'Invertébré marin et le milieu extérier." He then suggested: "... a l'état de nature, le plus grand nombre des organismes animaux a pour milieu interieur, ou point de vue minéral, le milieu marin lui-même."

Independently MacCallum (1910) observed: "The inorganic composition of these Medusæe, on this view, reflects to a certain extent the composition of the ocean water, not only of today but also of past, perhaps very remote, geological periods." Our bones not only provide unique challenges to contemporary cell biologists, but also bear witness to our evolution.

Lowenstam (1981) summarized "minerals formed by organisms": "four calcium carbonates, four calcium phosphates, two calcium oxalates, three calcium sulphates, calcium fluorite (all 14 of these calcium minerals with amorphous forms), as well as silica ($SiO_2 \cdot nH_2O$), five ferrous/ferric oxides, two pyrites, and manganese oxide. We have yet to understand how cells convert ions to stones. However, one can safely postulate that the mechanisms involved in regulating biomineralization and in avoiding intracellular precipitation of calcium salts preceded the evolution of the use of calcium for information transduction."

2

PHYSIOLOGICAL PROCESSES INVOLVING CALCIUM BINDING PROTEINS

As discussed in Chapter 1, calcium is unique among all of the inorganic metal ions and anions in its role as a secondary messenger in cell signaling. Although calcium binding proteins serve many functions and represent many homolog families, those that are most studied are involved in information transduction.

Nearly all normal intracellular and probably most extracellular processes involving calcium result in its interaction(s) with a protein(s) (reviewed by Kretsinger, 1976; Falke et al., 1994). There are few, if any, physiological interactions of calcium with nucleic acids, metabolites, or the carbohydrates of cell surfaces. One exception may be its interactions with the phospholipid headgroups of membranes. As discussed in Chapter 12, there are reasonable suggestions that a bridging Ca^{2+} ion may be coordinated by both phospholipids and annexins or C2 domains. The second exception may involve biominerals, such as shells or bones. All calcium biominerals that have been examined carefully contain specific proteins that are involved in the initial nucleation events and/or maintaining the stability of the biomineral. They may also contain small organic molecules. Calcium physiology involves calcium and protein interactions. In this chapter we summarize physiological processes that involve calcium; all involve calcium binding proteins.

Calcium Binding Proteins, By Eugene A. Permyakov and Robert H. Kretsinger

2.1. CALCIUM AS A SECONDARY MESSENGER

The concentration gradient of Ca^{2+} ions between extracellular and intracellular fluids far exceeds the gradients of Na^+, K^+, and Mg^{2+}. In extracellular fluids of metazoans, the concentration of free Ca^{2+} ion is about 2×10^{-3} M; within cells the free calcium concentration is about 10^{-8} M (Dipolo et al., 1976; Becker et al., 1980; O'Doherty et al., 1980; Fabiato, 1981). Such high concentration gradients across cell membranes are maintained by Ca-ATPase pumps. The energy invested in generating and maintaining this gradient is ultimately realized in information transduction.

Inside the cell, calcium is stored in multiple ways (reviewed by Mooren and Kinne, 1998). In vertebrate cells, calcium is usually stored, free and bound, in intracellular pools with exchange rates from seconds to hours or days. These slowly exchanging pools are localized in mitochondria or acidic compartments such as endosomes, lysosomes, and the trans-Golgi network. Rapidly exchanging calcium stores are found in the microsomal fraction, which consists mainly of *endoplasmic reticulum* (ER). The ER is a signaling organelle that controls a wide range of cellular processes, such as the entry and release of calcium, sterol biosynthesis, apoptosis, and the release of arachidonic acid (reviewed by Berridge, 2001, 2002). ER and its analog in muscle cells, *sarcoplasmic reticulum* (SR), are the main calcium reservoirs in most cells. The calcium storage capacity of ER depends on the intraluminal calcium binding proteins calsequestrin and calreticulin (Chapter 11). Calsequestrin is the main intraluminal component of SR in skeletal and cardiac muscle, whereas calreticulin occurs mainly in smooth muscle SR and nonmuscle ER.

Calcium is released from intracellular stores by two different routes: the inositol-1,4,5-triphosphate (IP_3) receptor–linked channel and the ryanodine receptor–linked channel. Binding of IP_3, which is produced by the G-protein-activated phospholipase C, triggers the release of calcium from internal stores. The ryanodine receptor is a cation-selective channel with high conductance values for both monovalent and divalent cations. Ryanodine, 0.01 to 10 μM, stimulates calcium release. Several proteins (i.e., annexin VI, S-100, calpain, and immunophilin FK506 binding protein) have been shown to be involved in the functioning of ryanodine receptors. These channels may also be opened by depolarization.

Plasma membrane calcium channels are classified according to their gating mechanisms. In response to an external stimulus, the concentration of low intracellular free calcium may increase 10- to 100-fold (O'Doherty et al., 1980; Blackmore et al., 1982; Rasmussen and Waisman, 1981). Calcium entry can be activated by voltage changes, by binding of some exogenous or endogenous transmitters, or simply by membrane stretch. The calcium flowing through these channels constitutes elementary events of calcium signaling. The changes in intracellular calcium concentration may be uniform throughout the entire cell or may be located in certain areas. These spatial and temporal patterns can vary depending on the nature of both the cell and the stimulus. Each cell also contains its own set of calcium binding proteins, for which the calcium dissociation constants, K_d(Ca), are within the range 10^{-6} to 10^{-8} M (reviewed by Heizmann and Hunziker, 1991). If these proteins bound with lower affinity, they would not bind calcium during stimulation; if with higher affinity, they would not release calcium during rest. The binding of calcium induces changes in the conformation of these proteins; this in turn changes the character of their interactions with other cell systems and triggers various cell processes; they provide the critical link in information transduction (reviewed by Ikura, 1996; Kretsinger, 1997).

Ca^{2+} ions in these systems are intracellular or second(ary) messengers (Kretsinger and Nelson, 1976; Rasmussen and Waisman, 1981). The extracellular stimulus, the first messenger, may be, for example, a hormone or neurotransmitter that interacts with receptors on the external surface of a cell. From the external surface of a cell the signal passes into the cell; this results in the entry of calcium from the extracellular fluid or its release from an internal store. At the same time, calcium itself may liberate calcium from these stores, thus adding one step to the signaling cascade, as if the liberated calcium were a "third" messenger. Calcium can act within milliseconds in highly localized regions, or it can act much more slowly as a global wave that spreads the signal throughout the cell (reviewed by Berridge, 2001, Berridge et al., 1999). Various pumps and exchangers are responsible for returning the elevated levels of calcium back to the resting state. Mitochondria also play a critical role in this process; they help the recovery process by taking calcium from the cytoplasm. Alterations in the ebb and flow of calcium through the mitochondria can lead to cell death. Figure 2.1 shows a calcium wave through a fertilizing maize egg.

Repetitive calcium spikes were first observed in agonist-stimulated hepatocytes and have now been observed in many cell types. Repetitive spikes arise from

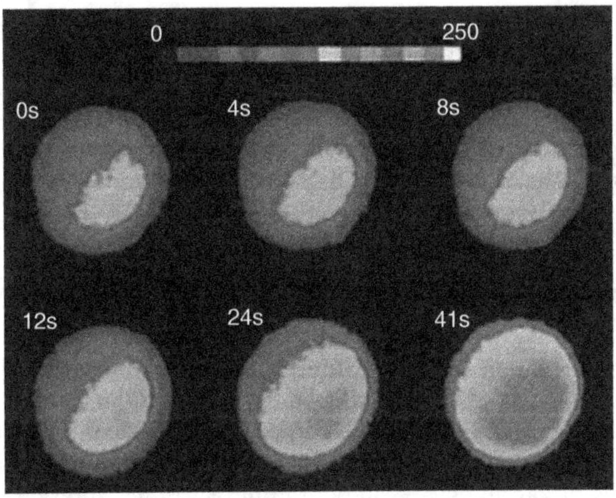

Figure 2.1. Calcium wave through a fertilized maize egg. (From Digonnet et al., 1997.) (*See insert for color representation of figure.*)

the periodic opening of plasma membrane channels (membrane oscillators) or from the periodic emptying of stores (cytosolic oscillators). Calcium waves can initiate at one point and spread across the cytoplasm (reviewed by Berridge, 2009). Sometimes, they form complex spirals. The propagation of the wave is based on positive feedback, in which calcium increases at the initiation site and diffuses to vicinal stores to activate them to release calcium, which in turn diffuses farther outward to trigger additional stores (calcium induced calcium release).

Slow (i.e., 0.2 to 2 $\mu m\ s^{-1}$) and ultraslow calcium waves have important morphogenic effects (reviewed by Jaffe, 2008). Both may be propagated by cycles in which the entry of calcium through the plasma membrane induces subsurface contraction. This contraction opens nearby stretch-sensitive channels. Calcium entry through these channels propagates the calcium wave.

The most commonly encountered form of calcium oscillations involves baseline transients or spikes. Usually, baseline spiking continues for at least a few cycles in the absence of extracellular calcium; thus, it represents cycles of discharge and reuptake of calcium by intracellular stores (reviewed by Putney and Bird, 2008). Woods et al. (1987) proposed a model in which the cellular level of IP_3 rises in spikes and the resulting changes in $[Ca]_{cyt}$ reflect these oscillations in IP_3. $InsP_3$ has two important functions in generating calcium oscillations (reviewed by Berridge, 2007): It releases calcium from the internal store, and it contributes to calcium entry. A hypothesis has been developed to describe a mechanism for calcium oscillations; it suggests that agonist concentration regulates oscillator frequency. The main idea is that the $InsP_3$ receptors are sensitized to release calcium periodically by cyclical fluctuations of calcium within the lumen of the endoplasmic reticulum. Each time a pulse of calcium is released, the luminal level of calcium declines and has to be replenished before the $InsP_3$ receptors are resensitized to deliver the next pulse of calcium. Variations in agonist-induced entry of external calcium, which can occur through different mechanisms, determine the variable rates of store loading responsible for adjusting the sensitivity of the $InsP_3$ receptors to produce the periodic pulses of calcium. The calcium oscillator is an effective analog-to-digital converter in that variations in the concentration of the external stimulus are translated into a change in oscillator frequency (Berridge, 2007).

The second proposal for the mechanism of baseline $[Ca]_{cyt}$ spiking involves fluctuations in $[Ca]_{cyt}$ while cellular levels of IP_3 are constant (Berridge, 1990). Such mechanisms require that the IP_3 receptor has the capacity to be regulated both positively and negatively by downstream signals, presumably by calcium itself. In the absence of extracellular calcium, oscillations are not maintained, indicating that some mechanism of calcium entry across the plasma membrane is required for their maintenance (Kawanishi et al., 1989).

Microdomains of calcium, which are formed at sites where calcium enters the cytoplasm either at the cell surface or at the internal stores, can combine to produce patterns of varying shape and duration (Berridge, 2006). These localized domains of calcium can regulate specific cellular processes in different regions of the cell. They are especially evident in neurons, where both pre- and postsynaptic events are controlled by highly localized pulses of calcium. The ability of single neurons to process enormous amounts of information depends on such miniaturization of the calcium-signaling system. Control of cardiac cell contraction and gene transcription provides another example of how the parallel processing of calcium signaling can occur through microdomains of intracellular calcium. Phasic smooth muscle cells (e.g., vas deferens, uterus, bladder) rely on membrane depolarization to drive calcium influx across the plasma membrane (Berridge, 2008). This depolarization can be induced by neurotransmitters or through the operation of a membrane oscillator. Many tonic smooth muscle cells (e.g., vascular airways, corpus cavernosum) are driven by a cytosolic calcium oscillator that generates periodic pulses of calcium. A similar oscillator is present in pacemaker cells such as the interstitial cells of Cajal and atypical smooth muscle cells that control other tonic smooth muscle cells (e.g., intestine, urethra, ureter). The changes in membrane potential induced by these cytosolic oscillators do not drive contraction directly but function to couple individual oscillators to provide this synchronization.

2.1.1. Contraction

Contraction refers to the movement of an entire cell relative to its environment or relative to another cell. *Transport* refers to the movement of organelles or molecular complexes within a cell; it depends on the expenditure of energy and does not rely solely on diffusion. Often, contraction and transport are interrelated.

Calcium plays a pivotal role in the regulation of contraction of all types of muscles. As in most cells, the free calcium concentration in a resting muscle cell is about 10^{-8} M. When an external signal stimulates the cell, the $[Ca]_{cyt}$ increases by one to two orders of magnitude. This messenger calcium is either released from the sarcoplasmic reticulum and/or flows down the concentration gradient from the extracellular space.

Within the SR, calcium is bound by calsequestrin, calnexin, and/or calreticulin (Chapter 11). These "sponges" permit the SR to sequester lots of calcium without increasing the concentration of the free Ca^{2+} ion to the level of precipitating calcium phosphate. In response to an electrical signal, the membrane of the SR becomes permeable to Ca^{2+} ions, and they diffuse into the sarcoplasm. The SR of all eukaryotic cells contains calmodulin, which is a calcium-dependent regulator of the activities of many enzyme systems (Chapter 11). Myosin is the main component of the thick filaments of myofibrils; its light chains bind calcium. Actin-based thin filaments bind the heterotrimer, troponin. The binding of calcium to troponin C (TnC) triggers the interaction of actin and myosin and subsequent contraction. Following contraction, calcium is pumped up a concentration gradient into the SR by the Ca^{2+}, Mg^{2+}-ATPase (Ca-ATPase) embedded in the membrane of the SR.

Myosin ATPase–Based Systems Force in biological systems is usually produced by motor proteins such as myosin, kinesin, and dynein; they convert chemical energy of ATP into mechanical energy (reviewed by Hackney, 1996). Myosin generates movement along actin filaments and functions in muscle contraction, organelle movement, and cytokinesis. ATP hydrolysis provides forces on the order of 5 to 10 pN and displacements between 5 and 10 nm (Svoboda et al., 1993; Finer et al., 1994). High resolution x-ray diffraction patterns of muscle fibrils reveal structural changes in contracting muscles that are interpreted in terms of molecular movements that underlie force generation (reviewed by Bershitsky et al., 2009). The initial stereo-specific locking of myosin heads onto the actin thin filaments is followed by tilting the orientation of the myosin head relative to its thick filament.

The myosin head has a large globular region known as the *catalytic domain*, which interacts with actin and ATP. A second region of the myosin head, known as the *light chain domain*, consists of a single α-helix, stabilized by two light chains that surround it. Both of these light chains consist of four EF-hand domains. The essential (enzymatic) light chains (ELCs) do not bind calcium, whereas the regulatory light chains (RLCs) bind one equivalent of calcium independent of their association with the heavy chains (Werber, 1978). The binding of calcium to RLCs triggers the contraction of scallop muscles (Kendrick-Jones and Jakes, 1977; Scholey et al., 1981). The role of the myosin light chains in these muscles is to switch on and off the binding of the cross-bridge to actin in response to the changing calcium levels within the muscle.

In both smooth and skeletal muscles, there are myosin light chain kinases (MLCKs) that catalyze the rapid incorporation of phosphate into a specific Ser of some light chains (reviewed by Stull et al., 1980). Dephosphorylation of these light chains is catalyzed by another enzyme, myosin light chain phosphatase. This class of light chains is referred to as the *phosphorylatable* or *P-light chains*. The phosphorylation of the myosin P-light chains controls contraction in smooth muscles. The calcium sensitivity of the MLCK is mediated by calmodulin. Calcium first binds to calmodulin; calci-calmodulin, in turn, binds to the inactive MLCK to form an active phosphotransferase. There is a close correlation between the degree of phosphorylation of the myosin light chains and tension development. The inhibition of MLCK results in dephosphorylation of the myosin light chains and inactivation of tension even in the presence of calcium. No such effects are known for striated muscle fibers (Kerrick et al., 1980).

Dynein ATPase–Based Systems Dynein and kinesin hydrolyze ATP and produce movements along microtubules. A conformational change in the motor domain of these proteins is proposed to produce a small power stroke that is amplified into a much larger movement by a long lever arm between the motor domain and the cargo binding domain (reviewed by Amos, 2008). Energy for these processes is supplied primarily by the hydrolysis of ATP (Hwang and Lang, 2009). Dynamic conformational changes in the motor domain can be regarded as controlling the flow of free energy to and from the surrounding heat reservoir. Multiple motor domains are organized in distinct ways to maintain motility under local physical constraints. A single dynein may run several millimeters along a microtubule in vitro, with typical velocities on the order of a few nanometers per second (reviewed by Serohijos et al., 2009). The average time between steps is a fraction of a second. Two dimeric, microtubule based motor proteins, kinesin-1 and cytoplasmic dynein, can take over 100 steps in vitro without detaching from the track. Such movements are essential for many different cellular and developmental functions, including organelle movement, localization of developmental determinants, mitosis, meiosis, and possibly long-range signaling in neurons. These movements allow them to shuttle attached cargoes over long distances, from a cell's center to its plasma membrane. Such continuous movement requires regulation and coordination to prevent premature microtubule dissociation and futile cycling of ATP that does not give rise to productive steps.

Dyneins consist of up to 15 different polypeptides, with a total mass 1 to 2 mDa. These complexes function as molecular motors and move the cargo to which they are attached toward the minus end of a microtubule (reviewed by Harrison and King, 2000). Dynein motor function is a property of its heavy chains; other subunits are involved in attachment to the appropriate cargo. The motor and cargo binding activities are tightly controlled by its light chains. The LC8 light chain is highly conserved and found in many enzyme systems. Like myosin, dynein and kinesin systems can be regulated via phosphorylation of light chains by a calmodulin regulated phosphorylase.

Spasmin-Based Systems Contractile systems based on myosin or dynein utilize the energy derived from the hydrolysis of ATP as an integral part of the contractile event. In contrast, the contraction of spasmin is based on a large change in conformation associated with the binding of calcium (Routledge, 1978). The cell expends ATP energy prior to contraction by pumping out Ca^{2+} ions to establish the calcium gradient that is involved in both information and energy transduction. Spasmin (Chapter 11) is the major component of a calcium-induced contractile filament, called the *spasmoneme*, found in vorticellid ciliates. Such filaments have not been observed in any organisms other than green algae.

2.1.2. Exocytosis

In simple exocytosis the membrane surrounding the exocytotic vesicle fuses with the cell membrane, and the contents of the vesicle diffuse into the extracellular environment. In a topological sense, endocytosis represents the reverse process, with special extracellular receptors and their catch ending up on the inner membrane of the endocytotic vesicle. In compound exocytosis, for which the adrenal medulla is a favorite example, multiple endocytotic vesicles fuse, mixing their contents, prior to exocytosis. The cytosol and the contents of these vesicles or of the extracellular space are always separated by a phospholipid bilayer. Calcium plays an important role in secretory cells, regulating both the biosynthesis of secretory proteins and the vesicular traffic through the Golgi apparatus in the process of formation of secretory granules. Microtubules provide directionality to the movement of the granules toward their site of secretion. Microtubule and granule associated proteins provide sites of anchorage for actin, another calcium binding protein, with movement being generated by a sliding filament mechanism, analogous to the contraction of skeletal muscle. The $[Ca^{2+}]$ of the vesicles is about 2.0 mM, as is the extracellular fluid. Further, some vesicles or the lumen of the endoplasmic reticulum, from which the vesicles pinched off, have many low-affinity, high-capacity calcium binding proteins, such as calsequestrin. For this reason the total cell content, per volume, of calcium may exceed the concentration of extracellular calcium, all the while maintaining $[Ca]_{cyt} < 3 \times 10^{-7}$ M.

The consensus model of glucose-induced insulin secretion involves the acceleration of metabolism, closure of ATP-sensitive K^+ channels, depolarization, influx of calcium, and increase in $[Ca]_{cyt}$ (reviewed by Henquin, 2009). In parallel, glucose activates a metabolic, amplifying pathway that does not raise $[Ca]_{cyt}$ further but augments the action of triggering calcium on exocytosis initially triggered by increased $[Ca]_{cyt}$. Initial insulin secretion is triggered by the rise in $[Ca]_{cyt}$ that occurs synchronously in all β-cells of every islet in response to a sudden increase in the glucose concentration. Its time course and duration are shaped by those of the calcium signal, and its amplitude is modulated by the magnitude of the rise in $[Ca]_{cyt}$ and, substantially, by amplifying mechanisms. During the second phase, synchronous calcium oscillations in all β-cells of an individual islet induce pulsatile insulin secretion, but these features of the signal and response are dampened in groups of intrinsically asynchronous islets. Glucose has hardly any influence on the amplitude of calcium oscillations and controls mainly the time course of a triggering signal. All the processes in which calcium participates are implemented via calcium binding proteins.

2.1.3. Metabolism

The properties of 70 subfamilies of EF-hand proteins are summarized in Chapter 11. Most are involved in the transduction of second messenger information; this ultimately involves covalent bond formation and/or cleavage. However, only a few are themselves enzymes; with the exception of aequorin (EC 1.13.12.5), these enzymes are chimeric proteins. The catalytic residues are in the non-EF-hand portions. Those with protein substrates include calcium-dependent protein kinase (EC 2.7.1.37), protein phosphatase (EC 3.1.3.16), and calpain (EC 3.4.122.17), an intracellular protease. Those with small-molecule substrates include diacylglycerol kinase (EC 2.7.1.107), glycerol-3-phosphate dehydrogenase, and phospholipase C (EC 3.1.4.3).

None of the annexins or synaptotagmins (Chapter 12) are themselves enzymes, nor have any been shown to interact directly with enzymes. Several phospholipases C and phospholipases A_2 are calcium regulated, as are calcium ATPases (Chapter 12). However, relatively few intracellular enzymes are directly regulated by calcium.

Many extracellular enzymes, including those in the blood-clotting cascade (Chapter 13), bind calcium and are inactive without calcium. However, their activities are not regulated by calcium.

2.1.4. Control of Gene Expression

Many pathways link calcium signaling to gene expression and especially to transcription (reviewed by Ikura et al., 2002). Calmodulin regulates various proteins, including calmodulin-dependent protein kinases, calcineurin, the nuclear factor of activated T-cells (NF-AT), and activator protein 1 (AP-1)—all of which are involved in the regulation of transcription. When calcium concentration increases within the cell, calci-calmodulin binds and activates its target kinases, resulting in transcriptional activation through phosphorylation of transcriptional factors. Calcineurin binds directly to the transcriptional complex NF-AT1; this enables calcineurin to dephosphorylate Ser's within the Ser–Pro repeats and the Ser-rich regions of NF-ATs. This prohibits interaction with the two nuclear localization sequences required for nuclear import. The activation of calcineurin follows calcium binding to the integral subunit calcineurin B (four EF-hands) and to calmodulin. S100B (two EF-hands) regulates the tumor suppressor protein, p53, which functions as a transcriptional activator of numerous genes (Méplan et al., 1999). Another calcium binding protein, calsenilin (DREAM or KChlP3, four EF-hands) is a calcium-regulated transcriptional repressor; it binds directly to DNA downstream from the promoter sequence.

2.2. CALCIUM BUFFERS

Any protein, or small molecule, in the cytosol that binds calcium with a log K_d(Ca) value of 6.0 to 8.0 is a calcium buffer. Most of these proteins (e.g., calmodulin) function directly in cell-signaling pathways. However, a few, such as parvalbumin, appear to shape, over time, the pulse of calcium (reviewed by Permyakov, 2007); others, such as intestinal calcium binding protein, are involved in facilitated diffusion (reviewed by Permyakov, 2009). Organelles such as mitochondia or SR serve similar buffering functions (reviewed by Mooren and Kinne, 1998; Carafoli et al., 2001; Carafoli, 2002).

2.3. CALCIUM PUMPS AND CHANNELS

2.3.1. Pumps

Calcium gradients across the plasma membrane and across the membranes of intracellular stores are generated and maintained by Ca-ATPase pumps or by Na^+/Ca^{2+} exchangers (reviewed by Stokes and Green, 2003). There are two different homolog families of Ca-ATPase pumps: those from the endoplasmic reticulum and those from the plasma membrane. Two equivalents of calcium are transported per equivalent of ATP hydrolyzed by the ER pump, and one equivalent is transported by the plasma membrane Ca-ATPase. The Na^+/Ca^{2+} exchanger has a stoichiometry of 3 Na^+ : 1 Ca^{2+}. Ca-ATPase of SR is a large multidomain protein consisting of a transmembrane domain, a phosphorylation domain, a nucleotide binding domain, and a transduction domain. The mechanism by which the energy of ATP is coupled to calcium transport includes large-scale conformational changes (Chapter 10).

Calcium signaling is seen as calcium waves, oscillations, or sparks within the cytosol. These vary in frequency, amplitude, and spatial distribution. Calcium-induced calcium release from ER-derived vesicles is an essential aspect of wave propagation; however, above a certain concentration, negative feedback causes a reduction in calcium concentration. These patterns vary with cell type and stimulation (Berridge, 2006, 2007, 2008, 2009).

2.3.2. Channels

Calcium channels are selectively permeable to Ca^{2+} ions. Voltage-gated calcium channels are found in excitable cells (e.g., muscle, glial cells, neurons) (reviewed by Cribbs, 2006; Kato et al., 2009). At resting membrane potential (-60 to -70 mV in neurones), voltage-gated calcium channels are normally closed. They are opened at depolarized membrane potentials (threshold potential about -55 mV). Activation of specific channels allows calcium entry into the cell, which, depending on the cell type, results in muscular contraction, excitation of neurons, up-regulation of gene expression, or exocytosis. High-voltage-gated calcium channels include the neural N-type channel blocked by ω-conotoxin, the R-type channel (resistant to the other blockers and toxins) involved in some processes in the brain, the closely related P/Q-type channel blocked by ω-agatoxins, and the dihydropyridine-sensitive L-type channels responsible for excitation–contraction coupling of skeletal, smooth, and cardiac muscle and for hormone secretion in endocrine cells.

Ligand-gated calcium channels are opened or closed in response to the binding of a chemical messenger (reviewed by Balla, 2009). Two different ligand-gated channels, inositol-1,4,5-triphosphate (IP_3) and ryanodine receptor linked, are involved in the release of calcium from intracellular vesicles. Binding of IP_3, which is produced by the G-protein-activated phospholipase C, triggers the release of calcium from internal stores.

2.4. MITOCHONDRIA

The mitochondrion and the chloroplast originated by endosymbiosis, the uptake of one bacterium by another (Margulis, 1975). The engulfed bacterium, surrounded by a layer of cell membrane invaginated from the host bacterium, survived and reproduced within its host. Over evolutionary time, many genes were moved to the cell nucleus or deleted from descendents of the engulfed bacterium, and it devolved absolutely on the host to perform most metabolic and regulatory functions. In turn, the engulfed bacterium performed some metabolic functions for the host. The mitochondrion specialized in oxidative phosphorylation; the chloroplast, in photosynthesis. Both organelles are enclosed within a double membrane.

The outer membrane, which faces the cytosol, retains similar composition and directionality with the cell membrane. The inner membrane resembles the inner membrane of the engulfed bacterium. The intermembrane space is topologically outside the cytosol and topologically outside the lumen, or matrix, of the mitochondrion. The inner mitochondrial membrane extrudes calcium to the intermembrane space, maintaining $pCa_{mito} \sim 7.5$. The calcium in the intermembrane space is inferred to leak into the cytosol, from which it is actively extruded into SR-derived vesicles or into the extracellular space.

Mitochondria can influence calcium signaling indirectly by changing the concentration of ATP, NAD(P)H, pyruvate, and reactive oxygen species; this, in turn, modulates components of the calcium-signaling machinery: that is, buffering, release from internal stores, influx from the extracellular solution, uptake into cellular organelles, and extrusion by plasma membrane–calcium pumps (reviewed by Walsh et al., 2009). Mitochondria can directly influence the calcium concentration in the cytosol of the cell by importing calcium via the mitochondrial Ca^{2+} uniporter or transporting calcium from the interior of the organelle into the cytosol by means of Na^+/Ca^{2+} or H^+/Ca^{2+} exchangers.

Several events are related to the increase in mitochondrial calcium, including regulation and activation of several calcium-dependent enzymes, such as phospholipases, proteases, and nucleases. Mitochondria and ER play pivotal roles in the maintenance of intracellular calcium homeostasis and regulation of cell death (reviewed by Smaili et al., 2009). During most cell-signaling events the matrix of the mitochondrion probably does not experience any increase in calcium concentration. However, it is possible that in exceptional physiological situations such as apoptosis (programmed cell death), the calcium concentration within the matrix rises significantly, thereby changing the activity of such enzymes as protein kinase C, calcineurin, calpains, and caspases (reviewed by Rizzuto et al., 2003). Isolated functional mitochondria certainly change their permeability, leading to a complete loss of potential and respiratory function when exposed to prolonged elevated levels of calcium. Similar effects may obtain in chloroplasts.

2.5. EUBACTERIA

The anatomy and physiology of Archae are not as well studied as are those of Eubacteria; however, most of the following generalizations probably apply equally well to both kingdoms (Eukaryotes being the third). Bacteria are surrounded by a single phospholipid bilayer membrane, although the lipid composition varies widely among bacteria and differs from that of eukaryotes. In gram (stain)-negative bacteria there is a second, complex membrane or wall that surrounds the inner or normal membrane. This outer membrane protects a gram-negative bacterium from osmotic shock and prevents the passage of some large molecules; however, it appears to be porous to small molecules and ions, such as Ca^{2+}. In most bacteria there are no internal membrane-bound vesicles or plastids; the DNA is not enclosed within a membrane. Bacteria extrude calcium to $[Ca]_{in} \sim 3 \times 10^{-8}$ M, as do eukaryotes. There is no evidence that bacteria use pulses of inflowing calcium to transmit information. Further, no calcium-dependent enzymes have been characterized. However, at least 16 bacterial proteins (Table 2.1) containing EF-hands have been found in several bacteria (reviewed by Michiels et al., 2002; Dominguez, 2004). Apart from calerythrin and calsymin, the functions of most of these EF-hand proteins from bacteria are hypothetical or unknown. Calerythrin might function as a calcium buffer or transporter (reviewed by Michiels et al., 2002); and calsymin might function as a tranducer of information between the bacterium *Rhizobium etli* and a host plant.

Note that pCa (or pH) 7.0 is difficult to characterize in a small volume: for example, within a bacterium or in a projection at the terminus of a dendrite $\sim 1.0\ \mu m$ in diameter $\rightarrow (10^{-6})^{-3}\ m^3$ volume.

$$
\begin{aligned}
1.0\ M &= 6.2 \times 10^{23}\ \text{atoms}\ 10^{-3}\ cm^{-3} && \text{definition of a mole in terms}\\
10^{-7}\ M &= 6.2 \times 10^{16}\ \text{atoms}\ 10^{-3}\ cm^{-3} && \text{of Avogadro's number}\\
10^{-7}\ M &= 6.2 \times 10^{19}\ \text{atoms}\ m^{-3}\\
10^{-7}\ M &= 6.2 \times 10^{1}\ \text{atoms}\ \mu m^{-3}
\end{aligned}
$$

TABLE 2.1. Bacterial Proteins Containing EF-Hands

Protein Name	Number of EF-Hands
Calsymin	6 or more
Calerythrin	4
SC6F11.09	4
SCJ33.05c	4
CC1180	4
SC10F4.20	3
CabA	3
Asp24	3
MLL5457	3
CC2752	3
CC2226	2
SC3D11.21	2
TAL_SYNY3	2
CC2193	2
MLR9645	2
AE0048271	2

In small volumes at pCa $\sim$ 8, pumps are looking for single ions; statistical fluctuations become important.

2.6. CALCIUM AND EXTRACELLULAR PROTEINS

Extracellular calcium binding proteins represent many homolog families. Unlike the intracellular calcium binding proteins, most of which function in information transduction, extracellular calcium binding proteins appear not to act as on/off switches because they are nearly constant at pCa $\sim$ 2.8. Several calcium binding proteins are associated with the extracellar matrix; these appear not to be enzymes but are stabilized by calcium. The second group, representing several different homolog families, has the characteristic posttranslational modification of Glu to γ-carboxyglutamic acid (Gla). Many of these are involved in blood clotting; a few, in biomineralization. The third group consists of many enzymes; all of which are stabilized by calcium, in some of which calcium in involved in the catalytic mechanism (Chapter 13).

2.6.1. Extracellular Matrix

The extracellular matrix is a network built up from a variety of proteins and proteoglycans (reviewed by Maurer and Hohenester, 1997). Its interaction with cells is mediated by cell adhesion molecules. Ca^{2+} ions are essential for maintenance of these structures; many matrix–matrix, cell–matrix, and cell–cell contacts involve calcium. Most extracellular matrix proteins are multidomain proteins assembled from a limited set of autonomous, independently folded domains. These extracellular calcium binding modules can be divided into two groups: Calcium can either be bound to a single domain or can mediate interactions between domains or between a domain and a sugar. Direct involvement of calcium in ligand binding is observed in the C-type lectin domain and the serum amyloid P component and also in the interaction of Gla (γ-carboxyglutamic acid) domains with phospholipid membranes. In some cases calcium acts indirectly, as when calcium binding induces a conformation that is required for enzymatic activity or a specific binding activity. Examples include matrix metalloproteinases, which are active only in the calcium-bound state, and BM-40, which binds collagens only in the presence of calcium. The independently folded domains of E-cadherin change their relative orientations upon binding calcium.

2.6.2. Blood Clotting Cascade

Blood coagulates very slowly in the absence of calcium: for example, when sequestered by the preservative citrate. The several vitamin K–dependent proteins involved in coagulation have K_d(Ca) values of 10^{-4} to 10^{-3} M (reviewed by Lorand, 2005; Smith, 2009). All of them contain from 10 to 12 Gla residues per molecule. The binding of calcium to these proteins induces their interaction with membranes; this is important in the coagulation process (Nelsestuen, 1984). Blood coagulation consists of one enzyme activating the next zymogen in the sequence. The culmination of this cascade is the conversion of fibrinogen to fibrin and the formation of blood clots. The cascade model of coagulation divides coagulation into extrinsic and intrinsic pathways. The factor VIIa/tissue factor complex of the extrinsic system is the major component of homeostasis in vivo. The intrinsic pathway (the contact system) does not play a physiological role in homeostasis. The cascade model has been replaced by a cell based model with three overlapping phases: initiation, amplification, and propagation (reviewed by Vine, 2009).

Although factors XI and XII of the intrinsic system are not involved in homeostasis, they may play a key role in abnormal homeostasis or thrombosis. The pathways of coagulation and inflammation are intertwined at numerous points. Thrombin and the pro-coagulant factors VIIa and Xa can activate members of the protease-activated receptor family, which play an important role in coagulation, inflammation, and vascular homeostasis. Factor V plays a crucial role in both the pro- and anticoagulant systems.

These Gla proteins are normally soluble in plasma and require acidic phospholipids for membrane binding. Since acidic phospholipids are rare or absent from the exterior surface of the plasma membrane, these proteins do not adhere significantly to intact cells in the circulation. Exposure of cytosolic membranes that are rich in acidic phospholipids provides an excellent surface for protein binding.

2.6.3. Other Extracellular Enzymes

Many extracellular enzymes bind calcium. This calcium is essential for stability and function; however, changes in $[Ca]_{ext}$ do not appear to be involved in the regulation of their activities. Among them are phospholipase A_2, staphylococcal nuclease, thermolysin, trypsin, chymotrypsin, calpain, and lipase. The binding of calcium to the enzymes might be functional in three ways. First, Ca^{2+} ions can stabilize an intermediate in the active site, as in phospholipase A_2 and staphylococcal nuclease. Second, calcium can stabilize the enzyme at high temperatures, as in thermolysin. Third, calcium can take part in the activation of a zymogen, as in trypsin, phospholipase A_2, and calpain.

2.7. BIOMINERALIZATION

Biomineralization refers to the formation by living cells of solids consisting solely or primarily of inorganic components, primarily calcium but also iron and magnesium (Lowenstam, 1981). The skeleton, teeth, and otoconia are normally the only mineralized tissues or organs in the human body. Physiological biomineralization occurs in collagenous as well as noncollagenous matrices. Usually, biominerals consist of amorphous compounds or small microcrystalline domains embedded in an amorphous matrix.

Skeletons are formed in a wide variety of shapes, sizes, and compositions of organic and mineral components (reviewed by Omelon et al., 2009; Wuthier, 1984). Many invertebrate skeletons are constructed from carbonate or silicate minerals, whereas vertebrate skeletons, which are continually rebuilt, repaired, and resorbed during growth and normal remodeling, are, instead, composed of a calcium phosphate mineral known as apatite. Silica biomineralization occurs, for example, in diatoms and silica sponges. In both calcification and silicification, genetically controlled organic macromolecules induce the formation of composite organic–inorganic nanoparticles, which behave as templates for subsequent assembly of the nanoparticles into micro- to macroarchitectures of complex pattern, and eventually, are mostly reabsorbed (reviewed by Bonucci, 2009). It is thought that the control of bone and calcifying cartilage mineralization is associated with phosphate-cleaving proteins. Skeletal mineralization is also associated with noncrystalline calcium- and phosphate-containing electron-dense granules that have been detected in vertebrate skeletal tissue prepared under nonaqueous conditions. Omelon et al. (2009) proposed that the electron-dense granules contain polyphosphates, which may play an important role in apatite biomineralization. The enzymatic formation (condensation) and destruction (hydrolytic degradation) of polyphosphates offer a simple mechanism for enzymatic control of phosphate accumulation and the relative saturation of apatite.

Many organisms first form a disordered mineral phase (reviewed by Weiner, 2008). Only when it is in place and has adopted its appropriate shape is it induced to crystallize. A unique group of proteins rich in Asp are involved in controlling mineral formation.

Although the matrix accounts for less than 5% w/w, it controls the synthesis of transient amorphous minerals and evolution to crystalline phases, the choice of the calcium carbonate polymorph (calcite vs. aragonite), and the organization of crystallites in complex shell microstructures (reviewed by Marin et al., 2008). Until recently, control of the biomineralization of molluskan shells was considered to consist of two antagonistic mechanisms: crystal nucleation and growth inhibition. Recent concepts and models incorporate a wide variety of shell proteins that have a broad range of pI values, from very acidic to very basic. The primary structures of several of these proteins are composed of different modules, suggesting that the proteins are multifunctional. Some of them exhibit enzymatic activities; others may be involved in cell signaling. The variety of ions existing in the mineralizing milieu may play an active role in regulating the precipitation process and nature of biogenic apatites (reviewed by Aoba et al., 2003).

Osteocalcin is a small, secreted bone protein that participates in the regulation of biomineralization through its large acidic and phosphorylated propeptide (Laizé et al., 2006). Osteopontin, bone sialoprotein, and bone acidic glycoprotein-75 are three acidic phosphoproteins that have been isolated from the mineralized phase of bone matrix, are synthesized by osteoblastic cells, and are generally restricted in their distribution to calcified tissues (Gorski, 1992; Hunter et al., 1994). These proteins are characterized by Asp plus Glu contents of 30 to 36%, and each contains multiple phosphoryl and sialyl groups. These proteins bind calcium and induce nucleation of calcium hydroxyapatite crystal formation.

The extracellular matrix protein asporin binds collagen type I. This binding is inhibited by recombinant asporin fragment LRR (leucine-rich repeat) 10 to 12 and by full-length decorin, but not by biglycan. Kalamajski et al. (2009) found that the poly-Asp domain binds calcium and regulates hydroxyapatite formation in vitro. Many other calcium binding proteins, yet to be fully characterized, take part in biomineralization.

Lowenstam (1981) in "Minerals Formed by Organisms" described four calcium phosphates, four calcium carbonates, two calcium oxalates, three calcium sulfates, one calcium fluorite, silica ($SiO_2 \cdot nH_2O$), five ferrous/ferric oxides, two pyrites (FeS_2), and manganese oxide that have been identified in the biosphere. The physical properties of these biominerals differ from those of their inorganic counterparts; even though the crystallographic unit cells are identical. These differences appear to reflect differences in domain sizes and orientations, which in turn reflect the incorporation of other cations and ions as well as proteins into the biomineral. The cellular and molecular mechanisms of biomineralization are poorly understood; however, it has been demonstrated, or inferred, that specific proteins are involved in the process and are incorporated into the final mineral (reviewed by Weiner, 2008).

The concentrations of salts of extracellular fluids of most metazoans are similar, but not identical, to those of contemporary oceans (Table 2.2). These serum concentrations are well regulated, especially those of [Ca^{2+}]; as McLean and Hastings (1935) remarked: In the context of human physiology, serum calcium is "... one of nature's physiological constants." In fact, there are variances in reported concentrations of all of these ions because it is difficult to distinguish free from total ions in the presence of other ions and ligands and because concentrations of both total and free ions vary with physiological conditions. LeGeros (1991) warned that commercial preparations of various calcium phosphates should be checked by (powder) x-ray diffraction and infrared absorption spectroscopy for both purity and crystal domain size.

The rough similarity of ocean and extracellular concentrations is assumed to reflect the environment of early cells (Table 2.3). Further, the intracellular concentrations of these salts are similar in both eukaryotes and prokaryotes.

TABLE 2.2. Concentrations of Free Ion (Sigel et al., 2005)

	Ocean	Serum	Cytosol (mM)
Na^+	470.0	144.0	12.0
K^+	10.0	4.22	139.0
Mg^{2+}	52.8	1.0	0.60
Ca^{2+}	10.3	1.25	0.00008
Cl^-	546.0	116.0	4.0

TABLE 2.3. Calcium Phosphate Crystal Forms

	K_{sp}	Space Group	a (Å)	b (Å)	c (Å)
Brushite, $CaHPO_4 \cdot H_2O$	$10^{-6.69}$ M^3	C2/c	5.18	15.18	6.2
Monetite, $CaHPO_4$	—	P1	6.91	6.63	6.99
OCP, $Ca_8H_2(PO_4)_6 \cdot 5H_2O$	$10^{-46.93}$ M^{16}	P1	19.87	9.63	6.87
Whitlock, TCP, $Ca_3(PO_4)_2$	$10^{-29.55}$ M^5	R3c	10.43	10.43	37.38
Hydroxyapatite, $Ca_5(PO_4)_3OH$	$10^{-56.47}$ M^9	P63/m	9.42	9.42	6.88

Calcium is the main exception to this generalization among the more abundant naturally occurring ions. Whereas ocean concentrations of the free Ca^{2+} ion are ~1.2 mM, the intracellular concentrations in both eukaryotes and prokaryotes are ~80 nM: that is, a gradient of $\sim 1.5 \times 10^4$. There is some variation in concentrations of free Ca^{2+} and free $CO_3{}^{2-}$. There are subtle and important variations in both $[Ca^{2+}]$ and $[CO_3{}^{2-}]$ with ocean depth and temperature and region of the ocean; however, most oceans are at or near saturation in calcium carbonate.

Most calcium salts are much less soluble than their potassium, sodium, or magnesium analogs. This is surely the reason that most, but not all, biominerals are calcium based (Lowenstein, 1981). Calcium phosphates form several salts: brushite ($CaHPO_4 \cdot H_2O$), octacalcium phosphate [$Ca_8(HPO_4)_2(PO_4)_4 \cdot 5H_2O$], whitlock [$Ca_3(PO_4)_2$], and least soluble of all at pH 7, hydroxyapatite [$Ca_5(PO_4)_3OH$] (Table 2.4). This crystallization process is complex in vitro. Ideally, the dissolution of five equivalents of Ca^{2+}, three of $PO_4{}^{3-}$, and one of OH^- would yield a constant ΔG until the solubility product ($K_s = [Ca^{2+}]^5 \cdot [PO_4^{3-}]^3 \cdot [OH] \sim 10^{-60}$ M^9) was reached, and then hydroxyapatite precipitate would form. In practice, a supersaturated solution is formed (Figure 2.2). The initial precipitate is brushite; this then "ripens" into hydroxyapatite. The relative stabilities vary with pH, as reflected in the incorporation of H^+ or OH^- ions in the crystal lattice. The extent and time of supersaturation are not well defined, nor is the time required for the transformation from brushite to hydroxyapatite. The process in vivo is even more complex. Trace amounts of fluoride, F^-, can substitute for a small fraction of OH^-, making the apatite even less soluble (and tooth enamel harder). Tribasic phosphate ($PO_4{}^{3-}$), as it appears in K_s, is only a small fraction of total phosphate, the important parameter for the organism. As illustrated in Figure 2.3a saturated solution of calcium phosphate contains much more total phosphate at pH 5 ($\sim 5 \times 10^{-3}$ M total phosphate at 10^{-4} $[Ca^{2+}]$) than it does at pH 7 ($\sim 7 \times 10^{-6}$ M total phosphate at 10^{-4} $[Ca^{2+}]$). This increased solubility in acidic conditions is exploited by organisms that form hydroxyapatite crystals (and underlies the admonition against drinking too many acidic beverages). Perhaps most important, all of these (calcium) biominerals contain specific calcium binding proteins that probably affect the

TABLE 2.4. Concentration of Tribasic Phosphate (PO_4^{3-}), at Saturation[a]

	pCa							
pOH	1	2	3	4	5	6	7	8
6			13			8		
7	16 (4.8×10^{-11})			11 (4.8×10^{-6})			6 (4.8×10^{-1})	
8		14			9			4
9	15.33 (1.4×10^{-6})		12	10.33 (1.4×10^{-1})		7		
10				10			5	

$p[PO_4^{3-}]$

$\Sigma[PO_4^{3-}] + [HPO_4^{2-}] + [H_2PO^{4-}] + [H_3PO_4]$

[a]Calculated from the solubility constant $K_s = [Ca^{2+}]^5 \cdot [PO_4^{3-}]^3 \cdot [OH] = 10^{-60}\ M^9$) for various combinations $[Ca^{2+}]$ and of [OH]. Note: pCa $= -\log[Ca^{2+}]$; pOH $= -\log[OH^-] = 14 -$ pH; $p[PO_4^{3-}] = -\log[PO_4^{3-}]$. The values in parentheses are the concentrations of total phosphate at the given values of pCa and pOH.

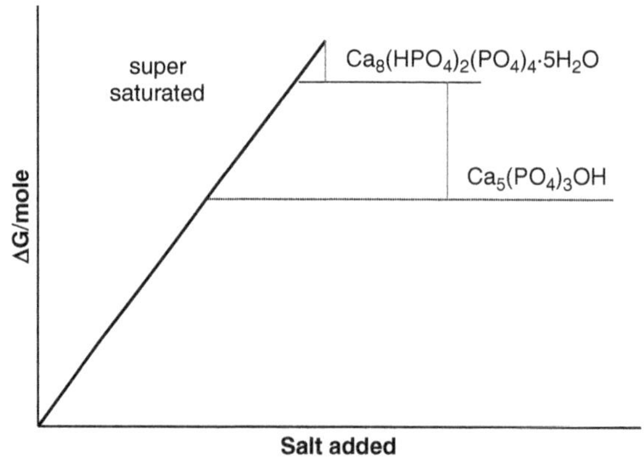

Figure 2.2. Representation of the change in free energy, ΔG, as a function of "salt" added, in this case Ca^{2+} and/or PO_4^{3-} and/or OH^- up to saturation (at a given temperature and pressure). Ideally, upon reaching saturation a precipitate (hydroxyapatite) would form with no further release or uptake of free energy. As indicated, many solutions can become supersaturated and semistable for a long time. When supersaturated calcium phosphate does precipitate, it can form several different crystal forms (e.g., octacalcium phosphate); these ultimately "ripen" to the most stable form, hydroxyapatite.

stabilities of supersaturated solutions as well as the domain sizes and physical characteristics of the resulting minerals.

The energy and information economy of most, if not all, cells is based on phosphorylated compounds, which, like phosphate itself, are poorly soluble in the presence of calcium. For instance, at pH 7 and an ocean level $[Ca^{2+}] \sim 10.0$ mM, total phosphate is limited to $\sim 10^{-10}$ M before precipitating. All cells, including

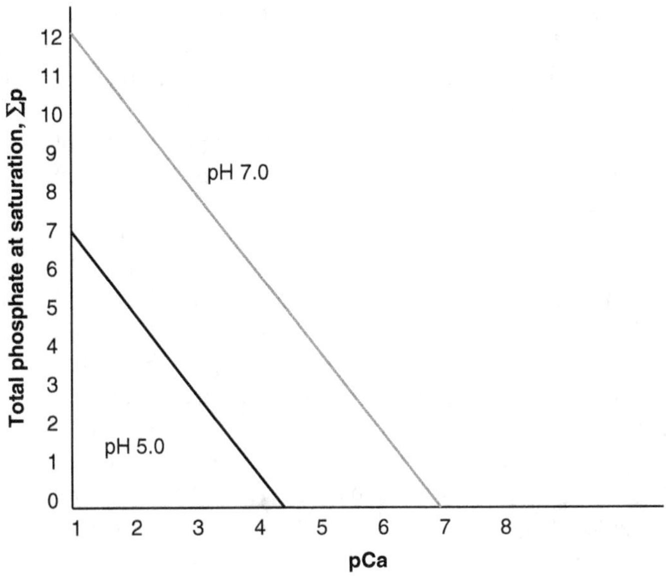

Figure 2.3. The concentration of total phosphate ($[PO_4^{3-}] + HPO_4^{2-}] + H_2PO^{4-}] + [H_3PO_4]$) is shown at pH 7.0 (blue line) and pH 5.0 (red line) as a function of pCa. (*See insert for color representation of figure.*)

eubacteria and archae, have evolved a mechanism for extruding calcium; this is essential to a phosphate-based economy (Kretsinger, 1990). Apparently, eukaryotes exploited the resulting $> 10^4$ concentration gradient to transmit information. Each pulse of messenger calcium, with $[Ca]_{cyt}$ rising briefly, and locally, from $\sim 10^{-7}$ to $\sim 10^{-5}$ M, risks precipitation and possible cell death. To what extent cells rely on the brief stability of supersaturation (Figure 2.2) is unknown.

Bacteria also extrude calcium. Calcium binding proteins, including many with EF-hands (Table 2.1), have been identified; however, to date, there is no solid evidence of calcium's functioning as a second messenger in prokaryotes (Permyakov and Kretsinger, 2008).

2.8. CALCIUM AND VIRUSES

Viruses perturb the calcium homeostasis of cells and cellular calcium binding proteins to their benefit in their replication cycles (reviewed by Zhou et al., 2009). Virus related elevations of cytosolic calcium levels allow increased viral protein expression (HIV-1, HSV-1/2), viral replication [HBx, enterovirus 2B, HTLV-1 p12(I), HHV-8, EBV], virion maturation (rotavirus), virus release (enterovirus 2B), and cell immortalization (EBV) (reviewed by Chami et al., 2006). Virus-induced decreased levels of cytosolic calcium have been found to be associated with the inhibition of immune cell functions (HIV-1 Tat, HHV-8 K15, EBV

LMP2A). Several viral proteins [HIV-1 Tat, HTLV-1 p13(II), HCV core, HBx, enterovirus 2B, HHV-8 K7] are able to modulate intracellular calcium signaling to control cell viability. The hepatitis C virus structural core and nonstructural NS5A proteins induce oxidative stress which is mediated by calcium alterations in human hepatocyte-derived cells. This might underlie the effects of both proteins in the pathogenesis of liver disorders associated with hepatitis C virus infection (Dionisio et al., 2009).

A search for EF-hand motifs in virus proteins resulted in the prediction, yet to be confirmed, of a total of 93 previously unrecognized calcium binding motifs (reviewed by Zhou et al., 2009). The presence of putative calcium binding motifs in viral proteins enlarges the spectrum of calcium–virus interactions. Calcium signaling may be altered during viral infection. The majority of viral calcium binding proteins reported are structural, including both coat and envelope proteins. In the former category, calcium is required to maintain the structural integrity and/or the proper assembly and disassembly of virions (turnip crinkle virus, tobacco mosaic virus, rotavirus, polyoma viruses, and HBV). Examples include the HIV-1 envelope glycoprotein, gp 160, which requires calcium for cell fusion. Calcium binding to the influenza B virus neuraminidase enhances both stability and enzymatic activity (reviewed by Zhou et al., 2009). Tobacco mosaic virus, a typical helical virus, has been shown to bind calcium with apparent affinities $< 100\ \mu M$ (Gallagher and Lauffer, 1983; Pattanayek et al., 1992). With the extracellular $[Ca^{2+}] \sim 1.0$ mM, these capsid proteins are in the calci form.

Many nonenveloped virus particles are stabilized by calcium ions bound in the interfaces between the protein subunits. These ions may have a role in the disassembly process within the cytosol. The small RNA phages of the Leviviridae family have a coat protein with a translational repressor activity. The crystal structure of phage PRR1 has been determined to 3.5-Å resolution (M. Persson et al., 2008). The structure shows a tentative binding site for a Ca^{2+} ion close to the quasi-threefold axis. The RNA binding surface used for repressor activity is mostly conserved. The structure does not show any significant differences between quasiequivalent subunits; this suggests that the assembly is not controlled by conformational switches as in many other simple viruses. Rotavirus outer-layer protein VP7 is a principal target of protective antibodies. The three well-ordered, 30-residue, N-terminal “arms” of each VP7 trimer grip the underlying trimer of VP6, an inner-capsid protein (Chen et al., 2009). Structural differences between free and particle-bound VP7 and between free and VP7-coated inner capsids may regulate mRNA transcription and release. The calcium-stabilized VP7 intratrimer contact region, which presents important neutralizing epitopes, is unaltered upon capsid binding. Removal of calcium ions results in dissociation of VP7 trimers into monomers, releasing VP7 from the virion, and initiation of penetration, inducing conformational changes in the other outer-layer protein, VP4. Aoki et al. (2009) reported the crystal structure at 3.4 Å resolution of VP7 bound with the Fab fragment of a neutralizing monoclonal antibody. The Fab binds across the outer surface of the intersubunit contact, which contains two calcium binding sites. The same region bears the epitopes of most neutralizing antibodies. The monovalent Fab is sufficient to neutralize infectivity. Aoki et al. (2009) proposed that neutralizing antibodies against VP7 act by stabilizing the trimer, thereby inhibiting the uncoating trigger for VP4 rearrangement.

3

COMPARISON OF THE Ca^{2+} ION WITH OTHER METAL CATIONS

Calcium (Latin *calc*, "lime") is the fifth most abundant element and the third most abundant metal in Earth's crust. The stable ion Ca^{2+}—$1s^2 2s^2 2p^6 3s^2 3p^6$—has the electron configuration of argon and two additional protons; it is less chemically reactive than the alkaline metals.

3.1. CALCIUM ISOTOPES

Calcium has four stable isotopes (^{40}Ca, ^{42}Ca, ^{43}Ca, and ^{44}Ca), plus two more isotopes (^{46}Ca and ^{48}Ca) that have such long half-lives that for all practical purposes they can be considered stable. It also has a radioactive isotope ^{41}Ca, which has a half-life of 103,000 years. Ninety-seven percent of naturally occurring calcium is ^{40}Ca.

3.2. CALCIUM IN THE ENVIRONMENT

Seawater contains approximately 400 ppm calcium, nearly all free Ca^{2+} ion ($\sim$10 mM). Rivers generally have 1 to 2 ppm calcium; however, in lime-rich areas they may have calcium concentrations as high as 100 ppm. Calcium makes up 3.5% of Earth's crust and occurs almost solely in inorganic compounds. It is obtained from carbonate minerals such as chalk, limestone, dolomite, and marble; it is widely distributed and quarried throughout the world.

Calcium Binding Proteins, By Eugene A. Permyakov and Robert H. Kretsinger

3.3. USES OF CALCIUM

Calcium carbonate, $CaCO_3$, is the most abundant calcium mineral. It occurs in two crystal forms: calcite, which is hexagonal, and aragonite, which is rhombohedral. It is used extensively in construction, either as marble (calcite) or as limestone aggregate for road building. $CaCO_3$ is the core ingredient of cement and mortar and is the starting material for the preparation of builder's lime. Calcite and Iceland spar are essentially pure carbonate forms; whereas, marble is impure and much more compact. Although calcium carbonate is poorly soluble in water, it is quite soluble if the water contains dissolved carbon dioxide and is slightly acidic. CaO (burnt lime, or quick lime) is used as an industrial dehydrating agent and in high intensity light arcs for its unusual spectral characteristics. Lime is an excellent absorbent for carbon dioxide. Calcium hydroxide, $Ca(OH)_2$ (slaked lime, hydrated lime, or pickling lime), has many applications in which the OH^- ion is necessary. It is used in making mortar, cements, paints, hard rubber products, and petrochemicals. Dolomite rock (also dolostone) is composed predominantly of the mineral dolomite, $CaMg(CO_3)_2$. Dolomite is used as an ornamental stone, a concrete aggregate, a source of magnesium oxide, and in the production of magnesium.

The sulfate hemihydrate ($CaSO_4 \cdot 0.5H_2O$) is known as plaster of paris, while the dihydrate ($CaSO_4 \cdot 2H_2O$) occurs naturally as gypsum. It constitutes a large portion of portland cement and is used to reduce the alkalinity of soils.

Calcium silicate, also known as orthosilicate, Ca_2SiO_4 ($2CaO \cdot SiO_2$), is one of a group of compounds obtained by reacting calcium oxide and silica in various ratios: for example, $3CaO{\cdot}SiO_2$, Ca_3SiO_5, $2CaO{\cdot}SiO_2$, Ca_2SiO_4, $3CaO{\cdot}2SiO_2$, $Ca_3Si_2O_7$, $CaO{\cdot}SiO_2$, and $CaSiO_3$. It is used as an antioxidant in the manufacture of steel as well as aluminum, beryllium, copper, lead, and magnesium alloys.

The calcium halogenides include phosphorescent fluoride (CaF_2), which is commonly used as to make windows for both infrared and ultraviolet spectroscopy, since it is transparent in these regions (about 0.15 to 9 μm) and has extremely weak birefringence. Calcium chloride ($CaCl_2$) is used as a brine for refrigeration plants, ice and dust control on roads, and in concrete. Because it is strongly hygroscopic, air or other gases may be channeled through a column of calcium chloride to remove moisture. Calcium sulfate ($CaSO_4$) is also used as a desiccant.

Calcium carbonate is widely used medicinally as an inexpensive dietary calcium supplement or antacid. It may be used as a phosphate binder for the treatment of hyperphosphatemia. It is also used in the pharmaceutical industry as an inert filler for tablets and other pharmaceuticals. Calcium carbonate is used in the production of toothpaste and in homeopathy as one of the constitutional remedies. Calcium hypochlorite, $Ca(ClO)_2$, is used to disinfect water and as a bleaching agent for cotton and linen. It is also used in the manufacture of chloroform. Although hydroxyapatite, $Ca_5(PO_4)_3OH$, is not a frequently encountered mineral, it is the main constituent of bone.

3.4. HEALTH EFFECTS OF CALCIUM

Calcium is the most abundant metal in the human body. It is the main constituent of bones and teeth, and it performs key metabolic functions. Calcium is commonly found in milk and milk products, as well as in vegetables, nuts, and beans. Gastric hydrochloric acid helps calcium absorption. The duodenum is the main location for absorption of calcium because farther down the small intestine the local environment becomes too alkaline. Maintaining a balanced blood calcium level is essential to life (Chapter 1); the serum content of [Ca^{2+}] is "... one of nature's physiological constants" (McLean and Hastings, 1935). A normal blood calcium level is about 10 mg per 100 mL of blood. Of that, approximately 5.5 mg is in ionic form, Ca^{2+}; about 4 mg is bound to carrier proteins; and about 0.5 mg is combined with phosphate or citrate. If there is not enough calcium in the diet to maintain sufficient amounts of calcium in the blood, the parathyroid glands will be activated to release more parathyroid hormone (PTH), which will then draw calcium out of the bones as well as increase intestinal absorption of available calcium (reviewed, e.g., by Goldsmith et al., 2009).

Toxicity from excess calcium is not common because the gastrointestinal tract normally limits the amount of calcium absorbed. Therefore, short term intake of large amounts of calcium does not generally produce any toxic effects aside from constipation and an increased risk of kidney stones. However, more severe toxicity can occur when excess calcium is ingested over long periods or when calcium is combined with increased amounts of vitamin D, which increases calcium absorption. Calcium toxicity is also sometimes found after excessive intravenous administration of calcium. Toxicity is manifested by abnormal deposition of calcium in tissues and by elevated blood calcium levels (hypercalcemia). Very high levels of calcium can result in appetite loss, nausea, vomiting, abdominal pain, confusion, seizures, and even coma.

Lack of calcium is one of the main causes of osteoporosis, a disease in which the bones become extremely porous, are subject to fracture, and heal slowly. This occurs especially in women following menopause. Adult men and premenopausal women need about 1000 mg of calcium per day to preserve their bones in normal condition. The recommended daily intake rises to 1500 mg for postmenopausal woman. The main sources of dietary calcium are dairy products, as well as nuts, green vegetables such as spinach, and pulses (*Pisum, Vigna, Vicia, Lens*). Calcium supplements are used, or are being investigated, for the treatment of hypertension, preeclampsia, cardiovascular disease, premenstrual syndrome, obesity, stroke, and colon cancer.

3.5. BIOLOGICALLY SIGNIFICANT METALS IN THE PERIODIC TABLE

Nearly one-fourth of the elements in the periodic table are found in living systems; most of these are essential for life. Six metals are present in the human body in

TABLE 3.1. Biologically Significant Metals

	Period											
Group	1	2	3	4	5	6	7	8	9	10	11	12
III	Na	Mg										
IV	K	Ca			V	Cr	Mn	Fe	Co	Ni	Cu	Zn
V						Mo						Cd
VI						W						

high concentrations: calcium (1700 g per 70 kg of body mass), potassium (250 g), sodium (70 g), magnesium (42 g), iron (5 g), and zinc (3 g). Other metals are present in quantities of less than 1 g per 70 kg of body mass.

The portion of the Mendeleev periodic table of elements shown in Table 3.1 includes 15 metals that play important roles in biological systems. Ten of these 15 biologically significant metals are located in the fourth row in the periodic table; sodium and magnesium are in the third row; and molybdenum and cadmium, and tungsten, are in the fifth and sixth rows, respectively. Their electronic structures are summarized in Table 3.2. These metals are considered in two groups:

TABLE 3.2. Electron Configurations and Ionic Radii of Some Biologically Significant Metals

	Element	Electron Configuration	Ionic Radius (Å)
Ne	neon	$1s^2 2s^2 2p^6$	
Na	sodium	$[Ne]3s^1$	1.02
Mg	magnesium	$[Ne]3s^2$	0.72
Ar	argon	$[Ne]3s^2 3p^6$	
K	potassium	$[Ar]4s^1$	1.38
Ca	calcium	$[Ar]4s^2$	0.99
Ca^{2+}	calcium ion	$1s^2 2s^2 2p^6 3s^2 3p^6$	0.99
V	vanadium	$[Ar]3d^3 4s^2$	0.59
Cr	chromium	$[Ar]3d^5 4s^1$	0.52
Mn	manganese	$[Ar]4s^2 3d^5$	0.46
Fe	iron	$[Ar]4s^2 3d^6$	0.645
Co	cobalt	$[Ar]4s^2 3d^7$	0.745
Ni	nickel	$[Ar]4s^2 3d^8$	0.69
Cu	copper	$[Ar]4s^1 3d^{10}$	0.73
Zn	zinc	$[Ar]4s^2 3d^{10}$	0.74
Kr	krypton	$[Ar]4s^2 3d^{10} 4p^6$	
Mo	molybdenum	$[Kr]5s^1 4d^5$	0.65
Cd	cadmium	$[Kr]4d^{10} 5s^2$	0.97
Xe	xenon	$[Kr]4d^{10} 5s^2 5p^6$	
W	tungsten	$[Xe]4f^{14} 5d^4 6s^2$	0.62

nontransition elements (Na, K, Mg, Ca, Zn, and Cd) and transition elements (V, Cr, Mn, Fe, Co, Ni, Cu, Mo, and W).

Atoms of alkali elements, sodium and potassium, possess one *s*-electron in the outer shell besides the electron structure of rare gas atoms; therefore, they are characterized by low ionization potentials (5.138 and 4.339 eV, respectively). Their ions have relatively large radii (1.02 and 1.38 Å, respectively). These atoms have very weak tendencies to form covalent bonds.

The ionization potential of the calcium atom ($1s^22s^22p^63s^23p^64s^2$) for the first electron is 6.1132 eV; the ionization potential for the second electron is 11.871 eV. In normal geological and biological conditions the Ca^+ ion does not exist. The completely filled octets of the Mg^{2+}, Ca^{2+}, and Sr^{2+} ions ($2s^22p^6$, $3s^23p^6$, and $4s^24p^6$, respectively) have no preferences with respect to the direction of bond formation and can be modeled by spheres with increasing radii (0.72, 0.99, and 1.35 Å) and decreasing charge density. The similar sizes of the ions of calcium and of the lanthanides (0.99 and 1.06 to 0.85 Å) allow the lanthanides to replace calcium in many binding sites; this is very useful for spectroscopic studies.

Transition elements (*d*-elements), significant for biological processes—V, Cr, Mn, Fe, Co, Ni, Cu, Mo, and W—have incompletely filled *d*-orbitals; Cu is actually $4s^13d^{10}$. They, including copper, are characterized by variable valences and formation of colored complexes and paramagnetic substances due to unpaired electrons. The transition elements often serve as catalysts of redox reactions.

Zinc and cadmium are not transition elements since they have no empty *d*-orbitals. Zn^{2+} ions are different from the other nontransition metal ions. While the radius of Zn^{2+} ion (0.74 Å) is close to the radius of the Mg^{2+} ion (0.72 Å), its ionization potentials are higher than those of calcium and magnesium. These high values of zinc ionization potentials (9.394 and 17.964 eV) are reflected in Zn^{2+}'s stronger tendency to form covalent bonds. It is bound by many proteins.

3.6. HYDRATION OF METAL IONS

Metal cations in aqueous solution are surrounded by water molecules oriented by the electric field of the ion; this creates charge–dipole interactions. The smaller the radius of an ion, the greater the charge density and the stronger its interactions with the dipole moment of water. At least two layers of water molecules are influenced by the central ion (Muñoz-Páez et al., 1995; Sakane et al., 1998). Some ions possess a rather rigid and stable first hydration shell that can have tetrahedral (Li^+) or octahedral (Mg^{2+}, Co^{2+}, Ni^{2+}) geometry. The Zn^{2+} ion can have both tetrahedral and octahedral coordination in the first hydration shell. The Mg^{2+} ion, with its 2+ charge and small radius, tightly orders six water molecules (the first or inner hydration shell) in an octahedral arrangement. The second, and perhaps third, layers of water are also organized by the charge of the ion and contribute to the overall hydration free energy of -455 kcal mol^{-1}. The K^+ ion is larger and has only a $+1$ charge; as a result, eight or nine water molecules pack around the ion in a less well-ordered manner; the hydration energy is -80 kcal mol^{-1}. Because the Mg^{2+} ion interacts strongly with six water molecules in $[Mg(H_2O)_6]^{2+}$, larger anions do not readily replace the water to give precipitates. The larger Ca^{2+} ion binds more strongly to those larger anions relative to water; hence, these anions displace water from calcium more readily. Large anions, $CO_3{}^{2-}$, and $PO_4{}^{3-}$, precipitate with calcium at lower metal ion concentrations than with magnesium. As a consequence, calcium has a lower concentration in seawater (~10 mM and ~50 mM for $[Ca^{2+}]$ and $[Mg^{2+}]$, respectively) and one can find deposits, both geochemical and biochemical, of $CaCO_3$, $CaSO_4$, and $Ca_3(PO_4)_2$, but not of the corresponding magnesium salts except mixed in the calcium salts: for example, dolomite $[CaMg(CO_3)_2]$. Calcium oxalate is insoluble; its crystals are found in plant tissue, but magnesium oxalate is soluble. Moreover, calcium tends to precipitate many polyanions, such as DNA, RNA, and some acidic proteins.

Calcium and monovalent metal cations, except Li^+, are characterized by a wide variety of geometries of the first hydration shell. The number of ion-associated water molecules and the distance between them and the central ion vary and increase with increasing ion size. Because the six oxygens of hexa-aquomagnesium are in optimal van der Waals contact and leave slowly, the magnesium association rate constant, k_{on}(Mg), for proteins is relatively low. In contrast, the waters of hepta-aquocalcium have more lateral mobility; thus, k_{on}(Ca) is much faster; this accounts for most of the difference in affinity of calcium binding proteins for calcium versus magnesium. The Ca^{2+} ion is heptahydrate and has a diffusion coefficient of 1.335×10^{-9} m^2 s^{-1} in water at 25°C (Ribeiroa et al., 2008) and a water exchange rate of 10^8 to 10^9 s^{-1} (Schwenk et al., 2001).

3.7. "HARD" AND "SOFT" METAL IONS

Both Ca^{2+} and Mg^{2+} are "hard" ions and prefer ligands of low polarizability (reviewed by Dudev and Lim, 2003, 2008). All Ca^{2+} and Mg^{2+} ion interactions are electrostatic with no covalent contribution. Mg^{2+} ions generally require more restricted orientations of suitable ligand groups. For Mg^{2+}, oxygen is the most preferred coordinating atom, followed by nitrogen. Calcium, like magnesium, prefers to bind to hard oxygen containing ligands but, generally, with lower free energy gain. By comparison, unlike Ca^{2+} and Mg^{2+}, Zn^{2+} and transition metal ions prefer "softer" polarizable ligands, such as the sulfur and nitrogen of Cys and His, although some of them are also coordinated by oxygens of Asp and Glu.

Although the favored coordination for the Mg^{2+} ion is sixfold octahedral, Ca^{2+} shows a greater diversity of coordination numbers, with seven and eight fold coordination the most common. Most of the crystal structures containing divalent calcium have coordination number 6, 7, or 8 (Katz et al., 1996). Bond distances between the Ca^{2+} ion and its ligands vary more than those of Mg^{2+}. The radius of the coordination sphere is significantly larger for calcium than for magnesium ions; bond distances to oxygen donor atoms typically range from 2.0 to 2.1 Å for Mg^{2+} and 2.3 to 2.6 Å for Ca^{2+}. Compared to Mg^{2+} ions, Ca^{2+} forms looser complexes of higher and variable coordination number, without directionality, and with variable bond lengths.

4

COMPLEXES OF CALCIUM AND OTHER CATIONS WITH COMPOUNDS OF LOW MOLECULAR WEIGHT

The structures, kinetics, thermodynamics, and solubilities of metalloorganic and inorganic complexes provide the background for understanding the special properties of calcium binding proteins and metalloproteins in general. In this chapter we survey these low molecular weight complexes and in Chapter 5 consider protein complexes of essential metals, other than calcium, and the biological interactions, often toxic, of nonessential metals.

4.1. CRYSTAL STRUCTURES OF COMPLEXES OF CALCIUM WITH LOW MOLECULAR WEIGHT COMPOUNDS

Examination of the crystal structures of calcium and small molecule complexes led to several conclusions (reviewed by Nelson and Kretsinger, 1976; Einspahr and Bugg, 1984; Katz et al., 1996). Nearly all of the ligands, excluding a few halides, are oxygen atoms. There are no Ca–Ca metal bonds, although a single carboxyl group can bridge two Ca^{2+} ions. Most of the Ca–O distances are 2.3 to 2.5 Å. In contrast, the (biological) transition metals—V, Cr, Mn, Fe, Co, Ni, Cu, and Zn—as well as Mo, Cd, and W (Chapter 3)—"soft metals"—are frequently complexed by nitrogen and sulfur atoms of organic compounds. Further, iron forms clusters with Fe–Fe bonds.

Many of these crystal structures of small-molecule complexes, as well as of calcium binding proteins, incorporate one to three waters of hydration in the primary coordination sphere. The Ca^{2+} ions tend to lie near the plane that

Calcium Binding Proteins, By Eugene A. Permyakov and Robert H. Kretsinger

relates the two hydrogen atoms of a coordinating water molecule. The acute angle between the dipole moment vector of the water and the Ca–O vector is generally between 0 and 60°, although a few Ca^{2+} ions lie at angles as high as 80°.

The coordination of calcium and magnesium can be understood in terms of Mg–O and Ca–O bond distances (2.0 and 2.3 Å) and of van der Waals radii (Mg ~ 0.70 Å, Ca ~1.00 Å, and O ~1.4 Å). The oxygens of hexa-aquomagnesium are at the vertices of an octahedron; the O–Mg–O angle is 90°. With Mg–O 2.0 Å, the O–O distance is 2.84 Å, optimal van der Waals packing. These are happy waters and are reluctant to leave; hence, a k_{off} value of 10^6 s^{-1} is slow. In contrast, the O–O distance in hexa-aquocalcium is $2.3 \times 2^{0.5} = 3.27$ Å. There is room for a seventh oxygen; it is usually accommodated in a pentagonal bipyramid. If the five oxygens about the circumference are in van der Waals contact, 2.8 Å, the Ca–O distance is $(2.8/2 \cdot \sin 36°) = 2.38$ Å. If the distance from Ca to the axial oxygen is 2.3 Å, the $O_{axial}–O_{equatorial}$ distance is $(2.3^2 + 2.38^2)^{0.5} =$ 3.28 Å. The equatorial oxygens can get a bit closer to the calcium by moving slightly off equator. Even so, the seven coordinating oxygens have a bit of room to roll about before coming into van der Waals contact. The k_{off} (10^9 s^{-1}) is much faster than for magnesium. In most protein–calcium complexes, the Ca^{2+} ion is seven-coordinate; Mg^{2+} is almost always six coordinate with oxygen.

In a square antiprism the eight oxygens come into van der Waals contact (2.8 Å) when still 2.3 to 2.4 Å from the calcium. Eight coordinate calcium sometimes occurs in organic complexes where the oxygens are less constrained than they would be in a protein.

The Mg^{2+} ion has a strong preference for sixfold (octahedral) coodination; this tends to restrict the Mg^{2+} ion to the dipole axis of water. In crystalline hydrates, Mg^{2+} ions are located within 0.2 Å of this axis. The average Mg–O distance is 2.07 Å.

Carboxylate oxygens of Asp and Glu, along with carbonyl oxygens of the main chain, are the primary ligands of calcium in proteins (reviewed by Einspahr and Bugg, 1984; Katz et al., 1996). Ca^{2+} ions in monodentate coordination can be located on any side of the C–O vector; the Ca–O–C angles are 120 to 150°, with the Ca^{2+} ions on the side directed to the second oxygen of the carboxylate group and within 140 to 160° on the other side of the C–O bond. In bidendate coordination both oxygens coordinate calcium; the Ca^{2+} ion lies near the plane of the carboxylate and near the line of the C–C bond. Ca^{2+} ions in bidentate coordination are restricted to a narrow region equidistant from the two carboxylate oxygens, with Ca–O–C angles of about 90°. Most Ca–O distances in such complexes are between 2.3 and 2.5 Å. These three types of interactions have slightly different mean Ca–O distances: 2.38 Å for monodentate ligands, 2.42 Å for bidentate, and 2.53 Å for carboxylate oxygen in combination with an α-ligand.

Mg^{2+} ions always interact with carboxyl groups in a monodentate manner. If a Mg^{2+} ion is bound by four monodentate carboxyl groups, it still has space for two waters or two other oxygen-containing groups.

In carbonyl interactions the Ca^{2+} ions tends to lie near the plane of the carbonyl group, but this tendency is not as pronounced as in the case of calcium carboxylate interactions. Most of the Ca^{2+} ions lie at Ca–O–C angles between 110 and 150°, but a number of examples lie outside this range. Cyclic peptide examples are tightly clustered in a region nearly collinear with the C–O bond. Most of the Ca–O distances lie between 2.30 and 2.45 Å. Averages of Ca–O distances for the unidentate, chelate, and cyclic peptide categories are 2.36, 2.42, and 2.30 Å, respectively. In general, calcium–carbonyl interactions lack the strong definition of geometrical preferences that was found for calcium–carboxylate interactions.

Ca^{2+} ions are bound to amino acids and peptides primarily through carboxylate and carbonyl groups, and calcium–ligand geometries in the complexes follow the general patterns described above. Phosphate, sulfate, and hydroxyl ions also interact with Ca^{2+} ions through oxygen atoms.

Unlike Mg^{2+} and Ca^{2+}, Zn^{2+} ions prefer "softer" ligands such as Cys and His, although they are also found coordinated to Asp and Glu side chains. The Cys side chains are deprotonated when bound to the metal. The Zn^{2+} ion is octahedrally bound to six water molecules in solution. At the same time, in both Zn finger proteins and enzymes, Zn^{2+} is usually tetrahedrally coordinated, but it can also adopt a five- or six coordinate geometry. The average Zn–ligand distances for a tetrahedral binding site are Zn–N (His) 2.07 to 2.09 Å, Zn–S (Cys) 2.21 to 2.35 Å, Zn–O (Asp/Glu) 1.95 to 2.04 Å, and Zn–O (water) 2.12 to 2.15 Å (reviewed by Dudev and Lim, 2003).

Kirberger et al. (2008) analyzed the calcium binding proteins from the Protein Data Bank to identify structural parameters associated with EF-hand and non-EF-hand calcium binding sites (Chapter 9). Non-EF-hand sites have lower coordination numbers (6 ± 2 vs. 7 ± 1), fewer protein ligands (4 ± 2 vs. 6 ± 1), and more water ligands (2 ± 2 vs. 1) than do EF-hand sites. The orders of ligand preference for non-EF-hands are H_2O (33.1%) > side-chain Asp (24.5%) > main-chain carbonyl (23.9%) > side-chain Glu (10.4%) and for EF-hand sites side-chain Asp (29.7%) > side-chain Glu (26.6%) > main-chain carbonyl (21.4%) > H_2O (13.3%). Fewer negative charges were observed in the non-EF-hand than in the EF-hand binding sites (1 ± 1 vs. 3 ± 1). Over 20% of non-EF-hand sites have formal charge values of zero, due to increased use of water and carbonyl oxygen ligands. The non-EF-hand sites have a broader range of ligand distances and bond angles than do EF-hand sites, possibly due to the highly conserved helix–loop–helix conformation of the EF-hand. Significant differences between ligand types (carbonyl, side chain, bidentate) show that angles associated with each type must be classified separately. The EF-hand side-chain Ca–O–C angles have an unusual bimodal distribution consistent with an Asp distribution that differs from the Gaussian model observed for non-EF-hand proteins.

4.2. DISSOCIATION CONSTANTS OF CALCIUM AND ANALOGS WITH SMALL COMPOUNDS

Simple compounds such as H_2O and NH_3 have a single donor atom (i.e., they function as monodentate ligands). More complex compounds or anions (e.g., NO^{3-}, $CO_3{}^{2-}$, $C_2O_4{}^{2-}$) can coordinate cations as both monodentate and bidentate ligands. Polydentate ligands coordinate cations using several donor atoms, and in the process do not lower the entropy of the system; this results in a drastic increase in the stabilities of the complexes. This effect is governed by both entropic and enthalpic factors. For example, the well-known chelator EDTA (ethylenediaminetetraacetic acid; $C_2H_4[N(CH_2COO)_2]_2{}^{4-}$) coordinates metal ions by six atoms (four oxygens and two nitrogens) and has very high affinity for Ca^{2+} and other di- and trivalent metal cations.

The most important thermodynamic characteristic of a coordination compound is its stability constant β_n, that is, the equilibrium constant of the reaction

$$M^{z+} + nL \leftrightarrow ML_n \tag{4.1}$$

$$\beta_n = \frac{[ML_n]}{[M][L]^n} \tag{4.2}$$

The stepwise stability constants are defined as

$$K_n = \frac{[ML_n]}{[ML_{n-1}][L]} \tag{4.3}$$

The stabilities of α-amino acids complexed with metal ions change in the series $Ca^{2+} < Mg^{2+} < Mn^{2+} < Fe^{2+} < Cd^{2+} < Co^{2+} < Zn^{2+} < Ni^{2+} < Cu^{2+} < Fe^{3+}$. Without constraints on the free metal ion concentrations in cells and on the steric demands of complicated ligands, calcium would not be bound by any organic molecules in cells. The concentrations of competing metal ions in the cytoplasm are reduced by binding to stronger donors so that the free ion levels are approximately (Williams, 2002)

Ion	Mn^{2+}	Fe^{2+}	Co^{2+}	Ni^{2+}	Cu^{2+}	Zn^{2+}	Cd^{2+}	Mg^{2+}	Ca^{2+}
$\log[M^{2+}_{free}]$	−7	−8	−9	−11	−15	−12	−15	−3	−8
Ionic radius Å	0.75	0.70	0.68	0.65	0.60	0.65	0.90	0.65	1.0

At these concentrations, none of these ions can bind to the calcium binding sites because the protein structures that hold calcium do not allow a collapse of ligand donor groups to give a smaller cavity than a radius of 1.0 Å, so chelation of the oxygen donors is relatively weak.

The coordination of the Ca^{2+} ion is ionic and spherical without significant directionality. To provide a good calcium site, one needs a nearly spherical

TABLE 4.1. Logarithms of the First Stability Constants for Complexes of Metal Cations with Organic Compounds

	Mg^{2+}	Ca^{2+}	Zn^{2+}	Gd^{3+}
Acetate	0.5	0.5	1.1	2.0
Lactate	0.9	1.1	1.9	2.9
Citrate	3.4	3.5	5.0	7.8
Oxydiacetate	1.8	4.4	3.6	5.4
Iminodiacetate	3.0	2.6	7.2	6.7
Alanine	2.0	1.2	4.6	
Aspartate	2.4	1.6	5.8	5.7
Nitrilotriacetate	5.5	6.4	10.7	11.4
EGTA	5.3	10.9	12.6	17.5
EDTA	8.8	10.6	16.4	17.4
Adenosin-5′-diphosphate	3.17	2.86	4.28	
Adenosine-5′-triphosphate	4.22	3.97	4.85	

Source: Data from Martell and Smith (1977).
Conditions: 25°C; 0.1 M ionic strength.

pocket of size appropriate to the coordination number with at least two negatively charged ligands. Table 4.1 contains logarithms of stability constants for some complexes of metal cations with organic substances.

The lanthanide stability constants are always higher than those of calcium. For this reason, one can expect that tripositive lanthanides will displace Ca^{2+} ions from protein binding sites; this agrees with experimental data.

Compared to magnesium, calcium prefers ligands without nitrogen donors (reviewed by Martin, 1984). For example, calcium binds more strongly to tridentate oxydiacetate, but magnesium binds strongly to iminodiacetate. Magnesium also binds to nitrogen donors in chlorophyll, but calcium binds neither. Calcium binds large multidentate anionic ligand groups with higher affinity than that for magnesium. For example, calcium binds 100 times more strongly to EGTA (ethylene glycol tetraacetic acid) than does magnesium since the Ca^{2+} ion fits the steric requirements of EGTA better than it fits the Mg^{2+} ion. In nucleotide phosphates, magnesium binds slightly more strongly than does calcium. Inside a cell, Mg^{2+} is associated with phosphates; whereas, Ca^{2+} is tightly bound by proteins.

4.3. SOLUBILITIES OF CALCIUM AND ANALOGS WITH SMALL COMPOUNDS

Solubilities of calcium and magnesium salts vary over a wide range. Calcium is naturally present in water. It may dissolve from rocks such as limestone, marble, calcite, dolomite, gypsum, fluorite, and apatite. Calcium carbonate has a solubility of 14 mg L^{-1}; it is five times higher at normal atmospheric CO_2 partial pressure. The solubility of calcium phosphate is 20 mg L^{-1}, calcium fluoride 16 mg L^{-1}, calcium chromate 170 g L^{-1}, and calcium hypochlorate 218 g L^{-1}. The solubilities of other calcium compounds lie within this range.

The mineral found in teeth and bone is an impure form of calcium hydroxyapatite, sometimes referred to as "biological apatite." However, a solution of calcium phosphate can give rise to a number of different salts that differ in their calcium/phosphate ratios. Table 4.2 shows the ideal forms of the calcium phosphate salts. Calcium hydroxyapatite is a naturally occurring form of calcium apatite with the formula $Ca_5(PO_4)_3OH$, but is usually written $Ca_{10}(PO_4)_6(OH)_2$ to denote that the crystal unit cell contains two equivalents. Seventy percent of bone consists of hydroxylapatite. Carbonated calcium-deficient hydroxylapatite is the main mineral of which dental enamel and dentin are comprised (Chapter 3).

Note that the large differences in solubility products are not always reflected in the actual solubilities of the various ions concerned. Impurities in biological apatite introduce significant stresses into the crystal structure, which make it much less stable. Biological apatite composition varies widely, and the apparent solubility of the source varies accordingly. The pH of seawater is about 8.4 and is close to saturation in $CaCO_3$. Many organisms in the sea make $CaCO_3$ shells quite easily (reviewed by Williams, 2002). Some freshwater organisms also make $CaCO_3$ shells; this means that the precipitation occurs due to elevated $[HCO_3^-]$, which is produced by the living system. The precipitation of phosphates in the sea is less probable than carbonates because of the low level of phosphate. Vertebrates have an extracellular pH ~7.0 and precipitate calcium phosphate $Ca(HPO_4)_2$ or $Ca_3(PO_4)_2$ before it is transformed into the more insoluble hydroxyapatite $[Ca_5(PO_4)_3OH]$.

TABLE 4.2. Solubility Products of Some Calcium Phosphate Salts

Salt	Ionic Composition	Solubility Product	[Me] at Saturation (M)
brushite	$Ca(HPO_4) \cdot 2H_2O$	2.32×10^{-7}	1.08×10^{-6}
calcium phosphate	$Ca_3(PO_4)_2$	2.83×10^{-30}	1.45×10^{-6}
magnesium phosphate	$Mg_3(PO_4)_2$	1×10^{-25}	3.9×10^{-6}
octacalcium phosphate	$Ca_4H(PO_4)_3$	2×10^{-49}	4.8×10^{-4}
hydroxyapatite	$Ca_5(PO_4)_3OH$	2.34×10^{-59}	4.36×10^{-7}
fluorapatite	$Ca_5(PO_4)_3F$	1.16×10^{-60}	3.47×10^{-7}
calcium carbonate	$CaCO_3$	3.8×10^{-9}	6.16×10^{-5}
magnesium carbonate	$MgCO_3$	3.5×10^{-8}	1.87×10^{-4}
calcium hydroxide	$Ca(OH)_2$	5.5×10^{-6}	1.11×10^{-2}
magnesium hydroxide	$Mg(OH)_2$	1.8×10^{-11}	1.65×10^{-4}
calcium oxalate hydrate	$CaC_2O_4 \cdot H_2O$	1.96×10^{-8}	1.4×10^{-4}
magnesium oxalate	MgC_2O_4	7×10^{-7}	8.37×10^{-4}
calcium sulfate	$CaSO_4$	9.1×10^{-6}	3.0×10^{-3}

Source: Data from McDowell et al. (1977).

Bone, which is a composite of polymer and crystals, can be redissolved by making the extracellular fluid acidic. Higher animals have special cells, osteoblasts, for such dissolution. These cells bind to bone, trapping a small aqueous volume between the bone and themselves. They then release acid into this volume to solubilize the phosphate and release enzymes to destroy the biopolymers found in bone.

5

STOICHIOMETRY, KINETICS, AND THERMODYNAMICS OF CALCIUM BINDING

One wants to know several fundamental characteristics about the binding of calcium, or of any ligand, to a protein:

1. Stoichiometry, affinity, and cooperativity of binding
2. Kinetics of binding
3. Partition of free energy of binding (ΔG) among enthalpy (ΔH) and entropy (ΔS)
4. Changes in structure associated with binding
5. Changes in function associated with binding

In this chapter we develop the theory and formalism of stoichiometry, affinity, cooperativity, kinetics, and thermodynamics. Chapter 6 covers the techniques used to measure calcium binding. Complexes of proteins with metals other than calcium are surveyed in Chapters 8 and 9. In Chapters 10 to 15 we characterize the basic structures, characteristics, and functions of the many homolog families of calcium binding proteins.

5.1. STOICHIOMETRY, AFFINITY, AND COOPERATIVITY OF BINDING

Two fundamental characteristics of any metal (ligand) binding protein are the number of metal ions bound (stoichiometry) and the affinity(s) of binding.

Calcium Binding Proteins, By Eugene A. Permyakov and Robert H. Kretsinger

Consider the reaction

$$P + Me \leftrightarrow PMe \tag{5.1}$$

in which P is a protein and Me is a metal ion. If one can determine experimentally any three of the following five concentrations: free protein, [P]; or [Me]; or [PMe]; or [P_0] (total protein concentration, [P] + [PMe]); or [Me_0] (total metal ion concentration, [Me] + [PMe]) (excepting the two triplets [P], [PMe], and [P_o] or [Me], [PMe], and [Me_o]), one can compute the dissociation constant:

$$K_d = \frac{[P][Me]}{[PMe]} (\text{molar}) \tag{5.2}$$

$$\nu = \frac{[PMe]}{[P_0]} = \frac{[Me]}{K_d + [Me]} \tag{5.3}$$

In the case of n independent, equivalent sites, equation (5.3) becomes

$$\nu = \frac{[PMe]}{[P_0]} = \frac{n[Me]}{K_d + [Me]} \tag{5.4}$$

or

$$\frac{\nu}{[Me]} = \frac{n}{K_d} - \frac{\nu}{K_d} \tag{5.5}$$

This is the well-known *Scatchard representation*. The plot of ν/[Me] against ν is called the *Scatchard plot*. The slope of the straight line of the Scatchard plot gives K_d, and the point of its intersection with the ordinate axis gives n. In the case of several independent binding sites with dissociation constants K_{di},

$$\nu = \sum \frac{n_i[Me]}{K_{di} + [Me]} \tag{5.6}$$

Ideally, one should measure activities, not concentrations; however, the difference between concentration and activity is usually within the experimental error of the measurements for these systems.

The change of apparent standard free energy upon the binding of the ith molecule of the ligand is

$$\Delta G_i^0 = \text{RT} \ln K_{di} \tag{5.7}$$

where R is the gas constant and T is the absolute temperature. Since

$$\Delta G_i^0 = \Delta H_i^0 - T \Delta S_i^0 \tag{5.8}$$

where ΔH_i^0 and ΔS_i^0 are standard enthalpy and entropy change, respectively. It follows that

$$\ln K_{di} = -\frac{\Delta H_i^0}{RT} + \frac{\Delta S_i^0}{R} \tag{5.9}$$

Differentiation of this expression yields the *van't Hoff equation:*

$$\frac{d \ln K_{di}}{dT} = \frac{\Delta H_i^0}{RT^2} \tag{5.10}$$

The values of calcium dissociation constants for various calcium binding proteins are within the range 10^{-9} to 10^{-4} M, depending on the type of the binding site. Intracellular calcium binding proteins (e.g., calmodulin, troponin C, parvalbumin) have the highest affinities for Ca^{2+} ions. Usually, their dissociation constants are comparable with intracellular calcium concentrations. The same generalization obtains for dissociation constants for magnesium; their values are comparable with intracellular concentrations of free Mg^{2+} ion. This same correspondence of concentration and affinity also obtains for sodium and potassium binding to these proteins. This means that several metal cations that occur in biological systems, the most important of which is magnesium, compete with calcium for well defined calcium binding sites. These balances of protein affinities and abundances of intracellur calcium and magnesium reflect the results of natural selection, not the affinities of small molecule ligands.

As noted in Chapters 3 and 4, the affinities of magnesium for many small, oxygen containing ligands exceeds that of calcium. In contrast, for most proteins, especially those in the EF-hand family, the affinity for calcium is three to five orders of magnitude higher than that for magnesium. This reflects the difficulty of wrapping a protein polymer around the Mg^{2+} ion in nearly rigid octahedral coordination relative to the more flexible pentagonal bipyramid of oxygens surrounding the Ca^{2+} ion and also the higher energy of interaction of Mg^{2+} ions with water molecules compared with Ca^{2+} ions.

If the protein has several interacting binding sites, the situation is complicated and the experimental challenge is more difficult. For n cooperative equivalent binding sites:

$$\mathrm{P} + n\mathrm{Me} \leftrightarrow \mathrm{PMe}_n \tag{5.11}$$

$$K_d = \frac{[\mathrm{P}][\mathrm{Me}]^n}{[\mathrm{PMe}_n]} \quad (\text{molar}^n) \tag{5.12}$$

$$\frac{\nu}{n} = \frac{[\mathrm{Me}]^n}{K_d^n + [\mathrm{Me}]^n} \tag{5.13}$$

In practice there is never absolute cooperativity, and experimental data reflecting cooperative binding are usually described by equations (5.12) and (5.13) with

Hill coefficient n_h instead of n: $1 \leq n_h \leq n$ in the case of positive cooperativity. When $n_h = n$ the system behaves fully cooperatively; when $n_h = 1$ there is no cooperativity. In negative cooperativity the binding of one Me reduces the affinity of the protein for subsequent Me's; n_h is less than 1.

Several complexities are encountered in practice. As discussed in Chapters 10 to 15, most calcium binding proteins have more than one calcium binding site. Although there is evidence of some cooperativity among these multiple sites, whatever cooperativity exists might vary for binding different cations. To the extent that cooperativity exists, there is not yet evidence of its physiological function.

Of greater concern, other cations might bind to sites on the protein that differ from calcium binding site(s). Such binding to an alternative site(s) might affect the affinity for calcium at its major site(s). These alternative situations are difficult to sort out, especially without knowledge of their various structures.

There are a few proteins with a single strong calcium binding site, such as α-lactalbumin, found in milk and as a component of lactose synthase. It is a favorite subject for evaluation of techniques and concepts.

There are many proteins with two strong calcium binding sites: for example, parvalbumins, calbindin 9_{kD}, and recoverin. The main difficulties in studying even two site calcium binding proteins are elucidation of the mechanism of the binding and determining whether the binding of one Ca^{2+} ion increases (positive cooperativity) or decreases (negative cooperativity) the affinity for calcium at the second site. In practice, it is difficult to determine the binding mechanism, even in the case of two binding sites. Consider the binding of Me to a protein P with two binding sites:

$$\begin{array}{ccc} & K_{11} & \\ \mathrm{P} & \leftrightarrow & \mathrm{MeP} \\ \updownarrow K_{12} & & \updownarrow K_{22} \\ \mathrm{PMe} & \leftrightarrow & \mathrm{MePMe} \\ & K_{21} & \end{array} \tag{5.14}$$

where K_{11}, K_{22}, K_{12}, and K_{21} are equilibrium dissociation constants for the two binding sites:

$$K_{11} = \frac{[\mathrm{P}][\mathrm{Me}]}{[\mathrm{MeP}]} \tag{5.15}$$

$$K_{22} = \frac{[\mathrm{MeP}][\mathrm{Me}]}{[\mathrm{MePMe}]} \tag{5.16}$$

$$K_{12} = \frac{[\mathrm{P}][\mathrm{Me}]}{[\mathrm{PMe}]} \tag{5.17}$$

$$K_{21} = \frac{[\mathrm{PMe}][\mathrm{Me}]}{[\mathrm{MePMe}]} \tag{5.18}$$

Concentrations of the reactants are related to each other by the equations of mass balance:

$$[\mathrm{P}] + [\mathrm{PMe}] + [\mathrm{MeP}] + [\mathrm{MePMe}] = [\mathrm{P_0}] \tag{5.19}$$

$$[\mathrm{Me}] + [\mathrm{PMe}] + [\mathrm{MeP}] + 2[\mathrm{MePMe}] = [\mathrm{Me_0}] \tag{5.20}$$

where $[\mathrm{P_0}]$ and $[\mathrm{Me_0}]$ are total protein and total metal concentrations.

The equilibrium binding constants are related by

$$K_{22}K_{21} = K_{12}K_{11} \tag{5.21}$$

$$\text{If} \quad K_{11} = K_{21}, \quad \text{then} \quad K_{12} = K_{22} \tag{5.22}$$

This means that the two calcium binding sites are independent [equation (5.22)] and the binding in each site occurs independent of the other site. One can then simplify:

$$K_{11} = K_{21} = K_1 \tag{5.23}$$

$$K_{12} = K_{22} = K_2 \tag{5.24}$$

If $K_2 \gg K_1$, the first site is filled (almost) completely before the second begins to fill. Scheme (5.14) simplifies to the sequential binding scheme

$$\mathrm{P} \overset{K_1}{\longleftrightarrow} \mathrm{MeP} \overset{K_2}{\longleftrightarrow} \mathrm{MePMe} \tag{5.25}$$

If K_{11} is similar to K_{12}, then K_{21} must be similar to K_{22}; one measures the apparent dissociation constants:

$$\frac{1}{K_{1\mathrm{app}}} = \frac{[\mathrm{MeP}] + [\mathrm{PMe}]}{[\mathrm{P}][\mathrm{Me}]} = \frac{1}{K_{11}} + \frac{1}{K_{12}} \tag{5.26}$$

$$\frac{1}{K_{2\mathrm{app}}} = \frac{[\mathrm{MePMe}]}{([\mathrm{MeP}] + [\mathrm{PMe}])[\mathrm{Me}]} = \frac{1}{K_{11} + K_{12}} \tag{5.27}$$

It is difficult to distinguish experimentally between schemes (5.14) ($K_{11} \neq K_{12} \neq K_{21} \neq K_{22}$) and (5.22) ($K_{11} = K_{21} = K_1$ and $K_{12} = K_{22} = K_2$) as well as between schemes (5.22) and (5.25) ($K_2 \gg K_1$).

Consider a protein P binding a compound Me with dissociation constant K_d. In the simplest case, one P binds one Me, as described previously:

$$\mathrm{P} + \mathrm{Me} \leftrightarrow \mathrm{PMe} \tag{5.1}$$

$$K_d = \frac{[\mathrm{P}][\mathrm{Me}]}{[\mathrm{PMe}]} \tag{5.28}$$

$$[\mathrm{P}] + [\mathrm{PMe}] = [\mathrm{P_0}] \tag{5.29}$$

$$[\mathrm{P}] = \frac{[\mathrm{P_0}]K_d}{K_d + [\mathrm{Me}]} \tag{5.30}$$

$$[\mathrm{PMe}] = \frac{[\mathrm{P_0}][\mathrm{Me}]}{K_d + [\mathrm{Me}]} \tag{5.31}$$

Combining equations, one obtains

$$[\mathrm{Me}]^2 + \{K_d + ([\mathrm{P_0}] - [\mathrm{Me_0}])\}[\mathrm{Me}] - [\mathrm{Me_0}]K_d = 0 \tag{5.32}$$

The solution of this quadratic equation is

$$[\mathrm{Me}] = \left\{\frac{[K_d + ([\mathrm{P_0}] - [\mathrm{Me_0}])]^2}{4 + [\mathrm{Me_0}]K_d}\right\}^{1/2} - \frac{K_d + ([\mathrm{P_0}] - [\mathrm{Me_0}])}{2} \tag{5.33}$$

Usually, to study the binding of Me to a protein by physical methods such as fluorescence spectroscopy, circular dichroism spectroscopy, nuclear magnetic resonance, and so on, the protein solution is titrated by small additions of Me, and physical parameters (e.g., fluorescence parameters of the protein) are measured after each addition. Fluorescence quantum yield (or fluorescence intensity at a fixed wavelength, Chapter 6) changes following the equation

$$q = \frac{q_{\mathrm{P}}[\mathrm{P}]}{[\mathrm{P_0}]} + \frac{q_{\mathrm{PMe}}[\mathrm{PMe}]}{[\mathrm{P_0}]} \tag{5.34}$$

in which q_{P} and q_{PMe} are fluorescence quantum yields of P and PMe, respectively. The theoretical dependence of q on concentration $[\mathrm{Me_0}]$ is

$$q = \frac{2q_{\mathrm{P}}K_d + q_{\mathrm{PMe}}(\{[K_d + ([\mathrm{P_0}] - [\mathrm{Me_0}])]^2/4 + [\mathrm{Me_0}]K_d\}^{1/2} - [K_d + ([\mathrm{P_0}] - [\mathrm{Me_0}])]/2)}{2K_d + \{[K_d + ([\mathrm{P_0}] - [\mathrm{Me_0}])]^2 + 4K_d[\mathrm{Me_0}]\}^{1/2} - [K_d + ([\mathrm{P_0}] - [\mathrm{Me_0}])]} \tag{5.35}$$

The shape of the curve $q([\mathrm{Me_0}])$ depends on the relationship between $[\mathrm{P_0}]$ and K_d (Figure 5.1). If $K_d \ll [\mathrm{P_0}]$, the dependence $q([\mathrm{Me_0}])$ is a straight line with an abrupt break (saturation) at $[\mathrm{Me_0}] = [\mathrm{P_0}]$. If $K_d = [\mathrm{P_0}]$, the curve $q([\mathrm{Me_0}])$ is far from saturation at $[\mathrm{Me_0}] = [\mathrm{P_0}]$. If $K_d \gg [\mathrm{P_0}]$, the curve is even less steep. The same relationships are valid in the more complicated case of a multisite protein.

It is evident that if $[\mathrm{P_0}] \gg K_d$, it is impossible to determine the binding constant from the $q([\mathrm{Me_0}])$ curve, but the stoichiometry of binding is seen clearly in this case. In contrast, if $[\mathrm{P_0}] \ll K_d$, it is difficult to determine stoichiometry (the

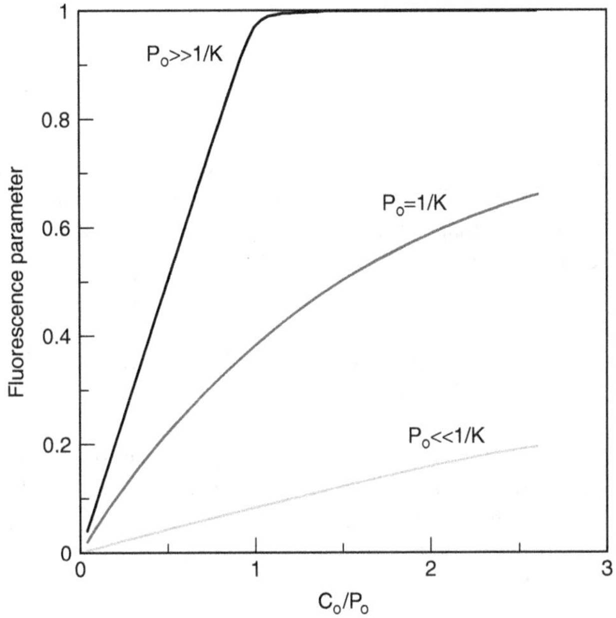

Figure 5.1. Model curves of spectrofluorimetric titration of a single-site protein by metal ions at different ratios of protein concentration and dissociation constant of the complex.

curve is very smooth), but the binding constant can be measured with sufficient accuracy. In the case of a very low K_d value, it is usually impossible to decrease $[P_0]$ so that it would be comparable with K_d because of the limited sensitivity of the physical instrument (e.g., spectrofluorimeter). One is then forced to carry out measurements under the conditions in which $[P_0] \gg K_d$. This does not permit determination of the binding constant.

In the opposite extreme, of a very high value of K_d, it is easy to measure its value if the binding stoichiometry is known. In this case the curve $q([Me_0])$ plateaus at very high $[Me_0]$ concentrations when $[Me_0] \gg K_d$ and the concentration of free [Me] is approximately equal to $[Me_0]$.

As mentioned above, the optimal situation is when $[P_0]$ is comparable to K_d. In this case one can measure both the stoichiometry and the binding constant. A convenient method of evaluation of K_d is computer fitting of the theoretical curve $q([Me_0])$ to experimental points by means of variation of K_d. Even in the absence of any other components in solution except for the protein and Ca^{2+} ions, complete characterization of the calcium binding properties of a protein is a challenging task.

5.2. KINETICS OF BINDING

It is important to know the association (k_{on}, $M^{-1}\ s^{-1}$) and dissociation (k_{off}, s^{-1}) rate constants when studying cell signaling. These on and off rates are related to the equilibrium constant by

$$K_d = \frac{k_{\text{off}}}{k_{\text{on}}} \tag{5.36}$$

For the stronger calcium binding sites, k_{on} is often very close to the diffusion-controlled limit, about $10^9\ M^{-1}\ s^{-1}$. For Mg^{2+} ions, k_{on} is usually several orders of magnitude lower than that for Ca^{2+} ions, while the k_{off} values for calcium and magnesium are close to one other.

5.3. PARTITION OF FREE ENERGY OF BINDING (ΔG) AMONG ENTHALPY (ΔH) AND ENTROPY (ΔS)

The thermal unfolding of most calcium binding proteins does not lead to the formation of a random coil structure (the main-chain [φ, ψ] and side-chain [χ's] dihedral angles are random, constrained only by van der Waals contacts). Instead, it results in the formation of a partially unfolded state with partially conserved metal binding sites. The affinity of this partially unfolded state(s) for calcium can be rather high and should be taken into consideration when evaluating in vitro experiments.

The situation may be additionally complicated since in the absence of bound cations, some metal binding proteins (e.g., α-lactalbumin and equine lysozyme) unfold at 20°C or even lower (Figure 5.2). In this case, the measurement of metal binding at room temperature gives only apparent metal binding constants and involves at least one intermediate state in the course of the measurements. Thus, the measurement of metal binding constants at a single fixed temperature may

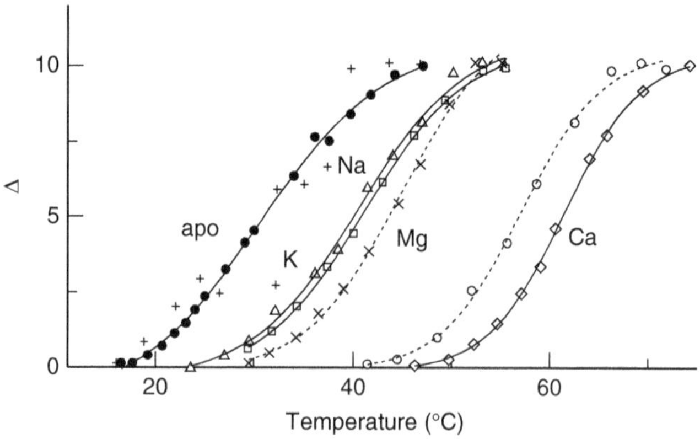

Figure 5.2. Thermal unfolding transition for various α-lactalbumin states (apo, Na^+-, K^+-, Mg^{2+}-, and Ca^{2+}-loaded) measured by intrinsic fluorescence. Δ is a fraction of conversion from the native to the thermally unfolded state.

lead to erroneous conclusions. To characterize the system completely, one should study metal binding at different temperatures and measure thermal unfolding of the protein in the presence of various metal concentrations. Given these data, one can construct a *phase diagram* using coordinates of free metal ion concentration and temperature (Permyakov and Permyakov, 2006).

Strictly speaking, the metal binding–induced structural or thermal transition in a protein cannot be considered as a classical phase transition, since certain distinctive features of phase transitions, such as the presence of a phase boundary, are absent in this case. Here, the term *phase transition* is applied to intramolecular transitions occurring within the protein molecule in response to metal association/dissociation or change of temperature. Nevertheless, in this case the phase diagram represents the most general characterization of the protein–metal system. It allows the easy visualization of regions of predominance of different protein states and the prediction of protein states under various experimental conditions. The phase diagram can be constructed relatively easily for proteins that possess a single metal binding site.

The equilibrium scheme of the binding of one metal ion (Me) to the protein molecule, taking into consideration equilibrium between native (P, PMe) and thermally changed (P*, P* Me) states of the protein, is

$$\begin{array}{ccc} & K_{\mathrm{Me}} & \\ & \mathrm{P} + \mathrm{Me} \leftrightarrow \mathrm{PMe} & \\ \gamma \ \downarrow\uparrow & & \downarrow\uparrow \ \beta \\ & \mathrm{P}^* + \mathrm{Me} \leftrightarrow \mathrm{P}^*\mathrm{Me} & \\ & K^*_{\mathrm{Me}} & \end{array} \tag{5.37}$$

in which K_{Me} and K^*_{Me} are intrinsic metal ion dissociation constants for the native and thermally denatured protein, respectively, and γ and β are equilibrium constants of the thermal denaturation of the protein in its apo and metal ion-bound forms, respectively:

$$K_{\mathrm{Me}} = \exp\left[\frac{(\Delta H_{\mathrm{Me}} - T\Delta S_{\mathrm{Me}})}{RT}\right] \tag{5.37}$$

$$K^*_{\mathrm{Me}} = \exp\left[\frac{(\Delta H^*_{\mathrm{Me}} - T\Delta S^*_{\mathrm{Me}})}{RT}\right] \tag{5.38}$$

$$\gamma = \exp\left[-\frac{(\Delta H_{\alpha} - T\Delta S_{\alpha})}{RT}\right] \tag{5.39}$$

$$\beta = \exp\left[-\frac{(\Delta H_{\alpha} - T\Delta S_{\beta})}{RT}\right] \tag{5.40}$$

ΔH_{Me}, ΔH^*_{Me} and ΔS_{Me}, ΔS^*_{Me} are enthalpy and entropy changes for the metal ion binding to native and thermally denatured protein. ΔH_{α}, ΔH_{β} and ΔS_{α}, ΔS_{β} are enthalpy and entropy changes for the thermal transitions in the apo and metal ion-bound protein. ΔH_{α} and ΔS_{β} can be determined using experiments studying the thermal denaturation of the apo-protein, while ΔH_β and ΔS_β can be determined from the thermal denaturation curve for the metal ion-bound protein.

The apparent metal ion dissociation constant determined from the fluorescence experiment is

$$\frac{1}{K_{app}} = \frac{[PMe] + [P^*Me]}{([P] + [P^*])[Me]} = \frac{1}{K_{Me}} \frac{1+\beta}{1+\gamma} \tag{5.41}$$

$$\frac{K^*_{Me}}{K_{Me}} = \frac{\gamma}{\beta} \tag{5.42}$$

Knowledge of all the thermodynamic parameters for metal cation binding allows calculation of the thermal denaturation curve of the protein in the presence of any given concentrations of the metal ion. More important, this knowledge allows computation of a phase diagram using the free calcium–temperature coordinates. Figure 5.3 shows such phase diagrams for bovine α-lactalbumin and equine lysozyme, proteins with a single strong calcium binding site (Permyakov et al., 2006). The phase diagram depicts both regions of predominance of separate protein states and the areas of transition between them, including lines of half-transitions in temperature (1) and calcium (2) scales. Thus, knowing the current temperature and pCa ($-\log[Ca^{2+}]$) values, the actual protein state can be assessed. Moreover, limits of curves 1 and 2 obviously demonstrate midtransition

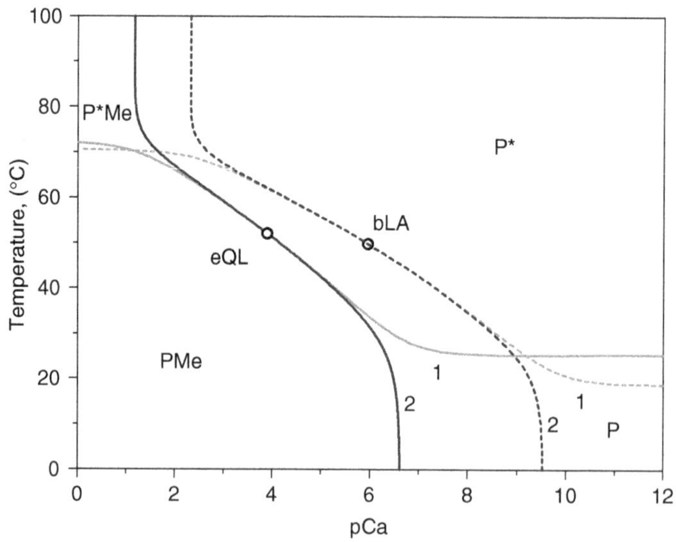

Figure 5.3. Phase diagram of equine lysozyme (eQL) and bovine α-lactalbumin (bLA) in the free calcium concentration–temperature space, calculated according to the four-states scheme (5.37). Curves 1 and 2 correspond to a half-transition for binding of calcium and thermal denaturation, respectively.

temperatures for apo- and calci-protein (curve 1, ordinate values) and calcium binding affinities of native and denatured states of protein (curve 2, abscissa values).

One is heartened by the expansion to proteins and improvement in techniques used to study calcium binding (Chapter 6). A great deal of empirical knowledge has been gathered, and the physical properties of these proteins have been related to their functions. Nonetheless, it is sobering to realize that given the sequence or even the structure of a protein, we cannot identify, with certainty, the site at which calcium might bind, and even if we can, we cannot estimate the binding affinity. Our ability to predict is poor; more disturbing, we do not really understand why.

6

EXPERIMENTAL METHODS USED TO STUDY CALCIUM BINDING TO PROTEINS

As noted in Chapter 1, bioassays and/or selective precipitation by oxalate provided the earliest assays of calcium concentrations: total and free ions. In Chapter 5 we developed three interrelated themes:

1. Stoichiometry, affinity, and cooperativity of calcium binding
2. Kinetics of binding
3. Partition of free energy of binding (ΔG) among enthalpy (ΔH) and entropy (ΔS)

In this chapter we describe the principal methods now used to study calcium binding and its effects on the binding protein. These methods are described or illustrated in terms of calcium binding; however, they are applicable to other metal–protein interactions (Chapters 7 and 8). All of these methods have associated systematic and random errors. It is essential that any determination be accompanied by controls and have some estimate of error. Chapters 10 to 15 characterize several homolog families of calcium binding proteins and show examples of their crystal and/or solution structures as determined by nuclear magnetic resonance (NMR).

6.1. RADIOACTIVITY

^{45}Ca has a half-life of 162.7 days. It emits β-particles with a maximum energy of 0.252 MeV; these can be counted quantitatively by liquid scintillation. These

Calcium Binding Proteins, By Eugene A. Permyakov and Robert H. Kretsinger

β-particles are 50% absorbed in about 50 cm of air and can be recorded by liquid detectors such as a solution of terphenyl in benzene. As in the case of atomic flame absorption spectroscopy, the radioactivity method is well suited to determining total calcium ($[Me_0] = [Me] + [PMe]$), or it can be used to measure $[Ca^{2+}]$ in a solution separated from the protein solution by a porous membrane in equilibrium dialysis. ^{45}Ca-labeled proteins can be measured semiquantitatively in gel electrophoresis experiments by direct measurement of β-particle emission by film darkening or by liquid scintillation counting of excised bands (reviewed by Skoog et al., 1998; Villarroya et al., 1999).

6.2. ION-SELECTIVE ELECTRODES

Compared to many other analytical techniques, ion-selective electrodes are relatively inexpensive, are effective over a wide range of concentrations, and enjoy a wide range of applications. An ion-selective electrode measures the potential of a specific ion in solution; the pH electrode is an ion-selective electrode for the H^+ ion (reviewed by Harris, 1999). This potential is measured against a stable reference electrode of constant potential. The potential difference between the two electrodes depends on the activity of the specific ion in solution. The relationship between the ionic concentration (activity) and the electrode potential is given by the *Nernst equation*:

$$E = E_0 + \frac{2.303RT}{nF} \log(A) \tag{6.1}$$

where E is the potential (in millivolts) between the measuring and reference electrodes, E_0 a constant that is characteristic of the particular ion-selective electrode–reference pair, R the gas constant, T the absolute temperature, n the charge on the ion (with sign), F the Faraday constant (96,500 C), and $\log(A)$ the logarithm of the activity of the measured ion, the activity being related to the concentration of the specific ion. Ion-selective electrodes have been developed for different ions. Such electrodes, long available for H^+, are commercially available for K^+, Na^+, and Ca^{2+} ions. They can measure $[Ca^{2+}]$ from 10^{-3} to $5\ 10^{-7}$ at 5 to 10% accuracy in volumes of less than 0.05 m.

A liquid-based ion-selective electrode uses a mobile carrier to transport the selected ion across a membrane impregnated with a liquid solution of the carrier. In the case of the Ca^{2+} ion electrode, the membrane is saturated with a hydrophobic calcium chelator (e.g., didecylphosphate dissolved in dioctylphenylphosphonate).

No electrode responds exclusively to one type of ion, and the selectivity coefficient gives the relative response of the electrode to different species with the same charge. The most serious interference with the Ca^{2+} electrode comes from Zn^{2+}, Fe^{2+}, Pb^{2+}, and Cu^{2+}, but high concentrations of Sr^{2+}, Mg^{2+}, Ba^{2+}, and Na^+ also interfere.

Micropipettes appropriate for intracellular use are now available. They are easily contaminated, and specificities vary. They should be calibrated with each use. Despite these problems, they are extremely valuable because in contrast to flame absorption spectroscopy and radioactivity measurements, they selectively measure free calcium as opposed to total calcium, and can do so in vivo. With the exception of calcium-specific dyes and fluors, these electrodes are the only technique that measures free calcium directly.

6.3. CALCIUM BUFFERS

Numerous small organic molecules bind calcium with moderate to high affinity and with reasonable selectivity relative to magnesium. Ethylenediaminetetraacetic acid (EDTA) and ethylene glycol tetraacetic acid (EGTA) (Figure 6.1) are well characterized and do not interact with (most) proteins (reviewed by Bers et al., 1994; Bers, 1996; Harris, 1999). When these atoms bind to a metal ion, the ligand atoms lose their protons. Frequently, the concentration of free Ca^{2+} ion is cited not from direct measurement (by the ion-specific electrode) but is calculated on the basis of the calcium dissociation constants of ED/GTA (Reitz and Pollack, 2003).

The most commonly used calcium chelator, EDTA, is hexaprotonic, with pK values of 0.0, 1.5, 2.0, 2.69, 6.13, and 10.37 (2.0, 2.65, 8.85, and 9.46 for EGTA). The first four pK values apply to carboxyl protons, and the last two are for the ammonium protons. The neutral acid is tetraprotic, H_4Y, and the commonly used reagent is the disodium salt $Na_2H_2Y \cdot 2H_2O$. Many of the EDTA–metal complexes adopt octahedral geometry. Some complexes of $EDTA^{4-}$ adopt more complex structures, due to the formation of an additional bond to water (i.e., seven-coordinate complexes, or to the displacement of one carboxylate ligand by water). EDTA forms especially strong complexes with Mn^{2+}, Cu^{2+}, Fe^{2+}, Pb^{2+}, and Co^{2+} (Table 6.1). EDTA is used to detoxify metal ions in chelation therapy (e.g., for treatment of mercury and lead poisoning).

BAPTA [1,2-bis(o-aminophenoxy)ethane-N, N, N', N'-tetraacetic acid] (Figure 6.1) is another calcium-specific polyaminocarboxylic acid. The presence of four carboxylic acid functional groups makes possible the binding of two Mg^{2+} ions. The extensive flexibility of the carboxylate ligands is critical to the coordination of metal ions. The dissociation constant (K_d) of BAPTA for calcium at pH 7.2 is 1.6×10^{-7} M.

Three precautions should be noted when working with calcium chelators. First, the affinities of most calcium buffers are pH dependent. Not only should the pH of the medium be measured, but one should also be aware that the addition of ED/GTA itself can change the pH of the medium, thereby affecting the reaction in question. Second, the addition of ED/GTA will consume some of the calcium in solution; the final pCa of the solution may be a half decade from the pK_d of the buffer. Finally, magnesium will also be buffered and to some extent consumed

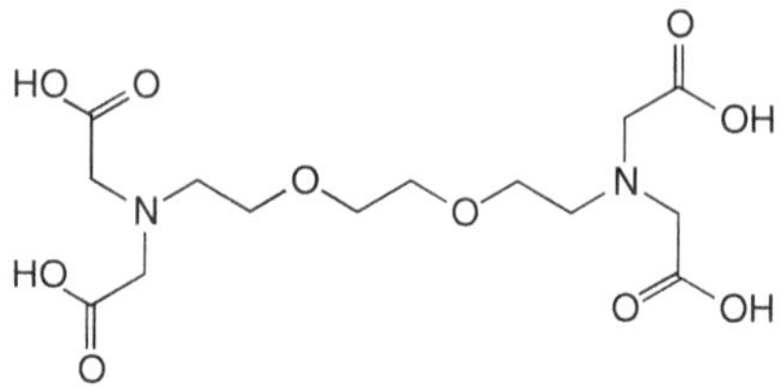

EGTA

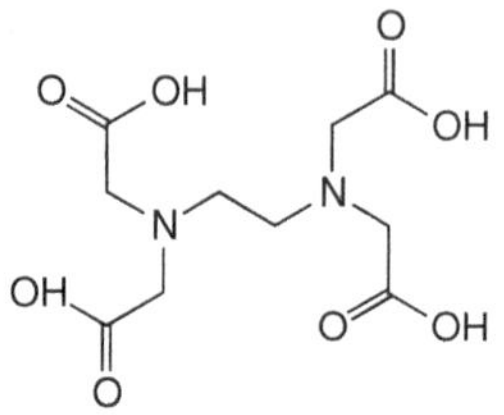

EDTA

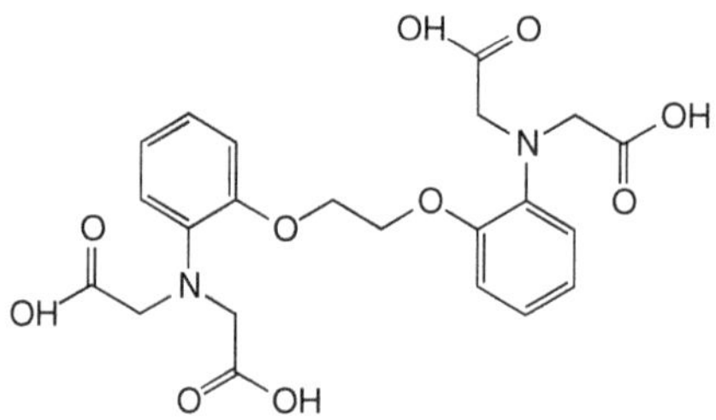

BAPTA

Figure 6.1. The structures of EGTA, EDTA, and BAPTA.

by the ED/GTA. Programs that calculate the final pCa, pMg, and pH of ED/GTA-buffered solutions are available (Reitz and Pollack, 2003).

6.4. DIALYSIS, EQUILIBRIUM, AND FLOW

In a simple dialysis experiment, two compartments are separated by a membrane that is permeable to small ions but not to proteins. The ions are usually labeled with radioactive isotopes for quantitative determinations, and ions bound to the macromolecule in the equilibrium state are determined directly from the difference between the free concentration in the dialysate and the total concentration in the protein solution. At equilibrium the free metal ion concentrations [Me] on both sides of the membrane should be equal; on the no-protein side, $[Me_{tot}]_{no} = [Me]_{prot}$. On the protein side, one can measure $[P_{tot}]_{prot}$ and $[Me_{tot}]_{prot}$. From $[Me]_{prot}$, $[P_{tot}]_{prot}$, and $[Me_{tot}]_{prot}$ one can calculate K_d if one assumes an n (stoichiometry). Measurements at several values of $[P_{tot}]$ should reveal n.

TABLE 6.1. Formation Constants for Metal–EDTA and Metal–EGTA Complexes[a]

Ion	Log K_{EDTA}	Log K_{EGTA}
Na^{+}	1.66	1.38
Mg^{2+}	8.79	5.2
Ca^{2+}	10.69	10.97
La^{3+}	15.50	15.79
Mn^{2+}	13.87	12.11
Fe^{2+}	14.32	11.92
Zn^{2+}	16.50	12.49

Source: Harris (1999).

[a]Equilibrium constant for the reaction $M^{n+} + Y^{4-} \leftrightarrow MY^{n-4}$ at 20°C and ionic strength 0.1 M.

Several cautions: Dialysis membranes should be washed thoroughly to remove calcium and other contaminants. Further, one should confirm that the same amount of protein is present in the solution at the end as at the beginning of the experiment since many proteins adhere to the membrane and/or pass through the membrane during extended exposure. The assumption that $[\text{Me}]_{\text{no}} = [\text{Me}]_{\text{prot}}$ is valid only to the extent that the osmotic pressure (total molar concentration of solutes) and ionic charges are balanced on both sides. In equilibrium dialysis, the actual concentrations of interacting species are measured after equilibrium has been established between two compartments separated by a semipermeable membrane. This is a relatively slow technique, making this approach less suitable when one of the components is labile. The chemical equilibrium between the ligand and the protein is usually reached within seconds or minutes.

The degree of binding or saturation fraction is $\nu = [\text{Me}]/[\text{P}]_0$. If the protein concentration, $[\text{P}]_0$, is known, a Scatchard plot can be used to determine the binding constants and the number of binding sites. In this case, values of ν should be plotted on the abscissa against $\nu/[\text{Me}]$ on the ordinate. If only one class of binding sites is present, the Scatchard plot results in a straight line with slope equal to $-K_a$ (association constant). The intercept on the abscissa give the value n (number of binding sites). If two classes of binding sites are involved, the plot takes the form of a hyberbola. In this case, the asymptotes have slopes equal to $-K_a$ for each class of site, and their intercepts on the abscissa give the two values for n.

In this method, the metal binding equilibrium, which is usually obtained within less than a second, has to be assessed after attainment of the diffusion equilibrium of metal ions across the membrane, which usually takes a much longer time, on the order of several hours.

Flow dialysis is a method for quickly measuring binding of ligand molecules that are able to diffuse across a semipermeable membrane. The system consists of an upper chamber containing the protein–ligand system under investigation, and a lower chamber, separated from the upper chamber by a semipermeable dialysis membrane, through which a mobile phase flows. Each solution in both of the chambers is mixed continuously with a magnetic stirring bar, and the concentration of free ligand in the protein solution is determined based on the rate of diffusion into the buffer chamber, which is proportional to the concentration of free ligand. The buffer chamber is connected to the reservoir and is flushed continuously with fresh buffer solution at a constant rate and the outlet is connected to a fraction collector to monitor the radioactivity in the effluent. When small amounts of the labeled metal ion in a small volume are added to the sample chamber, chemical equilibrium is usually attained within a fraction of a second and the free ions diffuse into the buffer chamber at a rate depending on the equilibrium concentration and the characteristics of the membrane. Under the constant flow rate of the buffer solution in the buffer chamber, the radioactivity in the buffer chamber becomes constant in several minutes when the steady state is reached; this is a measure of the concentration of the metal ion in the sample chamber. At a given flow rate, the concentration of metal ion measured in the mobile phase is proportional to the concentration of free metal ion in the upper chamber (Westerhoff et al., 1989; Porumb, 1994; Andre and Linse, 2002).

Usually, a relatively small amount of radioactive metal ion is added to a large pool of potential binding sites at protein concentrations in which practically all ligand is bound. By adding unlabeled ion, the bound labeled metal ion is replaced by the unlabeled ion and appears in the mobile phase. At the end, an excess of unlabeled metal ions is added to chase all labeled ions from the binding sites. The radioactivity measured in the mobile phase then reflects all of the radioactive metal ion being unbound, as in an experiment without protein. Evaluation of the concentration of the labeled metal ions in the mobile phase yields a reliable calcium binding curve in about an hour. The binding constants and the stoichiometry of calcium binding can be obtained by fitting the observed data to a standard curve (Andre and Linse, 2002).

6.5. PROTEOLYSIS

Proteolytics enzymes such as trypsin, chymotrypsin, pepsin, and their derivatives cleave at specific sites in a protein (reviewed by Righetti et al., 2004). The resulting mixture of peptides can be analyzed by gel electrophoresis, column chromatography, and/or mass spectroscopy. The relative reactivities of these sites may vary with calcium and/or target binding. The challenge is not so much identifying the sites and their reactivities but in relating these changes to calcium binding (reviewed by Permyakov, 2006).

6.6. DEUTERIUM EXCHANGE

In hydrogen–deuterium exchange a covalently bonded hydrogen atom is replaced by a deuterium atom, or vice versa. Proteins contain a number of hydrogens that can exchange with hydrogen in the surrounding solvent. If the normal H_2O solvent is changed to heavy water, D_2O, the protein gradually becomes deuterated. Hydrogen exchange is frequently used in conjunction with NMR (reviewed by Wüthrich, 1995; Hoofnagle et al., 2003; Zhu et al., 2004; Redfield, 2004). Ionization mass spectrometry can also be used to measure the hydrogen exchange (reviewed by Eyles and Kaltashov, 2004; Garcia et al., 2004).

Hydrogen–deuterium exchange gives information about the solvent accessibility of various parts of the protein and thus its tertiary structure. The rate of hydrogen exchange depends on hydrogen bonding, solvent accessibility, and pH. Folded proteins can have residues with hydrogen exchange rates as much as 10^9 times slower than the same residue that is solvent accessible. Local and global stability parameters can be derived from the hydrogen exchange rates. Calcium-induced changes of protein structure can be detected by the deuterium exchange method.

6.7. ISOTHERMAL TITRATION CALORIMETRY

Isothermal titration calorimetry (ITC) allows a study of the energetics of interaction of two molecules or ions (reviewed by Jelesarov and Bosshard, 1999; Leavitt and Freire, 2001; Cliff and Ladbury, 2003). When two molecules interact, heat is either generated or absorbed, and by measuring these interaction heats, dissociation constants (K_d), binding stoichiometry (n), and thermodynamic parameters, including enthalpy (ΔH) and entropy (ΔS), can be determined accurately. ITC is the only method for the direct experimental determination of ΔH. Carrying out the experiment at various temperatures allows determination of the heat capacity (ΔC_p) of the reaction.

An isothermal titration calorimeter has two identical cells made of a highly efficient thermal conducting material, surrounded by an adiabatic jacket. Sensitive thermopile/thermocouple circuits are used to detect temperature differences between the reference cell (filled with buffer or water) and the sample cell containing the macromolecule. The single titration experiment is performed at a constant temperature by titrating one binding partner (titrant) into a solution containing the other binding partner (titrand) in the sample cell of the calorimeter. After each addition of a small aliquot of titrant, the heat released or absorbed in the sample cell is measured with respect to a reference cell filled with solvent. The heat change is proportional to the electrical power required to maintain a constant small temperature difference between the sample and reference cells.

The experimental data can be plotted as the total heat accumulated up to injection i normalized to the total ligand concentration at step i or against the ratio of the total ligand concentration at step i to the total receptor concentration. In the integral mode, the total cumulative heat is plotted against the total ligand concentration to yield a hyperbolic saturation curve (Figure 6.2). Titration experiments are typically fast (approximately 1 h), and deconvolution analysis of the binding curves yields accurate values of K_d (in the range 10^{-2} to 10^{-8} M), n, ΔH, and

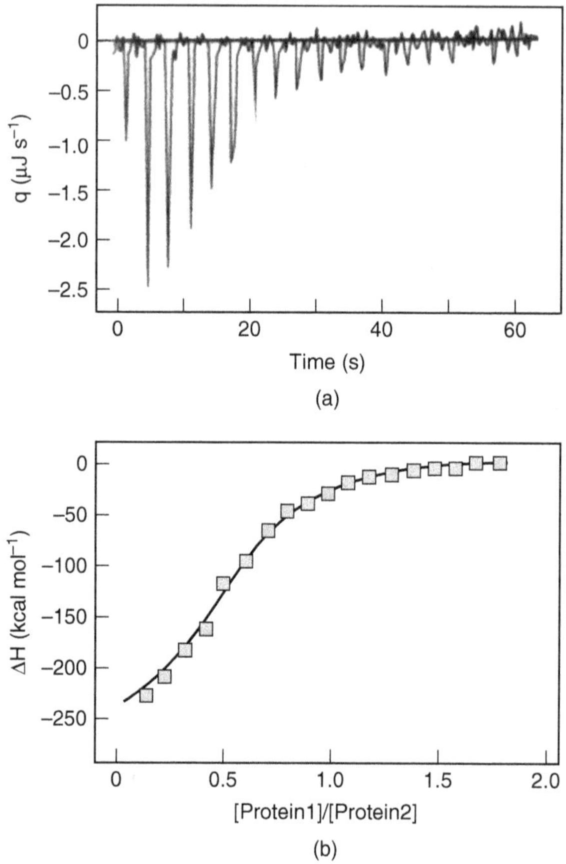

Figure 6.2. Isothermal titration of a protein solution 2 by a protein solution 1: (a) heat accumulated up in injection i vs. time; (b) total cumulative heat plotted against the total ligand concentration.

ΔS. In addition, kinetic data (for slow reactions) may be obtained in a batch injection experiment. The deconvolution has a firm thermodynamic foundation and avoids additional assumptions, as is necessary in an analysis of spectroscopic binding data.

For very tight binding reactions, optimal concentrations of the reagents are too low to yield measurable heat changes. It is for this reason that even with the most sensitive instruments presently available, a K_d value lower than 10^{-9} M (ΔG about -50 kJ mol^{-1} at room temperature) cannot be measured accurately. The method can be used to study the interactions of calcium with proteins. The isothermal titration calorimetry method can be used in combination with simultaneous (global) least squares analysis of titrations with calcium, with magnesium, with calcium in the presence of magnesium, or with either in the presence of a chelator (e.g., EDTA or EGTA) (Henzl et al., 2003).

6.8. DIFFERENTIAL SCANNING CALORIMETRY

Differential scanning calorimetry (DSC) is the most direct experimental technique available for studying the energetics of the thermally induced conformational transitions of biological macromolecules (reviewed by Plum and Breslauer, 1995; Sanchez-Ruiz, 1995; Shnyrov et al., 1997). DSC measures the heat capacity of a solution as a function of temperature. Scanning calorimeters usually have twin cells (a sample cell and a reference cell) placed in an adiabatic jacket and operated in a differential mode. The solution containing the solute is placed in the sample cell and an equal volume of solvent is placed in the reference cell.

The system is heated quasiadiabatically at a constant rate and a certain amount of electrical power is used to compensate for the temperature difference between the two cells. The power difference is a direct measure of the heat capacity difference between the solution and the solvent ΔC_p. When a protein solution is heated, the partial specific heat capacity of the protein $C_p(T)$ follows a curve reflecting the cooperative thermal unfolding of the protein (Figure 6.3). The area under the $C_p(T)$ profile represents the enthalpy of unfolding, the temperature of the maximum of the heat sorption curve provides the transition temperature, and the difference in the heat capacities of the native and unfolded states defines the temperature dependence of the enthalpy and entropy functions and, thus the temperature dependence of protein stability. In addition to these parameters, scanning calorimetry provides a direct estimate of the modes of protein unfolding. The sharpness of the heat capacity profile gives another characteristic of the

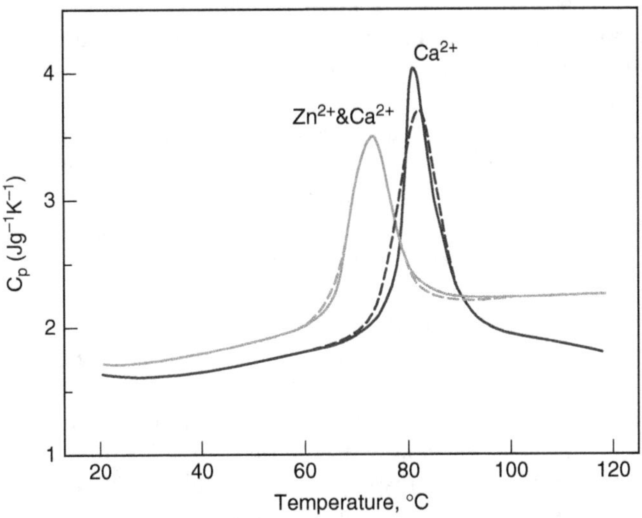

Figure 6.3. Heat sorption curves measured for calcium and for calcium- and zinc-loaded recoverin. Solid lines, experimental data; dotted lines, theoretical curves fitted to the experimental points according to the simple two-state scheme (Permyakov et al., 2003).

process observed, the effective enthalpy of transition, referred to as the *van't Hoff enthalpy* (Chapter 5). The ratio of these two enthalpies gives information about the mode of the transition observed. A ratio of 1 indicates that the transition observed is a two-state process, proceeding from the native to the unfolded state without a significant population of intermediates.

The DSC method is used to study the effects of metal ion binding on thermal transitions of proteins to get information about the character of the metal-induced changes in the protein structure.

6.9. MASS SPECTROSCOPY

Mass spectrometry consists of ionizing chemical compounds to generate charged molecules or fragments and then of measuring their mass/charge ratios. To measure the characteristics of individual ions, they are moved in external electric and magnetic fields. The mass spectrometer is a powerful detector for both quantitative and qualitative analysis (reviewed by Skoog et al., 1998; Heck and Van Den Heuvel, 2004; Page et al., 2004; X. Yan et al., 2004; Yates, 2004). It consists of three fundamental parts: the ionization source, the analyzer, and the detector. The sample under investigation has to be introduced into the ionization source of the instrument. Inside the ionization source the sample molecules are ionized, because ions are easier to manipulate than are neutral molecules. These ions go into the analyzer region of the mass spectrometer, where they are separated according to their mass (m)/charge (z) ratios (m/z). The ions separated are detected, and this signal is sent to a data system where the m/z ratios are stored together with their relative abundance for presentation as an m/z spectrum.

The ionization methods used for the majority of biochemical analyses in mass spectroscopy are electrospray ionization (ESI) and matrix-assisted laser desorption ionization (MALDI). ESI, one of the atmospheric pressure ionization (API) techniques, is well suited to the analysis of polar molecules, ranging from less than 100 Da to more than 1,000,000 Da, including proteins. In this method the sample is dissolved in a volatile polar solvent and pumped through a stainless steel capillary needle at a rate of a few microliters per minute. A high voltage of several kilovolts is applied to the tip of the capillary, which is situated within the ionization source of the mass spectrometer, and as a consequence of this strong electric field, the sample emerging from the tip is dispersed into an aerosol of highly charged droplets, a process that is aided by a coaxially introduced nebulizing gas flowing around the outside of the capillary. This gas, usually nitrogen, helps to direct the spray emerging from the capillary tip toward the mass spectrometer. Nanospray ionization is a low-flow-rate version of electrospray ionization. Using electrospray or nanospray ionization, an accuracy of 0.01% of the molecular mass is achievable. ESI is a soft ionization technique that is rapid and more sensitive than many other techniques used to characterize macromolecules. It is especially suitable for the study of proteins in their native state, as the solvent conditions for transferring and ionizing protein molecules from solution to the gas phase can be similar to those used for characterizing proteins in solution.

Standard "denaturing" ESI mass spectrometry conditions used to yield the best sensitivity in the determination of protein molecular weight are solutions of pH 2.0 to 4.0 in the positive ion mode and 8.0 to 10.0 in the negative ion mode. An organic co-solvent such as acetonitrile or methanol is often used to enhance sensitivity and signal stability. To study protein interactions by ESI mass spectrometry, nondenaturing conditions, including a volatile buffer such as ammonium acetate or ammonium carbonate close to neutral pH and low temperature, are used. However, these milder conditions lead to a significant decrease in sensitivity. Mass spectrometer parameters such as capillary temperature, ion mode, and voltage must be optimized for each system.

In the MALDI method, a laser beam (ultraviolet or infrared pulsed laser) serves as both a desorption and a ionization source. The matrix plays a key role in this technique by absorbing the light energy and causing a small part of the target substrate to vaporize. The sample is premixed with a highly absorbing matrix compound. The matrix transforms the laser energy into excitation energy for the sample; this leads to sputtering of analyte and matrix ions from the surface of the mixture. Most commercial MALDI mass spectrometers now have a pulsed nitrogen laser of wavelength 337 nm. Once the sample molecules are vaporized and ionized, they are transferred electrostatically into a mass spectrometer, where they are separated from the matrix ions and detected individually. The mass accuracy in this method depends on the type and performance of the analyzer of the mass spectrometer, but most modern instruments should be capable of measuring masses to within 0.01% up to 40 kDa.

A number of mass analyzers are currently available, the better known of which include quadrupoles, time-of-flight (TOF) spectrometers, magnetic sectors, and both Fourier transform and quadrupole ion traps. A quadrupole separator consists of four parallel metal rods to which are applied both a constant voltage and a radio-frequency oscillating voltage. The electric field deflects ions in complex trajectories as they migrate from the ionization chamber toward the detector; this field can be tuned to allow ions with only one particular m/z ratio to reach the detector. In a TOF mass spectrometer, sample molecules are converted to cations by electron impact. About 3000 to 20,000 times per second a grid voltage is applied to accelerate the ions through 3000 V, expelling them into the drift tube with a constant kinetic energy. Ions with the same kinetic energy but different values of m/z have different velocities: Light ions move faster and reach the detector sooner than do heavy ions. The mass spectrum shows detector current as a function of time. Using electrospray or nanospray ionization, a mass accuracy of within 0.01% of the molecular weight is achievable.

Both the range of molecular masses that can be resolved and the sensitivity and accuracy of their detections are quite remarkable. Often, one needs to know whether or not a protein(s) binds calcium and, if so, the stoichiometry of binding. In some cases one can compare the mass of an intact protein, only partially purified, with its mass after treatment with EGTA. Many metal ions are bound so strongly that they are not dissociated during ionization. Even a small fraction of the total mass with a shift in mass of 40 Da indicates a bound Ca^{2+} ion. Such experiments should be controlled by testing well-characterized calcium binding proteins under identical conditions.

Studies of calcium binding proteins in their native states using mass spectrometry can be grouped into three types: (1) stoichiometry and cooperativity of calcium binding determined by means of examination of the metal-bound species present in mass spectra and comparison to spectra of the metal-free protein; (2) conformational changes detected by a shift in the mass-to-charge (m/z) envelope or with the use of hydrogen–deuterium exchange; and (3) interactions of calcium binding proteins with other molecules not dissociated by the ionization process.

6.10. CALCIUM-SPECIFIC DYES AND FLUORS

Organic ligands whose spectral properties are changed by the coordination of calcium are widely used (reviewed by Roe et al., 1990; Simpson, 2002; Demaurex and Frieden, 2003). There are three general requirements for such calcium-sensitive dyes. First, the selectivity of the dye should be at least three decades relative to magnesium, sodium, and potassium—the ions most likely to compete and give a false signal. This selectivity should be four or five decades for use in vivo. The second consideration is that the signal from the dye be strong enough, and different enough in the calci vs. the apo form, so that the dye can be used in a small quantity and not function as a calcium buffer. Finally, if the dye is used in vivo, it should not be toxic.

The development of calcium dyes such as fura (Roe et al., 1990), a poly(aminocarboxylic acid), which can be injected into or targeted to the cytosol or even to the matrix of mitochondria, has revolutionized cellular calcium physiology. The most commonly used fluorescent calcium indicator is fura-2. The calcium binding properties of these indicators are formed by the presence of a tetracarboxylic acid core as found in the calcium chelator EGTA. Whereas the calcium binding of EGTA is highly pH dependent, the original calcium indicator, quin-2, and its successors were designed around an EGTA derivative, BAPTA [1,2-bis(*o*-aminophenoxy)ethane-N, N, N', N'-tetraacetic acid] (Figure 6.1), a largely pH-insensitive calcium chelator. Since the introduction of quin-2, fura-2, and indo-1, numerous other fluorescent calcium indicators have been synthesized, each with varying fluorescence characteristics and dissociation constants for calcium. The fundamental properties of these indicators are similar, in that the binding of the Ca^{2+} ion produces a wavelength shift in either the excitation or emission fluorescence spectra.

Single-wavelength calcium dyes show no calcium-induced changes in the position of absorption or emission maxima (Table 6.2). The intensity of their emission increases (with the exception of fura-red) when they bind calcium. This means that they are dependent on the intracellular dye concentration and other non-calcium-associated effects: for example, changes in membrane thickness.

TABLE 6.2. Spectral Properties of Single-Wavelength Calcium Dyes

		Absorption Maxima (nm)		Emission Maxima (nm)	
Indicator	K_d (nM)	Ca^{2+} Free	Ca^{2+} Bound	Ca^{2+} Free	Ca^{2+} Bound
Quin-2	115	352	332	492	498
Fluo-3	400	503	506	526	526
Calcium green-1	189	506	506	534	533
Calcium green-2	574	506	506	531	531
Calcium green-5N	3.3	506	506	531	531
Rhod-2	1.0	556	553	576	576
Calcium orange	328	554	555	575	576
Calcium crimson	205	588	588	611	611

The binding of calcium to ratiometric calcium dyes causes shifts in their absorption or emission spectral maxima (Table 6.3). Spectral parameters reflecting maximum position are not dependent on the intracellular concentration of the dye and can be used successfully for measurements of calcium concentration. The most commonly used dyes are excited by ultraviolet light; this can lead to such problems as the fluorescence of cell constituents or photolysis of photosensitive caged compounds. In terms of fluorescence properties, the indicators can be divided into two main groups: those that are excited by near-ultraviolet light at 340 to 380 nm (e.g., quin-2, fura-2, indo-1) and those that are excited by visible light at or above 450 nm (e.g., fluo-3, calcium green, rhod-2) (Figure 6.4). Most of the fluorophores for the visible indicators are fluoroscein and rhodamine derivatives.

Normally, the dye is derivatized as an uncharged ester, which is easily taken up by the cell, whose esterases cleave off the alcohol group; the resulting charged molecule does not diffuse out through the cell membrane. Once inside a cell, the hydrolyzed indicator should not escape easily; however, rapid decreases in signal intensities often occur during experiments. This may be due to photobleaching and transport of the indicators out of the cell.

Aequorin is an EF-hand photoprotein (Chapter 11) originally isolated from the hydromedusa, *Aequorea victoria*, and other marine organisms (reviewed by Alvarez and Montero, 2002; Vysotski and Lee, 2004). Aequorin contains the luminophore coelenterazine. The binding of Ca^{2+} ions to three EF-hands results

TABLE 6.3. Spectral Properties of Ratiometric Calcium Dyes

		Absorption Maxima (nm)		Emission Maxima (nm)	
Indicator	K_d (nM)	Ca^{2+} Free	Ca^{2+} Bound	Ca^{2+} Free	Ca^{2+} Bound
Fura-2	224	362	335	512	505
Fura red	133	472	436	645	640
Indo-1	250	349	331	485	410

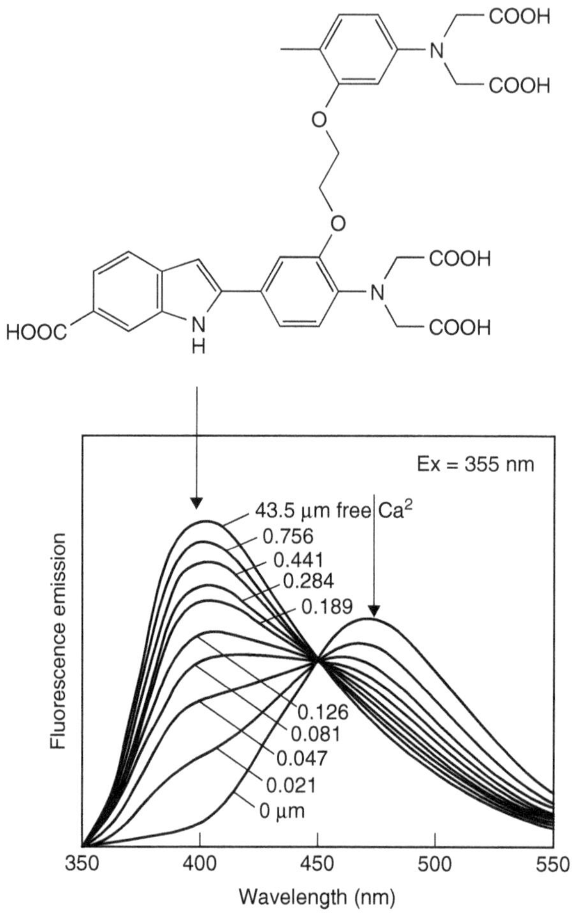

Figure 6.4. Structure and fluorescence spectra of indo-1 at different calcium concentrations. The single isosbestic point, marked by the arrow, indicates that each of the fluorescence curves can be represented as the partial sum of the curves of apo-indo-1 (peak, 482 nm) and of calci-indo-1 (peak, 398 nm). The ratio of fluorescence intensities at 405 and 450 nm can be used for calcium concentration measurements.

in changes of its structure and oxidation of coelenterazine to coelenteramide (Figure 6.5) a concomitant release of carbon dioxide and a quantum of blue light. The approximately third power dependence of aequorin's bioluminescence on [Ca^{2+}] allows the measurement of calcium from ~0.1 to > 100 μM. Unlike most fluorescent calcium indicators, calci-aequorin can be detected without illuminating the sample, thereby eliminating interference from autofluorescence and allowing simultaneous labeling with caged probes. The aequorin injected is depleted over time. The genes encoding aequorin and the enzymes required to synthesize coelenterazine can be incorporated into the genome of the cell under

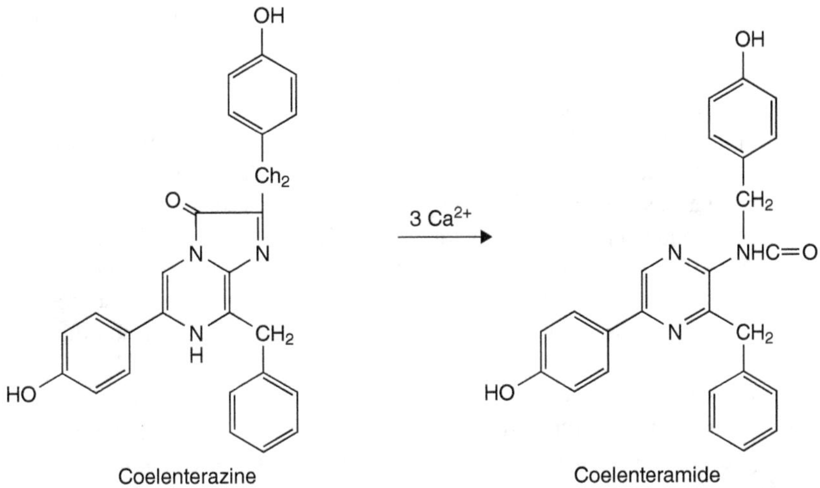

Figure 6.5. Oxidation of coelenterazine to coelenteramide in aequorin.

study, which then synthesizes its own calcium indicator (Alvarez and Montero, 2002).

Other fluorescent protein indicators for calcium have now been developed. They are based on fluorescence resonance energy transfer (FRET) between the blue or cyan emitting mutants of the green fluorescent protein (GFP) and the green or yellow emitting GFP mutants (Miyawaki et al., 1997). The fusion protein consists of two GFP mutants connected by calmodulin attached to a calmodulin binding peptide. When calcium binds to calmodulin, the complex binds to the calci-calmodulin binding peptide, bringing the GFP mutants sufficiently close for FRET to take place. Thus, when camelion-1 is excited at 380 nm, there is an increase in the 510/445 emission ratio on calcium binding. By using calmodulin mutants, this family of indicators should be able to monitor free calcium concentration in the range 10^{-8} to 10^{-2} M.

6.11. ATOMIC FLAME ABSORPTION SPECTROSCOPY

Atomic flame absorption spectroscopy is used to determine the concentration of total calcium in whatever ligand state in the presence of many other metals (reviewed by Jackson and Mahmood, 1994; D. C. Harris, 1999). The technique can be used to analyze the concentration of over 62 different metals in solution. In atomic absorption spectroscopy the radiation absorbed by the nonexcited atoms in the vapor state is measured. In emission spectroscopy, measurement is made of energy emitted when atoms in the excited state return to the ground state.

At the temperature of an air–acetylene flame (~2570 K), atoms of many elements exist mostly in the ground state. When a beam of light with the emission spectrum of the element that is to be determined passes through the flame, some of the ground-state atoms absorb energy of characteristic wavelengths (resonance lines) and undergo a transition to a higher-energy state. The unabsorbed radiation is passed through a monochrometer and detected by a photomultiplier. This method measures the intensity of characteristic absorption lines relative to the intensities of those lines in standard samples. Concentration determinations are made by comparing the observed absorption spectrum with a curve made on that instrument with standards of known concentration.

For some elements, such as sodium and potassium, an air–acetylene flame is hot enough not only to produce ground-state atoms but to transfer some of the atoms to an excited electronic state. The radiant energy emitted when the atoms return to the ground state is proportional to their concentration and is the basis of flame emission spectroscopy.

Ions or atoms in a sample must undergo desolvation and vaporization in a high-temperature source such as a flame or graphite furnace (atomization). Upon introduction, the sample solution is dispersed into a fine spray, the spray is then desolvated into salt particles in the flame, and the particles are subsequently vaporized into neutral atoms, ions, and molecules. All of these conversion processes occur in geometrically definable regions in the flame. The chamber is arranged in such a way that it gives an approximate 10 cm long path length. The height of the flame is controlled by means of the flow of the fuel mixture. For a low temperature flame (up to 2500 K), acetylene and air are used. A hot flame can be produced using acetylene and pure nitrous oxide (up to 3100 K). A hotter flame can be produced using acetylene and pure oxygen (up to 3400 K). Alternatively, a graphite tube is heated up to 3000 K by electrical current for atomization. This method requires less sample.

Excitation of atoms in atomic flame absorption spectroscopy is achieved by the light emitted by light sources with a line spectrum (line width 10^{-3} to 10^{-2} nm). Hollow cathode lamps are generally used for this purpose. Emission from these lamps arises due to the excitation of atoms at temperatures lower than that of the atomizer. The cathode of such a lamp is made of the element that is being measured. The lamp is filled with an inert gas, argon or neon, at low pressure. When about 600 V is applied between the anode and cathode, the filler gas is ionized and positive ions are accelerated toward the cathode. They strike the cathode and kick out metal atoms from the cathode. The free atoms are excited by collisions and then emit photons to return to the ground state. Lasers are also used in research instruments.

The atoms of a sample absorb this ultraviolet or visible light and make transitions to higher electronic energy levels. Unwanted lines are cut out by a monochromator. The emission is recorded by a photoelectrical multiplier. The amount of absorbed radiation at a characteristic wavelength is a quantitative measure of the concentration of the element to be analyzed.

A disadvantage of the atomic absorption technique is the nonlinearity of the calibration curves when absorbance becomes higher than 0.5 to 1.0. The relative standard deviations are between 0.3 and 1.0% for absorbances of 0.1 to 0.2.

Detection limits for flame atomic absorption spectroscopy vary enormously: from 1 to 5 ppb (e.g., Ca, Cd, Cr, Cu) to more than 1000 ppb (e.g., P). Some elements (e.g., B, C, Br) cannot be measured at all. Although the technique has often been replaced by ^{45}Ca determinations, it does circumvent the use of radioisotopes.

6.12. ABSORPTION SPECTROSCOPY

The polypeptide chain itself, free of ligands or posttranslational modifications, has low absorbance in the visible region. The side chains of Phe, Tyr, Cys, and especially, Trp have characteristic absorbances in the ultraviolet (UV) region (210 to 300 nm). To good approximation the UV absorption of a protein is a sum of its constituent amino acids. Changes in conformation of the protein, including those induced by calcium binding, usually induce little change in UV absorption of proteins. As a rule, such changes can be detected only in the differential measurement mode.

Infrared absorption of proteins is much more sensitive to changes in conformation. The absorption of infrared radiation excites vibrational transitions of molecules. In the mid- and far-infrared spectral regions this is generally the case when the frequencies of light and vibration are equal and when the molecular dipole moment changes during the vibration. Since vibrational frequency and probability of absorption depend on the strength and polarity of the vibrating bonds, they are influenced by intra- and intermolecular effects.

Certain bands are exquisitely sensitive to local changes in conformation. The peptide group has up to nine characteristic bands: amides A, B, I, II, VII (reviewed by Tamm and Tatulian, 1997; Barth and Zscherp, 2000, 2002; Jung, 2000; Pelton and McLean, 2000; Barth, 2007a,b). The amide A band (about 3500 cm^{-1}) and amide B (about 3100 cm^{-1}) originate from a Fermi resonance between the first overtone of amide II and the N–H stretching vibration. The bands of amides I and II are two major bands of the protein infrared spectrum. The amide I band (between 1600 and 1700 cm^{-1}) is associated mainly (70 to 85%) with the C=O stretching vibration and is directly related to the backbone conformation. Amide II results from the N–H bending vibration (40 to 60%) and from the C–N stretching vibration (18 to 40%). This band is sensitive to changes in conformation of the protein main chain. The hydrogen–deuterium exchange of amide protons can be monitored by the disappearence of the band characteristic of N–H bending near 1545 cm^{-1} (amide II) and the appearence of N–D absorption near 1455 cm^{-1} (amide II′). Amides III and IV are very complex bands, resulting from a mixture of several coordinate displacements. Out-of-plane motions are found in amides V, VI, and VIII. The shapes of the amide I band of globular proteins are characteristic of their secondary structures.

The determination of secondary structures in proteins can be carried out by means of Fourier transform infrared spectroscopy (FTIR) (reviewed by Cooper and Knutson, 1995; Jackson and Mantsch, 1995; Jung, 2000; Barth, 2007a,b). The availability of a high signal-to-noise ratio, measurement of digitized spectra, and the access to computers and software able to perform many operations on the spectra in a short time provide the very high potential of this method. The heart of a Fourier transform infrared spectrometer is its interferometer; the sample is usually placed between the output of the interferometer and the detector. The interferogram is the sum of contributions from all source wavelengths. Since the sample absorbs light of certain wavelengths, after passing the sample the interferogram contains the spectrum of the source minus the spectrum of the sample. An interferogram of a reference cell containing solvent is recorded and transformed into a spectrum. Then the interferogram of the sample is also recorded and transformed into a spectrum. The quotient of the second spectrum divided by the first is the infrared transmission spectrum of the sample. The interferogram is recorded at discrete intervals of time. The greater the number of the data points, the more time and memory are consumed by the Fourier transform computation. The mathematics of the Fourier transform dictate that the wavelength range of the spectrum is determined by how the interferogram is sampled: The closer the spacing between data points, the greater the wavelength range of the spectrum.

Curve fitting of amide I/I′ band profiles is often used to analyze underlying band components quantitatively. In the curve-fitting approach, the number of component bands estimated by Fourier self-deconvolution and derivative spectra, plus their approximated width, height, and shape, are used as input parameters in an iterative least-squares procedure that attempts to reproduce the measured amide I/I′ band profile by varying these parameters. When a reasonable fit is obtained, the fractional areas of the fitted components are taken as being directly proportional to the relative quantities of the structure elements they represent. The percentages of different secondary structure elements can then be estimated by adding the areas of all component bands assigned to each of these structures and expressing the sum as a fraction of the total amide I/I′ band area. The amide I′ bands in the range 1650 to 1658 cm^{-1} correspond to α-helices, and a component at approximatley 1645 cm^{-1} corresponds to irregular parts of polypeptide backbones. Turns are associated with various component bands between 1660 and 1690 cm^{-1}. One or more bands between 1620 and 1635 cm^{-1} can be attributed to β-sheets; the antiparallel component of the β-sheets can be identified by the presence of another weak band near 1675 to 1695 cm^{-1}.

The binding of metal cations can change the protein secondary structure; this can be detected by FTIR spectroscopy (Figure 6.6). When a metal ion interacts with only one oxygen atom of a COO^- group, the coordination structure is regarded as unidendate. In the bidentate coordination mode, the metal ion interacts equally with the two oxygen atoms of a COO^- group. In the bridging coordination mode, one metal ion binds to one of the two oxygens in a COO^- group and another metal ion to the other oxygen atom. The pseudobridging mode features a water molecule replacing one of the two ligands in the bridging coordination. Extensive infrared studies have been carried out on the relationship between COO^- stretching frequencies and coordination types (reviewed by Nara and Tanokura, 2008). The frequency of the COO^- antisymmetric stretch

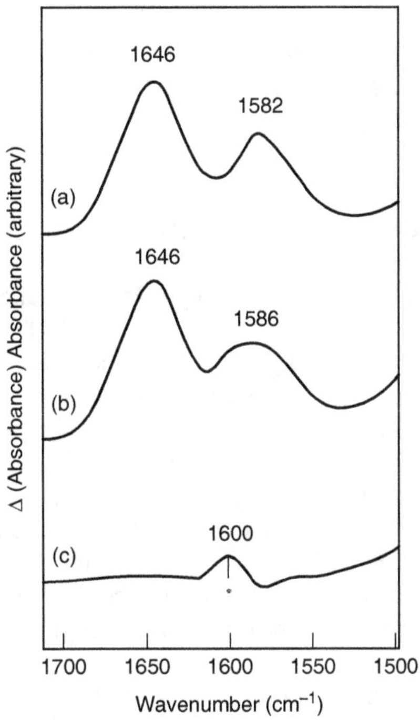

Figure 6.6. FTIR absorption spectra of (a) apo and (b) magnesium-bound E142Q mutant of Akazara troponin C. The difference spectrum (c) was obtained by subtracting (a) from (b). (From Nara and Tanokura, 2008.)

of the unidentate species is higher than that of the ionic (metal-free) species, which is in turn higher than that of the bidentate species. The reverse is the case for the COO^- symmetric stretch. As a result, the $\Delta\nu_{a-s}$ (frequency separation between the COO^- antisymmetric and symmetric stretching vibrations) values for unidentate, bridging, bidentate, and ionic species are in the following order: $\Delta\nu_{a-s}$(unidentate) > $\Delta\nu_{a-s}$(ionic) ~ $\Delta\nu_{a-s}$(bridging) > $\Delta\nu_{a-s}$(bidentate), where $\Delta\nu_{a-s}$(ionic) is approximately 160 to 170 cm^{-1}. These data can be used for characterizing the coordination of Ca^{2+} ions by carboxyl groups.

Many commonly employed buffers, such as phosphate, cacodylate, Tris-HCl, and HEPES, are acceptable for FTIR spectroscopy. Buffers containing carboxylic acid groups, such as acetate or carbonate buffers, have infrared absorption bands that overlap with those for the protein backbone. FTIR measurements of proteins in the presence of the calcium chelators EDTA or EGTA are complicated by the fact that the carboxylic groups of these chelators have strong infrared bands in the region 1570 to 1630 cm^{-1}. In addition, the spectral characteristics of these bands are influenced by calcium binding.

For FTIR measurements in H_2O solution, relatively high protein concentrations (>10 mg mL^{-1}) are required. Much lower protein concentrations (~1 mg mL^{-1}) are required to obtain high quality spectra of proteins in D_2O solution, because the latter measurements allow for the use of cells of much longer path length (3 to 8 μm for H_2O vs. 40 to 80 μm for D_2O). The aquisition of high quality infrared spectra requires a drastic reduction of the contributions of water vapor and CO_2 in the sample compartment of the spectrometer by purging the instrument with dry air or with nitrogen.

6.13. FLUORESCENCE SPECTROSCOPY

Luminescence is an emission of photons by electron excited states of molecules (reviewed by Lackowicz, 1999; Permyakov, 1993b, 2003). There are two types of luminescence: fluorescence and phosphorescence. Fluorescence is emitted due to radiation transitions between singlet states of a molecule, whereas phosphorescence arises due to radiation transitions between triplet and singlet states. Parameters of fluorescence reflect properties of excited states of fluorophores, characteristics of electron transitions in fluorophores, and interactions of fluorophores with their environment. Just the latter circumstance allows using aromatic amino acid residues and various dyes as fluorescent reporter groups in studies of proteins. Fluorescence is characterized by such parameters as fluorescence spectrum maximum, fluorescence quantum yield, fluorescence lifetime, and fluorescence anisotropy.

The *fluorescence quantum yield* is a ratio of the number of photons emitted from an excited state to the number of photons absorbed during transitions from the ground to the excited state by the same molecule per time unit. The fluorescence quantum yield is proportional to the area under the fluorescence spectrum. This parameter reflects the effectiveness of radiationless deactivation of excited states of the molecule. The fluorescence intensity at a fixed wavelength is proportional to the fluorescence quantum yield. *Fluorescence lifetime* is the time required by a population of N excited fluorophores to decrease exponentially to $N/e(e = 2.718)$ by losing excitation energy through fluorescence and other deactivation pathways. Fluorescence lifetime is inversely proportional to the effectiveness of radiationless processes in a molecule.

The position of the *fluorescence spectrum maximum* is used mostly in the case of structureless emission spectra. For fluorophores possessing dipole moments in both the ground and excited states, it reflects their interactions with the surrounding molecular ensemble and connected with the mobility of their polar environment. The shape of the fluorescence spectrum depends on the intensity of interactions between the fluorophore and its environment. If the interaction is weak, fluorescence spectra of some substances demonstrate vibrational structure, if the interaction is strong, fluorescence spectra become smooth.

Fluorescence anisotropy is defined as the ratio of the difference between the emission intensity parallel to the polarization of the electric vector of the exciting light ($I_{||}$) and that perpendicular to that vector ($I_{\perp}$) divided by the total intensity:

$$A = \frac{I_{||} - I_{\perp}}{I_{||} + 2I_{\perp}} \tag{6.2}$$

The anisotropy of emission (A) is related to the correlation time of the fluorophore (τ_c) through the Perrin equation: $A_0/A - 1 = \tau/\tau_c$, in which A_0 is the limiting anisotropy of the probe, which depends on the angle between the absorption and emission transition dipoles, and τ is the fluorescence lifetime. Measurements of fluorescence anisotropy can be used to obtain hydrodynamic information concerning macromolecules and macromolecular complexes.

Fluorescence resonance energy transfer (FRET) is the radiationless transfer of energy from an excited donor fluorophore to a suitable acceptor fluorophore, a physical process that depends on spectral overlap and proper dipole alignment of the two fluorophores. The transfer of excitation energy between the donor and the acceptor originates only upon the fulfillment of several conditions: (1) overlap of the absorption (excitation) spectrum of the acceptor with the emission spectrum of the donor, (2) the spatial proximity of the donor and the acceptor (usually up to several tens of angstroms), (3) a sufficiently high emission quantum yield of the donor, and (4) a favorable spatial orientation of the donor and acceptor. The fluorescence resonance energy transfer can be used as a molecular ruler to measure distances between the donor and acceptor. According to Förster, the efficiency of energy transfer, E, from the excited donor, D, to the nonexcited acceptor, A, located from D at a distance R_{DA} is determined by the equation

$$E = \frac{1}{1 + (R_{DA}/R_0)^6} \tag{6.3}$$

in which R_0 is the characteristic donor–acceptor distance, the Förster distance, which has a characteristic value for any given donor–acceptor pair.

The interaction of metal cations with a protein usually changes its conformation; this in turn can alter the environments of some of the side chains of Phe, Tyr, and Trp. These local changes often induce changes in the fluorescence parameters of these amino acids (fluorescence quantum yield, position of fluorescence spectrum maximum, fluorescence spectrum shape, and/or fluorescence lifetime) (reviewed by Permyakov, 1993, 2003). These changes can be used to determine the fraction of protein with bound cation(s) and sometimes the proximity of binding site(s) to the fluorescence side chain(s). Fluorescence changes reflect changes in the local environments of emitting amino acid residues. However, the parameters of the fluorescence cannot yet be predicted from the structure of the protein, nor the structure from the fluoresecence. The absence of change in fluorescence does not prove an absence of cation–protein interaction.

Tryptophan fluorescence (300 to 450 nm) provides rich information about proteins. Most proteins, however, possess several Trp residues, and the total protein emission yields information averaged over the entire protein structure. Since the indole group of Trp has a relatively large dipole moment in both the ground and excited states, it interacts extensively with polar and charged groups in its environment. This is reflected in changes of fluorescence. When a Trp is located in a rigid hydrophobic environment inside a protein molecule, its fluorescence spectrum displays a distinct vibrational structure with the main maximum at

308 nm (azurin is an example). Buried Trp's with some polar groups in their environments are characterized by a more structured spectra with the maximum at a slightly longer wavelength, 316 to 325 nm (e.g., L-asparaginase) (reviewed by Permyakov, 1993, 2003). The occurrence of more mobile polar groups in the environment of a buried Trp shifts its fluorescence maximum up to 330 to 335 nm (e.g., actin, chymotrypsin). The emission spectrum is structureless in this case. Trp's at the protein surface in contact with bound water molecules are characterized by emission spectra with maxima at 340 to 345 nm. Trp's in the environment of freely relaxing water molecules (totally unfolded proteins) have fluorescence spectra with maxima at 350 to 353 nm. Thus, the position of the fluorescence spectrum maximum reflects the relaxation properties of the polar environment of the chromophore; these, in turn, reflect the location of Trp's in a protein.

Fluorescence quantum yield and fluorescence lifetime of Trp's depend on the quenching properties of surrounding groups and their mobilities. These parameters can be used to monitor interactions of proteins with metal cations. Changes in the location of a Trp can both increase and decrease fluorescence quantum yield and fluorescence lifetime. The effect depends on specific features of the environment of the Trp.

Information about the accessibility of protein chromophores to solvent (and thus about the relative compactness of a protein) can be obtained from an analysis of dynamic quenching of intrinsic fluorescence by small molecules. The method of selective fluorescence quenching by external quenchers is used to determine the locations of chromophores in proteins. The method is based on the use of quenchers, which deactivate excited states of the chromophores during direct collisions. Such quenchers reduce the fluorescence of surface chromophores more effectively than the emission of buried chromophores. The collisional quenching is described by the *Stern–Volmer equation*:

$$\frac{I_0}{I} = (1 + K_{SV}[Q])e^{V[Q]} \tag{6.4}$$

in which I_0 and I are the fluorescence intensities in the absence and presence of quencher, K_{SV} is the dynamic quenching constant (Stern–Volmer constant), V is a static quenching constant, and [Q] is the concentration of the quencher.

Spectral probes and labels are widely used for the investigation and determination of proteins. Traditional luminescence probes include fluorescent derivatizing reagents, fluorescent probes, and chemiluminescence probes, all of which continue to be developed. Proteins have at least two functional groups where a derivatization may take place: the amino group and the carboxyl group. The carboxylic group at the C-terminus is less active and must be activated before derivatization; it is rarely used in protein-labeling procedures. On the contrary, the N-terminal amino group is easily derivatized. The reagents reacting with primary amino groups include fluorescein-5-isothiocyanate (FITC), 6-aminoquinolyl *N*-hydroxysuccinimidyl carbamate (6-AQC), fluorenylmethyloxycarbonyl chloride (FMOC-Cl), *o*-phthaldialdehyde (OPA), naphthalene-2,3-dialdehyde/cyanide

(NDA), 5-fluroylquinoline-3-carboxaldehyde (FQ), 3-(4-carboxybenzoyl)-2-quinolinecarboxaldehyde (CBQCA), and others. Almost all of the derivatizing reagents mentioned above react with primary or with both primary and secondary amines. Only 4-(*N*, *N*-dimethylamino-sulfonyl)-7-fluoro-2,1,3-benzoxadiazole (DBD-F) react specifically with secondary amines. Introducing a fluorophore to a thiol group is also a very common method to form a highly fluorescent derivative. Ammonium 7-fluoro-2,1,3-benzoxadiazole-4-sulfonate can be used for the derivatization of SH groups in proteins.

The dyes serving as noncovalent probes of proteins are almost all anionic dyes. These dyes can bind to the positively charged amino acid residues of proteins; therefore, pH is an important parameter in their use. Upon binding to proteins, the fluorescence intensity of the dyes may be enhanced or quenched. The enhancement of fluorescence of a dye comes mainly from a change in the microenvironment in which the dye exists. Very often, these probe reagents are nonfluorescent in water but highly fluorescent in apolar media. These dyes can bind to the hydrophobic regions of a protein through noncovalent binding, and their fluorescence yields are greatly enhanced. Typical probes of this type are naphthalene derivatives, Sypro dyes, and Nile red. The aromatic chromophore 1-anilino-8-naphthalene sulfonate (ANS) is feebly fluorescent in water, but its spectrum is blue-shifted and its intensity is increased dramatically in nonpolar solvents or when it binds to nonpolar sites of proteins. ANS is a widely used fluorescence probe, yet despite its popularity, significant questions remain concerning its binding selectivity for hydrophobic "patches" as well as the precise origin of its enhanced quantum yield when bound to such regions.

6.14. CIRCULAR DICHROIC AND OPTICAL ROTATORY DISPERSION SPECTROSCOPY

Linear polarized light can be considered as a superposition of opposite circular polarized light of equal amplitude and phase. Asymmetric chromophores or symmetric chromophores in asymmetric environments interact differently with right and left circularly polarized light, resulting in two related phenomena. *Circular dichroism* (CD) refers to the difference in absorption of right and left circularly polarized light at some wavelengths due to differences in extinction coefficients for the two polarized rays. At the same time, the right and left components of circularly polarized light travel through an optically active medium with different velocities due to the different indices of refraction for right and left circularly polarized light; this is called *optical rotation* or *circular birefringence*. The variation of optical rotation as a function of wavelength is called *optical rotary dispersion* (ORD).

The near-UV CD bands of proteins (deriving from Trp, Tyr, Phe, and Cys) reflect the tertiary and quaternary structure of the protein (reviewed by Kelly and Price, 2000; Greenfield, 2004; Sreerama and Woody, 2004). The CD bands in the far UV, which derive principally from peptide bond absorption, reflect the secondary structure of the protein (α-helix, β-strand, and coil). The most commonly used units are mean residue ellipticity (degrees · cm^2 $dmol^{-1}$) and the difference in molar extinction coefficients called *molar circular dichroism* or $\Delta\varepsilon$ (liters · mol^{-1} cm^{-1}). The molar ellipticity $[\Theta]$ is related to the difference in extinction coefficients by $[\Theta] = 3298(\Delta\varepsilon)$. Here $[\Theta]$ has the standard units of degrees · cm^2 $dmol^{-1}$ and the molar dichroism has the units degrees · dL mol^{-1} dm^{-1}.

Protein chromophores in CD spectroscopy can be divided into three classes: peptide bond, amino acid side chains, and prosthetic groups. The lowest-energy transition in the peptide chromophore is an $n \rightarrow p^*$ transition observed at 210 to 220 nm with very weak intensity. The $n \rightarrow p^*$ transition reflects α-helices as a small shoulder near 220 nm on the tail of a much stronger absorption band centered at 190 nm. This intense band, responsible for the majority of the peptide bond absorbance, is a $p \rightarrow p^*$ transition.

CD spectroscopy is especially powerful in monitoring conformational changes. In the region of 230 to 178 nm, one expects to observe effects of backbone conformational changes (secondary structure) while CD effects at longer wavelengths (>230 nm) show contributions from the aromatic chromophores, Phe, Tyr, and Trp. Since these CD effects are dependent on the environment, they reflect more global tertiary properties of the protein (reviewed by Kelly and Price, 2000; Greenfield, 2004; Sreerama and Woody, 2004). CD spectra between 180 and 260 nm can be analyzed to estimate the relative amounts of α-helix, β-strands, and coils, each of which gives rise to a characteristic shape and magnitude of the CD spectrum (Figure 6.7). The simplest method of determination of secondary structure content from CD data is to assume that a spectrum is a linear combination of CD spectra of each contributing secondary structure type weighted by its abundance in the polypeptide conformation. The major drawback

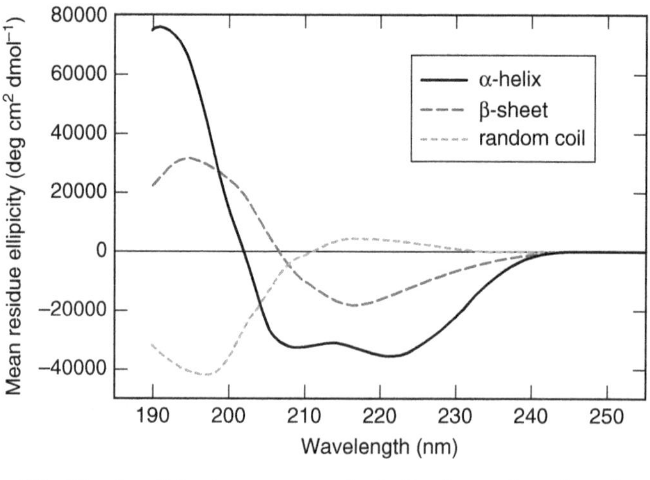

Figure 6.7. Circular dichroism spectra of α-helix, β-strand, and random coil.

of this approach is that there is no standard reference CD spectrum for "pure" secondary structures. To overcome this shortcoming, several methods have been developed that analyze the experimental CD spectra using a database of reference protein CD spectra containing known amounts of secondary structure. In one of these methods, single value decomposition is used to create orthogonal CD basis vectors from CD spectra of proteins with known secondary structure. Using the statistical technique of variable selection, unimportant variables are removed from an underdetermined system of equations, allowing a solution for the important ones. The great advantage of this and related techniques is that one makes no assumptions about the form of the CD from the individual secondary structural elements. CD spectroscopy is very useful in monitoring conformational changes induced by binding of metal ions.

Optically active material shows a different refractive index for right and left circular polarized light. As the light passes through the material, one component of the circularly polarized light is delayed with respect to the other. This delay, or phase shift, causes a change in the orientation of the linearly polarized light. As noted earlier, the difference in refractive index is the circular birefringence and optical rotation as a function of wavelength is optical rotary dispersion (ORD). The ORD curve changes sign at the absorption maximum. This occurs because the phase of the oscillations of the electrons becomes 180° out of phase from the incident light. This implies that the polarized light (right or left) that was retarded by the material at longer wavelengths now becomes advanced with respect to the other polarized direction at shorter wavelengths. Note that there is no optical rotation at the absorbance maximum because the light is now being absorbed. ORD spectra are dispersive (called the *Cotton effect* for a single band); whereas, circular dichroism spectra are absorptive. The two phenomena are related by the König–Kramers transform. CD is a more sensitive analytical technique than ORD because of its dispersive nature.

6.15. NUCLEAR MAGNETIC RESONANCE

Magnetic resonance arises from the interaction of the magnetic moment of an atomic nucleus (μ) with an external magnetic field (reviewed by Wüthrich, 1995; Pickford and Campbell, 2004; Tugarinov et al., 2004; Clarkson and Campbell, 2003; Dyson and Wright, 2004; Kumar and Yu, 2004; Opella and Marassi, 2004; Redfield, 2004). The nature of this magnetic moment is the quantum mechanical angular momentum (spin angular momentum) of the nuclei. The projection of this magnetic moment on an external magnetic field is described by

$$\mu_z = \frac{m\gamma h}{2\pi} \tag{6.5}$$

in which the constant γ is characteristic for each isotope and is called the *gyromagnetic ratio*. The magnetic quantum number m can be an integer between

$-I$ and $+I$. Thus, the external field causes a splitting of the energy levels. For spin-1/2 nuclei, two energy levels exist according to a parallel or antiparallel orientation of the magnetic moment with respect to the magnetic field. The energy of these levels is given by the classical formula for a magnetic dipole in a homogeneous magnetic field of strength B_0:

$$E = -\mu_z B_0 = \frac{-m\gamma h}{2\pi B_0} \tag{6.6}$$

The magnetic moment of each nucleus precesses around B_0. The frequency of this precession is the Larmor frequency (ω_0), which is equivalent to the resonance frequency of the nucleus and the energy difference between the two levels:

$$\frac{\gamma h}{2\pi B_0} = \Delta E = h\nu = \frac{h}{2\pi\omega_0} \tag{6.7}$$

$$\omega_0 = \gamma B_0 \tag{6.8}$$

The Larmor frequency depends on the gyromagnetic ratio and the strength of the magnetic field (i.e., it is different for each isotope). At a magnetic field of 18.7 T, the Larmor frequency of protons is 800 MHz. Transitions between different energy levels occur if the frequency of radiation is equivalent to the energy difference between the two levels.

In Fourier transform NMR the resonances are not measured one after another, but all nulei are excited at the same time by a radio frequency impulse. Normally, a radio emitter works at a fixed frequency ν_0. However, if the radiation is emitted as a very short pulse (a few microseconds) the impulse frequency becomes "uncertain." A short radio frequency impulse contains many frequencies in a broad band around ν_0 and thus excites the resonances of all spins in a sample at the same time.

The excited spins emit the absorbed radiation after the impulse. The signal emitted is a superposition of all excited frequencies and its evolution in time is recorded. The intensities of the several frequencies, which give the observed signal in their superposition, are calculated by Fourier transformation; this translates the time data into the frequency domain (Figure 6.8).

The electrons in a molecule surround the nuclei and create a small magnetic field, which shields the nuclei slightly from the external field. Therefore, the Larmor frequencies of different nuclei vary due to their different chemical environments ("chemical shift"). Different chemical groups give different chemical shifts; hence, the type of atom can be inferred from the NMR spectrum.

The introduction of multidimensional spectra and their Fourier transform have been major advances in NMR spectroscopy. The two-dimensional experiment is characterized by an indirect evolution time t_1 and a mixing sequence of pulses. Such an experiment includes the following steps: a disturbance of the equilibrium state of nulcei (preparation); their free precessing (evolution); doing something else (mixing); and detection of the result (detection).

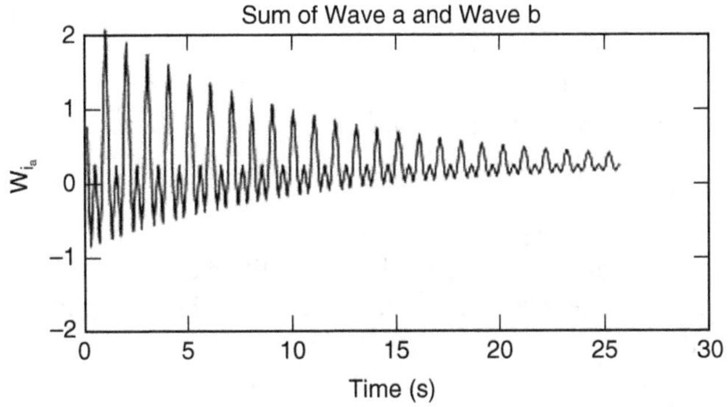

Fourier Transform: F := fft (W)

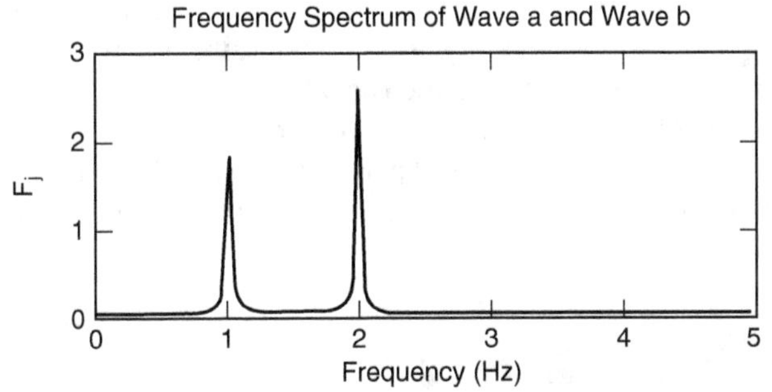

Figure 6.8. Evolution in time of the emitted signal, consisting of two hypothetical waves, in NMR and its Fourier transform into the frequency domain.

After preparation the spins precess freely for a given time, t_1. During this time the magnetization is labeled with the chemical shift of the first nucleus. During the mixing time the magnetization is then transferred from the first nucleus to a second nucleus. Mixing sequences utilize two mechanisms for magnetization transfer: scalar coupling or dipolar interaction [nuclear Overhauser effect (NOE)]. Data are collected at the end of the experiment (detection); during this time the magnetization is labeled with the chemical shift of the second nucleus. The cross signals originate from nuclei that exchanged magnetization during the mixing time and indicate an interaction of these two nuclei. Therefore, the cross signals contain the really important information related to two dimensional NMR spectra.

Nuclear Overhauser effect correlation spectroscopy (NOESY) is crucial for the determination of protein structure. It uses the dipolar interaction of spins [the nuclear Overhauser effect (NOE)] for the correlation of protons. The intensity of the NOE is to a first approximation proportional to r^{-6}, with r being the distance between the protons. The correlation between two protons depends on the distance between them. The NOESY experiment shows the correlation of all protons that are less than 5 Å apart. It correlates protons that are distant in sequence but nearby in space. This is the most important information for the determination of protein structures. The tertiary structures of many calcium binding proteins have been determined by NMR (see the Protein Data Bank).

The NMR active isotope ^{43}Ca is a quadrupolar ($I = 7/2$) nucleus; it produces broad peaks when bound to a protein. Isomorphous replacement of the Ca^{2+} ion with an ion with more favorable NMR properties often allows the resolution of NMR signals for individual sites. ^{113}Cd and ^{207}Pb both have $I = 1/2$ nuclei. They have been used successfully to characterize the calcium binding properties of several metalloproteins. The Cd^{2+} ion has a filled d-shell orbital and can form complexes with a variety of conformations and number of ligands. Cd^{2+} has an ionic radius (0.97 Å) similar to that of Ca^{2+}, 0.99 Å; whereas, Pb^{2+} is larger (1.20 Å). For both ^{113}Cd and ^{207}Pb, the chemical shift of the NMR signal is highly dependent on the ligand environment of the metal ion.

Solution NMR spectroscopy is used widely to determine the structure of proteins. In these experiments ^{2}H, ^{13}C, and ^{15}N isotopically labeled proteins are used in residue assignments. For modestly sized proteins, the assignments can be made by standard homonuclear ^{1}H two-dimensional methods. As the proteins exceed 10 kDa, the NMR spectra become more crowded with overlapping signals. With ^{13}C and ^{15}N labeling, heteronuclear experiments have allowed the spectra to be recorded in two, three, or four dimensions: thus increasing the resolution and decreasing the assignment ambiguities.

6.16. ELECTRON SPIN RESONANCE

Electron spin resonance (ESR) spectroscopy monitors unpaired electrons of free radicals or of transition metal ions in complexes with various ligands (reviewed by Poole and Farach, 1994). Since most stable molecules do not have unpaired electrons, this method is used less frequently than is the NMR method.

The physical principles underlying ESR are similar to those of NMR; however, instead of nuclear spins, electron spins are excited in ESR. Due to the great difference in masses of nuclei and electrons, weaker magnetic fields and higher frequencies are used in ESR than in NMR. For electrons in a magnetic field of 0.3 T, spin resonance occurs at about 10 GHz. Unpaired electrons interact with the external magnetic field; simultaneously, the system is perturbed by microwaves. If the strength of the magnetic field is increased gradually, at certain field strengths the orientation of the electron spin changes. This field strength is characteristic of a given electron environment. If an atom with which an unpaired electron is associated has a nonzero nuclear spin, its magnetic moment will affect the electron. This leads to the phenomenon of hyperfine coupling, splitting the ESR resonance signal into doublets, triplets, and so on. The g-factor and hyperfine coupling in an atom or molecule may not be the same for all orientations of an unpaired electron in an external magnetic field. The anisotropy depends on the electronic structure of the atom or molecule (e.g., free radical), so can provide information about the atomic or molecular orbital containing the unpaired electron.

Compounds with free radicals, called *spin labels*, can be attached to various groups in proteins to obtain information about chemical and physical characteristics of their environments. For example, the nitroxyl compound 2,2,6,6-tetramethyl-4-piperadon-1-oxyl, in which the unpaired electron is located on the NO group can be attached selectively to thiol groups of proteins. Spin labels should not change the native structure of labeled proteins if they are to provide information about the native structure.

ESR spectroscopy gives especially important information about iron, manganese, and copper containing proteins. Usually, one can measure ESR spectra of ions containing an odd number of unpaired electrons in the ground state.

6.17. SURFACE PLASMON RESONANCE

Surface plasmon resonance occurs when light is reflected off thin metal films. A fraction of the light energy incident at a sharply defined angle can interact with the delocalized electrons in the metal film (plasmon), thus reducing the intensity of the reflected light. Suitable metals include silver, gold, copper, and aluminum, with silver and gold used more commonly. The precise angle of incidence at which this occurs depends on a number of factors. The surface plasmon resonance device detects minute changes in the refractive index of the sensing surface in its immediate vicinity (reviewed by Lucosz, 1991, 1997; Fan et al., 2008; Piliarik et al., 2009). This resonance is based on the collective excitement of electrons (the surface plasmons) in a metal film on a substrate (such as glass), leading to total absorption of light at a particular angle of incidence that depends on the refractive indices of the materials on either side of the metal film. Resonance of a surface plasmon is excited at the metal–air interface when the angle of incidence of light is such that the evanescent component of its wave vector is equal to the wave vector of the propagating surface plasmon. Thus, the refractive index of the sensitizing layer and a thin layer immediately adjacent to it can be measured as a shift in the angle of the total absorption of light. One of the main factors is the refractive index near the backsurface of the metal film. Various target molecules immobilized on the surface interact with ligands in a mobile phase, leading to a change in surface plasmon resonance angle; this can be monitored in real time by detecting changes in the intensity of the reflected light, producing a sensorgram. The rates of change of the surface plasmon resonance signal can be used to obtain apparent rate constants for the association and dissociation phases of the reaction. The ratio of these values gives the apparent equilibrium constant of the reactants: one immobilized, the other in solution. The amplitude of the change in surface plasmon resonance signal is directly proportional to the mass being immobilized and can thus be interpreted crudely in terms of the stoichiometry of the interaction. Signals are easily obtained from submicrogram quantities of material. Since the surface plasmon resonance signal depends only on binding to the immobilized template, it is also possible to study binding events from molecules in extracts (i.e., it is not necessary to have highly purified components).

Surface plasmon resonance sensing has been demonstrated to be an exceedingly sensitive and quantitative method for studying the interactions of biopolymers with various ligands. These include other proteins, DNA, membrane fragments, and various ligands, including metal ions.

6.18. EXTENDED X-RAY ABSORPTION SPECTROSCOPY

X-ray absorption spectroscopy (XAS) includes both x-ray absorption near edge structure (XANES) and extended x-ray absorption fine structure (EXAFS). XAS is measurement of the x-ray absorption coefficient of a material. X-rays of a narrow bandwidth are directed to the sample, and the incident and transmitted x-ray intensity are recorded as the incident x-ray energy is incremented. When the incident x-ray energy matches the binding energy of an electron of an atom within the sample, the x-ray absorption by the sample increases dramatically, causing a drop in the x-ray intensity transmitted (Figure 6.9) (reviewed by Bergmann et al., 2001). This results in an absorption edge. Each element in the periodic table has a set of unique absorption edges, corresponding to different binding energies of its electrons. XAS spectra are most often collected at synchrotrons, because the the energy of the x-ray beam can be tuned or scanned. The normalized absorption spectra are often called *XANES spectra*.

X-ray absorption spectra are generally produced in the range 200 to 35,000 eV (0.35 to 61.99 Å). Below the absorption edge, the photons cannot excite the

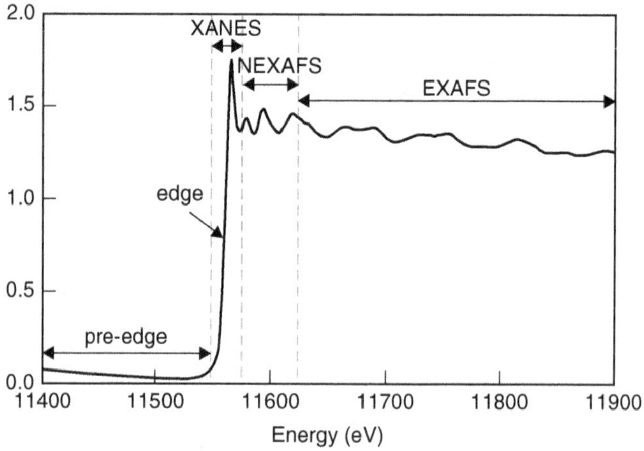

Figure 6.9. Model x-ray absorption spectrum (XAS), showing the absorption edge, the x-ray absorption near edge structure (XANES), and the extended x-ray absorption fine structure (EXAFS) regions.

electrons of the relevant atomic level and thus absorption is low; however, when the photon energy is sufficient to excite the electrons, a large increase in absorption occurs at the absorption edge. The resulting photoelectrons have a low kinetic energy and are backscattered by the atoms surrounding the emitting atom. The probability of backscattering is dependent on the energy of the photoelectrons. This backscattering of the photoelectron affects whether the x-ray photon is absorbed in the first place. The net result is a series of oscillations on the high-photon-energy side of the absorption edge (i.e., the EXAFS spectrum) (Figure 6.9). These oscillations can be used to determine the atomic number, distance, and coordination number of the atoms surrounding the element whose absorption edge is being examined. For calcium, EXAFS should distinguish six from seven from eight coordination in a lyophilized sample and give the mean Ca–O distance ± 0.02 Å (Yano and Yachandra, 2009).

6.19. SMALL ANGLE X-RAY SCATTERING

When a beam of neutrons or x-rays encounters a protein in solution, a small portion of the beam will be deflected or *scattered*. In small angle x-ray scattering (SAXS) the sample is irradiated by a highly collimated, monochromatic x-ray beam and the elastic scattering of x-rays (wavelength 1 to 2 Å) is recorded at low angles (typically, 0.1 to 10°). The angular dependence of the scattering is related to the shape of the protein. SAXS of a protein that is not interacting with its neighbors in solution can be computed as the spherically averaged Fourier transform of the structure of that protein. From a measurement of that transform, one cannot, even in principle, determine the detailed shape of the protein; however, if a protein is modeled as an ellipsoid of revolution, or a cylinder, or a dumbbell (two sphere joined by a rod), one can compare the intensity measured with that calculated (reviewed by Perkins et al., 1998; Doniach, 2001; Svergun and Koch, 2002). Although one treats the absolute determination of shape with caution, SAXS is exquisitely sensitive to a change in structure, such as that induced by binding calcium or a target. SAXS studies of the dumbbell shaped calmodulin and troponin C have been particularly fruitful (Fujisawa et al., 1989, 1990; Kataoka et al., 1991).

6.20. CRYSTALLOGRAPHY

Most of our understanding of protein structure, in the narrow sense of assigning Cartesian coordinates to the atoms of a protein or ligand, comes from x-ray crystallography and nuclear magnetic resonance (reviewed by Heinemann et al., 2001; Burley and Bonanno, 2002; L. Esposito et al., 2002; Lecomte et al., 2004). It is reassuring that results from crystallography and NMR are consistent with one another and with those from other techniques described above. Nonetheless, it is often difficult to relate the crystal structure of a protein to its thermodynamic, chemical, or functional properties.

Obtaining reasonable crystals has become the rate-limiting step in many investigations (reviewed by Wiencek, 1999). However, not all crystals are created equal. Usually, the content of solvent, or crystallization medium, varies from 27 to 78% of the crystal volume; the median volume per dalton of protein (the Matthews coefficient) is 2.52 $Å^3$ Da^{-1} (Kantardjieff and Rupp, 2003). All of the asymmetric units (a.s.u.) of the crystal have identical contents and there is usually one protein (subunit) per a.s.u. However, sometimes there is more than one subunit per a.s.u.; these subunits will have different intersubunit contacts. The same protein may also be crystallized from different solvents: for example, from polyethylene glycol and from ammonium sulfate. When these different subunits from the same or from different crystals are compared, they are found to be nearly identical; that is, neither crystal contacts nor solvent seem to alter the structure of proteins significantly. Nonetheless, if one is concerned with a subtle detail of structure at the surface of a protein, it is reassuring to see that detail in two different subunits within the asymmetric unit.

X-ray diffraction data are measured to a resolution usually cited in terms of the *Bragg equation*:

$$\sin\theta = \frac{\lambda}{2d} \tag{6.9}$$

in which 2θ is the angle relative to the direct beam at which the reflection is recorded, λ is the wavelength of x-ray used (usually between 1.0 and 2.0 Å), and d is the nominal spacing giving rise to that reflection, or constructive interference of diffracted intensity. For example, "1.5-Å data" would be measured at $\theta = 30°$ using $\lambda = 1.5$-Å x-radiation; 2.0-Å data would be measured at $\theta = \sim 16°$. Two point atoms (i.e., all electrons clustered at the nucleus), that are 2.0 Å apart could be resolved using 2.0 Å data. The functional resolution in protein crystallography is better than d. For example, a slight dip in electron density should be seen between single-bonded carbon atoms, 1.54 Å apart, using 1.5 Å data. The important point is that the positions of these carbon atoms in proteins are not (usually) refined independently as they are in crystal structures of small molecules. The atoms of a protein are constrained to honor (near) canonical bond lengths and angles; the structure of the protein is refined by varying the dihedral angles, φ and ψ, of the main chain and of the side chains (χ's). The effective resolution is about 0.05 Å for nonhydrogen protein atoms using 2.0 Å data and about 0.03 Å for 1.5 Å data. In contrast, the resolution for a Ca^{2+} ion (or for a water molecule) would be significantly lower, ~0.15 Å for 2.0 Å data since the distance to its oxygen ligands should *not* be constrained by canonical values during refinement. That is, the error in the Ca–O bond distance in a protein structure determined to 2.0 Å resolution would be about 0.2 Å.

The effective resolution of a protein crystal structure depends on several other factors. The refinement of the protein structure is evaluated by its residual, R, in which I_{obs} is the observed intensity of the diffracted x-ray and I_{calc} is the calculed intensity based on the calculated structure at that state of refinement. The sum is over all measured reflections:

$$R = \frac{\sum |I_{obs} - I_{calc}|}{\sum I_{obs}} \tag{6.10}$$

A well refined protein structure with 2.0 Å data should be refined to $R = 0.19$; with 1.5 Å data, $R = 0.17$; with 1.0 Å data, $R = 0.15$. By contrast, a small organic molecule determined to 1.0 Å should have $R \sim 0.07$. Proteins refine so poorly for several reasons. The solvent molecules in a small molecule are (almost) all well defined; only the first layer of waters about the surface of a protein is usually (partially) ordered. That may reflect only a tenth of the total solvent in the crystal. Usually, proteins having a lower solvent content refine better than do those with ~70% solvent.

Proteins in crystals undergo several types of motion or disorder. The entire protein, as a rigid unit, may be in slightly different orientation, at any instant in time, in different subunits, hence smearing the computed electron density associated with each atom. Again, proteins with a lower solvent content tend to have less disorder of orientation. Side chains, or loops of proteins, may assume alternate orientations in different proteins within one crystal. The electron density may show a superposition of these alternate densities. On occasion one is justified in refining, with appropriate weighting, two alternate conformations in a single-crystal structure. A similar but distinct effect results from a component, usually a side chain, of the structure oscillating about a single minimum. Finally, each atom vibrates about its mean position. The cumulative effects of the various oscillations and vibrations are approximated by a single *temperature* (B) *factor*; the higher the B factor, the greater the disorder. In a well-ordered protein whose diffraction data were measured from crystals maintained at liquid nitrogen boil-off temperature 77 K, the B values for main-chain atoms should be between 10 and 14, for side chains between 12 and 18. Helices have lower B values than do coil regions. If a stretch of residues has main chain B values over 20, there is probably some disorder in that region. The B assigned to a single Ca^{2+} ion should be less than 15 for 2.0 Å data.

A protein, refined with 2.0 Å data, that has $R > 0.22$ probably contains some significant errors in dihedral angles, even though the overall trace of the main chain is correct. Some portions of the protein, frequently N-terminal residues and loops that extend from the surface, are poorly defined or even absent in the electron density. If the protein model has been built into these regions, it will have high B values and be especially suspect. Even a protein with a good overall residual may contain local regions of disorder: hence the advisability of checking individual B factors, especially in any region of critical importance.

7

STRUCTURE AND EVOLUTION OF PROTEINS

In this chapter we summarize the concepts and terminology used to describe proteins in general. Because of the importance of cell signaling, calcium modulated proteins are cited to illustrate these concepts. Some terms used in the description of proteins have several meanings; this book honors the usage of this chapter.

7.1. DOMAIN

Domain has two distinct but usually concordant definitions: one based on structure, the other on evolution. Ideally, contiguous residues fold into a compact, stable, spherical structure. The crystal structures of most proteins have few cavities that would accommodate a sphere the size of a water molecule or xenon atom. In fact, proteins are dynamic; the crystal structure captures a time–space average. Further, nearly all proteins have several closely related structures, corresponding to various stages of their functional cycles. Most proteins have evolved to retain their structures at the ambient temperature of the host organism. Usually, a domain expressed alone or cut out of a full protein retains the structure and properties as seen in the intact parent protein; however, there are many examples of reduced stability and altered interactions of the excised domain, for example, the CD–EF lobe of parvalbumin, cut out of the intact protein (Chapter 10). Usually, the extreme axial ratio of a domain is less than 3 : 1. However, some very stable structures, such as coiled coils of α-helices or the collagen triple strand, are highly elongated. Other, nominally spherical domains

Calcium Binding Proteins, By Eugene A. Permyakov and Robert H. Kretsinger

of contiguous residues have helices protruding into solution or into a phospholipid bilayer, as for example in Ca-ATPase (Chapter 12).

From an evolutionary perspective, a *domain* is that uninterrupted stretch of residues recognized as a member of a homolog family. Homologous proteins are usually identified by the similarity of their amino acid sequences by BLAST (basic local alignment search tool; Altschul et al., 1990; http://blast.ncbi.nlm.nih.gov/Blast.cgi) or FASTA (http://www.ebi.ac.uk/Tools/fasta33/index.html). Although the percent identity between two sequences is often cited, this is a misleading metric. Thirty percent identity over 20 residues is not unusual; 30% identity over 200 is much more significant. The size of the database under consideration affects the probability that two (or more) sequences might have achieved that alignment score by chance. A Z-score of 5.0 indicates homology at 95% confidence; 6.0, 99% confidence (Miguel et al., 2002). Further, one should consider the similarity of residues. A Leu vs. Ile difference is common; Trp vs. Arg is rare. Small insertions or deletions (indels) carry a weight per event and per length when aligning sequences or scoring similarity (reviewed by Reese and Pearson, 2002; Tsigelny, 2002). If protein A is homologous to B and B is homologous to C, A is homologous to C. Homology within a group is inclusive; homology between groups is exclusive. However, in BLAST searches one inevitably finds a few alignments at the edge of significance. Further, using a different probe sequence chosen from the established homolog family will affect the number and significances of these outliers. Such analyses raise the question of the origins of each of the approximately 1000 recognized homolog families (reviewed by Sternberg, 2002; Tsigelny, 2002).

Often, domains in real proteins have a few extra residues at their N- and C-termini. If one chooses a reference domain from an established family and does a BLAST search against a database that contains homologs, the Z-scores of the homologs will vary slightly depending on how the N- and C-termini were trimmed: too tightly, the significance of the alignment is reduced; too much fluff, the significance is also lower. The Z-score of sequence comparison correlates strongly with the Z-score of structure comparison; a low Z-score homology is more easily recognized in structure comparison (Wood and Pearson, 1999). Many proteins consist of several domains of different (or the same) homolog families spliced together. If one protein, of domains A and B, is aligned with another(s) of domains A and C, the significance of the overall alignment will be poor; however, if the probe sequence from A and B is trimmed back to just domain A, the significance of its alignment with the A domain of A and C will be much higher. Characterization of the optimal domain boundaries can be an iterative process.

Ideally, the structural domain and the homolog domain are the same. Despite the numerous variations and exceptions, the concept of a protein domain is invaluable because it provides a solid reference for evaluating this enormous diversity.

7.2. STRUCTURE

There are several exceptions to this structural and evolutionary characterization of domains. Much of our understanding of protein structure comes from crystallography. Of necessity, only a single conformation is captured in the crystal; local *B* (temperature) factors (Chapter 6) give a rough sense of the disorder in that region of the protein. Not all stretches of residues in natural proteins have a single defined structure, or even a tight ensemble of similar structures, under the physiological conditions of their host organism. As much as 10% of the eukaryotic proteome is inherently disordered (reviewed by Uversky, 2002). These regions are greatly underrepresented in the Protein Data Bank since they are often removed to facilitate crystallization or, if present in the crystal, are not visible because of their disorder. These regions are not flawed; they have been selected by evolution, perhaps as flexible linkers between standard domains, perhaps to facilitate targeting by other proteins. Our ignorance is of little concern to nature. The vast majority of amino acid sequences, randomly assembled and in proportion to their occurrence in the proteome, are predicted to be inherently disordered (Uversky, 2002); a defined, stable tertiary structure is the product of natural selection.

Over 10% of the eukaryotic proteome consists of tandem repeats of units 1 to about 10 residues long (Ignoul and Eggermont, 2005; Mularoni, 2006; Dalby, 2009). Some of these repeats (e.g., the Gly–Pro–Pro of collagen) form stable structures, usually helical. Most are disordered. Low complexity regions, many of which consist of (quasi) tandem repeats, are edited out of protein sequences before they are subject to BLAST searches because they skew statistical analyses.

Many, if not most, protein monomers, synthesized as a single polypeptide chain, consist of more than one (homologous or nonhomologous) domain. Most of the change in conformation associated with binding an effector occurs in the interaction between domains of a monomer and between monomers of an oligomeric protein. Often, domains are linked by several flexible residues, not recognized as having evolved with either domain: for example, the linker between EF-hands 2 and 3 of calmodulin and troponin C (Chapter 11). Sometimes, domains provide an interface for dimerization, as do the fifth EF-hands of calpain (Chapter 11). Some monomers exchange, or swap, domains (Verdino et al., 2002) with another monomer as part of a functional cycle of dimerization and dissociation: for example, Phl p 7 (polcalcin) (Chapter 11).

Many proteins consist of several subunits or polypeptide chains. In a homodimer, both subunits have identical amino acid sequences; in a heterodimer, the two subunits differ in sequence. Whether, as discussed in the literature, a "protein" is monomeric, homo-oligomeric, or hetero-oligomeric often must be inferred from context. As noted for domains, often the contacts between monomers of an oligomeric protein change in response to binding of a substrate or of an effector.

The genes in eukaryotes that encode single polypeptide chains are interrupted by numerous introns; these, plus a few exons, may be spliced out in alternate combinations to yield proteins of different sequences, called *splice variants* (Lander et al., 2001). Further variation and complexity is generated by posttranslational modifications, of which hundreds are known (reviewed by Walsh, 2007): for example, myristylation of recoverin or phosphorylation of myosin light chain kinase (Chapter 11) as part of their functional cycles. Further, many proteins bind prosthetic groups or metal ions. These modifications further complicate the task of structure interpretation and prediction.

It is generally assumed that a polypeptide chain, as synthesized in the ribosome and subsequently "posttranslationally modified" assumes its minimum free energy conformation and is not trapped in a local minimum (reviewed by Baldwin, 2007); however, to assemble into their functional conformation, some proteins require chaperons, protected from proteolytic enzymes and in a reasonable time (reviewed by Macario and Conway de Macario, 2007).

Given the amino acid sequence of a protein and its posttranslational modifications, one should be able to predict its conformation: that is, assuming constant bond lengths and angles, predict all of the dihedral angles of main chain and side chains. This is the outstanding theoretical challenge of molecular biology. There are two broad categories of structure prediction algorithms: template based and *ab initio* or free (reviewed by Zhang, 2008).

The main chain [—(NH)—; (Cα—Cβ)—(C═O)—] of the target is aligned with, or threaded onto, that of an identified homolog target (Soding, 2005), or better as a control, onto several targets. The more similar the sequences of the target and of the template, the closer their tertiary structures (Wood and Pearson, 1999). The second step consists of adjusting the conformation of the main chain of the target to accommodate indels, or loops. Finally, the free dihedral angles (the φ's and ψ's of the main chain and χ's of the sides chains) are adjusted to minimize close van der Waals contacts and to optimize hydrogen bonding.

If the structures of no homologs are available, one turns to the much greater challenge of *ab initio* prediction. The (approximate) functions to compute the free energy of a protein in solution are known and generally accepted (reviewed by Rose et al., 2006); however, many algorithms add knowledge based terms. Computation of the minimum free energy is extremely time consuming and frequently settles into a local as opposed to a global minimum (Bradley et al., 2005; Moult et al., 2007; S. T. Wu et al., 2007). The starting structure for refinement must be quite near the final, "true" structure. For example, one may alter a few ligands near a metal binding site and compute with some confidence the conformation of the mutated structure (Chapter 16). However, one cannot assume that changes in conformation upon metal binding are restricted to side chains or even to the local main chain (Chapters 10 and 11). To generate the starting structure to predict an entire domain, one may overlap short peptide elements of sequence similar to, but not necessarily homologous with, the target (Bradley et al., 2005; Wu et al., 2007). This requires massive computing power and is limited in success. It seems that many important interactions that determine tertiary structure are distant in linear sequence (reviewed by Muñoz, 2008).

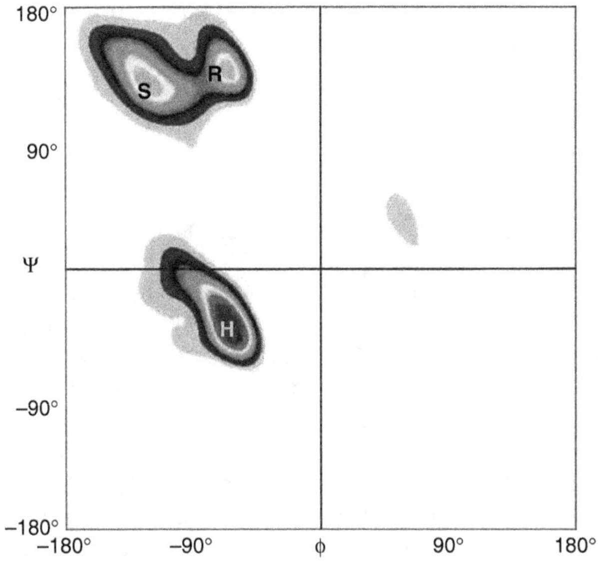

Figure 7.1. Ramachandran plot. The figure at the right top shows the regions with no bad van der Waals contacts, as originally proposed by Ramachrandran et al. (1963). Bond lengths and bond angles are assumed constant and identical for all 20 genetically encoded amino acids. Phi, ϕ, is the dihedral angle about the (N—H)—Cα bond; psi, ψ, is the dihedral angle about the Cα—(C═O) bond. The peptide bond, (C═O)—(N—H) is assumed to be planar with (C═O) and (N—H) trans. The figure at the left top shows the ϕ and ψ angles actually observed in a large sample of proteins in the PDB determined to high resolution. Three consecutive residues in the **H** region form an α-helix (or 3_{10} helix). Two consecutive residues in the **S** region form a β-strand, and three **R**'s, a polyproline helix. (Courtesy of T. Zhou and S. Hovmöller.)

Analyses of crystal structures of many proteins determined to high resolution indicate that the dihedral angles of the main chain are restricted to relatively small areas of ϕ,ψ space; to rough approximation they assume discrete values. α-Helices have a main-chain hydrogen-bonding pattern $NH_i \rightarrow CO_{i-4}$; and are recognized by having at least three contiguous residues in the H region of the ϕ,ψ (Ramachandran) plot (Figure 7.1). β-Strands usually, but not always, hydrogen-bond with other β-strands, in parallel or in antiparallel orientation, to form β-sheets. β-Strands are defined by having two contiguous residues in the S region of the ϕ,ψ plot. Three contiguous residues in the R region, which occurs infrequently, form a polyproline helix. In great simplification, all residues—except three contiguous in H, or two in S, or three in R—form "coils." Various algorithms assign H for "helix," S for "strand," and C for "coil" to known or predicted structures by slightly different but similar criteria (see, e.g., http://au.expasy.org/tools/).

These coil regions may be disordered but are seldom random in either structure or in sequence as passed through the filter of natural selection. They vary in length from one to scores of residues; the classification and terminology of coils is a subject of debate. The larger motions of proteins occur in coils; crystallographic temperature factors (Chapter 6) are usually higher in coils; and most insertions and deletions (indels) occur in coils.

7.3. EVOLUTION

Proteins may be classified by many criteria: metal affinity, function, cellular location; all have advantages and disadvantages. We first use evolution, as appropriate. A dendrogram (cladogram) shows the relationships among any set of entities that have measurable attributes: for example, the protein domains of a homolog family. A dendrogram provides a foundation for classification (e.g., organisms into species and genus or homologous proteins into subfamilies). Peripheral (terminal) nodes represent the individual domains under analysis; internal nodes are connected by three branches to other nodes, internal or peripheral. If there are n peripheral nodes (domains of different sequence), the dendrogram has $2n - 3$ branches; n branches to terminal nodes and $n - 3$ internal branches (Figure 7.2). The (sum of) branch lengths shows the distance (difference) between nodes; the more similar, the closer to one another. For n domains there are $n(n - 1)$ observed differences between domains but only $2n - 3$ degrees of freedom for the model (dendrogram) to accommodate the observations. For large data sets, finding the optimal topology and the optimal branch lengths is challenging; fortunately, programs (e.g., GenomeStudio's Protein Analysis Module) compute the (near) best "tree(s)" plus estimates of their uniqueness and fits to the data.

Some proteins have similar tertiary structures and hence are inferred to be homologous, even though the similarities of their sequences are not statistically significant. The dissimilarity in the tertiary structure of two homologous domains, measured by the root-mean-square deviation of optimal superposition of main-chain atoms, increases monotonically with decrease in the similarity of their sequences (Wood and Pearson, 1999; Zhang and Skolnick, 2005; Zhang, 2008). The extent of dissimilarity increases more rapidly for proteins with a high β-strand content relative to those with a high α-helix content. The structures of two homologs retain recognizable similarity even after the similarity of sequence has faded to statistical insignificance: hence the guideline "Structure preserves homology better than does sequence."

There are very many, but still limited, ways in which a few helices and strands can reasonably be packed together. That is, two small domains may have achieved similar structures as a result of convergent evolution; they are analogs, not homologs. Several protein classification schemes (e.g., SCOP, CATH, PFAM) use structural criteria, not sequence. Members of a resulting group are not necessarily homologous; however, they usually are. Some proteins have similar structures because of convergent evolution; they are analogs. The burden of proof rests with him who proposes homology.

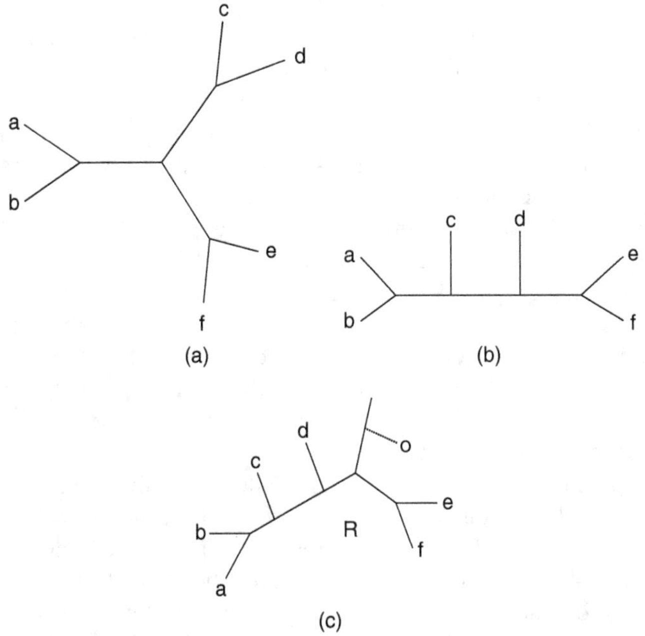

Figure 7.2. Examples of dendrograms, cladograms, or trees. Parts a and b show the two possible topologies for six terminal nodes, or domains (a to f). Both have four internal nodes and nine branches. There are $n \times (n-1)/2 = 15$ comparisons, or "differences," among the six domains. For a given topology and arrangement of terminal nodes, the 9 degrees of freedom, branch lengths, are adjusted to best fit the 15 observations. Part c shows a "rooted" dendrogram; the placement of this root (R) introduces an additional branch and requires either a postulated "oldest" branch or, better, an observed "outgroup" (O) to establish the oldest branch.

The number of networks or arrangement of terminal nodes within a given topology increases as follows:

n	2	3	4	5	6	7	8	9	10	
Networks	1	1	3	15	105	945	10,395	135,135	2,027,025	$(2n-5)!$
Rooted	1	3	15	105	945	10,395	135,135	2,027,025		$(2n-3)!$

Two (or more) similar species evolved from a common ancestor. Homologous proteins from these two species are related by the speciation event; they usually perform similar functions in the two species. These two homologous proteins are orthologs. A dendrogram of orthologs reflects, within statistical error, the evolutionary history of the species represented. The dendrograms of many sets of orthologous proteins from the same group of species should, within statistical error, have the same branching order and branch lengths. The more orthologs that are used in the calculation, the more reliable the resulting tree of the species.

Most domains are present in many different proteins within a single species. For example, *Homo sapiens* encodes $\sim 3 \times 10^4$ different proteins, excluding splice variants; these are built of $\sim 10^3$ different domains (reviewed by Mueller et al., 2007; Edwards, 2009). These multiple expressions of the same domain arose by gene duplication; they are paralogs. Dendrograms of paralogs do not reflect the evolution of their host organisms. One should assume paralogy until there is strong evidence for orthology.

Many proteins, single polypeptide chains, have evolved by gene duplication and splicing; they contain several domains and are called *multidomain* or *chimeric proteins* (Nakayama et al., 1992). These proteins may be homochimeric, such as calmodulin, which consists of four (homologous) EF-hand domains; others are heterochimeric, such as calciphosine, which consists of four EF-hands and an additional non-EF-hand domain that contains the binding site for calcium. Each domain of a chimeric protein has its own evolutionary history, often different from those of its fellow domains.

The concept of congruence is used to group the subfamilies of a homolog family. For example, if one generates a cladogram of all of the individual EF-hands of calmodulin, all of the domains 1 cluster together, as do domains 2, domains 3, and domains 4. The branching order within the four clusters is the same, within the error of statistical fluctuation. Domains 1 cluster closer to domains 3, and domains 2 cluster with domains 4. Further, the four EF-hands of troponins C, essential and regulatory light chains, nonvertebrate troponins, calmodulin-like leaf protein, squidulin, calctrin, Cal1, and pCAST cluster with the corresponding domains for calmodulin. These ten subfamilies are congruent; they are referred to as the *CTER group*. The most parsimonious interpretation is that these ten subfamilies evolved from a common four-domain precursor.

Correspondingly, the six four-EF-hand proteins—calcineurin B, secretagogin, p22, visinin (recoverin), calsenilin, DRE-antagonist modulator (DREAM), and sos3 gene product—form a congruent group (CPV) that is distinct from and not congruent with the CTER group (Bitto et al., 2009).

All of the S100 proteins, 26 of which have been identified in humans, are congruent with one another and with calbindin D_{9k}. They are also congruent with the EF-lobes of the heterochimeric proteins P26olf and profilaggrin (Chapter 11). The non-EF-hand domains of these proteins are not homologous to one another. The first EF-hand of the S100s binds calcium with moderate affinity, even though it has two residues inserted relative to the canonical calcium binding EF-hand.

Helices E and F are characterized by residues, n, with hydrophobic side chains on the inner aspects of the helices. In the canonical calcium binding EF-hand, the oxygens of side chains coordinating calcium can be assigned to the vertices—X, Y, Z—of an octahedron; both oxygens of the almost invariant Glu are assigned to vertex $-Z$ (Gifford et al., 2007). The residue at $-Y$ coordinates calcium with its carbonyl oxygen (=O). Two residues are inserted in domain 1 of S100, relative to the canonical EF-hand.

```
Canonical (S100, domain 2)
                    -------  Loop

  αααααααα E αααααα           ααααααα F αααααααα

                    1 1 1 1 1 1 1 1 1 1 2 2 2 2 2 2 2 2 2 2

  1 2 3 4 5 6 7 8 9 0 1 2 3 4 5 6 7 8 9 0 1 2 3 4 5 6 7 8 9

  E n * * n n * * n X * Y * Z G-Y I-X * *-Z n * * n n * *(n)

                              =0

S100, domain 1
                                            2 2 2 2 2 2 2 2 2

 1 2 3 4 5 6 7 8 9                          1 2 3 4 5 6 7 8 9

 E n * * n n * * n * * * E G D * * * L * K * E n * * n n * *(n)

                   X    Y  Z    -Y        -Z

                  =0   =0 =0    =0
```

The other EF-hand subfamilies, beyond the CTER, CPV, and S100 groups, do not show strong congruence to one another. A massive dendrogram generated with all, or a random sampling of all, EF-hands shows a strong tendency for the odd-numbered domains to group together and for the evens to group. Parvalbumin with three domains—AB, CD, and EF—also observes this pattern if one assumes that the first domain has been deleted and that $AB = 2, CD = 3$, and $EF = 4$. This general grouping of odd and even domains implies that the gene encoding an ur-EF-hand duplicated and fused to form to the original odd–even pair from which all EF-hands evolved.

To gain a full understanding of the changes and interactions of EF-hands, one should know the relative orientations of helix E and helix F of each domain. In addition, the interactions between the two EF-hands of a pair should be analyzed.

Yap et al. (1999) proposed a coordinate system for evaluating the relationship of helix E and helix F. The axis of helix E, of EF-hand 1 of apo-calmodulin, defines the z-axis. The $+x$-axis is defined by the N-end of helix F, residues 19 to 29. The helices E of other EF-hands are superimposed on this reference. When the structures of all known EF-hands are displayed in this common coordinate system, there is a broad distribution of F vectors; there is not a bimodal distribution corresponding to open and closed. However, for those specific EF-hands for which both a calci and an apo structure is available there is a consistent "opening," or separation of helices E and F upon binding calcium.

Kawasaki (unpublished) proposed a coordinate system for evaluating pairs of EF-hands. The z-axis is defined as the best fit to the midpoints of lines joining the 29 homologous residues of the two hands. The x-axis is initially defined as the line joining the centers of gravity of the 29 residues of the first (odd) EF-hand and the center of gravity of the second (even) EF-hand. This initial x-axis is slightly adjusted so that it passes through the z-axis and is perpendicular to it. Of greater interest, the yz-plane approximates the interaction surface between the odd and the even EF-hands. One can cleave apart the two EF-hands and view their projections on the yz-plane. Most revealing, one can see how the interactions of the two EF-hands change with calcium binding. There are several hydrophobic interactions between side chains of helices E1 and F2 and symmetrically between F1 and E2; also there are F1,F2 interactions in the apo lobe. Upon binding calcium, the E1,F2 and F1,E2 interactions shift slightly; however, the F1,F2 interactions are lost, as both EF-hands open. The details of these interactions vary with each EF-lobe.

8

PROTEIN COMPLEXES WITH METALS OTHER THAN CALCIUM

The theme of this book, calcium binding proteins, and the unique role of calcium in biology are brought into sharper focus by comparison with complexes of proteins with other metal ions. Na^+, K^+, Mg^{2+}, Zn^{2+}, and Cd^{2+}, in addition to Ca^{2+}, have only a single ionization state under biological conditions. In contrast, V(III/IV), Cr(III/VI), Mo(IV/V/VI), W(IV/V/VI), Mn(II/III/IV), Fe(II/III), Co(II/III), Ni(II/III), and Cu(I/II) are often involved in oxidation–reduction (redox) reactions. Iron, nickel, manganese, and molybdenum can be bound to organic ligands, most notably porphyrin and pterin, and the entire complex bound to a protein. Lead, mercury, cadmium, strontium, and so on, and some lanthanides are encountered in nature as contaminants or toxins; these elements, especially the lanthanides, may also be used to replace the naturally occurring metal ion for their valuable spectral properties (Chapter 6). Their mechanisms of toxicities and their interactions with proteins and with organic ligands in vivo are poorly understood.

The intracellular and, in metazoans, the extracellular concentrations of these essential cations are well regulated. Less well understood are the specificities and special characteristics of their respective sensors, channels, and pumps. The storage of metals (e.g., calcium in calsequestrin within the sarcoplasmic reticulum or iron in ferritin) assures both the availability and maintenance of physiological concentrations. Whether and how other metals are regulated and stored is a significant question.

These descriptions of "typical" cation sites lay the foundation for a discussion of exceptions and of more complex binding patterns. For example, some proteins,

Calcium Binding Proteins, By Eugene A. Permyakov and Robert H. Kretsinger

among them many calcium binding proteins, bind more than one type of metal ion, some at different sites within that protein and some by competition for a single site (Chapter 14).

The essential and nonessential metals discussed in Chapters 4 and 5 are sometimes encountered in nature as oxyanions: mono-, bi-, and polynuclear (e.g., Cr(VI) $2CrO_4^{2-} + 2H^+ \leftrightarrow Cr_2O_7^{2-} + H_2O$). Their formations, or dissociations, depend on the availability of oxygen and on pH. The complexes of these metals with proteins or ligands, as porphyrin or pterin, involve the metal stripped of its oxygen, or hydroxyl, ligand. How the oxyanion is converted to the (hydrated) cation in biological systems is poorly understood.

Investigations of these diverse metal–protein interactions has necessitated the development of techniques specific to each ion, some of which are discussed in Chapter 3. One can anticipate the development of more specific techniques for application in vivo. Some of the physical, chemical, and geological characteristics of these ions were summarized in Chapter 4; these are illustrated in the following summaries of these ions and their binding sites.

Five elements—carbon, nitrogen, oxygen, hydrogen, and phosphorus—are generally considered essential to life, as evolved here on Earth or imagined elsewhere. Any life-form would very probably engage in redox reactions by way of extracting energy from the environment and/or protecting itself from the environment; these reactions are much more easily realized by the several valence states available to sulfur (Amend et al., 2004), especially without recourse to redox metals. The halides—fluorine, chlorine, bromine, and iodine—are not essential to the origin of life; however, many organisms, *Homo sapiens* included, have evolved in the presence of these halides and have put (some of) them to use; they would die, or survive only poorly, without them; they have become essential.

Most organisms also require the essential metals, whose complexes with proteins are discussed in this chapter. The first group of six includes monovalent—Na^+ and K^+—and divalent cations—Mg^{2+}, Ca^{2+}, Zn^{2+}, and Cd^{2+}—that have only one valence state available under biological conditions. They are "hard" cations; they form purely electrostatic bonds and prefer oxygen (or halide) ligands (reviewed by Martin, 2002). They do not form oxyanions. The second group includes V(III/IV), Cr(III/VI), Mo(IV/V/VI), W(IV/V/VI), Mn(II/III/IV), Fe(II/III), Co(II/III), Ni(II/III), and Cu(I/II) (reviewed by Ma et al., 2009). They are often involved in redox reactions; they form partially covalent bonds and prefer nitrogen and sulfur ligands. Most of them, in their several valence states, form oxyanions. Not all organisms require all nine of these redox metals. However, once put to use, their absence would be highly detrimental; they, like the halides, have become essential. Several proteins bind to toxic metals (e.g., metallothioneins) with high specificity and supposedly are involved in detoxification; this might be the first step toward utilization and dependence.

The transport of these metals, both into and out of cells, often involves overlapping specificities. For example, the intestinal transporter for nonheme iron, divalent metal transporter 1 (DMT1), also mediates the transport of cadmium, manganese, and lead.

8.1. ESSENTIAL HARD CATIONS

8.1.1. Sodium Complexes

Sodium and potassium are necessary for maintenance of fluid balance in metazoans and for the function of nerves and muscles. The interactions between Na^+ or K^+ ions and their ligands are solely electrostatic (reviewed by Page and Di Cera, 2006). Ligand exchange rates (k_{ex}) of Me^+ are very high and allow rapid association and dissociation. Since the single positive charge is spread over a large volume, the charge density of any Me^+ is too low to enhance enzymatic catalysis,. However, Me^+ coordination can play an important role in rate enhancement (e.g., β-galactosidase and clotting proteases) and/or allosteric regulation of an enzyme (e.g., some kinases) (Page and Di Cera, 2006).

Coordination of a Me^+ ion in a folded polypeptide often involves carbonyl oxygen atoms donated from peptide bonds. The dipole moment of the carbonyl oxygen provides a strong electrostatic interaction with the ion. Recruitment of additional regions of the protein is mediated via hydrogen bonding through the amide hydrogen. Polypeptide segments can communicate through a bound Me^+ ion and its liganding carbonyl O atoms (Page and Di Cera, 2006). Sometimes, metal binding sites employ a negatively charged unidentate carboxyl or phosphoryl group.

The number of water molecules in the primary hydration shells of Me^+ ions (Li^+ 4, Na^+ 6, K^+ 7, Rb^+ 8) (Koneshan et al., 1998) reflect their radii and resulting charge densities (Chapter 3). The more tightly bound waters of Li^+ impose greater structure on its secondary hydration shell. The relative ease with which these waters can be displaced and the relative affinities of ligands for partially hydrated Na^+ and K^+ ions are essential to their specific transport.

Na^+ and K^+ ions are the metal ions most frequently found in biological systems. Na^+ ions are most abundant outside cells; whereas, K^+ ions are present in high concentration inside cells. Concentration gradients of these ions across the cell membrane provide an energy source for action potential generated by the opening of the Na^+ and K^+ channels, for ATP synthesis in some organisms, and for movements of various substances and other ions through the membranes by means of transporters.

LeuT transports Leu and Na^+ ions in the same direction across the cell membrane (Yamashita et al., 2005). It uses the energy of the Na^+ gradient to pump Leu into the cell (reviewed by Gouaux and MacKinnon, 2005). The crystal structure reveals a Leu and two completely dehydrated Na^+ ions bound deep inside the protein, partway across the membrane (Yamashita et al., 2005). One site contains six oxygen atoms in direct contact with the K^+ ion. Five of these oxygen atoms bear only a partial negative charge: the main-chain carbonyl, side-chain hydroxyl, and side chain amide, and one oxygen atom is from a carboxylate group. Five oxygens surrounding the Na^+ ion in the second site bear only a partial negative charge. The size of the binding-site cavity formed by the oxygen atoms is a good match to the Na^+ ion; the mean Na^+–O distance for both sites is 2.28 Å.

The sodium pump, also known as Na^+/K^+-ATPase, is responsible for establishing and maintaining electrochemical gradients in animal cells (reviewed by Jorgensen and Pedersen, 2001; Scheiner-Bobis, 2002; Rakowski and Sagar, 2003). Na^+/K^+-ATPase belongs to the P-type ATPases, a family of enzymes that become phosphorylated during transport by the γ-phosphate group of ATP at an Asp located within the highly conserved sequence DKTGS/T. Na^+/K^+-ATPase transforms the chemical energy of ATP to osmotic work and maintains electrochemical Na^+ and K^+ gradients across cell membranes. For every molecule of ATP hydrolyzed, three Na^+ ions from the intracellular space and two K^+ ions from the external medium are exchanged. Thus, the sodium pump contributes substantially to the maintenance of the membrane potential of the cell, provides the basis for neuronal communication, and contributes to the osmotic regulation of the cell volume. In addition, the electrochemical Na^+ gradient is the driving force behind secondary transport systems.

Na^+/K^+-ATPases are oligomeric and consist of α- and β-subunits. A third peptide, the γ-subunit, appears in some tissues to be involved in regulating the activity of the sodium pump and its interactions with Na^+ or K^+ ions. A number of isoforms of the α- and β-subunits have been isolated from various tissues of numerous species. It has clearly been demonstrated that the functioning of Na^+/K^+-ATPase requires the presence of both subunits. For example, renal [α1β1γ] Na^+/K^+-ATPase consists of three subunits, α with 1016 residues, β with 302, and γ with 55; it is the largest protein complex in the family of cation pump proteins.

The cytosolic loop (L4/5) between transmembrane helices M4 and M5 is essential to the function of the enzyme, because a series of residues within L4/5 are involved in ATP hydrolysis and enzyme function. The phosphorylation site is localized within this loop as a part of the sequence DKTGT/S that is highly conserved among all P-type ATPases. Na^+/K^+-ATPase is unique among cation pumps in its ability to bind either one molecule of free ATP or two K^+ ions per α1β1 unit with high affinity in the absence of other ligands.

All P-type ATPases hydrolyze ATP and occlude ions during the translocation process within the membrane-inserted segment of the protein. Through this process, the ionophore of every ion-transporting ATPase is accessible from only one side of the membrane at any given time. The Na^+/K^+-ATPase has two conformational states, E_1 and E_2 (reviewed by Scheiner-Bobis, 2002). These states are characterized by differences in their interactions with Na^+, K^+, ATP, and ouabain, and in their tryptic cleavage patterns. Relatively large E_1–E_2 conformational changes in the α subunit mediate long-range interactions between the ATP site and the cation sites in the membrane domain. In the first step of the reaction sequence, Na^+ ions and ATP bind with high affinity (K_d = 0.19 to 0.26 mM and 0.1 to 0.2 μM, respectively) to the E_1 conformation of the enzyme. This process is accompanied by phosphorylation of the Asp via the transfer of the γ-phosphate of ATP. Magnesium also takes part in this reaction. Thereafter, three Na^+ ions are occluded, while the enzyme remains in a phosphorylated state. Phosphorylation of the Na^+/K^+-ATPase causes a conformational change that brings about an alteration in the Na^+ binding site, allowing Na^+ to exit toward the extracellular side. One assumes that this conformational change occurs concomitantly with an expansion of the cation binding site (E_2 conformation of Na^+/K^+-ATPase), so that the larger K^+ ion can then be accommodated. After the E_2-P3Na^+ conformation is attained, the enzyme loses its affinity for Na^+ ($K_{0.5} = 14$ mM) and the affinity for K^+ is increased ($K_d \sim 0.1$ mM). Three Na^+ ions are then released to the extracellular medium and K^+ ions are taken up. The binding of K^+ to the enzyme induces a spontaneous dephosphorylation of the E_2–P conformation. The dephosphorylation of E_2–P leads to the occlusion of two K^+ ions and to the formation of E_2(2 K^+). Intracellular ATP increases the extent of the release of K^+ from the E_2(2 K^+) conformation and thereby also the return of the E_2(2 K^+) conformation to the E_1ATPNa conformation. The affinity of the E_2(2 K^+) conformation for ATP is very low, $K_{0.5} = 0.45$ mM.

Transitions between E_1 and E_2 forms of the protein are accompanied by large shifts of selectivity for Na^+ and K^+ ions, but little is known about the molecular basis for Na^+ vs. K^+ selectivity. Intramembrane Glu327 in transmembrane helix M4, Glu779 in M5, as well as Asp804 and Asp808 in M6, are essential for tight binding of K^+ and Na^+ (reviewed by Jorgensen and Pedersen, 2001). Asn324 and Glu327 in M4, Thr774, Asn776, and Glu779 in 771-YTLTSNIPEITP of M5 contribute to Na^+/K^+ selectivity.

Sweadner and Donnet (2001) noted the close structural similarity of the Na^+/K^+-ATPase and sarcoplasmic reticulum Ca^{2+}-ATPase (SERCA). They threaded the Na^+ pump sequence onto the known structure of SERCA to examine the topological location of specific residues. Although the overall sequence identity between SERCA and the Na^+ pump is about 27%, the identity is much higher in highly conserved regions of the molecule and in the α-helical transmembrane segments where the ion binding sites are located. It was concluded that the Na^+ pump and SERCA share the same fold. Ogawa and Toyoshima (2002) and Rakowski and Sagar (2003) tried to predict the detailed tertiary structure of Na^+/K^+-ATPase and the location and detailed coordination geometry of all three Na^+ binding sites and of both K^+ binding sites.

Enzymes activated by monovalent cations are abundant in plants and animals (reviewed by Di Cera, 2006). In general, enzymes requiring K^+, such as kinases and molecular chaperones, are also activated by NH_4^+ and Rb^+ but are not activated as well or at all by the larger cation, Cs^+, or by the smaller cations, Na^+ and Li^+. Enzymes requiring Na^+ such as β-galactosidase and clotting proteases are not activated as well by Li^+, or the larger cations K^+, Rb^+, and Cs^+. Because the concentration of Na^+ and K^+ is tightly controlled in vivo, monovalent cations do not function as regulators of enzyme activity. Rather, they facilitate substrate binding and catalysis by lowering energy barriers in the ground and/or transition states (reviewed by Page and Di Cera, 2006). Enzymes activated by Me^+ ions evolved to take advantage of the availability of Na^+ outside the cell and K^+ inside the cell to optimize their catalytic function. Indeed, a strong correlation exists between the preference for K^+ or Na^+ and the intracellular or extracellular localization of such enzymes.

Selective stabilization of one conformation of the enzyme through metal complexation may produce local and potentially long-range effects on the enzyme structure. The entropy of these solvent-accessible regions can affect the kinetic properties of the enzyme. Substrate binding to a stabilized metal–enzyme complex may be more favorable as the entropic penalty of ordering the enzyme to form the enzyme–substrate complex is paid by the previously bound ion. According to Page and Di Cera (2006), the variety of reported instances of weak metal activation of enzyme activity are possibly the result of global entropic effects rather than a specific Me^+ ion binding site interwoven into the catalytic process of the enzyme.

Alkali cations can affect the catalytic efficiency of enzymes. This is especially true when dealing with enzymes whose substrate bears a formal positive charge. For example, the combination of computational and biochemical experiments clearly showed that Li^+, Na^+, and K^+ may influence the ligand binding at the human acetylcholine esterase gorge (Petraglio et al., 2008). This enzyme is a serine hydrolase that plays a key role in terminating the nervous signal by hydrolyzing the neurotransmitter, acetylcholine.

A classification of monovalent cation–activated enzymes can be based on the selectivity of the effect, as established by kinetic studies, and the mechanism of activation, as shown from structural analysis. The mechanism of metal activation can be established unequivocally from crystal structures of the enzymes as cofactor-like (type I) or allosteric (type II) (Page and Di Cera, 2006). In the former case, substrate anchoring to the enzyme active site of the enzyme is mediated by Me^+ ions, often acting in tandem with a divalent cation such as Mg^{2+}. In such a mechanism, Me^+ ion coordination is absolutely required for catalysis or substrate recognition. In the latter, metal binding enhances enzyme activity through conformational transitions triggered upon binding to a site where the cation makes no direct contact with substrate. In this case, the Me^+ is not expected to be absolutely required for either binding or catalysis.

The strategy to anchor substrate to the active site is used by Na^+-activated type I enzymes fructose-1,6-bisphosphate aldolase (Hall et al., 1999) and tagatose-1,6-bisphosphate aldolase (Hall et al., 2002a), in which the tandem Na^+–Zn^{2+} is used (Page and Di Cera, 2006). Tagatose-1,6-bisphosphate aldolase (Figure 8.1) is unique in that it replaces a water molecule in the Na^+ coordination shell with the π-electrons of a Phe. In β-galactosidase from *Escherichia coli*, the interplay between Na^+ and Mg^{2+} is quite different from the partnership seen in kinases, with the Mg^{2+} ion binding far from substrate and Na^+ being in contact with the galactosyl 6-hydroxyl (Juers et al., 2000). β-Galactosidase hydrolyzes lactose to galactose and glucose. Its activity is greatly influenced by magnesium and enhanced by sodium or potassium. A Na^+ ion contacts the galactosyl 6-hydroxyl directly, and a Mg^{2+} ion binds distal to the substrate. In this case the Na^+ ion is

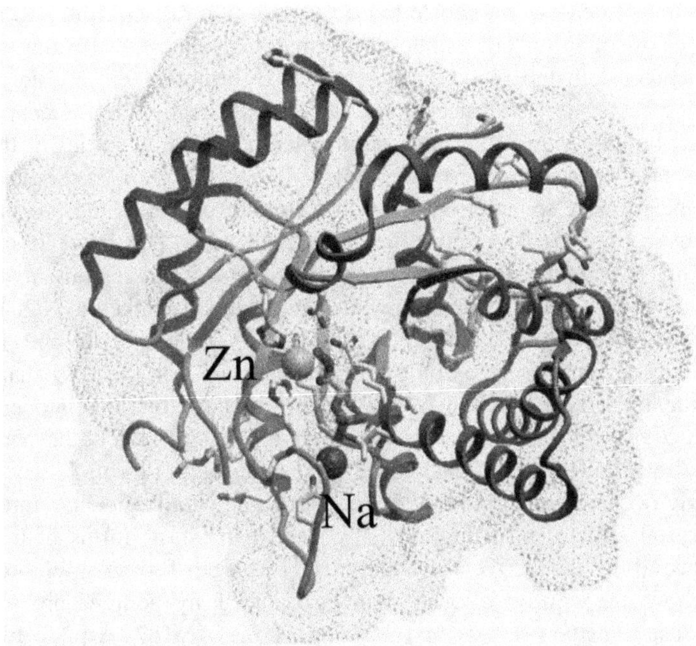

Figure 8.1. Structure of *Escherichia coli* tagatose-1,6-bisphosphate aldolase with bound Na^+ and Zn^{2+} ions complexed with phosphoglycolohydroxamic acid (PDB file 1GVF). (From Hall et al., 2002a.) (*See insert for color representation of figure*.)

coordinated by three protein atoms and two water molecules in the free enzyme, and lactose hydroxyl replaces one of the water molecules in the coordination shell. Change in the Na^+ coordination, which is induced by substrate binding, triggers a conformational transition of the 794–804 loop linked to the repositioning of Phe601, one of the Na^+ ligands (Page and Di Cera, 2006).

There are several examples of the Na^+-activated allosteric type II enzymes. Trp synthase is an $\alpha_2\beta_2$ tetramer with the subunits arranged in a linear $\alpha\beta\beta\alpha$ fashion (Woehl and Dunn, 1995). The α-subunit catalyzes cleavage of IGP to G3P and indole, which is then tunneled to a neighboring β-subunit that catalyzes condensation of indole with L-Ser to give L-Trp. Trp synthase requires sodium or potassium for optimal catalysis. Crystal structures of Trp synthase with bound Na^+ or K^+ ions show that the monovalent cation makes no contact with substrate or pyridoxalphosphate and binds to the β-subunit near the tunnel, enabling the indole intermediate to be shuttled to the active site for complexation with L-Ser. The tunnel is partially blocked by residues Phe280 and Tyr279 in the sodium form and is more open in the potassium form. The presence of Me^+ ion influences the nature of intramolecular interactions, including the salt bridge between Asp56 in the α-subunit and Lys167 in the β-subunit, which is critical for allosteric communication.

In clotting proteases, the allosteric effect of sodium arises from optimization of substrate recognition and subtle conformational changes affecting the catalytic machinery. In the blood-clotting protease thrombin, the allosteric effect of sodium affects the basic mechanism of substrate recognition. Occupancy of the sodium binding site in thrombin controls some of the activities of thrombin and is important for allostery (reviewed by Lane et al., 2005). Sodium binding near the primary specificity pocket of thrombin promotes the procoagulant, prothrombotic, and signaling functions of the enzyme. The effect is mediated allosterically by communication between the Na^+ site and regions involved in substrate recognition. The thrombin residues energetically linked to Na^+-induced allostery are Asp189, Glu217, Asp222, and Tyr225 (Pineda et al., 2004). All of these residues are in close proximity to the bound Na^+ ion. Asp189 shares with Asp221 the important function of transducing sodium binding into enhanced catalytic activity. Occupancy of its sodium binding site affects the activity of thrombin; nevertheless, since $[Na^+]$ in blood is constant (140 mM), the site will be similarly occupied under all but the most extreme pathological circumstances.

The crystal structures of the apo (slow) and natri (fast) forms of thrombin, free or bound to the active site inhibitor H-D–Phe–Pro–Arg–chloromethyl ketone, show the conformational changes induced by sodium binding. The slow → fast transition results in formation of the Arg187–Asp222 hydrogen bond, optimal orientation of Asp189 and Ser195 for substrate binding, and a significant shift of the side chain of Glu192 linked to a rearrangement of the network of water molecules that connect the bound Na^+ to Ser195 in the active site (Pineda et al., 2004). Na^+ binding turns Asp189 for correct orientation of the Arg side chain of substrate at the P1 position (Howard et al., 2006). This enables the enzyme to accomplish its procoagulant role in the blood. Long-range effects induced by Na^+ binding propagate through a network of buried water molecules up to the catalytic Ser195 located 15 Å away.

Among Me^+-activated enzymes, chymotrypsin-like proteases deserve special attention in view of the very important role played by Na^+ binding in their function and evolution (Page and Di Cera, 2006). This large family of enzymes catalyzes the hydrolytic cleavage of peptide bonds. Over 2000 proteases are included in the S1A family of peptidases, and they are most abundant within the animal kingdom.

Chymotrypsin cleaves peptides at the carboxyl side of Phe, Tyr, and Trp. In chymotrypsin, two β-barrels lie perpendicular to one another, in addition to a C-terminal α-helix, to comprise the chymotrypsin fold and form the basic unit of a number of key catalytic processes. Catalytic residues—His57, Asp102, and Ser195—are present in both β-barrels as well as two hot spots involved in protein–protein and protein–glycosaminoglycan interactions. A catalytic triad of residues acts to yield a Ser that engages the substrate through nucleophilic attack followed by hydrolysis. The majority of S1A family members are trypsin-like enzymes that cleave polypeptide chains on the C-terminal side of Arg or Lys residues.

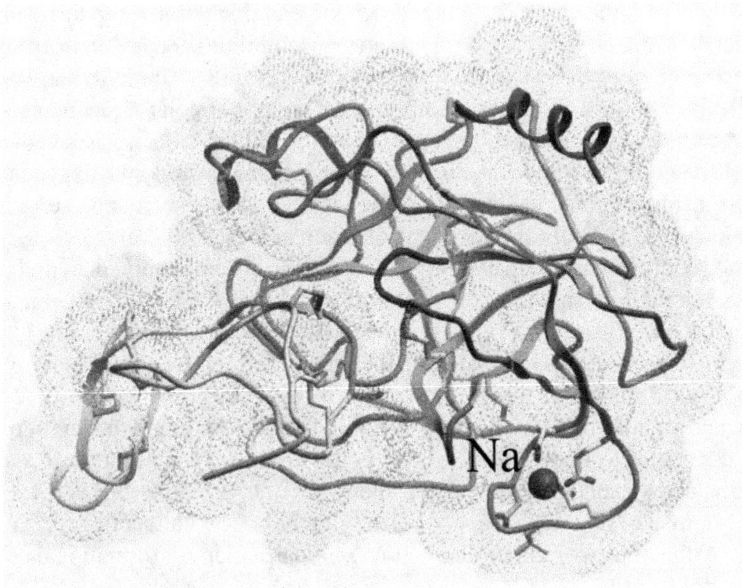

Figure 8.2. Structure of porcine trypsin with bound Na^+ ion (PDB file 1EJA). (From Rester et al., 2000.) (*See insert for color representation of figure.*)

The primary specificity pocket (S1 site) plays a dominant role in trypsin-catalyzed reactions. Na^+ coordination near the S1 binding pocket activates trypsinlike enzymes, and binding is linked allosterically with distant and distinct regions on the protease surface (reviewed by Krem and Di Cera, 2001). The Na^+ ion binding site lies immediately adjacent to Asp189 in the S1 pocket of the enzyme and establishes a direct link with substrate (Figure 8.2). Furthermore, sodium binding influences the extended substrate binding site of the enzyme. Tyr225 plays a crucial role in determining the sodium-dependent allosteric nature of serine proteases by allowing correct orientation of the backbone O atom of residue 224 (Guinto et al., 1999), which contributes to coordination of the Na^+ ion. The side chain of Tyr225 also secures the integrity of the water channel surrounding the primary specificity pocket required for correct substrate recognition. The allosteric core is assisted by another set of residues whose substitution with Ala reduces sodium affinity more than tenfold. These residues are Thr172, Tyr184, Arg187, Ser214, and Gly223; together with the residues of the allosteric core, they link the Na^+ binding site to the S3–S4 specificity pocket and the S1 specificity site.

8.1.2. Potassium Complexes

Channels are divided into voltage gated and ligand gated, depending on the activation mechanism. The K^+ ion channel is voltage gated. It conducts K^+ ions selectively across the cell membrane, down the electrochemical gradient. The K^+ ion channel is a fourfold symmetric tetramer resembling a teepee that surrounds a central pore (Figure 8.3) (Doyle et al., 1998; reviewed by Burdette and Lippard, 2003; Roux, 2005). The channel comprises a wide, nonpolar aqueous cavity on the intracellular side, leading up, on the extracellular side, to a narrow pore that is 12 Å long and lined exclusively by main-chain carbonyl oxygens. Four voltage-sensing regions, each made of helices S1 to S4, surround the pore and control its gates. The carbonyl oxygens of five residues, Thr–Val–Gly–Tyr–Gly, assume a conformation capable of coordinating a completely dehydrated K^+ but not the smaller Na^+ ions (Y. Zhou et al., 2001; reviewed by Gouaux and MacKinnon, 2005). In each of these sites, a K^+ ion is dehydrated and interacts with eight oxygen atoms. The size of the cavity formed by the selectivity filter sites is a good match to the K^+ ion, with a mean K^+–O distance of 2.84 Å. The greater number of oxygen atoms forming the K^+ binding sites (eight oxygen atoms) compared to that of the Na^+ sites (five or six oxygen atoms) is a simple geometric consequence of the larger radius of the K^+ ion; this allows more oxygen atoms to surround the ion. The selectivity for conducting K^+ ions is enabled by the ability of the protein to undergo particular conformational changes

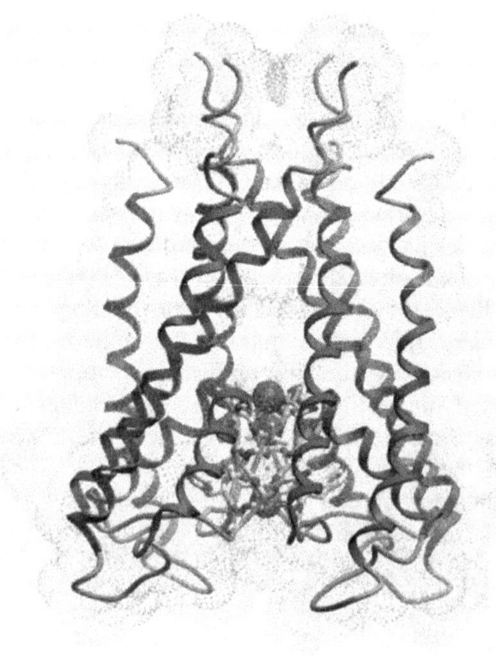

Figure 8.3. Structure of the tetrameric subunit of potassium channel with three K^+ ions inside the channel. The sequence Thr–Val–Gly–Tyr–Gly, which comprises the selectivity filter, is seen in each monomer (PDB file 1BL8). (From Doyle et al., 1998.) (*See insert for color representation of figure.*)

and thus form the correct coordination sphere for the dehydrated K^+ ion, in contrast to the (partially) hydrated Na^+ ion.

In most voltage-gated channels at resting potential (~ -70 mV in neurons), the activation gate is closed, and membrane depolarization causes a conformational change in the voltage-sensing domains that is transmitted to the pore domain and results in opening of the gate (reviewed by Tombola et al., 2006). It has been estimated that the conductance of a single channel drops at least 10^5 times, going from the open to the closed state and that the probability for a channel to be in the open state in resting conditions is less than 10^{-9}. The crystal structure of the Kv1.2 K^+-channel shows the four voltage-sensing homologous domains around the central pore. The domains are located at the corners of the square-shaped pore domain, and their interaction surface with the pore is rather small. As a result, a large portion of their perimeter is expected to face lipid. Based on these findings, Long et al. (2005) suggested that the voltage-sensing domains keep their position at the periphery of the channel floating in the membrane and that they interact only weakly with the pore domain. To make the opening of the activation gate voltage dependent, the gate must be controlled by a molecular sensor that detects the transmembrane potential in real time (reviewed by Tombola et al., 2006). This molecular sensor must contain charges, located in the membrane electric field, that change their position when the field changes. It is now well appreciated that these charges, called *gating charges*, reside in the voltage-sensing domains of voltage-gated channels. The S4 helix in each voltage-sensing domain contains four to eight positively charged residues, mostly Arg's, located at every third position. The S4 charges are responsible for most of the gating charge movement during activation (reviewed by Bezanilla, 2005). When the gating charges move across the electric field of the membrane as a result of a change in the applied membrane voltage, they generate an electric current called the *gating current*. The gating current is roughly two orders of magnitude smaller than the ionic current flowing through the open channel, and it is transient. It can be measured only when the number of channels in the membrane is high and when both the ionic current through the pore and the capacitive current required to charge and discharge the membrane are reduced or eliminated. Water-filled vestibules in the voltage-sensing domains focus the membrane electric field on the S4 Arg's. They move across pores connecting internal and external voltage-sensing domain vestibules. The coupling between the voltage sensor and gate may be allosteric (i.e., the activated conformation of the voltage sensor stabilizes the open state of the gate) or obligatory (i.e., the gate cannot open unless the voltage sensor is in the activated conformation). There are three major conceptual models of voltage sensing: the transporter model, the helical screw model, and the paddle model (reviewed by Tombola et al., 2006). They differ in the extent and nature of the transmembrane movement of the S4 helix and in terms of whether the S4 Arg's are exposed to the lipid hydrophobic core.

Diol and glycerol dehydratases provide the simplest example of type I enzymes. Diol dehydratase is a coenzyme B_{12}–dependent enzyme with an absolute requirement for potassium. In these proteins the K^+ ion is coordinated by five protein ligands and acts as a "bait" for the two hydroxyl oxygens from the substrate (Shibata et al., 1999; Di Cera, 2006; Page and Di Cera, 2006). In the substrate-free form, the enzyme retains K^+ in essentially the same coordination, yet replaces two substrate ligands with water molecules.

Enzymes involved in phosphoryl transfer reactions are the dominant group among monovalent cation–activated enzymes. In addition to K^+, these enzymes have an absolute requirement for a divalent cation, typically Mg^{2+}. The mechanism of the enzyme activation involves K^+ and Mg^{2+} ions acting in tandem to provide optimal docking for the phosphate group of the substrate into the protein active site to enable nucleophilic attack on the P_γ or transfer of phosphate groups. The clearest example of this mechanism comes from the ATP-driven folding machine, GroEL, which has the highest K^+ affinity ($K_d \sim 80$ μM) ever reported for a monovalent cation–activated enzyme (Viitanen et al., 1990). GroEL is an allosteric tetradecameric protein composed of two stacked heptamers that define a large central cavity when in complex with GroES. The activity of GroEL is influenced by magnesium and has an absolute requirement for potassium. The crystal structure of GroEL bound to ATP reveals Mg^{2+} and K^+ ions acting in tandem to assist binding of ATP to the protein. In this structure the K^+ ion is coordinated by one ATPα oxygen, the backbone oxygens of Thr30 and Lys51, and four waters (Wang and Boisvert, 2003). Nucleophilic attack on the Pα oxygen of ATP is mediated by Asp52. The K^+ ion fixes both groups in place through some of its ligating waters and is assisted by Mg^{2+}, which anchors all three P groups of ATP.

Some kinases belong to type II potassium-activated enzymes because in this case the K^+ ion does not contact ATP directly (reviewed by Di Cera, 2006). In this structure, potassium exerts its influence indirectly by perturbing the conformation of active site residues. In ribokinase (Andersson and Mowbray, 2002) and aminoimidazole riboside kinase (Zhang et al., 2004) the K^+–Mg^{2+} ion pair is broken by embracing the monovalent cation in a β-turn adjacent to the active site. Ribokinase catalyzes the phosphorylation of ribose using ATP and magnesium, with the ancillary requirement of potassium. Unlike other carbohydrate kinases, ribokinase is also activated by cesium to an extent comparable to potassium, but not at all by sodium or lithium. Ribokinase is a homodimer; each subunit is composed of two domains, a larger domain providing most of the binding interactions for substrate and a smaller domain that provides the dimer interface and a lid for the ribose.

8.1.3. Magnesium Complexes

The Mg^{2+} ion usually binds oxygen ligands rather than nitrogen or sulfur. It has a strong tendency to bind six ligands in a octahedral arrangement. Magnesium readily binds water; the hexa-aquo form, $Mg[H_2O]_6{}^{2+}$, is common in crystal structures. The high affinity of magnesium for water is partly due to the size of the cation; the six oxygens are in contact with each other (O–O is 2.9 Å), as well as with the Mg^{2+} (O–Mg^{2+} is 2.07 Å) (Chapter 3). Some other metal ions of similar size do not show this affinity for water. Molecular orbital calculations indicate that the energy penalty for changing the inner coordination number of Mg^{2+} from six is fairly high; $Mg[H_2O]_6{}^{2+}$ is 12.4 and 6.4 kcal mol^{-1} lower in energy than $Mg[H_2O]_4{}^{2+} \cdot 2H_2O$ and $Mg[H_2O]_5{}^{2+} \cdot H_2O$, respectively (reviewed by Glusker et al., 1999). Thus the inner coordination sphere of Mg^{2+} is fairly rigid and stable.

Within cells, magnesium plays a vital role in numerous regulatory processes. It is involved in the synthesis of DNA and RNA as well as in the maintenance of their conformation. Magnesium forms complexes with phospholipids and affects membrane fluidity and permeability. Intracellular magnesium is directly associated with processes as important as the secretion of hormones, including insulin and prolactin.

Many of the actions of magnesium reflect its role as a cofactor of a wide range of enzymes. It activates nearly all of the enzymes involved in the metabolism of phosphorylated compounds, as well as many enzymes in the glycolytic and tricarboxylic acid pathways. When bound to ATP in the active site of an enzyme, not all of the normal six coordination positions for ligands are filled by interaction with either the protein or by ATP. One or more water molecules remain coordinated with the Mg^{2+} ion. The function of its binding to the phosphoryl groups of ATP in many cases appears to be activation of the phosphate ester toward hydrolysis (reviewed by Maguire and Cowan, 2002). The relatively slow exchange rate of water in the hydration shell of Mg^{2+} may also play a role here, making it somewhat harder to lose a water molecule and therefore somewhat easier to allow formation of a structure containing a water molecule in a particular geometry.

Mandelate racemase binds a Mg^{2+} ion with three carboxyl groups (Figure 8.4). The mandelate displaces a water from the coordination sphere of the Mg^{2+} ion and binds it by means of its carboxylate group and a hydroxy group. The magnesium holds the substrate firmly in place so that the catalytic abstraction and addition of a hydrogen atom by His297 or Lys166 is effected precisely.

Topoisomerases play a central role among enzymes involved in processing the main chain of DNA (reviewed by Sissi and Palumbo, 2009). They participate in essential cellular processes such as DNA replication, transcription, and chromosome condensation, which require enzymes able to regulate the ensuing topological changes produced in the nucleic acid. Topoisomerases produce a topological change in a DNA chain by the unwinding or the supercoiling of the double helix, thereby releasing the torsional strain imposed by DNA processing. The topoisomerase-mediated cleavage process consists of a nucleophilic attack of a Tyr located in the catalytic pocket to a phosphodiester bond of the DNA backbone. The transphosphorylation reaction produces a covalent protein, nucleic acid linkage, and a free hydroxyl group at the split deoxyribose group. Topoisomerases often require cofactors for full catalytic activity. ATP regulates the conformational changes required for enzyme action through its binding and subsequent hydrolysis. Divalent metal ions, especially Mg^{2+}, are important, as they perform both structural and catalytic functions, besides acting in complex

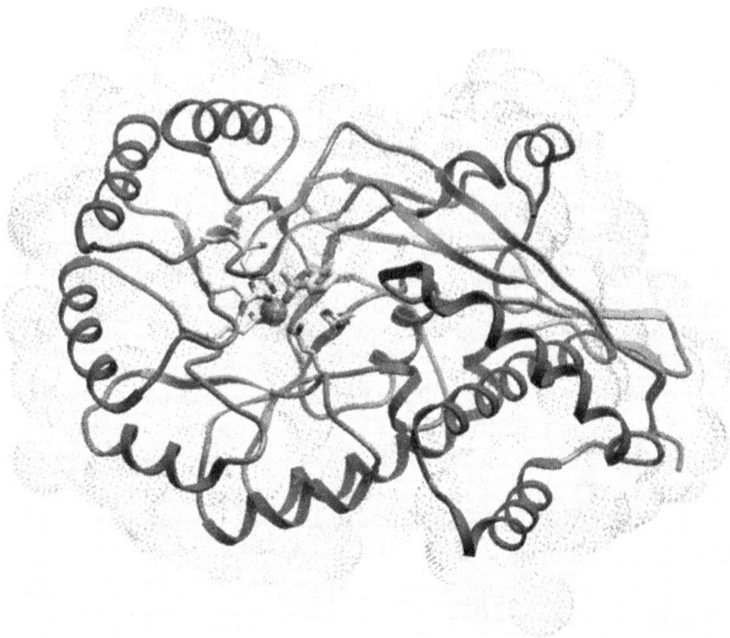

Figure 8.4. Structure of mandelate racemase from *Pseudomonas putida* with a bound Mg^{2+} ion (PDB file 1MDR). (From Landro et al., 1994.) (*See insert for color representation of figure.*)

with ATP. Topoisomerase I and topoisomerases II require magnesium to relax supercoiled (Stewart et al., 1996). They have a conserved domain of about 100 residues, called Toprim, that has an invariant Glu and a D × D motif (Figure 8.5) (reviewed by Sissi and Palumbo, 2009). The Glu is located in a sharp turn connecting a β-strand to an α-helix. The three acidic residues are near one another and are available for concerted interactions (Dong and Berger, 2007). Given its chemical nature and the requirement for magnesium, the triad motif has been proposed to represent the Me^{+} ion(s) binding element in the catalytic core. In the topoisomerase II family the D × D motif is extended to a D × D × D conserved pattern that offers an additional carboxylic site to generate a structural and electronic network for coordination and correct positioning of the catalytically relevant Mg^{2+} ion(s) (Figure 8.6).

The concentration of free Ca^{2+} ions in cells varies over several decades in tenths of seconds. Organisms evolved strategies for keeping cytoplasmic magnesium concentration nearly constant, $[Mg^{2+}]$ ~2.0 mM, despite fluctuations in magnesium concentration outside. The gram-negative bacterium *Salmonella enterica* sv. *typhimurium*, has three magnesium transporters—CorA, MgtA, and MgtB—and a regulatory system—PhoP/PhoQ—that respond to the extracytoplasmic levels of magnesium (reviewed by Chamnogpol and Groisman, 2002; Chamnongpol et al., 2003).

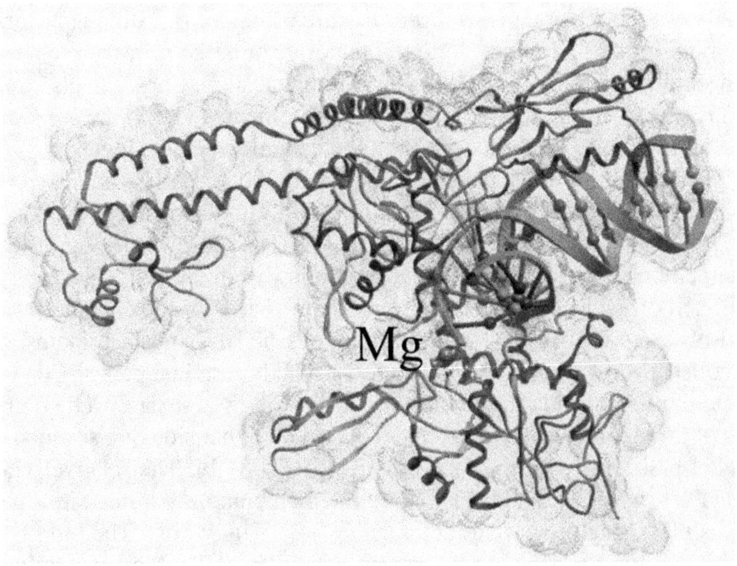

Figure 8.5. Structure of a fragment, 419 to 1177, of type IIA topoisomerase *Saccharomyces cerevisiae* with bound polydeoxyribonucleotide and a Mg^{2+} ion (PDB file 2RGR). (From Dong and Berger, 2007.) (*See insert for color representation of figure.*)

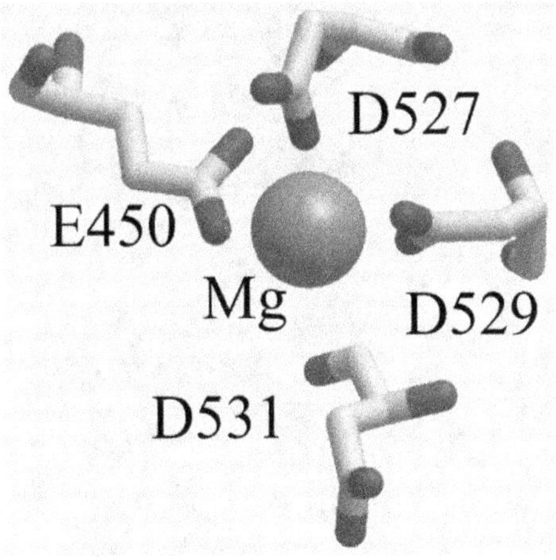

Figure 8.6. Magnesium binding site in type IIA topoisomerase *S. cerevisiae* (PDB file 2RGR). (From Dong and Berger, 2007.)

CorA is the major magnesium transporter in *Salmonella*; homologs have been identified in over 30 organisms from all three kingdoms. Although CorA can mediate both influx and efflux, influx is believed to be its primary physiological function (Smith and Maguire, 1998). Besides magnesium, CorA can also transport cobalt, nickel, and iron; although with reduced efficacy. CorA of *Thermatoga* is pentameric and funnel shaped, as seen in the crystal structure (Figure 8.7) (Lunin et al., 2006; Maguire, 2006; Payandeh and Pai, 2006). Each monomer contains two transmembrane helices; the pore is formed by the TM1 helix (residues 293 to 315), which is kinked at Pro303 and Gly312. The narrow opening of the pore is at the membrane interface; the wider mouth is within the cell. It is enclosed by the TM2 helix (residues 326 to 346); this "stem" helix extends about 100 Å into the cytosol, a record for the longest continuous helix found in proteins, except for the coiled coil of tropomyosin. The rest of the N-terminal part of CorA forms a separate domain outside the funnel. It consists of an $\alpha\beta\alpha$ sandwich comprising two sets of three α-helices ($\alpha_1\alpha_2\alpha_3$ and $\alpha_4\alpha_5\alpha_6$) on either side of a seven-stranded antiparallel β-sheet ($\beta_2 \uparrow \beta_1 \downarrow \beta_3 \uparrow \beta_7 \uparrow \beta_6 \downarrow \beta_5 \uparrow \beta_4 \downarrow$). The relatively long α_5 and α_6 helices, which extend back toward the membrane, are termed the "willow" helices, as they hang down like the branches of a willow tree. The loop between β_5 and β_6 extends back toward the membrane surface. The tips of this loop and

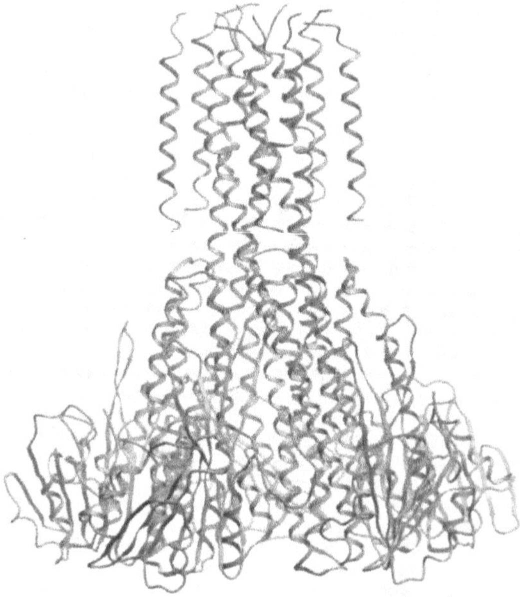

Figure 8.7. Structure of the magnesium transporter CorA (PDB file 2BBJ). (From Lunin et al., 2006.)

the tips of the willow helices have many Asp's and Glu's. This high concentration of negative charge is counterbalanced by a unique ring of Lys's, termed the *basic sphincter*. Asn314 at the extracellular membrane surface appears to completely block the entrance to the ion channel. The diameter of the channel varies from about 6 Å to about 2.5 Å (i.e., the channel is too narrow for a hydrated Mg^{2+} ion to pass). The side chain of Met302 constricts the channel to about 3.3 Å in the middle of the membrane; additional obstacles to ion movement are Leu294 and Met291. The side chains of these hydrophobic residues protrude into the interior of the channel, narrowing its diameter to 2.5 Å. Both the blocking by Asn314 and constriction by Met302 show that this is a closed state of CorA. It is assumed that the negatively charged willow helices can pull the positive charge of the basic sphincter away from the central axis of CorA at the level of Leu294 and Met291. This would allow the stalk helix to act as a lever to rotate or otherwise move the TM1α-helix, which forms an extension of the membrane. Together, these movements would have the combined effect of opening both the Leu294 barrier and the Met291–Asn314 barrier at both ends of the channel.

MgtA and MgtB are P-type ATPases that primarily mediate magnesium influx; they can also mediate nickel influx at a lower capacity.

PhoP/PhoQ is a two-component regulatory system that mediates adaptation to magnesium-limiting environments (reviewed by Chamnogpol and Groisman, 2002). Growth in micromolar concentrations of magnesium promotes transcription of PhoP-activated genes; whereas, growth in millimolar concentrations of magnesium represses transcription of PhoP-activated genes. Chamnogpol et al. (2003) demonstrated that residues G93, W97, H120, and T156 in PhoQ are required for a wild-type response to magnesium; they suggested that its binding to the periplasmic domain regulates several activities of the PhoQ protein.

In mammals, magnesium homeostasis is strictly controlled and depends on a balance between intestinal absorption and renal excretion (reviewed by Voets et al., 2004). The kidney provides the most sensitive control of magnesium balance. About 80% of the total serum magnesium is filtered through the glomerular membrane and then reabsorbed in consecutive segments of the nephron. The final excretion of magnesium to the urine is determined by the active reabsorption of magnesium in the distal convoluted renal tubules. Voets et al. (2004) demonstrated that TRPM6 contains a magnesium-permeable channel that is specifically localized to the apical membrane of magnesium-reabsorbing tubules in the kidney and the brush border membrane of the magnesium absorptive cells in the duodenum. The tight regulation of the TRPM6-induced current by intracellular magnesium provides a feedback mechanism of regulation of influx and implies that intracellular magnesium buffering and extrusion mechanisms have a strong impact on channel functioning. Parvalbumin and calbindin D_{28K}, which are coexpressed with TRPM6, might function as intracellular magnesium chelators in the distal convoluted renal tubule.

8.1.4. Zinc Complexes

About 10% of the human genome encodes zinc proteins, amounting to at least 3000 proteins; of these, 397 are hydrolases, 302 ligases, 167 transferases, 43 oxidoreductases, 24 lyases/isomerases, 957 transcription factors, 221 signaling proteins, and 141 transport/storage proteins (reviewed by Maret and Li, 2009). Fifty-three proteins contain structural metal sites; 19 proteins are involved in DNA repair, replication, and translation. In addition, there are 427 zinc finger proteins and 456 other zinc proteins of unknown function.

At least four types of zinc binding sites—structural, catalytic, co-catalytic, and protein interface—have been seen in crystal structures (reviewed by Auld, 2001). His, Glu, Asp, and Cys are the most common ligands.

In catalytic sites zinc generally forms complexes of any three of nitrogen, oxygen, or sulfur donors; His is most frequent, followed by Glu, then Asp and Cys. A characteristic feature of a catalytic zinc site is its open coordination sphere. The overall length of such sites can be as short as 11 residues, as is observed in the astacin superfamily of zinc proteases. A catalytic Zn^{2+} ion is located at the active site of the enzyme, where it participates directly in the catalytic mechanism, interacting directly with the substrate molecules undergoing reaction (reviewed by McCall et al., 2000; Auld, 2001).

The chemical properties of zinc in enzymes are largely attributed to its function as a relatively strong Lewis acid. Water is always a ligand of the catalytic zinc. The polarization or ionization of this activated water is brought about by a base from an active site residue. This provides a OH^- ion at neutral pH, and displacement of water or expansion of the coordination sphere results in Lewis acid catalysis by the Zn^{2+} ion. In the zinc hydrolases and lyases, such as the zinc proteases and carbonic anhydrases, the Zn^{2+} ion serves as a powerful electrophilic catalyst itself or by providing an activated water molecule for nucleophilic attack, polarization of the carbonyl of the scissile bond, and stabilization of the negative charge in the transition state.

Catalytic zinc is found in carboxypeptidases, alcohol dehydrogenases, astacin, β-lactamases, and carbonic anhydrases (reviewed by E. A. Permyakov, 2009). The carboxypeptidase family of exopeptidases and the thermolysin family of endopeptidases are examples of polarization assisted zinc water catalysis. Carboxypeptidases are exopeptidases that catalyze the hydrolysis of peptide bonds at the C-terminus of peptides and proteins (reviewed by Vendrell et al., 2000). Figure 8.8 shows the crystal structure of carboxypeptidase A; Zn^{2+} ion is coordinated by His69 (Nδ1), Glu72 (Oε1 and Oε2), His196 (Nδ1), and a water molecule. His69 and Glu72 are in a seven residue loop between a β-sheet and an α-helix; His196 is the last residue in a β-strand extending from residues 191 to 196 (Rees et al., 1983). The terminal carboxylate group of the peptide substrate is fixed by Asn144, Arg145, and Tyr248, while the carbonyl group of the scissile peptide bond is positioned near Glu270, Arg127, and the Zn^{2+} ion. The acyl pathway hypothesis proposes the existence of a covalent acyl–enzyme intermediate; in

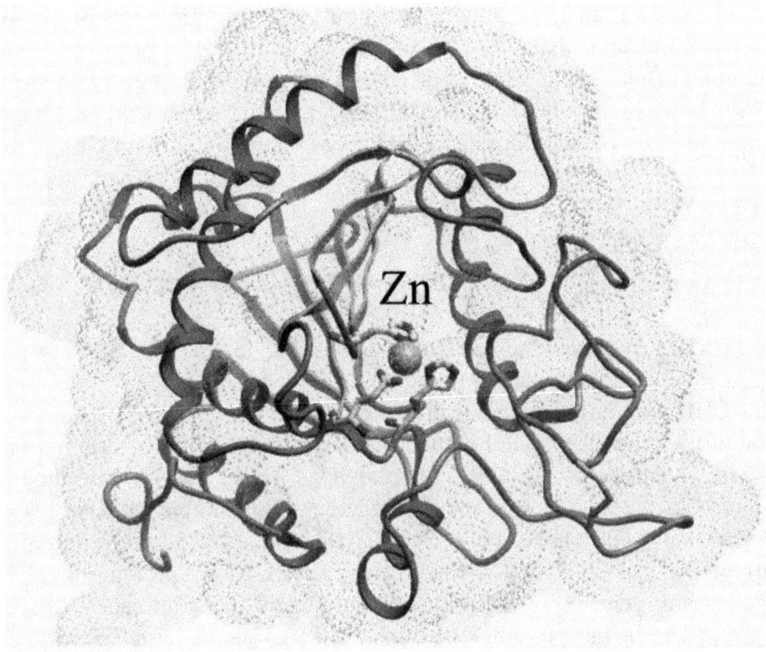

Figure 8.8. Structure of carboxypeptidase A with its catalytic zinc binding site (PDB file 5CPA). (From Rees et al., 1983.)

contrast, the promoted water pathway requires a simultaneous polarization of the substrate carbonyl group by the Zn^{2+} ion and a zinc promoted activation of a water molecule, which directly attacks the scissile peptide bond.

Many enzymes are sensitive to zinc inhibition; Maret and Li (2009) have characterized these inhibitory zinc sites. Phosphorylation signaling, mitochondrial respiration, and neurotransmission are examples where the biological importance of zinc inhibition has been recognized (Gazaryan et al., 2007; Hogstrand et al., 2009; Paoletti et al., 2009). The inhibition of some proteinases by zinc is believed to be physiologically relevant because these enzymes are secreted from cells together with zinc. Zinc inhibits carboxypeptidase A, a zinc proteinase, with a K_i of 0.5 μM (Larsen and Auld, 1989).

Structural zinc sites have four protein ligands and no bound waters. Cys is the preferred ligand in such sites; His is second most common (reviewed by Auld, 2001). The total number of residues spanning structural sites varies from 15 to 209, but most are in the range 20 to 40 residues. The role of these sites is to stabilize and maintain the structure of the protein. These sites can have indirect effects on enzyme activity by altering the chemical environment of the active center and/or influencing the alignment of active site residues for catalysis. Typically, zinc organizes small domains (about 20 amino acids), but when two or three zinc ions are used, much larger domains can be organized.

The zinc finger domain contains a structural Zn^{2+} ion. Zinc fingers use special domains in which side chains of Cys and His coordinate Zn^{2+} ions in tetrahedral configuration. There are at least 14 different classes of zinc fingers (reviewed by Eis, 1997; Laity et al., 2001; Pabo et al., 2001; Matthews and Sunde, 2002; Sri Krishna et al., 2003). The consensus sequences of the three main classes are:

CCHH: (Y,F)–X–C–$X_{2,4}$-C-X_3-F-X_5-L-X_2-H-X_3-H-X_{2-6}
CCHC: C–X_2–C–X_4-H-X_4-C
CCCC: C–X_2–C–X_{13}-C-X_2-C-X_{15-17}-C-X_5-C-X_9-C-X_2-C-X_4-C

X is a nonconserved residue. Isolated, single zinc fingers (25 to 35 residues) behave as independent structural units.

The CCHH zinc finger is characterized by the fold ββα. Two Cys's are located near the turn of the antiparallel β-sheet; two His's are in the C-terminal part of the α-helix (Narayan et al., 1997) (Figure 8.9). These ligands, coordinating a Zn^{2+} ion, hold the β-sheet and α-helix together. This simple fold is also stabilized by a small hydrophobic core. The CCHH zinc is bound with high affinity, $K_d(\text{Zn}) = 10^{-11}$ to 10^{-9} M (Michael et al., 1992). Many proteins involved in the regulation of gene expression contain zinc fingers. They interact with specific DNA sequences in the promoter and enhancer regions of target genes and are found in all eukaryotes and many prokaryotes.

Co-catalytic sites contain two or three Zn^{2+} ions in close proximity; two of the ions are bridged by the side chain of a single residue. Asp and His are the

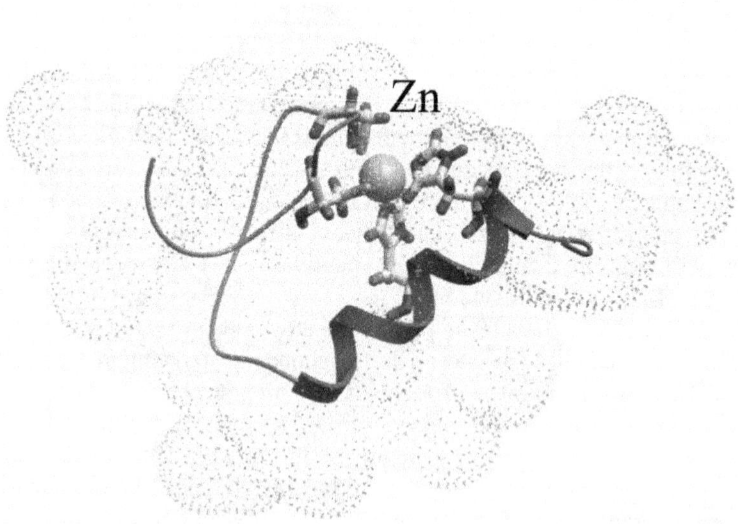

Figure 8.9. Structure of a CCHH zinc finger in transcription factor SP1F3 (PDB file 1SP1). (From Narayan et al., 1997.)

preferred residues for co-catalytic sites; Cys is not found in such bridging sites. In multimetal enzymes, several Zn^{2+} (or other metal) ions operate in concert to enhance catalysis (reviewed by McCall et al., 2000; Auld, 2001). The scaffolding of the zinc sites is important for the function and reactivity of the bound metal. The distance between the metal ions in these sites depends on the bridging amino acid. For Asp or Glu it is generally between 3 and 4 Å; for His the distance is about 6 Å, due to the width of the imidazole ring.

The bridging amino acids and water could have critical roles in catalysis. Their dissociation from either metal ion during catalysis could change the charge on the metal, promoting its action as a Lewis acid or allowing interaction with an electronegative atom of the substrate. Alternatively, the bridging ligand might participate transiently in the reaction as a nucleophile or general acid–base catalyst. The flexibility of the arm supplying the bridging ligands would be expected to influence the stability and reactivity of the two metal sites.

Only a few of the co-catalytic sites contain only zinc. Several contain Zn^{2+} ions in combination with $Cu^{+/2+}$, $Fe^{2+/3+}$, or Mg^{2+}. There is a Zn/Mg site in alkaline phosphatase and in lens aminopeptidase, a Fe(III)/Zn site in the purple acid phosphatase, and a Cu(II)/Zn site in the superoxide dismutase.

Superoxide (O_2^-) is generated in a number of cellular processes, including oxidative bursts from immune cells and as a by product of normal respiration. The Cu(II)/Zn superoxide dismutases (SOD) play a critical role in the physiological control of oxygen radicals by catalyzing the dismutation of the superoxide anion into molecular oxygen and hydrogen peroxide (reviewed by Cizewski Culotta et al., 2006). These proteins protect redox-sensitive cellular machinery from damage. There are crystal structures available for several eukaryotic and bacterial SODs (Strange et al., 2006) (Figure 8.10). The Cu^{2+} and Zn^{2+} ions bind to the protein in similar but chemically distinct environments. The Cu(II)/Zn site of SOD has a bridging His63, which must be deprotonated in order to bind both metals simultaneously. The zinc site is composed of Nδ1 nitrogens of His 63, His71, His80, and His83 and the Oδ1 oxygen of Asp81. The copper ligands are His46, His48, His63, and His120, the latter three coordinating by their Nε2 nitrogens and only the first, His46, by Nδ1. The copper-coordinating nitrogens form a distorted plane; water is a fifth ligand to complete a square pyramid. Zinc somehow facilitates the oxidation–reduction of copper during the catalytic cycle. Zinc may also be important for substrate specificity.

Zinc binding sites at protein interfaces are formed by residues from two proteins (reviewed by Auld, 2001). The resulting site usually has the coordination properties of a catalytic or structural zinc binding site. These interactions can lead to formation of homo-oligomers or link two different proteins. His, Glu, and Asp usually supply the ligands to these sites, but Cys containing interface sites are also found. The ligands are frequently contributed by β-strands. The resulting sites resemble catalytic sites as in γ-carbonic anhydrase, or structural sites as in the superantigen MHC class II complexes.

A feature common to all zinc sites is that the metal ion is surrounded by a shell of hydrophilic groups that is embedded within a larger shell of hydrophobic

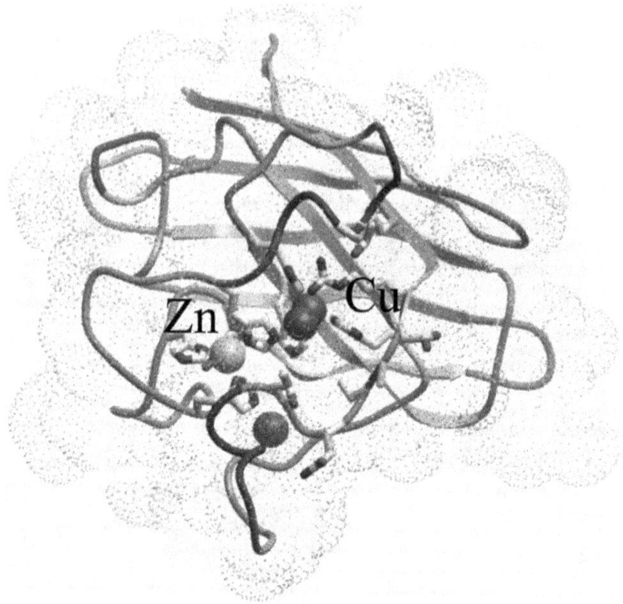

Figure 8.10. Structure of superoxide dismutase (SOD) with two Cu^{+} and two Zn^{2+} ions (PDB file 2C9V). (From Strange et al., 2006.) (*See insert for color representation of figure.*)

groups (reviewed by McCall et al., 2000). In addition, the side chains serving as zinc ligands often form hydrogen bonds with other residues, perhaps to preorder the zinc binding site and lower the entropic cost of binding the Zn^{2+} ion (Christianson, 1991). These interactions between the Zn^{2+} ion and ligands have been proposed to orient these ligands and enhance the electrostatic interaction between metal and ligand.

Despite the variation in their coordinations, these intracellular enzymes bind zinc very tightly (pK_d 10 to 12) (reviewed by Maret and Li, 2009). Zinc affinities of the cytosolic eukaryotic zinc proteins, carbonic anhydrase II, superoxide dismutase (SOD1), and sorbitol dehydrogenase are quite similar, even though these proteins have different numbers and types of ligands. Extracellular zinc proteins have a broader range of affinities.

For most cells, the total cellular zinc concentration is estimated to be 0.1 to 0.5 mM, similar to that of iron (Palmiter and Findley, 1995; Outten and O'Halloran, 2001; Krezel and Maret, 2006). Most of the zinc is bound to proteins with widely varying binding affinities. The concentration of free Zn^{2+} ion is only from 10^{-9} to 10^{-12} M (Peck and Ray, 1971; Simons, 1991). Elevated cellular zinc concentration is highly toxic; it can cause mitochondrial dysfunction, neuronal death, and brain diseases. Thirty to 40% of the cellular zinc is localized in the nucleus, 50% in the cytosol and cytosolic organelles, and the remainder is associated with membranes (Vallee and Falchuk, 1993).

Zinc transport or diffusion within the cytosol appears not to depend on specific metallo-chaperone proteins because too many would be required to supply all of the many zinc proteins. Yet the redistribution must occur under tight control to avoid nonspecific reactions and cytotoxic effects. This redistribution may involve protein–protein interactions without zinc ever being free (associative mechanism) and/or via free Zn^{2+} ions (dissociative mechanism) (reviewed by Maret and Li, 2009).

Zinc homeostasis in eukaryotic cells is controlled on the levels of uptake, intracellular sequestration in zinc storing vesicles ("zincosomes"), and nucleo-cytoplasmic distribution and elimination (reviewed by Beyersmann and Haase, 2001). The zinc binding protein metallothionein is involved in regulation of the cellular zinc level and the nuclear translocation of zinc during the cell cycle and differentiation. Metallothioneins are small (25 to 65 residues) Cys rich proteins with a low content of hydrophobic residues (reviewed by Henkel and Krebs, 2004). They were found throughout eukaryotes and bacteria. There is increasing evidence for the involvement of zinc in extracellular and intracellular signal transduction (reviewed by Permyakov and Kretsinger, 2009). Since zinc plays essential roles in many cell processes, organisms must maintain adequate intracellular zinc concentrations to support cell functioning even when extra-cellular or dietary zinc levels are low. To accomplish this regulation, cells have evolved efficient uptake systems to allow the accumulation of zinc (reviewed by Gaither and Eide, 2001). Once inside a eukaryotic cell, a portion of the zinc must be transported into intracellular organelles to serve as a cofactor for various zinc dependent enzymes and processes present in those compartments. Therefore, transporter proteins must be present in organelle membranes to facilitate this flux of zinc. It can also be stored in certain intracellular compartments when supplies are high and used later if zinc deficiency ensues.

The metal response element (MRE) binding transcription factor-1 (MTF-1) senses cellular Zn^{2+} ion concentrations in multicellular eukaryotes and activates the expression of proteins involved in zinc homeostatic mechanisms such as thionein. MTF-1 has six canonical zinc fingers at the N-terminus, which is the DNA binding region of the molecule (Radtke et al., 1993). In contrast, zinc finger domains are not used in sensing zinc ions in prokaryotic cells (reviewed by Maret and Li, 2009). Metal sensor proteins have been well studied in bacteria, especially members from three families: (1) ZntR from the MerR family, (2) Zur and FurB from the Fur family, and (3) SmtB and CzrA from the ArsR family. They all belong to the winged helix (helix–turn–helix) class of proteins with zinc binding sites at the interface of the subunits. As the extracellular Zn^{2+} ion concentrations rise, ZntR from *E. coli* binds zinc and turns on the production of a Zn^{2+} ion efflux pump, ZntA, which removes any excess zinc from the cell (reviewed by Maret and Li, 2009). The *znuABC* operon encodes a triplet of proteins, ZnuA (a periplasmic protein), ZnuB (a membrane protein), and ZnuC (an ATPase), and it is regulated by Zur (Patzer and Hantke, 2000). Sub-femtomolar ($<10^{-15}$ M) concentrations of free zinc ions trigger Zur repression of the *znuC* gene (Outten and O'Halloran, 2001; Outten et al., 2001). In the presence of excess Zn^{2+}

ions, the Zur dimer binds to the regulatory sequence located within the central *znu* operon and prevents the binding of RNA polymerase (Patzer and Hantke, 2000). When extracellular Zn^{2+} ion concentrations become critically low, Zur repression of the *znu* operon is lifted and the proteins involved in Zn^{2+} ion influx are expressed.

Escherichia coli ZntA, a member of the P1B-type ATPase transporter family, exports zinc across the inner membrane against a concentration gradient by utilizing the energy derived from ATP hydrolysis (reviewed by Maret and Li, 2009). The N-terminal fragment of ZntA (residues 46 to 118) from the cytoplasmic region contains a solvent-exposed zinc coordination site with sulfur donors of Cys59 and Cys62 and one or both carboxylate oxygens of Asp58 (Banci et al., 2002). Another zinc binding site, involving Cys392 and Cys394 and located in the transmembrane domain, is essential for transport because mutations of the ligands result in a protein that cannot catalyze metal ion–dependent ATP hydrolysis. Another *E. coli* zinc transporter is ZnuA, the periplasmic component of the ZnuABC complex. *E. coli* ZnuA (ZnuA-Ec) belongs to the ABC-type periplasmic ligand binding proteins. The crystal structure of ZnuA shows at least two zinc binding sites: the primary binding site (K_d < 20 nM) is tetrahedrally coordinated by His60, His143, His207, and one water molecule (B. R. Chandra et al., 2007). The second metal binding site involves His224 and several unidentified residues from the His rich loop.

Although zinc is essential for many cellular processes, excess zinc can be toxic (Koh et al., 1996). Therefore, while maintaining adequate levels of zinc for growth, cells must also control intracellular zinc levels when exposed to excessive zinc concentrations. Several mechanisms exist to detoxify excess zinc. One of them is the binding of zinc to cytoplasmic macromolecules (e.g., to metallothionein proteins); this may play such a detoxification role (Hamer, 1986). Zinc transporters can also take part in detoxification by facilitating intracellular sequestration within organelles or the efflux of zinc across the plasma membrane.

Alteration of zinc homeostasis in the brain may accompany neurological diseases such as Alzheimer's disease, Parkinson's disease, and amyotrophic lateral sclerosis, in which oxidative stress has been implicated as a cause (reviewed by Frederickson and Bush, 2001). Anomalous zinc content was found in amyloid plaques taken from Alzheimer's patients. Zinc has been shown to take part in the aggregation of peptides, forming senile plaques. Acute disturbance of zinc homeostasis following heart attacks, ischemia, and brain injury can induce neuronal disturbances.

8.1.5. Cadmium Complexes

Naturally occurring cadmium is composed of eight isotopes. Two are radioactive: ^{113}Cd (half-life is 8.7×10^{15} years) and ^{116}Cd (half-life is 2.9×10^{19} years). ^{106}Cd, ^{108}Cd, and ^{114}Cd are predicted to be radioactive, but their decay has not been observed, due to extremely long half-lives.

In humans, cadmium exposure, often from cigarette smoke, has been associated with cancers of the prostate, lungs, and testis (reviewed by Beyersmann and Hechtenberg, 1997; Waalkes, 2003; Bressler et al., 2004). Cadium has a long biological half-life of 17 to 30 years and accumulates throughout life. Cd^{3+} ions are taken up through calcium channels of the plasma membrane of various cell types and are accumulated intracellularly. The intestinal transporter for nonheme iron, divalent metal transporter 1 (DMT1), also mediates the transport of cadmium, manganese, and lead. The involvement of DMT1 in metal uptake may explain why an iron deficient diet is a risk factor for cadmium and lead poisoning (reviewed by Bressler et al., 2004). The uptake of manganese is markedly reduced in cadmium resistant cells (Himeno et al., 2002). The divalent metal transporter 1 (DMT1) is the only known mammalian transporter involved in the uptake of both cadmium and manganese.

Metallothionein coordinates zinc and copper in vivo. It also binds cadmium, cobalt, iron, mercury, bismuth, platinum, silver, and gold in vitro. Cys's are strictly conserved in metallothioneins and may be required for its metal binding properties. The presence of metallothionein within cells decreases cadmium toxicity (Klaassen et al., 1999).

Cadmium inhibits photosynthesis, respiration, and nitrogen metabolism as well as water and mineral uptake in plants (reviewed by Deckert, 2005). Heavy metal transport into plant cells is dependent on the presence of mycorrhizas, the binding properties of the cell wall, and root exudates (Hall, 2002). Cadmium is thought to enter plant cells through cation transporters in the plasma membrane with broad substrate specificity (Hall, 2002). These include ZIP proteins (IRT1 and ZNT1), natural resistance-associated macrophage protein (Nramp) and low-affinity cation transporter (LCT1) as well as by Ca^{2+} and K^{+} ion channels (reviewed by Deckert, 2005). In the cytosol, cadmium is bound by chelators such as phytochelatins and citric acid and then sequestered in the vacuole. In this process the participation of ATP binding cassette (ABC-type) transporters such as AtMRP, AtHMA4, and TcHMA4 is postulated as well as the involvement of cation diffusion facilitator (CDF). Clemens (2001) suggested that cadmium causes both the perturbation of intracellular calcium levels and interferes with calcium signaling, by binding to calmodulin. Cadmium has high affinity for Cys, Glu, Asp, and His side chains and often competes with zinc for a variety of important binding sites (e.g., in zinc fingers of transcription factors).

Carbonic anhydrase catalyzes the reversible hydration of carbon dioxide; it is a zinc enzyme in most organisms. In the oceans, where zinc is nearly depleted, diatoms (e.g., *Thalassiosira weissflogii*) use cadmium as a catalytic metal atom in carbonic anhydrase (Lane et al., 2005; Xu et al., 2008) (Figure 8.11). Despite low sequence similarity, cadmium carbonic anhydrase is a structural mimic of the functional β-carbonic anhydrase dimer, with striking similarity in the spatial organization of the active site residues. Cadmium carbonic anhydrase readily exchanges cadmium and zinc at its active site; this seems to be a unique adaptation to oceanic life that is explained by a stable opening of the metal coordinating site in the absence of metal. Hence, cadmium is tentatively regarded as essential, at

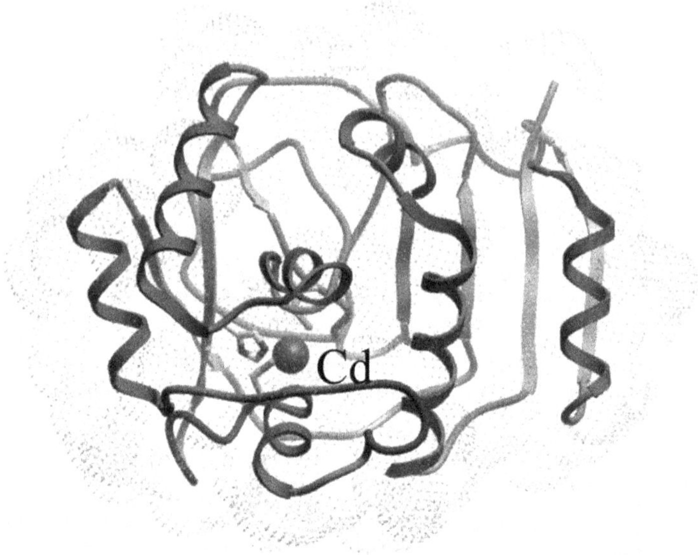

Figure 8.11. Structure of cadmium bound domain II of carbonic anhydrase from the marine diatom *Thalassiosira weissflogii* (PDB file 3BOB). (From Xu et al., 2008.)

least in one organism. As noted, this does not imply that it is essential to others; it may be toxic.

8.2. ESSENTIAL METALS WITH SEVERAL VALENCE STATES

Na^+, K^+, Mg^{2+}, Ca^{2+}, Zn^{2+}, and Cd^{2+} are hard cations and have only a single valence state under biological conditions. In contrast, V(III/IV), Cr(III/VI), Mo(IV/V/VI), W(IV/V/VI), Mn(II/III/IV), Fe(II/III), Co(II/III), Ni(II/III), and Cu(I/II) have several valence states and are often, but not always, involved in oxidation–reduction reactions. Their coordinations are often more complex than those of the six hard cations. They are usually bound by S, N, and O ligands in proteins. Sometimes it is the metal–organic ligand complex that is bound by the protein.

8.2.1. Vanadium Complexes

Both vanadium bromoperoxidases (V-BrPO), which are isolated mainly from marine algae, and vanadium chloroperoxidases (V-ClPO), which are isolated from certain terrestrial fungi, have been identified (reviewed by Butler et al., 1998). Many halogenated natural products have been isolated from marine organisms. These compounds range from volatile halogenated hydrocarbons (e.g., bromoform, chloroform), which are produced in very large quantities, to chiral halogenated terpenes, acetogenins, and indoles. The marine haloperoxidases are thought to be involved in the biosynthesis of these natural products. Protein-bound V(V) is required for activity in V-BrPO and V-C1PO. Phosphate can displace vanadium from V-BrPO, and the activity can be fully restored on addition of vanadate in the absence of phosphate.

The crystal structure (Figure 8.12) of the native form of *Curvularia inaequalis* V-C1PO shows that two four-helix bundles form its core V-ClPO (Messerschmidt et al., 1997). The other side of the broad channel is hydrophobic, containing Pro47, Pro211, Tyr350, Phe393, Pro395, Pro396, and Phe398. $Na_3(VO_4)$ and a small amount of azide (N_3^-) were added to the crystallization drop. The V(V) complex in the crystal structure is pentagonal bipyramid VO_3 (V–O, 1.65 Å), with N_3^- at one vertex and Nε of His496 (V–N, 1.96 Å) at the other. The 2− charge of the complex is partially compensated by a network of hydrogen bonds, with Lys353, Arg360, Arg390, and Ser402, as well as the amide nitrogen proton of Gly403, stabilizing the vanadate coordination to the protein.

In addition to the principal role of vanadium in phosphorylation and redox processes, this trace metal has been found in ascidians in remarkably high levels (Michibata et al., 2003; reviewed by Jelikić-Stankov et al., 2007). The suborder *Phlebobranchia* contains higher levels of vanadium than the *Stolidobranchia* contains. *Phlebobranchia* sequesters vanadium, as V(III), in specialized blood cells called vanadocytes (Smith, 1989). The extremely low pH of 1.9 found

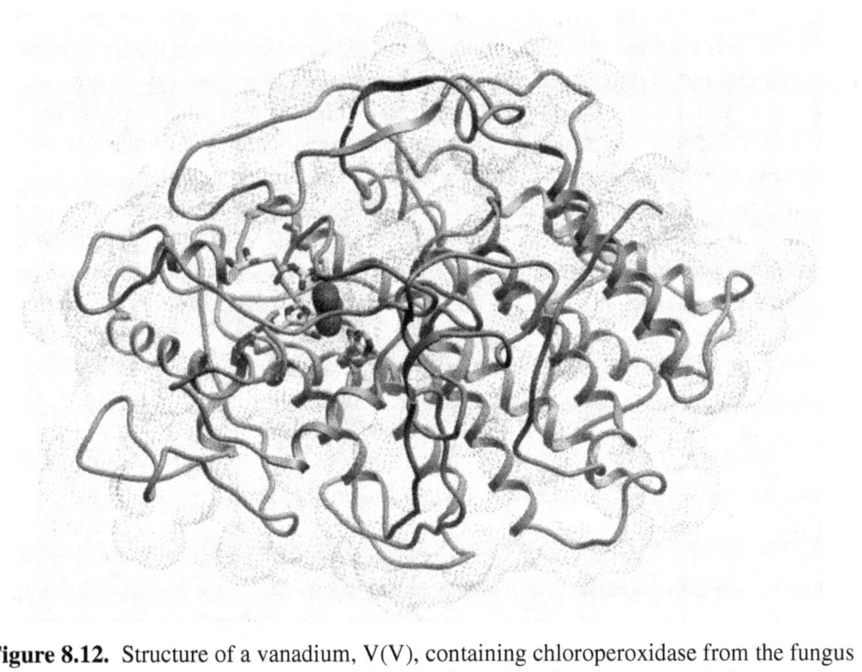

Figure 8.12. Structure of a vanadium, V(V), containing chloroperoxidase from the fungus *Curvularia inaequalis* (PDB file 1VNC). (From Messerschmidt et al., 1997.)

in vanadocytes suggests that protons, concentrated by an H^+-ATPase, might be linked energetically to the accumulation of vanadium. The concentration of vanadium (350 mM) in blood cells of *Ascidia gemmata* is 10^7 times higher than the concentration in seawater. Signet ring cells are the true vanadocytes in ascidians. Kanda et al. (1997) and Wuchiyama et al. (1997) identified vanadium binding Cys rich proteins (vanabins). Ueki et al. (2003a,b) found five different types of vanabin (vanabins 1 to 4 and vanabin P), very probably involved in vanadium accumulation processes as metallochaperones. Two independent but related vanabins, vanabin1 and vanabin2, bind 10 and 20 V^{4+} ions, respectively (Hamada et al., 2005), while an excess of Cu^{2+} ions was shown to inhibit in vitro binding of vanadium(IV) to the vanabins (Ueki et al., 2003a,b). The physiological roles of vanadium in ascidians remain unknown.

Vanadium compounds are important for their interaction at the level of enzymatic systems (reviewed by Jelikić-Stankov et al., 2007). They can act either as activators or inhibitors of some enzymes. Vanadium compounds also play an important role in the process of detoxification. Decavanadates are known especially for their remarkable bioactivity as antiviral agents, antioxidants, insulin mimetics, and so on. The possible general role of vanadium in all living organisms can be explained in terms of the structural analogy between vanadate and phosphate. Vanadate is a known inhibitor but also a stimulator of many phosphate-metabolizing enzymes. This includes the inhibition of a regulatory phosphatase, which is likely to lead to activation of protein kinase(s), whose activity is the key factor in the insulin mimetic action of vanadate. Further, vanadium also acts as an electron acceptor and hence as a trigger in the photocleavage of DNA molecules. Numerous vanadate species may interact with enzymatic systems. For example, the $H_2V_2O_7{}^{2-}$ anion influences enzymatic activity of hydrogenase, isomerase, and phosphatase, while $V_4O_{12}{}^{4-}$ inhibits both dehydrogenase and aldolase. $VO_3{}^{4-}$ at micromolar concentrations is an inhibitor of the glycolytic processes. $VO_3{}^{4-}$ is reduced to VO^{2+} within the cells.

Vanadium solutions contain different oligomeric vanadate species in equilibrium [e.g., monomeric (V_1), dimeric (V_2), tetrameric (V_4), and decameric (V_{10})] and sometimes with different conformations and states of protonation (reviewed by Jelikić-Stankov et al., 2007). After several hydrolytic and polymerization reactions, the vanadate ion, VO_2^+, in acidic solutions (pH $<$ 6.3) may form crystalline decavanadate $V_{10}O_{28}{}^{6-}$. The other polyoxovanadates, such as $V_2O_7{}^{2-}$ or $V_{14}O_{12}{}^{4-}$, can be obtained by varying the vanadium concentration and pH of the solution. An increase in pH leads to the formation of protonated decavanadate forms such as $HV_{10}O_{28}{}^{5-}$, $H_2V_{10}O_{28}{}^{4-}$, $H_3V_{10}O_{28}{}^{3-}$, and $H_4V_{10}O_{28}{}^{2-}$. The protons in the $V_{10}O_{28}{}^{6-}$ anion are always linked by O_2.

The uptake of vanadium by humans takes place primarily through foodstuffs (reviewed by Barceloux, 1999b). Foods rich in vanadium include mushrooms, shellfish, dill seed, parsley, and black pepper. In most of these cases, physiologically relevant forms of vanadium include vanadyl sulfate, sodium metavanadate, sodium orthovanadate, and vanadium pentoxide. Most of the vanadium ingested is transformed into the cationic vanadyl form in the stomach before being absorbed in the duodenum by an unknown mechanism (Hirano and Suzuki, 1996). After reaching the bloodstream, vanadate is converted into the vanadyl ion, although the "vanadate" form also exists. These vanadate ($H_2VO_4^-$) and vanadyl forms are rapidly transported by blood transferrin and by albumin and transferrin, respectively, to various tissues (Fantus et al., 1995). Vanadium has been reported to be incorporated in liver, kidney, brain, heart, muscles, and bone.

8.2.2. Group VIa

Chromium Complexes Chromium concentrations in the environment are low; chromium exists in various compounds and oxidation states. People can be exposed to chromium by breathing, drinking, and eating and through skin contact with chromium or chromium compounds (reviewed by Barceloux, 1999a). Inhalation is the most important route for occupational exposure, especially for people who work in the steel and textile industries. Inhalation of chromium compounds can result in ulceration and perforation of the mucous membranes of the nasal septum, irritation of the pharynx and larynx, and asthmatic bronchitis (reviewed by Mutti and Corradi, 2006). Adverse effects on the skin include ulcerations, dermatitis, and allergic skin reactions.

Cr(III) is poorly absorbed and hence not very toxic (reviewed by Barceloux, 1999a; Shrivastava et al., 2002). Cr(VI) is highly soluble and more readily absorbed (reviewed by Dayan and Paine, 2001; Shrivastava et al., 2002; Mutti and Corradi, 2006). Cr(III) is unable to enter cells, but Cr(VI) enters through membrane anionic transporters. Cr(VI) does not react with DNA, RNA, proteins, and lipids. However, intracellular Cr(VI) is metabolically reduced to Cr(III); it and the reductional intermediate, Cr(V), can form covalent bonds with macromolecules. Relatively stable oxoCr(V)-sialoglycoproteins and oxoCr(V)-carbohydrate complexes are generated when Cr(VI) reacts with the mucus and cells that line the respiratory tract (reviewed by Codd et al., 2003).

Glutathione and cysteine are the most important cofactors for the intracellular reduction of Cr(VI), but ascorbic acid, microsomes in the presence of NAD/NADH, microsomal cytochrome P450, mitochondria, and proteins such as hemoglobin and glutathione reductase may also be active in the reduction process.

Chromium is essential for proper lipid and carbohydrate metabolism in mammals; chromium deficiency leads to adult-onset diabetes and cardiovascular disease. Yamamoto et al. (1997) isolated and characterized a unique chromium binding oligopeptide, low-molecular-weight chromium binding substance (LMWCr), or chromodulin. Its molecular weight is ~1500 Da and it is comprised of only four types of amino acid: Gly, Cys, Glu, and Asp. Despite its low molecular weight, it binds a cluster of four Cr^{3+} ions. Chromodulin is widely distributed in mammals; it has been isolated and purified from livers of many mammals as well as from porcine kidney and bovine colostrum (reviewed by Vincent 2000a,b). The most important property of chromodulin is its ability to potentiate the effects of insulin on the conversion of glucose into carbon dioxide or lipid by isolated rat adipocytes (Davis and Vincent, 1997). This stimulation of insulin activation occurs without changing the concentration of insulin required for half maximal activity, suggesting that chromodulin plays an intrinsic role in the adipocytes. The stimulation is proportional to the chromium content of chromodulin (Yamamoto et al., 1989). No other naturally occurring chromium containing compound potentiates insulin action in this manner.

Chromodulin also activates a membrane phosphotyrosine phosphatase (Davis et al., 1996) and an insulin receptor tyrosine kinase (Davis et al., 1997). Vincent (2000a,b) suggested that in response to increases in blood sugar levels, insulin is released rapidly into the bloodstream, where it binds to an external α subunit of the transmembrane, insulin receptor, inducing a change in conformation of the receptor. The receptor autophosphorylates Tyr's on the internal part of its β-subunit, turning the receptor into an active kinase.

Chromodulin is stored in its apo form in both the cytosol and the nucleus of insulin-sensitive cells. Increases in plasma insulin result in the movement of Cr(III) from the blood to insulin-dependent cells (Morris et al., 1993). This process is mediated by a metal transport protein, probably transferrin. Apo-chromodulin binds the Cr^{3+} ion ($K_d \sim 10^{21}$ M). Cr_4-chromodulin can then bind to the insulin receptor, helping to maintain its active conformation and amplifying insulin signaling. Cr(III) is an essential nutrient for humans, and shortages may cause heart conditions, disruption of metabolism, and diabetes.

When blood concentrations of insulin diminish and receptor signaling is terminated, chromodulin may be eliminated from cells. This loss of chromodulin from cells is consistent with increased chromium(III) and chromomodulin in the urine after carbohydrate and sugar intake. The manner in which apo-chromodulin is replaced is unknown; presumably, it is synthesized as a pro protein, which is modified posttranslationally. How chromodulin levels are regulated remains unknown.

The addition of chromodulin to rat adipocytic membranes or isolated rat insulin receptor in the presence of 100 nM insulin results in an eightfold stimulation of protein tyrosine kinase activity. The dissociation constant for the binding of chromodulin to the insulin receptor is 250 to 875 pM. The site of activation appears to be located at or near the kinase active site. The addition of chromodulin to a fragment of the *b* subunit of the insulin receptor that contains the active site (but does not require insulin to be active) results in a similar stimulation of kinase activity (Davis et al., 1997). Chromomodulin also activates a membrane phosphotyrosine phosphatase (Davis et al., 1996).

Administration of chromium to mammals orally or by injection results in the appearance of Cr^{3+} ions in the iron transport protein, transferrin (reviewed by Vincent, 2000a,b). The addition of chromium to blood or blood plasma in vitro also results in the loading of transferrin with chromium, although albumin and some of its degradation products also bind chromium. In diabetics, in which blood chromium levels are reduced and urinary chromium losses are increased, this transport system may be overactive.

Chromium is important in altering the immune response, as shown by its effects on T- and B-lymphocytes, macrophages, and cytokine production, which may induce hypersensitivity reactions (reviewed by Shrivastava et al., 2002).

Cr(VI) reduction by bacteria has been proposed for bioremediation (reviewed by Cervantes et al., 2001). Chromium exists mainly in the oxyanion form, CrO_4^{2-}, and thus cannot be trapped by the anionic components of bacterial envelopes. However, cationic Cr(III) derivatives bind tightly to *Salmonella* lipopolysaccharides, *Bacillus subtilis* and *E. coli* cell walls, as well as to capsular polymers of *B. licheniformis*.

Cr(VI) may enter yeast cells via a nonspecific anion carrier, the permease system, which transports different anions such as sulfate and phosphate. Some chromate resistant mutants of *Neurospora crassa* have strongly reduced sulfate transport properties. Cr(VI) toxicity is due to its specific antagonism to sulfate uptake; whereas, Cr(III) toxicity results from antagonism with iron transport (Ramana and Sastry, 1994). Chromate resistance in fungi is related to modifications in ion uptake systems (reviewed by Cervantes et al., 2001).

Shewanella oneidensis MR-1 exposed to Cr(VI) form aseptate, nonmotile filaments that tend to aggregate. Transcriptome profiling and mass spectrometry based proteomic characterization revealed that the principal molecular response is the induction of prophage-related genes and their encoded products as well as a number of functionally undefined hypothetical genes that are located within the integrated phage regions of the MR-1 genome (Chourey et al., 2006).

Molybdenum Complexes Molybdenum and tungsten have equal atomic (1.40 Å) and ionic (0.68 Å) radii and similar electronegativities (1.3 and 1.4 for Mo and W, respectively). Both metals can be in various oxidation states (from II to VI), but only IV, V, and VI are relevant to biology.

Most organisms require molybdenum in trace amounts; however, an excess can be lethal. It is very abundant in the oceans as the oxyanion MoO_4^{2-}, and it is the only form available in soils. More than 50 enzymes are known to contain molybdenum; most of them occur in bacteria. Only six have been found in eukaryotes (reviewed by Mendel and Bittner, 2006). Uptake requires specific systems to scavenge molybdate in the presence of competing anions. High-affinity molybdate transporters in bacteria consist of three subunits and require ATP. Some bacteria have specific proteins that bind up to eight equivalents of molybdate. Eukaryotic molybdate transport and storage are still poorly understood.

Both molybdenum and tungsten can act as transducers between obligatory two- and one-electron oxidation–reduction systems. They can catalyze reactions such as the hydroxylation of carbon centers under more moderate conditions than are required by other systems. The physiological roles of these enzymes are fundamental and include the catalysis of key steps in carbon, nitrogen, and sulfur metabolism (reviewed by Hille, 2002; Mendel and Bittner, 2006).

Molybdenum and tungsten containing enzymes are characterized by rather complex architecture and contain multiple redox active sites. In order to be biologically active, molybdenum is usually complexed by a pterin compound, thereby forming the molybdenum cofactor (Moco) (Figure 8.13). Moco is synthesized by a conserved biosynthetic pathway that can be divided into four steps

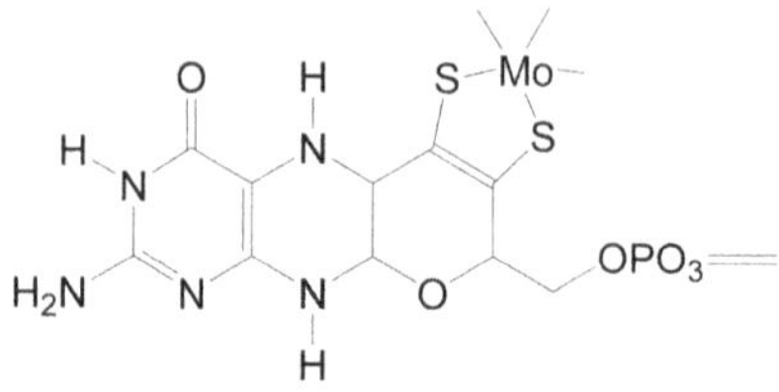

Figure 8.13. Structure of pterin–molybdenum cofactor (Moco).

according to the biosynthetic intermediates: precursor Z, molybdopterin, adenlyated molybdopterin, and Moco. In eukaryotes, six gene products are involved in Moco biosynthesis. The insertion of Moco into Mo-enzymes is not understood. Molybdenum is a constituent of the Fe–Mo cofactor in bacterial nitrogenase; this is the only known instance of its not being complexed with Moco.

For higher organisms a shortage of molybdenum in the diet or a compromised ability to use it leads to the loss of essential metabolic functions. All enzymes needing molybdenum lose their activity at the same time (reviewed by Mendel and Bittner, 2006).

In general, reactions catalyzed by Mo-enzymes are characterized by the transfer of an oxygen atom, ultimately derived from or incorporated into water, to or from a substrate in a two-electron redox reaction. Molybdenum-containing enzymes are divided into three families, each with a distinct active site structure and type of reaction catalyzed (reviewed by Hille, 2002). The first family, which contains xanthine oxidase from milk, has an L Mo(VI) OS(OH) core in the oxidized state, with one equivalent of the pterin cofactor (called L) coordinated to the metal. These enzymes typically catalyze the hydroxylation of carbon centers. The second family includes sulfite oxidase, typically isolated from avian or mammalian liver, and nitrate reductases, from plants that assimilate nitrate from the soil. Again, the oxidized metal center has a single equivalent of the pterin cofactor, but as part of an L Mo(VI) O_2(S-Cys) core, with the Cys provided by the protein. Members of this second family catalyze the transfer of an oxygen atom either to or from a lone pair of electrons on the substrate, respectively. The third family is diverse in both structure and function, but all members have two equivalents of the pterin cofactor (L) bound to the metal. The molybdenum coordination sphere is usually completed by a single oxygen atom and a sixth ligand in an L_2 Mo(VI) O(X) core. The sixth ligand, X, can be Ser (as in dimethyl sulfoxide reductase), Cys (e.g., in nitrate reductase), selenocysteine (e.g., formate dehydrogenase H), or hydroxide and/or a water molecule (e.g., arsenite oxidase). The reactions catalyzed by members of the latter family frequently involve the transfer of a single atom of oxygen, but dehydrogenation reactions also occur.

Tungsten Complexes Tungsten–sulfur bonds, such as those found in tungsten-containing enzymes, are more stable than their molybdenum counterparts (reviewed by Hille, 2002). Low-valent tungsten sulfides are also more soluble in aqueous solutions and thus would be more available in the anaerobic, highly reducing environment of the early Earth. Finally, the reduction potentials of tungsten containing complexes are lower than those that contain molybdenum and therefore are possibly more useful to early life forms.

In addition to the metal, the pterin cofactor is an integral part of the active site in both tungsten and molybdenum enzymes; this cofactor coordinates the metal via a dithiolate group. Tungsten containing enzymes, all of which come from bacterial or archaeal sources, fall into three groups (reviewed by Hille, 2002). Crystal structures are available only for members of the first group, which includes the aldehyde–ferredoxin and the formaldehyde–ferredoxin oxidoreductases from *Pyrococcus furiosus*. Members of this family catalyze the oxidation of aldehydes to carboxylic acids, with the reducing equivalents being transferred to a [4Fe–4S] ferredoxin. Their active sites have tungsten coordinated to two equivalents of the pterin cofactor, but no ligand contributed by the protein. The oxidized enzyme seems to have one W(VI)=O and one W(VI)=OH bond, analogous to the molybdenum center of arsenite oxidase; the reduced form probably has a single W(VI)–OH.

The tungsten-containing aldehyde–ferredoxin and formaldehyde–ferredoxin oxidoreductases both possess tungsten centers and [4Fe–4S] iron–sulfur clusters (Figure 8.14) (Hu et al., 1999; reviewed by Hille, 2002). These enzymes are composed of a single type of subunit; they form dimers and tetramers. Members of the formaldehyde–ferredoxin oxidoreductases are more complex, in that they have two to four types of subunits. The homodimeric aldehyde–ferredoxin oxidoreductase of *P. furiosus*, whose maximum growth temperature is 105°C, oxidizes a wide range of aliphatic and aromatic nonphosphorylated aldehydes (reviewed by Kletzin et al., 1995). The [4Fe–4S] cluster in each aldehyde ferredoxin oxidoreductase subunit, of $M_r \sim 67$ kDa, is liganded by Cys's at positions 288, 290, 295, and 494. It contains molybdopterin and one W, four or five Fe, one Mg, and two P atoms per subunit. Electron spin resonance analyses of the reduced enzyme indicated the presence of a single $[4Fe–4S]^+$ cluster with a spin = 3/2 ground state. The Cys494 forms a hydrogen bond to a ring nitrogen of one pterin, thus linking the two metal centers at their shortest distance. The two pterins are coordinated in part by two homologous DXXGL groups (positions 338 to 342 and 489 to 494), Arg76, Arg182 (which forms a polar hydrogen bond to the phosphate moiety of one of the pterins), Thr344, and Leu498. The carbonyl groups of Asn93 and Ala183 next to Arg182 coordinate a Mg^{2+} ion, which bridges the pterin phosphates, and Lys450. The side chains of Glu313 and His488 are in the vicinity of the substrate binding site and might participate in electron transfer reactions. Glu332 and His382 are part of a separate EXXH motif and coordinate the monomeric iron site.

In these enzymes, tungsten is bound similarly to molybdenum in Mo-enzymes, with hexa-substituted pterin (molybdopterin) (reviewed by L'vov et al., 2002). In contrast to Moco in Mo-enzymes, the pterin in W-cofactor is not bound with oligonucleotides (GMP, CMP, AMP, or IMP). The main feature of tungsten coordination in the active site of W-enzymes that probably provides their thermal

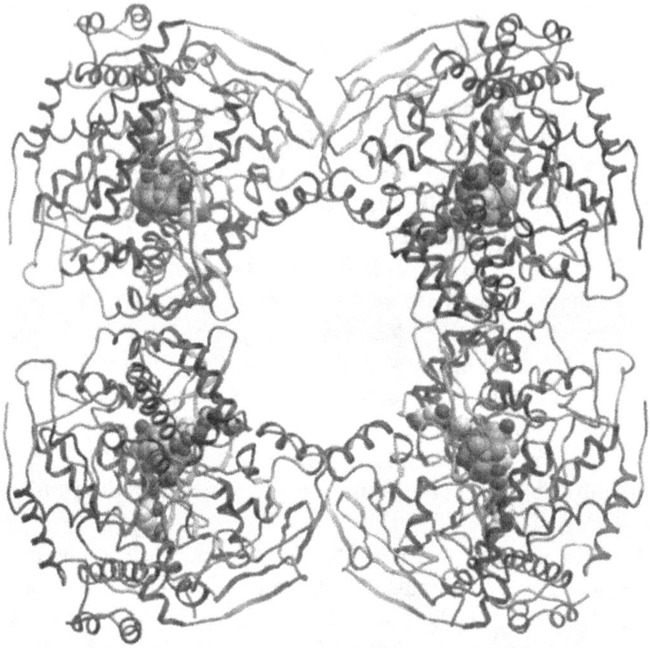

Figure 8.14. Structure of the tetrameric formaldehyde–ferredoxin oxidoreductase from *Pyrococcus furiosus* with iron–sulfur clusters and tungsten–pterin (PDB file 1B25). (From Hu et al., 1999.) (*See insert for color representation of figure.*)

stability is that every subunit contains two molecules of molybdopterin, which are coordinated by four sulfur ligands.

The second family of tungsten-containing enzymes consists of the formate and *N*-formylmethanofuran dehydrogenases, both of which function physiologically to reductively fix CO_2 into acetate and into *N*-formylmethanofuran, respectively (reviewed by Hille, 2002). They have amino acid sequence similarities to the molybdenum-containing formate dehydrogenase and dissimilatory nitrate reductase mentioned above, and probably have structures similar to these enzymes, with Cys and selenoCys, respectively, coordinating the metal in the oxidized enzyme in an L_2 W(VI) OX coordination sphere.

The third family of tungsten-containing enzymes consists of a single enzyme, acetylene hydratase from *Pelobacter acetylenicus*, which catalyzes the hydration of acetylene to acetaldehyde. Although this reaction is different from those catalyzed by members of the first group of tungsten-containing enzymes, the two families appear to be loosely related on the basis of sequence similarities and could have similar active site structures.

Molybdenum and tungsten are both transported into *E. coli* via the ABC transporter ModABC as $MoO_4{}^{2-}$ and $WO_4{}^{2-}$ oxyanions rather than as free ions (reviewed by Ma et al., 2009). ModA is the periplasmic SBP component of this transporter, and its structure has been determined in the presence of both molybdate and tungstate (Hu et al., 1997). ModA binds both molybdate and tungstate through seven hydrogen bonds between the protein and the anion and lacks direct metal–ligand bonds. These hydrogen bonds are derived from four main chain NH groups and three side chain OH groups (Ser12, Ser39, and Tyr120) that stabilize the oxyanion in an otherwise positively charged pocket.

8.2.3. Manganese Complexes

The ionic radius of the Mn^{2+} ion (0.80 Å) in aqueous solutions lies between that of Mg^{2+} (0.65 Å) and Ca^{2+} (0.99 Å) and is close to that of Fe^{2+} (0.76 Å) and several other transition metal ions. Manganese is quite electronegative and prefers "hard" ligands such as the oxygen atoms of Asp and Glu, as well as oxygens of water molecules (reviewed by Christianson and Cox, 1999). Sometimes, nitrogens of His coordinate manganese in metalloenzymes; whereas, sulfurs of Cys have not been found in the manganese enzymes. The coordination geometry of manganese is usually square pyramid or trigonal bipyramid with coordination number 5 or octahedral with coordination 6. Manganese is often interchangable with magnesium in proteins.

The participation of Mn^{2+} ions in the photosynthetic processes has been studied in most detail. The oxidation of H_2O to O_2 by the manganese-containing, oxygen-evolving complex in the chloroplasts, and the reduction of O_2 to H_2O by cytochrome *c* oxidase in mitochondria, form the important cycle of dioxygen metabolism that is essential to both plants and animals ($2H_2O \leftrightarrow O_2 + 4e^- + 4H^+$). Oxygen is relatively abundant in the atmosphere, primarily because of its constant regeneration by photosynthetic water oxidation by the oxygen-evolving complex. It consists of the Mn_4 cluster, Y_Z, and Ca^{2+} ion and Cl^- ion cofactors (reviewed by Robblee et al., 2001). The oxygen-evolving complex cycles through a series of intermediate states, S_i ($i = 0$ to 4), where i represents the number of oxidizing equivalents stored on the oxygen-evolving complex. This process is driven by the energy of four successive photons absorbed by the pigment P680 of the photosystem II (PS II) reaction center. The manganese complex in the oxygen-evolving complex couples the four-electron oxidation of water with the one-electron photochemical process occurring at the PS II reaction center. It acts as the locus of charge accumulation. In addition to Mn^{2+}, Ca^{2+} and Cl^- ions are essential cofactors that are required for activity. The precise location of the manganese complex is still unknown but seems to be located on the lumenal side of the membrane (reviewed by Nugent et al., 2001).

The structure of the manganese cluster has been studied extensively by x-ray spectroscopy (reviewed by Nugent et al., 2001; Robblee et al., 2001; Vrettos et al., 2001). These studies suggest a structure in the S_2 state similar to model compounds containing a Mn(III)–Mn(IV) dimer. The Mn–Mn distance in the cluster is 2.7 Å, with coordination by O and/or N ligands. The manganese cluster may consist of two nonequivalent μ-oxo-bridged Mn pairs. Some of the structures suggested for the manganese tetramer occupy a small volume; the four $Mn^{3+/4+}$

ions could require up to 24 ligands. If μ-oxo bridges supply eight ligands, then water, hydroxyl, or chloride side chains (charged and uncharged) and peptide carbonyl ligands could supply the rest. Other ligands, such as bicarbonate, may also be present. Extended x-ray absorption fine structure (EXAFS) data indicate that a Ca^{2+} ion binds close to the Mn_4 cluster. It can be replaced by cations of similar size, but only Sr^{2+} partially restores activity. Ca^{2+} and Cl^- ions are required for water oxidation and oxidation beyond the S_2-state.

Manganese is essential for several enzymes, including Mn-superoxide dismutase, Mn-catalase, and arginase (reviewed by Yocum and Pecoraro, 1999; Jakubovics and Jenkinson, 2001). Glycolysis cannot proceed fully without 3-phosphoglycerate mutase, which in several gram positive endospore forming bacteria is active only when associated with manganese.

Liver arginase is a 105-kDa homotrimer; each 35-kDa subunit contains a binuclear Mn(II) center that is critical for catalytic activity (reviewed by Ash, 2004; Cama et al., 2004). The overall fold of each subunit consists of a parallel eight stranded β-sheet flanked on both sides by numerous α-helices (Figure 8.15). Each arginase subunit contains a binuclear manganese cluster that is required for full catalytic activity (Figure 8.16). One of the Mn^{2+} ions can be dissociated reversibly, generating an enzyme retaining approximately half of its catalytic activity. The Mn^{2+}–Mn^{2+} cluster (separation = 3.3 Å) is situated on one edge of the central β-sheet. The Mn^{2+} ion that is deeper in the base of the active site

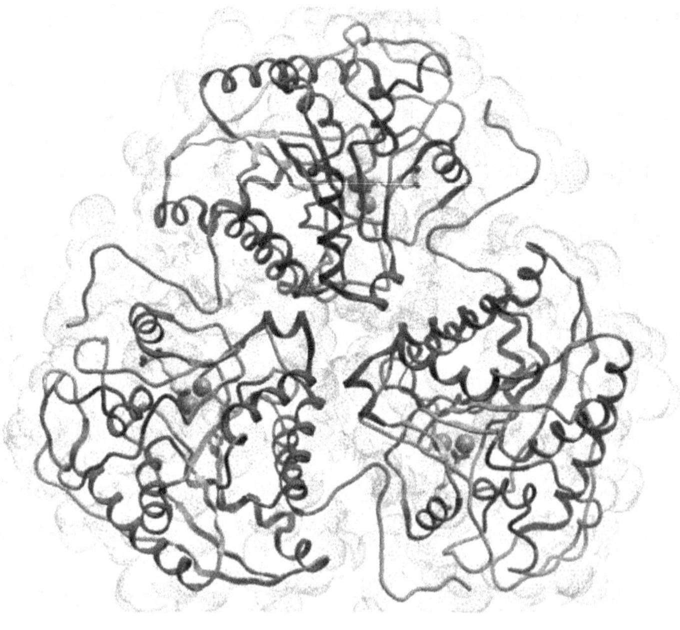

Figure 8.15. Structure of the arginase trimer, each complexed with a Mn^{2+}–Mn^{2+} ion cluster (PDB file 1T4P). (From Cama et al., 2004.)

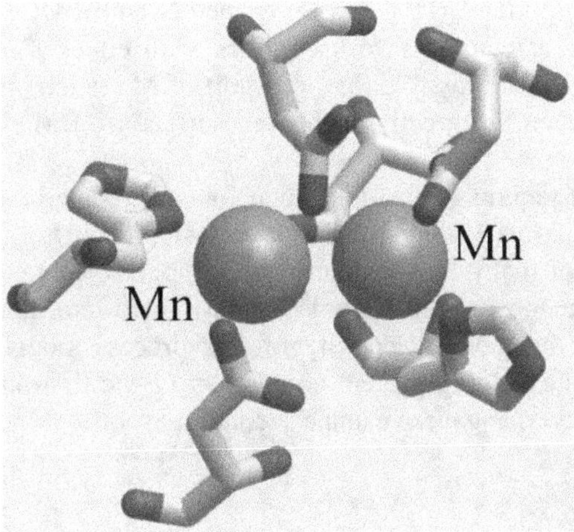

Figure 8.16. Structure of the Mn^{2+}–Mn^{2+} ion cluster in arginase (PDB file 1T4P). (From Cama et al., 2004.)

cleft is called Mn_A^{2+} and is coordinated by His101, Asp124, Asp128, Asp232, and a bridging OH^- ion with square pyramidal geometry (this OH^- ion also donates a hydrogen bond to Asp128). Mn_B^{2+} is coordinated by His126, Asp124, Asp232, Asp234, and a bridging OH^- ion in a distorted octahedron. In addition to the bridging OH^- ion, the Mn^{2+} ions are bridged by the carboxylates of Asp124 and Asp232. The Mn^{2+} ion activates arginase, although the divalent cation requirement for some arginases has been reported to be satisfied by Co^{2+} and Ni^{2+} and in some instances by Fe^{2+}, VO^{2+}, and Cd^{2+} (Ash, 2004). The mechanism of arginase proposed by Kanyo et al. (1996) includes the following steps: (1) a precatalytic binding side chain for arginine in which the side chain of Glu277 plays a role in substrate recognition; (2) attack of a nucleophilic metal-bridging OH^- ion; (3) formation of a neutral tetrahedral intermediate that is stabilized by the binuclear Mn(II) center; and (4) a possible role for His141 as a proton shuttle in mediating proton transfer between the active site and the bulk solvent.

Manganese is transported into cells via a mechanism similar to that for iron, as evidenced by the fact that both manganese and iron, as well as several other divalent transition metals, compete for uptake into a number of different cells (reviewed by Roth, 2006). There are two distinct but related mechanisms responsible for the transport of manganese and iron: a transferrin-dependent and a transferrin-independent pathway. In the former the Mn^{3+}–transferrin complex binds initially to the transferrin receptor on the cell surface in a manner similar to that for iron. After attachment of transferrin to the transferring receptor, the resulting endosomal vesicles are internalized and acidified by a H^+-ion ATPase pump, causing release of the metal from the transferrin–receptor complex. The Mn^{3+} ion released into the vesicle is presumably reduced to Mn^{2+}. Mn^{2+} is subsequently transported across the endosomal membrane via the transport protein, divalent metal transporter 1 (DMT1; also termed Nramp2, DCT1, or SLC11A2). Since Mn^{2+} forms a relatively weak complex with either a2-macroglobulin or serum albumin, it is also possible that transport of Mn^{+2} ions released from this complex can be taken up directly at the cell surface by DMT1, independent of transferrin.

Manganese is a required transition metal for most bacteria, and the evidence is compelling that Mn(II) homeostasis plays a significant role in the virulence and pathogenesis of many human microbial pathogens (reviewed by Ma et al., 2009). Like iron, manganese in bacteria is transported into the cytosol by ABC transporters, with the metal selectivity of this process thought to be dictated largely by the solute binding protein component of the transporter. In addition, many bacteria import manganese using a second transporter, MntH (Papp-Wallace and Maguire, 2006).

8.2.4. Iron Complexes

Iron is an abundant metal, being the fourth most plentiful element in the Earth's crust. It is essential for the growth of most organisms. The two stable oxidation states of iron, (II) (ferro) and (III) (ferri), easily interconvert; this flexibility makes iron a valuable redox mediator in biology. It can adopt high- or low-spin states in both ferrous and ferric forms, depending on its ligand environment. Several spin states of iron ions are catalysts in reactions with molecular oxygen. For these reasons, iron is considered to have been the ideal choice for incorporation into proteins during the evolution of early life and was probably the key constituent in the first prosthetic moieties.

Iron takes part in many important processes, such as respiration, DNA synthesis, nitrogen fixation, and photosynthesis. Its biological functionality is almost entirely dependent on its incorporation into proteins, as either a mono- or a binuclear species, or in a more complex form as part of iron–sulfur clusters or heme groups. Within proteins, iron can adopt the necessary redox potential (from 3300 to 700 mV), geometry, and spin state required for it to fulfill its designated biological function.

With the appearance of an oxygen-rich atmosphere on Earth, oxidation of the initial ferrous ion pool to the ferric form created two problems. Fe(II) activates dioxygen, with the general production of intermediate reactive species that might cause oxidative damage, thus preventing its use as a source of iron. In contrast, Fe(III) has a low solubility under physiological conditions; this obliges organisms to adopt efficient iron storage, transport, and usage mechanisms. The maximal solubility of iron in an oxidative environment such as extracellular fluids is limited by the solubility of $Fe(OH)_3$; at pH 8.0, $K_{sol} \sim 10^{-17}$ M, whereas Fe(II) is much more soluble, 10^{-1} M. To achieve effective iron homeostasis, organisms must balance their need to scavenge iron from their surroundings with the careful management of cellular levels to guard against iron induced toxicity (reviewed by Andrews et al., 2003).

Multiple processes are used to maintain iron homeostasis in diverse organisms. These include its absorption, storage, use, and excretion. In mammals, absorption of dietary iron is the only means known to regulate body content. Free iron in the intestine is reduced from Fe(III) to Fe(II) on the inner surface of enterocytes and is transported throughout the body by means of the divalent metal transporter DMT1 (reviewed by Arredondo and Nuñez, 2005). DMT1 transports iron into cells by an electrogenic mechanism that involves the cotransport of the Fe^{2+} ion and one proton. In enterocytes the iron can be stored in a complex with ferritin. Once internalized, iron traverses the intestinal mucosal cell, a process referred to as *transcytosis*. Heme absorbed from the diet is degraded by heme oxygenase, and the iron released enters the intracellular nonheme iron pool. Export of iron into the circulation requires the iron exporter ferroportin (FPN, also called IREG1 and MTP1) and a ferroxidase, to oxidize iron so that it can be bound by transferrin. Ferroportin is essential for iron export from enterocytes, macrophages, and hepatocytes. Ferroportin-1 is negatively regulated by the iron regulatory hormone hepcidin. Hepcidin controls the iron level by interacting directly with FPN1. This mechanism blocks the release of iron from macrophages, hepatocytes, and enterocytes. In the blood, iron is bound to transferrin and the complex passes through the portal circulation of liver. The liver is the major organ for iron storage. The major center of iron usage is the marrow, where it is used in heme synthesis.

Transferrin, which is synthesized in the liver, is a serum protein responsible for iron transport. The primary role of transferrin is to transport iron safely around the body to supply growing cells (reviewed by Gomme et al., 2005). Transferrins are a group of iron binding proteins that control the levels of iron in the body fluids of vertebrates due to their ability to bind two Fe^{3+} and two $CO_3{}^{2-}$ ions. Transferrin has several homologs or splicing variants, with more than 30 different forms detected to date. There are, however, three major isotypes: B, C, and D. Although transferrin can bind various metal cations, it has the highest affinity for ferric iron. Transferrin has two iron binding sites with dissociation constants on the order of 10^{-22} M for the Fe^{3+} ion (Aisen and Listowsky, 1980). Transferrin does not bind ferro-iron. Transferrin–Fe^{3+} interacts with a cell surface receptor. The entrance of the iron–transferrin–receptor complex into the cell is initiated by phosphorylation of the receptor by protein kinase C. After entrance, the iron is released.

The binding and release of iron by transferrin is controlled by several factors: for example, temperature and pH as well as chelator and ionic concentrations. Although a carbonate ion is essential for stabilizing the iron binding site, other anions, such as Cl^-, are also involved in maintaining the functionality of the binding site.

Transferrin has 679 residues and a molecular mass of ~79 kD (Hall et al., 2002b). It is stabilized by 19 intrachain disulfide bonds and has three carbohydrate side chains, of which two are N-linked (Asn413 and Asn611) and the third is O-linked (Ser32) (reviewed by Hirose, 2000; Gomme et al., 2005). Transferrin (Figure 8.17) consists of two homologous halves, the N-domain (336 residues) and the C-domain (343 residues), which are linked by a short spacer. There are

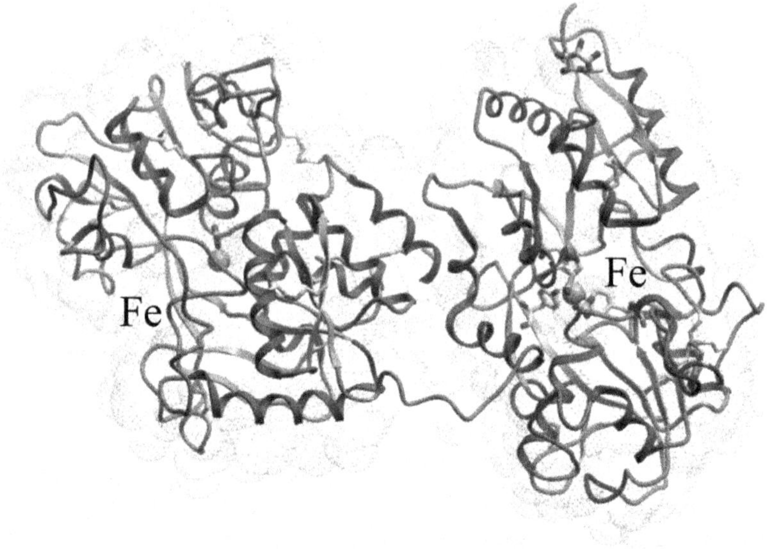

Figure 8.17. Structure of transferrin; one Fe^{2+} ion is bound in each of the two homologous domains (PDB file 1NO4). (From Guha Thakurta et al., 2003.)

subtle differences in the two domains in terms of structure, stability, and the ease of iron release, and there is some evidence that the iron status of one lobe can influence binding or release from the other.

Two features of its structure are of special importance (Hirose, 2000; Gomme et al., 2005). First, the iron binding cleft in transferrin is hydrophilic, with many polar side chains and 10 to 20 water molecules. Second, two antiparallel β-strands that connect the two domains at the back of the binding cleft contain a hinge that enables one domain to move relative to the other, opening or closing the cleft. Each domain contains two regions, each comprising a series of α-helices, which overlay a central β-sheet backbone. The iron binding site in both the N- and C-terminal domains has four conserved residues, including one Asp, two deprotonated Tyr's, and one His (in the N-terminal domain: Asp63, Tyr95, Tyr188, and His249). These residues are arranged in a distorted octahedron (Figure 8.18). The 3− charge of the ligands matches the 3+ charge of the metal ion. Moreover, the binding site requires two further oxygen molecules donated by a CO_3^{2-} ion to stabilize the iron atom. The surrounding amino acid residues (Gly65, Glu83, Tyr85, Arg124, Lys206, Ser248, and Lys296 in the N-terminal domain) are thought to further stabilize the iron binding site, and they have crucial roles in iron release. Upon uptake and release of iron, transferrins undergo a large conformational change: Domains 1 and 2 rotate as rigid bodies around a rotation axis that passes through the two antiparallel β-strands linking the domains (reviewed by Hirose, 2000; Baker et al., 2003; Gomme et al., 2005).

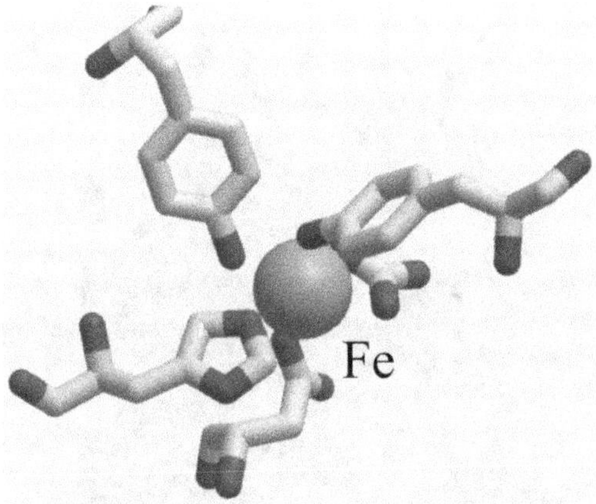

Figure 8.18. Structure of a Fe^{2+} ion binding site in transferrin (PDB file 1NO4). (From Guha Thakurta et al., 2003.)

Most of the intracellular iron is bound to ferritin and stored in the liver, skeletal muscles, and reticuloendotelial cells. If the capacity of ferritin in the cell is exceeded, iron is deposited near the iron–ferritin complexes in the cell (amorphous iron hemosiderin). Ferritins are a broad superfamily of iron storage proteins, widespread in both aerobic and anaerobic eukaryotes, eubacteria, and archae (reviewed by Carrondo, 2003; Liu and Thail, 2004). Ferritins are large protein cages (100 to 120 Å), controlling the reversible formation of iron-oxy minerals in a large cavity (50 to 80 Å). The function of ferritins is to supply cells with the necessary iron, leading to effective concentrations in living cells in the range 10^{-3} to 10^{-5} M. However, they can also be involved in cell redox stress resistance. Ferritins concentrate iron as a mineral within its cavity. Other types of bacterial ferritins use iron and oxygen to trap hydrogen peroxide or oxygen. Ferritins isolated from bacteria may contain heme b and are called bacterioferritins. A class of "mini-ferritins" from bacteria, also called *dps proteins*, are only half the size of ferritins. Deletion of ferritin genes increases oxidant sensitivity and is lethal under most conditions (Ferreira et al., 2000).

Ferritins and bacteroferritins have similar architectures; 24 subunits form a hollow sphere of external diameter ~120 Å and internal diameter ~80 Å (reviewed by (Theil, 2001, 2003, 2004; Carrondo, 2003; Theil et al., 2006) (Figure 8.19). The iron storage cavity has a diameter of ~80 Å and can accommodate up to 4500 Fe(III) ions as an inorganic complex core. Miniferritins have 12 subunits with 32 symmetry. Each ~20-kDa subunit folds as a four α-helix bundle capped by a shorter helix at the C-terminus. At the threefold axes there are channels that traverse the cavity. The subunits associate as in transferrin, and in bacterioferritins the hemes are located between pairs of subunits. Mammalian ferritins have two types of subunits, H and L chains (heavy and light), only 55% identical in sequence but mutually interchangeable in the fully assembled 24-mer.

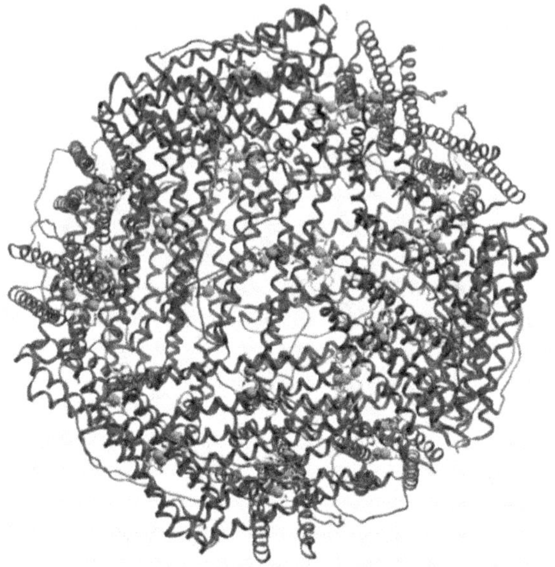

Figure 8.19. Structure of ferritin from the hyperthermophilic archaeal anaerobe *Pyrococcus furiosus* (PDP file 1JD7). (From Tatur et al., 2007.)

H-subunits have an active ferroxidase site, while L-subunits have a degenerate ferroxidase site. Protein mineralization surfaces are shared among ferritin and in other proteins that form biominerals.

Multiple functional sites are distributed symmetrically throughout ferritin protein cages. The functional sites include: (1) catalytic ferroxidase sites that couple two Fe(II) ions with oxygen, (2) pores where Fe(II) enter or leave the protein cage, and (3) the central cavity where the biomineral nucleates and iron is transferred reversibly between liquid and solid phases. The uptake of iron by (bacterio)ferritins involves an initial step of oxidation by molecular oxygen of an Fe^{2+} ion, through a binuclear di-iron center, the ferroxidase center (Le Brun et al., 1993, 1995). Fe-protein sites are located within each subunit, between pairs of subunits, among trimers of subunits, and on the cavity surface of the assembled molecules.

Two coordinated iron binding sites, A and B, are at the ferroxidase center (Liu and Theil, 2005). Site A is composed of Glu23, Glu58, and His61; site B consists of Glu58, Glu103, Gln137, and Asp140. These two iron sites are bridged by Glu58. The residues forming the di-iron center in bacterioferritin are Glu23, Glu56, Glu99, Glu132, His59, and His138. All of these residues are highly conserved in bacterioferritins. When an Fe(II) ion enters the ferritin protein pores, it reaches the ferroxidase site buried within each H-ferritin subunit within milliseconds. The ferroxidase site in ferritin oxidizes Fe(II) and forms diferroxo-mineral precursors (Fe^{3+}–O–Fe^{3+}) (Zhao et al., 2001). Hydrogen peroxide and diferric oxo/hydroxo mineral precursors are the products. Diferric peroxo intermediates form in milliseconds and decay in seconds. As filling of the cavity with mineral proceeds, oxidation occurs at both the protein catalytic sites and the mineral surface. Ferritin releases H_2O_2 and diferric-oxo mineral precursors, leaving behind an active site that is altered for a fairly long time, presumably to allow peroxide to diffuse away before binding the next Fe(II) ion.

In humans, about 70% of the total iron is bound to hemoglobin. Due to the system of storage and reuse, very little iron is excreted or required in the diet (1 to 2 mg per day). Excess iron either is not absorbed or is stored in enterocytes of the intestine. The entry of iron into the organism is regulated by hepcidin, a 25-residue peptide with a high Cys content, which inhibits the action of the iron transporters (DMT1 and Ireg1) in the membranes of the intestine.

Iron is complexed in enzymes and other proteins in addition to these iron storage proteins (reviewed by Theil, 2004; Arredondo and Nuñez, 2005; Wallander et al., 2006). It can be incorporated into proteins in the form of a heme, in various types of iron–sulfur clusters, in mixed metal centers (e.g., containing nickel), as di-iron, or in mononuclear iron centers. Hemoglobin is a tetrameric protein with molecular mass 64 kDa; it consists of two pairs of similar subunits (α and β), located approximately at the vertices of a tetrahedron (Figure 8.20) (reviewed by Linberg et al., 1998). Each subunit contains heme and is homologous to

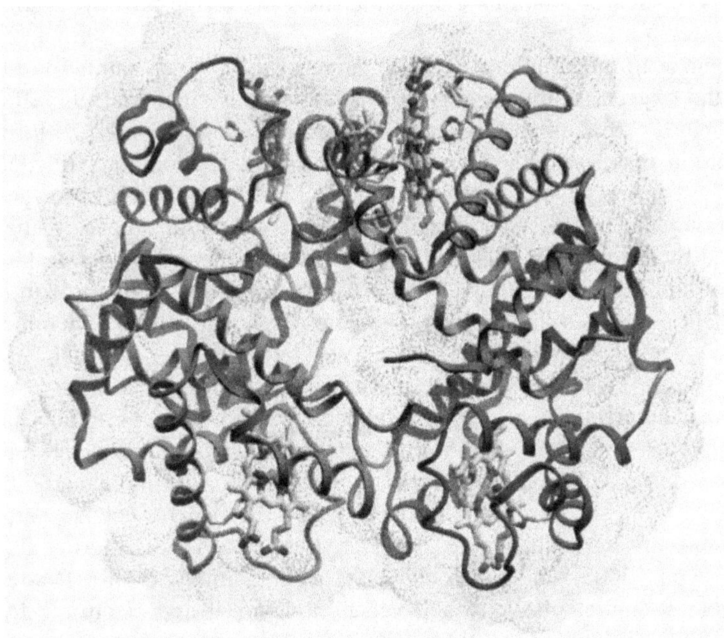

Figure 8.20. Structure of oxyhemoglobin; α- and β-subunits are blue and red, respectively (PDB file 1GZX). (From Paoli et al., 1996.) (*See insert for color representation of figure.*)

myoglobin. Deoxyhemoglobin is a high-spin, diamagnetic form of hemoglobin. The Fe(II) ion is 0.3 Å above the heme plane; its diameter is slightly larger than the hole in the center of the porphyrin ring. Triplet oxygen, the lowest-energy oxygen species, has two unpaired electrons in antibonding π^* molecular orbitals. The binding of oxygen results in the formation of low-spin Fe(II), which is slightly smaller then the high-spin Fe(II), and therefore it fits into the hole and can enter the plane of the porphyrin ring. This movement causes a change in the Fe–N(imidazole) bond; this in turn induces significant conformation changes in the hemoglobin. The result is a change in the quaternary structure of hemoglobin, reflected in a decrease of distances between heme groups upon oxygenation.

A proximal His anchors the heme group into the protein by the formation of a penta-coordinated high-spin ferrous complex, with the heme group itself providing the other four coordinating nitrogen ligands (reviewed by Brittain, 2005). Oxygen then binds at the vacant iron coordination site to form a pseudooctahedral, low-spin complex. In the absence of a well-defined protein pocket surrounding the heme group, such binding would immediately yield a collinear dioxygen complex and would lead rapidly to the formation of a nonfunctional ferric heme. The pocket in which the heme sits not only prevents such oxidative reactions but also modulates the strength of the iron–oxygen bond. The heme pocket prevents oxidative reactions primarily by steric means. The function of the "globin" in hemoglobins is to solubilize the heme group, slow the oxidation of Fe(II) to iron(III), and direct traffic for the binding of ligands; hemoglobin transports oxygen; it is not an enzyme, as are most of the proteins described in this chapter.

In contrast to the simple hyperbolic curve of the oxygen saturation of myoglobin, the oxygen saturation curve of hemoglobin is sigmoidal: the affinity of hemoglobin for small ligands (O_2 or CO for ferrohemoglobin and CN for methemoglobin) increases with the increase in ligand concentration; the Hill coefficient is 2.8. These properties make hemoglobin an efficient transporter of oxygen from the lungs to the tissues and of carbon dioxide from the tissues to the lungs (reviewed by Eaton et al., 1999). The reversible binding of oxygen to the heme groups of hemoglobin occurs with the active participation of the protein matrix. The protein provides a largely hydrophobic environment within which the heme resides.

Iron–sulfur [Fe–S] clusters are ubiquitous, evolutionary ancient prosthetic groups that are required to support fundamental life processes. The simplest Fe–S proteins contain a single iron tetrahedrally coordinated by four Cys thiolates. Clusters of iron are formed by the addition of inorganic sulfide to bridge iron atoms to form, for example, 2Fe–2S clusters, cubane and linear 3Fe–4S clusters, and cubane 4Fe–4S clusters (reviewed by Beinert et al., 1997; Johnson et al., 2005; Lukianova and Davis, 2005; Imlay, 2006). Due to their remarkable structural plasticity and versatile chemical and electronic features, [Fe–S] clusters participate in electron transfer, substrate binding and activation, storage of iron and sulfur, regulation of gene expression, and enzyme activity. The biochemical utility of clusters rests on two features: their ability to accept and donate electrons, and their tendency to bind the electron-rich oxygen and nitrogen atoms of organic substrates. Both features are influenced by the solvent exposure and electrostatic environment of the cluster. The ability to delocalize electron density over both Fe and S atoms makes [Fe–S] clusters ideally suited for their primary role in mediating biological electron transport. [Fe–S] clusters are major components in the photosynthetic and respiratory electron transport chains. They define the electron transport pathways in numerous membrane bound and soluble redox enzymes, and constitute the redox active centers in ferredoxins, one of the largest classes of mobile electron carriers in biology. Clusters involved in electron transfer contain [2Fe–2S], [3Fe–4S], [4Fe–4S], or [8Fe–7S] core units, with Cys generally completing the tetrahedral S coordination at each Fe site. Asp, His, Ser, or backbone amide nitrogens at a unique Fe site are occasionally encountered in clusters that function in electron transport, and these ligands are likely to play a role in modifying the redox potential, gating electron transport, or coupling proton and electron transport.

Ferredoxins are small, soluble iron–sulfur (Fe–S) proteins which serve as low-potential electron carriers (reviewed by Sticht and Rosch, 1998; Meyer, 2001). Ferredoxins take part in functioning of chloroplasts (reviewed by Dai et al., 2000; Vassiliev et al., 2001). All ferredoxins are homologous; the domain consists of a five-stranded β-sheet, two or three α-helices, and a long loop containing three or four Cys's for iron coordination (Figure 8.21). The β-strands for a mixed β-sheet represent two pairs of antiparallel two-stranded β-sheets connected by a complicated structure. All the plant-type ferredoxins have a conserved sequence $CX_4CX_2CX_{22-33}C$. The [2Fe–2S] cluster in ferredoxin is located near the surface of the protein. Two Fe^{2+} ions are coordinated tetrahedrally by sulfide ions and by Cys's.

Figure 8.22 shows the structure of hydrogenases containing Fe–S clusters. The Ni-containing hydrogenases are αβ heterodimers, with the large (α) and small (β) subunits having average masses of 60 and 30 kDa (reviewed by Vignais et al., 2001; Tamagnini et al., 2002). The crystal structures of the [Ni–Fe]-hydrogenases from *Desulfovibrio gigas* and *D. vulgaris* show that the two hydrogenase subunits interact very extensively through a large contact surface and form a heterodimer. The *D. gigas* hydrogenase is involved in periplasmic H_2 oxidation and is related phylogenetically to other bacterial-uptake hydrogenases.

The bimetallic Ni–Fe center of the active site of hydrogenase is located in the large subunit and is buried deep inside the protein. The small subunit contains up to three Fe–S clusters, which conduct electrons between the H_2-activating center and the physiological electron acceptor or donor of hydrogenase. The [4Fe–4S] cluster that is proximal to the active site is "essential" to H_2 activation in [Ni–Fe]-hydrogenases. Hydrophobic channels passing through both subunits have been identified by crystallographic analysis of xenon binding and molecular dynamics simulations of xenon and H_2 diffusion within the interior of the enzyme. Those channels linking the active site to the surface of the molecule were suggested to facilitate access of hydrogen to the active site.

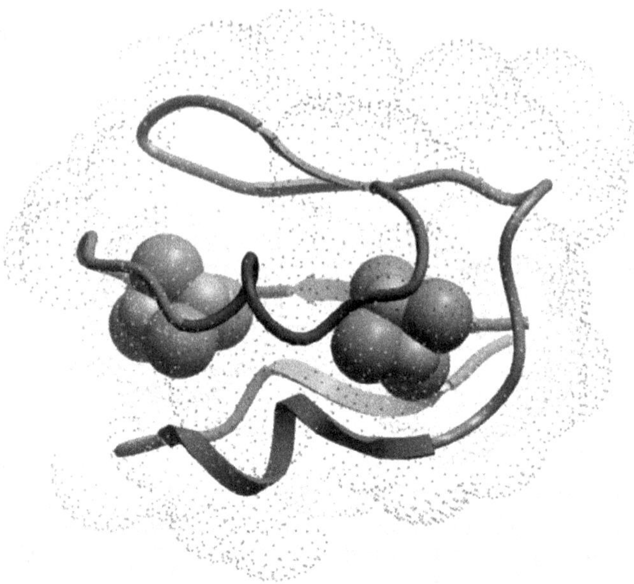

Figure 8.21. Structure of ferredoxin; note the two 4Fe–4S clusters (PDB file 1DUR). (From Moulis et al., 1996.)

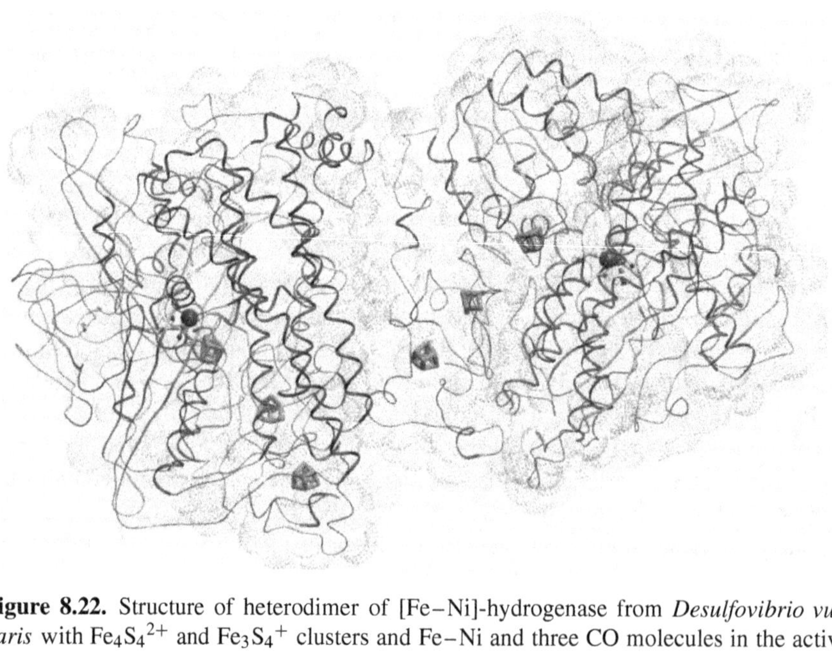

Figure 8.22. Structure of heterodimer of [Fe–Ni]-hydrogenase from *Desulfovibrio vulgaris* with $Fe_4S_4{}^{2+}$ and $Fe_3S_4{}^{+}$ clusters and Fe–Ni and three CO molecules in the active site (PDB file 1FRV). (From Ogata et al., 2002.) (*See insert for color representation of figure.*)

Most Fe–S proteins of eukaryotic cells are located within the mitochondria; these include aconitase of the citric acid cycle, subunits of the respiratory chain complexes I, II, and III, or biotin synthase (reviewed by Gerber and Lill, 2002). Up to 12 different Fe–S clusters are in the mitochondrial electron transport chain. Fe–S cluster biogenesis is crucial for maintaining mitochondrial iron homeostasis.

In vertebrates, cellular Fe levels are posttranscriptionally controlled by the activity of iron regulatory proteins (IRP1 and IRP2), cytosolic proteins that bind to structural elements called iron-responsive elements (IREs). Iron homoeostasis is controlled through several genes, an increasing number of which have been found to contain noncoding sequences, which are recognized at the mRNA level by IRP1 and IRP2 (reviewed by Cairo and Pietrangelo, 2000; Cairo et al., 2002).

Iron plays a very important role in bacteria. In order for iron to be used by a bacterium, it must be extracted from the environment or host by specialized uptake mechanisms. The major role of iron inside the bacterial cell is its involvement in enzymatic redox reactions. There are five strategies used by bacteria in the management of iron (reviewed by Andrews et al., 2003): (1) use of high-affinity iron transport proteins that enable iron to be scavenged, in various forms, from the surroundings; (2) use of intracellular iron stores to provide a source of iron that can be drawn upon when external supplies are limited; (3) employment of redox stress resistance systems (e.g., degradation of iron-induced reactive oxygen species and repair of redox stress-induced damage); (4) control of iron consumption by down-regulation of the expression of iron containing proteins under iron restricted conditions; and (5) use of an iron responsive regulatory system that coordinates the expression of the iron homeostatic machinery according to iron availability.

Iron acquisition in bacteria can occur either by direct uptake of iron containing proteins or by iron binding chelators called *siderophores* (reviewed by Ma et al., 2009). In gram negative bacteria, high-affinity outer membrane receptors first bind lactoferrin, ferritin, siderophores (ferrichrome, FhuA) (Ferguson et al., 1998), or low molecular weight iron chelators [e.g., ferric citrate (FecA)] (Ferguson et al., 2002), and mediate the transport of the iron complex through the outer membrane into the periplasm (reviewed by Wandersman and Delepelaire, 2004). These receptors are 22 stranded β-barrel proteins that contain extracellular loops that bind substrates and an N-terminal region or plug that folds into the barrel near the periplasmic surface. Transport across the outer membrane is coupled to the proton motive force of the cytoplasmic membrane via a periplasm spanning complex composed of TonB, ExbB, and ExbD. Once in the periplasm, the uptake of Fe(III) or Fe(III) chelates occurs through the transmembrane channel of ATP binding cassette (ABC) transporters (Davidson et al., 2008; Rees et al., 2009) in the plasma membrane in a process mediated by ATP hydrolysis of the cytosolic ATPase subunits. In gram-positive organisms, ABC transporters are found in the plasma membrane just beneath the cell wall. In both cases, the specificity of this transport is mediated by the solute binding protein (SBP) component of the transporter. The SBP is freely diffusible in the periplasm of gram negative bacteria but is anchored covalently to the plasma membrane in gram positive bacteria.

Crystallographic structures of six gram-negative Fe(III) binding SBPs have now been determined (reviewed by Krewulak and Vogel, 2008). Each employs at least four oxygen atoms to coordinate the Fe(III) atom and have a coordination number of 5 or 6. SBPs of *Neisseria gonorrheae* and *Haemophilis influenza* Fe(III) bind to two consecutive tyrosine residues, one glutamic acid, one histidine, a water molecule, and a phosphate ion (Bruns et al., 1997).

Many bacteria can also get Fe(III) from heme proteins and siderophores that the bacteria themselves secrete and then uptake these Fe(III) complexes directly into the cytosol before degrading them (reviewed by Z. Ma et al., 2009). Thus, the mechanism for transport of Fe(III) across the plasma membrane does not involve the direct coordination of Fe(III).

8.2.5. Cobalt Complexes

Cobalt has two naturally occurring oxidation states (II and III), but can exhibit oxidation states from −I to +IV. Although cobalt is encountered less frequently in metalloenzymes than in the other first-row transition metals, it is nevertheless an important cofactor in vitamin B_{12}–dependent enzymes.

Vitamin B_{12} contains cobalt in a substituted corrin macrocycle, a relative of porphyrin. The B_{12} coenzyme possesses an axial Co(III)-alkyl (5′-deoxyadenosine or methyl) group. These corrin–cobalt complexes are well studied; in contrast, only a few proteins containing noncorrin cobalt have been characterized. To date, eight noncorrin, cobalt containing enzymes (methionine aminopeptidase, prolidase, nitrile hydratase, glucose isomerase, methylmalonyl-CoA carboxytransferase, aldehyde decarbonylase, lysine-2,3-aminomutase, and bromoperoxidase) have been isolated and characterized (reviewed by Kobayashi and Shimizu, 1999).

Methionine aminopeptidase is a monomeric protein of 29 kDa consisting of 263 residues that has two Co^{2+} ions in its active site (Figure 8.23) (Lowther et al., 1999; reviewed by Kobayashi and Shimizu, 1999). The enzyme has several weak absorption bands between 550 and 700 nm, which arise from $d \rightarrow d$ transitions of the d^7 Co^{2+} ions. The two Co^{2+} ions are situated between two double antiparallel β-sheets that constitute part of the active site of the enzyme. The Co^{2+} ions are coordinated by Asp97, Asp108, His171, Glu204, and Glu238. These five residues are conserved in four other methionine aminopeptidases, with the exception of Glu235, which is Gln in the *Bacillus subtilis* enzyme. The coordination geometry of each cobalt site in methionine aminopeptidase is approximately octahedral, with the sixth position unoccupied in the crystal structure. The two Co^{2+} ions and six of the ligand atoms, other than the two oxygen atoms of Glu235 that bridge the two cobalts, are nearly coplanar. Solvent molecules may be bound to both Co^{2+} ions in place of the "absent" octahedral ligands. The distance between the Co^{2+} ions (2.9 Å) is somewhat larger than twice the covalent radius for

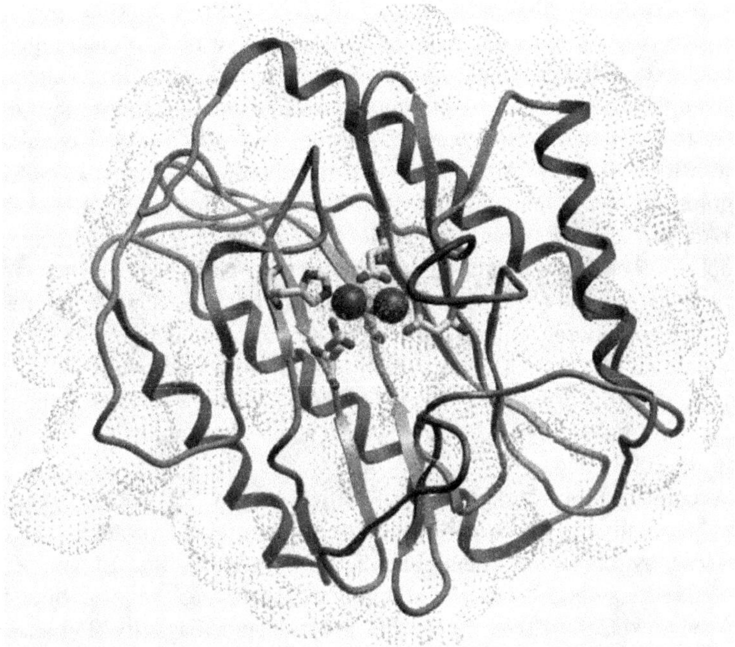

Figure 8.23. Structure of methionine aminopeptidase from *E. coli* with two bound Co^{2+} ions (PDB file 1C21). (From Lowther et al., 1999.)

cobalt (2.32 Å) and is similar to that between the two Zn^{2+} ions of leucine aminopeptidase (Zn^{2+}–Zn^{2+}: 2.9 Å).

8.2.6. Nickel Complexes

Nickel is an essential cofactor of enzymes found in eubacteria, archaebacteria, fungi, and plants (reviewed by Li and Zamble, 2009). Nickel and cobalt uptake in microorganisms is mediated by secondary transporters and by ATP-binding cassette (ABC) systems (reviewed by Eitinger et al., 2005). The best investigated ABC-type nickel permease is the NikABCDE system of *E. coli*, composed of a periplasmic binding protein (NikA), two integral membrane proteins (NikBC), and two ABC proteins (NikDE). Sequence determinations and phylogenetic analyses suggest that the Nik proteins are more closely related to oligopeptide transporters than to those that transport other types of metal ions (Navarro et al., 1993; Tam and Saier, 1993). On the basis of homology, the genes of the *nik* operon were designated as components of the expected core structure, which consists of two channel forming transmembrane proteins, NikB and NikC, and two cytoplasmic nucleotide binding proteins, NikD and NikE (Navarro et al., 1993). NikA is a soluble periplasmic binding protein that acts as a receptor for nickel and presumably delivers it to the NikBC pore. The first high resolution structure of NikA revealed a structure with the substrate binding pocket in a cleft between two globular domains that rotate with respect to each other about a flexible linker (Heddle et al., 2003). Although nickel binding causes some hinge closing in comparison to the apoprotein, the metal binding site remains accessible to solvent. It was suggested that NikA binds a nickel metallophore that is translocated across the periplasmic membrane, analogous to the Fe(III) siderophore uptake mechanisms. The affinity of NikA for Ni(II)–EDTA is only 30 μM (Addy et al., 2007), but it is not likely to be the physiological complex. In an attempt to identify the natural metallophore, holo-NikA was purified from the *E. coli* periplasm in the absence of EDTA (Cherrier et al., 2008). In the resulting structure, the compound bound to the metal was best modeled as butane-1,2,4-tricarboxylate (BTC), which contributes three carboxylate ligands to the square planar metal coordination site. A second nickel site on the surface of the protein with nickel coordinated by two histidines and four water molecules was revealed (Addy et al., 2007).

The secondary transporters are classified as nickel–cobalt transporters (NiCoTs), a family of prokaryotic and fungal membrane proteins, which are characterized by eight transmembrane helices (reviewed by Li and Zamble, 2009). These monomeric, single component permeases have a dissociation constant for nickel of ~10 to 20 nM but low transport capacity (Fulkerson and Mobley, 2002; reviewed by Li and Zamble, 2009).

Cellular nutrients for gram-negative bacteria must pass through an outer membrane prior to reaching the cytoplasmic membrane. Metal ions can diffuse through nonspecific transmembrane porins in the outer membrane, so it was assumed that nickel ions would get into the periplasm by this mechanism. However, there is now evidence that at least in some organisms, nickel uptake across the outer membrane occurs through a TonB-dependent transporter (TBDT) (Schauer et al., 2008). A TBDT receptor is essentially a barrel spanning the outer membrane with a plug at the bottom, and the interaction with TonB on the periplasmic side couples the opening of the plug and substrate passage to the electrochemical charge gradient of the inner membrane.

There are multiple enzymes that employ nickel as a cofactor to catalyze a diverse array of chemical reactions. In several of these enzymes, the nickel is embedded in an intricate, multinuclear cluster that includes modified amino acids and/or exogenous ligands (reviewed by Li and Zamble, 2009). Nickel is found in urease, carbon monoxide dehydrogenase, *S*-methyl coenzyme-M reductase, and hydrogenase (reviewed by Volbeda et al., 1996). It is an essential component of at least nine metalloenzymes involved in energy and nitrogen metabolism, in detoxification processes, and in pathogenesis. This list includes urease, [Ni–Fe]-hydrogenase, and Ni-superoxide dismutase.

The plant and fungal ureases are homo-oligomeric and the bacterial ureases are hetero-oligomeric (Mobley et al., 1995; Tange and Niwa, 1997). Each catalytic subunit of urease contains an active site with two Ni^{3+} ions that, in the case of *Klebsiella aerogenes* urease, are 3.5 Å apart (Jabri et al., 1995) (Figure 8.24).

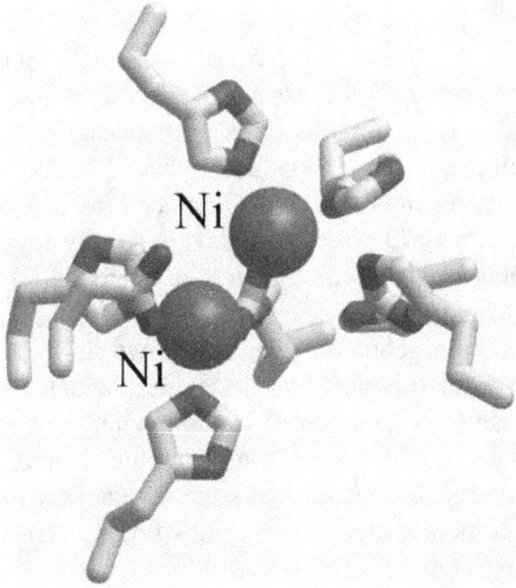

Figure 8.24. Structure of the di-nickel binding site in urease from *Klebsiella aerogenes* (PDB file 1FWJ). (From Pearson et al., 1997.)

One Ni^{3+} ion is bound to His246 and His272, while the second Ni^{3+} ion is bound to His134, His136, and Asp360.

Nickel and cobalt are necessary for cellular function and require strict regulation to hinder toxic side reactions or incorporation into non-native binding sites. The concentrations of "free" metal ions are under the control of transcription factors and appear to be exceedingly low in cells (Finney and O'Halloran, 2003). The MerR family of transcription factors responds to many metal ions, including Hg (MerR), Zn (ZntR), Cu (CueR), Pb (PbrR), Cd (CadR), and Co (CoaR), and to oxidative stress (SoxR) and drugs (BmrR). Proteins of this family have homologous N-terminal DNA binding domains, but different C-terminal sensor domains that enable them to bind specific coactivator molecules (reviewed by Barondeau and Getzoff, 2004).

As with other transition metals, the use of nickel ions is inherently complicated because they are also potentially toxic and can cause a plethora of cellular damage, so distribution and accumulation must be tightly controlled (reviewed by Li and Zamble, 2009). For this reason, organisms that require nickel also express a variety of proteins that specifically contribute to nickel homeostasis by assembling metallocenters, importing or exporting the metal, transporting nickel ions within the cell, or regulating the expression of the factors involved. Several exporters have been predicted and/or shown to pump nickel out of the cytoplasm (Nies, 2003) or even the periplasm (Grass et al., 2005), but in only a few cases are the transporters specific for nickel. An example is NreB from *Achromobacter xylosoxidans* 31A, a putative exporter from the major facilitator superfamily with 12 transmembrane segments and a histidine-rich C-terminus (Nies, 2003).

8.2.7. Copper Complexes

Copper is an essential element for all living organisms; copper binding proteins take part in various biological processes, from electron transfer to oxidation of various substrates. Many protein functions are associated with copper binding (reviewed by Nakamura and Go, 2005; Permyakov, 2009). Despite the usefulness of copper as a redox mediator, stray copper ions can be harmful and the distribution of copper ions in living organisms is strictly regulated.

Two oxidation states are usual for copper, Cu(I) and Cu(II) (reviewed by Belle et al., 2005). The $3d^9$ outer electronic configuration of Cu(II) does not have cubic symmetry; its coordination geometry is usually lower in symmetry with unequal bond lengths. The coordination numbers 4 (square planar), 5 (trigonal bipyramid or square pyramid), and 6 (octahedron) predominate. With coordination 6, the Jahn–Teller effect—geometrical distortion of nonlinear molecules with a degenerate, electronic ground state that removes the degeneracy—excludes a regular octahedron. Cu(II)–N bonds are often inert, while Cu(II)–O bonds are more labile.

Cu(I) prefers coordination numbers 2, 3, or 4 (tetrahedral geometry) and is stabilized by soft ligands. Coordination 5 (square pyramid) is known. Cu(I) is a closed-shell d^{10} transition metal ion and thus is diamagnetic. The disproportionation of Cu(I) is usual in aqueous solution, where the electrode potentials (vs. the normal hydrogen electrode) of the reactions

$$Cu^+(aq) + e^- \rightarrow Cu^0(aq) \qquad E^0 = 0.52\ V \tag{8.1}$$

$$Cu^{2+}(aq) + e^- \rightarrow Cu^+(aq) \qquad E^0 = 0.15\ V \tag{8.2}$$

are widely separated and drive

$$2Cu^+ \rightarrow Cu^0 + Cu^{2+} \qquad E^0 = 0.37\ V \qquad \text{disproportionation} \tag{8.3}$$

The equivalent reaction is not observed with iron.

Cu(I) catalyzes the Fenton reaction, as does Fe(II), with hydrogen peroxide (conversion of two molecules of hydrogen peroxide into two hydroxyl radicals and water in the presence of catalytic metal ion). The Cu(I) state exhibits the ability to bind and activate dioxygen, leading to the formation of $Cu–O_2$ intermediates such as peroxo, hydroperoxo, bis(l-oxo), or superoxo species.

Copper coordination sites in proteins are classified as type 1, type 2, or type 3, based on their structural and spectroscopic properties (Solomon et al., 1996). Type 1 (blue) sites shows maximum absorption around 610 nm, and type 3 sites show maximum absorption around 330 nm. Absorption at 610 nm is characterized by an extinction coefficient which is 100 times higher than those for model copper complexes. Strong visible absorption is characteristic of type 1 Cu sites when in the oxidized Cu(II) state. Type 1 and type 2 coppers are ESR (electron spin resonance) detectable, whereas the dinuclear type 3 coppers are ESR silent. The reduction potentials of most type 1 Cu sites are sufficiently low that the oxidized state is favored in the presence of ambient oxygen.

The type 1 copper binding site consists of two His's, one Cys, and one Met. The first three residues are essential for the blue copper binding site and tightly coordinate the Cu^{2+} ion in trigonal geometry, while the coordination of the fourth residue, an axial Met, is rather distant and weaker. This residue can be replaced with carbonyl oxygen of different amino acids, such as Leu or Phe.

The blue color of these proteins is due to charge transfer between the Cu and S atoms of the Cys ligand. Electron spin resonance spectra of the oxidized site show unusually low coupling constants (*A* values). The function of these sites is exclusively single electron transfer. Type 1 sites are found in small electron transfer proteins (cupredoxins) that ferry electrons between larger enzymes such as components of the denitrification pathway and photosynthesis. They are also found in the larger enzymes, such as nitrite reductase and multicopper oxidase, and function in intramolecular electron transfer to copper active sites. Within some of the large enzymes, a Cu_A site functions as an electron entry point. This site is an expansion of the type 1 site by a second Cu to form a metal–metal bond. Both type 1 and Cu_A sites are rigid and characterized by low reorganization energies to facilitate electron transfer.

Type 2 copper sites have a variety of amino acid ligands and geometries (reviewed by MacPherson and Murphy, 2007). Most type 2 sites are three to four coordinate, and one or more of the copper ligands are the imidazole side chains of His. The coordination sphere may be completed by Met, Glu, Gln, or Tyr. The absence of a thiol group results in weak visible absorption and thus no evident color. Electron spin resonance spectra of type 2 sites are characterized by a weaker signal with larger *A* values (isotropic hyperfine coupling constant) and have spectra clearly distinct from those of type 1. Coordination positions in type 2 Cu sites can be either vacant or occupied by exogenous ligands. These sites can be catalytically active by interacting directly with enzyme substrates. When molecular oxygen is the substrate, type 2 sites may function as oxidases, reducing oxygen to water or peroxide; as monooxygenases, in which one oxygen atom is inserted into the substrate and the other is reduced to water, or as dioxygenases, in which both oxygen atoms are incorporated into the substrate. In addition, type 2 sites are able to perform the dismutation of superoxide and reduce nitrite (NO_2^-) to nitric oxide (NO). More complex sites such as the Cu_A site mentioned above are constructed from multiple copper centers.

Type 3 copper sites consist of two antiferromagnetically coupled Cu atoms bridged by molecular oxygen or a hydroxyl (reviewed by MacPherson and Murphy, 2007). The type 3 pair plus a third, type 2 copper, is part of the trinuclear cluster in multicopper oxidases. A dinuclear type 3 site is found in hemocyanins, which function as oxygen carriers in invertebrates. There is a similar site in tyrosinases; however, these enzymes function as monooxygenases and activate oxygen for insertion into phenolic substrates.

Multicopper blue proteins include nitrite reductase and multicopper oxidases (reviewed by Permyakov, 2009). Multicopper blue proteins contain tandem repeats of a domain, which are homologous to the cupredoxin fold found in plastocyanin, azurin, pseudoazurin, rusticyanin, stellacyanin, and amicyanin. The cupredoxin domain, an eight-stranded Greek key β-barrel, was first observed in the small blue copper proteins, plastocyanin and azurin, and was subsequently observed in the related proteins pseudoazurin, amicyanin, and cucumber basic protein. The cupredoxin domain usually contains a type 1 copper binding site, which is responsible for the characteristic blue color of these proteins. The single domain cupredoxin fold proteins and multicopper blue proteins together are called *blue copper proteins* and the cupredoxin domain is called the *blue copper binding domain*.

Multicopper blue proteins often possess an interdomain binding site. Such sites usually have only His ligands. For example, in multicopper oxidases such as laccase and ceruloplasmin, the interdomain site is trinuclear, composed of one type 2 and two type 3 coppers coordinated by eight His's (four from each domain). Laccases are three domain multicopper blue proteins found in fungi and trees (reviewed by Baldrian, 2006). Laccases catalyze the reduction of oxygen to water accompanied by the oxidation of a substrate, typically a *p*-dihydroxyphenol or another phenolic compound. One of the functions of the enzyme is supposed to be the polymerization of lignans (phenol) to yield lignin (polyphenol), which forms the cell wall of plants, and to decompose the lignin in tree rotting by fungi. Crystal structures of several laccases have been determined (Figure 8.25). Laccases belong to the group of blue multicopper oxidases that catalyze a one electron oxidation concomitantly with the four-electron reduction of molecular oxygen to water (Solomon et al., 1996). The catalysis carried out by all members of this family employs different copper centers in the enzyme. All blue multicopper oxidases are characterized by the presence of at least one type 1 copper, together with at least three additional coppers: one type 2 and two type 3 Cu^+ ions, arranged in a trinuclear cluster (Figure 8.26). The substrates are oxidized by the type 1 copper and the extracted electrons are transferred, probably through a strongly conserved His–Cys–His tripeptide motif, to the type 2/type 3 site, where molecular oxygen is reduced to water (Messerschmidt, 1997). Interdomain sites of some laccases contain one type 2 copper atom and two type 3 copper atoms, coordinated by eight His's (four from each domain).

The function of the blue copper in multicopper blue proteins is to accept an electron from (oxidize) a substrate, and the function of the interdomain copper is to donate the electron to (reduce) another substrate. For example, nitrite reductase accepts an electron from electron transporter proteins such as azurin and pseudoazurin to its blue copper and reduces nitrite at the interdomain site. Multicopper oxidases oxidize a wide variety of substrates, such as phenol, methoxyphenol, aromatic amines, polyaromatic compounds, and metal ions, in the vicinity of the blue copper, and donate electrons to molecular oxygen to convert it to water at the interdomain site. The environment around the blue copper affects the electronic structure and the physical properties, such as the redox potential, of the blue copper. Blue copper proteins use several strategies to control the electronic environment of the copper ion. One is a modification of the loop region around

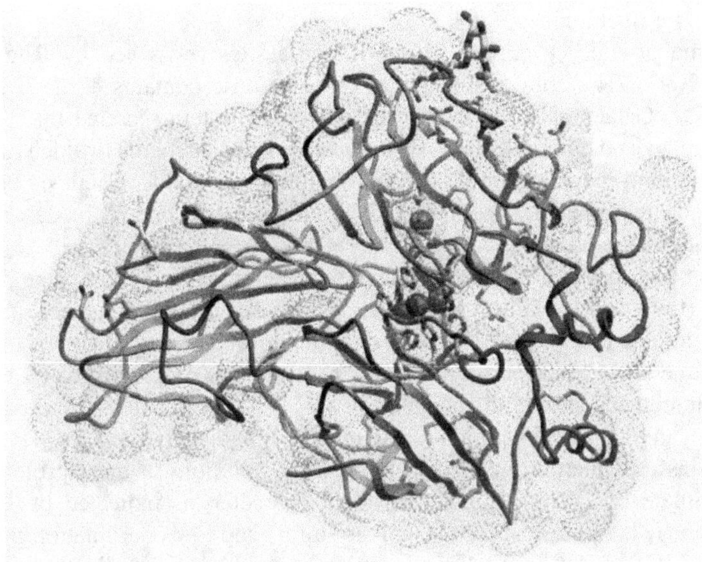

Figure 8.25. Structure of fungal laccase from *Trametes versicolor* (PDB file 1GIC). (From Piontek et al., 2002.) (*See insert for color representation of figure.*)

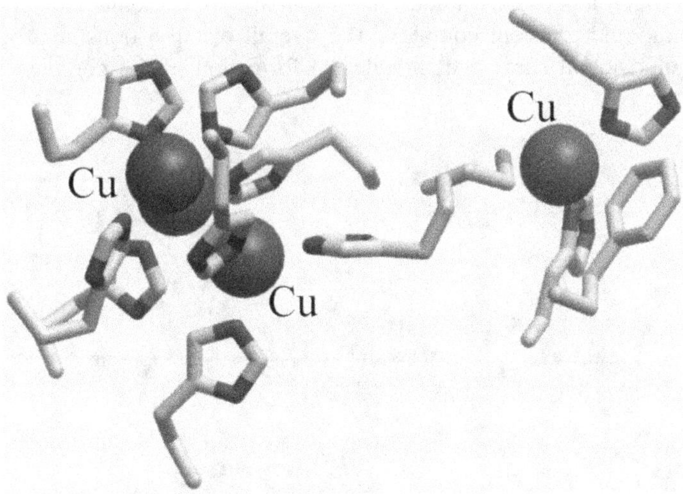

Figure 8.26. Four copper binding sites in fungal laccase from *T. versicolor* (PDB file 1GIC). (From Piontek et al., 2002.)

the metal binding residues. Another is to replace the axial ligand residue, which typically is a methionine.

Tyrosine provides a copper ligand in some sites (reviewed by Rogers and Dooley, 2003). The copper site of galactose oxidase contains Tyr497, His496, and His581. Galactose oxidase is an extracellular protein secreted by *Fusarium* spp.; it catalyzes the two-electron oxidation of primary alcohols to aldehydes with the concomitant reduction of dioxygen to hydrogen peroxide. Catalysis occurs at a mononuclear copper site, with Tyr272 providing the second redox center in the active state of the enzyme.

Several blue proteins with monocopper sites contain two His's, one Cys, and one Met (reviewed by Gray et al., 2003; Permyakov, 2009) (Figure 8.27). The distorted tetrahedral geometry of the coordination sphere of Cu(II) is reflected in its unique spectral properties. Moreover, this site is characterized by a higher redox potential and structural changes assisting electron transfer. The blue copper sites give an intense absorption band at 600 nm due to the $Cys^{-} \rightarrow Cu(II)$, charge transfer, which provides the blue color to solutions of these proteins. The S–Cu thiolate bond has a greater covalent character, as indicated by ESR. In general, Cu(I) favors soft ligands (such as sulfur) and low coordination numbers, whereas Cu(II) prefers harder ligands (oxygen) and higher coordination numbers. The cupredoxins (plastocyanins, azurin, stellacyanin, rustacyanin, amicyanin), members of the monocopper blue protein family, are 10- to 20-kDa soluble copper proteins that function as electron transfer agents, shuttling electrons, one by one, from a donor to an acceptor protein in the respiratory and photosynthetic chains of many bacteria and plants (reviewed by De Rienzo et al., 2004). Electron transfer between a cupredoxin and one of its partners is mediated by formation of a dynamic and transient complex. The overall electron transfer process is a sequence of reaction steps that include (1) formation of the complex between

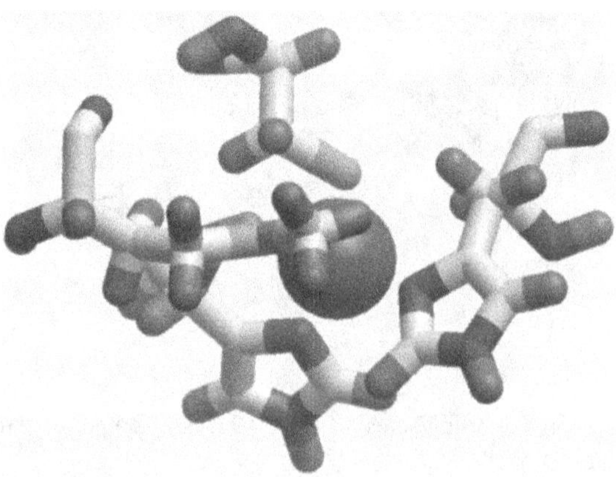

Figure 8.27. Copper binding site in plastocyanine; both His's coordinate with their Nδ atoms (PDB file 1JXD). (From Bertini et al., 2003.)

the two redox partners, (2) structural reorganization of the complex to bring the proteins into the right geometry for electron transfer, (3) the electron transfer itself, and (4) dissociation of the complex after the redox reaction has taken place.

In the cupredoxin-like structures, the copper is bound to the protein matrix by the Nδ atoms of two His's and the Sγ atom of a Cys thiolate arranged in a trigonal plane with a copper, ligand distance of $\sim$2 Å. In addition, the copper is coordinated by one or two weaker axial ligands ($\sim$3 Å), typically the Sγ atom of a Met and the carbonyl oxygen of Gly (in azurin) or the amide oxygen of Gln (in stellacyanin). The copper site has a rhombohedral geometry that is intermediate between that preferred by Cu(I) complexes (tetrahedral) and that preferred by Cu(II) complexes (tetragonal).

Small changes in the active site geometry of blue copper proteins can tune their redox properties. Conformational fluctuations of the protein structure can affect not only protein recognition and binding but also electron transfer properties. Small movements of atoms in protein regions far from the copper site can induce fluctuations of residues near the copper site and/or in regions involved in protein binding. These internal dynamic processes, which are essential to protein functionality, depend on structural features and can be influenced by mutation of the protein sequence.

Azurin (Figure 8.28) has one α-helix and eight β-strands that fold into a β-barrel structure in a double Greek key topology (reviewed by Wittung-Stafshede, 2004). The indole group of the single Trp in azurin is located in the center of the molecule. The environment of the indole is extremely rigid and nonpolar; this is reflected in the unusually blue spectrum of Trp (main maximum at 308 nm) fluorescence with distinct vibrational structure (Burstein et al., 1977). In azurin of *P. aeruginosa*, a redox active copper [Cu(II)/Cu(I)] is coordinated by His46, His117, Cys112, and two weaker axial ligands, sulfur of Met121 and oxygen of Gly45, in a trigonal bipyramid. The highly covalent nature of the copper–Cys bond is reflected in an intense absorption band with a maximum at 630 nm. It has been suggested that the polypeptide fold defines the metal site, leading to the rather unusual Cu(II) coordination in azurin as well as in other blue copper proteins. The overall tertiary structures of apo- and holo-azurin are identical (Nar et al., 1991, 1992). Copper remains bound to azurin even in the presence of guanidine hydrochloride (GuHCl), a strong denaturing agent. EXAFS experiments at high GuHCl concentrations established that copper [Cu(I)] in unfolded azurin is coordinated in a trigonal geometry to one thiolate (probably Cys112), one imidazole (perhaps His46 or His117), and a third, unknown ligand (reviewed by Wittung-Stafshede, 2004).

Binuclear Cu proteins play vital roles in O_2 binding and activation and can be classified into coupled and noncoupled binuclear sites based on the magnetic interaction between the two copper centers (reviewed by P. Chen and Solomon, 2004). Coupled binuclear Cu proteins include, for example, hemocyanin, tyrosinase, and catechol oxidase. These proteins have two copper centers strongly magnetically coupled through direct bridging ligands that provide a

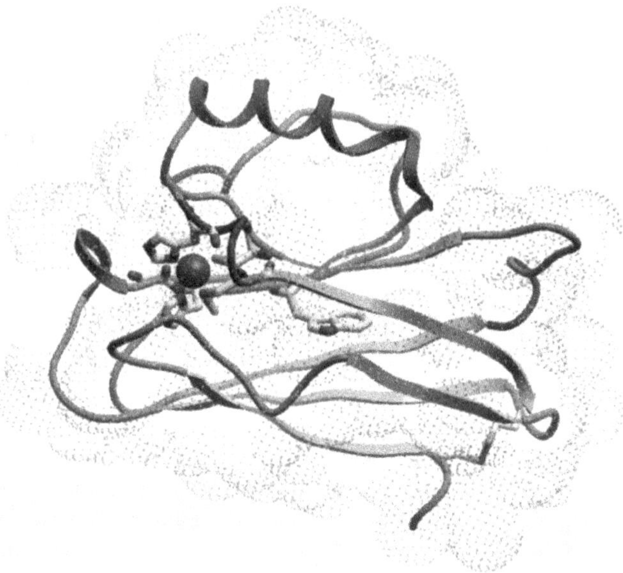

Figure 8.28. Structure of azurin with bound Cu^+ ion (PDB file 1JOI). (From Adman and Jensen, 1981.)

mechanism for the two-electron reduction of O_2. Noncoupled binuclear copper proteins include peptidylglycine α-hydroxylating monooxygenase and dopamine β-monooxygenase. These proteins have binuclear Cu active sites that are distant, that exhibit no exchange interaction, and that activate O_2 at a single Cu center to generate a reactive Cu(II)/O_2 species for H-atom abstraction from the C–H bond of substrates.

Cu_A centers contain [2Cu–2S] units similar to the [2Fe–2S] iron–sulfur clusters. The near planarity of the Cu_2S_2 core unit may be an important requirement for the access and the reversibility of the Cu(II)Cu(I)/Cu(I)Cu(II) couple. Binuclear Cu_A sites are found in cytochrome *c* oxidase (Iwata et al., 1995) and N_2O-reductase (Brown et al., 2000). Cytochrome *c* oxidase catalyzes the reduction of O_2 to H_2O during cellular respiration, while N_2O-reductase catalyzes the reduction of N_2O to N_2 during microbial denitrification. Binuclear Cu_A sites contain a dicopper unit with a delocalized mixed valence in its resting state. $Cu^{1.5+}$–$Cu^{1.5+}$ ions are 2.5 Å apart, both in a distorted tetrahedron coordination (Figure 8.29). The copper atoms are bridged by two Cys's and by two His's and in axial positions by Met and by carbonyl O of Trp in N_2O reductase, or of Glu in cytochrome *c* oxidase.

In adult humans, the net absorption of dietary copper is less than 1 mg per day (reviewed by Linder et al., 1998). Dietary copper joins endogenous copper flowing into the gastrointestinal tract through various digestive juices. Most of this copper returns to the circulation and to the tissues and organs (including

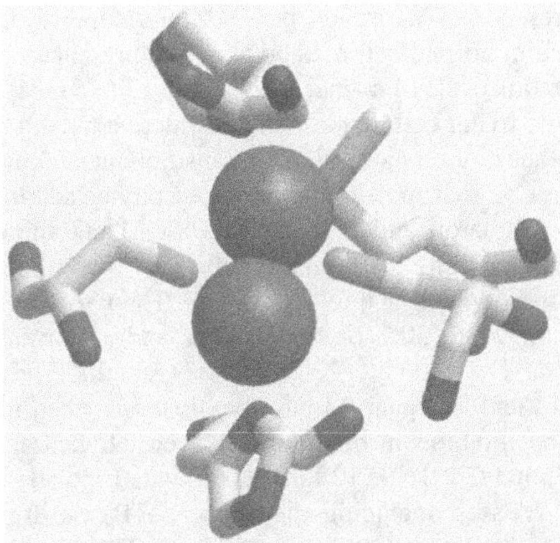

Figure 8.29. Binuclear copper binding site in cytochrome *c* oxidase; both His's coordinate with their Nδ atoms (PDB file 1OOC). (From Tsukihara et al., 1996.)

liver) that formed them. The flux of copper into and out of other organs of the body is much lower. Newly absorbed copper is transported to body tissues in two phases. In the first phase, copper goes from the intestine to the liver and kidney; in the second phase, copper usually goes from the liver (and perhaps also the kidney) to other organs. Under normal conditions, the concentration of copper in plasma is regulated by strong homeostatic mechanisms, yet to be characterizered.

Most of the changes observed in the concentration of plasma copper are associated with changes in a cuproprotein, ceruloplasmin. More than 75% of plasma copper is associated with ceruloplasmin. Ceruloplasmin plays a very important role in the second phase of copper transfer (Bento et al., 2005, 2007). Under normal dietary conditions, much of the copper entering the liver and kidney from the diet reemerges in the plasma on ceruloplasmin. Apart from blood, copper is also present in other body fluids, including fluids surrounding brain and the central nervous system, and amniotic fluid. Ceruloplasmin is present in these fluids and can take part in the copper transport.

Living cells need copper for normal functioning; however, excess copper is toxic. In humans, copper toxicosis is most often the result of genetic abnormalities and results in hepatic cirrhosis, hemolytic anemia, and degeneration of the basal ganglia. Due to its very high affinity for proteins, the characteristic half-life of copper dissociation from metalloenzymes ranges from several days to several weeks.

The redox interconversions of Cu(II) and Cu(I) are the basis of electron transfer reactions in copper enzymes. However, these same redox reactions contribute to the formation of hydroxyl radicals, which can cause oxidative damage of DNA, proteins, and lipids. Copper ions can initiate formation of hydroxyl radicals in solution in the presence of hydrogen peroxide and superoxide anion, both of which are produced in normal cell metabolism. The presence of reducing agents contributes to the redox cycle of copper. High levels (3 to 5 mM) of glutathione in the cytoplasm of the living cell increase the destructive potential of copper ions.

The principal sequestration molecules for transition metal ions, including Cu(I) and Cd(II), are the Cys-rich metallothioneins and phytochelatins as well as sulfide (reviewed by Dameron and Harrison, 1998). Metallothioneins are 25- to 62-residue, Cys rich proteins that contain very few hydrophobic residues. The metals are bound in polynuclear thiolate clusters. The Cys's in metallothioneins typically account for 20 to 30% of the residues and are arranged in repeating Cys–Cys, Cys–Xaa–Cys and Cys–Xaa–Xaa–Cys motifs. The mammalian protein has 62 residues and is organized into two domains when metals are bound. The mammalian metallothionein binds up to seven tetrahedral, four coordinate ions such as Zn(II) and Cd(II), or 12 three coordinate trigonal planar Cu(I) ions.

P-type, cation ATPases, including the copper ATPases, are involved in the movement or translocation of various metal ions. They are highly conserved from bacteria to humans (reviewed by Dameron and Harrison, 1998). The "P" designation comes from the covalent phosphorylation of a conserved Asp that is part of the reaction cycle. Defects in Menkes' ATPase translocase lead to a fatal copper deficiency disease in humans called *Menkes' syndrome*. This protein has the characteristic elements of a P-type ATPase: a phosphorylation domain, a phosphatase domain, an ATP binding site, and a transmembrane cation channel. The N-terminal regulatory region of Menkes' ATPase contains six repeated domains of about 70 residues, each of which contains a single conserved Met–Xaa–Cys–Xaa–Xaa–Cys motif. The Menkes protein functions to export excess intracellular copper and is postulated to be reversibly regulated by metals through the specialized copper binding subdomains in the N-terminus of the protein. This metallo-regulation couples the cellular export of copper to its intracellular concentration.

A number of effective methods for sequestering and detoxifying copper prevent its circulating freely inside a cell. Metalloenzymes need copper and are therefore faced with the challenge of acquiring their important metal cofactor in the absence of available copper. To overcome this dilemma, all eukaryotic organisms have evolved a family of intracellular copper binding proteins that help reserve a bioavailable pool of copper for the metalloenzymes, escort the metal to appropriate targets, and transfer the Cu^{2+} ion directly. These proteins have been collectively called *copper chaperones* (reviewed by Harrison et al., 1999; Huffman and O'Halloran, 2001; Rosenzweig, 2002). Copper chaperones can acquire the metal under conditions in which the metalloenzymes cannot then deliver and directly transfer copper to specific cellular targets. These molecules act to escort copper ions and protect them from copper-scavenging detoxification mechanisms. Encapsulation of copper in a solvent-proof complex with metallothionein is one of the mechanisms for limiting the interaction of copper with the cytoplasm; however, metallothionein does not transport copper to specific organelles and/or specific copper-containing proteins. There are three groups of copper chaperones: ATX1-like chaperones, copper chaperones for superoxide dismutase, and copper chaperones for cytochrome *c* oxidase. Copper chaperones for superoxide dismutase, known as CCS proteins, provide copper to eukaryotic antioxidant enzyme Cu,Zn-superoxide dismutase (reviewed by Harrison et al., 1999; Huffman and O'Halloran, 2001; Rosenzweig, 2002).

In contrast to iron, zinc, and manganese, the copper requirements are lower in most bacteria, with copper toxicity becoming acute at relatively low concentrations of cytosolic Cu(I) (reviewed by Ma et al., 2009). This explains the requirement for Cu(I) chaperones that traffic the metal in the cytosol and may also explain the extraordinarily high Cu(I) binding affinity of Cu(I) sensor proteins that up-regulate cytosolic Cu(I) efflux systems in response to Cu(I) stress.

8.3. CONCLUSIONS

Although the earliest life probably originated and evolved without metals, many were soon added to biology's repertoire. These essential metals are bound by proteins with reasonable selectivity for a specific ion, ion cluster, or ion–ligand complex. These metal–protein complexes perform a wide variety of functions, including catalysis, regulation, transport, and transduction. We know the structures of many of these complexes in vitro without a full understanding of their functions or selectivities in vivo. The generation and functions of fluctuations in $[Ca]_{cyt}$ are under intense investigation and are reasonably well understood (Chapters 2, 15, and 16). In contrast, much less is known about the intracellular storage and regulation of these other metals, except for sodium and potassium. Since several of these metals, especially magnesium and zinc, compete for calcium binding sites and/or bind to other sites on calcium binding proteins (Chapter 14), it is essential to understand their physiology and chemistry in order to have a full understanding of calcium binding proteins. The same concerns apply to metals such as lead and strontium, which are not (now) considered essential; however, they are frequently encountered by organisms as potential toxins (Chapter 7).

9

NONESSENTIAL METALS

This brief survey of nonessential metals is not meant as a review of their roles in either the biosphere or the geosphere. It focuses on four questions relevant to calcium binding proteins:

1. Every metal, essential or not, can harm an organism: "Alle Ding' sind Gift und nichts ohn' Gift; allein die Dosis macht, dass ein Ding kein Gift ist" (Paracelsus, ca. 1530) ("All things are poison and nothing is without poison, only the dose permits something not to be poisonous"). Can this potential toxicity be ascribed to the metals binding to specific proteins?
2. Is there a specific binding protein(s) for each nonessential metal, in competion with or distinct from those for the 15 essential metals and especially for calcium? For example, should MerR and PbrR be considered specific for binding mercury and lead, respectively?
3. To what extent, and how, is the intraorganismal and/or intracellular concentration of each metal, essential and nonessential, regulated?
4. How are nonessential metals used to study complexes of essential metals and proteins in general and especially, of calcium binding proteins?

9.1. ALKALI METALS (GROUP Ia)

There is a closer similarity between the elements of group I than in any other group of the periodic table. The alkali metals are strong reducing agents; their

Calcium Binding Proteins, By Eugene A. Permyakov and Robert H. Kretsinger

TABLE 9.1. Elements of Group I, the Alkali Metals

Metal	Symbol	Electron Configuration	Ionic Radius (Å)
Lithium	Li	[He]$2s^1$	0.68
Sodium	Na	[Ne]$3s^1$	0.98
Potassium	K	[Ar]$4s^1$	1.33
Rubidium[a]	Rb	[Kr]$5s^1$	1.48
Cesium[a]	Cs	[Xe]$6s^1$	1.67
Francium	Fr	[Rn]$7s^1$	

[a]Melting points: Rb, 39°C; Cs, 28°C.

reactivities increase from lithium to cesium. They can reduce oxygen, chlorine, ammonia, and hydrogen. The standard electrode potentials all lie between −2.7 and −3.0 V, indicating a strong tendency to form Me^+ cations in solution.

The ionic radii of the alkali metals are all much smaller than the corresponding atomic (van der Waals) radii (Table 9.1). This is because the atom contains one electron in an *s* level relatively far from the nucleus in a new quantum shell; when it is removed to form the ion, the remaining electrons are in levels closer to the nucleus. In addition, the increased, relative charge of the nucleus attracts the electrons toward the nucleus and decreases the size of the ion.

9.1.1. Lithium

Lithium (Li) is the lightest and least abundant metal; it accounts for only 0.0007% of the Earth's crust. Naturally occurring lithium is composed of two stable isotopes ^{6}Li (7.5% natural abundance) and ^{7}Li (92.5%). Seven lithium radioisotopes have been characterized, the most stable being ^{8}Li, with a half-life of 838 ms, and ^{9}Li, with a half-life of 179.3 ms. The five remaining radioisotopes have half-lives shorter than 9.6 ms.

A number of enzymes have been proposed as potential targets of lithium action, including inositol monophosphatase (IMPase), a family of related phosphomonoesterases, and the protein kinase glycogen synthase kinase-3 (GSK-3) (reviewed by Phiel and Klein, 2001). Inhibition of IMPase in the brain should prevent recycling of inositol, thus depleting free inositol that is required for phosphoinositol mediated intracellular signaling cascades. Indeed, extensive preclinical and some clinical studies have provided support for this hypothesis (reviewed by Atack, 2000). These potential targets are widely expressed, require metal ions for catalysis, and are generally inhibited by lithium in an uncompetitive manner, most likely by displacing a divalent cation, supposedly calcium.

Lithium and other mood stabilizers have significant effects on neurons in culture, including morphology, chemotaxis, and cell survival. At least some of these effects are due to inhibition of the enzyme glycogen synthase kinase-3 (GSK-3) (Klein and Melton, 1996). GSK-3 phosphorylates and deactivates glycogen synthase. It is a ubiquitous kinase, found in both neurons and glia, localized to the cytoplasm, nucleus, and mitochondria. Its two isoforms, α and β (51 and 47 kDa), have 97% sequence identity in their catalytic domains and generally have similar biological effects. Their constitutive activities arises from phosporylation of Tyr's 279 or 216 (α and β, respectively).

Long-term treatment with lithium protects primary cultures of rat brain neurons from glutamate-induced, NMDA receptor–mediated excito-toxicity (reviewed by Schou, 2001; Chuang, 2004). Lithium antagonizes glutamate induced activation of c-Jun-N-terminal kinase, p38 kinase, and AP1 binding; this has a major role in cytotoxicity and suppresses glutamate-induced loss of phosphorylated cAMP responsive element binding protein. It also induces the expression of brain derived neurotrophic factor (BDNF) and subsequent activation of TrkB, the receptor for BDNF, in cortical neurons. The activation of BDNF/TrkB signaling is essential for the neuroprotective effects of this drug. In addition, lithium stimulates the proliferation of neuroblasts in primary cultures of central nervous system neurons. Lithium also shows neuroprotective effects in rodent models of diseases.

Li^+/Na^+ exchange in erythrocytes is increased in patients with essential hypertension compared with normotensive controls (reviewed by Semplicini et al., 2003). Moreover, increased Li^+/Na^+ exchange has been found in some clinical conditions related to hypertension, such as overweight and diabetes. Among diabetic patients, Li^+/Na^+ exchange is particularly high in patients with nephropathy, hypertension, and microalbuminuria. Increased Li^+/Na^+ exchange is associated with severe and drug-resistant hypertension, insulin resistance, vascular and cardiac hypertrophy, hyperlipidemia, obesity, family history of hypertension, and of major cardiovascular accidents, suggesting that high Li^+/Na^+ exchange could be a biochemical marker for increased cardiovascular risk.

9.1.2. Rubidium

Rubidium (Rb) is a soft, silvery white metal; it is about the twenty-third most abundant element in the Earth's crust, roughly as abundant as zinc and more common than copper. Rubidium has 24 isotopes, with naturally occurring rubidium being composed of just two isotopes: ^{85}Rb (72.2% natural abundance) and radioactive ^{87}Rb (27.8%). ^{87}Rb has a half-life of 4.88×10^{10} years.

Rubidium is similar to potassium in physical and chemical characteristics but is more reactive. It readily substitutes for potassium in minerals. The Rb^+ ion has a hydrated ionic radius similar to that of K^+ (2.93 vs. 2.76 Å); this causes the biochemical similarity of the ions. Rubidium and cesium have not been found in biological systems and are considered moderately toxic. Rubidium reacts readily with skin moisture to form rubidium hydroxide, which causes chemical burns of eyes and skin.

Radioactive $^{86}Rb^+$ is used for studies of metal ion transport in various systems (Kanayama et al., 2005). Rubidium is also used to study the specificity of monovalent cations in proteins. Asp189 of thrombin defines its specificity for Arg side chains and participates indirectly in the coordination of Na^+ ions (Prasad et al., 2004). Substitution of Asp189 by Ala and Asn abolishes binding of monovalent cations; whereas, the Ser and Glu mutations change the monovalent cation preference from Na^+ to the smaller Li^+ or to the larger cation Rb^+, respectively. The kinetic properties of K^+ and Rb^+ with respect to Na^+/K^+ ATPase are very similar in cardiac and skeletal muscles, nerves, red cells, and other tissues.

Nonradioactive ^{87}Rb is often used for studies of K^+ transport in vivo (reviewed by Kupriyanov and Gruwel, 2005). The ^{87}Rb resonance consists of narrow and broad lines both centered at 150 Hz in water that comprise 40 and 60% of the total area, respectively.

Radioactive ^{86}Rb is used for studies of metal ion transport in various systems. For example, it was used to study axonal transport via the olfactory nerve pathway, from nasal cavity to the olfactory bulb, as well as to the olfactory cortex through the synaptic junctions in mice (Kanayama et al., 2005).

9.1.3. Cesium

Cesium (Cs) is a bright, silvery metal. It reacts explosively with cold water and reacts with ice at temperatures above -116° C. Cesium hydroxide is a strong base and attacks glass. Cesium reacts with the halogens to form a fluoride, chloride, bromide, and iodide. Cesium has at least 39 known isotopes; this is more than any other element except francium. The atomic masses of these isotopes range from 112 to 151. It has only one naturally occurring stable isotope, ^{133}Cs. Most of the other isotopes have half-lives from a few days to fractions of a second.

The nucleus of ^{133}Cs has spin 7/2 and its gyromagnetic ratio is roughly one-eighth that of ^{1}H. The relaxation of $^{133}Cs^+$ is dominated completely by the quadrupolar mechanism. Its nuclear quadrupole moment is very small, however, resulting in long relaxation times and narrow linewidths. $^{133}Cs^+$ and $^{87}Rb^+$ are both much more sensitive in NMR than is ^{39}K, whose NMR receptivity is roughly 100 times less than that of these two analogs. ^{133}Cs NMR is a valuable tool for noninvasively probing biological systems (reviewed by Goodman et al., 2005). One of the most useful properties of the Cs^+ ion in vivo is that its physical properties when it is present inside cells appear to be different from those when it is present outside cells. The intracellular $^{133}Cs^+$ signal is generally up-shifted by 1 to 2 ppm compared with the extracellular signal. Furthermore, extracellular Cs^+ generally has a ^{133}Cs T_1 relaxation time constant that is several times longer than that of intracellular Cs^+. This allows resolution of the magnetic resonance signal of ^{133}Cs in different tissue compartments on the basis of chemical shift and/or magnetic relaxation properties. This compartmental resolution applies not only to the intra- and extracellular spaces, but to subcellular compartments as well. ^{133}Cs NMR has been applied to evaluation of ion transport across membranes and the kinetic/chemical environment of the intracellular space in systems ranging from red blood cells to rat brain. CsCl density gradient centrifugation assumes little interaction with the proteins or with nucleic acids.

Cesium is accumulated by algae, fungi, yeasts, bacteria, cyanobacteria, and rhodococci (reviewed by Avery, 1995; Ivshina et al., 2002). Microbial Cs^+ (K^+) uptake is generally mediated by monovalent cation transport systems located on the plasma membrane. Microorganisms display large differences in their abilities to accumulate Cs^+. It appears to have an equal or greater affinity than K^+ for transport in some microorganisms. Microbial Cs^+ accumulation is markedly influenced by the presence of external cations (e.g., K^+, Na^+, $NH_4{}^+$ and H^+) and is generally accompanied by an approximate stoichiometric exchange for intracellular K^+. However, stimulation of growth of K^+-starved microbial cultures by Cs^+ is limited and it has been proposed that it is not the presence of Cs^+ in cells that inhibits their growth but, rather, the resulting loss of K^+. Most Cs^+ in microorganisms is maintained in soluble form in either the cytoplasm or vacuole (if present). Whereas Cs^+ is readily accumulated via monovalent cation transport systems and can replace intracellular K^+, Cs^+ apparently cannot substitute for K^+ in some or all of its essential biological functions. The precise intracellular target(s) for Cs^+-induced toxicity has yet to be clearly defined, although ribosomes become unstable in the presence of Cs^+, and Cs^+ is known to substitute poorly for K^+ in the activation of many K^+-requiring enzymes.

Environmental cesium accumulates in the intracellular space, primarily through action of the Na^+-K^+ pump (reviewed by Avery, 1995; Ivshina et al., 2002). Radioactive cesium, ^{137}Cs (30-year half-life) and ^{134}Cs (two year half-life), is transported throughout the body. Radioactive cesium is one of the most important artificial radionuclides produced by nuclear fission. Soil contamination with radioactive cesium has a long term impact due to its long half lives and its high biological availability (reviewed by Zhu and Smolders, (2000)).

9.1.4. Francium

Francium (Fr) does not occur in nature to any significant extent. There are 34 known isotopes of francium, ranging in atomic mass from 199 to 232. ^{223}Fr and ^{221}Fr are the only isotopes that occur in nature; the former is far more common. All known isotopes are radioactive and have short half-lives; 22 minutes is the longest. Trace amounts of francium are found in uranium and thorium ores, where ^{223}Fr continually forms and decays.

9.2. ALKALI EARTH METALS (GROUP IIa)

The alkali earth metals (Table 9.2) are all found in the Earth's crust, but not in elemental form since they are so reactive. Instead, they are widely distributed in minerals. Magnesium is found are carnellite, magnesite, and dolomite. Calcium is found in chalk, limestone, marble, gypsum, and anhydrite. Magnesium is the eighth most abundant element in the Earth's crust, and calcium the fifth.

In all of their compounds these metals have an oxidation state of II, and with few exceptions their compounds are ionic. Their electron configurations always have two electrons in an outer shell. These two electrons are relatively easy to remove; this results in the formation of Me^{2+}. However, removing the third electron is much more difficult, as reflected in their ionization energies.

TABLE 9.2. Elements of Group II, the Alkali Earth Metals

Metal	Symbol	Electron Configuration	Ionic Radius (Å)
Beryllium	Be	$[He]2s^2$	0.34
Magnesium	Mg	$[Ne]3s^2$	0.78
Calcium	Ca	$[Ar]4s^2$	0.99
Strontium	Sr	$[Kr]5s^2$	1.27
Barium	Ba	$[Xe]6s^2$	1.43
Radium	Ra	$[Rn]7s^2$	

9.2.1. Beryllium

Beryllium (Be) is a strong, extremely light, high-melting, silver-gray metal with a close-packed hexagonal crystal structure. Beryllium, $[He]2s^2$, has a small ionic radius, 0.34 Å. Its more common compounds, BeO and BeF_2, have partial covalent character, in contrast to magnesium $[Ne]2s^2$ and calcium $[Ar]2s^2$. Beryllium is tetrahedrally coordinated by various oxygen compounds [e.g., $Be_4O(O_2CR)_9$]. It is resistant to corrosion; it is stronger than steel, and because of its low density, about a third that of aluminum, it has found extensive use in industry.

Of beryllium's isotopes, only ^9Be is stable, and the others are relatively unstable or rare. It has a small neutron cross section, which makes it useful in the production of nuclear weapons and in sealed neutron sources. Beryllium is one of the most toxic elements. Coal fired power plants, industrial manufacturing, and nuclear weapons production have released beryllium to the environment (reviewed by T. P. Taylor et al., 2003).

Chronic berylliosis, a granulomatous lung disorder, occurs primarily among workers engaged in beryllium extraction, purification, and preparation of Be metal and its alloys (reviewed by Amicosante and Fontenot, 2006). The presence of circulating beryllium-specific $CD4^+$ T cells correlates directly with the severity of lymphocytic alveolitis (reviewed by Amicosante and Fontenot, 2006). Beryllium is listed as a class A EPA carcinogen (reviewed by Gordon and Bowser, 2003). It is not known how beryllium exerts its toxicity.

BeF_3^- serves as an analog of the (missing) γ-phosphate of GDP or ADP in studies of G proteins, F1-ATPase, nitrogenase, and some kinases (Chabre, 1990; Petsko, 2000). BeF_3^- can be bound to the bacterial response regulator CheY, as observed in its crystal structure (Guhaniyogi et al., 2006).

9.2.2. Strontium

Strontium (Sr) is a soft, silver–yellow, alkaline earth metal. Its physical and chemical properties are similar to those of calcium and barium; it has three allotropic crystal forms. It has 16 unstable isotopes; of greatest importance are ^{89}Sr (half-life of 50.5 days and ^{90}Sr (29.78 years). Strontium has four stable, naturally occurring isotopes: ^{84}Sr (0.56% occurrence), ^{86}Sr (9.86%), ^{87}Sr (7.0%), and ^{88}Sr (82.58%).

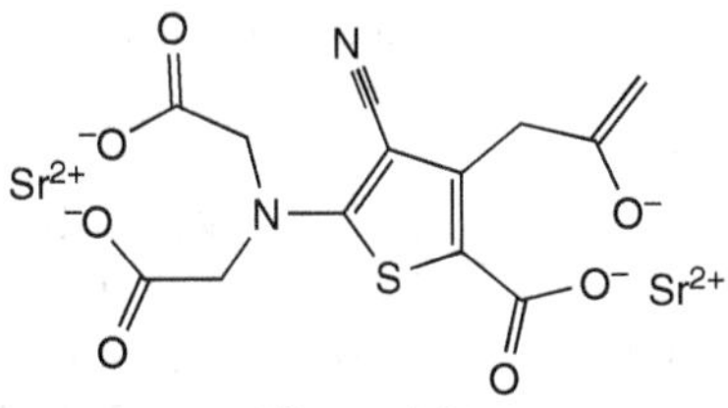

Figure 9.1. Structure of distrontium ranelate.

Strontium has a similar affinity to calcium for calcium binding proteins of the serum (reviewed by Nielsen, 2004). Strontium binds to the EF-hand sites in calmodulin (Lepsík and Field, 2007) and to the calcium binding sites of α-lactalbumin and serum albumin (Sandier et al., 1999).

Contraction of the rat portal vein induced by noradrenaline is lost when calcium in the bathing medium is replaced by strontium, which can enter through Ca^{2+} channels and be released from the sarcoplasmatic reticulum (Bonnevier et al., 2002). In isolated muscle tissue, 2 mM strontium has a calcium like action on hexose transport (Bihler et al., 1986). Calcium and strontium seem to share a common tubular transport path in renal tubules (Walser, 1969). Strontium in suspended renal proximal tubular cells inhibits PTH-dependent cyclic AMP production, as does calcium, at concentrations up to 10 mM (Mathias and Brown, 1991). The renal clearance of strontium is around three times that of calcium, perhaps due to smaller tubular reabsorption, which again might be due to the larger ionic radius of Sr^{2+}.

Administered strontium is deposited almost exclusively in bone. Strontium, sodium, and lead can be substituted in the Ca^{2+} sites of apatite (Chapter 3). The amount of strontium in the skeleton is only 3.5% by weight of its calcium content (reviewed by Nielsen, 2004).

Strontium ranelate (Figure 9.1) augments bone calcium in experimental animals and reduces the fracture rate in osteoporotic patients (reviewed by Nielsen, 2004; Marie, 2005; Simon, 2005). The rather unique action of strontium ranelate as an osteoblast agonist could be explained by the activation of a calcium sensing receptor (Marie et al., 2001). Such receptors have been identified in different cells of the body, including osteoblasts. Strontium ranelate may decrease bone resorption and increase bone formation, resulting in increased bone mass (Marie, 2005). In rat calvaria culture systems and rat osteoblastic cell cultures, strontium ranelate enhances preosteoblastic cell replication and increases collagen synthesis by osteoblasts (Canalis et al., 1996).

9.2.3. Barium

Barium (Ba) is the fourteenth most abundant element in the Earth's crust. Naturally occurring barium is a mix of seven stable isotopes, the most abundant being ^{138}Ba (71.7% natural occurrence). Twenty-two isotopes are known, but most of these are highly radioactive and have half-lives from several milliseconds to several days.

Barium salts are toxic in mammals. They are absorbed rapidly from the gastrointestinal tract and are deposited in the muscles, lungs, and bone. At low doses, barium acts as a muscle stimulant; doses of 0.8 mg in humans affect the nervous system, eventually leading to paralysis and death. When swallowed, barium coats the walls of the digestive tract; this allows the shape of the upper digestive tract to be imaged by x-rays (reviewed by C. J. Martin, 2004). The median dose and area product for barium enemas and meals can be 15.7 Gy $\cdot$ cm^2 and 4.8 Gy $\cdot$ cm^2 (radiation dose, Chapter 6), respectively. A reduction in dose can be achieved using newer x-ray equipment, such as low-dose pulsed fluoroscopy and digital imaging facilities.

Ba^{2+} and K^+ ions have similar radii. Eaton and Brodwick (1980), Armstrong and Taylor (1980), and Shieh et al. (1998), examined the interactions of Ba^{2+} with the permeant K^+ ion in Kir2.1 channels and found that extracellular Ba^{2+} blocking is relieved by intracellular K^+ through competition for the same binding site. Glu125 assists the entry of barium into the channel; ^{141}Thr stabilizes barium in its deep binding site (Alagem et al., 2001). Jiang and MacKinnon (2000) determined the crystal structures of several complexes of the KcsA K^+-channel. A Ba^{2+} ion is bound at a single location within the selectivity filter of the channel. The Ba^{2+} blocking site is on the internal aspect of the "inner ion" position, where an alkali metal cation is found in the absence of the blocking Ba^{2+} ion (Figure 9.2). Magnesium and spermine induce inward rectification by blocking the Kir2.1-channel; however, their binding sites are different.

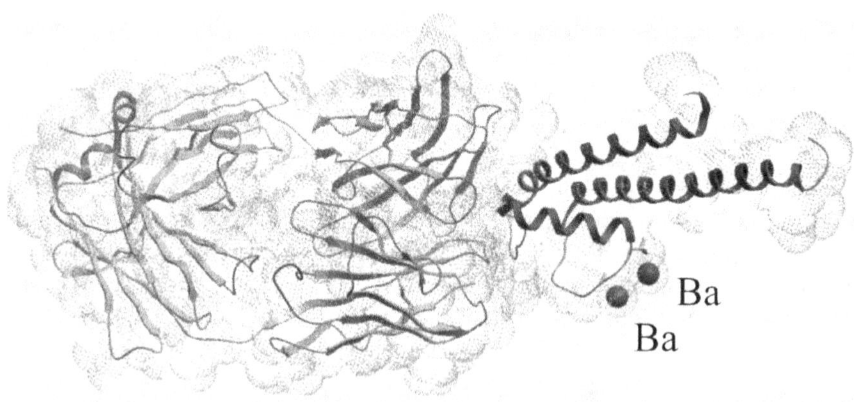

Figure 9.2. Structure of a part of the KcsA K^+-channel with two bound Ba^{2+} ions (PDB file 2ITD). (From Lockless et al., 2007.)

9.2.4. Radium

Radium (Ra) is brilliant white when freshly prepared, but blackens on exposure to air, probably due to formation of the nitride. It exhibits luminescence, as do its salts; it decomposes in water and is more volatile than barium. Radioactive isotopes include ^{222}Ra (half-life of 39.0 s), ^{223}Ra (11.43 days), ^{224}Ra (3.66 days), ^{225}Ra (14.9 days), ^{226}Ra (1600.0 years), and ^{228}Ra (5.76 years). Radium emits α-, β-, and γ-rays and when mixed with beryllium produces neutrons. The decay of ^{223}Ra produces radon, an invisible, odorless, radioactive gas (reviewed by Polpong and Bovornkitti, 1998). Radon emits α-particles and produces several solid radioactive products called *radon daughters*. Radon daughters can be inhaled; their α-radiation increases the risk of lung cancer.

Radium causes osteosarcomas and fibrosarcomas of bone, carcinomas of paranasal sinuses and mastoid air cells, and bone lesions, along with fractures in people with high exposures (reviewed by Stebbings, 2001). Various cancers are correlated with radium; however, multiple myeloma and excess lung cancer may have resulted from high γ-radiation and radon exposures rather than from internal radium.

The accumulation of radionuclides such as radium within the mineral phase of the bone matrix is believed to cause local irradiation only of those proliferating cells close to the bone surface. However, Atkinson et al. (2005) presented evidence for another pathway for the irradiation of target cells, mediated through the sequestration of heavy metal radionuclides by the intracellular iron storage protein ferritin. In vitro studies revealed the transfer of radionuclide from a ^{223}Ra–transferrin complex into immunoprecipitable cytosolic ferritin. In vivo studies confirmed the colocalization of incorporated ^{224}Ra and cellular iron stores.

9.3. GROUP IIIa

The 38 elements in groups 3 through 12 of the periodic table are called *transition metals*. As with all metals, the transition elements are both ductile and malleable and they conduct electricity and heat. Their valence electrons are present in more than one shell. This is the reason that they often exhibit several common oxidation states.

The first of four transition series includes scandium through copper (filling of the $3d$ subshell). Irregularities are observed for chromium and copper, because $3d^5 4s^1$ is more stable than $3d^4 4s^2$ and $3d^{10} 4s^1$ is more stable than $3d^9 4s^2$, respectively.

The second transition series includes yttrium through silver (filling of the $4d$ subshell). Irregularities are observed for niobium, which skips from $4d^2 5s^2$ to $4d^4 5s^1$, and palladium, which goes from $4d^8 5s^2$ to $4d^{10} 5s^1$.

The third includes lanthanum through gold (filling of the $5d$ subshell). Irregularities are observed for platinum, which skips from $5d^9 6s^2$ to $5d^{10} 6s^1$.

The fourth transition series is incomplete and includes actinium to elements 104 through 109 (filling of the $6d$ subshell).

The properties of transition metals are primarily dependent on the electronic configuration of the electrons in the outer shell and in the penultimate outer shell. The transition metals are listed in Table 9.3.

9.3.1. Scandium

Scandium (Sc) is scarce in nature; it is usually found in only two ores. Thortveitite is the primary source of scandium; uranium mill tailings are also an important source. There is one stable, naturally occurring isotope, ^{45}Sc, with a nuclear spin of 7/2. Thirteen radioisotopes have been characterized; the most stable are ^{46}Sc, half-life of 83.8 days; ^{47}Sc, 3.35 days; and ^{48}Sc, 43.7 hours. All of the remaining radioactive isotopes have half-lives of less than 4 hours, and the majority of these have half-lives of less than 2 minutes.

Scandium has no known biological function. Only trace amounts reach the food chain. Scadium is not highly toxic; although extended inhalation can cause lung embolisms.

9.3.2. Yttrium

Yttrium (Y) never occurs in nature as a free element. It is found in almost all rare earth minerals and in uranium ores. Natural yttrium is composed of only one isotope (^{89}Y). The most stable radioisotopes are ^{88}Y, half-life of 109.65 days, and ^{91}Y, 59.51 days. All the other isotopes have half-lives of less than a day except ^{87}Y, which has a half-life of 79.8 hours, and ^{90}Y, 64 hours. ^{90}Y emits β-particles and is used for the treatment of rheumatoid arthritis (reviewed by Heuft-Dorenbosch et al., 2000). The U.S. Food and Drug Administration approved the transarterial administration of ^{90}Y microspheres for liver-directed cancer therapy (reviewed by Murthy et al., 2005; Gulec and Fong, 2007).

9.3.3. Lathanum and Lanthanides

Lanthanides are silvery white, relatively soft metals that tarnish when exposed to air, forming their oxides. Their hardness increases somewhat with higher atomic number. Moving from left to right across the period, the radius of each lanthanide 3+ ion decreases steadily, called *lanthanide contraction*. They are very reactive and have high melting points and boiling points. The lanthanides usually exist as trivalent cations, in which case their electronic configurations are (Xe)$4f^n$, with n varying from 1 (Ce^{3+}) to 14 (Lu^{3+}). The transitions of the f-electrons are responsible for the interesting spectral properties of the lanthanides, such as long-lived luminescence and sharp absorption and emission lines.

In nature, lanthanides never exist as pure elements, only as sparsely distributed minerals (e.g., cerite, monazite and euxenite) or as admixtures in other ores. Lanthanides are termed *rare-earth elements*; nevertheless, they are not rare in nature because their levels in the Earth's crust are often equal to or higher than those of some physiologically significant elements, such as iodine, cobalt, and

TABLE 9.3. Transition metals

Z	Symbol	Neutral Atom	Positive Ion	Ionic Radius (Å)
21	Sc	[Ar]$3d^1 4s^2$	[Ar]$3d^1 4s^1$	0.745
22	Ti	[Ar]$3d^2 4s^2$	[Ar]$3d^2 4s^1$	0.605
23	V	[Ar]$3d^3 4s^2$	[Ar]$3d^4$	0.59
24	Cr	[Ar]$3d^5 4s^1$	[Ar]$3d^5$	0.52
25	Mn	[Ar]$3d^5 4s^2$	[Ar]$3d^5 4s^1$	0.46
26	Fe	[Ar]$3d^6 4s^2$	[Ar]$3d^6 4s^1$	0.645
27	Co	[Ar]$3d^7 4s^2$	[Ar]$3d^8$	0.745
28	Ni	[Ar]$3d^8 4s^2$	[Ar]$3d^9$	0.69
29	Cu	[Ar]$3d^{10} 4s^1$	[Ar]$3d^{10}$	0.73
30	Zn	[Ar]$3d^{10} 4s^2$	[Ar]$3d^{10} 4s^1$	0.74
⋮				
39	Y	[Kr]$4d^1 5s^2$	[Kr]$5s^2$	0.9
40	Zr	[Kr]$4d^2 5s^2$	[Kr]$4d^2 5s^1$	0.72
41	Nb	[Kr]$4d^4 5s^1$	[Kr]$4d^4$	0.69
42	Mo	[Kr]$4d^5 5s^1$	[Kr]$4d^5$	0.65
43	Tc	[Kr]$4d^5 5s^2$	[Kr]$4d^5 5s^1$	0.56
44	Ru	[Kr]$4d^7 5s^1$	[Kr]$4d^7$	0.68
45	Rh	[Kr]$4d^8 5s^1$	[Kr]$4d^8$	0.68
46	Pd	[Kr]$4d^{10}$	[Kr]$4d^9$	0.86
47	Ag	[Kr]$4d^{10} 5s^1$	[Kr]$4d^{10}$	1.26
48	Cd	[Kr]$4d^{10} 5s^2$	[Kr]$4\mathrm{d}^{10} 5s^1$	0.85
⋮				
57	La	[Xe]$5d^1 6s^2$	[Xe]$5d^2$	1.061
58	Ce	[Xe]$4f^1 5d^1 6s^2$	[Xe]$4f^1 5d^2$	1.034
59	Pr	[Xe]$4f^3 6s^2$	[Xe]$4f^3 6s^1$	1.013
60	Nd	[Xe]$4f^4 6s^2$	[Xe]$4f^4 6s^1$	0.995
61	Pm	[Xe]$4f^5 6s^2$	[Xe]$4f^5 6s^1$	0.979
62	Sm	[Xe]$4f^6 6s^2$	[Xe]$4f^6 6s^1$	0.964
63	Eu	[Xe]$4f^7 6s^2$	[Xe]$4f^7 6s^1$	0.947
64	Gd	[Xe]$4f^7 5d^1 6s^2$	[Xe]$4f^7 5d^1 6s^1$	0.938
65	Tb	[Xe]$4f^9 6s^2$	[Xe]$4f^9 6s^1$	0.923
66	Dy	[Xe]$4f^{10} 6s^2$	[Xe]$4f^{10} 6s^1$	0.912
67	Ho	[Xe]$4f^{11} 6s^2$	[Xe]$4f^{11} 6s^1$	0.901
68	Er	[Xe]$4f^{12} 6s^2$	[Xe]$4f^{12} 6s^1$	0.881
69	Tm	[Xe]$4f^{13} 6s^2$	[Xe]$4f^{13} 6s^1$	0.869
70	Yb	[Xe]$4f^{14} 6s^2$	[Xe]$4f^{14} 6s^1$	0.858
71	Lu	[Xe]$4f^{14} 5d^1 6s^2$	[Xe]$4f^{14} 6s^2$	0.848
72	Hf	[Xe]$4f^{14} 5d^2 6s^2$	[Xe]$4f^{14} 5d^1 6s^2$	0.71
73	Ta	[Xe]$4f^{14} 5d^3 6s^2$	[Xe]$4f^{14} 5d^3 6s^1$	0.64
74	W	[Xe]$4f^{14} 5d^4 6s^2$	[Xe]$4f^{14} 5d^4 6s^1$	0.62
75	Re	[Xe]$4f^{14} 5d^5 6s^2$	[Xe]$4f^{14} 5d^5 6s^1$	0.56

(*continued overleaf*)

TABLE 9.3. (*Continued*)

Z	Symbol	Neutral Atom	Positive Ion	Ionic Radius (Å)
76	Os	$[Xe]4f^{14}5d^6 6s^2$	$[Xe]4f^{14}5d^6 6s^1$	0.63
77	Ir	$[Xe]4f^{14}5d^7 6s^2$	$[Xe]4f^{14}5d^7 6s^1$	0.625
78	Pt	$[Xe]4f^{14}5d^9 6s^1$	$[Xe]4f^{14}5d^9$	0.625
79	Au	$[Xe]4f^{14}5d^{10}6s^1$	$[Xe]4f^{14}5d^{10}$	0.85
80	Hg	$[Xe]4f^{14}5d^{10}6s^2$	$[Xe]4f^{14}5d^{10}6s^1$	1.02
89	Ac	$[Rn]6d^1 7s^2$	$[Rn]7s^2$	1.119
90	Th	$[Rn]6d^2 7s^2$	$[Rn]6d^2 7s^1$	1.5
91	Pa	$[Rn]5f^2 6d^1 7s^2$	$[Rn]5f^2 7s^2$	0.78
92	U	$[Rn]5f^3 6d^1 7s^2$	$[Rn]5f^3 7s^2$	0.52

Source: http://environmentalchemistry.com/yogi/periodic/.

selenium. Cerium (68 ppm) and lanthanum (La; 32 ppm) are the most common. Lutetium and thulium are the rarest (about 0.5 ppm), while the concentrations of the remainder range from 1 to 9 ppm. Promethium is an artificial radioactive element with no stable isotopes.

Lanthanides have a very wide range of coordination numbers (generally, 6 to 12, but 2, 3, or 4 are common). Their coordination geometries are determined by ligand steric factors rather than crystal field effects. They form labile "ionic" complexes that undergo facile exchange of ligands. The $4f$ orbitals in the Ln^{3+} ions do not participate directly in bonding, being well shielded by the $5s^2$ and $5p^6$ orbitals. Their spectroscopic and magnetic properties are thus largely uninfluenced by the ligand. They prefer anionic ligands with donor atoms of rather high electronegativity (e.g., O, F). They readily form hydrated complexes (because of the high hydration energy of the small Ln^{3+} ion).

The combination of NMR, pseudocontact shifts induced by a site-specifically bound lanthanide ion, and prior knowledge of the three-dimensional structure of the lanthanide-labeled protein can be used to (1) assign NMR spectral lines, (2) identify the binding modes of low-molecular-weight compounds in a complex with proteins, and (3) characterize binding sites formed at protein–protein interfaces (reviewed by Pintacuda et al., 2007). Dynamic measurements by NMR of paramagnetic lanthanide complexes give information on the degrees of conformational freedom sampled by the various members of the EF-hand proteins, as well as on the time scales of their motions (reviewed by Capozzi et al., 2006).

Gadolinium(III) is the most used contrast agent in magnetic resonance imaging (MRI); however, other lanthanide ions (e.g., europium, dysprosium) and the 2+ oxidation state are also increasingly being investigated as alternatives to gadolinium(III) (reviewed by Bottrill et al., 2006). Many attempts have been made to design macromolecular–gadolinium complexes as blood contrast agents, as macromolecules do not readily diffuse across healthy vascular endothelium (reviewed by Mohs and Lu, 2007).

Lanthanide ions have been shown to affect the function of ligand gated ionic channels and R7G metabotropic receptors. La^{3+} and Gd^{3+} ions competitively block many types of voltage gated calcium channels in cells from many tissues and organs (reviewed by Pałasz and Czekaj, 2000). Gadolinium blocks stretch sensitive ionic channels in the sarcolemma of skeletal muscle fibers. The physiological effects of lanthanides observed at the cellular level result largely from the fact that their ionic radii are similar to those of calcium; this enables them to compete with calcium for the same binding sites. The ions with small radii are the strongest inhibitors. For the larger ions, the channel blocking potential varies in inverse proportion to the ionic radius. The potencies (0.1 to 1.0 μM) of lanthanides to block the voltage gated type T calcium channel are $Ho^{3+} > Yb^{3+} \geq Er^{3+} > Gd^{3+} > Nd^{3+} > Ce^{3+} > La^{3+}$.

The paramagnetism of lanthanide ions offers outstanding opportunities for fast determinations of the three dimensional structures of protein–ligand complexes by NMR spectroscopy. The combination of pseudocontact shifts induced by a site specifically bound lanthanide ion and prior knowledge of the three dimensional structure of the lanthanide labeled protein can be used to achieve (1) rapid assignments of NMR spectra, (2) structure determinations of protein–protein complexes, and (3) identification of the binding mode of low molecular weight compounds in complexes with proteins (reviewed by Pintacuda et al., 2007). Dynamic measurements by NMR, combined with calcium substitution by paramagnetic lanthanides, provide information on the intrinsic amplitude of the conformational degrees of freedom sampled by the various members of the EF-hand superfamily, as well as on the time scales of the motions (reviewed by Capozzi et al., 2006). NMR of lanthanide derivatives helps in capturing long time scale motions.

Gadolinium(III) remains the dominant material for contrast agent design in MRI but other lanthanide ions (e.g., europium, dysprosium) and other oxidation states (e.g., II) are also being investigated increasingly as alternatives to Gd(III) (reviewed by Bottrill et al., 2006). Many attempts have been initiated to design macromolecular Gd(III) complexes as magnetic resonance imaging contrast agents, as macromolecules do not readily diffuse across healthy vascular endothelium and remain intravascular (reviewed by Mohs and Lu, 2007). Although the use of Gd(III) is extremely efficacious in detecting and characterizing pathologic tissue, clinical development of these agents has been limited by concerns of potential toxicity from incomplete Gd(III) clearance. Recent innovative technologies, such as reversible protein binding contrast agents and biodegradable macromolecular contrast agents, may be valuable alternatives that combine the effective imaging characteristics of an intravascular contrast agent and the safety of clinically approved low-molecular-weight Gd(III) chelates.

A group of very closely related radioactive lanthanides and lanthanide, such as radioisotopes, is considered to be almost ideal (reviewed by Cutler et al., 2000). These radioisotopes have similar chemical properties, and all may be conjugated to biomolecules using a single chelator, DOTA (1,4,7,10-tetraazacyclododecane-1,4,7,10-tetraacetic acid) and its chemical analogs. They also provide a wide range of physical characteristics, such as half-lives and energies of β-particles, which can be chosen to match the biological properties of the conjugated biomolecule and the malignant tumor.

Naturally occurring lanthanum is composed of one stable isotope (^{139}La) (99.91% natural abundance) and one radioactive isotope (^{138}La). Thirty-eight radioisotopes have been characterized, with the most stable being ^{138}La, with a half-life of 1.05×10^{11} years, and ^{137}La, with a half-life of 6.0×10^{4} years.

Rats injected intravenously with chlorides of cerium, praseodymium, europium, dysprosium, ytterbium, and lutetium accumulate these compounds in the liver (over 78%), bones, and spleen. Cerium and praseodymium produce liver steatosis, jaundice, and increased levels of serum alanine aminotransferase and aspartate aminotransferase. The toxic effects of gadolinium include mineral deposits in capillaries (mainly in lungs and kidneys), necrosis in liver and spleen, mineralization of gastric mucosa without necrosis, thrombocytopenia, and prolonged prothrombin time (reviewed by Pałasz and Czekaj, 2000).

9.3.4. Actinium and Actinides

Actinides are the series of elements from thorium (90) to lawrencium (103). They all have two outer s-electrons ($7s^2$); increasing proton number corresponds to filling the $5f$ level. Although actinides are similar to lanthanids in that their electrons fill the $5f$ orbitals in order, their chemical properties are not uniform. Trivalent compounds are the most common. Other states exist; thorium, protactinium, uranium, and neptunium can assume the IV or higher oxidation states. Actinium (Ac) has no stable isotopes. Naturally occurring actinium is composed of one radioactive isotope, ^{227}Ac. Thirty-six radioisotopes have been characterized, with the most stable being ^{227}Ac, half-life of 21.772 yeard, ^{225}Ac, 10.0 days, and ^{226}Ac, 29.37 hours. All of the remaining radioactive isotopes have half-lives that are less than 10 hours, and the majority of these have half-lives that are less than one minute.

All the actinide elements are radioactive and very poisonous. Actinides that exist in nature in considerable amounts are thorium, protactinium, and uranium.

Ballangrud et al. (2004) suggested that ^{225}Ac-labeled trastuzumab, the humanized monoclonal antibody, directed against oncogene HER2/*neu*, may be a potent therapeutic agent against metastatic breast cancer cells exhibiting intermediate to high HER2/*neu* expression. ^{225}Ac, an α-emitter, is used to eradicate breast cancer metastases expressing variable levels of HER2/*neu*.

Interaction of the actinide cations in biologically active sites is only partially understood. Serum transferrin is reported to bind a wide variety of d-block transition metals, as well as actinides and lanthanides (Harris et al., 1981; Raymond et al., 1982). Furthermore, D. M. Taylor (1998) has suggested that the protein is able to stabilize the tetravalent state and forms stable $(M^{4+})_2$–transferrin complexes. For this reason, transferrin contamination by actinide cations is a critical issue of nuclear human toxicology.

Uranium Uranium (U) is a naturally occurring element with no stable isotopes. All uranium is radioactive and hence vanishing by radioactive decay, yet it is found in great quantity in the Earth's crust. The natural isotopes are ^{234}U, ^{235}U, and ^{238}U. In addition, other useful isotopes, such as ^{232}U, have been produced in mass quantity in breeder reactors. Uranium in small amounts exists in many rocks and soils.

There is an appreciable retention of incorporated uranium in skeleton, kidneys, and liver. In long term exposure, uranium accumulates mostly in the bone and kidneys (Leggett and Pellmar, 2003). The uranyl ion (UO_2^{2+}) is assumed to exchange with Ca^{2+} ions at the surfaces of bone mineral crystals but not to participate in crystal formation. The current biokinetic model of the International Commission on Radiological Protection suggests three compartments for uranium in human bone: bone surface, exchangeable bone volume, and nonexchangeable bone volume.

Biosorption utilizes various natural materials of biological origin, including bacteria, fungi, yeast, and algae, for removal of heavy metals (Pollmann et al., 2006). They can effectively sequester dissolved metal ions out of dilute, complex solutions quickly and completely. Among the promising biosorbents for heavy metal removal, *Saccharomyces cerevisiae* has received increasing attention, due to its unique nature for metal uptake compared with other fungi (reviewed by Wang and Chen, 2006). The biosorptive capacity of uranium by *S. cerevisiae* is usually between 150 and 300 mg U g^{-1} dry weight biomass. *Desulfovibrio vulgaris* can reduce the soluble oxidized form of uranium, U(VI), to insoluble U(IV) (Lovley et al., 1992, 1993). Using this reductive precipitation, one can effectively remove uranium from contaminated water. The list of bacteria known to reduce U(VI) is growing, yet a complete understanding of the biochemistry of this process in any one bacterium is lacking (reviewed by Wall and Krumholz, 2006).

Cell isolates of *Bacillus sphaericus JG-A12* from a uranium mining waste pile in Germany are able to accumulate high amounts of toxic metals such as U, Cu, Pb, Al, and Cd as well as precious metals. Some of these metals [i.e., U, Cu, Pd(II), Pt(II) and Au(III)] are also bound by the highly ordered paracrystalline proteinaceous surface layer (S-layer) that envelopes the cells of this strain. These special capabilities of the cells and the S-layer proteins of *B. sphaericus* JG-12 are valuable for the cleanup of uranium contaminated wastewaters, for the recovery of precious metals from electronic wastes, and for the production of metal nanoclusters. The S-layer of *B. sphaericus* JG-12 is able to bind up to 20 mg U g^{-1} protein.

The dramatic decrease in solubility accompanying the reduction of U(VI) to U(IV), producing the insoluble mineral uraninite, has been viewed as a potential mechanism for sequestration of environmental uranium contamination. Reductases specific for U(VI) have not been identified, and mutations affecting single genes do not eliminate U(VI) reduction; this suggests that multiple low redox potential electron carriers in a single bacterium may be involved in reducing uranium. It is possible that cells transfer a single electron to U(VI) and that U(IV) is generated through disproportionation.

Hainfeld (1992) described a method to deliver ^{235}U to tumors in which the isotope would then be fissioned by incident neutrons, producing localized lethal radiation sufficient for therapy. Apoferritin was loaded with an average of ~400^{235}U atoms per molecule. Fab antibody fragments were attached covalently to the uranium–ferritin; the immunoreactivity of the conjugate was 92% of that for antibody alone. Such biouranium constructs should provide significant advantages over boronated antibodies to meet the requirements for clinical neutron capture therapy.

Neptunium Most of the neptunium (Np) that is retained in the body deposits primarily in the bones; some is also retained in the liver. No health effects specific to exposure from neptunium have been observed in humans. Neptunium (IV) is especially interesting for its relative stability in physiological conditions and similarities with Pu(IV), Th(IV), and Fe(III). Combined x-ray absorption spectroscopy (XAS) and near-infrared absorption spectrometry (Chapter 6) were used to characterize a complex between Np(IV) and transferrin and the synergistic nitrilotriacetic acid anion (Llorens et al., 2005). The average Np(IV)–transferrin bond distance is 2.42 Å and is similar to the Ce(IV)–transferrin distance of 2.46 Å.

Plutonium ^{239}Pu emits α-particles, which do not penetrate the skin but can irradiate internal organs when plutonium is inhaled or ingested. It is released into the atmosphere primarily by atmospheric testing of nuclear weapons and by accidents at weapon production sites. Pu(IV), as well as most of the transition metal cations reported to be complexed by transferrin, are assumed to be located in iron sites (Harris, 1998).

9.4. GROUP IVa

9.4.1. Titanium

Titanuim (Ti) is the ninth most abundant element in the Earth's crust (0.63% by mass) and is present, bound with other elements, in most igneous rocks and in sediments derived from them. There are five naturally occurring stable isotopes: ^{46}Ti, ^{47}Ti, ^{48}Ti, ^{49}Ti, and ^{50}Ti, with ^{48}Ti being the most abundant (73.8%). Eleven radioisotopes have been characterized; the most stable are ^{44}Ti (half-life of 63 years), ^{45}Ti (184.8 minutes), ^{51}Ti (5.76 minutes), and ^{52}Ti (1.7 minutes). All of the remaining radioactive isotopes have half-lives of less than 33 seconds; the majority of these have half-lives of less than $\frac{1}{2}$ second. Titanium has no known biological function.

Ti(IV) is a potent inhibitor of trypsin but not of chymotrypsin (Duffy et al., 1998). The binding of titanium to trypsin involves a free carboxyl group at the bottom of the substrate binding pocket, resulting in a five-coordinate geometry of

$TiO(SO_4)(H_2O)$ (Duffy et al., 1998). Direct binding of octahedral or tetrahedral metal ion complexes is prevented by the inability of the enzyme active site of the enzyme to promote formation of a five coordinate transition state of the metal ion required for reaction.

Five coordinate titanyl sulfate completely inhibits the growth of *Escherichia coli, Salmonella typhimurium*, and *Pseudomonas aeruginosa*. Titanium is not poisonous; it is contained in some antitumor drugs (e.g., titanocene dichloride and budotitane) (reviewed by Meléndez, 2002; Caruso and Rossi, 2004). They do not show common side effects, such as emesis, alopecia, or bone marrow impairment of widely used cytostatic agents.

Titanium is used to make hip and knee replacements, pacemakers, bone plates, cranial plates, and dental implants (reviewed by Tengvall and Lundström, 1992). In aqueous environments, TiO_2 has low ion formation tendencies and low reactivity with macromolecules. Titanium does not facilitate reactive oxygen radical generation during inflammatory conditions as observed in vitro. The outermost layers of TiO_2 are in the Ti(IV) state, although according to electron spin resonance, Ti(III) is formed under atmospheric conditions.

9.4.2. Zirconium

Zirconium (Zr) is a very strong, malleable, ductile, lustrous, silver-gray metal. Its chemical and physical properties are similar to those of titanium. Naturally occurring zirconium is composed of four stable isotopes and one extremely long-lived radioisotope, ^{96}Zr, which decays via double beta decay with a half-life of 2.0×10^{19} years; it can also undergo single β-decay, which is not yet observed, but the theoretically predicted value of $T_{1/2}$ is 2.4×10^{20} years. The second most stable radioisotope is ^{93}Zr, which has a half-life of 1.53×10^6 years. Eighteen other radioisotopes have been observed. Most of these have half-lives of less than a day except for ^{95}Zr (64.0 days), ^{88}Zr (63.4 days), and ^{89}Zr (79.4 hours).

Zirconium and its salts generally have low toxicity (reviewed by Ghosh et al., 1992). Zr(IV) oxychloride and zirconium oxalate increase the thermal stability of collagen fibers by 8 to 10°C, supposedly by cross-linking (Nishad Fathima et al., 2003).

9.4.3. Hafnium

Hafnium has a very high neutron capture cross section; several isotopes of hafnium nuclei can absorb multiple neutrons. At least 34 isotopes of hafnium have been observed, ranging in mass number from 153 to 189. The five stable isotopes are in the range 176 to 180. The radioactive isotopes' half-lives range from only 400 ms for ^{153}Hf to 10^{15} years for the most stable, ^{174}Hf.

Hafnium has no known toxicity. Hafnium binds to both of the two specific binding sites of ovotransferrin in two slightly different configurations (Schwab et al., 1992). ^{181}Hf was used for time differential perturbed angular correlation (TDPAC) NMR study of serum transferrin (Then et al., 1986).

9.5. GROUP Va

9.5.1. Niobium

Niobium has physical and chemical properties similar to those of tantalum, and the two are therefore difficult to distinguish. Naturally occurring niobium consists of one stable isotope, ^{93}Nb. At least 32 radioisotopes have also been synthesized, ranging in atomic mass from 81 to 113. The most stable of these is ^{92}Nb, with a half-life of 34.7 million years. One of the least stable is ^{113}Nb, with an estimated half-life of 30 ms. Niobium and its compounds cause eye and skin irritation; however, there are no reports of toxicity.

9.5.2. Tantalum

Natural tantalum (Ta) consists of two isotopes: ^{180}Ta (0.012%) and ^{181}Ta (99.988%). ^{181}Ta is a stable isotope. It causes eye and skin irritation. Tantalum is widely used in protheses, especially in orthopedics; therefore, it potential interactions with various proteins, cells, and tissues have been studied.

9.6. GROUP VIIa

9.6.1. Technetium

The common oxidation states of technetium (Tc) are VII, V, and IV. Under oxidizing conditions, technetium(VII) exists as the pertechnetate ion, TcO_4^-. There are 22 reported isotopes of technetium, with masses ranging from 90 to 111. Technetium, as well as promethium, has no stable isotopes; all of its isotopes are radioactive. Technetium has three long-lived radioactive isotopes: ^{97}Tc (half-life of 2.6×10^6 years), ^{98}Tc (4.2×10^6 years), and ^{99}Tc (2.1×10^5 years). The most useful isotope of technetium is ^{99}Tc (9.01 hours); it is used in many medical tests because its half-life is short and it binds to many biologically active molecules. A technetium–somatostatin analog complex (^{99}Tc depreotide) has been used for the rapid, convenient, accurate, and cost-effective characterization of lung nodules with conventional gamma camera systems (Blum et al., 2002).

9.6.2. Rhenium

Rhenium (Re) is one of the rarest elements in the Earth's crust, with an average concentration of 1 ppb. Naturally occurring rhenium is 37.4% ^{185}Re, which is stable, and 62.6% ^{187}Re, which is unstable but has a very long half-life ($\sim$1010 years); that lifetime is affected by the charge state of rhenium atom. ^{188}Re is widely used for attachment to tumor-specific monoclonal antibodies for radioimmunotherapy (reviewed by Iznaga-Escobar, 1998; Jeong and Chung, 2003).

Rhenium interacts with His in proteins. Laser flash-quench methods have been used to generate Tyr and Trp radicals in rhenium modified *P. aeruginosa* azurins

(Miller et al., 2004). Electron tunneling from Cu(I) to Re(II) in Re(H107) azurin occurs in the microsecond range. This is much faster than Cu(I) to Ru(III) tunneling in Ru(H107) azurin, suggesting that a multistep mechanism might be involved. Although a Y108 radical can be generated by flash quenching a Re(H107)Me(II) (Me = Cu, Zn) protein, it is probably not an active intermediate in enhanced Cu(I) oxidation. Rather, the likely explanation is rapid conversion of Re(II)(H107) to deprotonated Re(I)(H107 radical), followed by electron tunneling from Cu(I) to the hole in the imidazole ligand.

9.7. GROUP VIII

9.7.1. Ruthenium

Ruthenium (Ru), together with rhodium, palladium, osmium, iridium, and platinum, are called the *platinum group metals*. Naturally occurring ruthenium is composed of seven stable isotopes. Additionally, 34 radioactive isotopes have been discovered. Of these radioisotopes, the most stable are ^{106}Ru with a half-life of 373.59 days; ^{103}Ru, 39.26 days; and ^{97}Ru, 2.9 days. Fifteen other radioisotopes have been characterized, with atomic weights ranging from 89.93 (^{90}Ru) to 114.93 (^{115}Ru).

Ruthenium is regarded as highly toxic and carcinogenic. The elemental form is seldom encountered. Ruthenium, rhodium, and platinum are widely used in antitumor complexes. Thousands of organic, inorganic octahedral, Ru(II) and Rh(III), and square planar, especially Pt(II), compounds have been developed as potential anticancer and diagnostic agents (reviewed by Wheate et al., 2007). The design and synthesis of new drugs is focused on bis-intercalators, which have two intercalating groups linked via a variety of ligands, and synergistic drugs, which combine the anticancer properties of intercalation with other functionalities, such as covalent binding or boron cages for radiation therapy. The development of metal based antitumor drugs has been stimulated partially by the clinical success of *cis*-diamminedichloroplatinum(II) (cisplatin) and its analogs and by the clinical trials of other platinum and ruthenium complexes with activity against resistant tumors and with reduced toxicity, including orally available platinum drugs (reviewed by Brabec, 2002).

9.7.2. Osmium

Osmium (Os) has seven naturally occurring isotopes, six of which are stable: ^{184}Os, ^{187}Os, ^{188}Os, ^{189}Os, ^{190}Os, and (most abundant) ^{192}Os. ^{186}Os undergoes α-decay with such a long half-life [$(2.0 \pm 1.1)\ 10^{15}$ years] that for practical purposes it can be considered stable. Osmium tetroxide, OsO_4, is a powerful highly toxic oxidizer. Concentrations in air as low as 10^{-7} g m^{-3} can cause lung congestion, skin damage, and severe eye damage (reviewed by Makarovsky et al., 2007).

9.7.3. Rhodium

Naturally occurring rhodium (Rh) is composed of only one isotope, ^{103}Rh. The most stable radioisotopes are ^{101}Rh with a half-life of 3.3 years; ^{102}Rh, 207 days; ^{102m}Rh, 2.9 years; and ^{99}Rh, 19.1 days.

There are few reports of humans being affected by rhodium; nevertheless, it should be regarded as highly toxic and carcinogenic. Various divalent rhodium complexes $Rh_2(L)_4$ (L denotes acetate, propionate, butyrate, trifluoroacetate, and trifluoroacetamidate) have been found to bind to nondefatted human serum albumin at molar ratios of about 8:1 (Espósito et al., 1999).

The interaction between $Rh(CO)_2$ (Rh acetylacetonate 1,5-cyclooctadiene) and serum albumin was characterized by means of matrix assisted laser desorption ionization time-of-flight mass spectrometry (MALDI-TOF MS), high-performance liquid chromatography/mass spectrometry (HPLC-MS), and scanning electron microscopy (Crobu et al., 2006). This biocatalyst consists of a tetrameric human serum albumin that binds up to 89 $Rh(CO)_2{}^+$ units.

9.7.4. Iridium

There are two natural isotopes of iridium (Ir) and many radioisotopes, the most stable radioisotope being ^{192}Ir (half-life of 73.83 days). ^{192}Ir β-decays into ^{192}Pt, whereas most of the other radioisotopes decay into osmium.

Some complexes of iridium possess antitumor activity (reviewed by Haiduc and Silvestru, 1989). The effects of square planar iridium (Ir acetylacetonate 1,5-cyclooctadiene) and rhodium (Rh acetylacetonate 1,5-cyclooctadiene) complexes and of octahedral ruthenium [*cis*-$RuCl_2$ $(DMSO)_4$; DMSO = dimethyl sulfoxide] have been compared with *cis*-dichlorodiamminoplatinum(II) (*cis*-PDD) (Sava et al., 1983). All the complexes tested similarly inhibit the growth of subcutaneous Lewis lung carcinoma and the development of spontaneous as well as of artificial metastases, with the exception of Ir acetylacetonate 1,5-cyclooctadiene, which is inactive on metastases. Radioactive ^{192}Ir is often used in interstitial radiation implants (brachytherapy) (reviewed by Brandt and Harney, 1989).

9.7.5. Palladium

Naturally occurring palladium (Pd) is composed of six isotopes. The most stable radioisotopes are ^{107}Pd (half-life of 9.5 million years; ^{103}Pd, 17 days; and ^{100}Pd, 3.63 days.

All palladium compounds are regarded as highly toxic and as carcinogenic (reviewed by Kielhorn et al., 2002). People with known nickel allergy may be especially susceptible. Palladium(II) complexes are used in cancer therapy (reviewed by Caires, 2007). Palladium(II) and platinum(II) complexes can be used for selective cleavage of peptides and proteins (Milović et al., 2003). In this procedure, the use of *cis*-$[Pt(en)(H_2O)_2]^{2+}$ is followed by the use of $[Pd(H_2O)_4]^{2+}$. In the peptide AcAla–Lys–Tyr–Gly–Gly–Met–Ala–Ala–Arg–Ala, the Pt(II) reagent cleaves the Met6–Ala7 peptide bond; whereas, the Pd(II) reagent cleaves the Gly4–Gly5 bond All cleavage reactions are regioselective and complete at pH 2.0 and 60°C. Each metal ion binds selectively to an anchoring side chain and then, as a Lewis acid, activates a proximal peptide bond toward hydrolysis by the solvent water. Both Pt(II) and Pd(II) reagents bind to the Met side chain; whereas, only the Pd(II) reagent binds to the His side chain under reaction conditions. Consequently, only Met's direct the cleavage by the Pt(II) reagent; whereas, both Met and His direct the cleavage by the Pd(II) reagent. Deliberately incomplete His-directed cleavage at many sites by the Pd(II) reagent produced numerous short fragments, suitable for protein identification by mass spectrometry.

9.7.6. Platinum

There are six naturally occurring isotopes: ^{194}Pt, which accounts for 33%, ^{195}Pt (34%), and ^{196}Pt (25%), ^{198}Pt (7%), ^{192}Pt (1%), and ^{190}Pt (0.01%). The latter is weakly radioactive, with a half-life of 7×10^{11} years.

Platinum containing antitumor drugs such as *cis*-diamminedichloroplatinum(II) (cisplatin) are widely used for the treatment of testicular, ovarian, and other cancers (reviewed by Bose, 2002). Cisplatin arrests the cell cycle at the G2 phase. The anticancer properties of these compounds are due to their covalent binding to DNA. Some of these platinum complexes with antitumor activity can interact weakly with serum albumin (Aleksenko et al., 2007) and with hemoglobin (Mandal et al., 2004).

9.8. GROUP Ib

9.8.1. Silver

Naturally occurring silver (Ag) is composed of two stable isotopes, ^{107}Ag (51.84% natural abundance) and ^{109}Ag. Twenty-eight radioisotopes have been characterized, the most stable being ^{105}Ag (half-life of 41.29 days), ^{111}Ag (7.45 days), and ^{112}Ag (3.13 hours). Soluble silver salts, especially $AgNO_3$, are lethal in doses of 2 g.

Silver has been used as an antimicrobial substance for thousands of years (reviewed by Silvestry-Rodriguez et al., 2007). It is used in topical application, in polymers of medical devices, and in water filter cartridges (reviewed by Atiyeh et al., 2007). Most bacterial resistances to toxic heavy metals result not from chemical detoxification, but from energy dependent ion efflux from the cell by ATPases or as chemiosmotic cation/proton antiporters. Silver resistance conferred by the *Salmonella* plasmid pMGH100 involves nine genes in three transcription units (reviewed by Silver, 2003). SilE is a small periplasmic protein that binds Ag^{+} ions specifically at the cell surface. When SilE is loaded with Ag(I), Cu(II), or Cd(II) and dialyzed, five Ag^{+} ions, but no copper or cadmium, are bound. SilE contains 10 His's; in contrast to many other metal binding proteins, such as metallothionein, SilE has no Cys. Binding of silver changes the structure of SilE from disordered to predominantly α-helical. Proton NMR spectroscopy demonstrates the specific binding of Ag(I) to the ten histidine imidazole N atoms by following perturbation of the proximal C2 and C4 carbon protons. There are four $HisX_6His$ motifs; the two imidazoles are inferred to be on the same surface of an α-helix, separated by two turns of the helix, and bridged by an Ag^+ ion (Gupta et al., 1999; reviewed by Silver, 2003). Silver compounds can be slowly absorbed by body tissues, causing a bluish or blackish skin pigmentation (argiria).

9.8.2. Gold

Gold (Au) has only one stable isotope, ^{197}Au, which is also its only naturally occurring isotope. Thirty-six radioisotopes have been synthesized, ranging in atomic mass from 169 to 205.

There is increasing documentation of allergic contact dermatitis and other effects from gold jewelry, gold dental restorations, and gold implants (reviewed by Eisler, 2004). Au(I) and Au(III) complexes form strong covalent attachments to targets. Au(III) is isoelectronic with Pt(II)-d^8 and usually forms square planar complexes. The square planar geometry of Pt(II) is important for its anticancer activity; Au(III) compounds can also be used and have the added advantage of decreased toxicity (reviewed by Kostova, 2006). In contrast to cisplatin, gold complexes target proteins but not DNA (reviewed by Marcon et al., 2003; Desoize, 2004; Talib et al., 2006; Milacic et al., 2008). Treatment of nude mice bearing human breast tumors with a Au(III) dithiocarbamate complex results in significant inhibition of tumor growth, associated with proteasome inhibition and massive apoptosis. To develop a novel clonable contrasting agent for electron microscopy, Mercogliano and DeRosier (2006, 2007) fused maltose binding protein (MBP) and metallothionein. This chimera can bind gold in proportion to the number of copies of metallothionein, with 10 to 20 gold atoms per metallothionein.

9.9. MERCURY (GROUP IIb)

There are seven stable isotopes of mercury (Hg), ^{202}Hg being the most abundant (29.86%). The longest-lived radioisotopes are ^{194}Hg (half-life of 444 years) and ^{203}Hg (49.612 days).

Mercury compounds, both organic and inorganic, are widely distributed and highly toxic to most organisms (Zalups, 2000). Genes encoding proteins of mercury resistance are widely distributed in eubacteria (reviewed by Barkay et al., 2003). The regulator, MerR, and the major detoxification enzyme, MerA, are each composed of discrete modules observed in paralogs. MerA reduces reactive ionic Hg(II) to volatile, relatively inert, monoatomic Hg(0) vapor. It is a cytosolic flavin disulfide oxidoreductase (homodimer of 120 kDa), which uses NADPH as a reductant (reviewed by Barkay et al., 2003). MerA has a flexible amino terminal domain, which is homologous to MerP, a small periplasmic mercury binding protein. MerT is a 116-residue membrane-bound protein involved in the uptake of Hg(II). It is predicted to have three transmembrane helices, the first of which has a Cys pair, which would lie within the first hydrophobic helix and perhaps be accessible from the periplasmic side. The second Cys pair is predicted to lie on the cytoplasmic face of the inner membrane between the second and third transmembrane helices. The Hg^{2+} ion may be transferred from the N-terminal, proximal Cys pair to form a di-coordinate protein complex and then be transferred to cytoplasmic, low-molecular-mass thiols and/or directly to MerA.

The genes encoding MerA and MerT are in a single operon under the control of MerR, a metal responsive regulator. Many HgR loci encode an additional enzyme, MerB, that degrades organomercurials by protonolysis, and one or more additional proteins apparently involved in transport.

The 72 amino acid periplasmic form of MerP is processed from a 91 residue precursor by removal of a typical Sec-type signal sequence. It functions as a monomer and binds a single Hg^{2+} ion via its two Cys's, at positions 14 and 17, which are key elements in defining a metal binding motif GMTCxxC found singly and as multiple repeats at the N-termini of P-type ATPases involved in the influx of beneficial transition metal cations or efflux of deleterious transition metal cations in prokaryotes and eukaryotes. The K_d value of fully reduced MerP with an Hg^{2+} ion is 2.7×10^{-5} M, and the K_d value for mercury bonding to thiolate anions is on the order of 10^{-15} to 10^{-20} M (reviewed by Rooney, 2007). In contrast, the affinity of mercury for oxygen or nitrogen ligands is about ten orders of magnitude lower.

9.10. GROUP IIIb

Groups 13 to 16 (the *p*-block) of the periodic table of elements include some metals, some metalloids, and some nonmetals. These groups are mixed since the transition from metal to nonmetal is gradual. The *p*-block elements have electron configuration s^2p^1 to s^2p^9. The *p*-block elements have the most diverse properties of all the other blocks of elements. *p*-Block elements may be solids, liquids, or gases. The reactivity of the *p*-block elements ranges from very reactive to very unreactive. In all these elements, whereas the *s*-orbitals are filled completely, their *p*-orbitals are incomplete. These are filled progressively by the addition of one electron from group 13 (ns^2np^1) to group 17 (ns^2np^5). In group 18 (ns^2np^6) both *s*- and *p*-orbitals are completely filled. Metals and metalloids of groups 13 to 16 are listed in Table 9.4.

9.10.1. Aluminum

Aluminium (Al) has nine isotopes, whose mass numbers range from 23 to 30. Only ^{27}Al (stable isotope) and ^{26}Al (radioactive isotope, half-life = 7.2×10^5 years) occur naturally. Aluminum, mostly as bauxite, comprises 9.8 mass % of the Earth's crust. Al^{3+}, as a "hard" trivalent metal ion, binds strongly to oxygen donor ligands such as citrate and phosphate (reviewed by Harris et al., 1996).

TABLE 9.4. Metals and Metalloids of the Groups 13 to 16

Element	Symbol	Electron Configuration	Ionic Radius (Å)
Aluminum	Al	$[Ne]3s^2p^1$	1.43
Gallium	Ga	$[Ar]3d^{10}4s^2p^1$	1.35
Indium	In	$[Kr]4d^{10}5s^2p^1$	1.67
Thallium	Tl	$[Xe]4f^{14}5d^{10}6s^2p^1$	1.70
Germanium	Ge	$[Ar]3d^{10}4s^2p^2$	1.22
Tin	Sn	$[Kr]4d^{10}5s^2p^2$	1.41
Lead	Pb	$[Xe]4f^{14}5d^{10}6s^2p^2$	1.46
Antimony	Sb	$[Kr]4d^{10}5s^2p^3$	1.40
Bismuth	Bi	$[Xe]4f^{14}5d^{10}6s^2p^3$	1.50
Polonium	Po	$[Xe]4f^{14}5d^{10}6s^2p^4$	1.68

Source: http://environmentalchemistry.com/yogi/periodic/.

The aqueous coordination chemistry of aluminum is complicated by the tendency of many Al complexes to hydrolyze and form polynuclear species, many of which are insoluble. Aluminum, $[Ne]3s^23p$, has a single oxide, alumina, Al_2O_3, but many polymorphs and hydrates. AlF_4^- is often leached from minerals by acid rain; it inhibits many enzymes. Aluminum compounds are moderately toxic. Aluminum in human serum is bound to the N-lobe site of transferrin (reviewed by Nagaoka and Maitani, 2005). It inhibits remodeling of bone, slowing both osteoblast and osteoclast activities and producing osteomalacia and adynamic bone disease and may play a role in Alzheimer's disease (reviewed by Jeffery et al., 1996; Yokel, 2000; Domingo, 2006; Savory et al., 2006). Parathyroid hormone levels are disrupted by aluminum. In hematopoietic tissue, aluminum causes a microcytic anemia, not reversible by iron.

9.10.2. Gallium

Gallium (Ga) has two naturally occurring isotopes, ^{69}Ga and ^{71}Ga. About a dozen radioactive isotopes are known. Gallium is trivalent ($3d^{10}$) in aqueous solution. The free hydrated ion, Ga^{3+}, hydrolyzes nearly completely at neutral pH, forming highly insoluble, amorphous $Ga(OH)_3$. The solution and coordination chemistries of Ga^{3+} are somewhat similar to those of Al^{3+} and In^{3+} and are very similar to those of Fe^{3+}.

Gallium inhibits the proliferation of some infectious microorganisms, including those causing syphilis, trypanosomiasis, and tuberculosis. This activity probably stems from its ability to enter microbes through their iron transport mechanisms, to disrupt their iron metabolism, and to interfere with synthesis of DNA and proteins (reviewed by Bernstein, 1998). A major exception appears to be the antiresorptive and anabolic effects on bone of gallium and its accumulation in remodeling bone.

Nearly all gallium in plasma is tightly bound to transferrin (Clausen et al., 1974; Vallabhajosula et al., 1980). Each domain of transferrin can independently bind one Fe^{3+} (or Ga^{3+}) ion, together with one carbonate or bicarbonate anion. Transferrin transports its bound metal into cells via the transferrin receptor, which binds most strongly to diferric transferrin, less strongly to monoferric transferrin, and weakly to apo-transferrin. The complex of metal-bearing transferrin and transferrin receptor is taken into the cell by endocytosis; the endosome is then acidified to release the metal, and the transferrin and its receptor are recycled. The interaction of gallium-loaded transferrin with the transferrin receptor occurs in a single very fast step with $K_d = 1.10 \pm 0.12$ μM and a second-order rate constant $k_d = (1.15 \pm 0.3) \times 10^{10}$ $M^{-1}s^{-1}$ (Chikh et al., 2007). Ga^{3+} ions can bind to the two metal sites of transferrin with binding constants log $K_1 = 20.3$ and log $K_2 = 19.3$ at normal plasma bicarbonate concentrations; these values are close to the iron–transferrin binding constants log $K_1 = 22.8$ and log $K_2 = 21.5$ under the same conditions (Harris and Pecoraro, 1983). Although the affinity of transferrin for the Fe^{3+} ion is approximately 400 times higher than for Ga^{3+}, the replacement of Ga^{3+} by Fe^{3+} is found to proceed very slowly, with an exchange half-life of 4.3 hours at 310 K (evaluated by ^{1}H NMR by Kubal et al., 1983).

Gallium binds even more avidly to the related protein lactoferrin, which can remove gallium from transferrin (W. R. Harris, 1986). Like transferrin, lactoferrin is a two-lobed protein ($M_r \sim 80,000$), which can bind two Fe^{3+} (or Ga^{3+}) ions (Levay and Viljoen, 1995) with extremely high affinity, K_d values of $10^{-21.4}$ and $10^{-20.6}$ M, 100 times more tightly than Ga-transferrin (Harris, 1986).

Gallium binds to a third iron binding protein, ferritin, a very large (440,000) nearly spherical protein that can hold as many as 4500 Fe^{3+} ions, situated mainly in a hydrated ferric oxide–hydroxide core (Crichton and Charloteaux-Wauters, 1987). Ferritin is used for iron storage in most cells and is particularly concentrated in the Kupffer cells of the liver. The transfer of gallium from transferrin or lactoferrin to ferritin appears to be mediated by adenosine triphosphate, pyrophosphate, and other phosphate containing compounds.

The Ga^{3+} ion can substitute for Fe^{3+} in some of the many iron-containing enzymes, such as ribonucleotide reductase (Narasimhan et al., 1992). Gallium is not, however, able to enter Fe(II) bearing proteins such as hemoglobin and cytochromes (Logan et al., 1981).

Acute exposure to $GaCl_3$ can cause throat irritation, difficulty breathing, and chest pain; its fumes can cause pulmonary edema and partial paralysis. Gallium has shown efficacy in the treatment of several apparently diverse disorders: (1) accelerated bone resorption, with or without elevated plasma calcium, (2) autoimmune disease and allograft rejection, (3) certain cancers, and (4) infectious diseases (reviewed by Bernstein, 1998). Gallium has antimicrobial activity; it can induce apoptosis and it inhibits proliferation of some infectious microorganisms, including those causing syphilis, trypanosomiasis, and tuberculosis. It is likely that the antimicrobial activity stems from its ability to enter microbes through their iron transport mechanisms, to disrupt their iron metabolism, and to interfere with their DNA and protein synthesis.

The dominant mechanism underlying most of the diverse activities of gallium is its ability to act as a chemically irreducible Fe(III) analog in a wide variety of systems (reviewed by Bernstein, 1998). A major exception appears to be the mechanism for gallium's accumulation in remodeling bone.

Special emphasis has been given to the development of peptides labeled with positron emitters. ^{68}Ga has a half-life of 68 minutes and decays by 89% through positron emission; it can be generated from an in-house generator (reviewed by Mäcke et al., 2005). Radiopeptides for ^{68}Ga labeling have been developed and tested for the targeting of somatostatin receptors, the melanocortin 1 receptor, and the bombesin receptor. The parent element, ^{68}Ge, is accelerator produced and decays with a half-life of 270.8 days by electron capture.

9.10.3. Indium

Indium (In) has two natural isotopes, ^{113}In and ^{115}In. ^{115}In is radioactive, with a half-life of 4.4×10^{14} years. A number of artificial radioactive isotopes of indium also exist.

The chemical behavior of indium resembles lead and silver in some situations and potassium and cesium in others. Despite the fact that group 13 metal ions (Al^{3+}, Ga^{3+}, In^{3+}, and $Tl^{+/3+}$) play no biological role, they are NMR-active nuclides that can be used in NMR spectroscopy of biologically relevant systems (reviewed by André and Mäcke, 2003). These metal ions are quadrupolar (with the exception of thallium); therefore, they are especially sensitive to ligand type and coordination geometry. The linewidths of the NMR signals of their complexes depend strongly on the symmetry of coordination.

Indium binds to transferrin (Zhang et al., 2004). The radioactive ^{111}In–annexin A5 conjugate is used for molecular imaging; ^{111}In has the advantage of having a half-life longer than that of ^{99m}Tc, also used for this purpose (reviewed by Virgolini et al., 2002; Boersma et al., 2005). Apoptotic cells externalize phosphatidylserine; annexin A5 binds to phosphatidylserine selectively and with high affinity. The high expression of somatostatin receptors (SSTR) on various tumor cells has provided the molecular basis for successful use of ^{111}In octreotide (somatostatin) analogs as tumor tracers in nuclear medicine (reviewed by Virgolini et al., 2002).

9.10.4. Thallium

Thallium (Tl) has 25 isotopes whose atomic masses range from 184 to 210.^{203}Tl and ^{205}Tl are the only stable isotopes; ^{204}Tl is the most stable radioisotope, with a half-life of 3.78 years. Thallium occurs in natural waters as the Tl^{+} ion; it is readily absorbed and highly toxic (reviewed by Peter and Viraraghavan, 2005). It binds to sulfhydryl groups of proteins, as do lead and gold.

Villeret et al. (1995) reported crystal structures of fructose-1,6-bisphosphatase complexed with K^+, with Tl^+, and with both Tl^+ and Li^+ (in which the enzyme is complexed with the substrate analog 2,5-anhydro-D-glucitol 1,6-bisphosphate). They concluded that Tl^+ or K^+ ions can substitute for Arg276 in the active site and polarize the 1-phosphate group, thus facilitating nucleophilic attack on the phosphorus center.

9.11. GROUP IVb

9.11.1. Germanium

Germanium (Ge) has five naturally occurring isotopes: ^{70}Ge, ^{72}Ge, ^{73}Ge, ^{74}Ge, and ^{76}Ge. ^{74}Ge is the most common, natural abundance of approximately 36%, ^{76}Ge is the least common, ~7%. Of these five, ^{76}Ge is very slightly radioactive, decaying by double β-decay with a half-life of 1.58×10^{21} years.

Germanium toxicity in humans is low (reviewed by Gerber and Léonard, 1997). Rubredoxin is a small electron transfer protein bearing a tetra-Cys-coordinated Fe(II)–Fe(III) redox couple. Ga^{3+}, as well as Zn^{2+} ions, can substitute into the rubredoxin active site to form a structure isomorphous with the Fe(II)–Fe(III) state. LeMaster et al. (2006) found that Ge^{4+} can also be exchanged into the rubredoxin active site to form the first reported Ge(IV) macromolecular complex. The tetracysteine metal coordination site of the rubredoxins from *Clostridium pasteurianum* and *Pyrococcus furiosus* are shown to bind the inorganic Ge^{4+} ion.

9.11.2. Tin

Tin (Sn) has the greatest number, 10, of stable isotopes; these include all those with atomic masses between 112 and 124, with the exception of 113, 121, and 123. The most abundant are ^{120}Sn (~32%), ^{118}Sn, and ^{116}Sn; the least abundant is ^{115}Sn. Metallic tin and inorganic compounds of tin are not very toxic. In contrast, many organotin compounds are used as biocides (reviewed by Rüdel, 2003).

9.11.3. Lead

Lead (Pb) has four stable isotopes, ^{204}Pb, ^{206}Pb, ^{207}Pb, and ^{208}Pb. ^{204}Pb is primordial on earth; ^{206}Pb, ^{207}Pb, and ^{208}Pb are formed from decay of uranium and thorium. The one common, radiogenic isotope, ^{202}Pb, has a half-life of approximately 53,000 years. Lead is one of four metals that have the most damaging effects on human health. It can enter the human body through the uptake of food, water, and air. The extensive use of leaded gasoline and other sources of lead has resulted in dramatically elevated blood lead levels.

Lead compounds are widely distributed and highly toxic—hence, the emphasis in reducing their industrial and commercial uses. Diets low in iron result in increased absorption of cadmium and lead, suggesting common mechanisms of Cd and Pb transport, probably involving divalent metal transporter 1 (DMT1) (reviewed by Ballatory, 2002; Bressler et al., 2004). ZntA and CadA are bacterial P-type ATPases that confers resistance to zinc, cadmium, and lead (reviewed by B. P. Rosen, 2002). P-type ATPases are a superfamily of enzymes that transport cations into or out of cells or intracellular compartments. They have common features: a conserved ATP binding domain, an Asp that is the site of phosphoenzyme formation, and a phosphatase domain. The soft metal ATPases have conserved features not present in other P-type ATPases, in particular, characteristic Cys- or His-rich metal binding motifs at the N-terminus and a Cys–Pro–Cys (or His) sequence in the sixth membrane-spanning helix. ZntA has the preference Pb(II) > Cd(II) ~ Zn(II) ~ Hg(II) (Sharma et al., 2000). Lead interacts with many other proteins. It is strongly inferred that its toxicity is based on its demonstrated interactions with many proteins in vitro; however, only a few of the specific interactions that are most harmful have yet to be identified (reviewed by Goering, 1993).

The intracellular concentration of lead appears to be determined largely by complexation with a group of low molecular weight proteins rich in Asp and Glu (reviewed by Fowler, 1998). These proteins are similar in many species, including vertebrates; the brain protein appears to be different from that found in the kidney. These proteins have a K_d value for lead of $\sim 10^{-8}$ M; they usually bind zinc normally. A lead binding protein of rat kidney is $\alpha_{2\mu}$-globulin, a member of the retinol binding protein family (Fowler and DuVal, 1991).

One of the best documented targets for lead is the second enzyme in the heme biosynthetic pathway, aminolevulinic acid dehydratase (ALAD) or porphobilinogen synthase (reviewed by Magyar et al., 2005). Inhibition of ALAD by lead is probably a major contributor to the anemia observed in persons with high levels of lead in the blood. ALAD contains an unusual Zn^{2+} ion, Cys_3 active site. X-ray absorption spectroscopy reveals that lead is three-coordinate with sulfur in ALAD (reviewed by Magyar et al., 2005). This indicates a trigonal pyramidal lead–S_3 coordination. In most small-molecule complexes, the coordination number of Pb is 5, 6, or 9. Lead avoids four-coordination in sulfur-rich sites, binding instead as trigonal pyramidal lead–S_3 or as lead–S_{5-9}. In structural zinc binding protein sites, lead binds in a three-coordinate mode, and in a geometry that is fundamentally different from the natural coordination of zinc in these sites, might explain some of the toxicity of lead. Shimoni-Livney et al. (1998) concluded that "all Pb(IV) structures in the CSD (Cambridge Structural Database) have a holodirected coordination geometry. Pb(II) compounds are hemidirected [all ligands are in one hemisphere] for low coordination numbers (2–5) and holodirected for high coordination numbers (9, 10), but for intermediate coordination numbers (6–8), examples of either type of stereochemistry are found."

Zawia et al. (2000) found that zinc fingers of transcription factors could be potential targets for perturbation by lead. Lead interferes with the DNA binding properties of transcription factors Sp1 and Egr-1, both in vivo and in vitro. Lead might also interfere with the DNA binding of a recombinant human Sp1 protein. The effects of lead on the DNA binding of the zinc finger protein transcription factor IIIA (TFIIIA) have been demonstrated. Alterations in its DNA binding correspond to changes in the expression of its target genes. The action of lead on Sp1, Egr-1, and TFIIIA suggests that it can also bind to other zinc finger proteins. Thus, by specifically targeting zinc finger proteins, Pb is able to produce multiple responses through its action on a common site that is present in enzymes, channels, and receptors.

The developing brain has been shown to be especially susceptible to the neurotoxic effects of lead (Reddy et al., 2007). Lead exposure significantly decreases the specific activity of acetylcholinesterase and increases the levels of acetylcholine in the synaptosomal fractions of cerebellum, hippocampus, and cerebral cortex.

9.12. GROUP Vb

9.12.1. Antimony

There are two naturally occurring isotopes of antimony (Sb), ^{121}Sb and ^{123}Sb. About 20 radioactive isotopes are also known. Antimony is a semimetallic chemical element that can exist in two forms. The metallic form is bright, silvery, hard, and brittle; the nonmetallic form is a gray powder.

Antimony is a potentially toxic trace element with no known biological function. The uptake of antimony and arsenic into *E. coli* is facilitated by the aqua-glyceroporin channel, GlpF. Transport of antimony is catalyzed by the ArsB carrier protein. Everted membrane vesicles accumulated Sb(III) with energy supplied by NADH oxidation; this mimics efflux from intact cells (Meng et al., 2004). ArsB is the most widespread determinant of arsenic resistance in bacteria and archaea. It is a member of the ion transporter superfamily, with 12 membrane spanning segments and a membrane topology that is similar to that of many other carrier proteins. ArsB is an antiporter that catalyzes metalloid–proton exchange. As(III) inhibits ArsB-mediated Sb(III) uptake; whereas, Sb(III) stimulates ArsB-mediated As(III) transport. It was proposed that the actual substrate of ArsB is a polymer of $(AsO)_n$, $(SbO)_n$, or a copolymer of the two metalloids (Meng et al., 2004).

The arsRDABC operon of *E. coli* plasmid R773 encodes the ArsAB ATPase, a metalloid pump that confers resistance by actively extruding Sb(III) and As(III) from cells (reviewed by Ruan et al., 2006). ArsA has two homologous halves, A1 and A2, connected by a short linker. Each half has a consensus nucleotide binding domain (NBD). In the presence or absence of the pump substrate, Sb(III) or As(III), both NBD1 and NBD2 hydrolyze ATP, with steady-state hydrolysis dominated by the activity of NBD1. The two NBDs are located at the interface between A1 and A2, in close proximity to each other. Ruan et al. (2006) showed that ArsA binds a single Sb^{3+} ion with high affinity only in the presence of Mg^{2+}-nucleotide. Mutations of Cys113 and Cys422 eliminate antimony binding to purified ArsA. C113A/C422A ArsA has basal ATPase activity similar to that of the wild type but lacks metalloid-stimulated activity. Cells expressing the mutant ArsA C113A/C422A pump can extrude metalloid with higher efficiency than ArsB alone, exhibiting intermediate resistance between cells with wild type ArsAB and cells with only ArsB. These results indicate that the basal activity of the ArsAB pump is sufficient for ATP-driven efflux of metalloid and that the MBD is an allosteric site, with metalloid binding increasing pump activity.

9.12.2. Bismuth

While bismuth (Bi) was traditionally regarded as the element with the heaviest stable isotope, ^{209}Bi, it had long been suspected to be unstable on theoretical grounds. This was finally demonstrated in 2003 when researchers at the Institut d'Astrophysique Spatiale in Orsay, France, measured the α-emission half-life of ^{209}Bi to be 1.9×10^{19} years.

Bismuth and its salts are one of the less toxic of the heavy metals. Bismuth salts are used for the treatment of various gastrointestinal disorders, including gastric and duodenal ulcers, dyspepsia, diarrhea, colitis, and *Helicobacter pylori* infections. Eight proteins from *H. pylori* are significantly up- or down-regulated in the presence of colloidal bismuth subcitrate (Ge et al., 2007). Using affinity chromatography with immobilized bismuth, the authors isolated and subsequently identified seven bismuth binding proteins from *H. pylori* cell extracts. The intracellular levels of four of these proteins (HspA, HspB, NapA, and TsaA) are influenced by the addition of colloidal bismuth subcitrate; this strongly suggests that they interact directly with bismuth. The other bismuth-interacting proteins identified are fumarase, the urease subunit, UreB, and a translational factor (Ef-Tu).

H. pylori infection causes an increase in circulating gastrin via a variety of mechanisms. Progastrin is produced by G-cells located within the gastric antrum and is processed to shorter peptides, such as Gly extended gastrin (Ggly) and amidated gastrin (Gamide). Ggly and progastrin induce proliferation and migration of various cell lines in vitro and proliferation of the colonic mucosa in vivo. Gastrins bind two Fe^{3+} ions with high affinity; they are essential for the biological activity of Ggly (Baldwin et al., 2001). Fluorescence and NMR spectroscopic data are consistent with the hypothesis that Bi^{3+} ions compete with Fe^{3+} for the Ggly binding sites (Pannequin et al., 2004).

Urease converts urea into ammonia and carbonic acid and is essential for colonization of the acidic environment of the stomach by *H. pylori*. Zhang et al. (2006) showed that three bismuth complexes inhibit urease. Bi(EDTA) and $Bi(Cys)_3$ are competitive inhibitors of jack bean urease, with K_i values of 1.74 and 1.84 mM, respectively; the antiulcer drug ranitidine bismuth citrate (RBC) is a noncompetitive inhibitor with a K_i value of 1.17 mM. The active C319A mutant of *Klebsiella aerogenes* displays a significantly reduced sensitivity toward inactivation by Bi(EDTA) compared with the wild-type enzyme, consistent with the binding of Bi^{3+} to the active site Cys319 as the mechanism of enzyme inactivation.

The uptake of Bi(III) by apo-lactoferrin is rapid and almost equal in both lobes (Zhang et al., 2001). The presence of ATP facilitates the release of Bi(III) from the Bi_2-lactoferrin complex when the pH is lowered. The Bi_2-lactoferrin complex blocks uptake of the radiolabeled ^{59}Fe-lactoferrin complex into rat IEC-6 cells. Apo-lactoferrin (but not apo-transferrin) is almost as effective as bismuth-loaded lactoferrin in blocking ^{59}Fe uptake. These results suggest that Bi(III)-loaded lactoferrin might be recognized by the lactoferrin receptor and be taken up into cells. Miquel et al. (2004) characterized the formation of the complex between transferrin and bismuth.

Bi(III) binds to *Staphylococcus aureus* pI258 CadC, a metal sensor protein that regulates expression of the cad operon, which encodes metal ion resistance proteins, and regulates the cad operator/promoter (O/P) binding negatively (Busenlehner et al., 2002). Bi(III) also binds strongly to metallothionein (MT) with a stoichiometry bismuth/metallothionein = 7 : 1 (Bi_7MT) and can readily displace Zn^{2+} and Cd^{2+} ions (Sun et al., 1999).

Sun et al. (1999) showed that bismuth binds strongly to metallothionein (MT) with a stoichiometry bismuth/metallothionein = 7 : 1 (Bi_7MT) and can readily displace zinc and cadmium. Even in strongly acidic solutions (pH 1) bismuth is still bound to the protein. Reactions of bismuth citrate with metallothionein are faster than those of $[Bi(EDTA)]_2$, and both exhibit biphasic kinetics. The EXAFS spectrum of Bi_7MT is very similar to that for the glutathione and *N*-acetyl-L-cysteine complexes $[Bi(GS)_3]$ and $[Bi(NAC)_3]$ with an inner coordination sphere of three sulfur atoms and average Bi–S distances of 2.55 Å. Some sites appear to contain additional short Bi–O bonds of 2.2 Å and longer Bi–S bonds of 3.1 Å. The Bi^{3+} ion sites in Bi_7MT are therefore highly distorted in compared with those of Zn^{2+} and Cd^{2+}.

9.13. POLONIUM (GROUP VIb)

Polonium (Po) has 27 known isotopes, all of which are radioactive. They have atomic masses ranging from 194 to 219.^{210}Po (half-life of 139.38 days) is the most widely available and is an α-emitter. ^{209}Po (half-life $\sim 10^3$ years) and ^{208}Po (half-life 2.9 years) can be made through the α, proton, or deuteron bombardment of lead or bismuth in a cyclotron. Polonium is found in tobacco smoke and its α-emission may be partially responsible for its carcinogenicity (reviewed by El-Aziza et al., 2005; Harrison et al., 2007).

9.14. CONCLUSIONS AND GENERALIZATIONS

All metals, essential or not, in high concentrations are toxic to most organisms. The responses of various organisms, humans being the best studied, consist of:

1. A transport or storage protein (e.g., Na^+, K^+ pump or transferrin) normally produced by the organism, perhaps to protect it from an excess of an essential metal, also serves to bind or buffer the nonessential metal.

2. No response. In its evolutionary history the organism was never challenged with the trace element, and/or if challenged, evolved no response.

Some specific proteins have been identified binding, in vitro or in vivo, a nonessential metal: for example, MerR for mercury and PbrR for lead. Researchers have yet to identify a single specific target associated with the pathological effects of a nonessential metal. Most responses to excess metals seem to be highly pleotropic.

Many metals, essential or not, are deposited in bone and/or teeth. It is assumed, but generally not established, that these metals, (II) valence or otherwise, substitute for a Ca^{2+} ion in hydroxyapatite, just as the F^- ion substitutes for OH^- in $Ca_5(PO_4)_3OH$. Further, it is not known whether these metal substitions occur directly by diffusion into the already formed apatite crystal or whether they must pass through the osteoblast and its poorly understood process of mineral formation.

We emphasize the tentative definition of "essential" metal. What is essential for one organism may not be for another. With over 10^7 different organisms in the biosphere, one is reluctant to call any metal absolutely nonessential. Further, the characterization of essential is conditional in at least two senses: essential only under special environmental conditions and/or essential only if another metal is unavailable.

10

PARVALBUMIN

Chapter 7 covered the evolution, classification, and domain patterns of EF-hand proteins. From this perspective, parvalbumin has a minor role. It is not a member of one of the larger subfamilies. By sequence analysis its CD domain clusters with odd EF-hands and its EF domains cluster with even. The AB domain more closely resembles the even grouping; apparently, parvalbumin evolved from a four-domain precursor and the first domain was subsequently deleted. Nor does it appear to be directly involved in cell signaling; rather, it functions primarily as a temporal buffer, or shaper, of the calcium pulse.

However, parvalbumin was the first EF-hand protein to have its crystal structure and its amino acid sequence determined. It has a canonical pair of EF-hands and it has been studied extensively (reviewed by Permyakov, 2006). Many of the techniques used to study calcium binding proteins in general and EF-hands in particular have been refined in investigations of parvalbumin. For these reasons it has become a paradigm; this entire chapter is devoted to its description.

10.1. STRUCTURE

10.1.1. Secondary and Tertiary Structures

The crystal structure of carp parvalbumin, pI 4.25, was determined by Nockolds et al. (1972) and refined at atomic resolution (Kretsinger and Nockolds, 1973; Moews and Kretsinger, 1975). Crystal and/or NMR structures of pike, hake,

Calcium Binding Proteins, By Eugene A. Permyakov and Robert H. Kretsinger

whiting, carp, shark, rat, and human parvalbumins are available for comparison (Ahmed et al., 1990). The crystal structure of pike (component pI 4.10) has been determined at 0.91 Å resolution (Declercq et al., 1991, 1999).

Parvalbumin contains six α-helices, A to F (Figure 10.1), which comprise 52 of its 108 residues. Its hydrophobic core is composed of the side chains seven Phe's, five Leu's, four Ile's, and three Val's. The surface has polar and charged groups exposed to solvent. The six helices are connected by loops, two of which, between helices C and D and between E and F, form calcium binding sites. Helix C, calcium binding loop CD, and helix D are related to helix E, the EF-loop, and helix F by an intramolecular approximately twofold axis.

One hundred fourteen ordered water molecules are seen in the crystal structure of silver hake major parvalbumin (Richardson et al., 2000). There are 50 water molecules within hydrogen bonding distance (<3.2 Å) of the main chain carbonyl oxygens, and 17 water molecules in contact with backbone amides.

The solution structure of a mutant (F102W) carp parvalbumin was determined using multidimensional NMR (Moncrieffe et al., 2000). It is very similar to the crystal structure of the wild type. The hydrophobic core is intact in the mutant protein, and Trp102 is buried in the hydrophobic core near Phe30, Phe47, Phe70, Phe85, and Phe66.

The structures of calcium binding proteins, just like those of other proteins, are subject to fast fluctuations. ^{13}C NMR data on structural dynamics show that

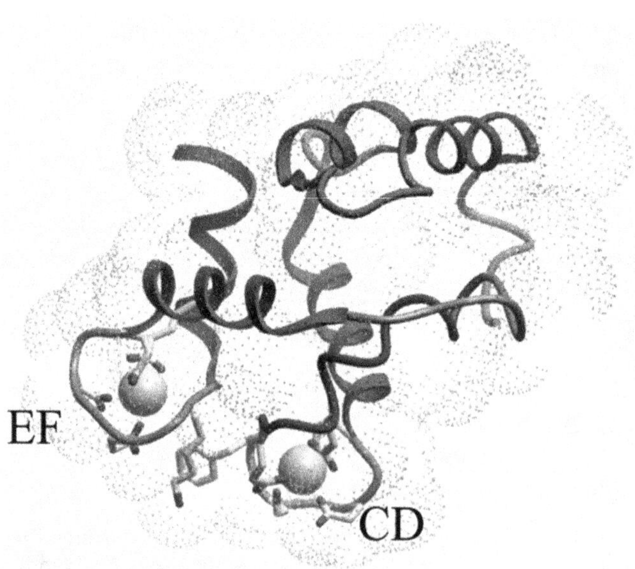

Figure 10.1. Structure of calci-parvalbumin pI 4.25 from carp (PDB file 4CPV). The EF-loop is to the left, the CD-loop to the right. Parvalbumin is viewed perpendicular to the approximate twofold axis relating the two EF-hands. Helices A and B are viewed side-on at the top of the figure. (From Kumar et al., 1990.)

α-carbons in parvalbumin are fixed rather rigidly, while the side chains possess considerable mobility on the nanosecond time scale (D. J. Nelson et al., 1976; Opella et al., 1974; Jardetzky, 1978). ^{15}N nuclear relaxation showed the extreme rigidity of the EF-hand domains and of the linker segment between them in rat parvalbumin (Baldellon et al., 1998). Both N- and C-termini of the protein are also restricted in their mobility; while the residues at relative position 2 in the calcium binding sites have enhanced mobility. Computer simulations of the molecular dynamics of parvalbumin in solution show changes in structure relative to that in the crystal; however, the helices remain largely intact. The average lifetime of hydrogen bonds in helices is several picoseconds. Parvalbumin is more rigid than most other EF-hand proteins; this is consistent with its not being directly involved in information transduction.

A characteristic feature of parvalbumin is a polar hydrogen bond between invariant Arg75 and Glu81. This strong dipole is isolated from solvent by the N-terminal segment of parvalbumin. Chemical modification of the single Arg75 in pike parvalbumin with 1,2-cyclohexandione causes a disorganization of a part of the α-helical regions and a pronounced decrease in affinity for calcium (Gosselin-Ray et al., 1973).

Parvalbumins are divided into two groups: α and β. The C-terminal helix in α-parvalbumin is one residue longer and pI < 5: pI(α) > 5, pI(β) < 5. Cys18 and Asp61 are typical for β-parvalbumins.

10.1.2. Calcium Binding

Parvalbumin binds two Ca^{2+} ions; both are seven-coordinate pentagonal bipyramid. Six amino acids provide ligands; the (almost) invariant Glu at 12(−Z) coordinates calcium with both oxygens of its carboxlyate. The six residues can be assigned to the vertices of an octahedron; the residues 3(+Y) and 9(−Y) correspond to the axis of the pentagonal bipyramid. The residue positions in the canonical EF-pattern are: 1(+X), 3(+Y), 5(+Z), 7(−X), 9(−Y), and 12(−Z) (Figure 10.2).

	1(+X)	3(+Y)	5(+Z)	7(−Y) (C=O)	9(−X)	12(−Z) (bident)
CD	D51	D53	S55	*F57*	E59	E62
EF	D90	D92	D94	*K96*	G98 H_2O	E101

E101D substitution at the EF-loop position 12(−Z) results in monodentate calcium coordination by Asp and dramatically reduces the affinity for calcium. Molecular dynamic simulations demonstrate that Asp is capable of attaining a suitable orientation for bidentate coordination, and only the inherent rigidity of the loop prevents bidentate coordination in the parvalbumin E101D mutant (Cates et al., 1999, 2002).

Substitution for Gly98 of Asp, Arg, Cys, or Ser into position 9(−X, "gateway" position) of the EF-loop reduces both calcium affinity and stability in silver hake parvalbumin (Elkins et al., 2001; Fahie et al., 2002). Energy calculations show

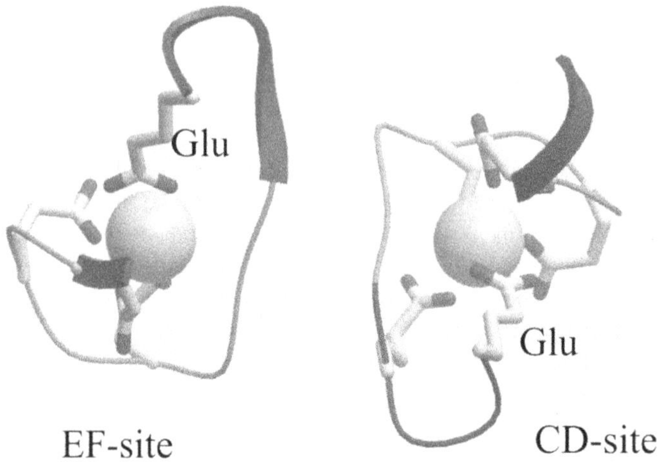

Figure 10.2. CD and EF calcium binding sites in carp parvalbumin pI 4.25 (PDB file 4CPV). Both Glu's at positions 12 (−Z) in the loops coordinate as bidentate ligands. (From Kumar et al., 1990.)

that it is easier to remove the Ca^{2+} ion with its coordinating water from the EF site than to remove the Ca^{2+} ion from the CD site after twisting around the Glu59 (Lockhart and Gray, 1987). Calcium binds sequentially to parvalbumin: P ↔ PCa ↔ CaPCa; the CD site is filled first (reviewed by Permyakov, 2006).

Even though parvalbumin is relatively rigid, the binding of calcium causes small changes over the entire molecule. Computer simulations showed that a new structure is obtained within approximately 60 ps after calcium binding or removal. In general, apo-parvalbumin becomes less compact, and the accessibility of many of its groups to solvent increases with calcium binding (reviewed by Permyakov, 1985, 1993a, 2006).

Permyakov et al. (1987a,b) used stopped-flow fluorescence to measure kinetics of calcium and magnesium dissociation from cod parvalbumin. In these experiments, metal-loaded protein was mixed with EDTA. All the dissociation kinetic curves for this parvalbumin measured in the temperature range from 10 to 30°C were found to be best fit by a sum of two exponential terms, which were identified as arising from the dissociation of calcium or of magnesium from the two binding sites of the protein. The calcium association rate constants for parvalbumin in the temperature range from 10 to 30°C are within the range 10^7 to 10^9 M^{-1} s^{-1} (i.e., they are close to the diffusion-controlled limit). Magnesium association rate constants for parvalbumin are several orders of magnitude lower. At the same time, the dissociation rate constants for calcium and magnesium are close to each other. The differences in calcium and magnesium binding constants are due to the differences in association rates.

True apo-parvalbumins (in the absence of both divalent and monovalent metal cations) of pike (both α- and β-isoforms, pI 5.0 and 4.2, respectively) are intrinsically disordered (Permyakov et al., 2008). Differential scanning calorimetric measurements of apo-parvalbumin revealed the absence of any thermally induced transitions with measurable denaturation enthalpy along with elevated specific heat capacity, implying the lack of rigid tertiary structure and exposure of hydrophobic protein groups to the solvent. Theoretical estimation of energetics of the charge–charge interactions in the parvalbumins indicated their pronounced destabilization upon calcium removal; this is consistent with sequence-based predictions of disordered regions (Permyakov et al., 2008). Overall, the theoretical and experimental data show that apo-parvalbumin is disordered and flexible.

Parvalbumin binds sodium and potassium (reviewed by Permyakov, 2006); hence, one should not use these counterions in studies of apo-parvalbumin. Na- or K-loaded forms of parvalbumin undergo a thermal transition at 30 to 40°C as measured by microcalorimetry, while apo-parvalbumin lacks a first-order thermal transition (Permyakov et al., 2008). Many works on supposedly "apo"-parvalbumins were carried out in millimolar disodium EGTA or EDTA (i.e., not on true apo forms).

A conformational rearrangement occurs upon calcium–magnesium exchange in the C-terminal EF-hand of parvalbumin (Declercq et al., 1991) (Figure 10.3). Glu101, with the relative position 12(−Z), is a bidentate ligand in the calci form but monodentate in the magnesi form; hence, the coordination number decreases from seven to six. The position of the cation within the loop is unchanged. A simple rotation of the Glu101 side chain around its C^{α}–C^{β} bond (χ_1 dihedral angle) allows this change of coordination number. In the CD site the homologous residue, Glu62, displays behavior similar to that of Glu101 upon calcium–magnesium exchange, based on NMR and FTIR experiments (Blancuzzi et al., 1993; Nara et al., 1994). Glu12 (−Z) is highly conserved in all EF-hands; it is replaced only by Asp and in only 8% of all those loops that bind calcium.

An additional, secondary calcium binding site in pike parvalbumin, pI 4.10, is formed by seven oxygens from Asp53, Glu59, and Asp61 and three water molecules and is located near the CD site (Declercq et al., 1988).

Most of the calcium dissociation constants (determined by fluorescence in the absence of any competing metal ions) in Table 10.1 are within the range 10^{-10} to 10^{-7} M^{-1}. Other methods give similar values of calcium association constants. The magnesium dissociation constants are mostly within the range 10^{-6} to 10^{-4} M^{-1}. The ratios between pK_d(Ca) and pK_d(Mg) are 10^{-5} to 10^{-3}.

Different isoforms of parvalbumin from the same animal may have very different binding constants. This is seen clearly in the case of carp parvalbumins, pI 4.25, 4.47, and 3.95, and pike parvalbumins, pI 5.0 and 4.2. These results suggest that in order to understand regulation of calcium flows in the sarcoplasm, one should know the exact cation affinities of different parvalbumins as well as their concentrations. The idea of a "mean" parvalbumin may be misleading. The selective advantage accruing to an animal with these different parvalbumins is not understood.

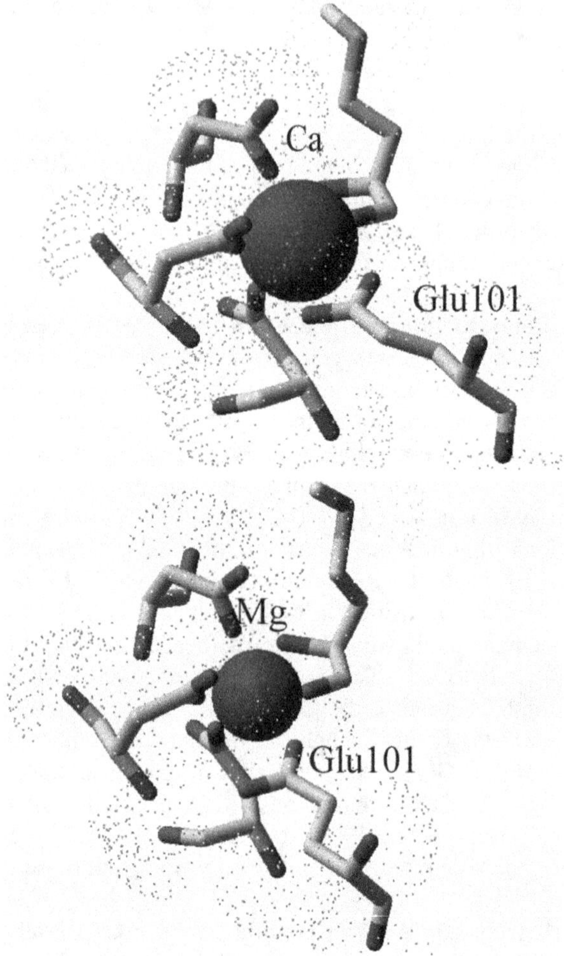

Figure 10.3. Comparison of the coordination of the Ca^{2+} and Mg^{2+} ions in the EF site of pike parvalbumin. The Glu, 12 (−Z), is bidentate coordinating calcium and monodentate with magnesium. (From Declercq et al., 1991.)

The CD and EF sites in most α-parvalbumins are similar. In β-parvalbumins, the calcium affinity of the CD site is lower, as is the case for pike parvalbumin pI 4.2 (Eberspach et al., 1988) and for toad parvalbumin. As seen in ^{15}N NMR relaxation data (Henzl et al., 2002), rat α-parvalbumin displays a higher value of the order parameter in helical regions. In contrast, rat β-parvalbumin has a higher-order parameter in the loop regions; one is located in the AB domain, in the vicinity of Pro21 and Pro26; another occurs in the CD-binding loop between residues 57 and 62. Interestingly, the sodium bound form has an average order parameter greater than that of the calci form.

TABLE 10.1. Equilibrium Parameters of Ca^{2+}, Mg^{2+}, Na^{+}, and K^{+} Binding to Parvalbumins[a]

	Ca^{2+}		Mg^{2+}		Na^{+}	K^{+}
Parvalbumin	$-\log K_1$	$-\log K_2$	$-\log K_1$	$-\log K_2$	$-\log K$	$-\log K$
Pike, *Esox lucius*						
pI 5.0	10.6	10.8	5.6	5.1	1.58	0.95
pI 4.2	10.5	5.3	5.4	3.6	1.55	1.28
Whiting, *Gadus merlangus*	7.7	6.9	3.6	4.3	1.40	0.95
Cod, *Gadus morhua narcocephalus*	9.0	9.7	5.8	4.8	1.70	1.04
Flatfish, *Limanda herzensteini*	9.2	9.9	4.9	5.1		
Yellow-fin flatfish, *Limand aspera*	9.9	10.2	6.0	5.4		
Alaska plaice, *Pleuronecte quadrituberculatus*	7.8	10.5	4.5	4.9		
Snakehead, *Parachanna obscura*	9.1	5.4	5.1	3.9		
Snakehead, *Ophiocephalus argus*	9.3	10.0	5.4	3.7		
Amur sleeper, *Percottus glehni dybowski*	10.3	10.2	3.7	3.3		
Saffron cod, *Eleginus gracialis*	9.3	9.3	3.9	2.9		
Carp, *Cyprinus carpio* pI 4.25	10.8	7.3				
pI 4.47	10.5	6.7				
pI 3.95	7.4	7.1				
Alaska pollock, *Theragra chalcograma*	9.7	9.4	4.8	4.0		

Source: Data from E. A. Permyakov et al. (2006).

[a]Conditions: pH 8, 20°C.

Many studies have addressed pairs of EF-hands (lobes), isolated EF-hand domains, and "fragments" of other proteins. One cannot anticipate, a priori, to what extent various properties of these fragments differ from those in situ. However, careful comparison of properties of a series of such fragments can suggest important interactions. Maximov et al. (1978) synthesized a 40 residue peptide that included a consensus EF-hand, as suggested by Pechére et al. (1973). This single, unpaired EF-hand possesses weak, but specific, calcium binding activity. Fragment 75–108, a single EF-hand, has a rather low affinity for calcium, the K_d(Ca) is 4.6 mM according to intrinsic fluorescence data and 1.2 mM according to dansyl fluorescence data (Medvedkin et al., 1987). This fragment does not contain Arg75 (or 74), which forms the polar hydrogen bond with Glu81 (or 80) in the native protein and plays a very important role in maintaining the native structure (Figure 10.4). Attachment of the missing Arg to the fragment increases the fragment's affinity for calcium by an order of magnitude. The attachment of Ala instead of Arg does not produce this increased affinity. This means that the interaction between Arg75 and Glu81 occurs even in such a short peptide and that this interaction makes the conformation more similar to that of the native protein. This is also reflected in an increase in ellipticity at 222 nm, which characterizes the helical content of the peptide. The binding of calcium to the 75–108 fragment of pike parvalbumin pI 4.2 induces very small changes in ellipticity at 220 nm. This is also the case for the EF-fragment 76–108 of silver hake parvalbumin (Revett et al., 1997). NMR and CD spectra indicate significant secondary structure promotion in the EF fragment in the presence of trivalent lanthanide cations.

Fragment 1–74, containing the CD, but not the EF-calcium binding site, also has rather low affinity for calcium, but it is still higher than that of fragment 75–108 (Medvedkin et al., 1987). The presence of the N-terminal region, 1–37 (with helices A and B), seems to support the intact conformation. Fragment 38–108 of pike parvalbumin pI 5.0, which differs from the intact protein by the absence of helices A and B, binds two Ca^{2+} ions with $K_d(Ca)$ 10^{-6} and 10^{-7} M. These values are almost two orders of magnitude weaker than those for the native intact protein. Circular dichroism indicates the low helical content of the apo state of the fragment. The binding of calcium results in a pronounced

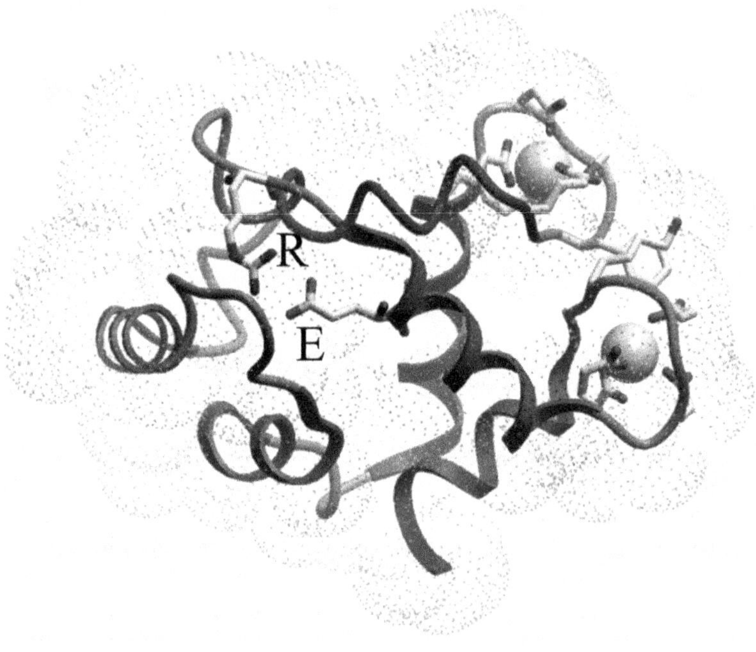

Figure 10.4. Arg75–Glu81 ionic hydrogen bond in carp parvalbumin pI 4.25 (PDB file 4CPV). (From Kumar et al., 1990.)

increase in helical content. In contrast, the binding of calcium to the intact protein causes a much less pronounced effect. Attachment of Trp or Met to the N-terminus of the 38–108 peptide decreases its affinity for calcium. The intact protein contains Met in this position. The crystal structure shows that the 38–108 fragment keeps the native parvalbumin fold, but is more compact, having a well-structured linker (Thepaut et al., 2001). The fragment has no stable apo form, has a lower affinity for calcium than does full-length parvalbumin, and does not bind magnesium. Structural differences in the hydrophobic core are responsible for lowering the calcium affinity of the truncated protein. This means that the entire protein structure, even those parts that do not take part directly in the coordination of the Ca^{2+} ion, is needed for normal, high-affinity calcium binding. Table 10.2 contains data on calcium binding for peptides of pike parvalbumin pI 5.0 (Medvedkin et al., 1987).

Thermal unfolding of metal loaded forms of five parvalbumins from northern pike (α- and β-isoforms), baltic cod and rat (α and β) have been studied by scanning calorimetry, circular dichroism, and fluorescence of the bis-ANS probe (S. E. Permyakov et al., 2009a). All of these calci-parvalbumins show complex denaturation behaviors, but only pike α and cod parvalbumins exhibit two distinct heat sorption peaks. The unfolding of pike α-parvalbumin is described by the model of two successive two state transitions; this suggests the presence of two thermodynamic domains. The first, lower temperature transition has midtemperature 90°C, while the second, high temperature transition occurs at 120°C. The intermediate state binds calcium and is characterized by largely preserved secondary structure and increased solvent exposure of hydrophobic groups. Magnesium and sodium loaded forms of pike α-parvalbumin show single two state transitions. This fact, along with the results of experiments on the effects of guanidinium chloride indicate that at least one of the calcium binding domains is essential for stabilization of the intermediate. The probability of individual residues being included in a fixed tertiary structure is indicated in the PONDR (predictors of natural disordered regions; see www.disprot.org/predictors.php) score. The lower the PONDR score of a specific residue, the higher the probability of its inclusion into the fixed tertiary structure of the protein. Such analyses show that the EF-hand is stabilized in parvalbumins exhibiting two distinct heat sorption peaks.

TABLE 10.2. Binding of Calcium to Fragments of Pike Parvalbumin[a]

Fragment	$-\log K_1$	$-\log K_2$ (M)
1–108	10.52	10.30
38–108	6.0	7.0
75–108	2.33	
Ala74–108	2.70	
Arg74–108	3.20	

[a] pI 5.0.

One approach to characterizing sites of loosened interactions between residues is to identify protein cavities, interior empty spaces that are not accessible to a solvent probe. Structural cavities can be calculated using the CASTp online service (http://sts-fw.bioengr.uic.edu/castp/) (Dundas et al., 2006). This analysis reveals that the interface between the CD-hand and the EF-hand of pike α-parvalbumin is much more loosely packed than in parvalbumins manifesting a single heat sorption peak (S. E. Permyakov et al., 2009a). The impairment of interactions between the CD and EF domains may cause a loss of structural interaction and the appearance of two separate thermodynamic domains. Overall, the mechanism of thermal unfolding of parvalbumin is switched by metal ions and ranges from the absence of observable first order transitions (apo state), to the two state or more complex mechanism, involving at least one intermediate. The loss of structural cooperativity seems to be due to loosened interactions between parvalbumin domains and calcium induced stabilization of the intermediate.

NMR analysis of rat β-parvalbumin in solution indicates that the transition from the calcium loaded to the sodium loaded state involves substantial conformational changes: reorientation of helices C, D, and E; reorganization of the hydrophobic core; reduced interdomain contact; and remodeling of the AB domain (Henzl and Tanner, 2007). In marked contrast to rat β-parvalbumin, the sodium and calcium loaded forms of rat α-isoform are quite similar (Henzl and Tanner, 2008). Significant structural differences appear to be confined to the loop regions of the molecule. The authors suggested that the α-isoform enjoys elevated divalent ion affinity because the metal ion binding events do not require major structural rearrangement and the concomitant sacrifice of binding energy. The six α-helices (A to F) are present in the apo-protein. The apparent content of α-helices in apo-parvalbumin is about 42% at 25°C (54% for the calci-parvalbumin) (Williams et al., 1986). Removal of both Ca^{2+} ions results in increased conformational fluctuations.

Calci-parvalbumin shows a remarkable stability against trypsin digestion (Cox et al., 1979). Even at trypsin/parvalbumin ratios as high as 1 : 40, perch parvalbumin II is unaffected by trypsin over a 24-hour incubation period at room temperature. Removal of calcium markedly increases parvalbumin susceptibility to proteolysis. The accessibility of Cys in parvalbumin to thiol reagents depends strongly on the binding of calcium. The single Cys of perch calci- or of magnesi-parvalbumin II reacts sluggishly with the thiol reagent Nbs_2; however, the reaction is completed within 30 minutes when calcium or magnesium is removed (Cox et al., 1979).

Fragments of parvalbumin preserve some affinity for each other. Permyakov et al. (1991b) and Henzl et al. (2004) observed the formation of a noncovalent complex between domain AB and the pair of domains CD–EF (dissociation constant 10^{-5} M for pike parvalbumin, 1.3×10^{-8} M for rat α-parvalbumin, and 1.1×10^{-6} M for rat β-parvalbumin in saturating calcium, accompanied by an increase in the affinity for calcium). These results indicate that the AB domain can modulate the calcium affinities of the CD and EF sites.

These experiments on parvalbumin fragments demonstrate that the entire protein is needed for high-affinity calcium binding, but some amino acid residues play an especially important role. For example, the removal of two C-terminal residues (Ala108 and Lys107) from carp parvalbumin pI 4.25 only slightly affects its affinity for calcium, but the removal of the next one or two residues (Val106 and Leu105) significantly decreases its affinity for metal cations (Corson et al., 1986). Helix F in carp parvalbumin appears to be bound to the protein fold through a series of hydrophobic contacts between its side chains (Val99, Phe102, Leu105, and Val106) and those of residues from helix B, the BC-linker, and helix C. Two features of parvalbumin are important for maintenance of its native structure: ionic interaction between the C-terminal carboxyl and the amino group of Lys27 and hydrophobic interaction of the internal part of the F-helix with residues of the remainder of the hydrophobic core. The removal of Lys107 and Ala108 produces a surrogate C-terminus, which is still well positioned to reestablish the polar hydrogen bond with Lys27. Hydrolysis of the Val106–Lys107 peptide bond also produces a C-terminal carboxyl group; however, its orientation does not permit bonding with Lys27.

Some cells contain high concentrations of various low-molecular-weight organic compounds, such as urea and methylamines (betaine and trimethylamine *n*-oxide, TMAO). Marine elasmobranchs are osmo-conformers that employ a combination of solutes to maintain extracellular and intracellular osmotic strength close to or slightly above that of their marine environment (Piermarini and Evans, 1998). Intracellular and extracellular osmoregulation in marine elasmobranchs involves high concentrations of urea (350 to 600 mM). Euryhaline elasmobranchs possess intracellular concentrations of urea between 140 and 200 mM and plasma concentrations of approximately 195 mM. Heffron and Moerland (2008) extracted parvalbumins from marine and freshwater Atlantic stingray *Dasyatis sabina*. Both populations have two isoforms, I (12.18 kDa) and II (11.96 kDa). Calcium dissociation constants of these proteins were determined in the presence and absence of physiological concentrations of urea, betaine, and TMAO by fluorescence spectroscopy using the fluorescent calcium indicator fluo-3. Parvalbumin I shows no significant changes in calcium binding from in situ muscle conditions, except in the presence of betaine. In contrast, the ability of parvalbumin II to bind calcium is increased relative to physiological conditions in the presence of each osmolyte independently. Thus, it appears that organic osmolytes have isoform specific effects on parvalbumin function.

10.2. FUNCTION

10.2.1. Temporal Calcium Buffering

The primary function of parvalbumin is assumed to be the control and/or modulation of the spatiotemporal parameters of intracellular calcium signals. These signals are short lived, especially in excitable cells such as neurons or muscle cells, in the range of a few to hundreds of milliseconds, and are called *calcium transients*. Various names are given to these calcium transients: *waves, spikes, sparks, puffs*, or *quarks*. Parvalbumin is inferred to function as a soluble relaxation factor in fast-twitch skeletal muscles, where it functions as a delayed calcium buffer. However, it has also been isolated from other tissues, including the central nervous system, testis, kidney, and several endocrine glands (reviewed by E. A. Permyakov, 2006).

The most reasonable hypothesis for the physiological function of parvalbumin in muscle cells is that its function is connected with regulation of contraction. It was proposed that parvalbumin serves as a soluble relaxing factor accelerating the relaxation phase in fast muscles (Haiech et al., 1979; Gillis et al., 1982). Parvalbumin can be detected in fast-contracting/relaxing muscle fibers of rodents starting 4 to 6 days after birth. The time period of the maximal increase in parvalbumin concentration coincides with the differentiation of the fast-twitch muscle function, consistent with its involvement in fast-twitch muscle function.

Since the concentration of free Mg^{2+} ion in a resting muscle cell is as high as 1 to 6 mM while free Ca^{2+} ion concentration is as low as 10^{-8} M, it was proposed that parvalbumin in resting muscle is mostly in the magnesi form. When calcium is released from sarcoplasmic reticulum, Mg^{2+} ions are replaced by Ca^{2+} ions. According to the hypothesis of Haiech et al. (1979), the magnesium binding properties of parvalbumin explain why high concentrations of parvalbumin do not prevent Ca^{2+} ions from reaching the calcium-specific sites of troponin C (Figure 10.5). A delay in the binding of calcium caused by magnesium dissociation allows calcium ions to reach the lower-affinity sites of troponin C, which are not occupied by magnesium. Subsequently, the divalent cation sites of parvalbumin will compete successfully for calcium and help to remove calcium from the lower-affinity sites of troponin C; this in turn causes muscle relaxation. Ca^{2+} ions bound to parvalbumin are then removed by the sarcoplasmic reticulum pump, causing muscle relaxation. That is, parvalbumin is too slow to affect the rapid rise in $[Ca^{2+}]_i$ occurring during the contraction phase, but increases the initial rate of $[Ca^{2+}]_i$ decay. Yet, under prolonged tetanic contraction, parvalbumin saturates with calcium and acts as a transient calcium source, extending the $[Ca^{2+}]_i$ decay and also the relaxation phase (Raymackers et al., 2000). Parvalbumin increases considerably the initial rate of $[Ca^{2+}]_i$ decay (the calcium binding phase), followed by a prolongation of the transient at later times (calcium releasing phase) yielding a biexponential $[Ca^{2+}]_i$ decay (Lee et al., 2000b).

Parvalbumin has been identified in heart tissue by immunohistochemistry of mouse, rat, chicken, rabbit, and pig and has been implicated in mediating relaxation in cardiac myocytes (Vongvatcharanon et al., 2008). All cardiac myocytes of each species have an identical pattern of parvalbumin immunoreactivity in their cytoplasm. Parvalbumin is found in the walls of all four cardiac chambers, with the left ventricle, in general, having the highest intensity. The intensity of immunoreactivity is correlated with the physical activity of the heart of each species.

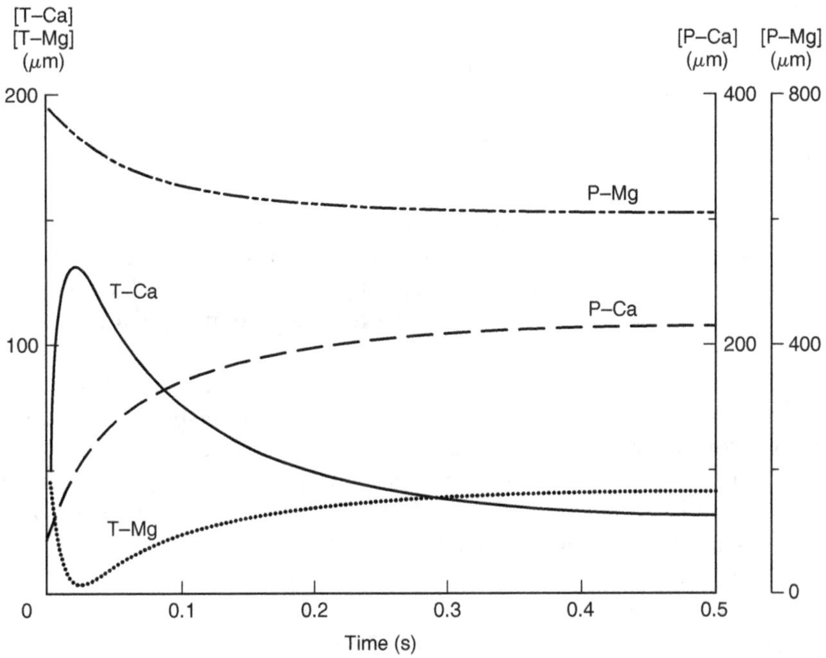

Figure 10.5. Calculated time course for the binding of calcium and of magnesium to the high affinity (P) sites of parvalbumin and to the lower affinity (T) sites of troponin, as in frog muscle. Initial calcium pulse of 200 μmol l^{-1} at $t = 0$. The model assumes no sequestration of calcium by the sarcoplasmic reticulum and P-site concentration of 840 μM. The calcium initially binds to the T-sites; over tens of contractions the calcium is 'transferred' to the P-sites. (From Gillis et al., 1982.)

10.2.2. Possible Function(s) in Cells Other Than Muscle

A biexponential decay in [Ca^{2+}] transients is seen in cultured parvalbumin containing hippocampal interneurons in vitro, but not in excitatory neurons devoid of parvalbumin expression (Lee et al., 2000a). Parvalbumin, acting as a calcium source, leads to a robust parvalbumin-dependent, delayed transmitter release at interneuron, interneuron synapses subsequent to presynaptic bursts of action potentials (Collin et al., 2005). Moreover, parvalbumin plays a major role in regulating the local inhibitory effects exerted by GABAergic interneurons on principal neurons.

Besides muscle and brain, parvalbumin is also expressed in the kidney. In the rat, parvalbumin expression occurs in part of the distal convoluted tubule, proximal collecting duct (Schneeberger and Heizmann, 1986), and in the early distal convoluted tubule (DCT1), which plays an essential role in NaCl reabsorption, where it colocalizes with the thiazide-sensitive Na^+/Cl^- cotransporter. A significant overlap exists between parvalbumin expression and the magnesium channel involved in renal magnesium absorption, TRPM6 (Voets et al., 2004).

10.2.3. Possible Extracellular Functions

Changes of parvalbumin level inside muscle cells, and even in serum, correlate with some muscle diseases. Jockusch et al. (1990) showed that parvalbumin is present in the serum of normal mice and that its level is indicative of the disease status of muscle. Elevated parvalbumin levels are found in mice with X-linked dystrophy (mdx) and reduced levels in myotonic (ADR) mice. Because myotonic mouse muscle is characterized by strongly reduced parvalbumin content, the reduced parvalbumin serum level in ADR mice indicates that serum parvalbumin is derived from skeletal muscle. It was suggested that serum parvalbumin in mdx mice, in which muscle parvalbumin content is close to normal, is a measure of the necrosis of fast muscle fibers.

Diastolic dysfunction is a characteristic feature of the aged mammalian heart. Impaired ventricular relaxation is one cause of diastolic dysfunction and is an important component of human heart failure. In about 40% of patients who suffer from heart failure, disease progression results specifically from a slowing of myocardial relaxation (diastolic dysfunction). In these patients, the heart relaxes too slowly after each contraction, thus compromising the refilling of its chambers with blood for the next contraction. No specific treatments currently exist. Genetic modification of intracellular calcium-handling proteins may help to correct this diastolic dysfunction. Since parvalbumin acts as a calcium sink and enhances relaxation in skeletal muscle, overexpression of parvalbumin in myocardium should increase cardiac relaxation in vivo as well as in vitro. Indeed, the expression of parvalbumin dramatically increases the rate of calcium sequestration and the relaxation rate in normal cardiac myocytes. Parvalbumin fully restores the relaxation rate in diseased myocytes isolated from an animal model of human diastolic dysfunction (Wahr et al., 1999). These authors concluded that gene therapy of parvalbumin may address the impaired calcium homeostasis and diastolic dysfunction without an increase in energy expenditure.

10.2.4. Allergen

Extensive studies of the cod *Gadus callarias* first identified parvalbumin as a fish allergen (Elsayed and Aas, 1971; Elsayed and Bennich, 1974, 1975; Ma et al., 2008). Most allergic reactions are caused by β-parvalbumins, which are considered as cross-reactive allergens in fish. The binding of calcium was found to be necessary for maintaining parvalbumin in a conformation that is able to bind IgE. Loss of calcium results in a change in conformation together with an associated loss of its IgE binding capacity. IgE antibodies against one parvalbumin will cross-react with parvalbumins from other fish species; this demonstrates the importance of parvalbumins as cross-reactive fish allergens and explains why allergic individuals exhibit clinical symptoms upon contact with various fish species. Parvalbumins from different fish species (e.g., cod, tuna, salmon) cross-react when tested with the sera of patients allergic to other fish (Bugajska-Schretter et al., 1998). After identification of parvalbumin as an allergen, calcium binding allergens were discovered in pollens of trees, grasses, and weeds and, subsequently, as autoallergens in humans (reviewed by Valenta et al., 1998).

Parvalbumins from fish are very stable; extremes in pH, temperature, or high concentrations of denaturing agents do not significantly alter their allergenicities (Elsayed and Aas, 1971; Elsayed and Bennich, 1975). Due to the very high stability of calci-parvalbumin, it can sensitize patients despite being cooked and exposed to the gastrointestinal tract.

The major isoform of northern pike muscle parvalbumin (pI 5.0, α-parvalbumin) exhibits microheterogeneity of amino acid sequence (Permyakov et al., 2009b). Mass-spectrometry analysis of an α-parvalbumin sample showed the presence of two components with a mass difference of 71 Da. Analysis of tryptic and endoproteinase Asp-N digests of α-parvalbumin by MALDI-TOF mass spectrometry revealed peptides corresponding to two different amino acid sequences. The sequence differences between variant proteins are limited to the AB-domain and include the substitutions K27A and L31K and an extra Leu residue between K11 and K12. Since the affected residues comprise a cluster on the surface of parvalbumin, involvement of the identified region in target recognition was suggested. The substitutions at positions 27 and 31 are located in the region of previously identified epitopes of parvalbumin relevant for parvalbumin-specific IgE and IgG binding; this suggests different immunoactivities of the variants. This microheterogeneity of parvalbumin is suggested to be important for physiological adaptation of the propulsive musculature to developmental and/or environmental requirements and may contribute to parvalbumin allergenicity.

IgE cross-reactivity to parvalbumin in five species consumed widely in the southern hemisphere was investigated (Beale et al., 2009). This study confirmed that pilchard parvalbumin is the major and most cross-reactive allergen in a group of subjects allergic to fish. Sera from several subjects displayed IgE reactivity to 24- and/or 38-kDa proteins in crude extracts of yellowtail, anchovy, and hake muscle. These were assumed to be dimers and trimers of parvalbumin due to their mobilities under electrophoresis compared to positions of dimeric and trimeric parvalbumin run as positive controls. Interestingly, some dimeric and other oligomeric forms of other proteins have been shown to be more allergenic than their monomeric forms.

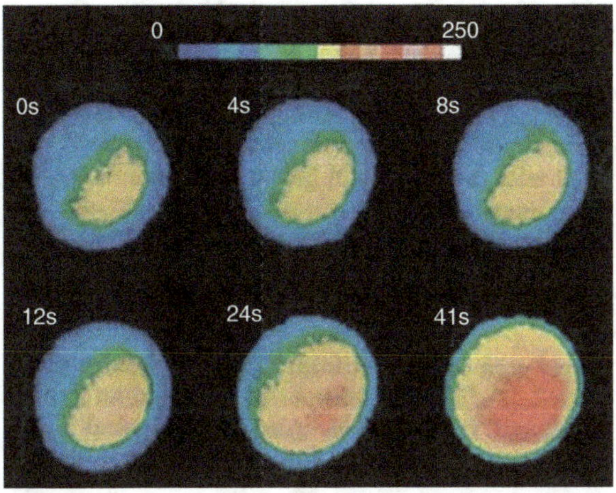

Figure 2.1. Calcium wave through a fertilized maize egg.

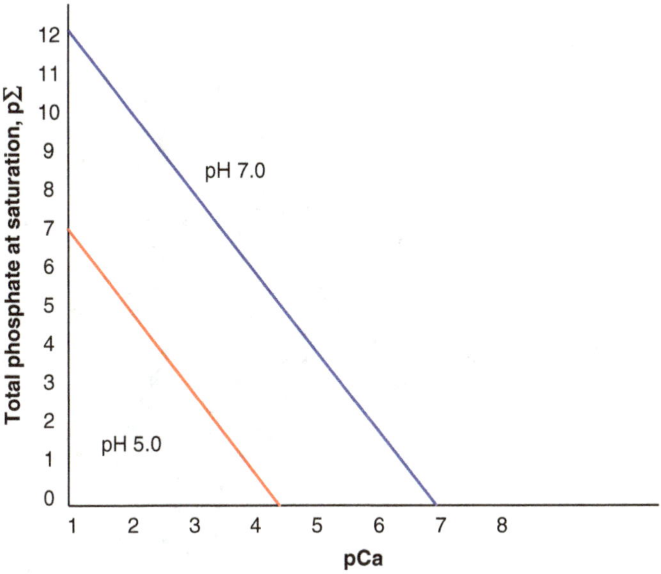

Figure 2.3. The concentration of total phosphate ($[PO_4^{3-}] + HPO_4^{2-}] + H_2PO^{4-}] +$ $[H_3PO_4]$) is shown at pH 7.0 (blue line) and pH 5.0 (red line) as a function of pCa.

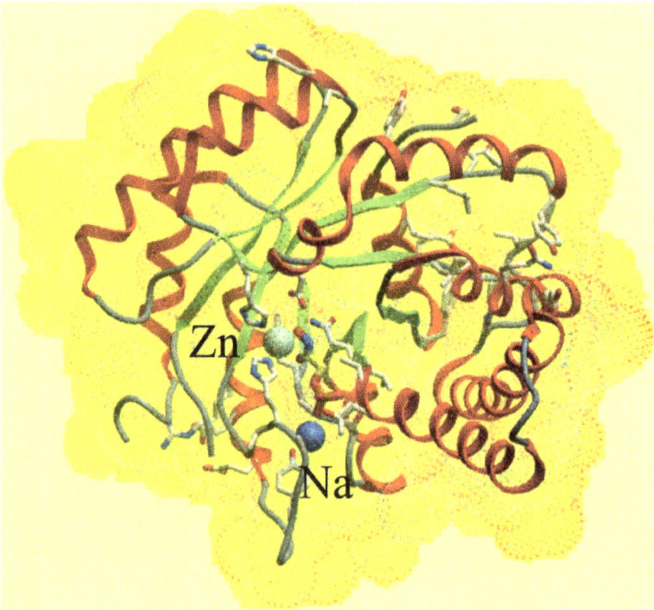

Figure 8.1. Structure of *Escherichia coli* tagatose-1,6-bisphosphate aldolase with bound Na^+ and Zn^{2+} ions complexed with phosphoglycolohydroxamic acid (PDB file 1GVF).

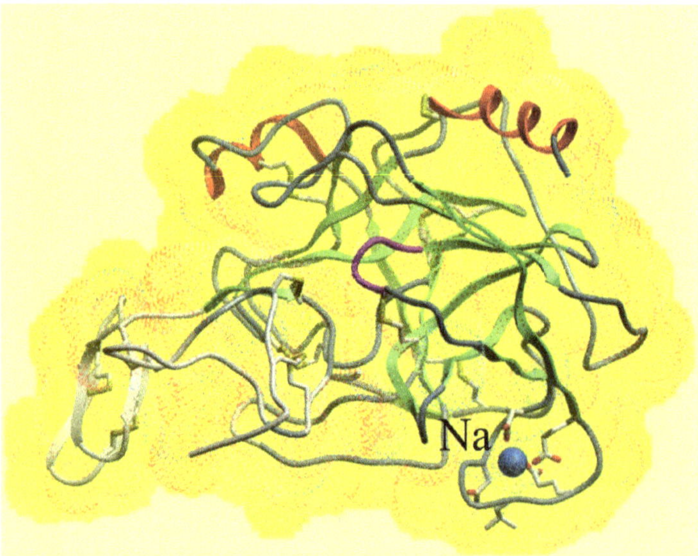

Figure 8.2. Structure of porcine trypsin with bound Na^+ ion (PDB file 1EJA).

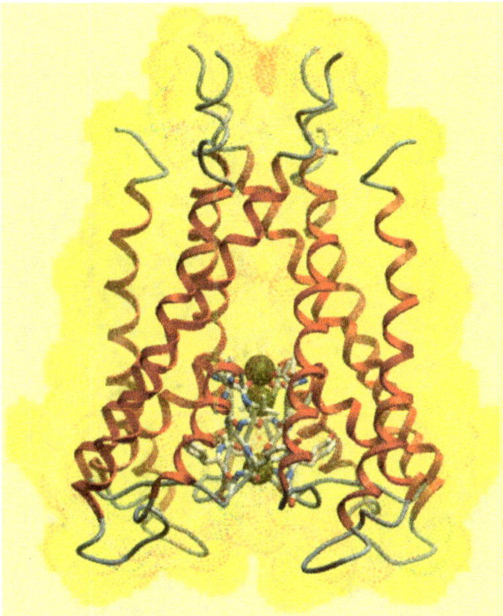

Figure 8.3. Structure of the tetrameric subunit of potassium channel with three K^+ ions inside the channel. The sequence Thr–Val–Gly–Tyr–Gly, which comprises the selectivity filter, is seen in each monomer (PDB file 1BL8).

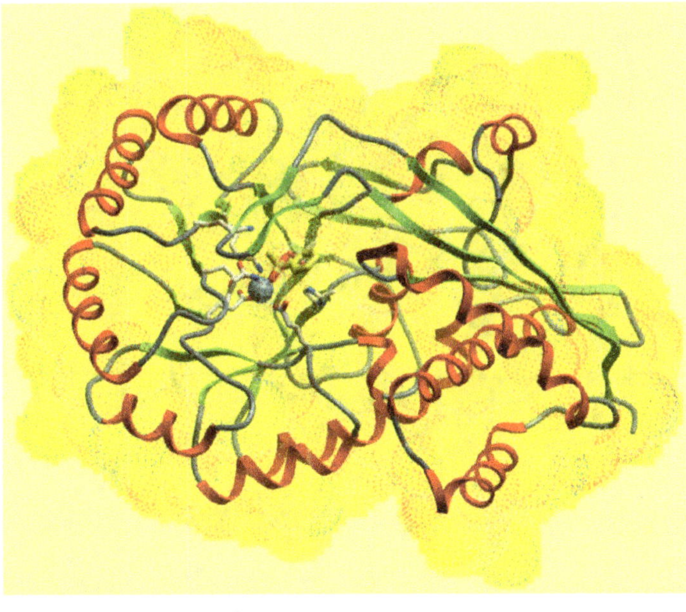

Figure 8.4. Structure of mandelate racemase from *Pseudomonas putida* with a bound Mg^{2+} ion (PDB file 1MDR).

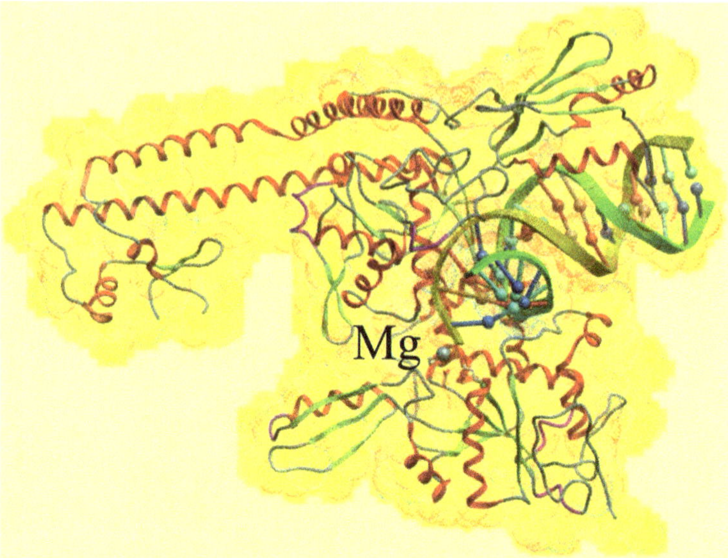

Figure 8.5. Structure of a fragment, 419 to 1177, of type IIA topoisomerase *Saccharomyces cerevisiae* with bound polydeoxyribonucleotide and a Mg^{2+} ion (PDB file 2RGR).

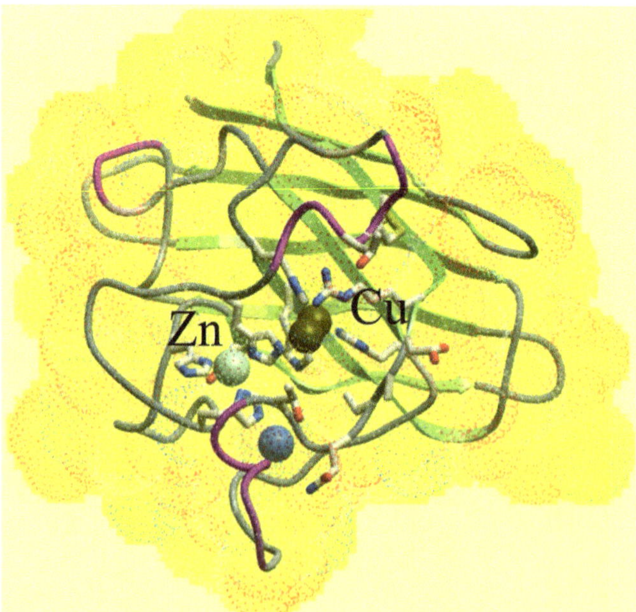

Figure 8.10. Structure of superoxide dismutase (SOD) with two Cu^{+} and two Zn^{2+} ions (PDB file 2C9V).

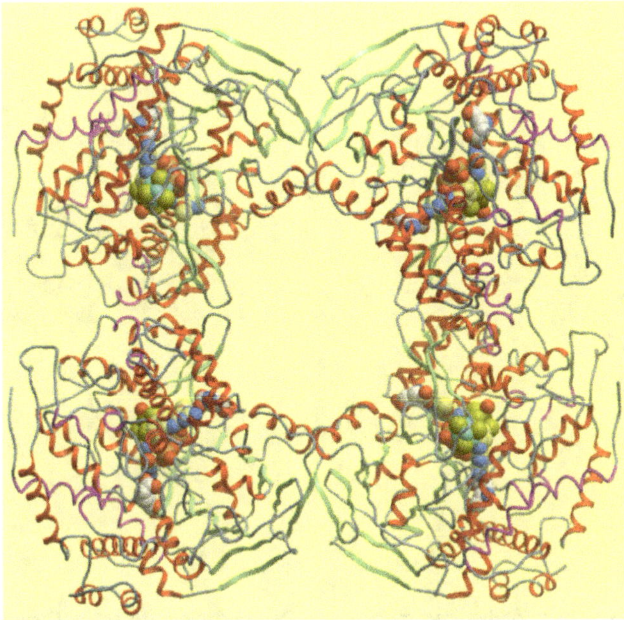

Figure 8.14. Structure of the tetrameric formaldehyde–ferredoxin oxidoreductase from *Pyrococcus furiosus* with iron–sulfur clusters and tungsten–pterin (PDB file 1B25).

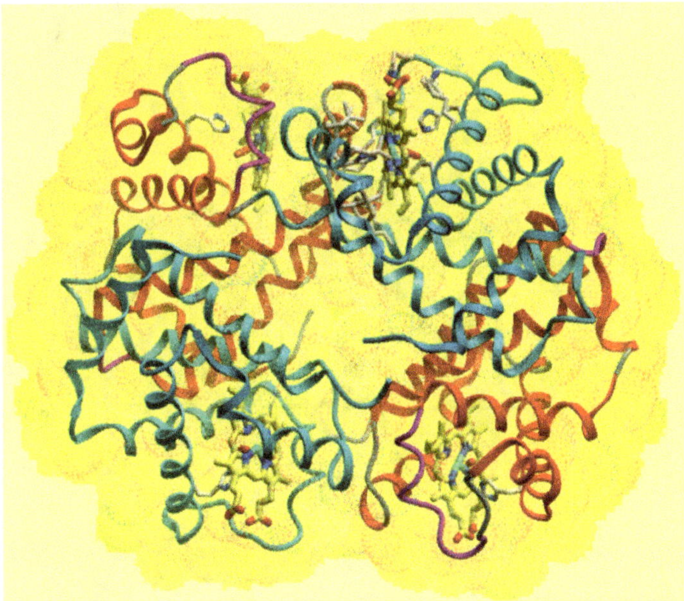

Figure 8.20. Structure of oxyhemoglobin; α- and β-subunits are blue and red, respectively (PDB file 1GZX).

Figure 8.22. Structure of heterodimer of [Fe–Ni]-hydrogenase from *Desulfovibrio vulgaris* with $Fe_4S_4^{2+}$ and $Fe_3S_4^{+}$ clusters and Fe–Ni and three 165I molecules in the active site (PDB file 1FRV).

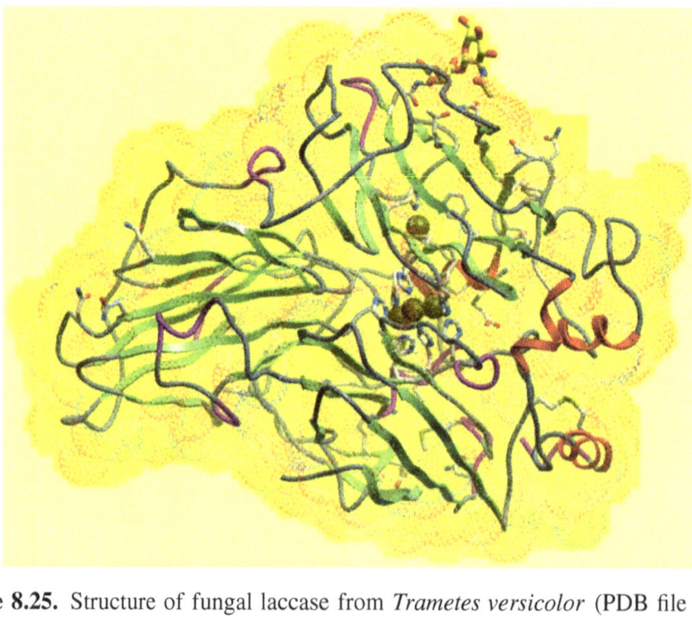

Figure 8.25. Structure of fungal laccase from *Trametes versicolor* (PDB file 1GIC).

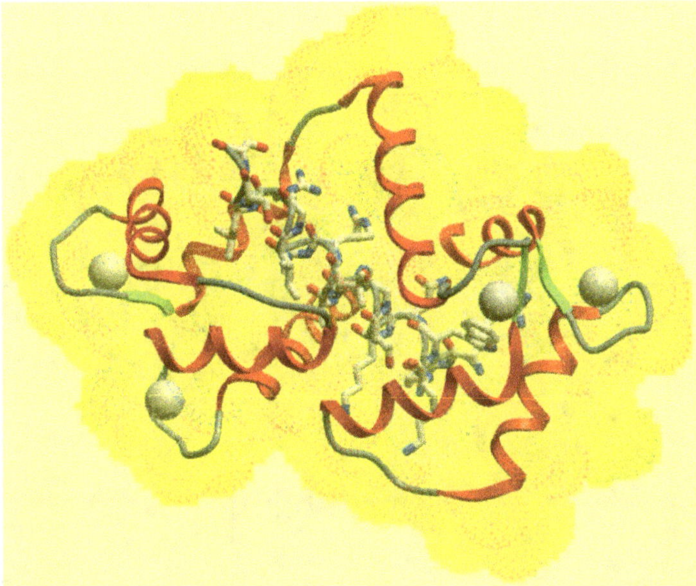

Figure 11.5. Structure of tetra-calci-calmodulin complexed with a calmodulin binding peptide of myosin light chain kinase (PDB file 1CDL). The peptide contacts both N- and C-lobes as well as the linker helix.

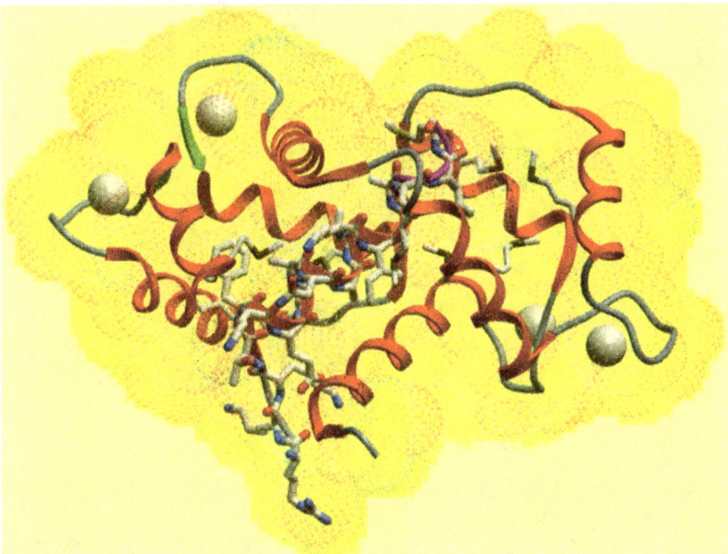

Figure 11.6. Structure of tetra-calci-calmodulin complexed with a fragment of NO-synthase (PDB file 1NIW).

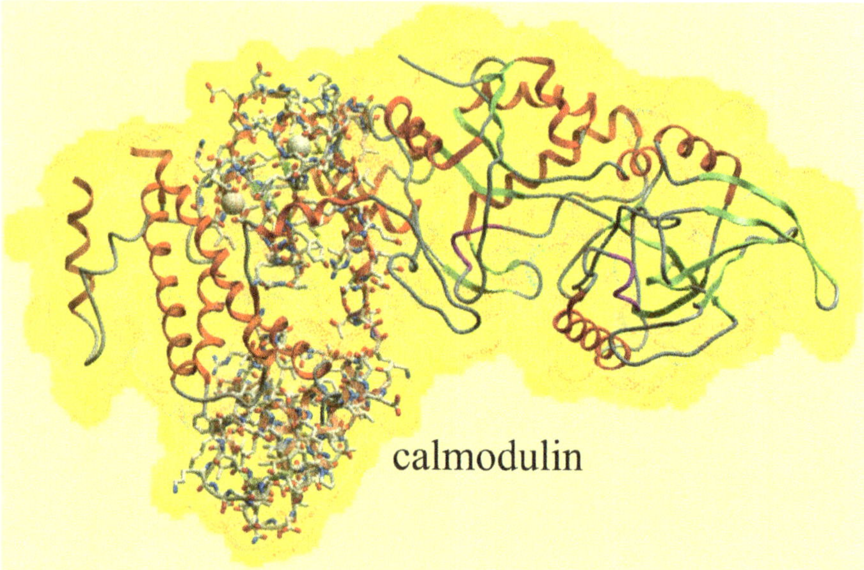

Figure 11.7. Di-calci-calmodulin complexed with *Bacillus anthracis* exotoxin (edema factor, adenylyl cyclase) (PDB file 1K93). Calmodulin is represented as a stick figure, the exotoxin as a ribbon model.

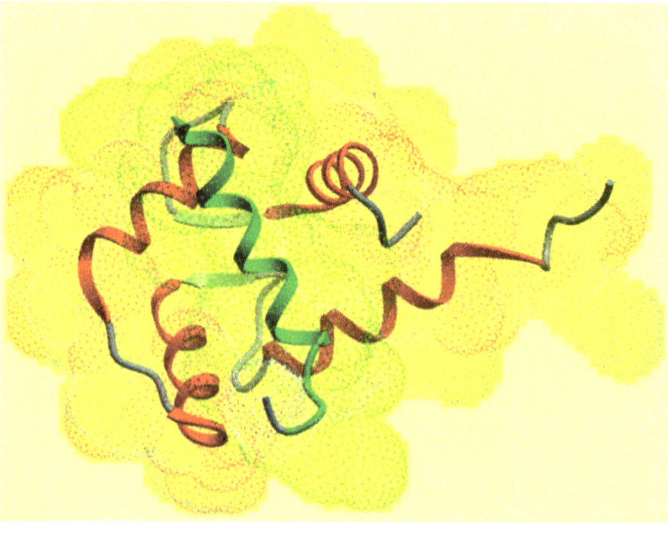

Figure 11.9. Structure of calcium-loaded C-terminal domain of caltractin in complex with CDC31P binding domain from KAR1P (PDB file 1OQP).

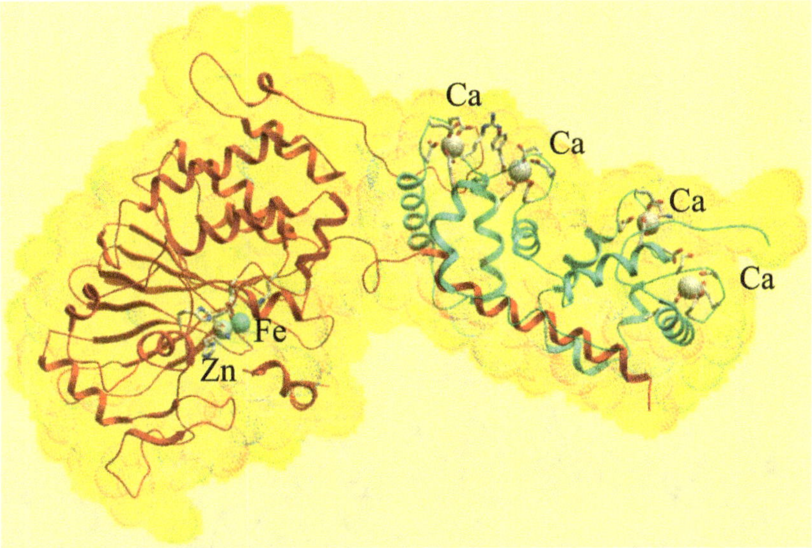

Figure 11.10. Structure of calcineurin heterodimer (PDB file 1AUI). Subunit B (green) has a Ca^{2+} ion in each of its four EF-hands. Subunit A has a Zn(II)–Fe(II) site.

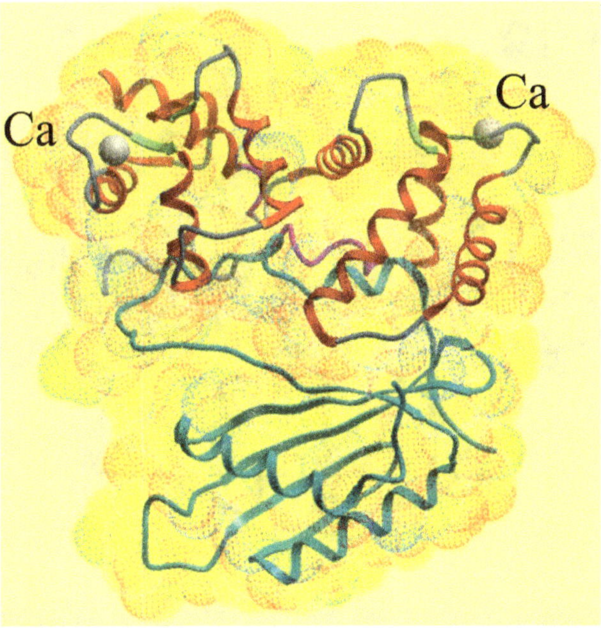

Figure 11.12. Structure of the C-terminal domain of the protein kinase AtSOS2 (blue) bound to AtSOS3 (red). Calcium is bound to EF-hands 1 and 4 (PDB file 2EHB).

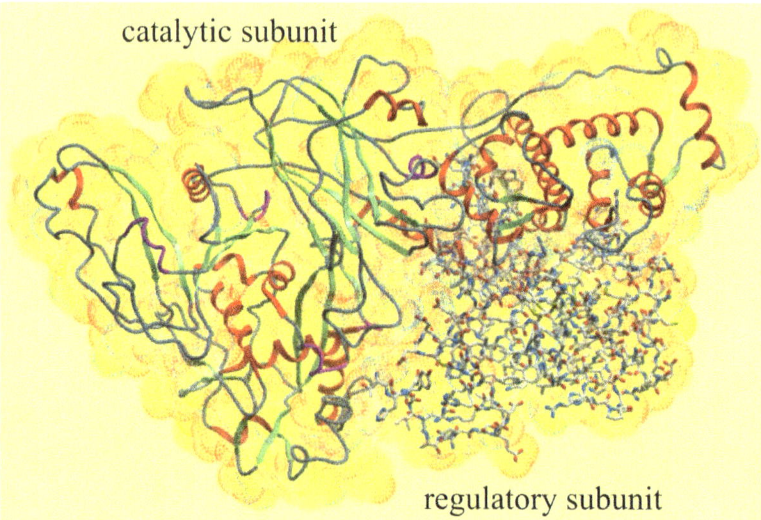

Figure 11.17. Structure of human m-calpain (PDP file 1KFU). The catalytic subunit is shown as a ribbon model; the regulatory subunit as a stick model.

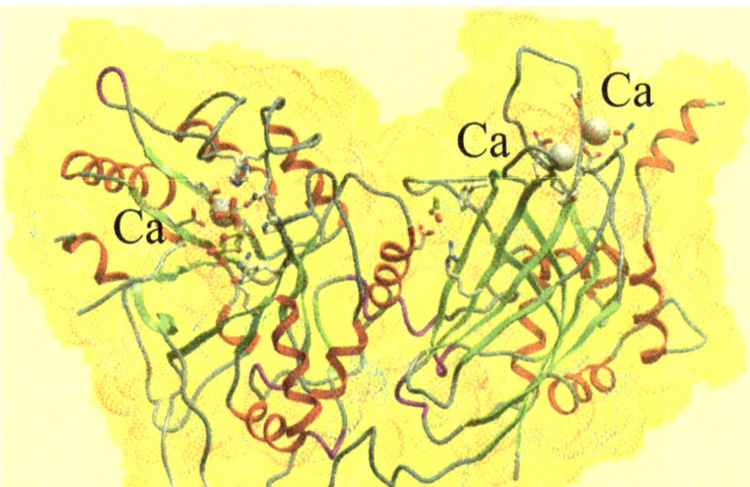

Figure 11.24. Crystal structure of a mammalian phosphoinositide-specific phospholipase C (PDB file 1DJW).

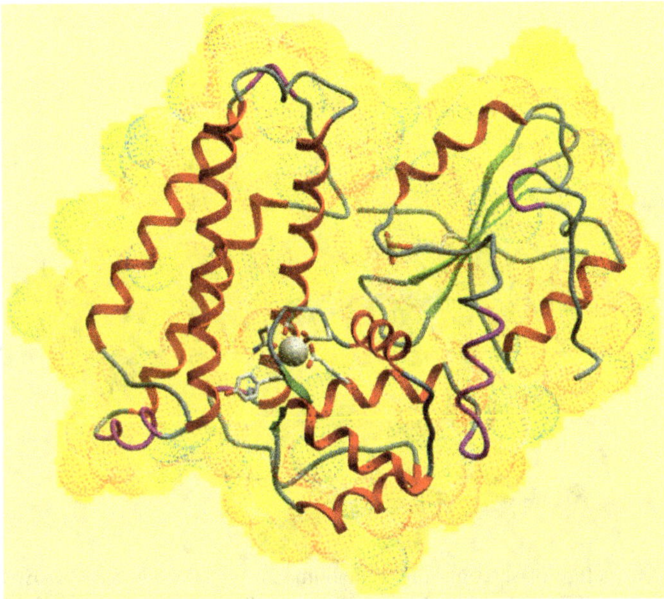

Figure 11.26. Structure of the amino-terminal domain of CBL complexed to its binding site on ZAP-70 kinase and Ca^{2+} (PDB file 2CBL). ZAP-70 kinase is viewed down the midline of its warped β-sheet.

```
S100A7     SNTQAERSIIGMIDMFHKYTRRDD
S100A1(      SELEKAVVALIDVFHQYSGREGDKHKLKKSELKELINNELSHFLEEIKEQ
S100A8     MLTELEKALNSIIDVYHKYSLIKGNFHAVYRDDLKKLLETECPQYIRKKG
S100A9      MSQMESSIETIINIFHQYSVRLGHYDTLIQKEFKQLVQKELPNFLKKQKK
S100A12      TKLEEHLEGIVNIFHQYSVRKGHFDTLSKGELKQLLTKELANTIKNIKDK
S100A3       MARPLEQAVA AIVCTFQEYA GRCGDKYKLC QAELKELLQK ELATWTPTEF


                      1111111111222222222
            12345678901234567890123456789
             n**nn**nX*Y*ZGy*x**zn**nn**n
S100A7     TNYLADVFEKKDKNEDKKIDFSEFLSLLGD IATDYHKQSHGAAPCSGGSQ
S100A1       EVVDKVMETLDSDGDGECDFQEFMAFVAMITTACHEFFEHE
S100A8          ADVWFKELDINTDGAVNFQEFLILVIKMGVAAHKKSHEESHKE
S100A9       NEAAINEIMEDLDTNVDKQLSFEEFIMLVARLTVASHEEMHNTAPPGQGH
S100A12      AVIDEIFQGLDANQDEQVDFQEFISLVAIALKAAHYHTHKE
S100A3     RECDYNKFMSVLDTNKDCEVDFVEYVRSLACLCLYCHEYFKDCPSEPPCSQ
```

Chapter 11, Page 285. Amino acid sequence.

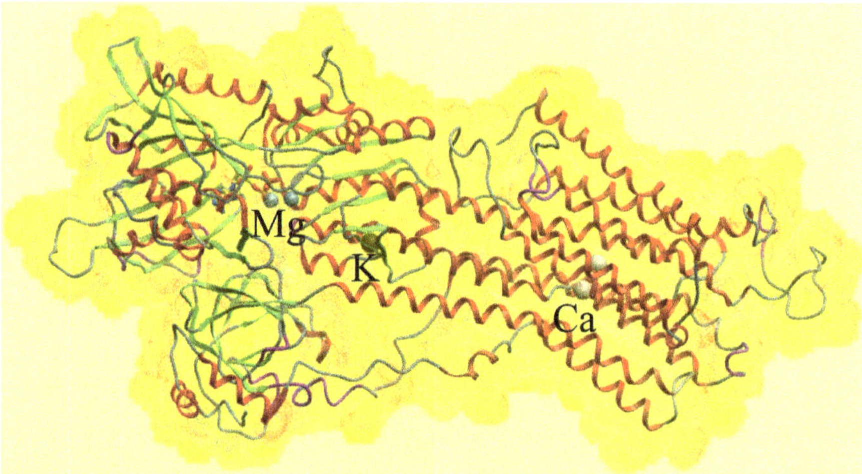

Figure 12.7. Structure of sarcoplasmic reticulum Ca^{2+}-ATPase in the Ca_2-E_1-ADP state (PDB file 1T5T). ATP is bound in the N domain, to the left. The transmembrane domain (helices M1 to M10) is to the right; the Ca^{2+} ion binding sites indicate the channel.

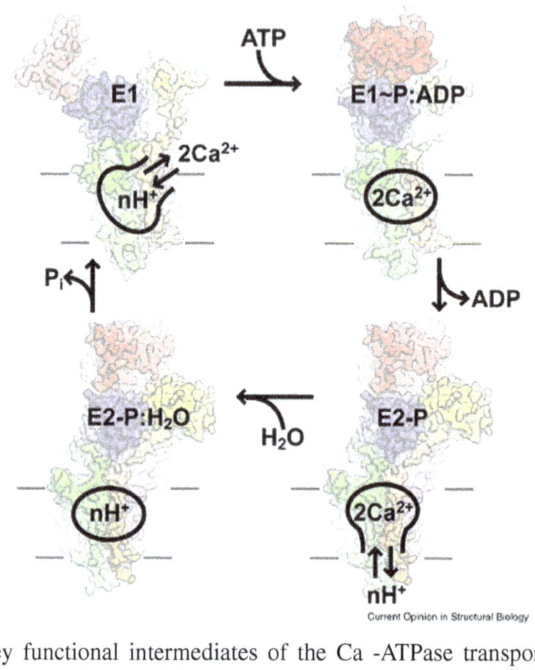

Figure 12.8. Key functional intermediates of the Ca -ATPase transport cycle and their structural correlates, as obtained from current structures determined by analysis of Ca^{2+}-ATPase crystals. Red, N domain; blue, P domain; yellow, A domain; purple, transmembrane segments 1 and 2 (M1–M2) with their cytosolic extensions; orange, M3–M4; green, M5–M10.

11

EF-HAND PROTEINS

The EF-hand proteins are discussed in this chapter in the context of their congruent subfamilies as summarized in Chapter 7—with one exception. Parvalbumin alone was reviewed in Chapter 10 since it has been the subject of so many chemical and physical studies that have provided techniques and precedents for investigation of other calcium binding proteins. These 70 subfamilies are only a fraction of the EF-hand proteins in the entire proteome. Some of their names, as well as subfamily and group designations, will change in the future as they have in the past. However, this sampling should provide a sense of the rich diversity of structure, function, and evolutionary history enjoyed by this 29-residue helix–loop–helix domain.

11.1. CTER (CALMODULIN, TROPONIN C, ESSENTIAL AND REGULATORY LIGHT CHAIN) SUBFAMILY

All CTER proteins evolved from a single precursor with four EF-hands Grabarek et al. (2006) discussed the structural basis for diversity of the EF-hand calcium binding proteins. All domains 1 within the CTER family resemble one another more closely than they do other domains within the same protein.

11.1.1. Calmodulin

Calmodulin is (inferred to be) present in all eukaryotic cells. It consists of four EF-hands and is involved in myriad signal transduction pathways. Many proteins involved in calcium signal transduction alter their activities in response to

Calcium Binding Proteins, By Eugene A. Permyakov and Robert H. Kretsinger

changes in levels of free Ca^{2+} ion, but are themselves not able to bind calcium. Some of these proteins use calmodulin as a sensor and mediator of the initial calcium signal. Calmodulin transduces the calcium signal by binding free Ca^{2+} ions to its N- and C-terminal EF-hand pairs; this causes a conformational change and enables calci-calmodulin to bind to the final targets of the calcium signaling pathway. The binding of calci-calmodulin to its target proteins alters their activities in a calcium-dependent manner. Calmodulin is involved in the regulation of cellular processes as diverse as platelet aggregation, cell–cell interactions, cell proliferation, smooth muscle contraction, neurosecretion, DNA repair, and secretion. It is also associated with the plasma membrane, the nuclear membrane, and the cytoplasmic surface of mitochondria and the endoplasmic reticulum. In the rat cerebellum, these structures contain much of the calmodulin antibody staining. At the synapse, the postsynaptic membrane contains abundant calmodulin. Most of these interactions occur only in the presence of calcium, but some of them take place in its absence (Jurado et al., 1999). Apo-calmodulin binds to other proteins and has specific and essential effects.

Calmodulin has been highly conserved during evolution, due to its specific interactions with at least 40 different targets (Table 11.1). In higher vertebrates, calmodulin is encoded by three distinct nonallelic genes (*CaM I, II, and III*) (reviewed by Kortvely and Gulya, 2004). The three calmodulin genes are collectively transcribed into at least eight mRNAs (4.2, 4.1, 1.7, and 1.0 kb for *CaM I*; a single, 1.4-kb transcript for *CaM II*; and 2.3, 1.9, and 0.9 kb for *CaM III*), by utilizing different polyadenylation signals. However, all the transcripts encode the identical sequence, 148 residues long.

Calmodulin constitutes at least 0.1% of the total protein present in cells and is expressed at even higher levels in rapidly growing cells, especially those undergoing cell division and differentiation (reviewed by Chin and Means, 2000; Kortvely and Gulya, 2004). The concentration of available calmodulin is lower than the sum of potential calmodulin binding partners. The diverse calmodulin-dependent activities are probably coupled by competition for free calmodulin.

Calmodulin is subject to diverse posttranslation modifications. These involve acetylation, trimethylation, carboxylmethylation, and proteolytic cleavage. Phosphorylation is by far the most common posttranslation modification, and calmodulin has many residues that can be phosphorylated by both serine/threonine kinases and receptor and nonreceptor protein tyrosine kinases. The roles of these modifications are not clear.

Calmodulin is found throughout the cytosol and nucleus of HeLa cells, although it is concentrated around the mitotic apparatus in cells undergoing mitosis, especially around the centrioles and the cytoplasmic furrow during cytokinesis. In response to a rise in calcium, calmodulin exhibits a complex pattern of cellular localization, including a significant redistribution from the cytosol to the nucleus. This stimulus-dependent movement of calmodulin to the nucleus and its activation have also been detected in neurons.

Calmodulin is a calcium sensor protein; hence, it must be able to detect and respond to a biologically relevant range of intracellular calcium concentrations.

TABLE 11.1. Some Well-Characterized Calci-calmodulin (1–5) and Apo-calmodulin (6–9) Binding Proteins

Group	Enzyme or Protein	Putative Function
1. Protein kinases	Phosphorylase b kinase	Regulate glycogen metabolism
	Myosin light chain kinase	Regulate smooth muscle contraction
	CaM kinase I	Multifunctional
	CaM kinase II	Multifunctional, regulate over 30 enzymes
	CaM kinase III	Translation?
	CaM kinase IV	Multifunctional
2. Phosphoprotein phosphatase second-messenger metabolism	Calcineurin (2b)	Metabolic regulation/cell cycle
	Type I adenylate cyclase	Produce cAMP
	cAMP phosphodiesterase	Degrade cAMP
	NO synthase	Produce NO
	Inositol 3-kinase	Produce IP_4
	Ca^{2+}-transport ATPase	Decrease cytosolic Ca^{2+}
	G protein coupled receptors	Various signaling
3. Cytoskeletal/ muscle	Dystrophin	Inhibit F-actin binding
	Spectrin/band 4.1	Inhibit F-actin binding
	Fodrin	Inhibit F-actin binding
	Caldesmon	Inhibit F-actin binding
	MARCKS protein	Inhibit F-actin binding
	Syntrophin	Inhibit dystrophin binding?
4. Metabolism	NAD kinase	Convert NAD to NADP
5. Other	Microtubules	Inhibit tubulin polymerization
6. Actin binding proteins	Brush border myosin I	Actin-activated ATPase, shedding membranes, twisting or moving the microvillus itself, moving microvillar membrane proteins and transporting vesicles
	Myr4	ATP-dependent binding to F-actin
	P190	Actin-activated ATPase
7. Cytoskeletal and membrane proteins	Neuromodulin	Reversible CaM storage, regulation of GTP binding to G_o, and regulation of phosphatidylinositol metabolism
	Neurogranin	Unidentified
	PEP-19	Unidentified
	Igloo	Unidentified

(*continued overleaf*)

TABLE 11.1. (*Continued*)

Group	Enzyme or Protein	Putative Function
	Syntrophin	Currently unknown; clustering Na^+ channels
	IQGAP	Unidentified
8. Enzymes	Phosphorylase b kinase Adenylyl cyclase	Regulation of glycogen metabolism cAMP production
	Inducible nitric oxide synthase	Nitric oxide synthesis
	Glutamate decarboxylase	Decarboxylation of glutamate to CO_2 and γ-aminobutyrate
	cGMP-dependent protein kinase	cGMP- and cAMP-dependent phosphorylation
9. Receptors and ion channels	Inositol 1,4,5-trisphosphate receptor	Inositol 1,4,5-trisphosphate binding
	SR Ca^{2+} release channel	Release of Ca^{2+} from the SR
	Cav1.4 L-type Ca^{2+} channels	Release of Ca^{2+}

Source: Data from Jurado et al. (1999).

Calmodulin fits this profile, as its affinity for calcium, $K_d(\text{Ca}) = 10^{-7}$ to 10^{-6} M, falls within the range of concentrations of free Ca^{2+} ion in the cytosols of most cells.

Calmodulin has four EF-hands, in two lobes, with an overal dumbbell shape. These globular lobes are connected by an eight-turn α-helix (Babu et al., 1985; Kretsinger et al., 1986; Wilson and Brunger, 2000) (Figure 11.1). The molecule is 65 Å long, and each lobe is 25 × 20 × 20 Å. Calmodulin contains seven α-helices (I to VII) comprising residues 7–19, 29–39, 46–55, 65–92, 102–112, 119–128, and 138–148. Its α-helical content is 63%. EF-hands 1 and 4 have canonical helix–loop–helix conformation and are similar to the two EF-hands in parvalbumin. The two lobes share 48% sequence identity; the backbones of the two lobes can be superimposed with a mean square deviation of ≈0.7 Å. Calcium binding loops are composed of the residues 20–31, 56–67, 93–104, and 129–140. EF-hands 2 and 3 are also canonical; however, the second helix of EF-hand 2 is continuous with an eight residue linker, which is in turn continuous with the first helix of EF-hand 3. The entire helix (IV) is $9 + 8 + 11 = 28$ residues long. Calmodulin is stabilized by multiple side-chain interactions between helices and by main chain hydrogen bonds of the antiparallel β-strands of each calcium binding loop. The conformation of the eight-residue linker between lobes 1 and 2 is delicately balanced between kinked and extended helices; the balance can be shifted by crystal packing. Fallon and Quiocho (2003) reported a 1.7-Å resolution crystal structure of calci-calmodulin that has an overall compact ellipsoidal conformation and shows a sharp bend at Ser81 in the linker helix and a more contracted lobe 1,2.

Figure 11.1. Structure of calci-calmodulin (PDB file 1CLL). The N-lobe (EF-hands 1 and 2) (left) is viewed perpendicular to its local, approximate twofold axis; the C-lobe (right) is viewed parallel to its axis. (From Chattopadhyaya et al., 1992.)

The calcium induced changes in calmodulin have been monitored by many physical methods (Burger et al., 1984; Abe et al., 1988; Trewhella et al., 1989; Kobayashi and Takada, 2006). The changes are distributed over the entire molecule but differ from the unfolding changes induced by denaturants. The helical content of calmodulin increases by only 15 to 20%. The largest change in helical content is caused by the binding of the first Ca^{2+} ion. The helical content change is complete after the binding of the third Ca^{2+} ion, but as indicated by Fourier transform infrared spectroscopy, most of the newly formed helices are distorted in comparison with the normal conformation. In contrast, the helices in the crystal structure of calci-calmodulin are not distorted.

Apo-calmodulin is overall a more compact structure than is calci-calmodulin, as seen in the crystal structure (a) and NMR spectra (b) (Figure 11.2) (reviewed by Zhang and Yuan, 1998 and Jurado et al., 1999). The major conformational change induced by the binding of calcium is a significant alteration of the relative orientations of the helices in each EF-hand; they change from 121–144° in apo-calmodulin to 86–116° in calci-calmodulin (Chapter 7). In apo-calmodulin the first helix of the N-lobe and the last helix of the C-lobe are roughly parallel to the central, linker helix. These terminal helices move away from the central helix in calci-calmodulin, and the central helix is (usually) not bent. In apo-calmodulin, the hydrophobic side chains pack together between α-helices in each lobe and form a compact, hydrophobic core. In contrast, in calci-calmodulin, these residues swing out, forming in both lobes, hydrophobic, concave patches that provide binding sites for the target enzymes. The rate of conformational change is faster in the C-lobe, ~490 μs, than in the N-lobe, 20 ms (Park et al., 2008).

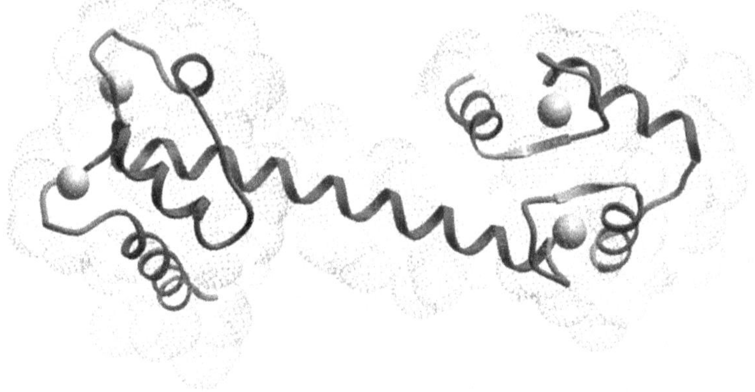

(a)

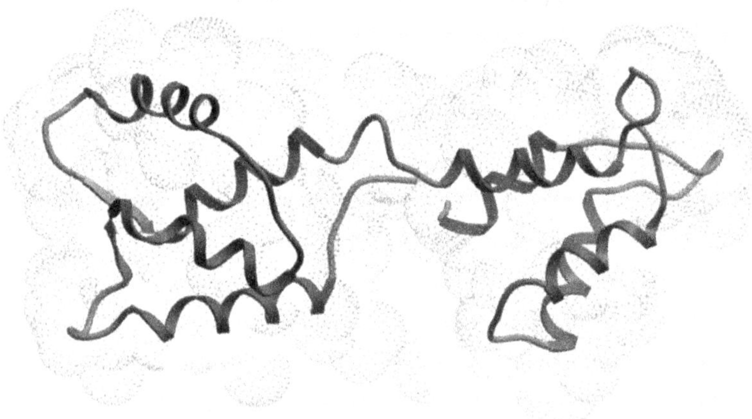

(b)

Figure 11.2. Calcium-induced changes in calmodulin (PDB files 1CLL and 1CFD): (a) calci-calmodulin; (b) apo-calmodulin. (From Wilson and Brunger, 2000; Kuboniwa et al., 1995.)

The removal of calcium changes the environment of many residues, in particular Tyr's, Phe's, and His's (Krebs and Carafoli, 1982) and increases the accessibility of calmodulin to proteolytic enzymes (Wall et al., 1981; Mackall and Klee, 1991). In apo-calmodulin the rates of trypsinolysis at Arg106 and Arg37 are rapid. In calci-calmodulin the bonds at Arg74, Lys75, and Lys77 in the central helix are cleaved more slowly; Arg37 and Arg106 are greatly protected; and the bonds at Lys13, Lys30, Arg86, Arg90, and Arg126 are not cleaved.

The two lobes of calmodulin are separated from each other by the long central helix; however, they still communicate with each other. Calcium binding to the high affinity sites 3 and 4 in the C-lobe alters the conformation of helix II in the N-lobe; even though, EF-hands 1 and 2 remain in the apo-form (Nakashima et al., 1999; Sorensen and Shea, 1998). This communication may arise from calcium induced disruption of interactions between the apo lobes. The interactions between the lobes of apo-calmodulin protect the N-lobe from proteolysis and increase its denaturation temperature by 10°C relative to the isolated N-lobe. Deletions or mutations within the linker helix do not abolish the ability of calmodulin to activate its targets (Persechini et al., 1991).

NMR data show that the observed α-helical conformation of the central helix is a consequence of crystal packing (Barbato et al., 1992). In solution the central helix is very flexible; therefore, the N- and C-lobes do not adopt a defined orientation relative to each but can explore various individual motions. Computer simulation of molecular dynamics of calmodulin in solution (Wriggers et al., 1998) showed that during a 3-ns simulation, the structure exhibits large conformational changes on the nanosecond time scale. The central helix, which unwinds locally upon binding of calmodulin to target proteins, bends and unwinds near Arg74; while the two lobes reorient with respect to each other and α-helices in the N-lobe rearrange to make the hydrophobic, peptide binding site more accessible. This was interpreted as a preparative step in the more extensive structural transition observed in the "flexible tether" region, 74–82, of the central helix upon complex formation.

Hertadi and Ikai (2002) measured the force required for mechanical unfolding of calmodulin using an atomic force microscope by stretching the protein from its N- and C-terminal residues. They measured force (nN) vs. extention (nm) curves for both apo- and calci-calmodulin. A very low force observed upon stretching of apo-calmodulin indicates a relatively high flexibility. In contrast, a relatively high force is needed for full extension of calci-calmodulin; this reflects a more rigid and more stable conformation.

Ion cyclotron resonance mass spectrometry, circular dichroism spectroscopy, and analytical ultracentrifugation indicate the formation of dimers of calmodulin (Lafitte et al., 1999). The use of denaturing solutions (1 : 1 acetonitile/water + 1% formic acid) favors the monomeric form. Whether these dimers are formed in vivo has yet to be established.

Magnesium binds sequentially to the two calcium binding loops in the N-lobe of calmodulin, with affinities such that nearly half of the loops are occupied by Mg^{2+} ions in resting eukaryotic cells (Malmendal et al., 1998). The microscopic Mg^{2+} dissociation constants calculated are $K_{d1}(\text{Mg}) = 8.5 \times 10^{-4}$ and $K_{d2}(\text{Mg}) = 2.0 \times 10^{-3}$ M. This magnesium binding seems to occur without ligation to the residue in the loop position 12(−Z). This position seems to be largely responsible for the major rearrangements induced by binding of the larger Ca^{2+} ion. Consequently, smaller magnesium-induced structural changes are indicated throughout the protein. The two calcium binding loops of the N-lobe have different magnesium binding characteristics. Ligands in loop 1 are better positioned for cation binding, resulting in higher affinity and slower dissociation kinetics than those for loop 2 ($k_{off} = 380\ s^{-1}$ at 37°C compared with ~10,000 s^{-1} at 25°C).

The cation binding parameters of calmodulin have been determined by many researchers, but their results were often inconsistent. However, most show K_d(Ca) in the range 10^{-6} to 10^{-8} M (Haiech et al., 1981; Linse et al., 1991). Magnesium and calcium compete for the same binding sites; reported K_d(Mg) values range from 10^{-3} to 10^{-5} M (Haiech et al., 1981; Kohse and Heilmeyer, 1981). Separated lobes of calmodulin retain the calcium binding properties of the intact molecule. The calcium affinity is sixfold higher for the C-lobe than for the N-lobe. Calcium dissociation from calmodulin occurs in two stages, with rate constants of about 10 s^{-1} and several hundreds of s^{-1} (Malencik et al., 1981; Suko et al., 1985; Martin et al., 1985).

Calmodulin also binds zinc (Permyakov et al., 1988a). Warren et al. (2007) determined the 1.3-Å crystal structure of the N-lobe with zinc bound. Zinc-calmodulin resembles apo-calmodulin despite zinc binding to both EF-hands. The coordination of zinc is different from that of calcium. The first Zn^{2+} ion is tetrahedrally coordinated by Asp22 (Y) and Asp24 (Z) of loop 1 and by two cacodylate molecules, the buffer used in the crystallization conditions. There is also one highly ordered water molecule in this site, coordinated by Asp20 (X) and by Thr26 (−Y) and normally involved in calcium binding. The second Zn^{2+} ion is also tetrahedrally coordinated by Asp 64 (−X) and Glu67 (−Z) of loop 2, as well as by Glu7 and Glu11 of the symmetry-related molecule. Thus, four of five residues of loop 2 normally involved in calcium binding are not liganded to the second Zn^{2+} ion. Comparisons of apo-calmodulin, calci-calmodulin, and cross-linked calci-calmodulin indicates that the zinc first binds to EF-hand 1. These data also suggest that metal ion coordination by Asp22 and Asp24 represents the initial step in the transition induced by metal binding. This initial step is followed by reordering of the N-terminal region of the helix exiting from this first binding loop.

The fourth EF-hand of yeast calmodulin does not bind calcium (Matsuura et al., 1991). Matsuura et al. (1993) created a chimera: Ala1–Ile130 of vertebrate and Asp131–Lys148 of yeast calmodulin. This chimera shows yeast properties, and its enzyme activation profiles are similar to those of yeast calmodulin. A single substitution, Q140E (−Z), in the chimera binds four equivalents of calcium and has vertebrate-type enzyme activation. Alternation of three residues in the fourth EF-hand of yeast calmodulin—substitutions S129D and Q140E and insertion of Ile130—result in binding four equivalents of calcium and vertebrate pattern of enzyme activation.

Calmodulin interacts with small organic compounds: for example derivatives of phenothiazine, trifluoperazine, and chlorpromazine (Levine and Weiss, 1977, 1978; Brostrom and Wolff, 1981). It has two classes of phenothiazine binding sites; the first has a low affinity for phenothiazines; it does not depend on calcium binding. The second binds two phenothiazines, with K_d values of 10^{-5} to 10^{-6} M, only in the presence of calcium. Both hydrophobic and electrostatic interactions are important in binding these organic compounds. For optimal binding they ought to have a bulky hydrophobic region formed by two aromatic rings and a basic nitrogen atom, positively charged at physiological pH, at a distance of not less than three carbon atoms from the hydrophobic rings.

The crystal structure of trifluoperazine bound to calmodulin shows only one trifluoperazine (Figure 11.3) (Cook et al., 1994). This interaction causes distortion of the linker helix and juxtaposition of the N- and C-lobes. The drug makes extensive contacts with residues in the C-lobe of calmodulin but only a few contacts with one residue in the N-lobe. Thirteen residues form 191 contacts (<5 Å) between calmodulin and trifluoperazine. Ten of the residues are hydrophobic; the other three are Glu, but they form only van der Waals contacts. The tricyclic ring of trifluoperazune lies in the hydrophobic pocket in the C-lobe. Vandonselaar et al. (1994) described the crystal structure of a calmodulin–trifluoperazine complex with four drug molecules complexed with calmodulin. Trifluoperazine 1 in their structure corresponds to the sole trifluoperazine in the C-lobe of calmodulin described by Cook et al. (1994).

The solution structure of calci-calmodulin complexed with its antagonist, *N*-(5-aminohexil)-5-chloro-1-naphthalene sulfonamide (W-7) was determined by multidimensional NMR spectroscopy (Figure 11.4) (Osawa et al., 1998). One molecule of W-7 is bound to each of the two lobes. In each lobe the W-7 chloronaphthalene ring interacts with the methyl groups of four Met's and with other

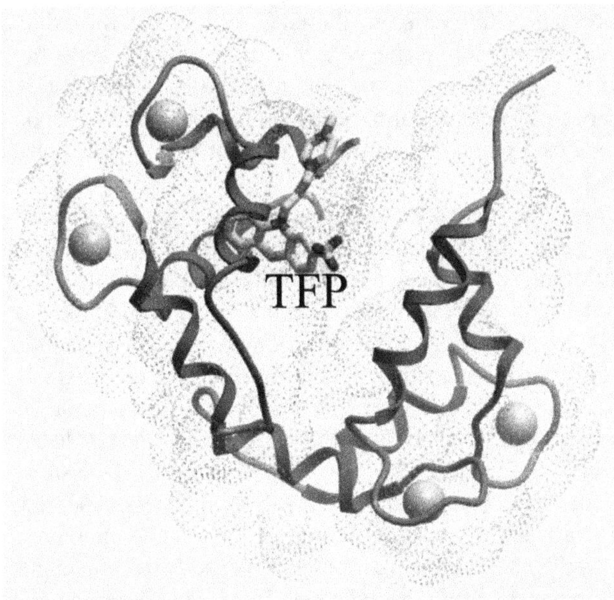

Figure 11.3. Structure of calci-calmodulin complexed with one molecule of trifluoperazine (PDB file 1CTR) The TFP is bound to the hydrophobic patch of the C-lobe. (From Cook et al., 1994.)

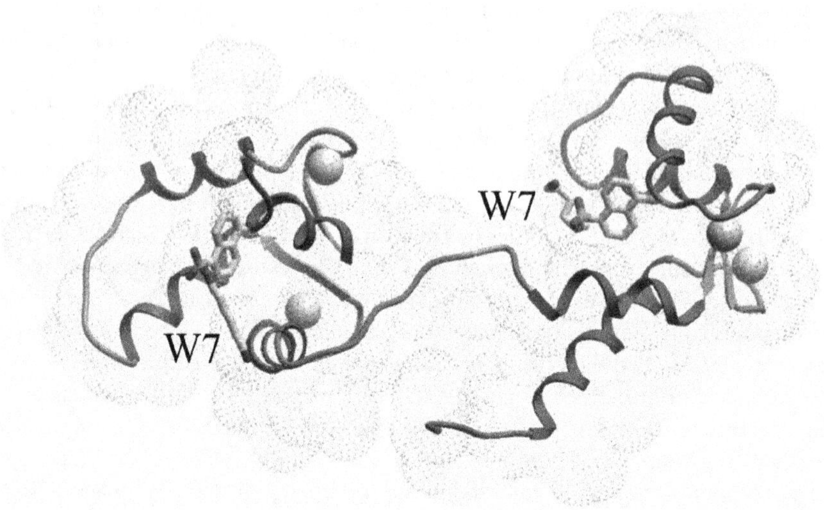

Figure 11.4. Structure of calci-calmodulin complexed with two molecules of *N*-(5-aminohexil)-5-chloro-1-naphthalene sulfonamide (W7); one W7 is bound to the hydrophobic patch of the N-lobe and one to the patch of the C-lobe (PDB file 1MUX) (From Osawa et al., 1998.)

aliphatic or aromatic side chains in the sites responsible for calmodulin–target interactions. The orientation of the W-7 ring differs significantly from that of the phenothiazine ring of trifluoperazine bound to the N- or the C-lobe. These two hydrophobic pockets of calmodulin can accommodate a variety of bulky aromatic rings; this provides a plausible structural basis for the diversity of targets bound by calmodulin.

Many other organic molecules—naphthalene sulfonamide derivatives, felodipine, nifidipine, dibucaine, tetracaine, medocaine, adreamicine, vinblastine, melatonin, and serotonin—bind to calci-calmodulin. The binding of all of these results in an inhibition of the calmodulin-dependent activation of some enzymes.

Calmodulin also binds to a number of synthetic and natural peptides mimicking calmodulin binding regions of their parent molecules. They share little sequence similarity to one another; however, they all have the potential to fold into a basic, amphiphilic α-helix. They have large hydrophobic residues in conserved positions, either 1–5–10 or 1–8–14, which point to one face of a presumed α-helix. The complexes of calci-calmodulin with short α-helical peptides are usually of ellipsoidal shape and much more compact than calci-calmodulin without the peptide (reviewed by Vetter and Leclerc, 2003). The peptide is engulfed by calci-calmodulin within a hydrophobic channel formed by bringing the N- and C-terminal lobes close together (Figure 11.5). The binding of such peptides is driven by hydrophobic interactions. The side chains of Met's, unusually abundant in calmodulin, play an important role in the binding of target peptides. Recent

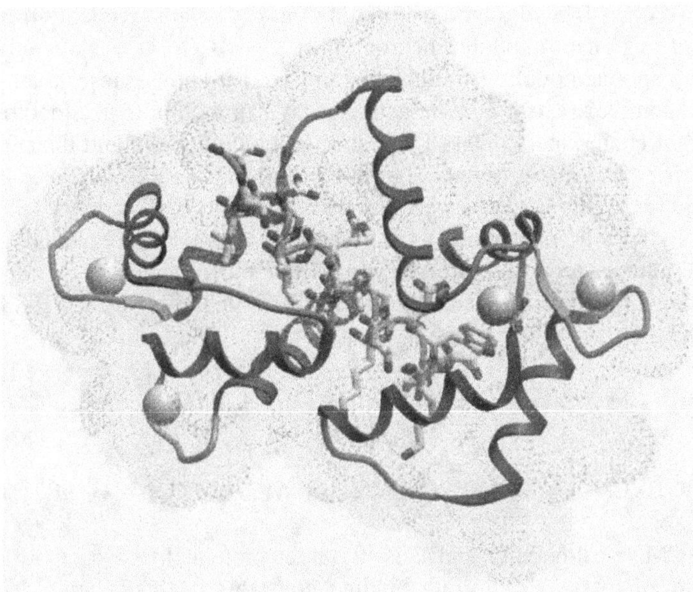

Figure 11.5. Structure of tetra-calci-calmodulin complexed with a calmodulin binding peptide of myosin light chain kinase (PDB file 1CDL). The peptide contacts both N- and C-lobes as well as the linker helix. (From Meador et al., 1993.) (*See insert for color representation of figure*.)

studies revealed the binding of a peptide from the calcium pump to only the C-lobe (Elshorst et al., 1999), dimer formation with fragments of the K^+-channel (Schumacher et al., 2001), and insertion of calmodulin between two domains of anthrax exotoxin (Drum et al., 2002).

The basic, amphipathic α-helices that bind calmodulin are classified according to the relative positions of hydrophobic residues that are known, or anticipated, to be in close contact with calmodulin upon binding. For example, melittin (26-residue peptide from bee venom) has a pattern of hydrophobic residues, 1–8–14. It binds to calci-calmodulin at a 1:1 molar ratio with K_d in the low-nanomolar range. In contrast, binding to apo-calmodulin is much weaker, with K_d about 10 μM. Newman et al. (2008) found that mutations in the C-lobe of *Paramecium* calmodulin alter its interaction with melittin; whereas, N-lobe mutations have no discernible effect. Melittin associates with both calmodulin lobe fragments (1–80) and (76–148), as well as with the C-lobe of full-length apo-calmodulin. In the presence of calcium, all of these interactions were again observed, in addition to which an association with the N-lobe of calci-calmodulin occurs.

Chin and Means (2000) divided calmodulin binding proteins into six classes according to their methods of regulation in the presence and in absence of calcium. Members of class A bind to both calci- and apo-calmodulin irreversibly. Calmodulin is considered a subunit of these proteins (e.g., phosphorylase kinase.

Members of class B bind to apo-calmodulin but dissociate reversibly from calci-calmodulin (e.g., neuromodulin, neurogranin). Class C form weak inactive complexes with apo-calmodulin. At high calcium concentrations, these target proteins form tight complexes with and are activated by calmodulin (e.g., smooth muscle myosin light chain kinase. Class D binds to calci-calmodulin and their functions are inhibited (e.g., kinases of the G-protein receptor). Class E is activated by calci-calmodulin (e.g., calmodulin-dependent protein kinases I, II, and IV). Class F are regulated by specific phosphorylation by another calmodulin-regulated kinase.

Several authors proposed the following scheme of interaction for calmodulin with its targets (Rasmussen and Weisman, 1981; Kilhoffer et al., 1983; Wang et al., 1983):

$$\begin{array}{ccccccccc}
\mathrm{CM} & \leftrightarrow & \mathrm{CM\cdot Ca} & \leftrightarrow & \mathrm{CM\cdot Ca_2} & \leftrightarrow & \mathrm{CM\cdot Ca_3} & \leftrightarrow & \mathrm{CM\cdot Ca_4} \\
\downharpoonleft\upharpoonright & & \downharpoonleft\upharpoonright & & \downharpoonleft\upharpoonright & & \downharpoonleft\upharpoonright & & \downharpoonleft\upharpoonright \\
\mathrm{E\cdot CM} & \leftrightarrow & \mathrm{E\cdot CM\cdot Ca} & \leftrightarrow & \mathrm{E\cdot CM\cdot Ca_2} & \leftrightarrow & \mathrm{E\cdot CM\cdot Ca_3} & \leftrightarrow & \mathrm{E\cdot CM\cdot Ca_4}
\end{array}$$

in which CM is calmodulin and E is its target protein. It is extremely difficult to measure all of the equilibrium binding constants in this scheme. Nevertheless, some of the constants were determined for interaction of calmodulin with phosphorylase b kinase and myosin light chain kinase (reviewed by Cox, 1988). Effective K_d values for interactions of calmodulin with most of the target proteins are within the range 10^{-9} to 10^{-7} M.

Calci-calmodulin binding domains in different target proteins show very little similarity in primary sequence. These binding sites are amphipathic α-helices, typically ~20 residues long, with basic and hydrophobic residues intermingled. In contrast, the binding sites for apo-calmodulin show greater similarity. A short peptide of about 20 residues that is responsible for binding of calci-calmodulin, called a *calmodulin binding region*, was found in many calmodulin regulated proteins and in other types of calmodulin binding proteins (Rhoads and Friedberg, 1997). Calmodulin binding regions of target proteins usually have a hydrophobic sequence containing basic amino acids, mostly Arg; the basic and hydrophobic residues of the region are normally spaced to produce an amphiphilic helix. An aromatic residue, frequently Trp, is almost always present in the N-terminal part of the region (Vorherr et al., 1990, 1993).

Several recognition motifs for calmodulin interaction have been found. According to Rhoads and Friedberg (1997), the calmodulin dependent calmodulin binding motifs can be grouped into two major classes, 1–8–14 and 1–5–10, based on the position of conserved hydrophobic residues. In contrast, Hoeflich and Ikura (2002) proposed that the calcium-dependent calmodulin binding motifs fall into three groups (1–14, 1–10, and 1–16), which are distinguished by their spacing of bulky hydrophobic and basic amino acids and are bracketed by aromatic residues near either end. The "IQ" calmodulin binding motif, IQXXXRGXXXR, appears in tandem repeats and binds multiple calmodulin molecules in a predominantly calcium-independent manner.

Calmodulin activated enzymes are inferred to have a regulatory domain containing a calmodulin binding site and a catalytic domain. In the absence of calmodulin, the regulatory domain inhibits catalysis, but the binding of calmodulin removes the internal inhibition. Calmodulin binding/autoinhibitory domains regulate enzymes through a fine balance between internal binding, causing inhibition, and external calmodulin binding, producing activation (James et al., 1995). Calmodulin-activated enzymes all have cationic, amphipatic regions that are similar in these characteristics to melittin or mastoparan. Calmodulin binding enzymes are also predicted to contain at least one site that binds cationic, amphipatic regions of a peptide sequence. Jarett and Madhavan (1991) found such regions in calcineurin, myosin light chain kinase, phosphorylase b kinase, phosphodiesterase, and NAD kinase. They showed that these enzymes bind biotinylated melittin and an anti-melittin antibody.

Myosin light chain kinase associates with fluorescently labeled calmodulin at a rate of $2.8 \times 10^7\ M^{-1}s^{-1}$ and dissociates at $0.031\ s^{-1}$ (Kasturi et al., 1993). Caldesmon, a target of both calmodulin and S100, associates with calmodulin at $5.3 \times 10^8\ M^{-1}s^{-1}$ and dissociates at $57\ s^{-1}$.

Some targets of calmodulin serve functions that appear contradictory (J. D. Johnson and Mills, 1986). For example, calmodulin stimulates adenylate cyclase to increase cellular cAMP levels, and at the same time it stimulates phosphodiesterase to decrease cellular cAMP levels. One of the most challenging questions in calmodulin research is how this single protein can regulate scores of different proteins with any degree of selectivity or specificity. Compartmentalization of calmodulin appears unlikely since calmodulin is a soluble cytoplasmic protein found in practically all cellular organelles. That different calci-calmodulin species activate different target proteins also seems to be unlikely, since it now appears that once calmodulin has three bound Ca^{2+} ions, it can bind and activate most target proteins. Drug interactions with calmodulin provided evidence that calmodulin is an allosteric protein with respect to its drug/protein binding sites (Johnson and Mills, 1986). Its interactions with some of its target proteins may be mediated by ligand binding to allosteric sites. If endogenous regulatory ligands bind to an allosteric regulatory site on calmodulin, they might allosterically potentiate its interaction with one protein over another, resulting in selective activation. Another possible mechanism for selective activation of one target protein over another is a ligand-induced conformational change in a target protein; this would alter its affinity for calmodulin. For example, the binding of calmodulin to myosin light chain kinase increases its affinity for its substrate, myosin light chain. This also means that substrate binding to myosin light chain kinase increases its affinity for calmodulin. Thus, calmodulin binding to this enzyme increases its affinity for substrate, and substrate binding to this target protein increases its affinity for calmodulin.

Cox (1988) suggested, based on initial results, that calmodulin contains only a single region, including the linker helix, recognized by its various targets and that these calmodulin binding domains are encoded by a single exon. Recent results indicate more complex patterns. For example, use of a label selection technique with trace acetylated calmodulin demonstrates that the face of the central helix containing Lys75 is involved in interaction with myosin light chain kinase and also suggests that an additional contact occurs near loop 1 (Jackson et al., 1987). A photoaffinity labeling study of the interaction of calmodulin with the plasma membrane calcium pump (Vorherr et al., 1990) showed that only the C-lobe of calmodulin has high affinity for the 20-residue peptide, corresponding to the calmodulin binding domain of the ATPase (K_d in the nanomolar range).

Nitric oxide synthase interacts with both the central linker and the C-terminus of calmodulin as seen in the crystal structure (Aoyagi et al., 2003) (Figure 11.6). An α-helix of the enzyme binds the linker helix through Met76 in antiparallel orientation through extensive hydrophobic interactions. The Phe496 side chain of the enzyme orients the C-lobe, allowing Met144 and Met145 of calmodulin to interact extensively with the N-terminal half of the peptide.

Calmodulin modulates its targets in several ways. The edema factor, a calmodulin-activated adenylyl cyclase, is important in the pathogenesis of anthrax. This exotoxin enters host cells through a transporter (protective antigen) produced by the pathogen. Activated as an adenylyl cyclase by calmodulin, it produces cAMP from ATP in an uncontrolled fashion, and to this extent impairs the host immune defenses. In the crystal structure of the complex, calmodulin is inserted between two domains of edema factor instead of being bound to an exposed helix or an external loop (Figure 11.7) (Drum et al., 2002; reviewed

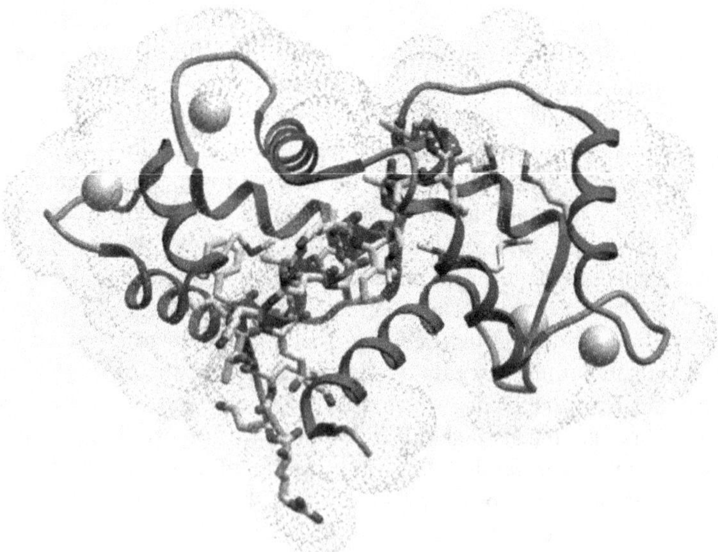

Figure 11.6. Structure of tetra-calci-calmodulin complexed with a fragment of NO-synthase (PDB file 1NIW). (From Aoyagi et al., 2003.) (*See insert for color representation of figure.*)

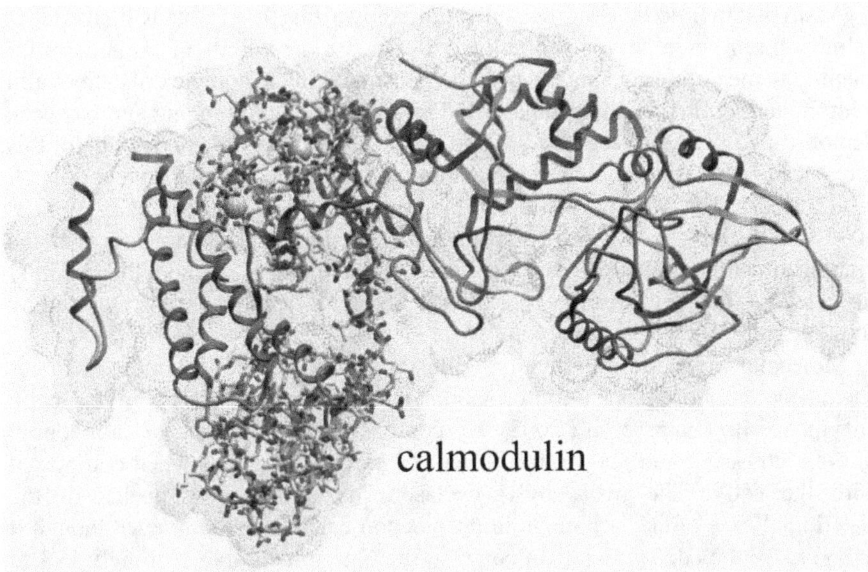

Figure 11.7. Di-calci-calmodulin complexed with *Bacillus anthracis* exotoxin (edema factor, adenylyl cyclase) (PDB file 1K93). Calmodulin is represented as a stick figure, the exotoxin as a ribbon model. (From Drum et al., 2002.) (*See insert for color representation of figure.*)

by Vetter and Leclerc, 2003). The bound calmodulin is elongated and its cental helix is partially unwound. Only the C-lobe has calcium bound; 53 residues of calmodulin interact with 63 residues of the exotoxin. Such extensive interactions cause large changes in relative orientation of the two domains of the exotoxin. Upon calmodulin binding, the C-terminal domain of edema factor moves by 15 Å and turns by 30° relative to its N-domain. Moreover, several short β-sheets, α-helices, and loops change conformations. The hydrophobic region, typical for calci-calmodulin, is exposed and forms a peptide binding channel, which interacts with an α-helix of the edema factor. This is a positively charged amphiphilic helix containing hydrophobic residues in positions 1, 5, and 10.

The binding of calmodulin to edema factor increases its catalytic activity 1000-fold. The activation of the enzyme is not achieved by changing the architecture of the active site dramatically, or by removing an autoinhibitory domain as in other calmodulin-activated enzymes (reviewed by Vetter and Leclerc, 2003). Calmodulin does not bind near the active site. The structures of the N-terminal domain of edema factor alone and bound to calmodulin can be superimposed and show no large conformational changes as the result of calmodulin binding. Instead, several loops near the active site become stabilized. This relatively minor change around the active site enhances the catalytic activity through better substrate binding and positioning in the active site.

A simple thermodynamic consideration shows that if a protein binds calci-calmodulin preferentially, this binding will increase the affinity of calmodulin for calcium in the complex. This has been demonstrated in the complex of calmodulin with inducible nitric oxide synthase (iNOS). A calmodulin binding site has been demonstrated between residues 504 and 532 of iNOS. A synthetic peptide of this sequence binds calmodulin, and the binding is highly calcium dependent (Anagli et al., 1995). The affinity of the peptide for calmodulin in 0.1 mM calcium is 1000-fold greater than it is in the absence of calcium. This suggests that the binding of calmodulin to these sequences could increase calmodulin's affinity for calcium 1000-fold, from ~11 mM (see above) to about 11 nM (reviewed by Jurado et al., 1999).

Molecular dynamics simulations illustrate how dicalci-calmodulin forces edema factor into an open conformation, thereby forming its adenylyl cyclase enzymatic site (Laine et al., 2008). By contrast, the linker helix of calmodulin, in both tetra-calci and apo forms, bends and partially unwinds when complexed with the active site of adenylyl cyclase. The hydrophobic pocket of the N-terminal lobe of calci-calmodulin opens upon calcium binding; even though it is bound by the edema factor. In contrast, the C-terminal lobe is strongly locked in the open conformation by the edema factor, and the removal of calcium induces a collapse of the catalytic site of the edema factor.

Vogel (1994) and Zhang and Yuan (1998) summarized features of calmodulin which allow it to be a universal and versatile calcium regulator: (1) (modest) positive cooperativity of calcium binding to all four binding sites of calmodulin in the presence of target proteins; (2) high Met content of the two hydrophobic patches on the two lobes of calmodulin, which creates a flexible and pliable, yet sticky interaction surface that does not place high demands on the specificity of the interaction; and (3) intrinsic flexibility of the central linker region that allows the two lobes of calmodulin to slide over the surface of the bound α-helix to find the most favorable orientation.

In vitro biochemical reactions are usually studied in dilute solution. The cytoplasm of eukaryotic cells is crowded with mobile and immobile macromolecules. Such crowded conditions exert volume exclusion and other entropic forces that often affect chemical equilibria and reaction rates (reviewed by Zhou et al., 2008). Homouz et al. (2009) used fluorescence spectroscopy to measure the distance between N- and C-lobes of calmodulin in the absence and presence of Ficoll 70. At the same time, they used coarse-grained molecular simulations that adequately probe large-scale fluctuations of the structure of calmodulin in a spatiotemporally complex system, providing complementary analysis to the experimental findings. They found that crowding stabilizes several different compact conformations and suggested that both EF-hands in the C-lobe are flexible and can be thought of as a switch, whereas those in the N-lobe are stiff, analogous to a rheostat. New combinatorial signaling properties may arise from the differential plasticity of the two lobes of calmodulin in the presence of crowding. The interpretation of these results is that crowding changes the free energy barrier between the minima corresponding to different compact conformations of calmodulin. As a consequence, there is a shift in the distribution of substates to those more likely to bind calcium and/or target enzymes.

Apo-calmodulin also binds to some targets, but it binds them differently, utilizing different binding motifs such as the IQ motif and noncontiguous binding sites (reviewed by Jurado et al., 1999). The apo-calmodulin binding targets are a diverse group of at least 15 (Table 11.1) proteins, including enzymes and actin-binding proteins, as well as cytoskeletal and other membrane proteins, including receptors and ion channels. Calci- and apo-calmodulin may bind to the same or overlapping regions of a target: for example, a skeletal muscle calcium release channel (Moore et al., 1999). Much of calmodulin is bound in a calcium-independent manner to membranes within the cell, and the proportion-bound changes with cell growth and density, suggesting that it may be a storage form. When binding sequences are localized and peptides synthesized, these peptides are bound preferentially by calci-calmodulin. This observation suggests that binding to apo-calmodulin utilizes many of the same types of protein–protein interactions that account for calci-calmodulin binding (reviewed by Jurado et al., 1999).

Calmodulin is highly enriched in brain neurons. Identified neuronal calmodulin binding proteins include adenyl cyclase VII, calcineurin A, calci-calmodulin kinase II, calci-calmodulin kinase kinase, neuronal nitric oxide synthase, and various calcium ion channels. Calmodulin is involved in controlling synaptic vesicle recruitment via activation of calci-calmodilin-dependent protein kinases and the phosphorylation of the synapsin proteins on a synaptic vesicle (reviewed by Burgoyne and Weiss, 2001). This phosphorylation leads to the dissociation of synapsins from the vesicle, release of synaptic vesicles from a cytoskeletal attachment, and vesicle movement to the presynaptic plasma membrane. Calmodulin dependent kinases are abundant in the brain and thymus. They are involved in a cascade that consists of three calci-calmodulin dependent kinases: calmodulin kinase kinase (CaMKK), calmodulin kinase I (CaMKI), and calmodulin kinase IV (CaMKIV) (Soderling, 1997). CaMKK acts as an upstream activator of CaMKI and CaMKIV by phosphorylating Thr177 in CaMKI and Thr196 in CaMKIV. This results in the increased efficiency of CaMKI and CaMKIV in the phosphorylation of various protein substrates, including MAP kinases. CaMKI and CaMKIV are localized predominantly in the cytoplasm; CaMKIV and CaMKK can also be localized in the nucleus. One of the well-characterized functions of nuclear CaMKIV involves the regulation of transcription through the phosphorylation of CREB, the cAMP-regulated transcriptional activator that activates transcription of target genes in part through direct interactions with the coactivator calcium binding protein, p300.

Calmodulin takes part in many events of calcium dependent modulation of neuronal function, including postsynaptic changes during synaptic plasticity, alternation of gene expression, calcium dependent inactivation of voltage gated Ca^{2+} channels, activation of calcium dependent potassium channels, modulation of a glutamate receptor, and metabolic function (reviewed by Burgoyne and Weiss, 2001).

11.1.2. Troponin C

Troponin is the regulatory complex of the myofibrillar thin filament; it plays a critical role in regulating excitation–contraction coupling in muscles. It is composed of three subunits: troponin C (TnC), the 18-kDa calcium binding subunit; troponin I (TnI), the 23-kDa inhibitory subunit; and troponin T (TnT), the 35-kDa subunit that attaches troponin to tropomyosin and to the myofibrillar thin filament. TnI prevents contraction in the absence of calci-TnC. In the contraction–relaxation cycle, calcium binds to the two EF-hands in the N-lobe of TnC. The binding of calcium to TnC triggers a change in the orientation of the troponin–tropomyosin complex; this may either remove a steric hindrance to the approach of the myosin head to actin or may allow the release of inorganic phosphate from the actin myosin head ADP $\cdot P_i$ complex. Such thin filament activation, in turn, facilitates cross-bridge cycling, myofibrillar activation, and contraction of the muscle. The intrinsic length–tension properties of myofibrills are mediated primarily through calcium-responsive activation of the thin filament.

The troponin subunits are encoded by single copy genes in the human genome. There are at least two genes for TnC in tissues of higher animals (reviewed by Filatov et al., 1999). One gene codes the isoform of fast skeletal muscle fibers; the second gene encodes the isoform found in slow skeletal muscle and heart. All isoforms of TnC have low isoelectric points, and they are similar in sequence.

TnC, like calmodulin, has a dumbbell shape (Figure 11.8) (Herzberg and James, 1985; Satyshur et al., 1988). Turkey skeletal muscle di-calci-TnC is 75 Å long, with 67% of its residues in helices. The two pairs of domains are separated by a nine-turn α-helix with no contacts between the lobes. The distance between the centers of mass of the two lobes is about 40 Å. The N-terminal lobe includes the first 85 amino acid residues; the C-lobe comprises residues 97 to 162. Like calmodulin, TnC has four calcium binding sites: two in each lobe. Although the crystals were obtained from solutions containing excess calcium, only the two high-affinity sites of the C-lobe (EF-hands 3 and 5) are occupied by calcium. Loops 1 and 2 are the low-affinity calcium binding sites. A long helix spanning residues Phe75–Asp106 connects the two lobes. The helix is slightly bent, such that the angle between the D and E helix axes is about 10°. The helices in the two calci-EF-hands of the C-lobe are canonical. In contrast, the four helices of the apo-N-lobe form a nonideal, antiparallel helix bundle.

	1(+X)	3(+Y)	5(+Z)	7(−Y) (C=O)	9(−X)	12(−Z) (bident)
EF-1	D30	D32	G34=O	*D*36	S38	E41
EF-2	D66	D68	S70	*T*72	D74	E77
EF-3	D106	S108	N110	*F*112	D114	E117
EF-4	D142	N144	D146	*R*148	D150	E153

The Glu's conserved in position 12(−Z) of the calcium binding loops donate two oxygen atoms of their carboxyl groups to coordinate the Ca^{2+} ion.

Figure 11.8. Structure of di-calci-troponin C from skeletal muscle (PDB file 1TOP). The N-lobe is apo; the C-lobe, viewed parallel to its local twofold axis, has two bound Ca^{2+} ions. (From Satyshur et al., 1994.)

TnC from skeletal muscles has four calcium binding sites; whereas, in cardiac troponin, C EF-hand 1 does not bind calcium. It is possible to make this first EF-hand bind calcium by deletion of Val28 and conversion of amino acids 29 to 32 to those found at the homologous positions in EF-hand 1 of skeletal TnC (Putkey et al., 1988, 1991).

The overall solution structure of calci-TnC is dumbbell shaped (Sia et al., 1997). The N-lobe of calci-cardiac TnC is significantly more compact than the C-lobe of calci-skeletal TnC. In particular, helix B of EF-hand I is in the closed conformation with interhelical (helices A and B) angle of 142°.

The solvent exposed, central (linker) helix, a rare feature in protein structures, is at the edge of conformational stability and has exactly the sensitivity required to respond to energy changes produced by calcium binding (Schutt, 1985). The bend in the central helix in TnC occurs at Gly92; calmodulin has no Gly in this position. The central helix is unwound at the center and bent by 90° in the complex of TnC with the N-terminal fragment of troponin I (residues 1 to 47) (Vassilyev et al., 1998). As a result, TnC in the complex has a compact conformation allowing direct interactions between its N- and C-lobes.

Calmodulin lacks the N-terminal α-helical arm that preceeds helix A of the first EF-hand of TnC, and its central helix is one turn shorter than its counterpart in TnC, due to the absence of residues 88 to 90 (Gulati et al., 1993). Deleting both of these regions in TnC leads to the appearance of calmodulin-like regulation as tested by smooth muscle contractility and by the activation of phosphodiesterase. Surprisingly, the calcium binding capacity of the mutant and the effect on maximally calcium activated tension of skinned rabbit psoas muscle fibers are both conserved. When the TnC region ^{85}EDAKG^{90} of the central helix is replaced by calmodulin-specific DTD, it generates an even more effective calmodulin mimic, whether or not the N-helix is also retained. Gulati et al. (1995) next mutated

Arg11 in rabbit skeletal TnC to Ala because the interactions of Arg11 with distal residues in the N-lobe seem to link the N-terminal helix to the rest of the structure. This mutant activates phosphodiesterase to 50% of the level of activation by calmodulin. If the ^{91}KGK^{93} triplet is also deleted, phosphodiesterase activation increases up to ~80% of that of calmodulin. Both mutants retain near complete TnC activity. Computer simulations of the molecular dynamics of tetra-calci-calmodulin, tetra-calci-TnC and tetra-calci-R11A-TnC show two types of structural changes, both of which are considered to have functional roles in calmodulin: (1) formation of a more globular form and (2) reorientation of the calcium binding lobes around the central helix. Both the wild type and the R11A mutant are compacted, but differently, so that reorientation of the N- and C-lobes was found only in the simulated structure of the mutant.

The two calcium binding loops of EF-hands 3 and 4 in TnC are canonical (Herzberg and James, 1985). This specific arrangement guarantees proper location of the calcium ligands. It consists of two reversive turns on each side of the loop and four Asx turns (a cyclic hydrogen-bonded structure, including oxygen of the side chain of residue n and amide nitrogen of the main chain of residue $n+2$). The fourth coordination positions in both loops of TnC are occupied by water molecules, which are located within the distance of a hydrogen bond from Asp. This results in indirect interaction between the Ca^{2+} ion and negatively charged carboxylate. The same framework is characteristic of the loops in parvalbumin and calbindin despite the variability of side chains in equivalent positions. The locations of the Ca^{2+} ion and of the coordinating water molecule are conserved in all these cases.

Calcium release causes a decrease in the helical content in TnC from 50% to 34% (Johnson and Potter, 1978); the environments of many residues change drastically (Gillis et al., 2000, 2003). Proton NMR indicates that in the apo-state its hydrophobic cores are conserved and there is essentially no unfolding of the protein (Evans et al., 1980). The removal of calcium changes the hydrophobic contacts of the side chains of many Phe's in TnC and makes all Tyr side chains in TnC from skeletal and cardiac muscles more accessible to water; however, this accessibility is lower than that for low-molecular-mass compounds. Calcium removal increases the accessibility of Cys's, especially Cys84, in cardiac TnC, to chemical reagents (Ingraham and Hodges, 1988). Even surface Lys's change their reactivities upon calcium removal. For example, Lys52 in TnC becomes more reactive to acetic anhydride in apo-TnC (Hitchcock, 1981).

Small-angle x-ray scattering (Fujisawa et al., 1989; 1990) shows that the two lobes of TnC move closer to each other and that the molecule shrinks along its long axis upon calcium binding in the absence of magnesium. When calcium binds to TnC, its radius of gyration decreases from 27.8 Å to 24.3 Å. The solution structure of calci-TnC is essentially the same as that of magnesi-TnC. The effect of calcium binding to the low-affinity sites in the presence of magnesium is quite different. Scattering experiments on tryptic fragments of TnC showed that the C-lobe shrinks, with the radius of gyration changing from 17.0 Å to 14.9 Å, while the N-lobe swells from 13.9 Å to 15.0 Å upon magnesium binding.

The model proposed by Herzberg, Molt, and James (HMJ model) (Herzberg et al., 1986). predicts changes in the C_α coordinates in several regions of the N-lobe upon calcium binding to EF-hands I and II of TnC. The changes are predicted to be minimal for helices N, A, and D, but substantial for helices B and C and the linker region between these two helices. The HMJ model predicts calcium-induced exposure of a short patch of hydrophobic residues in helix B. According to the model, the N-lobe of skeletal TnC has a closed conformation in the absence of bound calcium and an open conformation in its presence. Houdusse et al. (1997) determined the crystal structures of di-calci-TnC and of tetra-calci-TnC from rabbit at 2.0 Å resolution. The change in conformation of the N-terminal lobe is close to that predicted by the HMJ model. The linker helix connecting the N- and C-lobes is more stabile in the di-calci than in the tetra-calci state. A small number of differences in the sequences of the two lobes accounts for the fact that the C-lobe is stable only in the open, calcium-bound state; whereas the N-lobe can switch between two stable states.

Calcium dissociation constants for TnC from skeletal and cardiac muscles have been measured by many authors, and these data are in rather good agreement. K_d(Ca) values for the two strong sites of the C-lobe are between 10^{-8} to 10^{-6} M for both skeletal and cardiac troponin C. K_d values for the two weak sites of skeletal and one of cardiac TnC are within 10^{-6} to 10^{-4} M. K_d(Mg) values for the two strong sites are within 10^{-4} to 10^{-3} M. The weak sites in TnC are considered to be regulatory; the binding of calcium triggers muscle contraction (Szczesna et al., 1996).

The Glu41Ala mutant has 100-fold reduced calcium affinity in site I and 10-fold reduction in site II (Li et al., 1997). Asp65Ala does not bind calcium at site II and is functionally inactive (Brito et al., 1991).

Cleavage of skeletal TnC into lobe 1,2 and lobe 3,4 does not change its calcium binding parameters. This implies that the interaction, if any, of the two lobes does not affect their respective affinities for calcium. ^{1}H NMR spectra (Findlay and Sykes, 1993; Findlay et al., 1994) indicate that the secondary structure and global fold of the apo-N-lobe (residues 12 to 87) turkey skeletal muscle TnC are very similar to those expected of the corresponding region in its crystal structure (Gagne et al., 1995). The overall structure of the apo-N-lobe appears not to be affected by the loss of the N-helix, which precedes the first EF-hand.

Despite some contradictions in the literature concerning divalent cation dissociation kinetics in TnC, one can draw the following conclusions. The calcium dissociation rate constant for the two low-affinity sites and the rate constant for a dissociation-induced conformational change are comparable with each other and are within the range from tens to hundreds of s^{-1}. Calcium dissociation from high-affinity binding sites causes a process with a rate constant of several s^{-1} (i.e., approximately the same as for parvalbumin). It is reasonable to assume that this process reflects the dissociation of the cations.

It is important to know the conformation of TnC in the troponin complex (TnC, TnI, and TnT) (Gagne et al., 1997). Small-angle neutron and x-ray scattering showed that chicken skeletal muscle TnC in the troponin complex adopts a dumbbell conformation in both the calci and apo states, in contrast to the bent conformation found in the human cardiac TnC crystal structure (King et al., 2005). The rigid central helix of calci-TnC binds to the inhibitory segment of TnI (Vinogradova et al., 2005). The removal of calcium causes the central helix of TnC to collapse and to release the inhibitory segment of TnI; this changes the conformation from an extended loop in the presence of calcium to a short α-helix in its absence. Neutron scattering showed that TnC within thin filaments adopts a conformation consistent with the extended dumbbell structure; this is different from the structures found in the crystals of various troponin complexes (Matsumoto et al., 2004).

The interaction between TnC and TnI involves residues 51–62, 84–100 (N-lobe) and 127–138 (C-lobe) of TnC and residues 96–103 and 104–115 of TnI (M. Chandra et al., 1994; T. Kobayashi et al., 1995). The region of interaction with TnT is possibly located in the N-lobe of troponin C. The binding of calcium to TnC changes its structure, and this change is transferred to TnI and then to tropomyosin.

SAXS and neutron scattering data indicate that TnC in complex with TnI has an extended dumbbell shape with a radius of gyration 23.9 ± 0.5 Å and a maximum linear dimension of about 72 Å (Olah et al., 1994), similar to the values obtained from the crystal structure of TnC. TnI is even more extended than is TnC with radius of gyration 41 ± 2 Å and a maximum length of about 118 Å. The centers of mass for both components of the complex are approximately coincident (<10 Å separation), as are their long axes. Olah and Trewhella (1994) reported a model structure for calci-TnC-TnI derived from small-angle x-ray and neutron scattering data using a Monte Carlo method. In their model, TnI is a spiral that wraps around TnC, which adopts a dumbbell conformation. The TnI spiral has the approximate dimensions of an α-helix and winds through the hydrophobic "cups" in both lobes of TnC.

Incorporation of TnC into the troponin complex increases the calcium affinities of both strong and weak sites by an order of magnitude. The binding of calcium to sites 1 and 2 of TnC in the complex seems to be cooperative.

Information about interactions of TnC and TnI with TnT is rather limited, partly due to the relative insolubility of the entire troponin heterotrimer. The N-terminal portion, residues 159 to 221, of the T2 fragment (159–259) of TnT has been proposed to be the major calci-TnC binding region, whereas the C-terminal portion of this fragment (206–258) is considered as the secondary binding region for TnC (Tanokura et al., 1983). It was suggested that TnT, in association with TnI, may participate in the signal transduction of calcium binding by TnC to tropomyosin. The region around residue 155 probably acts as a linker between troponin and actin–tropomyosin in this signal transmission.

11.1.3. Essential Light Chains of Myosin

Myosin, as found in various muscles, consists of two heavy chains and of two essential (enzymatic) light chains and two regulatory light chains. The heavy chain has a large, globular catalytic head, which contains the sites for interactions with actin and with ATP. A second region of the myosin head, known as the *light chain binding domain*, consists of a single α-helix, which is stabilized by the two light chains that surround it. Both essential and regulatory light chains consist of four EF-hands; both are congruent with calmodulin and troponin C in that their EF-hands 1 resemble one another more closely than they resemble other EF-hands within the same light chain.

None of the EF-hands of essential light chains bind calcium. Only EF-hand 1 of the regulatory chain binds calcium; this is independent of its association with the heavy chain (Werber, 1978). The binding of calcium to the light chains triggers the contraction of scallop muscles (Kendrick-Jones and Jakes, 1977; Scholey et al., 1981). The role of the myosin light chain in these muscles is to switch on and off the binding of the myosin head to actin in response to the changing calcium levels within the muscle.

The essential (enzymatic) light chain of myosin is important for structural stability of the α-helical lever arm domain of the myosin head, but its function in the contraction of striated muscle is poorly understood (reviewed by Hernandez et al., 2007). Two essential light chain isoforms are expressed in fast skeletal muscle, a long isoform and its N-terminal ~40-amino acid shorter counterpart; whereas, only the long essential light chain is observed in the heart. The positively charged N-terminus of the long essential light chain makes direct contact with the negatively charged C-terminus of actin during the contraction of striated muscle (reviewed by Hernandez et al., 2007). The essential light chain is bound to the lever arm of the myosin head and, together with the regulatory light chain, stabilizes the 85-Å α-helical region of the myosin heavy chain. Whereas the functional significance of the myosin lever arm has been well documented and described, the specific roles of essential light chain and/or regulatory light chain in force development and the regulation of muscle contraction have yet to be elucidated.

The essential light chains of vertebrates do not bind calcium; however, scallop essential light chains binds one Ca^{2+} ion in EF-hand 3 or 4 (Kwon et al., 1990). The affinity for calcium is increased when scallop essential light chain is complexed with a fragment of the heavy chain and with the regulatory light chain.

Both essential and regulatory light chains bind to two successive IQ motifs of the heavy chain, located at the junction of its head and tail regions (reviewed by Schaub and Heizman, 2008). Although EF-hands 1 and 2 form a lobe, as do EF-hands 3 and 4, both light chains are elongated; their C-termini are near the head–tail hinge of the heavy chain. Their N-termini are near the actin binding site of the myosin head (Kendrick-Jones and Jakes, 1977). Both essential light chains (MLC1V, 21.9 kDa, 195-amino acid residues) and regulatory light chains (MLC2V, 18.8 kDa, 166-amino acid residues) of heart ventricle are also expressed in slow skeletal muscle; while the heart atria have their own isoforms (MLC1A and MLC2A) (reviewed by Schaub et al., 1998).

The IQ motif (23 to 25 residues) with a consensus core, IQxxxRGxxxR (where x is any amino acid), is repeated tandemly (two to six times) in the heavy chains of many myosins. This motif confers on these proteins the capacity to bind myosin light chains or calmodulin (Houdusse and Cohen, 1995). Crystal structures show that the lobes of the light chains adopt three distinct conformations when bound to the heavy chain. Two of these conformations correspond to the "open" and "closed" forms that the lobe assumes when divalent cations are bound or absent, respectively (Chapter 7). The third state is an unusual "semiopen" conformation with no metal bound. This conformation may depend not only on divalent cation binding but also on the precise pattern of residues in the target peptide.

Myosin light chain kinases that catalyze the rapid incorporation of phosphate into a specific Ser residue of some light chains are found in both smooth and skeletal muscle (reviewed by Stull et al., 1980; Holroyde et al., 1979; Takashima, 2009). Dephosphorylation of these light chains is catalyzed by myosin light chain phosphatase. These targeted light chains are called *phosphorylatable* or *P-light chains*. Phosphorylation of these P-light chains is an effective mechanism for control of contraction in smooth muscles. The calcium sensitivity of the myosin light chain kinase is mediated by calmodulin. Calcium first binds to calmodulin, which, in turn, binds to the inactive myosin light chain kinase to form a catalytically active phosphotransferase (reviewed by Somlyo and Somlyo, 2003). Experimental data on smooth (intestinal or arterial) muscles show that there is a close correlation between the degree of phosphorylation of the myosin light chains and tension development. Inhibition of the light chain kinase results in dephosphorylation of the myosin light chains and inactivation of tension even in the presence of calcium. No such effects were revealed for striated muscle fibers (Kerrick et al., 1980).

Actomyosin activity in secretory tissue is also regulated by changes in the phosphorylation state of the myosin light chains. Secretory activity is increased by calci-calmodulin interaction with myosin light chain kinase; however, it is decreased by phosphorylation by cAMP-dependent kinase or protein kinase C (Trifaro et al., 1985; Umekawa et al., 1985).

The functional significance of myosin light chain phosphorylation in skeletal and cardiac muscles is not clear. In smooth muscles the phosphorylation of myosin light chains is one of the regulatory mechanisms of muscle contraction (Bremel, 1974; Sommerville and Hartshorne, 1986). Calci-calmodulin also interacts with caldesmon, which is located on the thin filaments of smooth muscles and interacts with actin and tropomyosin (Sommerville and Hartshorne, 1986; Marston et al., 1988; Medvedeva et al., 2001). This binding results in elimination of the inhibition of the actomyosin interaction. In smooth muscles there are at least two mechanisms of calcium regulation of contraction, one via thick and one via thin filaments; calmodulin is a key element in both of them.

Smooth muscle cell contraction requires an increase in $[Ca^{2+}]_{cyt}$. Agonist-induced increases in calcium were observed to occur over the entire cell, followed by a lower, sustained plateau of calcium. However, more recently, with higher-resolution microscopy, these caclium increases have been observed to occur as a series of calcium oscillations (Perez and Sanderson, 2005; reviewed by Sanderson et al., 2008). These calcium oscillations have dynamic spatial characteristics and commonly occur as calcium waves that propagate along the full length of the smooth muscle cells. These waves often originate at one end of the cell. The mechanisms by which calcium is repetitively released and sequestered appear to vary between smooth muscle cells of different species or from different locations within tissues. There are two basic release mechanisms, based on either the inositol triphosphate receptor (IP_3R) or the ryanodine receptor (RyR) with calcium-induced calcium release occurring in both.

In one of the mechanisms the binding of IP_3 opens the IP_3R to allow a calcium efflux from the sarcoplasmic reticulum (Liu and Farley, 1996a,b). The resulting increase in $[Ca^{2+}]_{cyt}$ promotes the binding of calcium to the IP_3R, and this increases the probability that they will open with the result that the $[Ca^{2+}]_{cyt}$ continues to rise, which, in turn, initiates the propagation of a calcium wave by stimulating neighboring IP_3Rs. However, the binding of additional calcium to a second site on the IP_3R reduces its open probability, and this, together with a localized depletion in $[Ca^{2+}]_{cyt}$ of the sarcoplasmic reticulum, leads to termination of the calcium release. In an alternate scheme, calcium oscillations are proposed to be mainly mediated by calcium-induced calcium release via the ryanodine receptor RyR (Prakash et al., 1997).

These elemental calcium signals are usually mediated by small groups of receptors and are termed calcium *sparks* or *puffs* if mediated predominantly by RyRs or IP_3Rs, respectively (reviewed by Sanderson et al., 2008). Calcium sparks were initially observed in cardiac cells; whereas calcium puffs have been well documented in oocytes. Calcium sparks often appear to occur spontaneously or in response to low concentrations of caffeine, whereas puffs are usually invoked by low concentrations of agonist or cytosolic IP_3.

11.1.4. Regulatory Light Chains of Myosin

Cardiac regulatory light chain, MLC2V, binds calcium with $K_d \sim 10^{-7}$ M and magnesium with $K_d \sim 10^{-5}$ M; thus, the two ions compete with one another for binding to the light chain (Werber, 1978; reviewed by Schaub and Heizman, 2008). Magnesium may be bound during relaxation at low calcium levels, while on sustained increase of cytoplasmic calcium the magnesium would be replaced (Watterson et al., 1979; Holroyde et al., 1979). On the basis of the affinities for calcium and for magnesium, this EF-hand of MLC2V should always be occupied by one of these two ions. It could, however, also function as a relaxation site, as proposed for parvalbumin, by taking up the activating calcium from cardiac TnC with some delay because of the slow exchange rate of magnesium.

Regulatory light chains from both smooth and skeletal muscles interact with melittin in a calcium-dependent manner with $K_d \sim 2 \times 10^{-8}$ M (Malencik and Anderson, 1983a).

11.1.5. Other CTER Proteins

The following proteins are definitely conguent with the CTER subfamily; their initial names may change as more is known of their functions.

Troponin, Nonvertebrate (TPNV) Troponin is found in various nonvertebrates; for example, lobster has three isoforms (Garone et al., 1991). It is closely related to vertebrate troponin C. However, it clusters distinct from TnC, and its mode of function remains unknown. The interlobe linkers of CaM, TnC, ELC, and RLC are bent when those proteins are bound to their targets. In contrast, the 2,3 linker of TPNV is predicted to be strongly helical and hence not to bend.

Calmodulin Like Leaf Protein (CLAT) The calmodulin like leaf protein from *Arabidopsis thaliana* (CLAT) differs from calmodulin in that it has a 47-residue domain of unknown homology at its C-terminus (Lingn and Zielinski, 1993). Its linker between lobe 1,2 and lobe 3,4 is TYSEK, in contrast to the MKDTDSEE of CaM.

Chiasson et al. (2005) characterized a tomato gene, *APR134*, encoding a CaM-related protein that is induced in disease resistant leaves in response to attack by *Pseudomonas syringae* pv. *tomato*. They found that suppression of *APR134* expression in tomato (*Solanum lycopersicum*), using virus induced gene silencing (VIGS), compromises the plant's immune response. *APR134*-like genes from *Arabidopsis*, termed *CML42* and *CML43*, were isolated to investigate whether they serve a similar role. Gene expression analysis revealed that *CML43* is rapidly induced in disease-resistant *Arabidopsis* leaves following inoculation with *P. syringae*. Recombinant APR134, CML42, and CML43 proteins all bind calcium in vitro. These data support a role for CML43 and APR134 as important mediators of calcium signals in the plant's immune response to bacterial pathogens.

Squidulin Squidulin is identical to vertebrate calmodulin in 101 of its 149 positions. However, its Lys115 is not methylated; Lys115 of CaM is trimethylated. Further, it has a Pro at the second residue of the 2,3 liner, which is helical in CaM when it is not bound to a target. Squidulin copurifies with the 235-kDa myosin of squid axoplasm and is inferred to function as a myosin light chain (Bearer et al., 1996; Head, 1996).

Caltractin (Centrin) Caltractin (also known as centrin) is an essential component of the centrosome, which mediates chromosome segregation during mitosis (reviewed by Veeraraghavan et al., 2002; Friedberg, 2006). Caltractin is essential to proper cellular division because it regulates the cell cycle–dependent duplication and segregation of the microtubule organizing centers (reviewed by Sheehan et al., 2006). In lower organisms the protein is found in homologous structures (e.g., the microtubule organizing center in algae and the spindle pole body in yeast. Other functions of caltractin include (1) initiation of flagellar excision in *Chlamydomonas reinhardtii* through a fiber based microtubule severing mechanism, (2) formation of part of the human heterotrimeric DNA damage recognition complex required for global genome nucleotide excision repair, (3) modulation of homologous recombination and nucleotide excision repair in *Arabidopsis*, and (4) involvement with the nuclear mRNA export machinery in yeast. The *CDC31* d gene product, a yeast form of caltractin, interacts with the *KAR1* product. *KARI* encodes an essential component of the yeast spindle pole body that is required for karyogamy and duplication of the spindle pole body (Biggins and Ros, 1994; Belhadri, 1995). Caltractin is essential to proper cellular division because it regulates the cell cycle–dependent duplication and segregation of the microtubule organizing centers (MTOCs). The distribution and function of caltractin is restricted mostly to basal bodies (e.g., in rods of the retina), axonemes, flagella, cilia, and centrosomes.

In mammalian retinal photoreceptor cells, four caltractin isoforms are expressed as prominent components in the cilium that connects the light sensitive outer segment compartment with the metabolically active inner segment (Wolfrum et al., 2002). Caltractin isoform 1 is specifically localized in the nonmotile cilium (reviewed by J. H. Park et al., 2006). This connecting cilium links the light sensitive outer segment with the biosynthetic-active inner segment of the photoreceptor cell. All intracellular exchanges between these compartments have to occur through this cilium. Calcium activated caltractin isoforms assemble into protein complexes with the visual heterotrimeric G-protein transducin, which is thought to regulate light driven movements of transducin between photoreceptor cell compartments (Pulvermüller et al., 2002; Trojan et al., 2008a,b). This interaction is mediated by binding to the βγ-dimer of the heterotrimeric G-protein. These interactions of caltractin with transducin are reciprocally regulated via site-specific phosphorylations mediated by the protein kinase CK2. Caltractins are differentially phosphorylated during photoreceptor dark adaptation. CK2 and ciliary caltractins colocalize in the photoreceptor cilium. Direct binding of CK2 and caltractins to ciliary microtubules may spatially integrate the enzyme substrate specificity in the cilium. Phosphorylated caltractin has a lower calcium affinity and resulting reduced affinity for transducin.

Caltractins are 167 to 188 residues long; they have about 50% sequence identity with calmodulin. Caltractin has two lobes connected by a central linker; each of its four EF-hands can bind calcium (reviewed by Park et al., 2006; Thompson et al., 2006). There are 40 conserved residues within caltractins that are not shared by calmodulins (Levy et al., 1996). Unlike calmodulin, caltractin has a positively charged, 23-residue (or longer) N-tail and conserved aromatic residues near both termini. The crystal structure of mouse centrin 1 (MmCen1-L) resembles the overall structure of di-calci-troponin C. Residues W24 and R25 of the N-tail of MmCen1-L bind to its C-lobe in a conformation resembleing a target–protein interaction. Caltractin has two calcium binding sites with normal calcium affinity, ($K_d = 1.2 \times 10^{-6}$ M); it is distinguished from calmodulin by having two sites with lower affinity ($K_d = 1.6 \times 10^{-4}$ M). EF-hand 4 in *C. reinhardtii* caltractin binds calcium approximately 100 times more strongly than does EF-hand 3 (Hu et al., 2004). In the absence of calcium, the protein occupies a mixture of closed conformations. Binding of a single Ca^{2+} ion in site 4 promotes an open conformation. However, even at saturating levels of calcium, there remains a distribution between closed and open conformations. The open conformation is stabilized by the presence of the target peptide from yeast Kar1p (239–257). This is reflected in the enhancement of the calcium affinity in this site by more than an order of magnitude.

The α-helical content of caltractin is 60% and 53% in the presence and absence of calcium or magnesium, respectively (Pastrana-Rios et al., 2002). The β-strand content is 12% and 36%, while the random coil component is 7% and 13.5% in the presence and absence of calcium. Changes in the side chain band in FTIR spectra are due mostly to the monodentate coordination of calcium by Asp. A shift of approximately 4 cm^{-1} (for the COO^- antisymmetric stretch in Asp) from 1565 to 1569 cm^{-1} is observed for apo-caltractin and calci-caltractin, respectively. Thermally dependent conformational transitions occur at 37°C for apo- and at 37 and at 45°C for calci-centrin (Weber et al., 1994; Pastrana-Rios et al., 2002).

Titration of the isolated N- and C-lobes of caltractin and of intact caltractin with a peptide fragment from Kar1p shows that the two lobes interact with the peptide in a calcium dependent manner, with the C-lobe binding much more strongly than the N-lobe (Veeraraghavan et al., 2002). The C-lobe of caltractin mediates its calcium dependent interaction with Kar1p and Kic1p, essential components of the spindle pole body of yeast (Sullivan et al., 1998). Hu et al. (2004) suggested that the C-lobe of caltractin serves as a constitutive anchor to target proteins, and the N-lobe serves as the sensor of calcium signals. The structure of the N-lobe of *Chlamydomonas rheinhardtii* caltractin was determined in the presence of calcium by solution NMR spectroscopy (Sheehan et al., 2006). It has an open conformation typical of EF-hand calcium sensors. NMR titration of the N-lobe with a fragment of the caltractin target, Sfi1, reveals binding of the peptide to a discrete site on the protein; this supports the proposal that the N-lobe of caltractin serves as a calcium sensor.

Both SC-HsCen2 (T94-Y172), the C-lobe, and LC-HsCen2 (M84-Y172) are highly disordered in the apo-state but become more structured (although not conformationally homogeneous) in the presence of calcium, depending on whether potassium or sodium is in the buffer (Matei et al., 2003). The solution structure of LC-HsCen2 reveals an open two EF-hand structure, similar to the conformation of related calcium saturated regulatory domains. The N-terminal helix segment (F86–T94) lies over the exposed hydrophobic cavity. This unusual intramolecular interaction increases the calcium affinity considerably.

Yang et al. (2006) reported the solution structures of two human centrin isoforms, HsCen1 and HsCen2. The N-lobe of HsCen2 has a compact core consisting of four almost antiparallel α-helices in closed conformation and a short antiparallel β-sheet, very similar to the apo-state structure of other calcium regulatory EF-hand lobes. The first 25 residues show a highly irregular and dynamic structure. Calcium titration of the apo-N-terminal domains of HsCen1 and HsCen2, monitored by NMR spectroscopy, revealed a very weak affinity (10^{-3} to 10^{-2} M), suggesting that the cellular role of this lobe is not calcium dependent. Isothermal calorimetric titrations showed that an 18-residue peptide, derived from the N-tail, has a significant affinity ($\sim 10^{-5}$ M) for the isolated C-lobe, suggesting an active role in the self-assembly of centrin.

The high-resolution solution structure of the complex between the calci-C-lobe of *C. reinhardtii* caltractin and a 19-residue peptide fragment comprising the putative cdc31p-binding region of Kar1p was determined by multidimensional heteronuclear NMR spectroscopy (Hu and Chazin, 2003). Formation of the complex is calcium dependent and is stabilized by extensive interactions between caltractin and three key hydrophobic anchors (Trp10, Leu13, and Leu14) in the peptide as well as favorable electrostatic interactions at the protein–peptide interface (Figure 11.9).

Hu et al. (2004) found that site 4 in *C. reinhardtii* centrin binds calcium approximately 100-fold more strongly than does site 3. Apo-centrin occupies a mixture of closed conformations. Binding of calcium to site 4 alters the equilibrium radically, promoting occupancy of an open conformation. However, an exchange between closed and open conformations remains even at saturating levels of calcium. The population of the open conformation is substantially stabilized by the presence of the target peptide Kar1p-(239–257), to a point where a single Ca^{2+} ion bound in site 4 shifts the equilibrium to the open conformation. This is reflected in the increase of the calcium affinity in this site by more than an order of magnitude.

Cal-1 Gene Product The *cal-1* gene was isolated from the nemotode, *Caenorhabditis elegans*, by hybridization with a human calmodulin cDNA probe (Salvato et al., 1986). CAL has four EF-hands in its 161 residues. It is distinct from CAM.

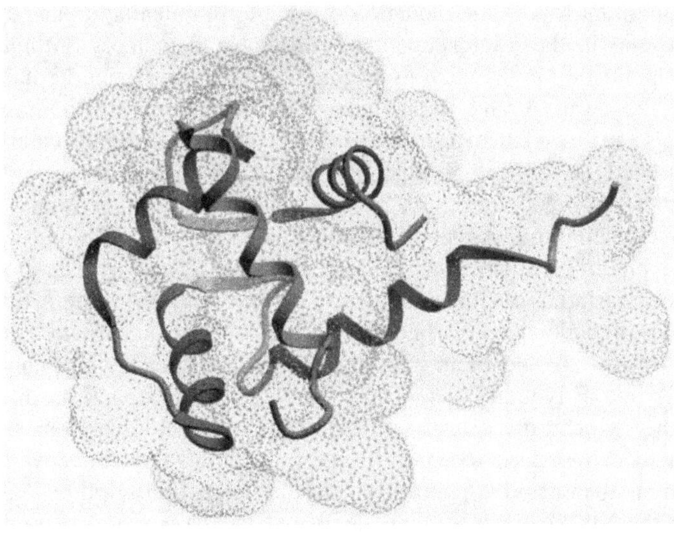

Figure 11.9. Structure of calcium loaded C-terminal domain of caltractin in complex with CDC31P binding domain from KAR1P (PDB file 1OQP). (From Hu and Chasin, 2003.) (*See insert for color representation of figure*.)

pCAST Gene Product The *pCAST* clone from *Solanum tuberosum* (potato) encodes four EF-hands preceded by 38 residues (Gellatly and Lefebvre, 1993). Its 2,3 linker is 23 residues long. It is not in the same subfamily with PMAT from *Arabidopsis*, which is not congruent with CTER; however, both contain $(Ser)_5$ in their N-terminal domains.

11.2. CPR (CALCINEURIN B, P22, RECOVERIN) SUBFAMILY

The seven members of the CPR group all contain four EF-hands; however, CPR is not congruent with CTER. Three of the seven are heterochimeric; these non-EF-hand domains are not homologous to one another.

11.2.1. Calcineurin B

Calcineurin is a Thr/Ser phosphatase regulated by calcium and calmodulin (reviewed by Guerini, 1997). The name *calcineurin* reflects its calcium binding properties and localization to neuronal tissues (Klee et al., 1979). It is the only protein phosphatase dependent on calcium and on calmodulin for its activity and is a common intracellular transducer of calcium signaling. Changes in the concentration of free Ca^{2+} ion, within the cytosol, control gene expression in many types of cells by calmodulin-dependent activation of calcineurin (reviewed by Bandyopadhyay et al., 2004). Calcineurin is present in high concentrations in the brain and in lymphocytes. In the brain, calcineurin is found in high concentration in neurons, in particular in those of the neurostriatum and cerebellum. It plays an important role in the induction of long-term potentiation and long-term depression, and in the establishment of learning and memory. Although it is expressed abundantly in the nervous system, calcineurin is found in other tissues as well. Calcineurin A and B subunits have been observed in most tissues of mammals (reviewed by Rusnak and Mertz, 2000). In brain, calcineurin is highly enriched in the cytoplasmic and microsomal fractions as well as in synaptosomes. It predominates in the cytoplasm and synaptosomal cytosol. There is overwhelming evidence for calcineurin in the nucleus along with other calmodulin binding proteins such as casein kinase-2 and myosin light chain kinase. Calcineurin has also been shown to be associated with the cytoskeleton.

Numerous functions have been identified for calcineurin in mammals (reviewed by Rusnak and Mertz, 2000). It plays a role in the programmed cell death of T- and B-lymphocytes. It also plays a role in apoptosis in neuronal cells via the cytochrome c/caspase-3 pathway. In T-cell hybridomas, binding of antigen to cell surface receptors triggers calcium entry, thereby activating calcineurin. It then removes phosphate groups from transcriptional regulatory proteins of the NF-AT (nuclear factor–activated T-cell) family that are necessary for T-cell proliferation (Liu et al., 1991; Clipstone and Crabtree, 1992; O'Keefe et al., 1992). It induces apoptosis through Bcl_2 (the prototype for a family of mammalian genes that govern mitochondrial outer membrane permeabilization and can be either pro- or antiapoptotic) and modulates calcium release by interacting directly with the Ins(1,4,5)P_3R (IP_3 receptor) and with the ryanodine receptor. In other cell types, calcineurin regulates calcium pumps and exchangers to maintain calcium homeostasis. Calcineurin activation is also essential for other calcium-regulated processes, such as chemotaxis of neutrophils and pathogenesis of hypertrophic cardiomyopathy (reviewed by Bandyopadhyay et al., 2004). In *Saccharomyces cerevisiae*, calcineurin is required for cation homeostasis and regulates cell wall biosynthesis by Tcn1/Crz1, a transcription factor distantly related to NF-AT. Tcn1/Crz1 translocates to the nucleus and causes gene activation. Calcineurin takes part in calcium sparklets produced by calcium influx via L-type channels in smooth muscle (Santana and Navedo, 2009). Targeting of protein kinase Cα by the scaffolding protein AKAP150 to specific sarcolemmal domains is required for the activation of persistent calcium sparklets. Calcineurin, which is also associated with AKAP150, reduces the actions of protein kinase Cα on sparklets.

Regardless of its source, calcineurin always consists of two subunits: catalytic calmodulin binding subunit A (60 kDa) and calcium binding regulatory subunit B (19 kDa) (reviewed by Klee et al., 1998) (Figure 11.10). The two subunit structure is conserved from yeast to man and is essential for activity, as are the amino acid sequences of both the catalytic and regulatory domains of calcineurin A. The subunits are tightly bound to each other and can be dissociated only under denaturing conditions. The A subunit alone has only very low activity; the B subunit is essential for the high specific activity of the phosphatase.

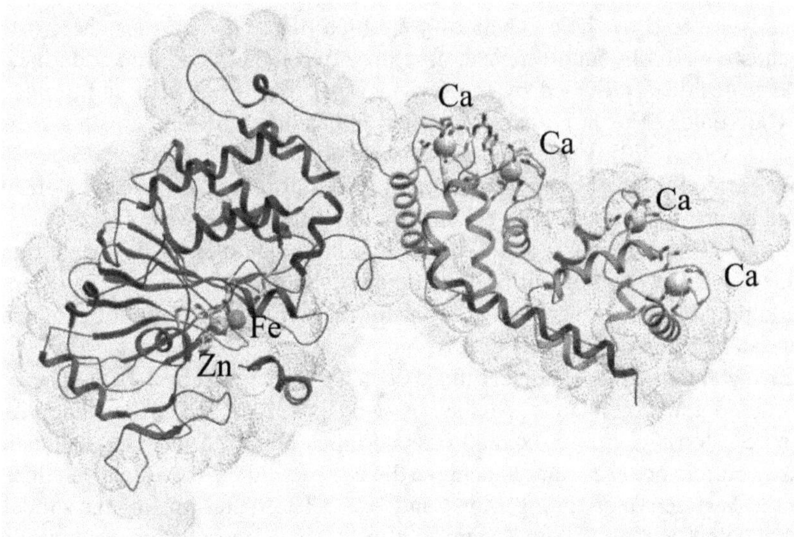

Figure 11.10. Structure of calcineurin heterodimer (PDB file 1AUI). Subunit B (green) has a Ca^{2+} ion in each of its four EF-hands. Subunit A has a Zn(II)–Fe(II) site. (From Kissinger et al., 1995.) (*See insert for color representation of figure.*)

The B subunit, 168 residues, consists of four EF-hands. It is highly conserved throughout evolution, with mammalian calcineurin B showing 86% sequence identity with *Drosophila* calcineurin B and 54% identity with calcineurin B from *S. cerevisiae* (reviewed by Rusnak and Mertz, 2000). Its dumbbell structure, determined by multidimensional NMR, is similar to those of calmodulin and TnC (Anglister et al., 1994). In comparison to calmodulin, helices B and C are shorter while helix G is considerably longer. Moreover, calcineurin B does not have a single long central helix; rather, helices D and E are separated by a flexible six-residue linker. The central linker of calcineurin B makes a sharp bend at Gly85 to place the N- and C-lobes on the same side of the target peptide. All four EF-loops bind calcium (Kissinger et al., 1995; Griffith et al., 1995). One binding site has high affinity for calcium [K_d(Ca) $\sim 10^{-7}$ M], and three sites have K_d(Ca) in the micromolar range (Kakalis et al., 1995). Equally conserved from yeast to humans is the myristoylation of the N-terminal glycine of subunit B. This modification has been conserved throughout evolution from yeast to mammals, suggesting possible physiological role. The myristoylated protein exhibits increased thermal stability ($\sim 12^\circ$C) relative to the nonmyristolyated protein (Kennedy et al., 1996). The N-terminal myristoyl group of calcineurin B is situated in a hydrophobic cleft between two amphipathic helices that are integral components of two EF-hands.

The A subunit has five domains (reviewed by Klee et al., 1998; Kissinger et al., 1995; Griffith et al., 1995). The N-terminal domain may be involved in the modulation of catalytic activity. It interacts with the B subunit, with the catalytic core, a region of 350 amino acids with high sequence similarity to other protein phosphatases, with the B subunit binding domain, and with the inhibitory domain. The catalytic domain is a β-sandwich that consists of a sheet of six β-strands covered by three α-helices plus three β-strands and of a sheet of five β-strands covered by an all helical structure. Two metal ions, iron and zinc, are bound to residues provided by the two faces of the β-sandwich and define the catalytic center. The last β-sheet extends into a five-turn amphipathic α-helix (residues 350 to 370) whose top face, completely nonpolar, is covered by a 33-Å groove formed by the N- and C-lobes and the C-terminal β-strand of calcineurin B. Calcineurin B interacts with calcineurin A by linkers between helix 1 and the calcium binding loops 1, 3, and 4, the central helix, and the C-tail. Interaction of residues 14 to 23 of calcineurin A with the C-lobe of calcineurin B may provide the additional binding energy to account for the very high affinity of calcineurin B for calcineurin A ($K_d \sim 10^{-13}$ M).

The metal ions in the dinuclear metal center are separated by 3.14 Å. The Zn^{2+} ion is coordinated by ligands provided by two His's, Asn, and an Asp that bridges to the Fe^{3+} ion. One more Asp and His coordinate the Fe^{3+} ion. Divalent cations in assay buffers are necessary to achieve the high activities of purified calcineurin, with the best activators being Mn^{2+} and Ni^{2+}. The metal ions of the dinuclear center could function in numerous ways to catalyze phosphate ester hydrolysis (reviewed by Rusnak and Mertz, 2000). The Lewis acidity of the metal ion(s) could serve to activate a solvent molecule, a well-known mechanism in several metalloenzymes such as carbonic anhydrase. Alternatively, a metal-coordinated solvent molecule could serve as a general acid to donate a proton to the leaving group, as has been proposed for inorganic pyrophosphatase. Another possible role for the metal ions could be to orient the substrate for inline attack.

Purified calcineurin from bovine brain contains up to 0.6 equivalent of phosphate (King and Huang, 1984), suggesting that it may be phosphorylated in vivo. Calcineurin can be phosphorylated by protein kinase C, casein kinase I, and casein kinase II in vitro (reviewed by Rusnak and Mertz, 2000). The site of phosphorylation by casein kinase II has been determined to be the Ser near the C-terminus of the calmodulin binding domain (Hashimoto and Soderling, 1989).

The binding of calcium directly to calcineurin has a small but significant influence on the activity of the phosphatase, but the main effect of calcium is mediated by calmodulin (Stemmer and Klee, 1994). The dependence of calcineurin activity on calcium is the result of the concerted action of calmodulin, which increases the turnover rate of the enzyme and modulates its response to calcium transients and of calcineurin B; this in turn decreases the K_m value of the enzyme for its substrate. Calmodulin increases $V_{\max}$ without changing K_m. The calcium-dependent activation of calcineurin by calmodulin is highly cooperative (Hill coefficient of 2.8 to 3.0). Conversely, the affinity of calmodulin for calcium is increased by more than two orders of magnitude in the presence of a peptide corresponding to the calmodulin binding domain of calcineurin A.

Calciuneurin B, unlike calmodulin, interacts with calcineurin A in the absence of calcium. Calcium binding to calcineurin B stimulates native calcineurin up to only 10% of the maximum activity achieved with calmodulin. Calcineurin B and calmodulin share 35% sequence identity, but the two proteins bind to different sites on the A subunit and are not interchangeable (Klee et al., 1988).

Rusnak and Mertz (2000) proposed a model for the catalytic mechanism of calcineurin. His151 should act both as a general base and a general acid. In the first stage, substrate binds to the active site, and His151 deprotonates an iron-bound water molecule to form hydroxide. In the second stage, the bond cleavage to the leaving group occurs to a greater extent than bond formation to the hydroxyl group in the transition state. The Zn^{2+} ion may aid in neutralization of the negative charge on the leaving group in the transition state. The leaving group is protonated by His151 and then leaves the active site. In the third stage the phosphate bridges the two metal ions. In the fourth stage a water molecule displaces bound phosphate, and the enzyme is ready for another round of turnover. Another substrate then enters the active site and displaces the bridging water molecule, which then becomes a terminal water molecule that must be deprotonated to a hydroxide molecule.

Calcineurin is regulated in vivo by changes in $[Ca^{2+}]_{cyt}$. In a resting cell calcineurin is unable to bind calmodulin, and the enzyme exists in an inactive form. In signaling pathways that lead to a rise in intracellular $[Ca^{2+}]$ calcium binding to calmodulin results in a conformational change, thereby allowing it to bind to calcineurin and activate its phosphatase activity (Klee and Cohen, 1980). Calcium binding to calcineurin B also appears to play a role (Stemmer and Klee, 1994). Calcineurin activity has also been shown to be affected by phospholipids, with either activation or inhibition, depending on the phospholipid and substrate used (Politno and King, 1990). Calcineurin is susceptible to redox regulation in vitro (Yu et al., 1997), a process that may also occur in vivo. Because the Fe(II)–Fe(III) form of calcineurin can lose activity by oxidation to the Fe(III) state, it may be that calcineurin exists in oxidation-sensitive Fe(II) and an oxidation inert (Fe^{3+}–Zn^{2+}) state in vivo (Rusnak and Mertz, 2000).

The enzymatic function of calcineurin is uniquely sensitive to the macrolides cyclosporine (CsA) and FK506 (tacrolimus) (reviewed by Stie and Fox, 2008). These agents interact with calcineurin through the immunophilins cyclophilin A and FK506 binding protein 12 (FKBP12), respectively. CsA-cyclophilin A and FK506-FKBP12 are structurally unrelated complexes that interact at distinct but overlapping sites on calcineurin (Liu et al., 1991). The macrolides bind to macrophilin 12 to form a complex capable of blocking calcineurin, which dephosphorylates, and thereby activates, the cytoplasmic subunit of nuclear factor of activated T-cells (NF-AT) (Liu et al., 1991; Clipstone and Crabtree, 1992; O'Keefe et al., 1992). Thus, in the presence of calcineurin inhibitors, NF-AT cannot enter the nucleus and interact with the promoter regions of many cytokine genes. In addition to T-cells, calcineurin inhibitors have also been shown to inhibit the activation of a number of other cell types in the skin immune system, including eosinophils, basophils, and Langerhans cells. These drugs are highly effective in treating atopic dermatitis, even in monotherapy, and they provide rapid, progressive, and sustained improvement (reviewed by Sárdy et al., 2009). Calcineurin inhibitors are well tolerated for long-term treatment, and there are virtually no contraindications when using these medications on the face, eyelids, flexural skin, or mucous membranes. Although calcineurin inhibitors are immunomodulators, no significantly increased incidence of infections can be detected during therapy. Currently, the calcineurin inhibitors cyclosporine and tacrolimus remain the cornerstones of immunosuppressive regimens, regardless of their well-described side effects, including nephrotoxicity. Tacrolimus and cyclosporine are widely used to prevent the rejection or graft versus host disease after transplantations.

11.2.2. p22

p22 has four EF-hands and is N-myristoylated. Lin and Barber (1996) reported that p22 inhibits GTPase-stimulated Na^+/H^+ exchange. The N-myristoylate group and the calcium-induced change in conformation are essential to the function of p22 (Barroso et al., 1996). It is required for constitutive, exocytotic membrane traffic in rat liver and modulates the organization and dynamics of the microtubule cytoskeleton in a calcium-independent manner and affects ER network assembly in a calcium-dependent manner (Andrade et al., 2004).

11.2.3. Recoverin

The neuronal calcium sensor (NCS) proteins (recoverin, frequenin, VILIP, Kv-channel interacting protein, guanylate cyclase activating protein) have restricted expression in neurons, neuroendocrine cells, and retinal photoreceptors, suggesting that they have specialized roles in these cell types (reviewed by Burgoyne and Weiss, 2001; Philippov and Koch, 2006).

Two members of the family (recoverin and guanylate cyclase activating protein) have established roles in the regulation of phototransduction. Recoverin, or visinin, regulates guanylate cyclase activity in the retinal rod outer segments of vertebrates. In the dark, cationic channels are kept open by bound cGMP. Calcium enters through these channels; this influx is matched by efflux through a N^+/K^+, Ca^{2+} exchanger. Light activates an enzyme cascade that stimulates cGMP hydrolysis, leading to channel closure; this blocks calcium influx but not its efflux. The dark level of cGMP is maintained by a balance between its synthesis by a guanylate cyclase and its hydrolysis by a cGMP-phosphodiesterase (reviewed by Ebrey and Koutalos, 2001). A steady concentration of cytoplasmic calcium is maintained at about 3×10^{-7} M; this inhibits the guanylate cyclase. Absorbed photons isomerize the chromophore of the light-sensitive pigment, rhodopsin. Its isomerization leads to the formation of an active photochemical intermediate. This intermediate activates a G-protein, transducin, which in turn stimulates the hydrolytic activity of the cGMP-phosphodiesterase. As a result, the cytosolic cGMP concentration falls, and the cGMP-gated channels close. Free $[Ca^{2+}]$ within the cell drops within 0.5 s (Oberwinkler and Stavenga, 2000) of light stimulation. Recovery from light excitation involves the inactivation of each of the active intermediates of the cGMP cascade; the activated rhodopsin molecules are inactivated through phosphorylation by rhodopsin kinase and subsequent capping by a 48-kDa protein, arrestin. Transducin hydrolyzes bound GTP to GDP. The inactivation of transducin ends the stimulation of the cGMP phosphodiesterase, bringing the rate of hydrolysis of cGMP down to its original level.

Recoverin is present in both rod and cone photoreceptors and was named *recoverin* because it promotes recovery of the dark state. Hence, recoverin makes the desensitization of rhodopsin sensitive to calcium. The shortened lifetime of photoexcited rhodopsin at low levels of calcium may promote visual recovery and contribute to the adaptation to background light. Recoverin prolongs the photo response by blocking the phosphorylation of photoexcited rhodopsin by rhodopsin kinase.

The recoverin subfamily includes neuronal calcium sensors such as VILIP (Lenz et al., 1992), neurocalcin (Okazaki et al., 1992), frequenin (Pongs et al., 1993), hippocalcin (Kobayashi et al., 1992; Takamatsu et al., 1996), neuronal calcium sensor 1 (Nef et al., 1995), and calsenin/DREAM/KChIP3 (Osawa et al., 2001). All of the members of this subfamily have myristoylated N-termini and four EF-hands. They all contain a Cys–Pro sequence in the first EF-hand that prevents calcium ion binding by this EF-hand. The cellular function of many of the neuronal proteins of this family remains unknown.

Unlike other members of this family, neurocalcins are differentially distributed. They are expressed mainly in the central nervous system, spinal cord, retina, inner ear, and olfactory epithelium; they are also found in the zonaglomerulosa of the adrenal gland (reviewed by Vijay-Kumar and Kumar, 2002). Both the unmyristoylated and myristoylated forms of neurocalcin bind three equivalents of calcium at saturating concentrations. Unmyristoylated neurocalcin binds calcium with higher affinity than the myristoylated form; however, calcium binding by the myristoylated form is cooperative (Hill coefficient: 1.8). The high degree of conservation of hydrophobic residues among all members of this family of proteins, including neurocalcin, suggests that these residues are important in the calcium induced extrusion of the myristoyl group, and that the mechanism of the calcium–myristoyl switch is similar in all family members from yeast to human.

In the crystal structure (Flaherty et al., 1993) of nonmyristoylated recoverin a single Sm^{3+} ion is bound in place of Ca^{2+} in EF-hand 2. The hydrophobic patches of lobe 1,2 and lobe 3,4 are exposed and oriented so that their two clefts cannot enfold a target helix as seen in calmodulin and troponin C. NMR studies (Ames et al., 1995b) have shown that the myristoyl side chain is buried between the two lobes in the apo form but is exposed following calcium binding to EF-hands 2 and 3.

Recoverin has four EF-hands (Dizhur et al., 1991; Stryer, 1991). Recoverin is myristoylated at its N-terminus. Upon binding calcium, recoverin undergoes a conformational change that allows it to associate with membranes in a manner that requires N-myristoyl participation. NMR data indicate that the myristoyl group is in contact with residues in the hydrophobic cluster in apo-recoverin and that it is exposed to solution in calci-recoverin (Hughes et al., 1995; Ames et al., 1995a,b; Tanaka et al., 1995).

The crystal structure of myristoylated calci-recoverin has been determined (Flaherty et al., 1993) and the solution strucuture of unmyristoylated apo-recoverin has been determined by NMR spectroscopy (Ames et al., 1994, 1997; Tanaka et al., 1995, 1998). The four EF-hands of recoverin are arranged in a compact globule, in contrast with the dumbbell shape of calmodulin and of TnC (Figure 11.11). Only EF-hands 2 and 3 bind calcium; EF-hands 1 and 4 do not. The hydrophobic side chains of Leu28, Trp31, and Tyr32 form a cluster that interacts with the end (C1 to C8) of the myristoyl group, whereas residues Phe49, Phe56, Tyr86, Val87, and Leu90 interact with the other end (C9 to C14). The myristoyl group is in a slightly bent conformation.

The binding of calcium to recoverin increases the content of α-helices and decreases β-strands and coils (Johnson et al., 1997; Ames et al., 1995a,b). The NMR data indicate that the myristoyl group makes close contact with a number of aromatic residues in apo-recoverin, whereas the myristoyl group makes no obsevable contacts with the protein in calci-recoverin [i.e., the binding of calcium induces the extrusion of its myristoyl group into the solvent (calcium myrisoyl switch)]. This is accompanied by a relative rotation of 45° of the two lobes and by solvent exposure of many hydrophobic residues. Calcium binding leads to the exposure of the covalently attached myristoyl group to solvent. This is accomplished by rotation of a "hinge" at Gly42, which moves helix I of the first EF-hand, pulling the N-terminal helix away from the rest of the protein and exposing the myristoyl group. The binding of calcium to recoverin results in transfer of some of its Trp side chains from the interior to the surface; this

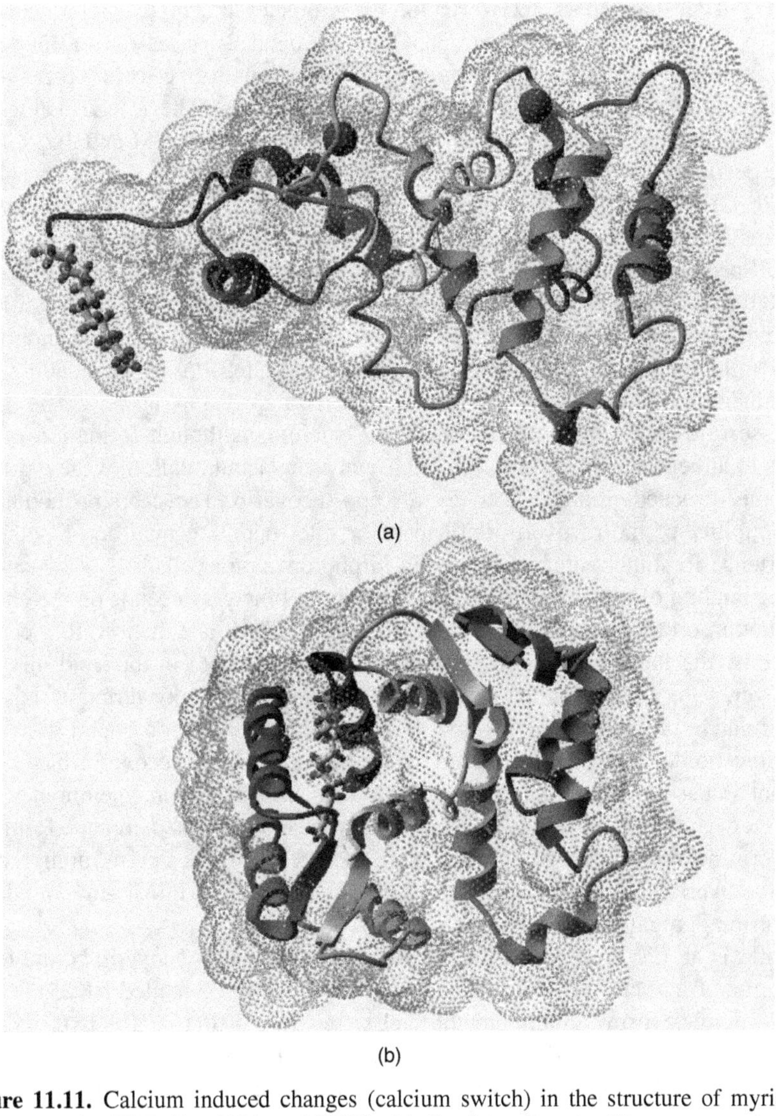

Figure 11.11. Calcium induced changes (calcium switch) in the structure of myristoylated recoverin (PDB files 1JSA and 1IKU): (a) di-calci-recoverin; (b) apo-recoverin. The myristoyl group attached to the N-terminus is extruded upon calcium binding. (From Ames et al., 1997; Tanaka et al., 1995.)

is reflected in a calcium induced red shift of the fluorescence of these Trp's (Permyakov et al., 2000a).

Unmyristoylated recoverin binds two Ca^{2+} ions with no cooperativity (Hill coefficient = 1.0). In contrast, myristoylated recoverin binds two equivalents of calcium with a Hill coefficient of 1.75 (Ames et al., 1995). The substitution,

E121Q, in EF-hand 3 of recoverin totally abolishes the high calcium affinity of recoverin; while the mutation, E85Q, in EF-hand 2 causes only a moderate decrease in calcium affinity. The binding of Ca^{2+} ions to recoverin seems to be a sequential process, with the EF-hand 3 being filled first. The K_d(Ca) values for sites 3 and 2 from spectrofluorimetric titration are 2.7×10^{-7} M and 3.2×10^{-6} M, respectively (S. E. Permyakov et al., 2000a).

Like all members of the neuronal calcium sensor (NCS) family, recoverin contains a conserved cysteine (Cys38) in EF-hand 1, which does not bind calcium. Spectrophotometric titration of calci-recoverin gave 7.6 for the pK_a of Cys38 thiol, suggesting partial deprotonation of the thiol under in vivo conditions. The ability of recoverin to form a disulfide dimer and thiol-oxidized monomer under mild oxidizing conditions was found using SDS-PAGE in reducing and nonreducing conditions and Ellman's test (S. E. Permyakov et al., 2007). Both processes are reversible and modulated by calcium. Although formation of the disulfide dimer takes place only for calci-recoverin, accumulation of the oxidized monomer proceeds more effectively for apo-recoverin. The calcium-modulated susceptibility of the recoverin thiol to reversible oxidation may be of potential importance for functioning of recoverin in photoreceptor cells.

The binding of myristoylated recoverin to lipid bilayers depends on the charge distribution on the phospholipids (Senin et al., 2007). It is enhanced to a certain degree by the inclusion of phosphatidylserine (up to 60%) in the lipid mixture. However, a recoverin mutant that lacks the charged carboxy terminus has the same relative binding affinity as does wild type recoverin when bound to neutral or acidic lipids. Instead, the charged carboxy terminus of recoverin has a significant impact on the biphasic dissociation of recoverin from membranes. On the other hand, the nonmyristoylated wild-type and truncated mutant forms of recoverin do not bind to lipid bilayers. In the presence of calcium, unmyristoylated recoverin is monomeric and globular in solution, while N-myristoylated recoverin aggregates (Kataoka et al., 1993).

Ames et al. (2006) reported the NMR structure of calci-recoverin bound to an N-terminal fragment of rhodopsin kinase (residues 1 to 25, called RK25). RK25 in the complex forms an amphipathic helix (residues 4 to 16). The hydrophobic face of the RK25 helix (Val9, Val10, Ala11, Ala14, and Phe15) interacts with an exposed hydrophobic groove on the surface of recoverin lined by the side chains of Trp31, Phe35, Phe49, Ile52, Tyr53, Phe56, Phe57, Tyr86, and Leu90. The residues of recoverin that contact RK25 are highly conserved, suggesting a similar target binding site in all neuronal calcium sensor proteins. Site-specific mutagenesis and deletion analyses confirm that the hydrophobic residues at the interface are necessary and sufficient for binding. The recoverin–RK25 complex exhibits calcium-induced binding to rhodopsin immobilized on concanavalin-A resin. It was proposed that calci-recoverin is bound between rhodopsin and rhodopsin kinase in a ternary complex on rod outer segment disk membranes, thereby blocking rhodopsin kinase interaction with rhodopsin at high calcium concentration (Ames et al., 2006).

Using competing synthetic peptides and site directed mutagenesis, Tachibanaki et al. (2000) established that a series of residues in the N-terminal region of the frog recoverin, S-modulin, are required for interaction with rhodopsin kinase. These residues (Phe22, Glu26, Phe55, and Thr92) are normally buried in apo-recoverin and are exposed on calcium binding.

Recoverin can be an autoantigen in some cancers. Cancer associated retinopathy (CAR) patients present with a sudden loss of vision and diminished a- and b-waves of the electroretinogram suggestive of photoreceptor cell degeneration with no signs of optic nerve damage or inflammation (Subramanian and Polans, 2004). Although CAR is often associated with small cell carcinoma of the lung, additional types of cancer have been associated with the disease, including epithelial ovarian carcinoma, endothelial carcinoma, and breast cancer. Recoverin has been identified as the CAR autoantigen in several studies. Recoverin was subsequently localized by immunohistochemical methods to retinal photoreceptor cells and a small subset of bipolar cells. Because CAR spares the inner retina while causing photoreceptor cells to degenerate, the cellular localization of recoverin corresponds to the pathological sites. Recoverin antibodies can cross the blood–retina barrier in animal models, enter living retinal cells, and induce apoptosis specifically in photoreceptor cells. Additional studies have demonstrated that the apoptotic event induced by recoverin involves caspases 3 and 9 via the intrinsic mitochondrial pathway (reviewed by Subraminian and Polans, 2004).

11.2.4. Other CPR Proteins

Calsenilin Calsenilin binds both presenilin 1 and presenilin 2 in a calcium-dependent manner. Most early onset familial Alzheimer cases are caused by mutations in presenilin. Calsenilin may mediate the role of presenilin in apoptosis and β-amyloid formation (Buxbaum et al., 1998). Calsenilin is 251 residues long and contains four EF-hands. No homolog for its first 80 residues has been identified. The sequences of calsenilin and of DREAM are quite similar and all domains cluster together. They are tentatively placed in different subfamilies because of apparent differences in function.

DREAM (DRE Antagonist Modulator) Downstream regulatory element antagonist modulator (DREAM), calsenilin, and KchIP3 are expressed in the mammalian brain and are closely related to recoverin (Buxbaum et al., 1998). DREAM is 284 residues long and contains four EF-hands. The 110 residues preceding the first EF-hand may be responsible for binding to DNA; however, no homolog is recognized. The first and second EF-hands have unusual distributions of potential calcium binding ligands: X, Y, Z, −X, −Z; Asn, Cys, Thr, Asp, Thr and Asp, Asp, Asn, His, Asp. The third and fourth EF-hands are canonical. The C-terminal 190 residues of DREAM are about 40% identical to recoverin; however, unlike recoverin, DREAM contains an extra stretch of residues at its N-terminus and lacks a myristoylation consensus sequence. DREAM binds three Ca^{2+} ions to EF-2, EF-3, and EF-4. EF-2 binds calcium with relatively low affinity [K_d(Ca),

50 μM]; under physiological conditions magnesium might occupy the second EF-hand site. Dynamic light scattering and size-exclusion chromatography studies revealed that apo-DREAM exists as a tetramer at protein concentrations higher than 20 μM, while calci-DREAM forms a stable dimer in solution at low protein concentrations and forms a tetramer at concentrations higher than 200 μM (Osawa et al., 2001).

DREAM binds to the DRE silencer element of the prodynorphin gene (Carrión et al., 1999; Osawa et al., 2001). DRE silences transcription of the prodynorphin ad *c-fos* genes. "Upon stimulation by Ca^{2+}, DREAM's ability to bind to the DRE and its repressor function are prevented" (Carrión et al., 1999; Fontán-Lozano et al., 2009). Tetrameric apo-DREAM binds to four molecules of DRE at saturation. One DRE binds tightly to the tetrameric protein with a dissociation constant of 75 nM, and the other three bind more weakly with dissociation constant 700 nM. No significant DNA binding was observed for calci-DREAM. Isothermal titration calorimetry studies revealed that DRE binding to DREAM is entropically driven with $\Delta H = +25$ kcal mol^{-1}. DNA binding studies performed on a series of deletion mutants of DREAM showed that the C-terminal residues, 65 to 256, possess all the essential features to serve as a calcum-regulated transcriptional repressor.

DREAM regulates gene transcription and Kv potassium channels in neurons and has also been claimed to interact with presenilins, which are involved in the generation of β-amyloid and in the regulation of the calcium content in the endoplasmic reticulum (reviewed by Fedrizzi et al., 2008). The overexpression of DREAM has transcriptional and posttranscriptional effects. The transient coexpression of DREAM and presenilin potentiates the decrease of endoplasmic reticulum calcium observed in cells that overexpress presenilin. This could be due to a direct effect of DREAM on presenilin, as the two proteins interact in a calcium-independent fashion. The expression of calsenilin suppresses the ATP-induced increase in $[Ca^{2+}]_{cyt}$. DREAM is distributed throughout the cytoplasm of SK-N-BE2(C), as seen in immunohistochemical and subcellular fractionation analyses of Jurkat and HeLa cells (Woo et al., 2008).

DREAM is translocated into the nucleus, following increases in $[Ca]_{cyt}$. Eighteen-month-old *dream*(-/-) mice exhibit marked analgesia and markedly enhanced learning and synaptic plasticity related to improved cognition. DREAM functions as a negative regulator of the key memory factor CREB in a calcium-dependent manner; loss of DREAM facilitates cAMP responsive element binding (CREB) protein-dependent transcription during learning. Moreover, loss of DREAM protects mice from brain degeneration in aging. These data indicate that DREAM is a key regulator of memory and brain aging. DREAM is critically important for sensing pain (Cheng et al., 2002). DREAM-knockout mice exhibit a reduced pain response, probably caused by the constitutive activation of the dynorphin-selective κ-opiate receptor.

Rivas et al. (2009) showed that DREAM modulates TSH receptor activity in the thyroid gland through direct protein–protein interaction. In transgenic mice, DREAM overexpression provokes a marked enlargement of the thyroid gland. Increased levels of DREAM were observed in human multinodular goiters, suggesting a novel etiopathogenic mechanism in nodular development in humans.

Calcium/Calmodulin Protein Kinase The calcium/calmodulin protein kinase (CCaMK) of *Lilium sp.* is expressed preferentially in anthers (Patil et al., 1995). It is 520 residues long and consists of a serine/threonine protein kinase domain, a calmodulin binding region, and four EF-hands. EF-hands 2 and 3 are crucial for calcium induced conformational changes. Deletion mutants lacking one, two, or all three calcium-binding EF-hands showed that all three are necessary for the full calcium/calmodulin-dependent kinase activity. As each EF-hand is deleted, there is a gradual reduction in kinase activity from 100% to 4%. The des 1–322 mutant lacks the four EF-hands and the calmodulin binding domain is constitutively active, indicating the presence of an autoinhibitory domain around the calmodulin binding domain (Ramachandiran et al., 1997). The calmodulin binding region of CCaMK is similar to the calmodulin binding region of the α-subunit of multifunctional Ca^{2+}/calmodulin-dependent protein kinase (CaMKII). Autophosphorylation of CCaMK increases calcium/calmodulin-dependent protein kinase activity about fivefold, whereas it does not affect its calcium-independent activity.

The importance of calcium ions in coupling physiological responses to external and developmental signals in plants has been well documented (reviewed by Harmon, 2003). Cytoplasmic calcium brings about responses by interacting with target proteins, many of which contain EF-hand calcium binding motifs. In plants there are at least five classes of protein kinases, all of which are in the CDPK/SnRK family, that either contain EF-hands within their structure or interact with proteins that contain EF-hands. Calcium dependent protein kinases (CDPKs) and calcium/calmodulin activated protein kinases (CCaMKs) both contain EF-hands in their C-terminal domains and are activated by the binding of calcium.

Calcineurin B Like (CBL) Sensors Plants lack calcineurin B–activated kinases and neuronal calcium sensors. However, land plants have evolved at least 25 calcineurin B like (CBL) sensors and tens of target kinases, each of which has multiple targets. There is not a one-to-one correspondence but a highly pleiotropic matrix of interactions that somehow allow different cells to fashion an initial external signal into temporally and spatially specific response(s). These CBLs have four EF-hands, the first of which has 14 residues in its calcium binding loop instead 12; the other three are canonical. Many of these CBLs are myristoylated or acetylated; this helps in bringing their partner kinases to membrane targets (reviewed by Batistic and Kudla, 2009; Luan, 2009; Ishitani et al., 2000).

The protein kinase salt overly sensitive 2 (SOS2) phosphorylates the Na^+/H^+ antiporter SOS1 to bring about cellular ion homeostasis under conditions of salt stress in *Arabidopsis* (Liu and Zhu, 1997). SOS2 is activated by calci-SOS3, a CBL, and by autophosphorylation at Ser-228. It is activated by calci-SOS3 and by autophosphorylation at Ser-228 (Fujii and Zhu, 2009). In the crystal

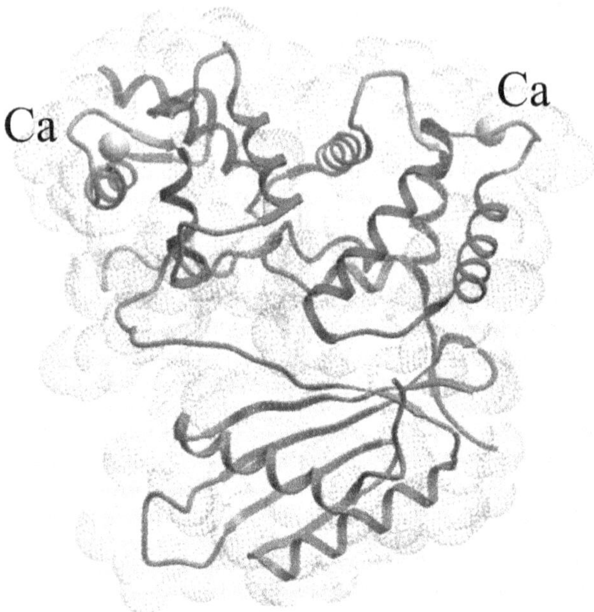

Figure 11.12. Structure of the C-terminal domain of the protein kinase AtSOS2 (blue) bound to AtSOS3 (red). Calcium is bound to EF-hands 1 and 4 (PDB file 2EHB). (From Sánchez-Barrena et al., 2007.) (*See insert for color representation of figure.*)

structure (Figure 11.12) of di-calci-SOS3 complexed with residues 306 to 430 of SOS2, the Phe–Ile–Ser–Leu (FISL) binding motif fits into a cleft formed by the open forms of both lobes 1,2 and 3,4 (Sánchez-Barrena et al., 2007). N-terminal myristoylation of SOS3 in vivo facilitates its interaction with the plasma membranes and with SOS2. One anticipates that other CBLs will have similar but subtlly different modes of binding to their respective targets.

11.3. S100 SUBFAMILY

S100 proteins, of which there are at least 26 members, are expressed in various cells and tissues of vertebrates exclusively (reviewed by Donato, 2001). These proteins were called S100 because of their solubility in a 100% saturated solution of ammonium sulfate. They are numbered sequentially (Schafer et al., 1995); older names are indicated for clarification. Many functions have been proposed for S100s and several human disorders, such as cancer, neurodegenerative diseases, cardiomyopathies, inflammations, diabetes, and allergies, are correlated with their altered expressions (reviewed by Heizmann and Cox, 1998). It is assumed that the functions of the S100s are associated with their abilities to bind calcium, to dimerize, and to interact specifically with target proteins.

Calbindin D_{9k} is congruent with the S100s; it functions as a monomer. The S100 lobe of two EF-hands is also found within the sequences of three large proteins: profillaggrin, trychohyalin, and repetin (Lee et al., 1993; Presland et al., 1995; Krieg et al., 1997; Huber et al., 2005). The EF-lobes of these chimeric proteins are congruent with the S100 proteins.

11.3.1. S100

S100s are involved in many biological processes (reviewed by Donato, 2001; Schaub and Heizmann, 2008; Donato et al., 2009; Eckert et al., 2004). Enzymes inhibited by S100s include casein kinase I and II by S100A8 and S100A9; twitchin kinase, fructose-1,6-bisphosphate aldolase, phosphoglucomutase, membrane-bound guanylate cyclase, phospholipase A_2, and actomyosin ATPase. S100s interact with microtubules, microfilaments, tropomyosin, myosin, the transcription factors MyoD and E12. They are involved in calcium homeostasis as well as cell proliferation and differentiation.

S100s are released into the extracellular space by unknown mechanisms. Extracellular S100s stimulate neuronal survival and/or differentiation and astrocyte proliferation. They cause neuronal death via apoptosis, and stimulate (in some cases) or inhibit (in other cases) the activity of inflammatory cells. A cell surface receptor, RAGE, for S100A12 and S100B was identified on inflammatory cells and neurons. S100A10 might play a role in regulation of the extrinsic pathway of blood coagulation by binding to plasminogen. The variety of intracellular target proteins of S100s and the cell specificities of expression of S100s indicates their roles in the regulation of effector proteins as well as signaling pathways and cellular functions.

Most S100 genes are located in a cluster on human chromosome, 1q21, a region frequently rearranged in human cancers (reviewed by Heizmann et al., 2007). The functional diversity of S100s is achieved by their specific cell and tissue expression patterns, structural variations, different binding properties for calcium, zinc, and copper, as well as by their ability to form homo- and hetero-oligomerss. Twenty-one S100s (S100A1 to S100A18, trichohylin, filaggrin, and repetin) have genes at chromosome locus 1q21; while other S100 genes are found at chromosome loci 4p16 (S100P), 5q14 (S100Z), 21q22 (S100B), and Xp22 (S100G). Fourteen S100 genes are located within the epidermal differentiation complex on human chromosome 1q21. Thirteen S100s (S100A2, S100A3, S100A4, S100A6, S100A7, S100A8, S100A9, S100A10, S100A11, S100A12, S100A15, S100B, and S100P) are expressed in normal and/or diseased epidermis (Eckert et al., 2004). Upon calcium binding, S100s interact with target proteins to regulate cell function; they undergo changes in subcellular distribution in response to extracellular stimuli (Table 11.1). They also function as chemotactic agents and may play a role in the pathogenesis of epidermal disease, as selected S100s are markedly overexpressed in psoriasis, wound healing, skin cancer, inflammation, cellular stress, and other epidermal disorders.

TABLE 11.2. Calcium-Dependent Interactions of S100 Proteins

Protein	Target	Function
S100A1	Aldolase C	Stimulation of aldolase C activity
	Annexin A5	Unknown
	Annexin A6	Regulation of calcium flux and IF assembly
	CacyBP/SIP1	Regulation of CacyBP/SIP
	Caldesmon	Decrease in inhibition of actomyosin by caldesmon
	CapZ (TRTK-12)	Modulation of actin organization
	Desmin	Inhibition of desmin intermediate filament assembly
	F-actin	Regulation of actin filament polymerization
	Fructose-1,6-bisphosphate aldolase	Regulation of energy metabolism
	GFAP	Inhibition of GFAP assembly
	Microtubules/tubulin	Regulation of microtubule dynamics
	MyOD	Inhibition of MyoD phosphorylation and DNA binding
	NDR	Activation of NDR kinase activity
	p53	Disruption of tubulin, S100A1 complex formation
	Phosphoglucomutase	Inhibition of phosphoglucomutase activity
	RAGE	Promotion of cell survival
	Ryanodine receptor	Regulation of ryanodine and cardiac contractility
	SERCA2a and PLB	Regulation of EC-coupling in the heart
	Synapsin I	Regulation of catalytic activity of synapsins
	Titin (PEVK domain)	Inhibition of the actin–or nebulin–PEVK interaction
	Twitchin kinase	Activation of twitchin kinase
S100A2 (S100L)	ΔNp63	Downstream mediation of ΔNp63
	p53	Activation of p53 transcriptional activity
	Tropomyosin	Modulation of the actin cytoskeleton organization

TABLE 11.2. (*Continued*)

Protein	Target	Function
S100A4 (placental calcium binding protein)	CCN3	Modulation of S100A4 affinity to calcium and function
	F-actin	Unknown
	MetAP2	Regulation of MetAP2
	Myosin heavy chain II-A	Regulation of cytoskeletal dynamics
	Myosin heavy chain II-B	Regulation of tropomyosin, actin association
	Tropomyosin isoform2	Increase affinity of S100A4 to calcium
	p37	Enhancement of p53-dependent apoptosis in tumors
	p53	
S100A6 (calcyclin)	Annexin A11	Regulation of annexin A11 function
	Annexin A2	
	Annexin A5	
	Annexin A6	
	CacyBP	
	CacyBP/SIP	Regulation of CacyBP/SIP ubiquitinylation complex
	Caldesmon	
	Fetuin (biotinylated) GADPH	Unknown
	Lysozyme	Unknown
	Tropomyosin	Regulation of smooth muscle contraction
	Sgt1	Regulation of protein ubiquitination via Stg1
S100A8/S100A9 (MRP-8 to MRP-14)	Cytochrome b_{558}	Activation of cytochrome b_{558}
	Glycosaminoglycans	Localization of MRPs to endothelium
	S100A8/S100A9 (tetramer)	Increasing affinity for calcium
S100A10 (p11)	Phospholipase A2	Anti-inflammatory activity
S100A11	Actin	Regulation of actin-activated myosin ATPase
	Actomyosin ATPase	Unknown
	Annexin A1	Targeting and membrane cross-linking
S100A12 (calgranulin C)	Annexin A5	
	CacyBP/SIP	Degradation of α/β-catenin
	Paramyosin	Development of keratitis
	RAGE	Inflammatory processes
	S100A12 (hexamer)	Receptor signaling

(*continued overleaf*)

TABLE 11.2. (*Continued*)

Protein	Target	Function
S100B	AHNAK	Regulation of calcium homoeostasis
	Aldolase C	Stimulation of aldolase C activity
	Annexin A6	Regulation of calcium flux and IF assembly
	Annexin II	Unknown
	CacyBP/SIP	Regulation of ubiquitination
	Caldesmon	Decrease in inhibition of actomyosin by caldesmon
	CapZ (TRTK-12)	Regulation of actin filament extension
	FtsZ	
	GFAB	Assembly of intermediate filaments
	Guanylate cyclase	Activation of guanylate cyclase
	Intermediate filaments	Regulation of IF assembly and disassembly
	IQGAP1	Membrane rearrangement
	MAG	Regulation of glial cell cytoskeleton
	MARCKS	Unknown
	Microtubules	Regulation of microtubule dynamics
	NDR	Modulation of NDR kinase activity
	Neurogranin	Unknown
	Neuromodulin	Inhibition of neuromodulin phosphorylation by PKC
	p53	Inhibition of p53 function
	Phosphoglucomutase	Stimulation of phosphoglucomutase activity
	RAGE	Promotion of cell survival
	Tau	Inhibition of tau phosphorylation by protein kinase II
S100P	CacyBP	
	Dormant ezrin	Regulation of ezrin ability to bind actin
	Melittin	
	RAGE	Stimulation of cell proliferation and survival via RAGE
	S100PBPR	Involvement in early pancreatic cancer
	Sgt1	

Source: Data from Donato (2001), Santamaria-Kisiel et al. (2006).

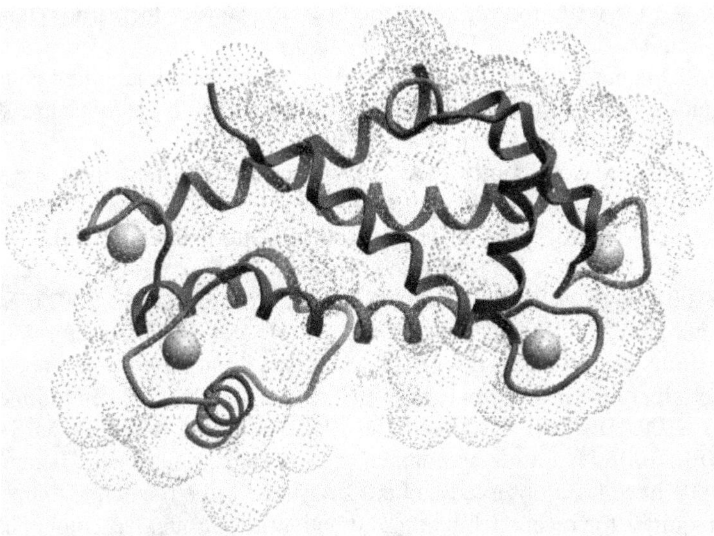

Figure 11.13. Structure of the tetra-calci-S100A12 dimer (PDB file 1E8A). The two monomers form a four-helix bundle with their two helices A and two helices D. (From Moroz et al., 2001.)

```
Canonical (S100, domain 2)
                     -------  Loop

 αααααααα E αααααα            ααααααα F αααααααα

                     1 1 1 1 1 1 1 1 1 1 2 2 2 2 2 2 2 2 2 2

 1 2 3 4 5 6 7 8 9 0 1 2 3 4 5 6 7 8 9 0 1 2 3 4 5 6 7 8 9

 E n * * n n * * n X * Y * Z G-Y I-X * *-Z n * * n n * *(n)

                               =0

S100, domain 1

                                       2 2 2 2 2 2 2 2 2

1 2 3 4 5 6 7 8 9                      1 2 3 4 5 6 7 8 9

E n * * n n * * n * * * E G D * * * L * K * E n * * n n * *(n)

                  X     Y   Z   -Y       -Z

                 =0    =0  =0   =0
```

S100s consist of two EF-hands; the overall fold of the monomer has four helices and three loops (Figure 11.13). The C-terminal EF-hand is canonical and binds calcium with relatively high affinity, $K_d \approx 10$ to 50 μM. The coordinating ligands are Asp61 (X), Asn63 (Y), Asp65 (Z), oxygen of the main-chain carbonyl of Glu67 (−Y), a water oxygen (−X), and two oxygens of the carboxyl group of Glu72 (−Z). The N-terminal EF-loop has two more residues (14 instead of 12) than does the canonical EF-hand. Several of these calcium liganding residues are not canonical. The N-terminal EF-loop in S100A7 and both EF-loops in S100A10 do not bind calcium (Brodersen et al., 1998).

Most S100s exist as homodimers, in which the two monomers are related by a twofold axis. The crystal structures of S100B (Matsumura et al., 1998; Smith and Shaw, 1998), holo- and apo-S100A6 (Otterbein et al., 2002), S100A7 (Brodersen et al., 1998), S100A8 (Ishikawa et al., 2000), S100A10 (Rety et al., 1999), S100A11 (Rety et al., 2000), S100A12 (Moroz et al., 2001), S100A9 (Itou et al., 2002), S100P (Zhang et al., 2003), and NMR structures of apo-S100A6 (Potts et al., 1995), holo-S100A6 (Sastry et al., 1998), and apo-S100B (Drohat et al., 1996) have been determined. Some S100s form heterodimers: for example, S100A1/S100B, S100A8/S100A9, S100B/S100A6, S100A1/S100A4, and S100B/S100A11. Larger assemblies (i.e., hexamers, tetramers, octamers) of some S100s have been observed and are suggested to be the active extracellular species required for receptor binding and activation through receptor multimerization (reviewed by Heizmann et al., 2007). Calbindin D_{9k}, a member of the S100 subfamily, is a monomer.

Dimerization of S100s is important for their biological activities (reviewed by Donato, 2001; Santamaria-Kisiel et al., 2006). It is assumed that one S100 monomer does not bind strongly enough to target proteins. Upon calcium binding, each S100 monomer assumes a more open form to accommodate a target protein. Moreover, the S100 dimer can bind target proteins on opposite sides (Donato, 2001). Calcium binding to S100s, with the exception of S100A10, results in a $\sim 40^{\circ}$ shift in the relative position of helix 3; this exposes a broad hydrophobic patch that enables the S100s to interact with a variety of target proteins.

In S100A6 (calcyclin), the binding of calcium results in a change in position and orientation of helix 3, the first helix of the second EF-hand; the rest of the protein, including the dimer interface, remains virtually unchanged, as determined by heteronuclear solution NMR (Mäler et al., 2002; Sastry et al., 1998). As seen in the crystal structue (Otterbein et al., 2002), calcium binding to S100A6 causes a 86° reorientation of helix C and marked changes in the positioning and structure of helix B, the loop between helices B and C, and the C-terminal end of helix D. The combination of these structural changes results in a dramatic change in the global shape and charge distribution of the S100A6 dimer, leading to exposure of two target binding sites, related by the rotation axis. The apo state of S100A6 is more compact ($40 \times 29 \times 43$ Å^3) than the calcium-bound state ($42 \times 31 \times 51$ Å^3). The total surface exposed to the solvent is smaller in the apo state (9136 Å^2) than in the calci state (9549 Å^2). The solvent contacting interface of the calci dimer is less charged and in some regions is more hydrophobic than that of the apo form. The calcium affinity of some S100s is not very high. For example, Mani and Kay (1983) reported $K_d(\text{Ca}) = 3.0 \times 10^{-5}$ M for S100a.

Several S100s bind zinc and copper at sites distinct from their EF-loops (Ogoma et al., 1992). The Zn^{2+} ion binding site of human S100A7 (psoriasin) is formed by three His's (H17, H86, H90) and Asp24 (Figure 11.14) (Brodersen et al., 1998, 1999). Sequence comparison suggests that this zinc site is present in a number of the other S100s. It is inferred that the effects reported for the binding of zinc to various S100s derive from binding at the H17, H86, H90, and D24 sites; however, it is possible that other sites are present in some S100s.

```
S100A7     SNTQAERSIIGMIDMFHKYTRRDD
S100A1(      SELEKAVVALIDVFHQYSGREGDKHKLKKSELKELINNELSHFLEEIKEQ
S100A8     MLTELEKALNSIIDVYHKYSLIKGNFHAVYRDDLKKLLETECPQYIRKKG
S100A9      MSQMESSIETIINIFHQYSVRLGHYDTLIQKEFKQLVQKELPNFLKKQKK
S100A12      TKLEEHLEGIVNIFHQYSVRKGHFDTLSKGELKQLLTKELANTIKNIKDK
S100A3       MARPLEQAVA AIVCTFQEYA GRCGDKYKLC QAELKELLQK ELATWTPTEF

                        1111111111222222222
                1234567890123456789012345678 9
                 n**nn**nX*Y*ZGy*x**zn**nn**n
S100A7      TNYLADVFEKKDKNEDKKIDFSEFLSLLGD IATDYHKQSHGAAPCSGGSQ
S100A1       EVVDKVMETLDSDGDGECDFQEFMAFVAMITTACHEFFEHE
S100A8          ADVWFKELDINTDGAVNFQEFLILVIKMGVAAHKKSHEESHKE
S100A9       NEAAINEIMEDLDTNVDKQLSFEEFIMLVARLTVASHEEMHNTAPPGQGH
S100A12      AVIDEIFQGLDANQDEQVDFQEFISLVAIALKAAHYHTHKE
S100A3     RECDYNKFMSVLDTNKDCEVDFVEYVRSLACLCLYCHEYFKDCPSEPPCSQ
```

(*See insert for color representation of the amino acid sequence.*)

S100A7 coordinates zinc with His17, Asp24, His86, and His90 (Figure 11.4). One infers that S100s A1β, A8, A9, and A12 also bind zinc with similarly spaced His's. In contrast, S100A3 has 10 Cys's (yellow), several of which are surely involved in coordinating zinc at both similar and different sites. In S100B, His15, His25, Cys84, His85, and perhaps His90 are involved in coordinating zinc in a site that is very similar to that observed in the crystal structure of S100A7; this was confirmed by NMR spectroscopy and site directed mutagenesis (Wilder et al., 2003).

The Cys–His motif is conserved in at least five S100 proteins (S100B, S100A3, S100A11, S100A14, and S100P), but only three of the five are known to bind zinc (S100B, S100A3, and S100P), and of these three, S100P binds very weakly ($K_d \sim 1$ mM). Asn can be a ligand for the Zn^{2+} ion; S100A1, S100A2, S100A4, S100A10, and S100Z have Cys and Asn in the C-terminal loop; it

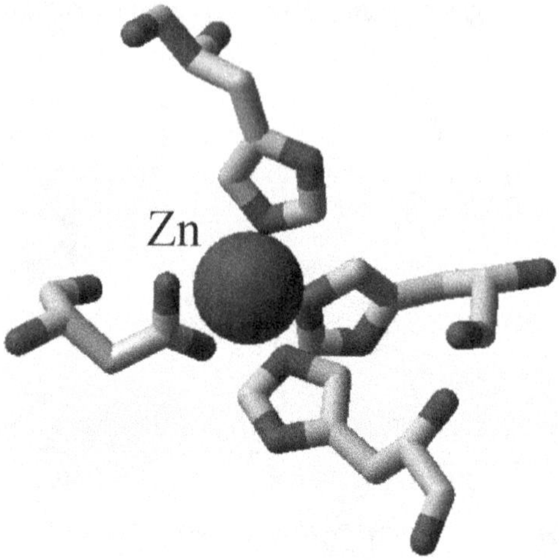

Figure 11.14. Zinc binding site in S100A7 (psoriasin) (PDB file 2PSR), formed by both oxygens of the carboxyl group of Asp24 and the three Nε's of His17, His86, and His90. (From Brodersen et al., 1998, 1999.)

might bind zinc in a manner similar to that of the Cys–His sequence. Of these, S100A1, S100A2, and S100A4 bind zinc. In helix A the critically important zinc ligand, His15, is conserved in 12 out of 17 S100s. At least 17 S100s have a Cys, His, Asp, or Glu between Glu21 and His25. Sixteen out of 17 S100s have one or more Cys's, His's, Asp's, or Glu's in the C-terminus.

The binding of zinc not only causes a structural change near the zinc binding site but also affects other properties of S100. Zinc binding increases the affinity for calcium and decreases the antagonistic effect of potassium on calcium binding (Baudier et al., 1983). Zinc binding causes helix D to extend by one full turn when compared to calcium-bound S100B. This change in secondary structure probably contributes to the increased affinity of S100B for target peptides in the presence of zinc (Wilder et al., 2003). As evidenced by fluorimetric titration, in the absence of zinc, S100A8 (calgranulin) binds one equivalent of calcium with K_d(Ca) of about 2×10^{-4} M; whereas, zinc–S100A8 binds two equivalents per monomer with K_d(Ca) values of $\sim 3 \times 10^{-6}$ and $\sim 2 \times 10^{-3}$ M (Dell'Angelica et al., 1994).

Zinc binding induces a large conformational change in S100A3, perturbing the hydrophobic interface between two S100A3 subunits (Fritz et al., 1998). S100A3 binds calcium with low affinity ($K_d = 0.5$ to 30×10^{-3} M) but binds zinc with exceptionally high affinity (two sites per monomer with $K_d = 0.5$ to 8×10^{-6} M). This high affinity for zinc is attributed to the high Cys content of S100A3 and involves sites other than H17, H86, H90, and D24 (Föhr et al., 1995). Zinc binding induces a large conformational change in S100A3, perturbing the hydrophobic interface between two S100A3 subunits (Fritz et al., 2002).

S100A12 binds copper at the predicted zinc/copper binding site, which is located close to the surface of the protein. Copper-mediated generation of reactive oxygen species by S100A12 was suggested as its function in the host–parasite response (Moroz et al., 2001). S100A5 binds one equivalent of zinc and two of copper per monomer (Schäfer et al., 2000). The zinc and copper binding sites in S100A5 seem to be different from those in other S100s. In light of the tertiary structure of S100s, it was suggested that in each subunit of the zinc site is located at the opposite side of the EF-hands and that the two copper binding sites possibly share ligands of the EF-hands.

Each S100 interacts in a calcium dependent manner with a number of target proteins, leading to specific physiological responses. Furthermore, some of the S100s target the same molecules as does calmodulin [e.g., cytoskeletal proteins, microtubule associated proteins (MAPs), caldesmon, the enzymes adenyl cyclase and glycogen phosphorylase, nuclear kinase, and the cell cycle associated proteins, neuromodulin, and p53].

Heizmann and Cox (1998) suggested that the biological activity of S100 in some cases is regulated by zinc and/or copper rather than by calcium. Zinc was identified as an essential component in the binding of S100β to MAP–tau (Yu and Fraser, 2001). S100β is overexpressed in Alzheimer's disease and hyperphosphorylated tau constitutes the primary component of neurofibrillary tangles. The S100β–tau interaction may be disrupted by hyperphosphorylation and/or imbalances in zinc metabolism; this may contribute to the neurite dystrophy associated with Alzheimer's disease.

S100A1 is highly abundant in the human heart (cardiomyocytes), although it is also found in lower amounts in skeletal muscle, brain, and kidney (reviewed by Schaub and Heizmann, 2008). It associates with the sarcolemma, junctional, and longitudinal SR, sarcomere, intercalated disks, and mitochondria of ventricular cardiomyocytes. At the molecular level, S100A1 interacts in a calcium-dependent manner with the cardiac isoforms of sarcoplasmic calcium release channel RYR2, calcium pump SERCA2A, phospholamban, titin, and the mitochondrial F1-ATP synthase in complex V of the respiratory chain. These target proteins are important regulators of cardiac excitation, contraction coupling, and energy homeostasis.

Stimulation of sarcoplasmic calcium release channel RYR2 by S100A1 increases systolic performance through enhanced calcium-induced calcium release (CICR) from the sarcoplasmic reticulum. On the other hand, S100A1 reduces the calcium spark frequency, thus reducing the calcium leak from the sarcoplasmic reticulum.

S100A1 stimulates calcium uptake into the sarcoplasmic reticulum not only by direct interaction and stimulation of the calcium pump, SERCA2A, but also by combining with phospholamban and thus derepressing its inhibitory effect on SERCA2A. Both effects together significantly hasten relaxation by rapid removal of cytoplasmic calcium after contraction.

The N-terminus of the giant protein titin (~3000 kDa) is at the Z-disk and runs near the actin filaments in the I-band (nonoverlap region) and along the myosin filaments in the A-band (actin and myosin filaments overlapping) up to the middle of the sarcomere in the M-band. In the extensible region of the I-band (between the Z-disk and the tips of the myosin filaments), cardiac titin contains a cardiac specific splice element, either N2B or N2BA, followed by the elastic PEVK (with over 70% Pro, Glu, Val, and Lys) domain. Both N2B and PEVK elements interact with the actin filaments in the I-band near the Z-disk. This interaction with actin contributes to passive tension and resists filament sliding during contraction. S100A1 can specifically bind to the N2B (or N2BA) and PEVK regions of titin in a calcium dependent manner. Binding of S100A1 to the extensible region releases titin from the actin filaments, thereby reducing passive tension. This action of S100A1 facilitates contraction. S100A1 also interacts in a calcium-dependent manner with the α- and β-chains of the F1-ATPase at the inner mitochondrial membrane, thereby stimulating production of ATP.

Over 50% of normal left ventricular S100A1 is required for cardiac adaptation to acute hemodynamic stress (reviewed by Schaub and Heizmann, 2008). On the other hand, transgenic overexpression of S100A1 provides chronically improved contractile function without concomitant development of cardiac hypertrophy as occurs under long term β-adrenergic overstimulation. Thus, S100A1 induced cytoplasmic calcium fluxes seem to bypass the effects of chronically elevated calcium levels, leading to hypertrophy as occurring not only with β-adrenergic stimulation, but also under increased activity of angiotensin-II, endothelin-1, or prostaglandin PGE2. Defective mobilization of calcium from intracellular stores may be responsible for the impaired inotropic reserve in hearts deficient in S100A1. Indeed, a failing human myocardium displays a marked loss of S100A1 expression at both the transcriptional and translational levels.

S100B is used as a marker of brain damage (reviewed by Korfias et al., 2006; Kleindienst et al., 2007). Over the past decade, numerous studies have reported a positive correlation of S100B serum levels with a poor outcome following traumatic brain injury. As an indicator for accurate determination of the extent of the brain damage, S100B meets the following requirements: (1) origin in the central nervous system with no contribution from extracerebral sources; (2) passive release from damaged neurons and/or glial cells without any active release; (3) lack of specific effects on neurons and/or glial cells interfering with the initial injury; and (4) unlimited passage through the blood–brain barrier. S100B can be measured in the arterial and venous serum; it is not affected by hemolysis and remains stable for several hours without the need for immediate analysis. Its long half life makes measurements crucial in emergency and intensive care settings. S100B can be monitored by several immunoassays in biological fluids (i.e., cerebrospinal fluid, blood, amniotic fluid, and urine) from fetuses and newborns at high risk of perinatal brain damage (Michetti and Gazzolo, 2002). S100B in biological fluids increases at an early stage when standard monitoring procedures are still silent in the study populations that later developed brain damage.

Altered S100 levels are associated with a broad range of diseases, including cardiomyopathy, inflammatory and immune disorders, neurodegenerative disorders, and cancer. There is increasing evidence that altered expressions of S100 are seen in many cancers, including those of breast, lung, bladder, kidney, thyroid, stomach, prostate, and mouth (reviewed by Yao et al., 2007; Salama et al., 2008).

Breast tumors exhibit altered levels of S100A2, S100A4, and S100A7 during tumorigenesis and/or progression (Emberley et al., 2004). Although the precise role of S100 in carcinogenesis is poorly understood, it seems that formation of homo- and heterodimers, binding of calcium, and interaction with effector molecules are correlated with the development and progression of many cancers. S100s are commonly up-regulated in tumors, and this often correlates with tumor progression. S100A2, S100A11, and S100A9 have been documented as tumor suppressors in some cancers but as tumor promoters in others (Emberley et al., 2004). This demonstrates the complexity and variability of their functions. Although the precise roles of S100s in cancer has yet to be discovered, many members of this protein family are associated with promoting metastases through interactions with matrix metalloproteinases or by acting as chemoattractants. There is also evidence that some members can regulate transcription factors such as p53. S100B is used in monitoring of malignant melanoma. In animal models of bladder cancer, several S100s are differentially expressed relative to normal urothelium. In human bladder cancer, overexpression of S100A4, S100A8, or S100A11 correlates with stage progression, invasion, metastasis, and poor survival.

Several proteins have been identified as S100A4 targets, including liprin β1, methionine aminopeptidase, the p53 tumor suppressor protein, and proteins involved in cytoskeletal rearrangement and cell motility, such as F-actin, tropomyosin, and the heavy chain of nonmuscle myosin II (reviewed by Garrett et al., 2006). In all of these cases, the interaction of these targets with S100A4 is calcium dependent and thus links the cellular functions of these proteins to changes in the intracellular calcium concentration. Transgenic mice that overexpress S100A4 in the mammary epithelium are phenotypically indistinguishable from wild type mice, demonstrating that S100A4 itself is not tumorigenic; however, transgenic mouse models of breast cancer have shown that S100A4 expression correlates with metastasis. Increased expression of S100A4 mRNA has recently been found in proliferating rheumatoid arthritis synovial fibroblasts and synovial tissues from patients with rheumatoid arthritis (Senolt et al., 2006). Synovial hyperplasia in rheumatoid arthritis consists of inflammatory cells and activated synovial lining cells; this contributes to the progressive destruction of the joints during the disease. Since several phenomena are similar between rheumatoid arthritis and malignant tumors, it is possible that S100A4 contributes to the invasive and tumorlike behavior of rheumatoid arthritis synovium. Extracellular S100A4 has been demonstrated to contribute to angiogenesis and the increased production of matrix-degrading enzymes by both endothelial and tumor cells. S100A4 might be responsible for TCRγδ T-cell-mediated lysis and negative regulation of matrix mineralization.

11.3.2. Calbindin D_{9k}

The name *calbindin* was initially applied to two different EF-hand proteins (Szebenyi et al., 1981; Dalgarno et al., 1983; Szebenyi and Moffat, 1986). In intestine and kidney their syntheses are dependent on vitamin D. Vitamin D maintains calcium homeostasis and is required for bone development and maintenance. Calbindin D_{9k} is a member of the S100 subfamily (Kawasaki et al., 1998; reviewed by Schwaller, 2008). It is found mainly in the mammalian brain, intestine, kidney, placenta, teeth, and yolk sac. In rat kidney, calbindin D_{9k} is expressed in the loops of Henle and the distal convoluted tubule. In the collecting duct, calbindin D_{9k} is restricted to the intercalated cells. These proteins appear to be calcium buffers in these cells or may be involved in calcium translocation across cells (reviewed by Choi et al., 2005). Calbindin D_{28k} has six EF-hand domains, as do calretinin and secretagogin (see Section 11.5.3). It is found not only in intestine, but also in brain, endocrine pancreatic cells, and endocrine pancreatic tumoral cells (Pochet et al., 1989).

Calcium is absorbed in the mammalian small intestine by two general mechanisms: a transcellular active transport process, located largely in the duodenum and upper jejunum, and a passive, paracellular process that functions throughout the length of the intestine (reviewed by Bronner, 2003). The transcellular process involves three major steps: entry across the brush border, mediated by a complex termed CaT1; facilitated diffusion, mediated largely by the cytosolic calbindin D_{9k}; and extrusion at the serosal surface, mediated largely by Ca-ATPase. When calcium intake is low, transcellular calcium transport accounts for a substantial fraction of the calcium absorbed. When calcium intake is high, transcellular transport accounts for only a minor portion of the calcium absorbed because of down regulation of CaT1 and calbindin D_{9k}. Biosynthesis of calbindin D_{9k} is fully dependent and CaT1 function is approximately 90% dependent on vitamin D. The presence of calbindin D_{9k} facilitates the diffusion of calcium across the cell without raising the concentration of free Ca^{2+} ions. Calbindin D_{9k} may also stimulate the basolateral calcium pumping ATPases. Interestingly, calbindin D_{9k} null mice are normal; this indicates that other calcium transporter genes can compensate for the lack of calbindin D_{9k} (reviewed by Choi and Jeung, 2008). The calbindin D_{9k} gene is regulated at the transcriptional or posttranscriptional level by 1,25-dihydroxyvitamin D3, a hormonal form of vitamin D (Darwish and DeLuca, 1992; reviewed by Choi et al., 2005). This hormonal form causes a parallel increase in calbindin D_{9k} mRNA and intestinal absorption of calcium in rats. In addition, uterine calbindin D_{9k} may be involved in controlling myometrial activity related to the intracellular level of calcium, but the exact role of calbindin D_{9k} in the uterus is still under investigation.

Calbindin D_{9k} is 75 residues long and consists of two EF-hands (Figure 11.15). Interactions among its four helices are mostly hydrophobic; however, it possesses several essential hydrogen bonds, especially within the calcium binding loops. The NMR structure of calbindin D_{9k} in solution corresponds well to its crystal structure (Drakenberg et al., 1989). The only difference is a short turn of helix in the loop between helices 2 and 3 in the crystal structure; this is disordered in solution.

The C-terminal EF-hand of calbindin D_{9k} is canonical. The N-terminal EF-hand has two residues inserted, as do other S100s. Residence times of the two

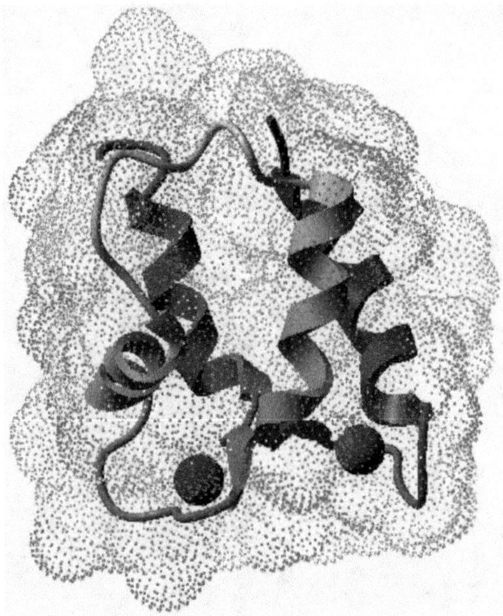

Figure 11.15. Structure of calbindin D_{9k} (PDB file 1B1G) viewed perpendicular to its approximate twofold axis. (From Kördel et al., 1997.)

water molecules that ligate Ca^{2+} ions are in the range 5 ns to 7 μs, much longer than for calcium-coordinated water in bulk solution (Denisov and Halle, 1995).

The N-terminal EF-loop can be converted into a canonical EF-loop by several point mutations (Johansson et al., 1993). Calbindin D_{9k} binds Ca^{2+} without significant changes in conformation (Skelton et al., 1995; Ababou and Desjarlais, 2001). The binding of calcium causes the largest changes in the middle of helix D and in the packing of helix C into the remainder of calbindin D_{9k} (Figure 11.16).

The binding of calcium changes the backbone dynamics of calbindin D_{9k} (Akke et al., 1993; Malmendal et al., 1998). Its two calcium binding loops respond differently to the binding of metal ions. The entire N-terminal EF-hand is rigid and allows calcium binding with only minor rearrangements. The structural and dynamical properties of the entire EF-hand are the major determinants of loop flexibility in this system. ^{15}N NMR spectroscopy showed that high-frequency fluctuations (10^9 to 10^{12} s^{-1}) of the N-terminal calcium binding loop are not affected by calcium binding, whereas residues G57, D58, G59, and E60 in the C-terminal calcium binding loop have significantly lower-order parameters in the apo state than in the metal-bound states. The dynamical responses of the four helices to the binding of ions are much smaller than that for the second calcium binding loop, with the strongest effect on helix C.

Calbindin D_{9k} has a high affinity for calcium [$K_d(Ca) \sim 10^{-7}$ M] (Shelling and Sykes, 1985). The sequence of calcium binding is still not clear. Proton NMR

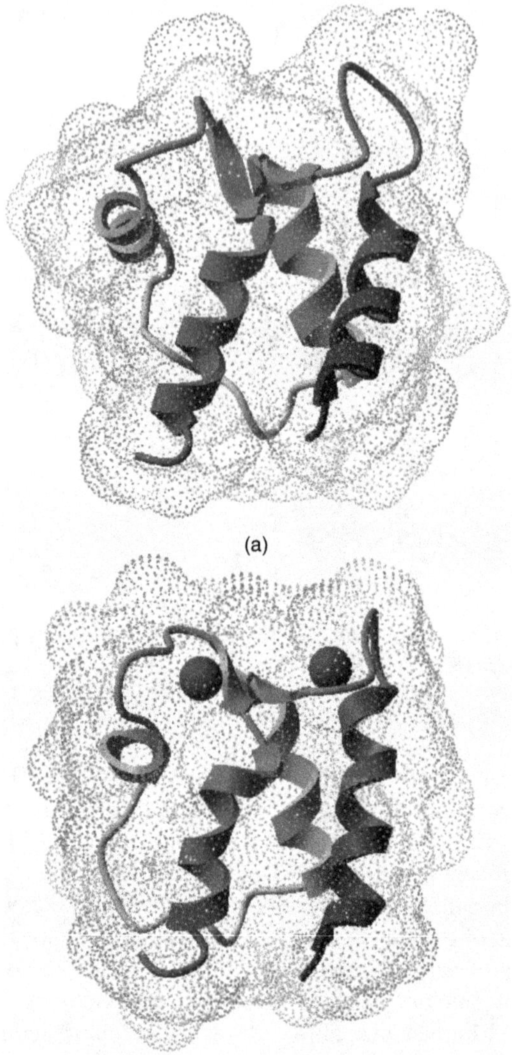

(a)

(b)

Figure 11.16. Structures of (a) apo- and (b) di-calci-calbindin D_{9k} (PDB files 1CLB and 1B1G). Both are viewed perpendicular to their local twofold axes. (From Skelton et al., 1995; Kördel et al., 1997.)

spectra are best fit by a scheme of random binding to two independent sites with the same association constants. Other data indicate an interaction between the two binding sites of calbindin D_{9k} (Hofmann et al., 1988; Mäler et al., 2000); the binding of a Ca^{2+} ion to site 1 results in a reorganization of the protein structure, reduces the flexibility of the protein on the picosecond to nanosecond time scale and significantly slows the rate of unfolding events leading to amide proton exchange. The reorganization associated with the first binding step lowers the free energy penalty for reorganization required for binding the second Ca^{2+} ion and drives the conformation and dynamics of the protein far toward the di-calci state, thereby facilitating binding of the second ion (Wimberly et al., 1995).

Mutations in the hydrophobic core of calbindin D_{9k} exert large effects on its stability; these are evident from the midpoints in urea induced unfolding (Kragelund et al., 1998; Julenius et al., 1998). This parameter varies from 1.8 M urea for L23G up to 6.6 M for V70L mutations. Both increases and decreases in the side chain surface areas cause quantitatively equivalent effects on the stability. These differences in free energy of unfolding correlate well with the changes in the surface areas of the side chains that are caused by the mutations. Despite the large effects of the mutations on the difference in free energy of unfolding and cooperativity, the structures of the mutants in the native form remain practically unchanged, as evidenced by circular dichroism, NMR, and fluorescence spectroscopy. For example, detailed NMR investigations of V61A showed that this mutation causes only minimal perturbations in the immediate vicinity of residue 61. The substitution of Ala or of Gly for bulky residues in the center of the protein core causes a reduction in calcium affinity and an increase in calcium dissociation rate. Small effects, both increases and decreases in stability, were revealed for substitutions involving residues located far from the calcium binding sites and toward the outer part of the hydrophobic core. F66W has a 25-fold increase in overall affinity for calcium and a sixfold reduction in calcium dissociation rate. The hydrophobic core in calbindin D_{9k} promotes calcium binding both by contributing to the preformation of the calcium binding sites in the apo state and by preferentially stabilizing the calcium-bound state. Its calcium binding properties are functions of its entire structure, and can be fine tuned by adjustments at the opposite end of the protein.

The isolated N-terminal EF-hand of calbindin D_{9k} tends to dimerize in the calci state (Julenius et al., 2002). The C-terminal EF-hand shows a much smaller tendency to dimerize; this may be related to its larger net negative charge. The calcium affinities of both EF-hands are similar and in the range $K_d = 10^{-4.6}$ to $10^{-5.3}$ M.

Forsén et al. (1988) studied the kinetics of calcium dissociation from the two sites of calbindin D_{9k} and its mutants using fluorescence of the calcium chelator quin 2. Wild-type and several mutants show only a single dissociation process with rate constants 2.8 to 3.8 s^{-1}. Substitutions and deletions of amino acid residues in the calcium binding loop of site I affect primarily the calcium dissociation rate for this site and only slightly influence the dynamic properties of site II. The calcium association rate in all cases is close to the diffusion-controlled limit ($\sim 10^9$ M^{-1} s^{-1}).

11.3.3. P26olf (Dicalcin)

P26olf (dicalcin) was prepared from the olfactory epithelium of *Rana catesberiana* (Grobler and Hurley, 1998) and from *Xenopus laevis* eggs (Miwa et al.,

2007). P26olf contains four EF-hands. Both EF-lobes resemble S100 in that the first and third EF-hands have two residues inserted in their calcium binding loops; EF-hands 2 and 4 are canonical. The first two EF-hands closely resemble the second two, reflecting a recent gene duplication. The 36-residue linker between EF-hands 2 and 3 is 60% identical to the C-terminal 20 residues of a rabbit S100. Tanaka et al. (1999) built a model of p26olf based on multiple sequence alignments and on the NMR structures of dimeric S100B(ββ) in the apo state. This predicted structure of the p26olf monomer adopts a folding pattern remarkably similar to that of dimeric S100B(ββ). The circular dichroism spectral data support the structure predicted.

P26olf binds four equivalents of calcium with K_d(Ca) 2.4 μM (Miwa et al., 1998, 2001). At 100 mM potassium p26olf binds calcium with K_d(Ca) 22 μM and a Hill coefficient of 2.0 (Miwa and Kawamura, 2003). P26olf immunoreactivity is observed in the cilia, dendritic knobs, and dendrites of frog olfactory receptor cells. It binds to a β-adrenergic receptor kinase like protein in a calcium dependent manner). P26olf appears to function in olfactory transduction or adaptation (Miwa et al., 2001).

11.3.4. Fused Gene Family

The fused gene family comprises profilaggrin, trichohyalin, repetin, hornerin, the profilaggrin related protein, and cornulin. These proteins are 500 to 1000 residues long; all have an S100 fold at the N-terminus, followed by many copies of a 60-residue domain. They interact with keratin based intermediate filaments and are especially abundant in the epidermis. For example, cornulin is encoded by human gene *c1orf10* on 1q21 (near gene for S100A11). Its S100 lobe is most similar to S100A8. It is especially abundant in esophagus, scalp, and foreskin.

11.4. PENTA-EF-HAND SUBFAMILY

The penta-EF-hand proteins were inititially thought to contain only four EF-hands (reviewed by Maki et al., 2002). However, the crystal structure of the small subunit of calpain (calpain dVI) revealed a fifth EF-hand, not easily recognized by comparison of sequences (Blanchard et al., 1997; Lin et al., 1997). This fifth EF-hand has 11 (not 12) residues in its loop and an Ala in place of the canonical Asp in position X. The fifth EF-hand forms part of the dimer interface by pairing with the corresponding fifth EF-hand of the second monomer. Penta-EF-hand proteins have several common features: (1) dimerization through unpaired C-terminal EF-hands, (2) possession of hydrophobic Gly/Pro-rich N-terminal regions, and (3) calcium-dependent translocation to membranes.

11.4.1. Group II

The mammalian penta-EF-hands are considered in two groups II (calpain, calcium-dependent proteases, sorcin, and grancalcin) and I (ALG-2 and peflin). Group I genes are also found in invertebrates, plants, fungi, and protists.

Calpain Calpains are cytosolic cysteine (papain-like) proteases whose enzymatic activities depend on calcium (reviewed by Sorimachi et al., 1997; Sorimachi and Suzuki, 2001). Calpains are the only known enzymes that combine in one molecule protease activity with a dependence on calcium binding to EF-hands. Calpain is not specific for residues or sequences but cleaves bonds between domains. As a consequence, calpain hydrolyzes substrate proteins in a limited manner, and large fragments retaining intact domains are produced by hydrolysis. There are more than 12 calpain isoforms. Calpains participate in integrin-mediated cell migration, cytoskeletal remodeling, cell differentiation, and apoptosis. Overactivation of calpain as well as mutations abrogating calpain activity have been implicated in muscular dystrophy, cardiac and cerebral ischemia, platelet aggregation, neurodegenerative diseases, rheumatoid arthritis, cataract formation, and Alzheimer's disease.

The m- and μ-calpains are the best characterized. Both are heterodimers composed of a large 78- to 80-kDa catalytic subunit and a common small 29-kDa regulatory subunit. The large subunit consists of four domains (dI to dIV), while the small subunit has two domains (dV and dVI). The crystal structure (Figure 11.17) of the full length human m-calpain in the apo form shows that domain dI comprises a single α-helix in a cavity of the dVI domain (Strobl et al., 2000). Domain dII contains the catalytic site and can be divided into two subdomains, IIa and IIb, containing the active sites Cys105 and His262/Asn286, respectively. The active site Cys105 in IIa is >10 Å from other catalytic triad residues (His262 and Asn286) on IIb, indicating that IIa and IIb are slightly too separated and open to constitute a functional catalytic triad in the apo form. Domain dIII consists of eight β-strands with topology similar to that of the C2 domain (Chapter 10) found in some proteins, including protein kinase C and phospholipase C, which are known to interact with calcium and phospholipids (Rizo and Südhof, 1998). This domain may be responsible for the calcium-dependent binding of calpain to membranes. Domains dIV and dVI both contain five EF-hands. The fifth EF-hands of dIV and dVI do not bind calcium; however, they do provide the heterodimerization interface between the large and small subunits of calpain.

Mini-calpain is a recombinant protein that comprises the catalytic domain IIa–IIb, which is active only upon binding calcium. Its crystal structure revealed two non-EF-hand calcium binding sites (Moldoveanu et al., 2002). The first site, located in the dIIa subdomain, involves residues Val99, Gly101, and Asp106 in the loop preceding α-helix 3 and Glu185, which is located in the loop connected to the N-terminus of the core α-helix 5. The second site is located in the dIIb subdomain; the Ca^{2+} ion is coordinated by four acidic residues: Glu302 and Asp309 in the loop containing Trp298, another residue in the active site, and Asp331, Glu333, and Met329 in the loop between α-helices 8 and 11. S-Ca bonds are very unusual; their significance is unknown. The active site of calcium bound μ-calpain markedly resembles the active site in other cysteine proteases, such as papain and cathepsins. The location and spacing between the critical active site

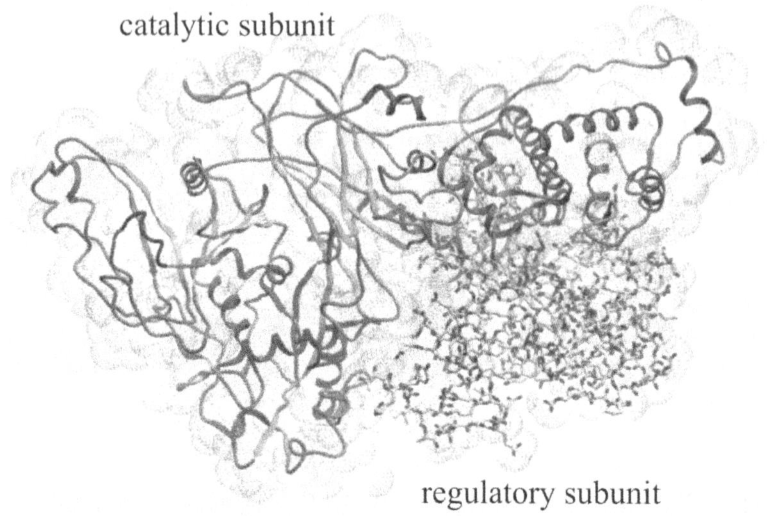

Figure 11.17. Structure of human m-calpain (PDP file 1KFU). The catalytic subunit is shown as a ribbon model; the regulatory subunit as a stick model. (From Strobl et al., 2000.) (*See insert for color representation of figure.*)

residues, Cys115, His272, Asn296, Gln109, and Trp298, correspond well with the corresponding residues in papain.

Moldoveanu et al. (2002) suggested a two step mechanism for calcium activation of calpain. The first step is the release of constraints imposed by the circular arrangement of the domains: calcium binding results in subtle conformational changes in domains dIV and dVI, leading to the abrogation of the interaction between the N-terminal α-helix of dI and the second EF-hand of dVI. The second step is the realignment of the active site cleft caused by the cooperative binding of calcium to dIIa and dIIb.

Purified calpains display a low affinity for calcium in vitro. The concentration giving half-maximal activity, $[Ca^{2+}]_{1/2}$, ranges from 2 to 75 μM for μ-calpain to 200 to 1000 μM for m-calpain (reviewed by Friedrich, 2004). Since such high calcium concentrations do not exist in intact living cells (normal range 50 to 300 nM), some ancillary mechanism or factor should enhance the affinity of calpains for calcium. Phospholipids, in particular phosphatidylinositides, reduce the concentration of calcium required for activation. It is also possible that calpain anchored to the cell membrane in the vicinity of calcium channels may be in the presence of much higher local calcium concentrations than in the cytoplasmic environment. Friedrich (2004) suggested that the high $[Ca^{2+}]_{1/2}$ value has probably been adjusted during evolution to allow fractions of the calpain pool to be activated in some cytoplasmic subcompartments, at the same time serving as a safety device protecting the cell from random calcium noise.

Both μ- and m-calpains are inhibited by reagents such as E64, leupeptin, and *N*-Ac–Leu–Leu–norleucine that react with the active site Cys (reviewed by Suzuki et al., 2004). These inhibitors also react with other cysteine proteases and are not specific for calpain. The nonpeptidic inhibitor PD150606 inhibits calpain by interacting with the EF-hands.

Within the cell, proteases are synthesized as zymogens, or inactive precursors, to prevent inappropriate or premature degradation of substrates. Zymogen conversion to the active enzyme is triggered by a variety of factors, such as a change in pH. Even though calpain has some zymogenlike characteristics, it is usually not referred to as a zymogen, as it clearly does not function as do most zymogens.

Calpains are ubiquitously distributed, along with their endogeneous inhibitor, calpastatin. The calcium-dependent proteolytic system, which consists of calpain and calpastatin, is activated by an increase in the intracellular concentration of calcium and then kept under control by the inhibitory effects of calpastatin. Calpastatin is a 120-kDa protein that is widely distributed in most mammalian tissues. It contains four inhibitory domains, homologous to one another, each having three conserved regions, termed A, B, and C, and an N-terminal domain, termed L. Calpastatin, the endogenous inhibitor of calpain, is an intrinsically unstructured protein proposed to undergo folding transitions upon binding to the enzyme. Calpastatin was suggested to have little secondary structure under physiological conditions, but the amino acid sequences of the conserved regions A and C suggest their ability to form amphipatic α-helices. Each of the conserved regions has been shown to bind preferentially to separate domains of calpain. A and C bind to the dIV and dVI domains, while B binds dII (Takano et al., 1995).

The calpain–calpastatin interaction is calcium dependent and is thought to be similar to that of calmodulin, in which hydrophobic pockets are exposed upon calcium binding. Unlike calmodulin, calpain has hydrophobic areas exposed even in the absence of calcium; however, calcium binding exposes additional hydrophobic pockets (Yang et al., 1994). DIC19 is a 19-residue peptide corresponding to one of the three interacting domains of calpastatin; DIC19 interacts with domain VI of calpain. In the crystal structure of DIC19 bound to domain VI of calpain (Todd et al., 2003), 11 residues of DIC19 form a helix. The hydrophobic residues of DIC19 are buried, while the polar and charged residues are in contact with the solvent or interact with the neighboring charged residues. Upon binding DIC19, the movement of E_1 and E_2 causes the formation of a hydrophobic pocket in which Leu606 binds. This hydrophobic pocket is lined with several bulky aromatic residues (Phe99, Phe162, and Trp166) as well as with other hydrophobic residues (Leu102, Leu106, Ileu121, and Val125).

There are several correlations between calpain gene mutations and human diseases (reviewed by Suzuki et al., 2004). Cloning of genes associated with diseases has revealed the association of the calpain 3 gene (CAPN3) with limb girdle muscular dystrophy type 2A (Richard et al., 1995) and the calpain 10 gene (*CAPN10*) with type 2 diabetes (Hirokawa et al., 2000).

Calpain directly regulates calcineurin activity through proteolysis in glutamate-stimulated neurons in culture and in vivo (reviewed by Wu et al., 2007). This calpain mediated proteolytic cleavage of calcineurin increases phosphatase activity; this, in turn, promotes caspase-mediated neuronal cell death. Activation of the calpain–calcineurin pathway could contribute to calcium dependent disorders, especially those associated with Alzheimer's disease and myocardial hypertrophy.

Sorcin Sorcin (soluble resistance-related, calcium binding protein) is another member of the penta-EF-hand family (Figure 11.18). Like many members of this family, sorcin binds more strongly to membranes in the presence of calcium. It binds to target proteins on these membranes. Sorcin is thought to participate in different calcium-triggered biochemical cascades since different target proteins have been identified in the different cell types where sorcin is expressed constitutively. The best characterized function of sorcin is its association with the cardiac ryanodine receptor RyR (Meyers et al., 1995a). In response to the small calcium release triggered by voltage controlled L-type calcium channels, RyRs mediate the release of larger amounts of calcium from the sarcoplasmic reticulum ("calcium induced calcium release") and trigger muscle contraction. In the presence of micromolar concentrations of calcium, sorcin binds reversibly to RyRs and inhibits calcium release (Lokuta et al., 1997). This provides rapid attenuation of the RyR response and prevents excessive depletion of calcium stores. Other targets of sorcin have been identified as L-type calcium channels in cardiac and skeletal muscle, annexin VII in adrenal medulla, and presenilin 2 in brain. The

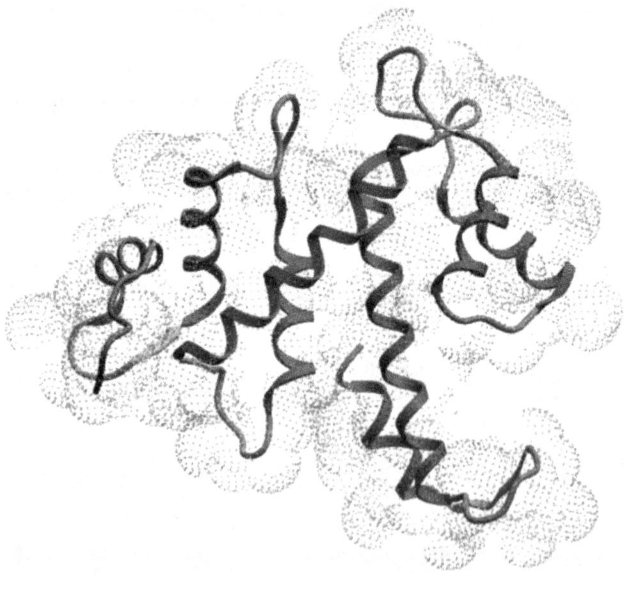

Figure 11.18. Structure of human apo-sorcin (PDP file 1JUO). (From Xie et al., 2001.)

N-terminal region of sorcin is required to interact with the membrane-localized annexin VII in a calcium-dependent manner (Verzili et al., 2000). Since sorcin is found in a wide variety of cell types, it probably has other functions in addition to its established role in cardiac muscle contraction.

The sorcin monomer (22 kDa) consists of two EF-lobes, as do all the penta-EF-hand proteins. It has a flexible, hydrophobic, Gly-rich N-tail. The crystal structures of the sorcin calcium binding lobes (residues 33 to 198) and the full-length protein have been determined in the apo form (Xie et al., 2001; Ilari et al., 2002). Apart from an extended N-terminal portion, sorcin has a globular shape. The protein consists of eight α-helices (A to H) in five EF-hands (EF1 to EF5). Helix D is common to EF2 and EF3, while helix G is common to EF4 and EF5.

Sorcin binds two Ca^{2+} ions per monomer with high affinity [K_d(Ca)$\sim$1 μM; Meyers et al. 1995b]. Sedimentation equilibrium analysis shows that sorcin forms dimers in the absence of calcium. Addition of calcium induces aggregation (Meyers et al., 1995b). The binding of calcium to sorcin weakens the interactions between its N- and C-lobes and leads to their reorientation. This exposes large hydrophobic regions, facilitating calcium-dependent binding to target proteins at or near membranes; this, in turn, results in a dramatic decrease in the solubility of sorcin and all other proteins of this family (Ilari et al., 2002).

Grancalcin Grancalcin, another member of the penta-EF-hand family, is expressed at high levels in neutrophils and may play a role in regulation of adherence and migration of neutrophils (Teahan et al., 1992). Grancalcin displays calcium-dependent binding to the granules and plasma membrane of neutrophiles. The crystal structure of human grancalcin was determined at 1.9-Å resolution in the absence of calcium and at 2.5-Å resolution in the presence of calcium (Jia et al., 2000). It contains eight α-helices and two short stretches of two stranded β-sheets between the loops of paired EF-hands. Grancalcin dimerizes by means of association of its EF-hands 5 in a manner similar to that observed in calpain. Only one Ca^{2+} ion is present per dimer under conditions of crystallization that included $CaCl_2$. It binds to EF-hand 3 in one molecule; this site in the second remains unoccupied. Hence, it was suggested that calcium binding induces rather small conformational changes and is subject to subtle crystal packing contacts.

11.4.2. Group I (ALG-2 and Peflin)

ALG-2 (encoded by apoptosis-linked gene 2) is 191 residues long and has five EF-hands (Lo et al., 1999; Jia et al., 2001). It interacts with Alix and TSG101, proteins involved in endosomal sorting and budding. It is involved in apoptosis and is required either downstream or independently of caspases for apoptosis (Vito et al., 1996). ALG-2 has been identified in the nucleus as well as cytoplasm of Jurkat and breast cancer cells.

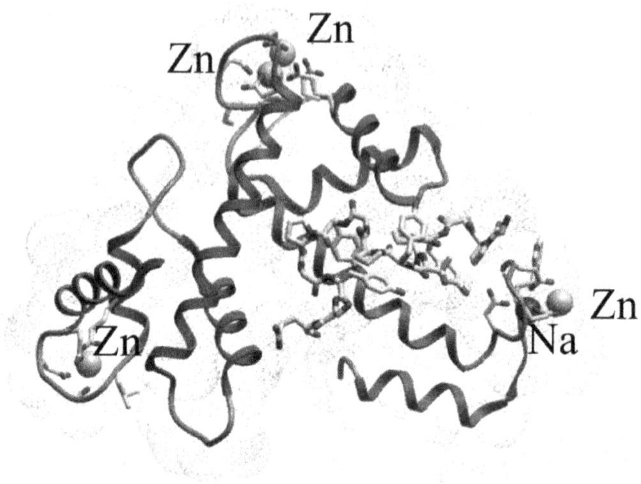

Figure 11.19. Crystal structure of zinc form of des3-23ALG-2 (peflin), complexed with Alix ABS peptide (PDB file 2ZNE). (From H. Suzuki et al., 2008.)

Several crystal structures are available (Suzuki et al., 2008). Zn^{2+} can replace the three Ca^{2+} ions normally bound in EF-hands 1, 3, and 5. Zinc is coordinated by the same canonical ligands, at similar metal–oxygen distances as is calcium. Zinc-ALG-2 has higher temperature factors and is less stable. ALG-2 dimerizes using its fifth EF-hand, as do other penta-EF-hand proteins. The target peptide, Alix799–804, binds in a groove between lobes 1,2 and 3,4 that consists of two pockets (Figure 11.19). Based on these structures, Sukuki et al. (2008) proposed a model in which the binding of calcium to EF-hand 3 enables the side chain of Arg125 in the 3,4 loop to move, thereby exposing the first hydrophobic pocket, which binds the GPPYP motif of Alix. This GPPYP is also involved in the binding of Alix to the 55-kDa centrosome protein.

The penta-EF-hand region of peflin (30 kDa) is most similar to the corresponding domain of ALG-2, particularly in the EF1 and EF3 regions (Kitaura et al., 1999). It has a longer N-terminal hydrophobic domain than that of any other member of the group I penta-EF-hand subfamily; it contains nine nonapeptide (A/PPGGPYGGP) repeats. Peflin is expressed in various nonadherent and adherent cultured human cell lines; however, its function is unknown.

11.5. PROTEINS WITH SIX EF-HANDS

The congruence of the CTER, CRP, S100, and penta-EF-hand groups is well established. The grouping of calbindin D_{28K}, calretinin, and tri-(Eps15 homolog) domain, CREC, and *Plasmodium falciparum* surface protein is a bit more heuristic. The relationship of lobe 1,2 relative to the pair of lobes 3,4 and 5,6 in calbindin D_{28k} and calretinin differs; however, this may reflect different states of ligation. Structures of the latter three are not available; their congruence is tenuous.

11.5.1. Calbindin D_{28k}

Calbindin D_{28k} is widely distributed and is involved in many essential physiological processes. It seems to act as both calcium buffer and calcium sensor. It is expressed in brain, kidney, bone, pancreas, and many other tissues (Pochet et al., 1989). Its concentration in auditory neurons approaches 2 mM (Oberholtzer et al., 1988).

Calbindin D_{28k} is 261 residues long and contains six EF-hands, packed into one globular structure (Figure 11.20); however, it binds only four equivalents of calcium (Gross et al., 1993; Veenstra et al., 1995, 1997; Berggård et al., 2002; Vanbelle et al., 2005; Kojetin et al., 2006). A high-affinity site ($K_d \approx 10^{-8}$ M) is located in EF-hand 1; while domains 3, 4, and 5 bind calcium less tightly. Domains 1, 3, and 5 more closely resemble domains 2, 4, and 6. Interdomains 1,2 and 3,4 and 5,6 are similar. Calbindin D_{28k} evolved by two gene duplication and fusion events from a two-domain precursor.

Experiments with calbindin D_{28k} mutants lacking various EF-hands showed that EF-hands 2 and 6 are not essential for calcium binding to the protein and that

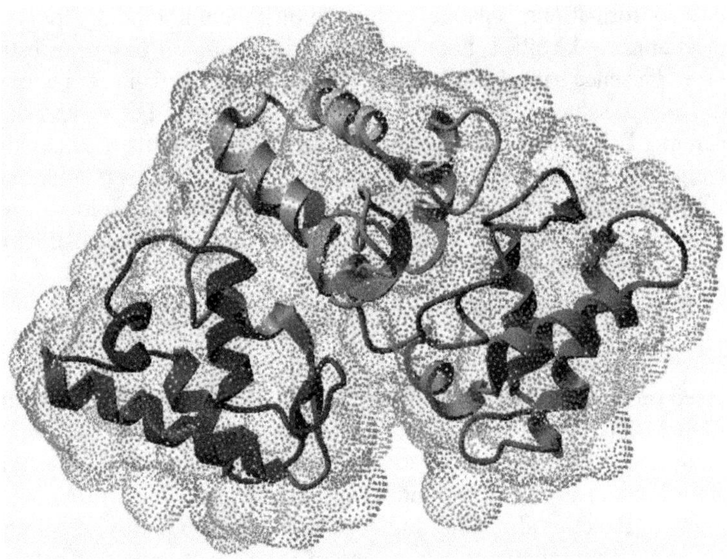

Figure 11.20. Structure of calci-calbindin D_{28k} (PDB file 2G9B). The three lobes (six EF-hands) are packed together. The Ca^{2+} ions are not indicated in 2G9B as determined by NMR spectroscopy. (From Kojetin et al., 2006.)

EF hands 1, 4, and 5 bind terbium. Four macroscopic K_d(Ca) values measured at low ionic strength are 2.6, 8.3, 12.3, and 58.8 μM. Magnesium binds to the same sites as does calcium with effective K_d(Mg) = 0.7 mM in 0.15 M KCl. The on-rate of calcium binding of calbindin D_{28k} under physiological conditions inside a cell is 1×10^7 to 8×10^7 M^{-1} s^{-1} (Schwaller, 2003). Thus, in activated neurons, the early phase of calcium transients are affected by calbindin D_{28k}. An immobile fraction of calbindin D_{28k} is observed in spines and dendrites, but not in axons of Purkinje cells. This immobilization involves approximately 20% of calbindin D_{28k} in spines and dendrites. It lasts for several seconds and can be increased by stimulation of climbing fibers at either 1 or 20 Hz. The binding partner in Purkinje cell spines and dendrites is myoinositol monophosphatase (IMPase), a key enzyme of the inositol-1,4,5-triphosphate signaling cascade. Additional reported calbindin D_{28k} binding proteins include caspase-3,3′, 5′-cyclic nucleotide phosphodiesterase, plasma membrane ATPase, and the L-type Ca^{2+} channel subunit CANAC1C (reviewed by Schwaller, 2008).

The calcium-induced conformational change in calbindin D_{28k} is readily detected by several spectroscopic methods (Berggård et al., 2000). ANS fluorescence experiments revealed that calbindin D_{28k} adopts distinct hydrophobic states upon the binding of calcium (Kojetin et al., 2006). The nature of the transition between apo- and calci-calbindin D_{28k} is pH dependent. At pH 6.5 apo-calbindin D_{28k} is notably more hydrophobic than is calcium-loaded calbindin D_{28k}, but this difference decreases with increasing pH. The calcium-induced change in the exposed hydrophobic surface of calbindin D_{28k} is considerably less pronounced than that in calmodulin. Interestingly, despite the fact that a substantial conformational change occurs upon calcium binding, only minor structural changes take place upon magnesium binding. It has been estimated that in the presence of physiological salt concentrations in a resting cell, calbindin D_{28k} is saturated to 40 to 75% with magnesium but to less than 9% with calcium. This means that upon calcium activation of a cell, calbindin D_{28k} undergoes significant calcium induced structural changes. Calbindin D_{28k} has a diffusion coefficient of 20 mm^2 s^{-1}, approximately twofold slower than parvalbumin measured under identical conditions and slower than dextrans of similar molecular weight (Schmidt et al., 2005).

By use of radioblots, isothermal titration calorimetry, and competition with a fluorescent zinc chelator, Bauer et al. (2008) found that calbindin D_{28k} binds Zn^{2+} ions to three rather strong sites with dissociation constants in the low micromolar range. Based on spectroscopic investigations, they concluded that zinc–calbindin D_{28k} is structurally distinct from calci-calbindin D_{28k} and that the two forms are different and connected by negative allosteric interaction between the zinc and calcium binding events. The binding of zinc changes the hydrophobicity of calbindin D_{28k} (Bauer et al., 2008) and correlates with the ability of calbindin to activate myoinositol monophosphatase, one of the known targets of calbindin.

Under nonreducing conditions, Cys94 and Cys100 in calbindin D_{28k} are in both reduced and oxidized forms, in the latter case in an intramolecular disulfide bond (Vanbelle et al., 2005). In contrast, Cys's 187, 219, and 257 are not involved in disulfide bonds. Both the reduced and oxidized forms of the protein bind four Ca^{2+} ions with high affinity in a parallel manner and with positive cooperativity. One assumes that no disulfide bonds are formed within the cytosol.

Many functions for calbindin D_{28k} have been proposed. These include a role in calcium resorption in the kidney, modulation of insulin production, and secretion in pancreatic β cells, and neuroprotection against excitotoxicity (reviewed by Schwaller, 2008). The importance of calbindin D_{28k} in cerebellar Purkinje cells and cerebellar function is seen in a mouse strain, from which the protein is selectively eliminated in Purkinje cells (Barski et al., 2002). These experiments demonstrated a contribution of calbindin D_{28k} in the modulation of calcium signals in Purkinje cell dendrites and spines. The absence of calbindin D_{28k} also affects short-term plasticity, evidenced either in cortical multipolar bursting cells or at the mossy fiber–CA3 pyramidal cell synapse in the hippocampus (Blatow et al., 2003). The interacting binding partner of calbindin D_{28k} in Purkinje cell spines and dendrites is myoinositol monophosphatase (IMPase), a key enzyme of the inositol-1,4,5-triphosphate signaling cascade. In vitro calbindin D_{28k} associates with IMPase (Berggård et al., 2002) as well as with Ran-binding protein (RanBP) M (Lutz et al., 2003). Additional reported calbindin D_{28k} binding partners include caspase-3 (Bellido et al., 2000), 3′, 5′-cyclic nucleotide phosphodiesterase (Reisner et al., 1992), and plasma membrane ATPase (Morgan et al., 1986).

The L-type Ca^{2+} channel subunit CANAC1C (Christakos et al., 2007) binds calbindin D_{28k}, and in the kidney calbindin D_{28k} binds to TRPV5 and modulates its activity (Lambers et al., 2006). The binding peptide SKSIKNLEP of procaspase-3 is highly similar to the reported IMPase peptide, suggesting that both bind to the same site on calbindin D_{28k} (Kojetin et al., 2006). These studies indicate that calbindin D_{28k} has characteristics of a Ca^{2+} sensor (Airaksinen et al., 1997); however, the Ca^{2+} dependency of the interaction between calbindin D_{28k} and binding partners has not been demonstrated directly.

The peptide regions of RanBP M, procaspase-3, and IMPase interact with a surface of calbindin D_{28k} consisting primarily of the N-terminal part, the first helix of EF-hand domain 3, the second helix of domain 4, and residues in the linker regions EF2–EF3 and EF4–EF5 (Kojetin et al., 2006).

11.5.2. Calretinin

Calretinin and calbindin D_{28k} have 58% sequence identity (Celio, 1996; Schwaller, 1996). Even so, they have different domain characteristics (Palczewska et al., 2003). Calretinin is 269 to 271 (depending on species) residues long; five of its six EF-hands bind calcium. Calretinin has four high-affinity calcium binding sites with positive cooperativity (Hill coefficient 1.3) and apparent dissociation constant 1.5 μM and one low-affinity site with a dissociation constant of 0.5 mM (Schwaller et al., 1997). Calcium binding causes very little change in the secondary structure of the protein. However, the trypsinolysis pattern of calretinin is markedly dependent on the presence or absence of calcium. Calretinin possesses calcium-dependent hydrophobic surfaces, but in contrast to calmodulin, apo-calretinin also displays a significant hydrophobic surface. EF-lobes 3,4 and 5,6 form relatively tight structures; while the linker between lobes is exposed to solvent and accessible to proteolytic enzymes.

The central nervous system contains many calcium binding proteins. Those most extensively studied include calretinin, calbindin D_{28k}, parvalbumin, S100, calmodulin, calcineurin, and synaptotagmin. Most of these proteins are found in neurons. Calretinin interacts with cytoskeleton components in a calcium-dependent manner (Marilley and Schwaller, 2000).

Calretinin, calbindin D_{28k}, and parvalbumin are useful neuronal markers for a variety of functional brain regions and cell types. Calbindin and S100h are present in glia. The functions of these proteins remain unknown; they are assumed to be buffering, transport of calcium, protection against calcium overload, and regulation of various enzyme systems (reviewed by Heizmann, 1993). It is assumed that neurons containing certain intracellular calcium binding proteins may have a greater capacity to buffer calcium and therefore would be more resistant to degeneration.

Calretinin, calbindin D_{28k}, and parvalbumin show specific developmental and distribution patterns (reviewed by Hof et al., 1999; Schwaller et al., 2002; Bastianelli, 2003). Calretinin is abundant in the olfactory bulb and auditory pathways. Most Purkinje cells express calbindin D_{28k} and parvalbumin, whereas basket, stellate, and Golgi cells express parvalbumin only. Calbindin D_{28k} and parvalbumin are present throughout the axon, soma, dendrites, and spines of Purkinje cells. Granule, lugaro, and unipolar brush cells present the opposite immunoreactivity profile, most of them being calretinin positive while lacking calbindin D_{28k} and parvalbumin.

Calretinin, calbindin D_{28k}, and parvalbumin are found in morphologically distinct classes of inhibitory interneurons as well as in some pyramidal neurons in the mammalian neocortex (reviewed by Hof et al., 1999). There is a fundamental similarity among the mammalian species investigated so far, in terms of the distribution of calretinin, calbindin D_{28k}, and parvalbumin across the depth of the neocortex. Calretinin and calbindin D_{28k} immunoreactive neurons are predominant in layers II and III, but are present across all cortical layers, whereas parvalbumin immunoreactive neurons are more prevalent in the middle and lower cortical layers. These different neuronal populations have well-defined regional and laminar distribution, neurochemical characteristics and synaptic connections, and each of these cell types displays a particular developmental sequence. Calbindin D_{28k} appears early, shortly after cessation of mitosis when neurons become ready to start migration and differentiation; parvalbumin is expressed later, in parallel with an increase in neuronal activity. The other calcium binding proteins are detected later (reviewed by Hof et al., 1999). Analysis of different brain regions suggests that these proteins are involved in regulating calcium pools critical for synaptic plasticity. Evidence of a major role for any of these three proteins as an endogenous neuroprotectant has yet to be found.

11.5.3. Eps15 Homology Domain

The gene product of *eps15* (EPS15) encodes a protein of 897 residues. The first 282 residues contain a candidate phosphorylation site as well as six EF-hands, two of which are inferred to bind calcium. EPS15 is phosphorylated on tyrosine following EGFR activation by epidemal growth factor in vivo, or directly in vitro (Takemasa et al., 1990). EPS15 exhibits transforming activity (Fazioli et al., 1993). The Reps 1 protein binds specifically certain linear and cyclic peptides. The spatial relationship of the three EF-lobes is not known (Kim et al., 2001a,b).

11.5.4. CREC Family (Reticulocalbin)

The CREC (Ca^{2+}-binding protein of 45 kDa, reticulocalbin, ER Ca^{2+}-binding protein of 55 kDa, and calumenin) family are encoded by five genes, RCN1, RCN2, RCN3, SDF4, and CALU, encoding reticulocalbin-1, ER Ca^{2+} binding protein of 55 kDa (ERC-55), reticulocalbin-3, Ca^{2+} binding protein of 45 kDa (Cab45), and calumenin. Alternative splicing increases the number of gene products (reviewed by Honoré, 2009). Reticulocalbin 1 (RCN1) is a marker for several cancers (non-small cell lung, liver, colorectal, and breast). Its surface expression on bone marrow endothelial cells is regulated by proinflammatory tumor necrosis factor (Cooper et al., 2008). ERC-55 transport is up-regulated in absence epilepsy and in multiple sclerosis. The expression levels of calumenin is altered in various cancers. The CREC proteins are involved in calcium-dependent processes in secretory pathways, especially those involved in γ-carboxylation of coagulation factors. They have six EF-hands.

11.5.5. *Plasmodium falciparum* Surface Protein

Plasmodium falciparum surface protein (PFS) has six EF-hands and other sorts of domains at both N- and C-termini (La Greca et al., 1997). The C-terminal IDEL motif indicates that PFS is found within the endoplasmic reticulum (Saijo et al., 1997).

11.5.6. Calsymin

Calsymin is found in the gram-negative *Rhizobium etli*, a bacterium that lives on the roots of its leguminous host (**11.9.2.**). Calsymin has six EF-hands in three lobes, all apparently competent to bind calcium (Xi et al., 2000).

11.6. PROTEINS WITH EIGHT AND 12 EF-HANDS

The following EF-hand proteins do not show significant congruence with other subfamilies. Some of their domains most closely resemble other EF-hands within the same protein; this implied self-duplication occurred recently in evolutionary history.

11.6.1. *Lytechinus pictus* SPEC-Resembling Protein

Lytechinus pictus SPEC-resembling protein (LPS) has eight EF-hands and evolved by a recent gene duplication and fusion from a four domain precursor, possibly common to the precursor of SPEC. The distribution of introns in the two halves of LPS are similar to one another and to that in SPEC. Further, interdomains 2,3 and 6,7 of LPS are similar to interdomain 2,3 of SPEC (Linse et al., 1997). The function of SPEC is not known.

11.6.2. EF12

A 542-residue protein of unknown function has been isolated from *Caenorhabditis elegans* (Xiang et al., 1991). It has 12 EF-hands, 11 of which are (near) canonical. The fourth EF-hand is highly diverged; however, of all the other 11 domains, it most closely resembles EF12 #2. The 11 interdomain spacings are all six residues; this pattern strengthens the interpretation that the 41 residues $(6 + 29 + 6)$ residues between EF-hands 3 and 5 harbor an aberrant domain. The six odd domains cluster together and are congruent with the six even domains. Further, lobes 1,2 and 7,8; 3,4 and 9,10; as well as 5,6 and 11,12 most closely resemble one another, reflecting four cycles of gene duplication and fusion. The *O. vulvulus* and the *C. elegans* EF12's align well, with no deletions over the 414 $(12 \times 29 + 11 \times 6)$ residues encompassing the 12 EF-hands but show many deletions in the N- and C-terminal domains. The six EF-lobes might be arranged helically and undergo a cooperative change in conformation upon binding calcium.

11.7. PROTEINS WITH FOUR EF-HANDS

11.7.1. *Tetrahymena pyriformis* Calcium Binding Protein

Tetrahymena pyriformis calcium binding protein (TCBP)-25 is localized in the whole cell cortex and around both the migratory and stationary pronuclei at the pronuclear exchange stage during conjugation (Nakagawa et al., 2008). It is 218 residues long and contains four EF-hands, the latter three of which are inferred to bind calcium. The isoforms TCBP-23 and TCBP-25 are 35% identical in sequence; however, their encoding DNAs are 49% identical (Numata et al,, 2000). Lobe 1,2 most closely resembles lobe 3,4; this reflects a recent gene duplication and fusion.

11.7.2. CBP

Dictyostelium discoideum expresses at least eight four EF-hand proteins (CBP1 to 8). CBP1 first appears just prior to cell aggregation and is then present at relatively constant levels throughout development. In a two hybrid screen, CBP1 binds in a calcium dependent manner to protovillin and toEF-1α, both actin-binding proteins (Dharamsi et al., 2000). CBP3 expression is up-regulated under the control of the actin 15 promoter and down regulated by RNA interference in *D. discoideum*. Using fluorescence microscopy, C. H. Lee et al. (2005) also showd that CBP3 associates with the actin cytoskeleton during development.

11.7.3. Calcyphosine

Calcyphosine was initially identified as a major phosphorylated substrate for cyclic AMP-dependent protein kinase after stimulation of dog thyroid cells with thyrotropin (Lecocq et al., 1979). Its synthesis is enhanced by thyrotropin and cyclic AMP analogs that trigger cell proliferation and maintain expression of the differentiated thyrocyte phenotype. Its synthesis is decreased by epidermal growth factor and phorbol 12-myristate 13-acetate, which also trigger cell proliferation but repress differentiation (Lecocq et al., 1990). Another member of this subfamily has been found in lobster and crayfish muscle (Engman et al., 1989). Although the exact function of calcyphosine remains unknown, it is implicated in the cross signaling between cAMP and calcium, phosphatidylinositol cascades to coordinate cellular proliferation, and differentiation in the thyroid. It is highly conserved among canine, rabbit, bovine, and human species, but is absent from mouse and five other rodents (Clement et al., 1997). The expression of calcyphosine in pediatric primitive neuroectodermal tumors and ependymomas is abberent (de Bont et al., 2007). It is considered to be tumor specific and may be a marker of a new subgroup of ependymomas and a potential drug target for therapy in pediatric brain tumors.

Calcyphosine contains two pairs of EF-hands, as seen in its crystal structure, determined to 2.65 Å resolution (Dong et al., 2008). It shares a similar overall topology with calmodulin; however, there are striking differences in both the N- and C-lobes, interdomain linkers, and EF-hand 4. Calcyphosine is composed of 11 α-helices and four short β-strands and folded into two globular lobes, connected by linker residues 88 to 91 (Figure 11.21). It changes its conformation in a calcium-dependent manner. All four Ca^{2+} ions are in typical pentagonal bipyramidal coordination through the following groups:

	1 (+X)	3 (+Y)	5 (+Z)	7 (−*Y* (*C*=*O*))	9 (−X)	12 (−Z) (bident)
EF-1	D34	N36	S38	*S40*	H_2O	E45
EF-2	D70	N72	S74	*T76*	H_2O	E81
EF-3	D106	S108	D110	*V112*	H_2O	D117
EF-4	D149	H_2O	D154	*Q156*	H_2O	E161

A fifth Ca^{2+} ion is found in a position not typically associated with EF-hand domains.

The eight α-helices of the four EF-hands are labeled sequentially B&C, D&E, F&G, and I&J. Calcyphosine has three additional helices: the N-terminal helix A, the C-terminal helix K, and helix H within the unusually long linker region

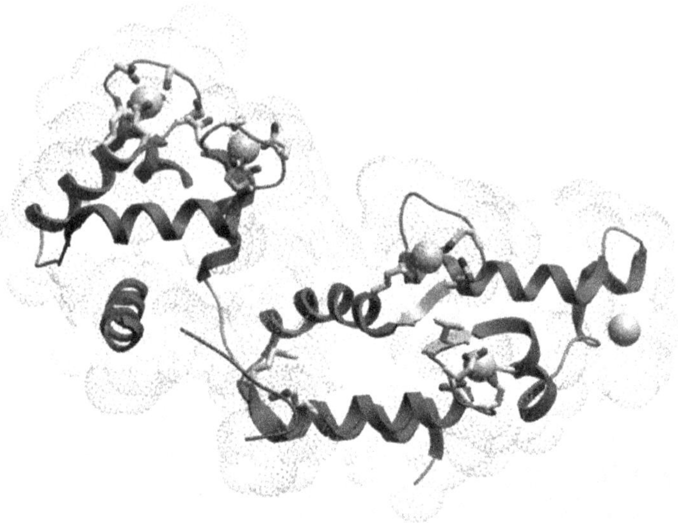

Figure 11.21. Structure of calcyphosine with five bound Ca^{2+} ions, one to each of its four EF-hands and one to the non-EF-hand domain (PDB file 3E3R). (From Dong et al., 2008.)

between helices G and I. Such a long linker is found in the neurocalcin and calpain subfamilies.

The N-lobe is slightly contracted and the C-lobe is slightly more open compared with those of other pairs of EF-hands. This is reflected in the separation of the two pairs of Ca^{2+} ions, 11.1 Å and 12.2 Å. The C-lobe of calcyphosine possesses a large hydrophobic pocket in the presence of Ca^{2+} ions that might be implicated in ligand binding, while its N-terminal hydrophobic pocket is partially shielded by the additional terminal helix A. Helix A is tightly packed on top of the hydrophobic pocket through hydrophobic interactions between residues Met1, Val4, Met8, and Leu11 and those in the hydrophobic pocket.

11.7.4. *Strongylocentrotus purpuratus* Ectodermal Calcium Binding Protein

Strongylocentrotus purpuratus ectodermal calcium binding protein (SPEC) mRNAs begin to accumulate 20 hours after fertilization in aboral cells. It has four EF-hands, at least two of which, 2 and 3, bind calcium. The function of SPEC is not known; however, it was suggested to resemble calcyphosine (Sauter et al., 1995).

11.7.5. Calflagin

Calflagin (1F8, TB17, calcimedin) cDNAs were prepared from highly redundant mRNAs from *Trypanosoma cruzei* (Lee et al., 1990) and from *T. brucei* (Wu et al., 1992; Kobayashi et al., 1990). Calcimedin is located in the flagellum. It has four EF-hands, all of which are inferred to bind calcium. It is modified by the addition of myristate or palmitate at its N-terminus. Fluorescence titration revealed two classes of calcium binding sites in unmyristoylated calflagin, $K_d = 3.0 \times 10^{-7}$ and 5.3×10^{-5} M. (Pinto et al., 2003).

11.7.6. Sarcoplasm Calcium Binding Protein

Sarcoplasm calcium binding protein (SARC) is found both as a monomer and as a dimer; it appears to function as a calcium buffer within the lumen of sarcoplasmic reticulum. Neither the sequences of its four EF-hands nor the three linkers closely resemble other EF-hands or linkers.

The main-chain traces from the crystal structures of SARC from the sandworm *Nereis diversicolor* (Cook et al., 1991) and from the amphioxus *Branchiostoma lanceolatum* (Cook et al., 1993) are almost superimposable, even though they share only 22% overall identity. The overall structure of the protein is highly compact and globular with a predominant hydrophobic core (Figure 11.22) (Vijay-Kumar and Cook, 1992). The protein is mainly α-helical (58%) with eight helices (A to H). A characteristic feature of the structure is a tight turn in the linker helix connecting EF-hands 2 and 3; this turn brings the two EF-lobes together. It has a central hydrophobic core of 20 residues, most of which are aromatic. A hydrophobic tail at the C-terminus adds to the structural stability

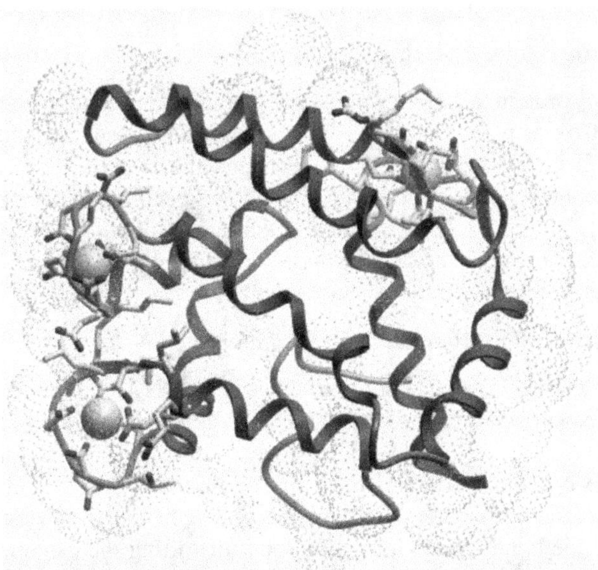

Figure 11.22. Structure of sarcoplasmic calcium binding protein from *Nereis* (PDP file 2SCP). It is viewed perpendicular to the local, approximate twofold axis relating EF-hands 3 and 4. EF-hand 2 does not bind calcium. (From Vijay-Kumar and Cook, 1992.)

by packing against a hydrophobic pocket created by the folding of the N- and C-lobes.

The apo-SARC of *Nereis* is highly disordered (Prêcheur et al., 1996; Christova et al., 2000). Addition of the first equivalent of calcium causes a disorder-to-order transition, bringing the protein to a stable conformation, close to the native, tri-calci state. The protein binds three equivalents of calcium with high affinity, $K_d(\text{Ca}) = 1.2$, 0.38, and 6.6 μM (Cox and Stein, 1981).

11.7.7. *Hra32*

The cDNA clone, *Hra32*, corresponds to a RNA transcript that accumulates in *Phaseolus vulgaris* (jack bean) during a hypersensitive reaction (Jakobek et al., 1999). The encoded protein, H32, is161 residues long and has four EF-hands, all of which are predicted to bind calcium.

11.7.8. EFH5

EFH5 of trypanosomes is encoded by the *TbEFH5* gene, which is transcribed as a polycistronic messenger. It follows the CAM gene *C* by 111 bp and precedes *ubiquitin-EP52/1* by 116 bp. It has four EF-hands, two of which might bind calcium. *TbEFH5* is inferred to be under the control of a distant upstream promoter in *T. brucei, T. cruzi*, and *L. tarentolae* (Wong et al., 1993; Aijioka and Swindle, 1993).

11.7.9. Calcium Vector Protein

Calcium vector protein (CVP) of *Amphioxus* is inferred to bind to a target protein. In a model (Cox et al., 1990) of CVP based on the crystal structures of CaM and of TNC, a target α-helix is bound between the two EF-lobes of CVP and the disulfide bond linking α -helices E1 and F2 is accommodated with no steric interference. EF-lobe 1,2 does not bind calcium. EF-lobe 3,4 is canonical and shows significant changes in solution upon binding calcium (Théret et al., 2001).

11.7.10. *PM129* Clone from *Arabidopsis thaliana*

The *PM129* clone from *A. thaliana* (PMAT) cDNA, enriched for plasma membrane–associated proteins, contains a single continuous open reading frame (Bartling et al., 1993). The 37 residues before the first of four EF-hands may comprise a distinct domain. Each of the linkers—1,2; 2,3; 3,4—is eight residues long. There are three Cys's in the fourth EF-hand and one in the 3.4 linker; a disulfide bond can be formed without distorting the canonical EF-hand. Domains 1 and 2 of PMAT most closely resemble one another, and domain 4 closely resembles domains 1 and 3 of CVP.

11.7.11. Calcium and Integrin Binding Protein

Calcium and integrin binding protein (CIB1 or calmyrin) is a ubiquitous, multifunctional regulatory protein that is involved in platelet aggregation by binding the cytoplasmic tail of the IIb subunit of the platelet-specific integrin $\alpha_{II}\beta_{b3}$ (Naik et al., 1997). It has 191 residues and contains four EF-hands. No homolog has been identified for the first 50 residues.

CIB1 was crystallized in the presence of reduced glutathione (GSH). CIB1 forms a dimer with GSH bound to one of the two monomers, near the free thiol of Cys35 (Blamey et al., 2005). The conformation of Cys35, His31, and Ser48 resembles a Cys-type protein photphatase. However, Yamniuk et al. (2009) found neither dimerization, GSH binding, nor phosphatase activity in solution. Although the dimer form of CIB1 may not be physiologically relevant, the hydrophobic interface may be involved in binding of other signaling partners. The C-terminal lobe is canonical.

Yamniuk et al. (2009) showed that EF-hands 1 and 2 do not bind calcium; EF-3 has low affinity and EF-4 higher affinity for calcium. The D127E mutant, which restores the canonical Glu at −Z of EF-3, has an increased affinity for calcium and a decreased affinity for integrin $\alpha_{II}\beta_{b3}$.

11.7.12. Spasmin

The stalked ciliated protozoan *Vorticella* has a stalk with a highly contractile cytoskeleton, consisting of spasmonemes and myonemes. The major component of these contractile organelles is the calcium binding protein(s) called *spasmin*. It contracts vigorously upon binding calcium without the concomitant hydrolysis of ATP. The energy for contraction is invested in the generation of the calcium gradient (Amos, 1971). It binds calcium even in the presence of 6 M urea (Routledge, 1978). Spasmins from *V. convallaria* (Maciejewski et al., 1999) and *Zoothamnium arbuscula* (Itabashi et al., 2003) have been cloned and sequenced. There are two isoforms in *Z. arbuscula*, both encoded by a single exon. They are 177 residues long and consist of four EF-hands, the first of which appears to bind calcium stongly. Spasmin is most closely related to centrin; how the binding of this single equivalent of calcium causes such a massive change in conformation of the entire organelle remains unknown.

11.7.13. Aequorin

Aequorin (EC 1.13.12.5 renilla luciferin 2-monooxygenase) is a calcium-dependent photoprotein that oxidizes luciferin (coelenterazine) and thereby produces the luminescence of the marine coelenterate, *Aequoria victoria* (Prasher et al., 1987; Kumar et al., 1990). In the anthozoan *Renilla reniformis*, calcium-induced bioluminescence involves two proteins, luciferase and luciferin binding protein, (LBP), which has four EF-hands but lacks the luciferase activity of aequorin. Aequorin has only a few residues N- and C-terminal to the four EF-hands; its luciferase activity appears to be associated with its aberrant second domain.

The subfamily of EF-hand photoproteins is very unusual. It includes calcium-activated aequorin from *Aequorea victoria*, obelin from the hydroid *Obelia*, and some other photoproteins (*halistaurin, phialidin, clytin, mitrocomin*) (reviewed by Cambell, 1988; Vysotski and Lee, 2004; Vysotski et al., 2006). The photoproteins emit photons upon binding calcium. The bioluminescence surely benefits the host; however, the nature of that benefit is not clear.

Aequorin is 189 residues long. It contains a prosthetic group, coelenterazine, a substituted imidozolopyrazine, linked covalently to apo-aequorin through a peroxide bond. The binding of calcium to the three competent sites triggers a conformational change resulting in the oxidation of coelenterazine to coelenteramide and the emission of a quantum of blue light ($\sim$470 nm). In principle, the binding of calcium is not strictly essential for bioluminescence of photoproteins because alone they give off a very low level of light emission called *calcium independent luminescence*; however, the light intensity is increased up to a million fold on the addition of calcium.

There are five distinct conformations of aequorin and obelin controlled by the binding of various ligands: apo-protein, no coelenterazine (state I); apo-protein, with coelenterazine (state II); calci-protein, with coelenteramide (state III); or apo-protein, with coelenteramide (state IV), and calci-protein, no coelenterazine (state V). Only the tertiary structures of states II and IV are known.

The crystal structures of calci-aequorin and calci-obelin were determined at resolutions 1.7 and 2.2 Å, respectively (Liu et al., 2000; Deng et al., 2005). The protein has four EF-hands and a hydrophobic coelenterazine binding site (Figure 11.23). As expected from the similarity of their amino acid sequences, aequorin and obelin have similar tertiary structures. They consist of two sets of four helices A to D in the N-terminal lobe and E to H in the C-lobe. Ca^{2+} ions are bound in each of EF-hand loops 1, 3, and 4 in both proteins. Each Ca^{2+} ion is coordinated in canonical pentagonal bipyramidal configuration. Mg^{2+} ions bind to only EF-loops 1 and 3 (Ohashi et al., 2005). EF-hand 3 binds magnesium with higher affinity than does EF-hand 1. Only EF-hand 3 seems to be occupied by magnesium under physiological conditions.

Deng et al. (2005) proposed that the binding of calcium to loop 1 and optimization of the pentagonal, bipyramidal geometry results in a "twist" of EF-hand 1 around a pivot point reflected in changes in hydrogen bond distances between the main-chain atoms of Ile37 and Ile83. These changes pull helix A in the direction of the N terminus of the protein, which results in a displacement of helix H and the C terminus. Binding of only one Ca^{2+} ion to loop 1 may be sufficient to trigger bioluminescence. Similarly, the binding of calcium to loop 4 and optimization of the pentagonal, bipyramidal geometry produces a "twist" of EF-hand 4 around a pivot point reflected in changes in hydrogen bond distances between the main-chain atoms of Ile130 and Leu166. This induces a small change of interhelical angle between helices H and G. The displacement of His175 in

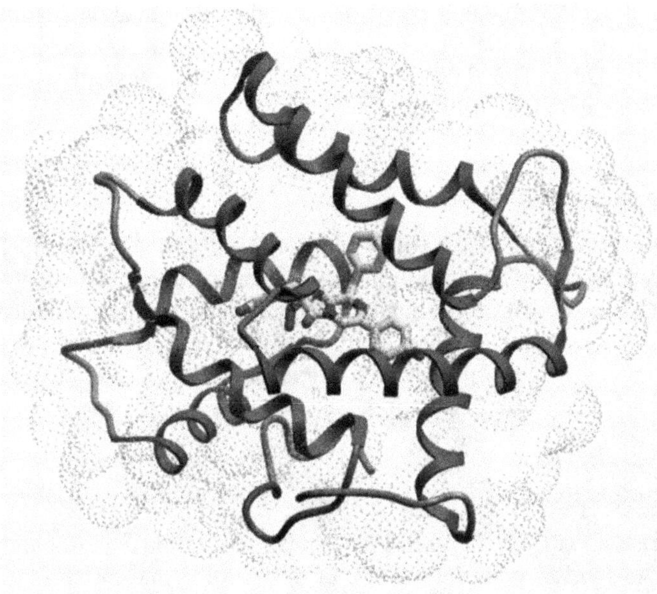

Figure 11.23. Structure of apo-aequorin (PDB file 1EJ3). The coelenterazine molecule is located in the center of the protein. (From Head et al., 2000.)

helix H is a crucial step for triggering the bioluminescence. The displacement of helix G produces a rearrangement of helix F; this, in turn, adjusts loop 3, increasing its affinity for calcium. The binding of calcium in this loop completes the rearrangements of the helices E and F and leads to the final reaction of the bioluminescence.

Calcium regulated photoproteins are sensitive to an increase in calcium concentration beginning with 0.3 to 0.5 μM. Obelin and aequorin are used to measure the intracellular concentration of free Ca^{2+} ions in the ranges 0.5 to 100 μM and 0.3 to 20 μM, respectively. Obelin and aequorin can be microinjected into cells or incorporated into liposomes that are fused with the targeted cell. One can also transduce, then express the gene encoding aequorin in the targeted cell (reviewed by Vysotski and Lee, 2004; Vysotski et al., 2006). It is also possible to target photoproteins to various cell compartments using specific amino acid signal sequences (molecular address) attached to the protein by genetic engineering (Brini et al., 1999).

11.7.14. *Plasmodium falciparum* Protein Kinase

Plasmodium falciparum protein kinase (PFCPK) (EC 2.7.1.37) is encoded by the *PfCPK* gene; PFCPK is 524 residues long. Residues 56 to 325 encoded a region homologous to the b subunit of Ca/CAM-dependent protein kinase II with which it shares ~40% sequence identity (Hardin et al., 1988). A pseudosubstrate sequence and a CAM binding motif are found at the junction region between the kinase domain and the four EF-hands. A membrane ATPase is one of the substrates of CDPK. The *PfCPK* gene contain four introns, all within the EF-hands region (Zhao et al., 1993).

11.7.15. Protein Phosphatase

Protein phosphatase (EC 3.1.3.16) is encoded by the *rdgC* gene of *Drosophila melanogaster*; it has a phosphoprotein phosphatase domain at its N-terminus (Steele et al., 1992). Three EF-hands are easily recognized and indicated as 1, 3, and 4 (Jakobek et al., 1999). Nominal EF-hand 2, occupying 56 residues between domains 1 and 3, may be a far-diverged EF-hand paired with EF-hand 1, an example of guilt by association.

11.7.16. Phospholipase C

Phospholipase C (PLC) (EC 3.1.4.3) hydrolyzes various phosphatidylinositol phosphates to inositol phosphate and diacylglycerol. It has an absolute requirement for calcium and is activated by guanine nucleotide binding proteins (PLC-β), by receptor tyrosine kinases (PLC-γ), or by transglutaminase II (PLC-δ). PLC-ζ is specific to mammalian sperm and is injected into fertilized eggs. It is responsible for the initiation of calcium oscillations, which are due to Ca^{2+} release from the endoplasmic reticulum mainly through the IP3 receptor. All four vertebrate isoforms have four EF-hands at the N-terminus, a plekstrin homology domain, and a C-terminal C2 domain (Essen et al., 1996; André, 1996) (Chapter 12). None of the four EF-hands have canonical calcium binding loops nor is calcium bound to these loops in the crystal structure of mammalian phosphoinositide-specific phospholipase C (Essen et al., 1996) (Figure 11.24) PLC-ζ is fully active at 100 nM [Ca^{2+}]. Deletion of the first two EF-hands of PLC-ζ reduces enzyme activity; EF-hand 3 is responsible for calcium sensitivity. Deletion of all four EF-hands or of the C2 domain abolishes enzyme activity (Kouchi et al., 2005). Deletion of any of the EF-hands or of the C2 domain render sperm incapable of initiating calcium oscillation in eggs (Nomikos et al., 2005).

11.7.17. LAV1

LAV1 cDNA encodes a 355-residue protein. Different isoforms are expressed in amoebae and in plasmodia of *Physarum polycephalum* (Laroche et al., 1989). EF-hands 2 and 4 of LAV most closely resemble one another. Although LAV appears not to be congruent with any other EF-hand proteins, its interdomains 1,2 and 3,4 resemble 1,2 and 3,4 of CTER. Linker 2,3 consists of a sole Leu. This places severe constraints on the spatial relationship between lobe 1,2 and lobe 3,4. The function of LAV is not known, and no homolog has been identified for the ~200 residues at its N-terminus.

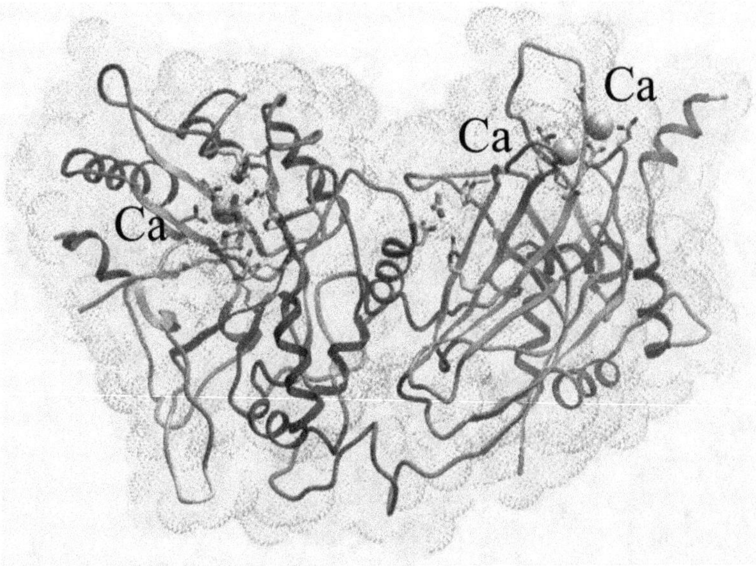

Figure 11.24. Crystal structure of a mammalian phosphoinositide-specific phospholipase C (PDB file 1DJW). (From Essen et al., 1996.) (*See insert for color representation of figure*.)

11.8. PROTEINS WITH TWO EF-HANDS

11.8.1. Phl p 7 (Polcalcin)

Members of a novel family of two EF-hand proteins, Phl p 7 (polcalcin), originating from pollens of a variety of plants, such as trees, grasses, and weeds, are potent allergens (Valenta et al., 1998). Phl p 7 is expressed in pollens but not in other plant tissues. Due to the strong IgE cross-reactivity within the two EF-hand allergen family, allergic patients are polysensitized to pollens of various plants. IgE antibodies of patients allergic to two EF-hand allergens recognize preferentially the calcium-bound form of the protein, whereas, the apo form has a strongly reduced IgE binding capacity.

Verdino et al. (2002) reported the first three dimensional structure of a representative Phl p 7 (polcalcin) in the calcium bound form. Polcalcin adopts an extended conformation (Figure 11.25). Each monomer consists of two EF-hands ((helices A, B, C, and D) connected by a three-residue linker, much shorter than the linkers in most other EF-hand lobes. This favors an extended conformation of the two EF-hands and precludes their self-pairing. In the C-terminal to the second EF-hand in Phl p 7 there is an 11-amino acid distorted helix (Z-helix). Due to a sharp kink at residues 67/68, this helix protrudes from the C-terminal EF-hand at $\sim 100°$. The Z-helix, whose sequence is highly conserved throughout the two EF-hand pollen allergens, is unusually hydrophobic.

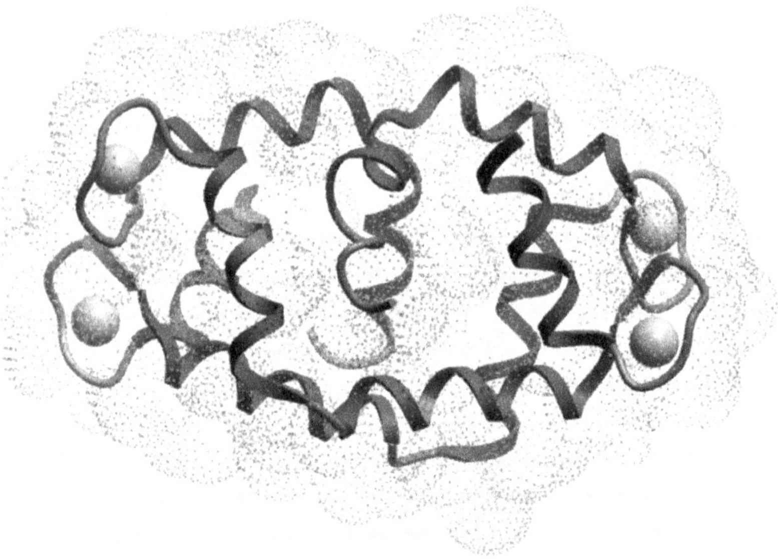

Figure 11.25. Ribbon model of tetra-calci-Phl p 7 (polcalcin) (PDB file 1K9U). The local, approximate twofold axes of both lobes are near coincident. (From Verdino et al., 2002.)

Two monomers dimerize in a head-to-tail arrangement with domain swapped EF-hand pairing: EF-hand I of one monomer with EF-hand II of the other. Both Z-helices intertwine with helix C of the opposite monomer. The helices F of both the N- and C-terminal EF-hands of each monomer come into close contact; they are related by an approximate twofold rotational axis, which passes close to the hinge regions on one side and at the kink regions on the opposite side of the monomer.

The overall structure of the Phl p 7 dimer is barrel shaped, with approximate dimensions 35 × 45 × 35 Å. The calcium coordinating loops form the top and the bottom of the barrel; the helices E and F form the upper and lower sidewalls. The Ca^{2+} ion is enclosed by a 12 amino acid loop and coordinated by seven oxygen atoms of Asp or Asn, a peptide carbonyl, a water molecule, and a bidentate Glu at the vertices of a pentagonal bipyramid.

The C-terminal Z helices form an equatorial belt. The extended conformation and the intertwined head-to-tail arrangement of the protein monomers lead to the formation of an extended cavity within the interior of the dimer. This cavity is lined with hydrophobic side chains and provides an enclosed ligand binding site. The cavity occupies a volume of 800 $Å^3$ in the center of the protein dimer.

Calcium binding acts as a conformational switch between an open and a closed dimeric form of Phl p 7. These findings are relevant to understanding lipid and calcium dependent pollen tube growth. Furthermore, the structure of Phl p 7 allows for the rational development of vaccines for treatment of sensitized, allergic patients (Verdino et al., 2002).

As judged by isothermal titration calorimetry, Phl p 7 binds calcium cooperatively with stepwise K_d(Ca) values of 5.8×10^{-7} and 5.8×10^{-7} M (Henzl et al., 2008). In contrast, magnesium binding is sequential with K_d(Mg) values of 3.6×10^{-5} and 5.9×10^{-3} M. Spectral data indicate that calcium binding provokes a major conformational change that does not occur upon magnesium binding. In phosphate buffered saline containing EDTA, the apo-protein undergoes two-state denaturation with a transition at 78.4°C. In the presence of 0.02 mM Ca^{2+}, the transition midtemperature exceeds 120°C. Analytical ultracentrifugation indicates that the protein is monomeric in neutral solution at concentrations exceeding 1.0 mM in both the apo and calci states.

Most EF-hands occur in odd–even pairs separated by 5 to 20 residues (Chapter 7). In turn, the two EF-hands interact with one another and are related by an approximate twofold axis. Three exceptions are EF-hand 1 of parvalbumin (helix A, AB loop, helix B). The fifth EF-hands of the penta-EF-hand subfamily interact with one another to form homo- or heterodimers. The two EF-hands of Phl p 7 are swapped (odd-1 & even-2 and even-1 & odd-2) to form a homodimer.

The crystal structure of recombinant Che a 3 (rChe a 3), the polcalcin from the weed *Chenopodium album*, contains an α-helical fold that is essentially identical with that of the two EF-hand allergens from birch pollen, Bet v 4, and timothy grass pollen, Phl p 7 (Verdino et al., 2008). The major timothy grass (*Phleum pratense*) allergen Phl p 3 was isolated from an extract of timothy grass pollen. The solution structure of Phl p 3 as determined by NMR spectroscopy reveals that the protein consists of a core of hydrophobic side chains from two β-sheets of five and four antiparallel β-strands, respectively (Schweimer et al., 2008). This structure is very similar to that of Phl p 2 and strongly resembles the known conformation of the carboxy-terminal domain of Phl p 1, the major difference being the loop orientations.

Pollen germination and pollen tube growth depend critically on calcium. The exact biological functions of the pollen specific two EF-hand proteins are unknown to date, but due to their calcium binding ability they were suggested to be involved in the control of calcium metabolism in pollen germination and pollen tube growth. The hydrophobic cavity of Phl p 7 suggests a ligand binding function rather than a mere calcium-buffering function, corresponding well to the observation that Phl p 7 and its homologous proteins are expressed tissue specifically in pollens (Verdino et al., 2002).

11.8.2. α-Actinin and α-Spectrin

α-Actinin cross links actin filaments noncovalently. It is a homodimer whose subunits assemble in an antiparallel fashion to form a rodlike structure (Taylor and Taylor, 1993). The difference between nonmuscle and muscle forms is that the cross linking activity of the former is completely inhibited by micro-molar calcium; whereas, the activity of the muscle form is not calcium sensititve. Both forms have two EF-hands at their C-termini (reviewed by Viel, 1999).

α-Actinin plays important structural and regulatory roles in cytoskeleton organization and in muscle contraction. It bundles actin filaments in multiple cytoskeleton frameworks (reviewed by Sjöblom et al., 2008). In nonmuscle cells, α-actinin is found along the actin filaments in adhesion sites. In striated, cardiac, and smooth muscle cells, it is localized at the Z-disk and analogous dense bodies, where it forms a latticelike structure and stabilizes the muscle contractile apparatus. Besides binding to actin filaments, α-actinin associates with numerous other cytoskeletal and signaling molecules, cytoplasmic domains of transmembrane receptors, and ion channels.

α-Actinin is composed of three domains: an N-terminal domain with actin binding sites, four "spectrin-like" repeats each 122 residues long, and a C-terminal lobe that contains two EF-hands. It assembles into rod shaped, antiparallel homodimers with a subunit mass of 94 to 115 kDa and a high α-helical content. Wenegieme et al. (1994) found in α-actinin three or four high-affinity calcium binding sites per dimer, $K_d(\mathrm{Ca}) = 6.36\ \mu\mathrm{M}$, and many low-affinity calcium binding sites. Binding of calcium to the EF-hands regulates the activity of nonmuscle α-actinin; whereas, muscle α-actinins cross-link actin filaments in a calcium-insensitive manner.

α-Spectrin was first described in the erythrocyte, in which it reinforces the cell membrane by cross linking ankyrin and protein 4.1. Nonerythroid αII-spectrin and its close homolog, α-fodrin, form elongated double stranded filaments. Each strand consists of segments of triple-stranded α-helices that are formed by the strand going forward, back, forward. The two EF-hands at the C-terminus impart a calcium-sensitive (Lundberg et al., 1997) interaction with actin and other cytosolic proteins (L. J. Brown et al., 1994). The first domains of α-actinin and of α-spectrin are most closely related; their second EF-hands are also quite similar to one another. Portions of their non-EF-hand regions appear to be related.

The spectrin family is characterized by spectrin repeats, actin binding domains, and EF-hands. Spectrins are involved in actin bundling and membrane anchoring; they are found throughout eukaryotes. Spectrin is a component of the cytoskeleton. It forms pentagonal or hexagonal clusters and covers cytosolic surfaces of the plasma membranes of many cell types. It plays an important role in the maintenance of plasma membrane integrity and in the structure of the cytoskeleton. The spectrin repeats share several features with the α-actinin rod domain. The rod domain forms a rigid connection between two actin binding domains positioned at the two ends of the α-actinin dimer. The exact distance and rigidity are important for organizing the muscle Z-line and maintaining its architecture during muscle contraction. The spectrin repeats of α-actinin make tight antiparallel, homodimer contacts. The spectrin repeats are important interaction sites for multiple structural and signaling proteins.

Spectrin is a heterodimer or tetramer consisting of α and β subunits (reviewed by Viel and Branton, 1996). Intersubunit interactions between the ends of spectrin chains integrate spectrin tetramers into a network by means of associations with protein 4.1, actin, and others. Much of the length of spectrin consists of tandemly repeated motifs, each of which contains 100 to 120 amino acids.

The nonrepetitive segment of α-spectrin consists of EF-hands (Trave et al., 1995). In α-actinin, the four repeats (R1 to R4) of the rod domain are involved in homodimer formation. α-Actinin repeats R1 and R2 are homologous to β-spectrin repeats β2 and β3, respectively, whereas repeats R3 and R4 are homologous to α-spectrin repeats α20 and α21, respectively. Spectrin can be split into N- and C-terminal lobes, which fold independently. Only the N-lobe binds calcium. The homologous domains in α-actinin, and possibly in dystrophin, may share the same structure. However, only nonmuscle α-actinins appear to have one or two EF-hands competent to bind calcium.

Spectrin has two EF-hands in its C-lobe and binds calcium with low affinity, K_d(Ca) 150 to 400 μM (Trave et al., 1995a,b; Lundberg et al., 1997). The binding appears to be sequential. The equine isoforms of spectrin have much higher calcium affinity with K_d(Ca) in the nanomolar range (Wallis et al., 1993). Calcium binding causes a redistribution of hydrophobic interactions within EF-hand 1, resulting in an opening of the structure. These conformational changes may modify the loop structure between EF-hand 1 and EF-hand 2; this plays an important role in interchain binding at the C-terminal of the spectrin subunits (Viel and Branton, 1994).

Spectrin repeats form three-helix bundles; these occur in a large number of diverse proteins, either as single copies or in tandem arrangements of multiple repeats. They can serve structural purposes by coordination of cytoskeletal interactions with high spatial precision as well as a "switchboard" for interactions with multiple proteins with a regulatory role.

11.8.3. Calmodulin-Related Gene Product

Calmodulin-related gene product is encoded by the gene T+, which contains a transposonlike human repeat element, *THE1*, in the 3′ untranslated region of its message (Deka et al., 1988). The first EF-hands of calmodulin related gene product and nucleobindin are most closely related to one another.

11.8.4. Calsensin

Calsensin is expressed in some of the peripheral neurons that fasciculate in a single axon tract of the leech *Haemopsis* sp. It has 83 residues, and both EF-hands bind calcium (Briggs et al., 1995). Calsensin is a monomer in solution. The residues in helix E of the second EF-hand undergo slow conformational exchange, indicating that the motion of this helix is associated with calcium binding (Venkitaramani et al., 2005). The backbone dynamics as measured by ^{15}N relaxation rates and heteronuclear NOEs correlate well with the tertiary structure.

Calsensin copurifies with a 200-kDa protein, yet to be characterized. Calsensin binds to the C-terminal tail of the leech homolog of protein phosphatase 4 regulatory subunit 2 (PP4-R2) in a calcium-dependent manner (Venkitaramani et al., 2006). The binding groove is formed by hydrophobic residues from helix E of the first EF-hand and helix F of the second.

11.8.5. Groovin

Groovin (kakapo) is encoded by *grv* (*kak*) and is expressed in tendon cells of *D. melanogaster*. Groovin is essential for muscle-dependent tendon cell differentiation (Strumpf and Volk, 1998). It contains two EF-hands, both inferred to bind calcium.

11.8.6. Diacylglycerol Kinase

Diacylglycerol kinase (DGK EC 2.7.1.107) has two EF-hands that span residues 114 to 202. The domain at the N-terminus is not similar to other known domains. The region (203 to 735) at the C-terminus contains an ATP binding site and Cys-rich zinc finger domains similar to those found in protein kinase C (Yamada et al., 1997). Isoforms α, β, and γ have K_d(Ca) values 3.0×10^{-7}, 3×10^{-8}, and 3×10^{-8} M, respectively. DGK modulates the balance between the two diacyly glycerol (DAG) and phosphatidic acid (PA), by phosphorylating DAG to yield PA; 10 mammalian DGK isozymes have been identified (Sakane et al., 2007).

11.8.7. Glycerol-3-Phosphate Dehydrogenase

FAD-dependent mitochondrial glycerol-3-phosphate dehydrogenase (GcPDH, EC 1.1.99.5) is located on the outer surface of the inner mitochondrial membrane. Both mammalian GcPDH and FAD isocitrate dehydrogenase are homologous to the yeast and eubacterial FAD glycerol-3-phosphate dehydrogenases; both have two EF-hands at their C-terminii (Autieri, 1996).

GcPDH, pyruvate dehydrogenase, NAD-isocitrate dehydrogenase, and oxoglutarate dehydrogenase are located within the mitochondria and are regulated by changes in calcium concentration in the matrix (reviewed by Denton, 2009). FAD-glycerol phosphate dehydrogenase is located on the outer surface of the inner mitochondrial membrane and is influenced by changes in $[Ca]_{cyt}$. The activation of these enzymes is important in the stimulation of the respiratory chain and hence ATP supply under conditions of increased ATP demand.

Mitochondrial GcPDH is most abundant in testis and pancreatic islets in the rat. At a physiologic concentration of glycerol phosphate (75 μM), half-maximal activity of Triton X-100–solubilized testis GcPD is achieved in the presence of 0.1 to 0.25 μM free calcium (MacDonald and Brown, 1996). Calcium (10^{-6} to 10^{-5} M) lowers the K_m value of GcPDH from 3.2 mM glycerol phosphate (testis) and 2.5 mM glycerol phosphate (islet) to 0.4 mM glycerol phosphate. This indicates that cytosolic calcium can regulate GcPDH activity. Since simultaneous oscillations in electrical activity, cytosolic calcium, glycolysis, and insulin release occur in pancreatic β-cells, GcPDH activity might also fluctuate and allow the glycerol phosphate shuttle to participate in glycolytic oscillations.

11.8.8. Fimbrin

Fimbrin is an actin filament bundling protein from chicken. It has two EF-hands at its N-terminus and four 125-residue actin binding domains, homologous to those of α-actinin, at its C-terminus (de Arruda et al., 1990). The human homologs are I-, L-, and T-plastin, whose genes are on chromosomes 3, 13, and X (Lin et al., 1994). Even though fimbrin bears some similarity to α-actinin, their EF-lobes are not closely related.

11.8.9. Ras Guanyl Nucleotide–Releasing Protein

Ras guanyl nucleotide releasing protein (GRP) is 795 residues long and contains a homolog of a REM domain, a CDC25 domain, two EF-hands, and a diacylglycerol binding domain (Ebinu et al., 1998). The two EF-hands are canonical and are predicted to bind calcium. Helices E1 and F2 appear conventional. However, there are only 15 residues for helices F1 and E2. One or both of these helices is truncated and the relationship between the two EF-hands is unclear. Rat fibroblasts expressing the *RasGRP* encoding gene are transformed. Addition of a DAG analog causes sustained activation of Ras-Erk signaling and changes in cell morphology. RasGRP is activated by phorbol esters (Lorenzo et al., 2000).

11.8.10. Polycystin-2

The *PKD2* gene product (polycystin-2) is 968 residues long and has two EF-hands, residues 680 to 796. It has six membrane-spanning helices and is a cation channel regulated by calcium. The K_d(Ca) values are 55 and 179 μm. In the absence of calcium, polycystin-2 forms a dimer; it is dissociated and inactivated by calcium. Mutations in this gene in humans segregate with polycystic kidney disease (Schumann et al., 2009). The homolog, PKD1, constitutes a voltage-activated calcium (and sodium) channel in *C. elegans*.

11.8.11. Ryanodine Receptor Protein

The ryanodine receptor (RYR) functions as a channel for release of calcium from endoplasmic reticulum of muscle and brain. The isoforms are ~5000 residues long and contain two EF-hands, the second of which is inferred to bind calcium. The channels are open in the presence of $\sim 10^{-6}$ M [Ca^{2+}] (calcium-induced calcium release) and closed at $\sim 10^{-3}$ M [Ca^{2+}]. The two EF-hands have low calcium affinity and have been suggested to function in closing the channels (Xiong et al., 1998).

11.8.12. CBL

CBL binds to phosphorylated tyrosine residues and functions as a negative regulator of many signaling pathways. The crystal structure (Meng et al., 1999)

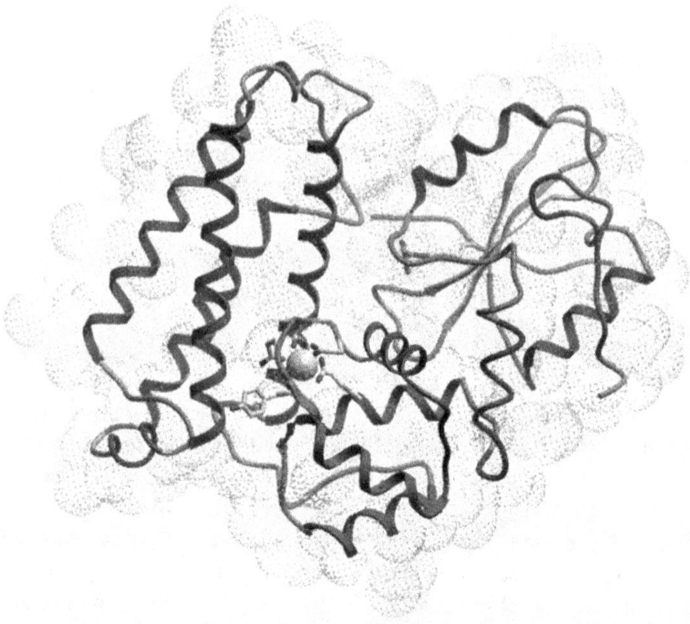

Figure 11.26. Structure of the amino-terminal domain of CBL complexed to its binding site on ZAP-70 kinase and Ca^{2+} (PDB file 2CBL). ZAP-70 kinase is viewed down the midline of its warped β-sheet. (From Meng et al., 1999.) (*See insert for color representation of figure.*)

of CBL, minus residues 1 to 24, alone and in complex with a phosphotyrosine decapeptide that represents its binding site in ZAP-70 shows a pair of EF-hands wedged between a four-helix bundle (4HB) and a divergent SH2 domain (Figure 11.26). The coordination of calcium is unique. Although the residue at −X is Ser, the −X ligand comes from the carboxylate group of a Glu from the fourth helix of the 4HB. Although calcium is required for phosphopeptide binding, it interacts with the SH2 domain, not with the pair of EF-hands. The EF-lobe may position the 4HB relative to the SH2 domain.

11.8.13. Nucleobindin

Nucleobindin, also known as calnuc, is found in the cytoplasm, where it binds to Gαi3, a G protein-coupled receptor on the cytosolic surface of Golgi membranes (Weiss et al., 2001). It is widely expressed in cells and tissues (Miura et al., 1992, 1996; Wendel et al., 1995; Cooper et al., 2008) and is well conserved from flies to humans (Lin et al., 1998). Kanuru et al. (2009) found that nucleobindin functions as a downstream effector for the G-protein α-subunit. They showed that calcium binds with an affinity of 7 μM and causes structural changes. Although magnesium binds to nucleobindin with very weak affinity, the structural changes that it causes are enhanced further by calcium binding. The nuclear EF-hand acidic (NEFA) gene product is closely related to nucleobindin; its function is unknown (Den et al., 1998).

Nucleobindin is 455 residues long. It has a signal sequence at its N-terminus, followed by a 200-residue domain of unknown homology, two canonical EF-hands, then by seven heptad repeats as found in a coiled coil α-helices or leucine zippers. The calcium affinity of nucleobindin is governed by the conformational state of the G-protein. Kanuru et al. (2009) proposed that nucleobindin binds more calcium upon binding to Giα. Isothermal titration calorimetry shows that nucleobindin and the G-protein associate with an affinity of 13 nM. They suggested that nucleobindin maintains calcium homeostasis in the cell.

11.8.14. Allograft Inflammatory Factor-1

Allograft inflammatory factor-1 is "... a cytokine inducible, tissue specific transcript transiently expressed in response to vascular trauma" (Hohenester et al., 1996). It has 143 residues with two EF-hands. Homologs of the first 50 residues have not been identified. The first domains of allograft inflamatory factor-1 and of glycerol-3-phosphate dehydrogenase are quite similar. The arrangement of possible calcium binding ligands (S * G * G * * * S * * D) is unusual.

11.8.15. BM-40 (Osteonectin)

BM-40 (osteonectin or SPARC, secreted protein acidic and rich in cysteine) is an extracellular glycoprotein 304 residues long. It contains a follistatinlike domain as well as two EF-hands, both of which bind calcium in the crystal structure. It is the only known secreted extracellur EF-hand protein. Parvalbumin and several S100s, as well as some annexins, are found extracellularly under physiological condtions. These do not have the leader sequences characteristic of secreted proteins; how they are exported from their normal residences in the cytosol is not known. The structure and functions of BM-40 are discussed in Chapter 13 with other extracellular calcium binding proteins.

11.9. EF-HAND PROTEINS IN BACTERIA AND VIRUSES

Calcium is involved in many bacterial processes, including cell division, pathogenesis, motility, chemotaxis, and maintaining competence (reviewed by Norris et al., 1991; Onek and Smith, 1992; Michiels et al., 2002). For example, in *E. coli*, repellents cause an increase in intracellular calcium concentration and provoke tumbling of the bacteria, whereas in the presence of attractants, calcium is reduced and bacteria move smoothly (Watkins et al., 1995).

Calcium homeostasis has been demonstrated in some bacteria, including *E. coli* and *B. subtilis* (Gangola and Rosen, 1987; Herbaud et al., 1998; Jones et al., 1999). The concentration of free Ca^{2+} ion in unstarved *E. coli* cells is maintained at 90 ± 10 nM, irrespective of the calcium concentration in the extracellular medium (Gangola and Rosen, 1987). In starved cells $[Ca^{2+}]_{cyt}$ varies from 0.2 to 0.7 μM as the loading calcium concentrations vary from 10 μM to 10 mM. Addition of glucose lowers $[Ca^{2+}]_{cyt}$ to 90 nM. Bacteria do not sequester Ca^{2+} ions in intracellular stores such as the endoplasmic reticulum of eukaryotes. However, their cell membranes and cell walls are rich in calcium; specific proteins or phospholipids might provide calcium to the cytoplasm. Several bacteria have been shown to possess both primary ATP-dependent Ca^{2+} pumps and secondary Ca^{2+} transporters to remove Ca^{2+} from the cell (Norris et al., 1996). The latter systems are antiporters using proton or sodium electrochemical gradients.

11.9.1. Calerythrin

Calerythrin, from the gram-positive bacterium *Saccharopolyspora erythraea*, is 20 kD and consists of four EF-hands, three of which (1, 2, and 4) bind calcium (Swan et al., 1987). NMR spectroscopy revealed a turn that brings the two EF-hand lobes into a compact globular structure similar to that of sarcoplasmic calcium binding proteins (Tossavainen et al., 2003); hence, the suggestion that it might function as buffer or transporter of calcium rather than as a regulator (Aitio et al., 1999).

In the absence of bound metal ions, calerythrin is disordered. Calcium first binds cooperatively to EF-hands 3 and 4 and then to EF-hand 1, which is paired with an atypical, non-calcium binding EF-hand (Aitio et al., 2001). Apo-calerythrin is in equilibrium between ordered and less ordered states with slow exchange kinetics at low temperatures, as observed by NMR. As the temperature is raised, signal dispersion in NMR spectra is reduced and the intensity of near-UV CD bands decreases. Yet far-UV CD spectra indicate only a small decrease in secondary structure, and SAXS data show that no significant change occurs in the overall size and shape of the protein. Thus, at elevated temperatures, the equilibrium is shifted toward a state with the characteristics of a molten globule. The fully structured state is reached by the addition of calcium.

11.9.2. Other EF-Hand Proteins in Bacteria

More then 15 EF-hand proteins have been identified in prokaryotes: SC6F11.09, SCJ33.05c, CC1180, SC10F4.20, CabA, Asp24, MLL5457, and others (reviewed by Michiels et al., 2002). Most of them are 70 to 184 residues long, have a high Phe/Tyr ratio, and are acidic (pI 4.2 to 5.0) or basic (pI 9.6 to 10.3); none have been identified in *Archae*. Calcium binding protein from *Saccharopolyspora eythraea* (CMSE) was the first protein identified from eubacteria (Swan et al., 1989). Its function remains unknown.

Calsymin was found in the gram-negative bacterium *Rhizobium etli* (Xi et al., 2000). It has six predicted EF-hands in three lobes with molecular mass 30 kDa, pI 3.6. *R. etli* secretes calsymin without N-terminal cleavage of the protein.

R. etli induces the formation of nodules on the roots of *Phaseolus vulgaris*, and within these structures the bacteria are present inside plant cells surrounded by a plant derived membrane. The gene encoding calsymin *casA* is highly expressed during colonization and infection of *R. etli*. Expression of *casA* is controlled by a repressor protein, termed CasR, belonging to the TetR family of regulatory proteins. Mutation of the *casA* gene affects the development of bacteroids during symbiosis and symbiotic nitrogen fixation. It was suggested that calsymin could function as a transducer of information between the bacteria and the plant. These various bacterial proteins are inferred to bind calcium; however, their involvement in signal transduction has yet to be established (Permyakov and Kretsinger, 2009).

11.9.3. MSV Virus

The entire genome of the pox virus, Entomopoxvirinae, of the North American migratory grasshopper *Melanopolus sanguinipes* is 236 kb and contains 267 open reading frames, of which 107 are similar to genes described previously (Afonso et al., 1999). One gene encodes MSV, 140 residues. It has four EF-hands, the first two of which are inferred to bind calcium. It was the first EF-hand protein identified from a virus. Zhou et al., (2009) reviewed “viral calciomics” in terms of the disturbance of host homeostasis, virus host interactions dependent on host calcium binding proteins, and binding of calcium by viral proteins. There are many examples of viral coat proteins that bind calcium with various distributions of contiguous and noncontiguous ligands. However, the identification of EF-hands remains tenuous and awaits structural determinations.

12

CYTOSOLIC CALCIUM BINDING PROTEINS LACKING EF-HANDS

Chapters 9, 10, and 11 cover the classification, evolution, structures, energetics, and functions of EF-hand proteins. Seventy distinct subfamilies have been identified, several of which (e.g., parvalbumin, calmodulin, S100) have been studied extensively. Those techniques and results have provided precedents for other less well studied EF-hand proteins and for calcium binding proteins in other homolog families. Several generalizations are helpful; however, they all have exceptions:

1. Calcium binding proteins in the cytosol are subject to calcium modulation by calcium functioning as a secondary message. However, several proteins, notably parvalbumin and proteins within the endoplasmic reticulum, serve primarily or solely a calcium buffering or modulating role.
2. Calcium-modulated proteins are found only, or mainly, in the cytosol. To be involved in information transduction, they should have K_d(Ca) values of 10^{-5} to 10^{-7} M, the range of concentration of free Ca^{2+} ion in the cytosol, $[Ca]_{cyt}$, observed during pulses of messenger calcium. However, there are several non-EF-hand calcium binding proteins discussed in this chapter with $K_d(Ca) \sim 10^{-4}$ M. Does their calcium binding have other physiological roles and/or are there local domains, perhaps next to plasma or endoplasmic membranes, where $[Ca]_{cyt}$ rises to 10^{-4} M briefly?
3. Many extracellular proteins bind calcium, supposedly for its role in substrate binding and/or protein stabilization. They should not be calcium modulated, because the extracellular concentration of free Ca^{2+} ion, $[Ca]_{ext}$,

Calcium Binding Proteins, By Eugene A. Permyakov and Robert H. Kretsinger

is constant, $\sim 10^{-2.8}$ M. However, there is strong evidence that $[Ca]_{ext}$ varies over time in microenvironments and some extracelluar calcium binding proteins may also be involved in information transduction (Permyakov and Kretsinger, 2009).

4. EF-hand proteins are intracellular. However, parvalbumin and S100 are also found in extracellular fluids; even though, their genes do not encode the leader sequence, usually found in protein destined for the lumen of the endoplasmic reticulum or for export from the cell. Osteonectin (BM-40), discussed in Chapters 9 and 11, has a pair of EF-hands, even though it is an extracellular matrix protein. Other generalizations, and exceptions, can be expected; stay tuned.

12.1. ANNEXINS

The name *annexin* is derived from the Greek *annex*, meaning "bring/hold together" and was chosen to describe the principal property of all or at least nearly all annexins: the binding to and possibly holding together of certain biological structures, in particular membranes (Gerke and Moss, 2002). Initially, annexins received diverse and unrelated names referring to their biochemical properties. These included synexin, chromobindins, calcimedins, lipocortins, and calpactins. Intensive biochemical work, protein and cDNA sequencing, as well as gene cloning led to the realization that all such proteins identified shared key biochemical properties as well as gene structure and sequence features. Hence, the concept of a novel multigene family arisen by gene duplication was developed and the common name *annexin* was introduced to solve the terminology tangle (Crumpton and Dedman, 1990). The 16 subfamilies common to animals are now called A1 to A16. They are ubiquitous, intracellular, calcium and phospholipid binding proteins whose many inferred functions have yet to be confirmed (reviewed by Moss, 1997; Gerke and Moss, 2002; Moss and Morgan, 2004). Most eukaryotes have $\sim$20 annexin genes; even the primitive protist, *Giardia*, has at least seven; however, annexins are absent from fungi and prokaryotes (Gerke and Moss, 2002). Any single cell type expresses a range of annexins, but no single annexin is expressed in all cells, implying that regulation of annexin gene expression is tightly controlled (Gerke and Moss, 2002). Most annexins are encoded by 12 to 15 exons, the variation depending in large part on the length of the N-terminal tails. For several annexins, particularly those with long N-tails, alternative splicing adds to the diversity of annexin isoforms; this may in turn amplify functional variability within the family as a whole. Annexin A6, with eight domains, is encoded by 26 exons and is the largest annexin gene, extending over $\sim$60 kb (Smith et al., 1994).

Plant annexins are a smaller and less diverse group; they comprise a distinct monophyletic group (reviewed by Mortimer et al., 2008; Talukdar et al., 2009). Their molecular weights range from 32 to 42 kDa and although sharing a common evolutionary precursor, they differ structurally from their animal counterparts. Plant annexins have, on average, a larger surface area than mammalian annexins (Clark et al., 2001). This is due to extra grooves and clefts, perhaps suggesting a wide range of interaction partners and a broad range of roles within the cell. However, in contrast to their animal counterparts, the N-tails of all known plant annexins are less than 10 residues long. Sequence alignment with mammalian annexins revealed that the phosphatidylserine-binding site is only poorly conserved in plant annexins. They have been found in all monocot and dicot plants tested to date. The majority of plant annexins are found in the cytosol, but some can be found associated with the plasma membrane, endomembranes, or the nuclear envelope. Despite divergence in amino acid sequence, phosphatidylserine binding activity is conserved in at least some plant annexins, including those from maize, bell pepper, and cotton.

Annexins consist of a core of four homologous annexin domains and of a highly variable N-terminal tail. The core is inferred to have evolved by two cycles of gene duplication and fusion; domains 1 and 3 and domains 2 and 4 resemble one another more closely. No examples of proteins having only a single or only a pair of annexin domains are known. Annexin A6, and only A6, has eight domains; it evolved by another cycle of gene duplication and fusion (Figure 12.1) (Huber et al., 1992; Kaneko et al., 1997; Musat et al., 1997; Avila-Sakar et al., 1998, 2000). Annexins fall into three major groups: first, A7, A11, and A13; second, A4, A5, and A8; third, A1, A2, A3, with A9 and A10 as somewhat distant members (Gerke and Moss, 2002). Annexin A6 is more difficult to categorize, because the N-terminal half is more closely related to the A4, A5, A8 group and the C-half to the A1, A2, A3 group.

Each annexin domain consists of five α-helices; A, B, D, and E form a right-handed four-helix bundle. The four domains pack into a flattened trapezoid; the axes of the four four-helix bundles are perpendicular to the surface of the trapezoid. Annexin domains 4 and 1 are related to domains 2 and 3 by a local twofold axis, also perpendicular to the plane of the trapezoid. It has a slightly concave surface on which the amino and carboxyl termini of each domain come into close apposition and a convex surface on which the $(4 \times 2 = 8)$ potential calcium binding loops, AB and DE, are located. Ca^{2+} ions that bind to this convex surface are inferred to cross-link carbonyl and carboxyl groups of (some of) the eight AB and CD loops with phosphoryl groups of membrane phospholipids. Some annexins may also interact with the hydrocarbon chains of membrane lipids (reviewed by Moss and Morgan, 2004). The concave side of annexin faces the cytoplasm and may interact with other proteins. The N-tails of the annexins sometimes fold back on this concave surface, reaching to the central pore in the trapezoid formed by packing of the four domains.

Calcium binding loops are in domains 2, 3, and 4 of annexin A2 and in domains 1, 2, and 4 of annexin A5 (Jost et al., 1992). Annexin A6, with eight domains, has six calcium binding sites: five in AB loops and one in a DE loop (Avila-Sakar et al., 1998) (Figure 12.1). The AB calcium binding sites in annexins are similar to those of phospholipase A_2. Typically, three calcium ligands are

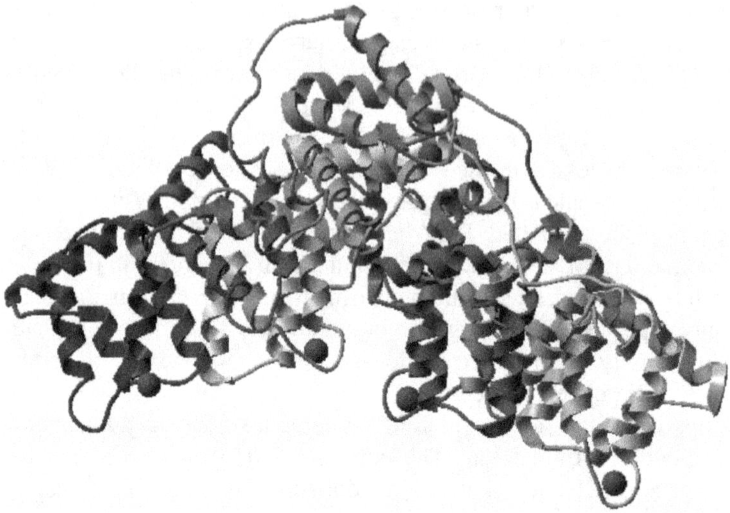

Figure 12.1. Structure of annexin A6 (PDB file 1AVC). The two flattened trapezoids (domains 1 to 4 and 5 to 8) are both seen approximately edge on. The three Ca^{2+} ions in each trapezoid are indicated as spheres; they are all on the inferred membrane binding convex surface. (From Avila-Sakar et al., 1998.)

carbonyl oxygens and two are from a bidentate carboxylate. This carboxylate is about 40 residues away from the AB loop and is called the *capping residue*. The remaining coordinating oxygens are usually contributed by water molecules. The calcium sites of the DE loop also typically have three carbonyl ligands and a bidentate carboxylate, which is closer to the carbonyls in the linear sequence. Annexins have lower affinity for calcium than do most EF-hands; therefore, they are inferred to be modulated by changes in calcium concentration within limited regions of the cytosol.

The N-terminal tails contain highly variable sequences 35 to 60 residues long; they are of low complexity and are inherently disordered; only portions of them are seen in crystal structures of annexins. The N-tails of annexins A1 and A2 are expelled from the core domain on calcium binding. These N-tails are sites of interactions with various targets, such as S-100, F-actin, profilin, and tissue plasmin activator. They are often phosphorylated by distinct protein kinases in different annexins. The N-terminal tails of annexins are located on the concave side of the trapezoid, probably facing the cytosol. This high variability in sequence and phosphorylation accounts for much of the target specificity of annexins.

The calcium dependent membrane aggregation of annexins A1, A2, and A4 is strongly inhibited by phosphorylation of the N-tail (Wang and Creutz, 1992); whereas, for annexin A7, aggregation is activated (Caohuy and Pollard, 2001).

Annexin A1 is 346 residues long (reviewed by D'Acquisto et al., 2008). The N-tail contains several putative Ser and Thr phosphorylation sites as well as consensus sequences for glycosylation (Asn43–Ser45) and transglutamination.

Annexin A1 also has several possible sites in the N-terminal domain for proteolysis; this profoundly modifies its physical and biological properties. A fragment of A1 whose N-terminus has been truncated is commonly found in fluids associated with inflammation. The N-tail is folded onto the concave surface in the region of the central pore; however, in the presence of calcium the N-tail can "flip" out (Rosengarth et al., 2001a). Six calcium binding sites of differing affinities on the convex aspect mediate the binding of annexins to negatively charged phospholipids (Rosengarth et al., 2001b).

Annexins are generally cytosolic proteins; they occur both in soluble forms and associated with components of the cytoskeleton or with proteins that mediate interactions between the cell and the extracellular matrix (reviewed by Moss and Morgan, 2004). Annexins interact with membranes in a calcium-dependent manner. This binding is reversible, and removal of calcium leads to the liberation of annexins from the phospholipid matrix. They provide a link between calcium signaling and membrane functions (reviewed by Rescher and Gerke, 2004; Gerke et al., 2005). By forming networks on the membrane surface, annexins organize membrane domains and membrane recruitment platforms for proteins with which they interact, such as cytoskeleton (F-actin, spectrin), endosomes, liposomes, and other vesicular structures (Gerke et al., 2005). A2 and A11 have been found in the nucleus.

Annexins not only bind to the cytosolic surfaces of membranes as single molecules but can also form lateral assemblies (reviewed by Gerke et al., 2005; Bitto and Cho, 1998, 1999). For example, annexin A5 forms two dimensional crystals on phosphatidylserine bilayers, with annexin trimers representing the principal building block (Voges et al., 1994). Annexins A1 and A2 associate into different types of oligomers. Such assemblies of annexins might stabilize selected plasma membrane structures and change membrane curvature and therefore cell shape. Annexins A1, A2, A4, A6, and A7 can aggregate membranes in a calcium-dependent manner.

It has been suggested that release of intracellular calcium induced by several stimuli triggers recruitment of annexins; the $[Ca]_{cyt}$ required for membrane translocation differs between different annexins (reviewed by Raynal and Pollard, 1994; Gerke and Moss, 2002). Thus, depending on the mode of calcium release, its location, and the spatiotemporal pattern of the calcium signal, different annexins can probably be recruited independently from one another to their respective target membranes (reviewed by Rescher and Gerke, 2004). Some annexins can interact with membranes in the absence of calcium. Moreover, some annexins, such as annexin A9, annexin A10, and the *Caenorhabditis elegans* annexin, NEX4, lack high affinity calcium binding sites; this indicates that they are probably not affected by fluctuations in cytosolic calcium.

Many cytosolic proteins, including members of the EF-hand family, bind selectively to the N-terminal tails of annexins. Several S100 proteins and annexins interact in calcium-dependent and calcium independent manners, and form complexes that exhibit biological activities (reviewed by Miwa et al., 2008; Rintala-Dempsey et al., 2008). When an S100 binds calcium, it undergoes a reorientation of its helices, exposing a hydrophobic patch capable of interacting with its targets, including the N-tails of annexins. For example, annexins A1 and A2 interact with S100A11 and S100A10, annexin A7 binds to sorcin, and annexin A11 binds to S100A6. Crystal structures of complexes of annexins and S100s have revealed the binding surfaces and conformations of the N-tails of annexin (Rintala-Dempsey et al., 2008). However, other S100–annexin interactions, such as those between S100A11 and annexin A6, or between dicalcin and annexins A1, A2, and A5, appear to be more complicated, involving the annexin core region, perhaps in concert with the N-tail. The diversity of these interactions indicates that multiple forms of recognition exist between S100s and annexins. S100–annexin interactions have been suggested to play a role in membrane fusion events by the bridging together of two annexin proteins, bound to phospholipid membranes, by an S100.

Annexin A1 (macrocortin, renocortin, lipomodulin, lipocortin-1), is 37 kDa, binds calcium and phospholipid, and is a strong inhibitor of glucocorticoid-induced eicosanoid synthesis as well as of phospholipase A2 (reviewed by Lim and Pervaiz, 2007). Annexin A1 is involved in a variety of inflammatory pathways, in cell proliferation machinery, in the regulation of cell death signaling, in phagocytic clearance of apoptosing cells, and in carcinogenesis. Annexin A1 has a powerful suppressive effect on the innate immune system (reviewed by D'Acquisto et al., 2008), acting on polymorphonuclear leukocytes, monocyte macrophages, and mast cells as well as on many other cell types to inhibit cell trafficking as well as for the generation or release of mediators. Annexin A1 has been implicated in the apoptotic cell "eat me" signal and ensuing phagocytosis, and evidence is accumulating to support a role in the resolution phase of inflammation. Specific "eat me" signals on apoptotic cells serve as markers for phagocytes to recognize and ingest them. Annexin A1 is recruited to the phosphatidylserine rich region of apoptotic cell surface in a caspase mediated mechanism that involves the release of intracellular calcium (reviewed by Lim and Pervaiz, 2007).

Endocytic transport of the epidermal growth factor receptor and its localization at the plasma membrane is important for the formation of compartment specific signaling complexes. It is regulated by scaffolding and targeting proteins, including calcium effector proteins, such as calmodulin and annexins, in particular annexins A1, A2, A6, and A8 (Grewal and Enrich, 2009). Since these annexins show differences in their expression patterns, subcellular localization, and mode of action, they probably contribute differentially and cooperate in the fine tuning of epidermal growth factor receptor activity. In support of this hypothesis, current literature suggests these annexins to be involved in different steps that control the endocytic transport and signaling of the epidermal growth factor receptor.

Annexin A1 plays a critical role in effecting the negative feedback effects of glucocorticoids on the release of corticotrophin (ACTH) and its hypothalamic releasing hormones and is particularly pertinent to the early onset actions of the steroids that are mediated via a nongenomic mechanism (reviewed by Buckingham et al., 2006). In rheumatoid arthritis, an impaired induction of annexin A1 by glucocorticoids in monocytes (Morand et al., 1994) as well as a lower capacity of annexin A1 binding to monocytes and neutrophils is observed (Goulding et al., 1992). Annexin A1 is present in arthritic synovium and mediates the effects of exogenous glucocorticoids on neutrophil migration in carrageenan-induced acute arthritis.

Annexin A1 is found in many cells and tissues (lung, bone marrow, intestine) at concentrations below 50 ng mL^{-1}, with the highest levels in the seminal fluid (150 μg mL^{-1}) (reviewed by Lim and Pervaiz, 2007). It accounts for 2 to 4% of the total cytosolic protein and is found in gelatinase granules in neutrophils. Annexin A1 is secreted on cellular adhesion to the endothelium. On release, annexin A1 is thought to bind to its receptor, which brings about cell detachment, and inhibits transmigration of leukocytes, thereby blocking the inflammatory response. It is undetectable in plasma.

The myocardium contains annexins A1, A2, A4, A5, A6, and A7; annexins A5 and A6 are the most abundant (reviewed by Camors et al., 2005). They can be coexpressed in the same cell type but with their own localization. In the myocardium, annexin A2 is localized in intramyocardial capillaries, extracellular matrix, and in endothelial cells of the coronary arteries. It is undetectable in ventricular and atrial myocytes. It has been reported to play a central role in the control of plasmin mediated processes. While annexin A2 is involved in regulation of fibrin homeostasis, alterations in expression and activity of annexins A5, A6, and A7 have been associated with regulation of calcium in the heart, but the target of each annexin has not yet been identified. In human cardiomyocytes, annexin A4 is a cytosolic protein whose expression increases during heart failure. Moreover, the punctated longitudinal staining additionally observed in the nonfailing atria was absent in the failing tissue, demonstrating loss of annexin A4 organization in the failing hearts. Annexin A5 is mainly localized in cardiomyocytes; it is one of the most abundant annexins in rat and human myocardium (3.7 μg mg^{-1} of total proteins). It could be relocated to interstitial tissue in ischemic and failing hearts, or it could be externalized and exhibit a pro-apoptotic effect in cardiomyocytes. Annexin A6 is the most abundant annexin in the heart (13.5 μg mg^{-1} of total proteins of human heart) and has been localized in various cell types, including myocytes. Overexpression of annexin A6 causes alterations in the contraction cycle, leading to dilated cardiomyopathy; whereas, knockout has been found to induce faster changes in calcium transient and increased contractility, suggesting a negative inotropic role for annexin A6. Annexin A7 (synexin) is expressed in heart and skeletal muscle. In annexin A7 null mutant mice, a decrease in the force–frequency relationship was observed in adult cardiomyocytes, consistent with regulation of $[Ca]_{cyt}$.

Ductal organs and the digestive system contain annexins A1, A2, A4, and A5, are expressed in the (reviewed by Markoff and Gerke, 2005). These annexins are well represented in the kidney, together with the long isoform of annexin A13, annexin A13b. The functional relevance of annexins expressed in the kidney appears to be linked to several physiological properties important for epithelial cells. First, annexins are required for membrane organization and membrane transport events required for the establishment and maintenance of epithelial polarity. Second, there is accumulating evidence of an association of annexins with ion channels, as membrane guiding, auxiliary proteins, or modulators of channel activity. Last, some annexins seem to work as extracellular autocrine modulators of receptor function under different physiological conditions.

Recent studies have shown that S100–annexin complexes have a role in the differentiation of gonad cells and neurological disorders, such as depression (reviewed by Miwa et al., 2008). These complexes regulate the organization of membranes and vesicles, and thereby may participate in the appropriate disposition of membrane-associated proteins, including ion channels and/or receptors.

There is no evidence that loss, mutation, translocation, or amplification of any annexin gene leads to a known human disease (reviewed by Hayes and Moss, 2004). However, there is good evidence that in certain clinical conditions, including cardiovascular disease and cancer, changes in levels of expression or localization may correlate with the pathological consequences and sequelae of disease (reviewed by Hayes and Moss, 2004).

One of the most prominent characteristics of apoptosis is the externalization of phosphatidylserine (PS), a plasma cell membrane phospholipid. In healthy cells, phosphatidylserine is present on the inner leaflet of the plasma cell membrane. Annexin A5 has a strong affinity for phosphatidylserine with dissociation constant in the nanomolar range. Through the coupling of annexin A5 to contrast agents, visualization of apoptotic cell death in vivo in animal models and in patients has become feasible (reviewed by Laufer et al., 2008). Annexin A5 imaging has proven to be a suitable imaging biomarker for the evaluation of cell death–modifying compounds and plaque-stabilizing strategies. Moreover, it was shown that externalization of phosphatidyl serine not only occurs in apoptosis but is also observed in activated macrophages and stressed cells. In addition, it has been shown that annexin A5 not only binds to externalized phosphatidyl serine, but is also internalized through an annexin A5–specific mechanism; this indicates that annexin A5 imaging is not only valuable for apoptosis detection, but can also be used to visualize inflammation and cell stress.

Some annexins have been found at the cell surface or in extracellur spaces, despite the absence of secretory signal peptide. Some parvalbumins and S100s are also found extracellularly, and both also lack secretory peptides.

12.2. C2-DOMAIN PROTEINS

The C2 domain was originally identified as the second of four domains (Cl to C4) in the α, β, and γ isoforms of mammalian Ca^{2+}-dependent protein kinase C (Coussens et al., 1986). The C2 domain is one of the more common and widely distributed calcium binding motifs. It exists in both calcium binding and nonbinding forms. Both interact with membranes and with multiple other proteins (reviewed by Rizo and Südhof, 1998). About 100 different C2-domain sequences are listed in current data banks. The C2 domain is widely distributed in eukaryotes but rare or nonexistent in prokaryotes, where calcium signaling is less widely used as a second messenger (reviewed by Nalefski and Falke, 1996). Single and multiple copies of C2 domains have been identified in eukaryotic signaling proteins that interact with cellular membranes and mediate a broad array of critical intracellular processes, including the generation of lipid second messengers (cytoplasmic phospholipase A_2, phosphoinositide specific phospholipase C, phosphatidylinositol 3-kinases, etc.), membrane trafficking (synaptotagmins, rabphillin-3, etc.), and activation of GTPases. Although the calcium binding properties of many C2-domain proteins remain to be examined, the large number of C2-domain proteins in the vertebrate genome makes it likely that this domain represents the second most common calcium binding motif after the EF-hand. The C2 domain and the EF-hand motif of the calmodulin superfamily are the two most frequently occurring Ca^{2+} sensors: for example, the *C. elegans, D. melanogaster*, and *H. sapiens* genomes contain at least 61, 57, and 126 C2 domain proteins, respectively, while the same genomes contain at least 58, 92, and 144 EF-hand proteins, respectively (Schultz et al. 1998).

C2 domains are about 130 residues long; they fold autonomously and form a compact β-sandwich composed of two four stranded β-sheets (Figure 12.2). Three loops at the top of the domain and four at the bottom connect the eight β-strands. Calcium binding occurs at the top three loops. The calcium binding

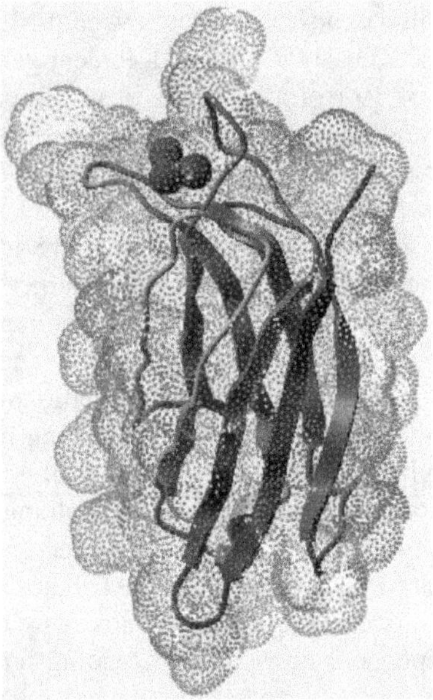

Figure 12.2. Structure of the C2A domain of synaptotagmin I (PDB file 1BYN). Three Ca^{2+} ions are bound at the top edge of the β-sandwich, which consists of two β-sheets, four strands each. (From Shao et al., 1998.)

sites are formed primarily by Asp side chains that serve as bidentate ligands for two or three Ca^{2+} ions.

Although considerable functional diversity exists, most C2 domains are activated by calcium binding and then dock to a specific membrane. The C2 domains of protein kinase Cα (PKCα) and cytosolic phospholipase A2α (cPLA$_2$α), for example, are known to dock to different membrane surfaces during an intracellular calcium signal (reviewed by Corbin et al., 2007). They discriminate two different mechanisms of C2-domain directed intracellular targeting: messenger-activated target affinity (MATA) and target-activated messenger affinity (TAMA).

The C2 domains of PKCα and cPLA$_2$α both utilize the TAMA mechanism, in which the calcium affinity of the C2 domain is too low to be activated by calcium signals in most regions of the cell. Only when the C2 domain nears its target membrane, which provides a high local concentration of target lipid, is the effective calcium affinity increased by the coupled binding equilibrium to a level that enables substantial calcium activation and target docking.

12.2.1. Synaptotagmin

Synaptotagmin has an established postdocking role in secretion and also appears to be a key component of the endocytosis machinery. Current evidence is consistent with the idea that synaptotagmin is a major calcium sensor that triggers the release of neurotransmitters. Synaptotagmin 1 was identified as p65 protein in a monoclonal antibody screen for synaptic proteins and proposed as a potential calcium sensor for regulated exocytosis (Matthew et al., 1981). Twelve additional synaptotagmins were subsequently discovered. Recent genomic analyses indicate that this family consists of 19 isoforms, some of which exhibit alternative splicing. Synaptotagmins 1 and 2 function as calcium sensors in synaptic vesicle exocytosis. Less is known about the other synaptotagmins; although, many of them are abundantly coexpressed with synaptotagmins 1 and 2 in the brain. The most abundant of these, synaptotagmins 3 and 7, are localized on the plasma membrane opposite synaptic vesicles and exhibit distinct calcium affinities, suggesting that plasma membrane and vesicular synaptotagmins may function as complementary calcium sensors in exocytosis with a range of calcium affinities. The presence of distinct synaptotagmins on the membranes of synaptic vesicles and on active zones of the membranes that fuse during neurotransmitter release suggests a potential explanation for the existence of multiple synaptotagmins (Südhof, 2002). It was proposed that at least synaptotagmins 1, 2, 3, 6, and 7 perform complementary functions in calcium triggered exocytosis; whereby, calcium binding to each class of synaptotagmins contributes differently to triggering exocytosis. As calcium sensors, all synaptotagmins share the same calcium cooperativity and calcium dependent phospholipid binding but have distinct calcium binding properties. Vesicular synaptotagmins have a lower calcium affinity and are more important in fast synaptic exocytosis than in endocrine exocytosis, most of which is much slower. Plasma membrane synaptotagmins, in contrast, have a higher calcium affinity and may be more important for endocrine exocytosis.

With these distinct properties, the combination and relative abundance of various synaptotagmins could contribute to shaping the characteristic calcium responses of synapses.

Synaptotagmins have been detected in tissues outside the nervous system (reviewed by Tucker and Chapman, 2002). For example, synaptotagmin XIII has been found in sperm heads (localized to acrosomes) and in the kidney. Similarly, synaptotagmin VII has been detected not only in brain, but also in heart, lung, and spleen.

Synaptotagmin is essential for fast synaptic vesicle exocytosis and is probably the main calcium receptor in this process (reviewed by Rizo and Südhof, 1998; Schiavo et al., 1998; Bai and Chapman, 2004). It binds calcium- and phosphatidylserine-containing membranes synergistically and is localized to synaptic vesicles and large dense core secretory vesicles. Synaptotagmin consists of a small intraluminal N-terminal domain, a single transmembrane segment, and a Cys-rich cytosolic region (Sutton et al., 1995). The N-terminal domain is highly variable between isoforms and is glycosylated, while the cytosolic region is palmitoylated in some isoforms. The large cytosolic region of synaptotagmin contains two C2 domains. The first of them, C2A, is connected to the Cys-rich region and the membrane-spanning domain by a highly charged segment and to the second C2 domain, C2B, by a short hinge region (Figure 12.3). C2A may function in neurotransmitter release through its calcium-dependent interaction with syntaxin and phospholipids.

A cluster of three calcium binding sites (Ca1 to Ca3) is located at the top of the three loops of the β-sandwich of C2A. They are formed by three carbonyl

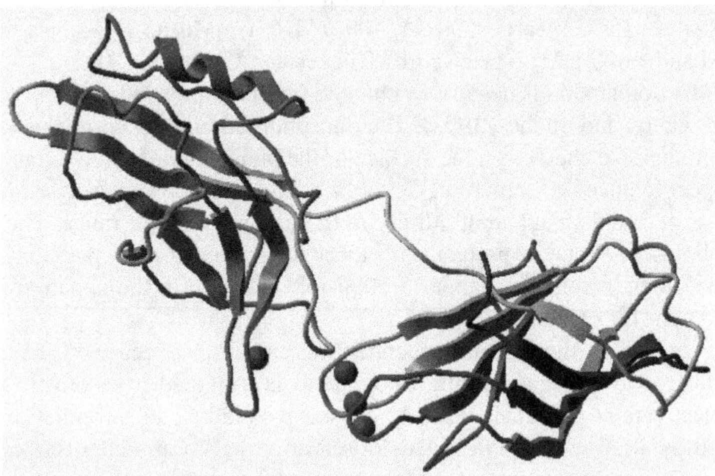

Figure 12.3. Ribbon model of C2A and C2B domains of synaptotagmin III with four bound Mg^{2+} ions (PDB file 1DQV). Three Mg^{2+} ions are bound at the "top" end of C2A and one at the "top" of C2B. (From Sutton et al., 1995.)

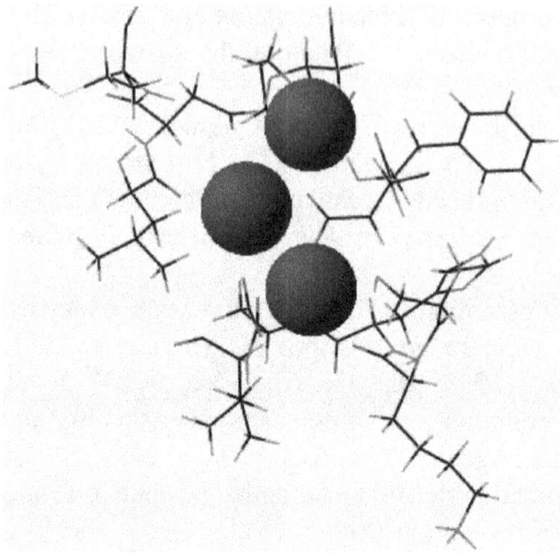

Figure 12.4. The three divalent cation binding sites in the C2A domain of synaptotagmin I (PDB file 1BYN). (From Shao et al., 1998.)

groups and the side chains of five Asp's and one Ser (Figure 12.4). The Ca1 and Ca2 sites have six ligands each. The coordination spheres of the bound Ca^{2+} ions are incomplete as observed in crystal structures, especially for the Ca3 site; this is reflected in the low calcium affinity of this site. The Ca1–Ca2, Ca1–Ca3, and Ca2–Ca3 distances are 3.2, 6.0, and 4.1 Å, respectively. Calcium binding to all three sites is necessary for syntaxin 1 and phospholipid binding. In the absence of synaptotagmin binding partners, calcium binds to these sites with a low affinity (the K_d values for C2A are ~60 μM, 400 μM, >1 mM; the K_d values for C2B are ~400 and ~600 μM) (reviewed by Tucker and Chapman, 2002).

The C2A domain binds negatively charged phospholipids independent of head-group (Davletov and Südhof, 1993). The phospholipid binding correlates with the density of negative charges on the surface of the phospholipids bilayer rather than with a specific chemical structure. Calcium binds C2A cooperatively with a Hill coefficient of 3 and an apparent affinity in the low-micromolar range. The bound phospholipids or syntaxin probably fill unsatisfied coordination positions of the three Ca^{2+} ions, resulting in about a 1000-fold increase in the apparent affinity of the domain for calcium.

NMR analysis of the calcium dependent interaction between the C2A domain and syntaxin showed that binding to syntaxin is triggered by a drastic change in the electrostatic potential of C2A, caused by binding of several Ca^{2+} ions (reviewed by Rizo et al., 2002). This observation led to an electrostatic switch model of neurotransmitter release, whereby repulsion between negatively charged surfaces of the C2A domain and syntaxin prevents exocytosis in the absence of calcium. Upon calcium influx, calcium binding to C2A reverses its electrostatic potential, inducing binding to syntaxin and initiating exocytosis.

Similar to C2A, the C2B domain is a compact β-sandwich formed by two four-stranded β-sheets with loops emerging from the top and bottom. C2B binds two Ca^{2+} ions (Ubach et al., 1998). C2B does not exhibit the same calcium-dependent phospholipid binding properties as does C2A. C2Bs of synaptotagmins I and II mediate the calcium dependent self association of synaptotagmins. C2As and C2Bs of most synaptotagmins bind calcium and are specialized for different calcium dependent activities (Ubach et al., 1998).

Calcium binding does not induce substantial changes in the conformation of these C2 domains but does cause rotations of some side chains of synaptotagmin I (Shao et al., 1998). NMR data indicate that the calcium binding region in synaptotagmin I is flexible in the absence of calcium and is stabilized after calcium binding. Calcium binding increases the denaturation temperature as well as the resistance of the C2A domain to proteolysis (Davletov and Südhof, 1994). Rizo and Südhof (1998) suggested that calcium causes a major change in the electrostatic potential of the C2A domain that may be important for regulating interactions with target molecules. The region around the calcium binding sites of C2A was shown to be responsible for calcium-dependent binding to syntaxin (Shao et al., 1997). This region contains a cluster of Asp's that coordinate calcium and a ring of basic amino acids surrounding it. The binding to syntaxin, which is negatively charged, could therefore be driven by the change in electrostatic potential caused by calcium binding. This means that synaptotagmin I acts as an electrostatic switch in neurotransmitter release. The binding of phospholipids by the C2A domain is also explained by this model. Phospholipid binding is inhibited by high ionic strength and correlates with the density of negative charges on the surface of the phospholipids bilayer.

Studies on mice, *C. elegans*, and *Drosophila* demonstrated that synaptotagmin I is essential for rapid and synchronous, calcium-triggered release of neurotransmitters. SNARE (soluble N-ethylmaleimide sensitive factor attachment protein receptor) proteins are involved in exocytosis at the presynpatic plasma membrane. SNAREs play a role in docking and fusion of synaptic vesicles to the active zone, as well as in the calcium triggering step itself, most likely in combination with synaptotagmin (reviewed by Brunger, 2005). Vesicles are docked, primed, and ready to fuse at the presynaptic nerve terminal before the arrival of the calcium signal. In the docked state of vesicle and plasma membranes, SNAREs are assembled into a complex, although they do not attain a fully fusogenic state. The role of this protein complex is to keep the SNAREs in a prefusion state (reviewed by Carr and Munson, 2007). Complexin might perform this function by binding to partly or fully assembled SNARE complexes, while preventing their final, fusogenic conformation. Complexins are largely unstructured in solution and bind tightly to the SNARE complex, forming an α-helix that interacts in an antiparallel fashion with the synaptobrevin and syntaxin-1 SNARE motifs. Different SNARE domains are involved in different processes, such as regulation, docking, and fusion. SNAREs exhibit multiple conformational and oliogomeric states. SNARE core domains undergo progressive disorder to order transitions upon interactions with other proteins. Complexin–SNARE interactions are calcium insensitive, suggesting that a calcium-sensing protein must also be involved; synaptotagmin is a likely candidate for this role. Current models indicate that exocytosis is regulated by synaptotagmin I and mediated by SNARE proteins (reviewed by Bai and Chapman, 2004; Carr and Munson, 2007; Rizo and Rosenmund, 2008). Indeed, at physiological calcium concentrations, and in the presence of phospholipids, synaptotagmin seems to bind more tightly to SNARE complexes and displace complexin (Tang et al., 2006). At the same time, complexin might not need to be fully released from the SNARE complexes, as Schaub et al. (2006) observed binding of both synaptotagmin and complexin to SNARE-containing liposomes in the presence of calcium. This competition model is also consistent with recent data showing that the binding site on the SNARE complex for the calcium–synaptotagmin–phospholipid complex might partly overlap the complexin binding site (Dai et al., 2007). This model predicts that the displacement of complexin by calci-synaptotagmin would free the SNARE complexes for membrane fusion.

Neuronal exocytosis is among the fastest processes in cell biology. The lag time between increases in presynaptic [Ca^{2+}] and postsynaptic responses can be as short at 60 to 200 μs in some synapses (Sabatini et al., 1996). Up to 5200 vesicles are secreted in 10 ms in certain synapses that process sensory information (Sun and Wu, 2001). The sensitivity of the Ca^{2+} sensor for synaptic exocytosis has been determined for only two synapses. In goldfish (*Carassius auratus*) retinal bipolar neurons, the threshold for detectable exocytosis is $\geq$ 20 μM [Ca^{2+}], and half-maximal rates of secretion occur at 194 μM Ca^{2+} (Heidelberger et al., 1994), while in the rat calyx of Held synapse the threshold for exocytosis is $\leq$ 1 μM [Ca^{2+}] (Bollmann et al., 2000). One possible explanation for this difference is the differential expression of Ca^{2+} sensors with low and high affinities for Ca^{2+}, respectively. More is known concerning the calcium requirements for the fusion of large dense core vesicles, which appear to have low calcium thresholds for release and in many cases exhibit half-maximal rates of release in the range 1 to 20 μM [Ca^{2+}] (reviewed by Tucker and Chapman, 2002).

Most synaptotagmins bind to syntaxin 1 in a calcium dependent manner (Li et al., 1995), suggesting an intersection of the mechanism of fusion (via the SNARE protein syntaxin) with the calcium dependent regulation of fusion (via synaptotagmin 1). Binding of synaptotagmin 1 to syntaxin, SNAP-25 (another SNARE protein), and assembled SNARE complexes has been reported (Li et al., 1995; Kee and Scheller, 1996). Synaptotagmin binding sites on syntaxin have been localized on the N-terminal Habc domain and the C-terminal SNARE motif.

In most vertebrate neurons, synaptic transmission is mediated by members of the Ca_v2 calcium channel family, in particular N- and P/Q-type channels. The α 1 subunits of both N- and P/Q-type calcium channels contain a long intracellular linker region connecting domains II and III, which associates with key synaptic proteins such as syntaxin 1, SNAP-25, synaptotagmin 1, and cysteine string protein (CSP), and has been termed the *synaptic protein interaction site* (reviewed by Zamponi, 2003). The stability of the complexes is critically dependent on intracellular calcium concentration, thus allowing for a dynamic rearrangement of calcium channel, SNARE protein complexes in response to calcium influx.

12.2.2. Phospholipases C

The physiological effects of many extracellular stimuli are mediated by receptor promoted activation of phospholipase C (PLC) and resulting activation of inositol lipid signaling pathways. In response to many extracellular stimuli, such as hormones, neurotransmitters, antigens, and growth factors, phospholipases C catalyze the hydrolysis of phosphatidylinositol (4,5)-bisphosphate, thereby generating two second messengers, inositol 1,4,5-trisphosphate and diacylglycerol. Eleven phospholipase C isozymes encoded by different genes have been identified in mammals and, on the basis of their structure and sequence relationships, have been classified into five families designated PLCβ (1 to 4), PLCγ (1 and 2), PLCδ (1, 3, and 4), PLCε (1), and PLCζ (1) (reviewed by Cockcroft, 2006). In 2005, four groups, using genomic data mining, independently identified an entirely new family of PLCs—η1 and η2—in mammals.

Mammalian phosphoinositide-specific phospholipases C produce two second messengers, D-myo-inositol 1,4,5-trisphosphate (1,4,5-IP_3) and *sn*-1,2-diacylglycerol (DAG), by hydrolyzing phospholipid phosphatidylinositol 4,5-bisphosphate. The second messenger, 1,4,5-IP_3, induces an increase of intracellular calcium levels by binding to 1,4,5-IP_3 receptors; whereas, DAG activates protein kinase C.

All PLCs contain the catalytic X and Y domains, in addition to other regulatory domains, including C2 and the EF-hand. The crystal structure of the C2 domain of phosphoinositide-specific phospholipase C is similar to those of synaptotagmins (Figure 12.5) (Essen et al., 1997). The C2 domains from both proteins share an antiparallel, eight-stranded β-sandwich architecture but differ topologically by a circular permutation of the strands. Despite this topological difference, the calcium binding regions of both C2 domains reside on the same end of the β-sandwich.

Agonist binding to the hepta-helical G-protein-coupled receptors leads to the activation of PLCβ's1 to 4 through direct activation by either the Gqα or Gβγ subunits of heterotrimeric G-proteins (Rebecchi and Pentyala, 2000). Many extracellular messenger molecules, such as bradykinin, angiotensin-2, acetycholine, and vasopressin, bind to receptors coupled to Gq and activate PLCβ isoforms.

Stimulation of receptor tyrosine kinases by growth factors such as nerve growth factor leads to the activation of PLCγ1 and 2. The single transmembrane domain of receptor tyrosine kinase dimerizes upon ligand binding, then phosphorylates itself; this leads to docking of various effector molecules to the receptor, including PLCγ. Finally, the tyrosin kinase domain of the receptor also phosphorylates PLCγ, which in addition to docking is also important for activation of this enzyme (Rebecchi and Pentyala, 2000).

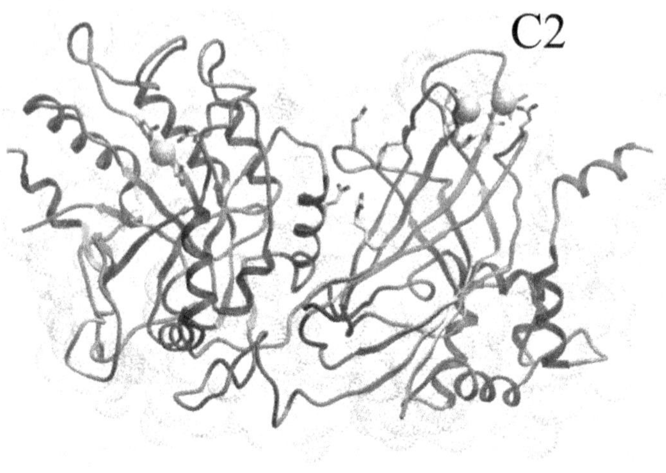

Figure 12.5. Crystal structure of phosphoinositide-specific phospholipase C (PDB file 1DJI). (From Essen et al., 1997.)

The third group consists of PLCδ1, 3, and 4 (reviewed by Rebecchi and Pentyala, 2000; Rohacs et al., 2008). PLCδ2 is a bovine homolog of the human/mouse PLCδ4. Increased cytoplasmic calcium activates these PLCδ's. Thus, these PLCδ's are thought to serve as signal amplifiers after PLCβ or PLCγ activation. All PLC isoforms require some calcium for activity, but PLCδ's are the most sensitive to changes of calcium in the physiological range. PKCs β, γ, and δ have no other known activators, unlike the PKCs ε and ζ.

Crystal structures of PLCδ1 (756 residues, 85 kDa) (Essen et al., 1996, 1997) reveal a four-module organization of the enzyme comprising the amino terminal PH domain, the EF-hand domain, catalytic domain, and the carboxyl-terminal C2 domain. Based on these structures, it was proposed that PH and C2 domains, both of which are well characterized membrane targeting domains, are involved in the membrane targeting of PLCδ1. The PH domain is a β-barrel-like structure that is present in many membrane binding proteins. The essential role of the PH domain in the membrane targeting of PLCδ1 has been demonstrated experimentally both in vitro and in vivo. The PH domain of PLCδ1 is capable of anchoring the protein to the membrane by specifically binding to phosphatidylinositol (4,5)-bisphosphate in the membrane, and the competitive binding of the PH domain to soluble IP3 can induce the membrane dissociation of PLCδ1. The role of the C2 domain in PLCδ1 catalysis remains unclear. Beta and γ isozymes have additional domains responsible for specific interactions with G-protein subunits or tyrosine kinases. Besides the catalytic TIM barrel, catalytic activity strictly requires the C-terminal helix of the EF-hand and an intact C2 domain. The crystal structure of phospholipase Cδ1 revealed that three loops of this region (CBR1, residues 643 to 653; CBR2, 675 to 680; and CBR3, 706 to 714) are involved in binding divalent ions. CBR1 and CBR3 undergo conformational changes during metal binding (Essen et al., 1996, 1997).

PLCε is activated by the small G-protein, ras (Bunney and Katan, 2006). PLCζ is injected from the sperm into the oocyte (sperm factor), where it induces calcium oscillations (Swann et al., 2006). PLCη1 and PLCη2 are the most recently cloned PLC isoforms. Their regulation is not well understood, but they may be even more sensitive to calcium than are the PLCδ's (Stewart et al., 2007). Two PLC homologs with no apparent PLC activity (PLCL1 and PLCL2) have also been identified (Takenaka et al., 2003).

The initiation of the calcium wave at fertilization in invertebrates can most generally be explained by a mechanism in which an unidentified, activating substance enters the egg from the sperm after fusion, activating the egg by stimulating PLC activation through a src family kinase pathway. In mammals this initiation involves the diffusion of a sperm specific PLC from sperm to egg after fusion (reviewed by Whitaker, 2006). A major source of the participating calcium is the endoplasmic reticulum, where calcium accumulation is driven by a SERCA pump. Calcium waves comprise the summation of elementary release events from either $InsP_3$ or ryanodine receptors. Calcium released from a receptor or cluster of receptors can diffuse to neighboring receptors, inducing further calcium release. In mammals the fertilization calcium wave is carried by the $InsP_3$ receptor; in fish and echinoderm eggs there appears to be a substantial contribution from the ryanodine receptors. PLC activation is central to the initiation of the fertilization calcium transient. In echinoderms and ascidians, PLCγ is activated by a src-like kinase; in mammals, PLCζ is activated by many factors, including cytokines, growth factors, mitogens, and calcium ions.

PLCζ is the smallest known PLC isozyme. It is composed of four EF-hands in the N-terminus (Chapter 11), X and Y catalytic domains, and a C2 domain in the C-terminus common to other PLC isozymes; however, it lacks a N-terminal pleckstrin homology (PH) domain (Saunders et al., 2002). PLCζ is similar to PLCδ1 (38% identity and 49% similarity in 647 amino acid residues of PLCζ); although the PH domain is present in PLCδ1 but absent in PLCζ. Increase in $[Ca^{2+}]_{cyt}$ causes successive activation of calmodulin-dependent kinase II and E3 ubiquitin ligase; this leads to proteolysis of ubiquitinated cyclin B1 and inactivation of metaphase promoting factor (Cdk1/cyclin B1 complex) and results in the release of eggs from meiotic arrest. PLCζ is very sensitive to $[Ca^{2+}]$ in vitro. Injection of PLCζ-encoding RNA or recombinant PLCζ into mouse eggs induces calcium oscillations as well as nuclear translocation. This is possibly related to cell cycle–dependent regulation of calcium oscillations (reviewed by Miyazaki and Ito, 2006; Swann and Yu, 2008).

The transient receptor potential, vanilloid type 1 (TRPV1) channels are involved in both thermosensation and nociception. Calcium flowing through TRPV1 activates PLC and the resulting depletion of phosphatidylinositol 4,5-bisphosphate (PIP2) was proposed to play a role in capsaicin-induced desensitization of these channels (reviewed by Rohacs et al., 2008).

Gas gangrene is an infectious disease caused by a group of anaerobic spore-forming bacteria. α-Toxin is a PLC that hydrolyzes phosphatidylcholine in eukaryotic cell membranes. The key determinant in gas gangrene is a calcium binding protein, whose C-terminal domain has a fold that is similar to the eukaryotic C2 domains (Naylor et al., 1999). The position of the three calcium binding sites at the top of the domain and their function to facilitate membrane binding is similar to those for the C2 domains. The K_d(Ca) values for these sites are rather high: 175 to 250 μM.

12.2.3. Phospholipase A_2

The phospholipase A_2 (PLA_2) family consists of several nonhomologous groups of enzymes that catalyze the hydrolysis of the sn-2 ester bond in a variety of phospholipids. The products of this reaction, a free fatty acid and a lysophospholipid, have many different physiological roles.

The crystal structure of the C2 domain of phospholipase A_2 reveals a β-sandwich with the same topology as the C2 domains in phosphoinositide-specific phospholipase Cδ1 (Perisic et al., 1998). Two clusters of exposed hydrophobic residues surround two adjacent calcium binding sites. This region, together with some basic residues, appears to constitute the membrane binding site. It is assumed that the interaction of the C2 domain with membranes is analogous to the hydrophobic electrostatic switch, which modulates reversible membrane binding of several myristoylated proteins. In this case the electrostatic switch is calcium, which changes the electrostatic properties of the surface of the C2 domain (i.e., neutralizes a cluster of negative charges and enables membrane binding). Another protein of this family, rabphilin 3A, which has two C2 domains, has an N-terminal extension containing a zinc finger domain.

The C2 domains from phospholipase A_2, PKC, and Nedd4 bind phospholipids at micromolar calcium concentrations, as do the C2A domains of synaptotagmins. The C2 domain from phospholipase A_2 interacts with neutral phospholipids; in contrast, the C2A domains of synaptotagmins and the C2 domain of PKC preferentially bind to negatively charged phospholipids (reviewed by Rizo and Südhof, 1998). Furthermore, although C2 from PKC is similar to the synaptotagmin C2A domains, it does not bind to syntaxin 1 as a function of calcium. Therefore, even among C2 domains that share calcium dependent phospholipid binding, there are functional distinctions that may be important for their biological roles.

Full activation of phospholipase A_2 requires calcium binding to the C2 domain and phosphorylation of Ser's (reviewed by Hirabayashi et al., 2004). The calcium binding induces translocation of phospholipase $A_2α$ from the cytosol to the perinuclear membranes. Ser phosphorylation is mediated by mitogen-activated protein kinases (MAPK), Ca^{2+}/calmodulin-dependent protein kinase II, and MAPK-interacting kinase Mnk1. Interaction with proteins and lipids, which include vimentin, annexins, NADPH oxidase, phosphatidylcholine, phosphatidylinositol 4,5-bisphosphate (PIP_2), and ceramide-1-phosphate, can also modulate the activity of phospholipase $A_2α$.

Among the fatty acids released by PLA_2, arachidonic acid is of particular biological importance, because it is subsequently converted to prostanoids and leukotrienes by cyclo-oxygenases (COX) and lipoxygenases (LOX), respectively (reviewed by Taketo and Sonoshita, 2002). Arachidonic acid derived from $cPLA_2\alpha$ is cytotoxic. The intracellular levels of arachidonic acid are regulated through distinct and nonoverlapping mechanisms. In resting cells, low concentrations of arachidonic acid are tightly maintained by basal levels of rapid catabolism, and also by membrane phospholipid recycling via the actions of the arachidonyl-CoA transferase (CoA-T) (Lands cycle). On the other hand, stimulus-induced release of arachidonic acid by $cPLA_2\alpha$ results in its rapid metabolism by induced COX-2, thereby limiting the intracellular arachidonic acid pools. Free arachidonic acid may also stimulate apoptosis through activation of sphingomyelinase. Alternatively, oxidized metabolites generated from arachidonic acid by LOX may induce apoptosis. In some cells induced to undergo apoptosis, arachidonic acid is released concomitant with loss of cell viability, caspase activation, and DNA fragmentation. Such effects appear to be mediated by activation of cytosolic PLA_2 and/or calcium-independent PLA_2. In addition to cytosolic PLA_2 and calcium-independent PLA_2, secretory PLA_2's are involved in apoptosis.

There are five main types of PLA_2: the cytosolic $cPLA_2$'s, the secreted $sPLA_2$'s, the calcium-independent $iPLA_2$'s, the PAF acetylhydrolases, and the lysosomal PLA_2's. There are more than 19 different isoforms of PLA_2 in mammals. The group IV phospholipase A_2 family is comprised of six cytosolic phospholipases A_2 ($cPLA_2\alpha$, $cPLA_2\beta$, $cPLA_2\gamma$, $cPLA_2\delta$, $cPLA_2\varepsilon$, and $cPLA_2\zeta$) (reviewed by Ghosh et al., 2006). They are most similar to PLA, PLB, and lysophospholipases of filamentous fungi, particularly in regions containing conserved residues involved in catalysis.

$cPLA_2$ is 85 kDa and has no sequence similarity to any other phospholipase A_2. It is widely expressed in mammalian cells and mediates the production of functionally diverse lipid products in response to extracellular stimuli. It has PLA_2 and lysophospholipase activities and is the only PLA_2 that is specific for phospholipid-containing arachidonic acid. Because of its role in initiating agonist-induced release of arachidonic acid for the production of eicosanoids, $cPLA_2\alpha$ takes part in regulating normal and pathological processes in a variety of tissues. Stimulation of cells with agents that mobilize intracellular calcium and/or promote the phosphorylation of $cPLA_2\alpha$ leads to (1) translocation of the enzyme from cytosol to endoplasmic reticulum, Golgi apparatus, and perinuclear membranes, where it associates with the arachidonic acid in close proximity to downstream enzymes that make eicosanoids; and (2) the change in conformation induced by phosphorylation increases phospholipid affinity and arachidonic acid release. Its N-terminal C2 domain is responsible for calcium dependent translocation of the enzyme to membranes for hydrolysis of substrates resident in the membrane.

Phagocytic cells contain $cPLA_2$ as well as several types of $sPLA_2$, all of which produce inflammatory lipid mediators (reviewed by Levy, 2006). By modulating $cPLA_2\alpha$ expression in a variety of phagocytes, it was found that $cPLA_2\alpha$ plays a major role in the formation of eicosanoids. In addition, $cPLA_2\alpha$ also regulates the activation of NADPH oxidase. The dual subcellular localization of cPLA$_2\alpha$ in different compartments, in the plasma membranes and in the nucleus, suggests a molecular mechanism for the participation of cPLA$_2\alpha$ in different processes: stimulation of NADPH oxidase and formation of eicosanoids in the same cells.

Ischemic stroke is caused by obstruction of blood flow to the brain, resulting in energy failure that initiates a complex series of metabolic events, ultimately causing neuronal death. One such critical metabolic event is the activation of phospholipase A$_2$, resulting in hydrolysis of membrane phospholipids and release of free fatty acids, including arachidonic acid, a metabolic precursor for important cell signaling eicosanoids (reviewed by Sun et al., 2004; Muralikrishna Adibhatla and Hatcher, 2006; Garcia-Garcia et al., 2009). Cardiolipin hydrolysis by mitochondrial sPLA$_2$ disrupts its respiratory chain and increases production of reactive oxygen species. Oxidative metabolism of arachidonic acid also generates reactive oxygen species. These two processes contribute to formation of lipid peroxides, which degrade to reactive aldehyde products (malondialdehyde, 4-hydroxynonenal, and acrolein) that covalently bind to proteins/nucleic acids, altering their function and causing cellular damage. Activation of PLA$_2$ in cerebral ischemia has been shown, while other studies have separately demonstrated increased lipid peroxidation.

Several studies demonstrate that PLA$_2$ regulates growth and signaling in several cell types. Most tumor cells contain elevated levels of PLA$_2$, and increased production of eicosanoids has been implicated in cell growth (reviewed by Sun et al., 2004; Nakanishi and Rosenberg, 2006). Using specific PLA$_2$ inhibitors, van Rossum et al. (2002) demonstrated the involvement of cPLA$_2$ activity in cell cycle progression, especially from G1 to S phase in neuroblastoma (N2A) cells. Cytosolic PLA$_2\alpha$ has a high selectivity for liberating arachidonic acid that is subsequently metabolized by a panel of downstream enzymes for eicosanoid production. Although concentrations of free arachidonic acid are maintained at low levels in resting cells, alterations in arachidonic acid production, often resulting from deregulation of cPLA$_2\alpha$ activity, are observed in transformed cells (reviewed by Nakanishi and Rosenberg, 2006). Blaine et al. (2001) have shown that Ras induces expression of cPLA$_2\alpha$ in non-small cell lung cancer cells (NSCLC) cells via direct binding of the transcription factors, Sp1 and c-jun, to the cPLA$_2\alpha$ promoter. In human skin-derived carcinoma cells (A431), the translocation and phosphorylation of cPLA$_2\alpha$ can be accomplished by another target of c-jun, epidermal growth factor receptor (EGFR) (Sato et al., 1997).

Eicosanoid products of the cyclooxygenase (COX) and lipoxygenase (LOX) pathways are important mediators of the proliferation of prostate cancer cells in culture and regulate tumor vascularization and metastasis in animal models. Pharmacological agents that block either COX or LOX products effectively reduce the size of prostate cancer xenografts. PLA$_2$'s, which regulate the provision of arachidonic acid to both COX- and LOX-derived eicosanoids, also regulate the growth of prostate cancer cells and tumors, with one enzyme, secreted PLA$_2$-IIA, being increased in prostate cancer tissues (reviewed by Dong et al., 2006). Annexins A1 and A2, known inhibitors of cytosolic PLA$_2\alpha$ activity, are absent in prostate cancer tissues. Dong et al. (2006) proposed that PLA_2 function is deregulated by aberrant up-regulation of secreted enzymes and down-regulation of endogenous inhibitors of cytosolic PLA_2 activity in prostate cancer and that this deregulation contributes to the pathogenesis of prostate cancer. Thus, in addition to COX and LOX enzymes, PLA_2 enzymes represent important targets for the treatment of prostate cancer.

PLA_2's from snake venom induce a wide spectrum of pharmacological effects, despite similarity in primary, secondary, and tertiary structures and common catalytic properties (reviewed by Kini, 2005). The strong anticoagulant PLA_2 CM-IV from cobra (*Naja nigricollis*) venom inhibits both the extrinsic tenase and prothrombinase complexes; whereas, the weak anticoagulant PLA_2's (CM-I and CM-II) inhibit only the extrinsic tenase complex. PLA_2 CM-IV binds to factor Xa and interferes in its interaction with factor Va and formation of the prothrombinase complex. In contrast, PLA_2 CM-I and CM-II do not affect the formation of the prothrombinase complex. In addition, PLA_2 CM-IV inhibits the extrinsic tenase complex by a combination of enzymatic and nonenzymatic mechanisms; while, PLA_2s CM-I and CM-II inhibit only by enzymatic mechanisms. These functional differences explain the disparity in the anticoagulant potency of *N. nigricollis* PLA_2 enzymes. Similarly, human secretory enzyme binds to factor Xa and inhibits the prothrombinase complex. The anticoagulant region of PLA_2 enzymes was predicted to be between 54 and 77 residues long. This region is basic in the strong anticoagulant PLA_2s and neutral or negatively charged in weak and nonanticoagulant enzymes. These studies strongly support the target model, which suggests that protein–protein interaction rather than protein–phospholipid interaction determines the pharmacological specificity of PLA_2 enzymes.

12.2.4. Protein Kinase C

Protein kinase C (PKC) belongs to the serine and threonine kinase family (reviewed by Yonezawa et al., 2009). At least 10 PKC isoforms have been identified and subdivided into three groups: classical (α, βI, βII, and γ), novel (δ, ε, θ, and η), and atypical (ζ and ι/λ). Two calcium-insensitive isoforms, PKC δ and ε, have received special attention as promising targets for new drugs. PKCs play a multifaceted role in cellular responses in a range of tissues. PKCα, PKCε, PKCζ, and PKCλ/ι preferentially function to promote cell proliferation and survival, while PKCδ is an important regulator of apoptosis (reviewed by Reyland, 2009). The essential role of PKCs in both cell survival and apoptosis suggests that specific isoforms may function as molecular sensors, promoting cell survival or cell death, depending on environmental cues. Given their central roles in cell and tissue homeostasis, it is not surprising that the expression or activity of some of these kinases is altered in human diseases, especially cancer.

The classical PKCs (α, βI, βII, γ) are activated by calcium and diacylglycerol. Novel PKCs (ε, δ, θ, η) respond only to diacylglycerol (reviewed by Kheifets and Mochly-Rosen, 2007). The atypical PKCs (η, λ/ι) do not respond to either of these second messengers.

Calcium binding increases the affinities of PKCs (α, βI, βII, γ) for phosphatidylserine at the cell membrane (reviewed by Kheifets and Mochly-Rosen, 2007). This, in turn, enables PKC to bind diacylglycerol that is found at low abundance in the membrane. The binding of its two regulatory domains to the membrane releases the autoinhibitory pseudosubstrate site (Ψ substrate) from the active site in the catalytic domain, producing conformational changes that leave the catalytic domain accessible to substrate binding and phosphorylation. Upon activation, PKCs translocate from the soluble fraction to cellular membranes, where they bind to anionic phospholipids and are localized to diverse subcellular sites by binding to receptors for activated C kinase (RACKs), which anchor them near a subset of protein substrates and away from others. Many of the PKC isozymes are expressed in the same cells and respond to the same activators; however, they translocate to different intracellular sites, to mediate unique and sometimes even opposing functions.

A class of phosphoinositol-specific PKCs with a novel domain structure, PKCη, has been found in mammals. This class consists of two isoforms, PKCη1 and PKCη2 (reviewed by Stewart et al., 2007). Both enzymes hydrolyze phosphatidylinositol 4,5-bisphosphate and are more sensitive to calcium than are other PLC isozymes and are likely to mediate G-protein-coupled receptor (GPCR) signaling pathways. Both enzymes are expressed in neuron-enriched regions and are abundant in the brain.

The two C1 domains within the regulatory domain of the classical and novel PKCs bind diacylglycerol as well as the tumor promoters, phorbol esters. Each C1 domain in both classical and novel PKCs consists of two tandem, Cys-rich zinc fingers, A and B, that depend on zinc ions for both proper folding and function (Fukuda et al., 1999); the atypical PKCs has only one zinc finger. While both C1 domains are thought to be oriented for potential membrane interaction and can bind diacylglycerol and phorbol esters, only one of the C1 domains usually binds to the membrane. In addition to diacylglycerol binding, the C1 domain has been implicated in PKC targeting to subcellular sites through both lipid and protein interactions.

The C2 domains of PKCs (α, βI, βII, and γ) bind two or three Ca^{2+} ions. This induces both electrostatic and conformational changes; it also enables binding of phosphatidylserine and the penetration of C2 into the membrane (Medkova and Cho, 1998). This calcium-dependent binding of C2 to membranes is highly specific to 1,2-*sn*-phosphatidyl-L-serine. The C2 domain affects the affinity of the C1 domain for activators; removal of C2 increases the sensitivity of PKC to diacylglycerol and phorbol esters (reviewed by Corbalan-Garcia et al., 1999; Kheifets and Mochly-Rosen, 2007).

The structure of PKCδ is similar to those of the classical PKCs, even though δ does not bind calcium. The regulatory domain (C1A, C1B, and C2) interacts with the membrane surface; the catalytic domain is in the cytosol (Figure 12.6).

Two models have been proposed for docking the C2 domain of PKCα to membranes. In the parallel model, the membrane docking surface includes the calcium binding loops and an anion binding site on β-strands 3 and 4, such that

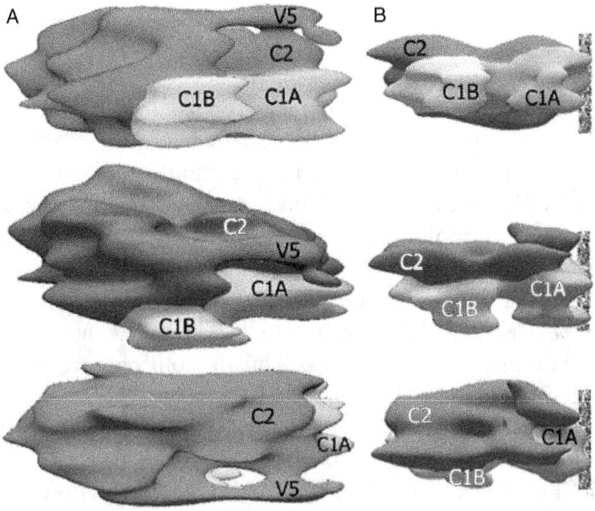

Figure 12.6. Low-resolution structure of protein kinase Cδ, A, and of its regulatory domain, B, as determined by electron microscopy of two-dimensional crystals. The right-hand side of PKC and of its regulatory domain is inferred to bind to the membrane surface. Both molecules are seen in three views as rotated about an axis perpendicular to the membrane surface. (From Solodukhin et al., 2007.)

the β-strands are oriented parallel to the membrane surface. In the perpendicular model, the docking surface is localized to the Ca^{2+} binding loops and the β-strands are oriented perpendicular to the membrane surface. Fluorescence and EPR data (Kohout et al., 2003) indicate that the first and the third calcium binding loops insert into the lipid headgroup region of the membrane and that the anion binding site on β-strands 3 and 4 is located near the headgroups. The β-strands are inferred to tilt parallel to the membrane surface. The crystal structure of the C2 domain of PKCα shows the binding of two Ca^{2+} ions and of one 1,2-dicaproyl-*sn*-phosphatidyl-L-serine molecule that was coordinated directly to one of the Ca^{2+} ions (Ochoa et al., 2002). An interaction between Ca^{2+} ions and phosphate groups or phospholipid molecules in the C2 domain of PKCα is inferred from the specificity and spatial organization of the binding sites in the domain.

PKCs are processed by a series of ordered phosphorylation events to mature into the catalytically competent species (reviewed by Newton, 2003). Maturation is initiated by phosphorylation of Thr in the activation loop by the upstream kinase, phosphoinositide-dependent kinase-1(PDK-1). It then requires a mammalian target of rapamycin complex 2 (mTORC2) and its own intrinsic kinase activity for phosphorylation/autophosphorylation at two carboxyl-terminal sites, the hydrophobic and turn motifs (reviewed by Newton, 2003; Gallegos and Newton, 2008).

The recognition of protein kinase C (PKC) as the long-sought-after receptor for the tumor promoting phorbol esters established the possible role of PKC in carcinogenesis and as a potentially important target for cancer therapeutics (Nishizuka, 1992).

PKC has been found to be involved in gene expression, the control of cytoskeleton, membrane and secretagogue dependent signal transduction for secretion of acid. Gastric carcinoma and adenocarcinoma cells often show dysregulated PKC-dependent cell signal transduction compared to normal gastric cells (reviewed by Fährmann, 2008). Moreover, PKC was the first known target of tumor promoting phorbol esters. These findings support PKC as a potential chemotherapy target in gastric cancer.

PKCs are important in normal keratinocyte biology and in skin diseases, especially skin cancer (Denning, 2004). Keratinocyte apoptosis induced by ultraviolet (UV) radiation is a major protective mechanism from photocarcinogenesis. The constitutively active catalytic domain of PKCδ is an apoptotic effector generated by caspase-3 cleavage of full-length PKCδ in response to a wide variety of apoptotic stimuli, including UV radiation (Sitailo et al., 2004). The activated PKCδ catalytic domain triggers the redistribution and activation of Bax that can directly induce cytochrome *c* release. PKCδ also activates upstream components of the death effector pathway to ensure apoptosis. The cleavage and activation of PKCδ are critical components of UV-induced apoptosis in human keratinocytes. Inactivation of PKCδ can promote the survival of keratinocytes exposed to UV radiation (D'Costa and Denning, 2005).

PKCα and PKCδ promote apoptosis in androgen dependent prostate cancer cells (reviewed by Teicher, 2006). Atypical PKCι is highly expressed in human non-small cell lung carcinoma (NSCLC) lines, whereas PKCζ, is undetectable in these cells. PKCL is encoded by a critical lung cancer gene; it activates a Rac1 → Pak → Mek1,2 → Erk1,2 signaling pathway required for transformed growth. PKC is directly involved in modulating apoptosis hematopoietic neoplasms.

PKCs are activated in response to various pathogenic stimuli in the diabetic state. Hyperglycemia is the predominant stimulus that induces the activation of distinct PKC isoforms within a cell, each mediating specific functions, probably through differential subcellular localization (reviewed by Meier et al., 2009).

12.3. CALCIUM ATPases

Lipid bilayer membranes surround all cells and organelles, forming barriers that limit the free exchange of polar solutes. A wide variety of proteins responsible for controlling the diffusion or active transport of ions and nutrients are inserted into these membranes. The P-type ATPases, also known as E_1–E_2 ATPases, are a large group of evolutionarily related ion pumps that are found in bacteria, archaea, and eukaryotes. They autophosphorylate a conserved Asp. They are α-helical bundle primary transporters; they all appear to interconvert between at least two different conformations, denoted by E_1 and E_2. They are encoded by five main gene families (I, II, III, IV, and V) (Palmgren and Axelsen, 1998). Those that share specificity for Ca^{2+}, K^+, and Na^+ ions group together in a single clade and are designated as P-type II ATPases. They include five subfamilies: A, B, C, D, and E; also known as SERCA (sarco/endoplasmic reticulum Ca^{2+}-ATPase), PMCA (plasma membrane Ca^{2+} ATPase), NK/HK (Na^+,K^+- and H^+,K^+-ATPase), ENA (P-type ATPases which are able to extrude Na^+, Li^+, and K^+ and encoded by *ENA* genes in *S. cerevisiae*; the name given by Rodriguez-Navarro et al. (1994), that is, exitus natru, for the Latin words meaning "exit sodium") and ACU (P-type ATPases which mediate high-affinity Na^+ and K^+ ion uptake and are encoded by *ACU* genes in fungi), respectively.

Calcium pumps, or Ca-ATPases, are located in the plasma membrane or in the membranes of endoplasmic or sarcoplasmic reticulum. They create calcium gradients, which are used in a variety of signaling systems mediated by gated calcium channels. The central event in the activity of this family of P-type ATPases is the formation of an acid stable aspartyl phosphate intermediate (reviewed by Stokes and Green, 2003; Inesi et al., 2008). This event is initiated by cooperative binding of two cytoplasmic Ca^{2+} ions to transport sites. The energy of this intermediate is used to induce a conformational change that closes the ion gate from the cytoplasm, reduces the affinity of these transport sites for Ca^{2+} ions, and opens the ion gate toward the lumenal, or extracellular, side of the membrane. After releasing calcium, protons bind to the transport sites and the aspartyl phosphate is hydrolyzed to complete the cycle. Active calcium transport was postulated to be achieved by changing the enzyme structure from the state in which calcium binding sites have high affinity and face the cytoplasm (E_1 state) to the state in which calcium binding sites have low affinity and face the lumen of the sarcoplasmic reticulum (E_2 state).

There are various types of ATPases, which can differ in function (ATP synthesis and/or hydrolysis), structure (F-, V-, and A-ATPases contain rotary motors) and in the type of ions they transport. P-ATPases (sometime known as E_1–E_2 ATPases) are found in bacteria and in eukaryotic plasma membranes and organelles. P-ATPases transport a variety of different compounds, including ions and phospholipids, across a membrane using ATP hydrolysis for energy. There are many different classes of P-ATPases, each of which transports a specific type of ion: H^+, Na^+, K^+, Mg^{2+}, Ca^{2+}, Zn^{2+}, Co^{2+}, Ni^{2+}, Cu^+, or Cu^{2+}.

The best characterized member of the P-type, ion-translocating ATPase superfamily is Ca^{2+}-ATPase from rabbit skeletal muscle SR. It is an integral membrane protein of 110 kDa; it can transport two Ca^{2+} ions per one ATP hydrolyzed against a concentration gradient across the membrane. Toyoshima et al. (2002, 2003) determined the structure of Ca^{2+}-ATPase in various conformations, including $E_1 \cdot Ca_2$. The protein consists of four basic domains (Figure 12.7). The transmembrane domain is almost entirely helical (10 transmembrane helices M1 to M10) and has short loops on the lumenal and cytoplasmic surfaces. Four of the transmembrane helices extend into the cytoplasm to form a stalk. The three cytoplasmic domains are built from two large cytoplasmic loops between transmembrane helices M2/M3 and M4/M5. The M4/M5 loop forms the phosphorylation (P) domain and the nucleotide binding (N) domain, which is inserted

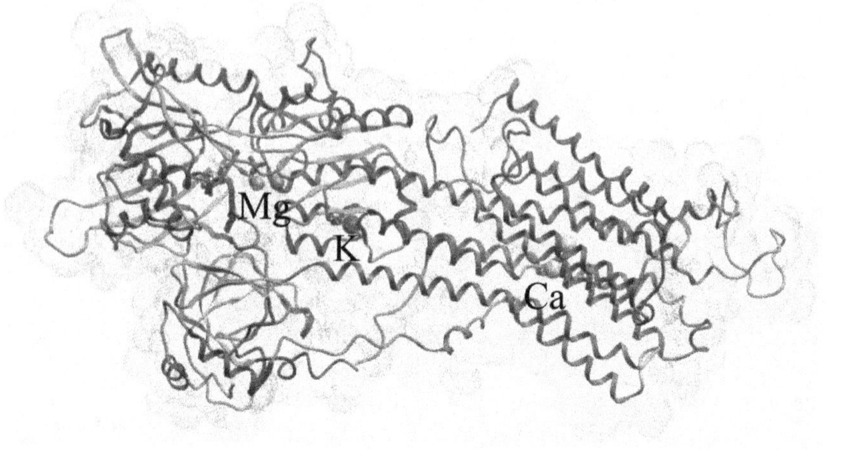

Figure 12.7. Structure of sarcoplasmic reticulum Ca^{2+}-ATPase in the Ca_2-E_1-ADP state (PDB file 1T5T). ATP is bound in the N domain, to the left. The transmembrane domain (helices M1 to M10) is to the right; the Ca^{2+} ion binding sites indicate the channel. (From Sorensen et al., 2001.) (*See insert for color representation of figure.*)

within the P domain. The third cytoplasmic loop, forming the transduction or anchor (A) domain, includes the smaller M2/M3 loop as well as the N-terminus.

The transmembrane domain contains two calcium binding sites. In the E_1 conformation the site cooperatively binds two Ca^{2+} ions from the cytoplasm. Residues on M4, M5, M6, and M8 helices provide the oxygen ligands for calcium. Asn768 and Glu771 on M5, Thr799 and Asp800 on M6, and Glu908 on M8 bind a Ca^{2+} ion in the first binding site. Main chain carbonyl oxygens along M4 and side-chain oxygens from Glu on M4 and Asn796 and Asp800 on M6 bind calcium in the second binding site. A highly conserved sequence motif ($PEGL^{311}$) on M4 is located in the middle of the second site and probably confers cooperativity of the binding. It is supposed that the binding of the first Ca^{2+} ion to M5/M6/M8 somehow induces a favorable conformation of M4 to facilitate the binding of the second Ca^{2+} ion.

Both calcium binding sites are located near the cytoplasmic side. Site II is closer to the cytoplasmic surface than is site I (Toyoshima et al., 2002). The binding of the two Ca^{2+} ions seems to be sequential, the first Ca^{2+} ion binding in site I. Toyoshima et al. (2002) suggested that Ca^{2+} ions enter the binding cavity via Glu309 and the locking of its side chain conformation by transmembrane helices causes "occlusion" of the bound Ca^{2+} ions. Calcium dissociation constants for these sites are in the micromolar range.

"Snapshots" of the SR Ca^{2+}-ATPase in the Ca^{2+}-bound state, the Ca^{2+}-free, thapsigargin-inhibited state, and the calcium-free, vanadate-inhibited state (i.e., in different states of the Ca^{2+}-transport cycle) reveal the impressive breadth and complexity of conformational change (reviewed by Young and Stokes, 2004;

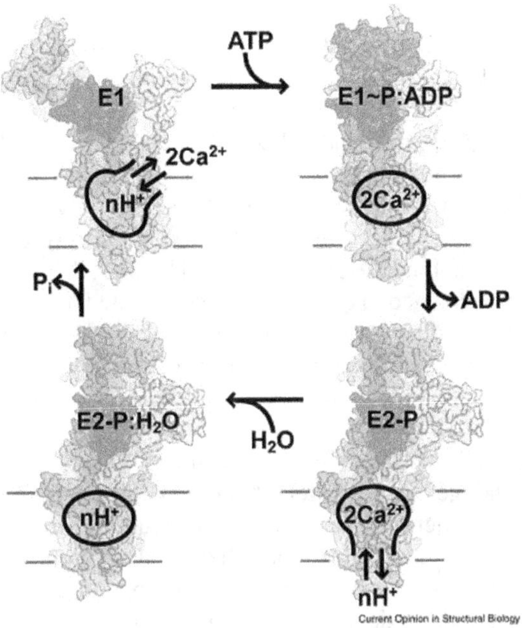

Figure 12.8. Key functional intermediates of the Ca -ATPase transport cycle and their structural correlates, as obtained from current structures determined by analysis of Ca^{2+}-ATPase crystals. Red, N domain; blue, P domain; yellow, A domain; purple, transmembrane segments 1 and 2 (M1–M2) with their cytosolic extensions; orange, M3–M4; green, M5–M10. (From Møller et al., 2005.) (*See insert for color representation of figure.*)

Møller et al., 2005) (Figure 12.8). The transition from the E_1Ca_2 to the E_2 conformation includes pronounced rigid body movements of the three cytoplasmic domains: a 110° rotation of the A domain about an axis normal to the membrane, a 30° rotation of the P domain with respect to the membrane plane, and a further 50° rotation of the N domain relative to the P domain (Toyoshima and Nomura, 2002). The structures within the three cytoplasmic domains are largely unchanged. The domain movements couple the energy of ATP hydrolysis to the transport of Ca^{2+} ions across the membrane. In contrast to the movements of the cytoplasmic domains, the transmembrane domain undergoes extensive deformations along most of its helices. The loss of calcium causes helix M6 to unwind, resulting in a 90° rotation of side chains Asn796, Thr799, and Asp800. The M4 helix shifts down, into the membrane, by almost 5 Å, and the side chain of Glu309 rotates completely away from the site to face helix M1. A bend in M1 pulls Glu58 out of the site; Glu771 and Glu908 in M5 and M8 do not change much, although Asn768 rotates 30° toward M4. The calcium sites face the cytoplasm in both E_1 and E_2 states.

In the Ca^{2+}-ATPase transport cycle, the A domain forms at least four distinct interfaces with the N and P domains, at first controlling phosphate transfer from

ATP and later controlling the stability of the phosphoenzyme. At least two structural elements couple ATP hydrolysis to calcium transport by coordinating the respective sites in the cytoplasmic and transmembrane domains. First, M4 and M5 helices provide rigid lever arms between the calcium binding sites and the P domain. Second, the A domain and helices M1 to M3 appear to be responsive to one another, forming a link between the calcium binding sites and the P domain.

The Ca^{2+}-ATPase pump has to discriminate between Ca^{2+} and Na^+, K^+, and Mg^{2+} ions (reviewed by Gouaux and Mackinnon, 2005). An obvious difference between the Ca^{2+} ion binding sites and the N^+ and K^+ sites is the greater importance of fully charged oxygen atoms contributed by Glu and Asp side chains for calcium coordination. A higher charge density is apparently required to compensate for the dehydration of a divalent cation. Part of the selectivity for the Ca^{2+} ion derives from its being seven coordinate, as opposed to six for the Mg^{2+} ion.

Phospholamban, a 52 residue protein spanning the sarcoplasmic reticulum membrane, is an endogenous inhibitor of sarco(endo)plasmic reticulum calcium ATPase (SERCA), lowering the apparent calcium affinity of the ATPase (reviewed by Traaseth et al., 2008). The relief of SERCA inhibition is achieved by phosphorylation of phospholamban at Ser16 by protein kinase A and/or Thr17 by Ca^{2+}/calmodulin-dependent protein kinase (Wegener et al., 1989). Sarcolipin, a single-pass membrane protein, has a primary sequence homologous to that of the transmembrane domain of phospholamban (Wawrzynow et al., 1992). Sarcolipin also binds to SERCA through intramembrane interactions, impeding calcium translocation. In vitro experiments have shown that sarcolipin can be phosphorylated at Thr5 when cotransfected with serine/threonine kinase 16 (STK16) (Gramolini et al., 2006). It was suggested that phosphorylation of phospholamban and sarcolipin constitute driving forces for calcium reuptake into cardiac sarcoplasmic reticulum (Traaseth et al., 2008).

12.4. CALCIUM BINDING PROTEINS OF THE ENDOPLASMIC RETICULUM

The endoplasmic reticulum is intracellular; however, the lumen of the endoplasmic reticulum (ER) is topologically extracytosolic. Proteins that exit the endoplasmic reticulum are, for the most part, properly folded and assembled, owing to the coordinated activities of several folding enzymes, molecular chaperones, and a rigorous quality control system that retains and disposes of misfolded proteins. Folding enzymes include at least 17 members of the protein disulfide isomerase family that catalyze the formation and interchange of disulfide bonds, peptide prolyl isomerases that catalyze *cis-trans* isomerization of peptidyl proline bonds, and components of the Asn-linked glycosylation system.

Calnexin and calreticulin are the most intensively studied chaperones of the ER because of their unusual modes of substrate recognition, their intimate relationship with the Asn-linked glycosylation system, and the diversity of functions attributed to them. They are lectins that interact with newly synthesized glycoproteins that have undergone partial trimming of their core N-linked oligosacharides. They serve simultaneously as molecular chaperones in the endoplasmic reticulum (reviewed by Helenius et al., 1997).

12.4.1. Calsequestrin

Calsequestrin, sarcoplasmic reticulum (SR) calcium binding protein, is found within the SR of skeletal and cardiac muscles but does not bind to the membrane. The total calcium concentration in SR is as high as 50 mM, but a large portion of this calcium is bound to calsequestrin. Due to calsequestrin, which acts as a luminal buffering system, the concentration of free Ca^{2+} ion within the lumen of the SR can be maintained below the inhibitory level of the calcium pump (1 mM). For the fastest muscles, a limiting step in the contraction–relaxation cycle is pumping of calcium into the SR. In this process, calsequestrin plays a key role through buffering the calcium levels within the lumen of the SR. Calsequestrin is not a passive calcium buffer, but rather, plays an active role in regulating calcium levels and facilitating its release from the lumen of the sarcoplasmic reticulum. Calsequestrin contains no transmembrane segments and is therefore believed to be located within the lumen of the SR.

The crystal structure of rabbit skeletal calsequestrin consists of three homologous domains, each with a thioredoxin fold, which is a five-strand β-sheet sandwiched by four α-helices (Wang et al., 1998b). Each domain has a hydrophobic core with acidic residues on the exterior, forming electronegative surfaces. The connecting loops and the secondary structural elements that fill the interdomain spaces contain mostly acidic residues. Cations are required to stabilize the acidic center of calsequestrin.

Calsequestrin (44 kDa) is a glycoprotein with an enormous number of calcium binding sites (30 to 50 mol mol^{-1}) with rather low affinity [K_d(Ca) ~1 mM] (Tada et al., 1978; Heilmann and Spamer, 1996). Over 30% of its residues are Asp or Glu; its isoelectric point is 3.75. Consistent with its calcium buffering function, calsequestrin (e.g., of rabbit skeletal muscle), has a total of 110 carboxyl groups and an excess of 80 carboxyl side chain groups over the sum of the positively charged groups. The amino terminal half of calsequestrin is slightly negative; whereas, over 35% of the residues in the carboxyl terminal half are Asp or Glu. Weak cooperativity of calcium binding was observed with the soluble form of calsequestrin; however, strong cooperativity accompanies the polymerization of calsequestrin into its insoluble forms (Krause et al., 1991). The aggregates are easily dissociated by potassium (Slupsky et al., 1987; Park et al., 2003).

The cardiac and fast muscle forms of calsequestrin differ mainly in their C-terminii. The cardiac form has an extended C-terminus with 71% acidic residues. Rabbit skeletal muscle calsequestrin has 31 calcium binding sites with a mean K_d(Ca) value of 0.79 mM in the absence of magnesium, and 23 sites with a K_d(Ca) of 0.88 mM in the presence of 3 mM magnesium (Krause et al., 1991). Calcium and magnesium do not bind to the same sites in calsequestrin.

The binding of calcium makes calsequestrin insoluble. In the absence of calcium, calsequestrin is in almost random coil conformation with α-helical content 11%. The binding of calcium increases its helical content up to 20% and changes its shape from elongated (Stokes radius 45 Å) to much more compact (35 Å) (Cozens and Reithmeir, 1984). Calcium, but not magnesium, blocks the binding of calsequestrin to a 26-kDa protein of the junctional sarcoplasmic reticulum. Based on this finding, Mitchell et al. (1988) proposed that calsequestrin is not only a calcium buffer but is also involved in the regulation of calcium release from sarcoplasmic reticulum. It was suggested that the calcium-dependent conformational change of calsequestrin causes a change in the shape of SR membrane proteins. Calsequestrin controls the ryanodine receptor channel in a phosphorylation-dependent fashion. It is also possible that calsequestrin interacts with the 95-kDa protein triadin, which contains a region of basic amino acids in its luminal domain. This luminal domain interacts with calsequestrin in a calcium-dependent manner; therefore, triadin might anchor calsequestrin to the junctional region of the SR (reviewed by Berchtold et al., 2000).

Calsequestrin has several functions in the lumen of the sarcoplasmic reticulum in addition to its well recognized role as a calcium buffer (reviewed by Beard et al., 2004). First, it is a luminal regulator of ryanodine receptor activity. In the presence of triadin and junctin, calsequestrin maximally inhibits the Ca^{2+} release channel when the free [Ca^{2+}] in the lumen of the SR is 1 mM. This inhibition is relieved when the [Ca^{2+}] changes, either because of small changes in the conformation of calsequestrin and/or its dissociation from the junctional face membrane. These changes in the association of calsequestrin with the ryanodine receptor amplify the direct effects of luminal [Ca^{2+}] on the activity of the ryanodine receptor. Additional functions for calsequestrin are connected with the kinase activity of the protein, its thioredoxinlike structure, and its influence over store operated calcium entry.

Calsequestrin plays a different role in the heart and skeletal muscle, enhancing calcium release in the heart but depressing release in skeletal muscle (reviewed by Beard et al., 2009). The contribution of cardiac calsequestrin to SR calcium storage and release during excitation–contraction coupling is largely dispensable (reviewed by Knollmann, 2009). The primary role of cardiac calsequestrin appears to be protection against premature calcium release and resulting arrhythmias. Furthermore, both cardiac calsequestrin and triadin are important for the structural organization of the SR.

Catecholaminergic polymorphic ventricular tachycardia (CPVT) is a familial arrhythmogenic disorder characterized by the onset of ventricular tachycardia (VT) during stress. Two forms have been described, an autosomal dominant (CPVT-1) resulting from mutations of ryanodin receptor and an autosomal recessive (CPVT-2) resulting from mutations of calsequestrin (Lahat et al., 2001; reviewed by Venetucci and Eisner, 2008). Several calsequestrin mutations that cause CPVT have been identified, and their effects on calsequestrin function and calcium handling have been studied either by overexpression or in transgenic mice. In a homozygous knock in mouse model, the D307H mutation decreased calsequestrin levels and increase calreticulin, another SR calcium buffering protein. The decreased calsequestrin increases the incidence of diastolic calcium release, delayed after depolarizations and ventricular tachycardia after β-adrenergic stimulation. This was ascribed to a reduced inhibitory action of calsequestrin on the ryanodin receptor.

12.4.2. Calnexin

Calnexin, calreticulin, and ERp57 (a glycoprotein-specific thiol-disulfide oxidoreductase) are endoplasmic chaperones that interact with partially folded glycoproteins and determine whether the proteins are to be released from the endoplasmic reticulum, or alternatively, whether they are to be sent to the proteosome for degradation (reviewed by Bedard et al., 2005). Accumulation of misfolded protein in the ER leads to activation of genes responsible for the expression of ER chaperones. When accumulation of misfolded protein becomes toxic, apoptosis is triggered, possibly with membrane kinase, IRE1, involved in signaling via caspase-12 (reviewed by Groenendyk and Michalak, 2005).

In mammalian cells, a large number of secretory proteins, both membrane bound and soluble, are Asn N-glycosylated (reviewed by Dejgaard et al., 2004). Proteins entering the ER are N-glycosylated by the oligosaccharyltransferase as they emerge from the translocon. Removal of the outermost and following glucoses by the successive action of glucosidase I and glucosidase II exposes the $Glc_1Man_9GlcNAc_2$ epitope. This structure is then recognized by two ER resident lectins (calnexin and calreticulin) that specifically bind monoglucosylated polymannose glycans (reviewed by Caramelo and Parodi, 2008). This is followed by removal of the innermost glucose by glucosidase I, thus liberating the glycoprotein from the lectin anchor. The protein-linked glycan is then reglucosylated by the soluble ER enzyme UDP-Glc:glycoprotein glucosyltransferase only if the protein displays a nonnative tertiary structure, as this enzyme behaves as a conformational sensor. Cycles of calnexin/calreticulin-glycoprotein binding and liberation, catalyzed by the opposing activities of UDP-Glc:glycoprotein glucosyltransferase and glucosidase II, are terminated once glycoproteins attain their native structures. Glucose-free glycoproteins then continue their transit through the secretory pathway. Alternatively, permanently misfolded glycoproteins may then be transported to the cytosol for proteasomal degradation.

Calnexin is a nonglycosylated transmembrane protein with an extracellular N-terminus and cytoplasmic C-terminus, called type I (65 kDa, 573 amino acid residues). Its substrate binding domain is located in the lumen of the ER. The structure of the luminal portion of calnexin (Schrag et al., 2001) has two domains: a globular β-sandwich that resembles legume lectins, and an extended 140-Å arm consisting of two β-strands folded into a hairpin. Each β-strand is composed of four tandemly repeated Pro-rich repeats (P domains). The 89-residue cytoplasmic tail of calnexin is phosphorylated and carries a C-terminal RKPRRE sequence that serves as a signal for ER localization. The luminal domain of calnexin (Figure 12.9) binds substrates, calcium, and Mg-ATP (Ou et al., 1995). The

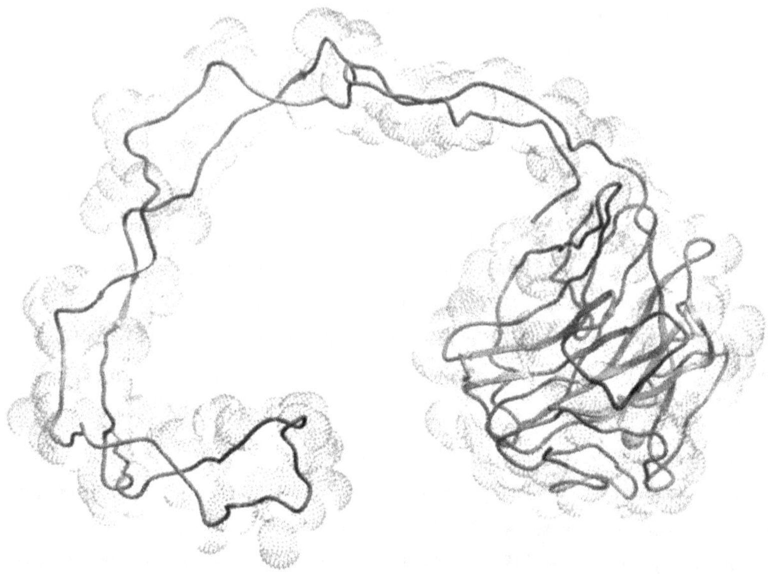

Figure 12.9. Structure of the luminal domain of calnexin (PDB file 1JHN). The 89-residue cytoplasmic tail (left) serves as a signal for localization to the endoplasmic reticulum. (From Schrag et al., 2001.)

subdomain of calnexin containing the four repeats that bind calcium with the highest affinity was isolated (Tjoelker et al., 1994). The sequence represents a conserved motif for the high affinity binding of calcium.

Extensive mapping studies informed by the crystal structure of calnexin have demonstrated that the oligosaccharide binding site is in a cleft on the surface of the globular domain (Schrag et al., 2001). This site recognizes not only the terminal glucose residue of the $Glc_1Man_9GlcNAc_2$ oligosaccharide but also three underlying mannose residues. This terminal glucose is coordinated by six residues within the lectin site, and mutation of any one of these in calnexin results in a dramatic loss of lectin function.

12.4.3. Calreticulin

Calreticulin is an ER luminal calcium buffering chaperone (reviewed by Michalak et al., 2009). It is involved in the regulation of intracellular calcium homoeostasis and the calcium storage capacity of the endoplasmic reticulum. The protein affects store operated calcium influx and influences calcium dependent transcriptional pathways during embryonic development. Calreticulin is also involved in the folding of newly synthesized proteins and glycoproteins and, together with calnexin and ERp57, constitutes the *calreticulin–calnexin cycle*, which is responsible for folding and quality control of newly synthesized glycoproteins. The goal of these mechanisms is to prevent the expression and secretion of misfolded proteins. Protein misfolding can be detrimental to the cell and contributes to the pathology in several inherited disorders (e.g., cystic fibrosis, familial hypercholesterolemia, diabetes insipidus).

Calreticulin has been implicated in many functions both inside and outside the ER. Calreticulin deficiency is lethal, due to impaired cardiac development. Overexpression in developing and postnatal heart leads to bradycardia, complete heart block, and sudden death (reviewed by Michalak et al., 2004). Calreticulin is a key upstream regulator of calcineurin in the calcium signaling cascade (Lynch and Michalak, 2003).

Calreticulin (46 kDa), the soluble homolog of calnexin, is a high capacity calcium binding protein found in almost every cell of higher organisms. Calreticulum and the luminal domain of calnexin are homologous; together they comprise an ER chaperone system that ensures the proper folding and quality control of newly synthesized glycoproteins. Calreticulin has at least three structural and functional domains (reviewed by Helenius et al., 1997; Krause and Michalak, 1997). The protein has an N-terminal signal sequence and its N-terminal domain (residues 1 to 180) is the most conserved domain among calreticulins from different species. The solution structure of calreticulin P-domain (residues 181 to 280) is characterized by an extended hairpin fold, with the N- and C-termini in close proximity. It is rich in Pro and contains two sets of three sequence repeats. It contains a zinc binding site, consisting of His25, His82, His128, and His153 and can bind ATP (Baksh et al., 1995). This extended structure is stabilized by three antiparallel β-sheets, with the β-strands comprising residues 189–192 and 276–279; 206–209 and 262–265; and 223–226 and 248–251 (Ellgaard et al., 2001). The two strands of the global hairpin interact not only through formation of the three short β-sheets but also by amino acid side-chain interactions in the three hydrophobic clusters. Each cluster involves two highly conserved Trp's, one from each strand of the hairpin; the three clusters share a similar arrangement of the side chains. The two Trp rings are packed against the aliphatic region of Pro and Lys side chains located three and four residues apart from one of the Trp's. The periodicity of the amino acid sequence is reflected in a repeating pattern of interactions across the hairpin. Within the global hairpin fold there are three well-ordered subdomains (residues 219–258, 189–209, and 262–284), connected by poorly ordered linkers.

The C-domain of calreticulin is similar to the C-terminal domain of calsequestrin. Calreticulin binds calcium with high capacity (>25 equivalents of calcium per protein) but low affinity [K_d(Ca) $\sim$2 mM]. The C-domain terminates with the ER retrieval signal, KDEL. Calreticulin appears to function as calcium storage or buffer in the lumen of the ER. In addition, it might participate in the coordination of protein synthesis, gene expression, and calcium signaling (reviewed by Krause and Michalak, 1997).

Calnexin and calreticulin are molecular chaperons that bind to many glycoproteins in the endoplasmic reticulum. Binding of glycoproteins to calnexin and calreticulin is a transient step during early maturation. This binding requires the presence of one or more monoglucosylated N-linked glycans. In one of the models proposed (Helenius et al., 1997), calnexin and calreticulin function as lectins; whereas, glucosyltransferase monitors the folding status. Whether the binding to the chaperones also involves protein–protein interactions is not resolved (reviewed by Williams, 2006). Calnexin and calreticulin interact with a wide variety of proteins, including soluble secretory proteins, complex membrane receptors and ion channels, lysosomal hydrolases, and extracellular matrix components. They play a role in the biosynthesis of viral glycoproteins, including those of HIV-1, hepatitis C virus, and influenza virus. Of the $\sim$100 newly synthesized glycoproteins that associate with calnexin and calreticulin, some interact with only one chaperone, whereas others may bind both either simultaneously or sequentially (reviewed by Williams, 2006).

13

EXTRACELLULAR CALCIUM BINDING PROTEINS

The cytosolic calcium binding protein of Chapters 10, 11, and 12 were discussed within the context of cell signaling. In great oversimplification, these proteins bind calcium functioning as a secondary messenger and in their calci forms are active; in their apo forms, inactive. This implies that their K_d(Ca) values are in the range 10^{-6} to 10^{-8} M; the range over which cellular calcium oscillates.

Of the many exceptions to this valid generalization, some proteins, such as actin, appear to bind calcium with high affinity and not release it during cell quiescence. Calcium is thought to stabilize actin, rain or shine. Correspondingly, extracellular calcium binding proteins are thought to use calcium for stabilization, configuration of active sites, and/or bridges to other extracellular proteins or carbohydrates.

Extracellular calcium is usually maintained at about 1.4×10^{-3} M free Ca^{2+} ion. This has been regarded as a physiological constant for vertebrates and other metazoans (McLean and Hastings, 1935). However, there is increasing evidence that [Ca_{out}] within microdomains varies in a regulated way. Some extracellular calcium binding proteins may be calcium modulated and involved in information transfer (reviewed by Hofer, 2005; Permyakov and Kretsinger, 2009). In humans and many other animals, the parathyroid gland, bones, kidneys, and intestine all cooperate to balance the uptake, excretion, and recycling of calcium. Despite the constant concentration of serum calcium, there are local fluctuations in the restricted volume of interstitial fluids bathing cells (reviewed by Hofer, 2005). Cells become major sources and sinks for calcium during cytoplasmic signaling events, owing to the activation of calcium export (e.g., by the plasma membrane

Calcium Binding Proteins, By Eugene A. Permyakov and Robert H. Kretsinger

Ca^{2+}-ATPase) and calcium entry (e.g., by store operated channels) across the plasma membrane. Calcium sensors include the well characterized extracellular Ca^{2+}-sensing receptor, a G-protein-coupled receptor originally isolated from the parathyroid gland (revewed by Hofer and Brown, 2003). In addition, other receptors, channels, and membrane proteins, such as gap junction hemichannels, metabotropic glutamate receptors, HERG K^+ channels, and the receptor Notch, are all sensitive to external [Ca] fluctuations (reviewed by Hofer, 2005). Interestingly, Han et al. (2003) showed that increased external [Ca] also causes intracellular calcium signaling in *Arabidopsis* guard cells. Emerging evidence indicates that [Ca] in the local microenvironment outside the cell undergoes alterations potentially sufficient to exert biological actions through these sensor proteins.

Calcium plays a key role in the blood-clotting process. Many enzymes involved in the blood-clotting cascade have several posttranslationally modified γ-carboxyglutamate (Gla) residues in an N-terminal Gla domain, which specifically bind calcium. In addition to the Gla domains, many blood clotting factors, as well as other extracellular proteins, have epidermal growth factor like (EGF) domains, some of which bind calcium. Many other extracellular proteins also bind calcium, in which the Ca^{2+} ion usually plays a structural and/or stabilizing role. These include some of the C-type lysozymes and the related α-lactalbumin, a protein involved in lactose biosynthesis, the pancreatic enzyme deoxyribonuclease I, and the bacterial protease, subtilisin. Intracellular proteolytic enzymes bind calcium as well, but in this case the role of the Ca^{2+} ion is often more complex. In the absence of calcium, the active sites of these enzymes are not functional, requiring calcium as the activator of the protease rather than a more conventional mechanism of zymogen activation, such as limited cleavage. Many actin filament-severing proteins also bind calcium; these include gelsolin, villin, and adseverin. These proteins sever actin filaments and cap the high affinity end of these filaments upon binding calcium.

13.1. α-LACTALBUMIN

α-Lactalbumin is a calcium binding protein from milk that is homologous to the c-type lysozyme of chicken. Either c-type or g (goose)-type lysozymes occur in vertebrates; invertebrates generally have i-type or sometimes also c-type lysozymes. In contrast to most lysozymes, echidna milk, horse milk, and pigeon egg white lysozymes all bind calcium. The crystal structures of α-lactalbumins from several sources are known (Stuart et al., 1986; Acharya et al., 1989, 1990, 1991, 1994; reviewed by Permyakov, 2005). The calcium binding domain in α-lactalbumin is not homologous to the EF-hand domain; however, its structure shares several properties.

The crystal structure of α-lactalbumin is similar to that of lysozyme (Acharya et al., 1989, 1990) (Figure 13–1). Both have two clusters of aromatic residues (Acharya et al., 1991; Alexandrescu et al., 1992). A single strong calcium binding site in α-lactalbumin consists of seven oxygen ligands, which belong to carboxylates of Asp82, Asp87 and Asp88, carbonyls of Lys79 and Asp84, and two water molecules. The coordinating ligands form a pentagonal bipyramid. Ca–O distances range from 2.3 to 2.5 Å. The calcium binding loop in α-lactalbumin is flanked by two α-helices. The conformation of the main chain in the calcium binding region of α-lactalbumin is similar to that of hen egg white lysozyme, but the side chains in this region of α-lactalbumin and lysozyme are different.

In contrast to the EF-hand proteins, which have no disulfide bonds, bovine α-lactalbumin has four disulfide bonds, Cys6–Cys120, Cys28–Cys111, Cys61–Cys77, and Cys73–Cys91; these contribute to its stability. One of them, Cys6–Cys120, has an unusually high reactivity to thiol reagents (Kuwajima et al., 1990; Ikeguchi et al., 1992). In the calcium-bound native state, the Cys6–Cys120 and Cys28–Cys111 disulfides are reduced sequentially, reflecting their relative accessibilities (Ewbank and Creighton, 1993a,b).

α-Lactalbumin is the modifier protein in lactose synthesis, which occurs in the lactating mammary gland. α-Lactalbumin complexes with galactosyl transferase and alters the substrate specificity of the enzyme to favor glucose as the acceptor molecule. α-Lactalbumin is a major source of amino acids in milk. Some forms of α-lactalbumin can induce apoptosis in tumor cells (Hakansson et al., 1995; Svensson et al., 1999; Köhler et al., 1999, 2001) and some of its peptides are bactericidal (Pelligrini et al., 1999; Hakansson et al., 2000).

Prolonged exposure of either calci- or apo- human α-lactalbumin to ultraviolet light (270 to 290 nm, 1 mW cm^{-2}, for 2 to 4 h) results in the reduction of the Cys61–Cys77 and Cys73–Cys91 disulfide bridges (Permyakov et al., 2003). In goat α-lactalbumin similar UV-illumination results in disruption of the Cys6–Cys120 and Cys73–Cys91 bonds (Vanhooren et al., 2002). The breakage of the disulfide bridges occurs as a consequence of energy transfer from nearby Trp's.

The binding of calcium to α-lactalbumin at room and higher temperatures causes pronounced changes mostly in tertiary, but not secondary, structure; this is clearly seen from both fluorescence (Permyakov et al., 1981, 1985) and circular dichroism (Anderson et al., 1997) data. Calcium binding results in both a blue shift of Trp fluorescence and a decrease in fluorescence quantum yield. The magnitude of the spectral changes depends on temperature; minimal changes are observed at temperatures below 10°C (transition between native apo and calci states), while the maximal changes are seen above 40°C (transition between the thermally denatured apo state and the native calci state). Time resolved fluorescence measurements demonstrate that the calcium-induced effects are due to changes in the environment of all emitting Trp's (four in bovine α-lactalbumin and three in human α-lactalbumin) (Ostrovsky et al., 1988). The pronounced fluorescence changes can be used for evaluation of the calcium association constant, which is very high and usually cannot be obtained from direct calcium titrations but can be calculated more accurately from titrations of calci-α-lactalbumin with a strong calcium chelator (Permyakov et al., 1981). Both the apo and calci forms of α-lactalbumin adopt nearly the same conformation at low temperatures, as evidenced by similarity in their circular dichroism and proton NMR spectra (Kuwajima et al., 1986). Removal of calcium from the protein enhances its sensitivity to pH and ionic conditions due to noncompensated negative charge–charge interactions at the cation binding site; this reduces its overall stability significantly (Griko and Remeta, 1999).

Chrysina et al. (2000) determined crystal structures of apo and calci forms of bovine α-lactalbumin at 2.2 Å resolution (Figure 13.1). Global conformations of both forms of the protein are similar. The removal of calcium has only minor effects on the structure of the metal binding site, and the largest structural change

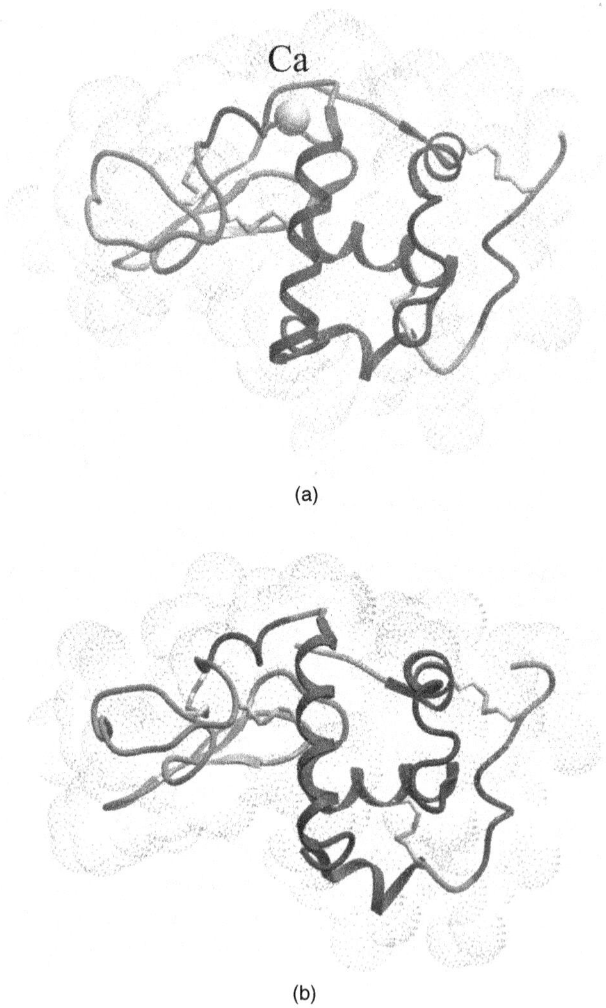

Figure 13.1. Structures of (a) calci- (with indicated Ca^{2+} ion) and (b) apo- α-lactalbumin (PDB files 1F6S and 1F6R). (From Chrysina et al., 2000.)

is observed in the cleft on the opposite side of the molecule. Tyr103 is shifted toward the interior of the cleft, and water mediated interactions with Gln54 and Asn56 replace the direct hydrogen bonds. The changes result in increased separation of the α and β domains, solvent exposure of a buried solvent molecule near the calcium binding site, and the replacement of inter- and intralobe hydrogen bonds of Tyr103 by interactions with newly immobilized water molecules.

Overall, the calcium binding loop in apo-α-lactalbumin has a slightly greater solvent accessibility and significantly increased temperature factors (crystallographic B-factors) compared with the corresponding loop in calci-α-lactalbumin. In apo-α-lactalbumin, the side chain of Asp87 is reoriented, but there is no other major structural change. Despite the minor structural changes in the calcium binding site of the apo-protein, it is reasonable to suggest that the larger change in the cleft region is propagated from the calcium binding loop. Charge repulsion between the carboxylates of the five Asp's in this region is a probable source of this change; the reorientation of the side chain of Asp87 and a slight expansion of the loop appear to reduce the negative charge density.

Small structural changes cause pronounced changes in the stability of the protein. Comparison of the B-factors in apo-α-lactalbumin and calci-α-lactalbumin indicates that the removal of calcium results in an increase in mobility in main chain and side chain atoms in the calcium binding site and C-terminal region but a decrease in mobility in the β-domain.

In the crystal structure, apo-α-lactalbumin is actually not in genuine apo state since it is probably stabilized in the native conformation by the high ionic strength of the crystallization medium. The structural changes caused by the removal of calcium are more pronounced in low-ionic-strength solution. The calcium free form of recombinant α-lactalbumin was studied in the presence of 0.5 M NaCl at pH 7.1. Taking into consideration the ability of α-lactalbumin to bind sodium, it is evident that in these conditions the protein is different from the pure apo form obtained in the absence of divalent and large concentrations of monovalent cations. Moreover, the choice of 20°C for these experiments was not appropriate, since at this temperature both native and thermally denatured forms of both apo- and sodium-loaded α-lactalbumin are present.

Calci-α-lactalbumin also binds other physiologically significant cations, such as Mg^{2+}, Mn^{2+}, Na^{+}, and K^{+}, which can compete with Ca^{2+} for the same binding site (Permyakov et al., 1981, 1985). They induce similar, albeit smaller structural changes in α-lactalbumin. Table 13.1 contains apparent equilibrium binding constants of some metal ions for bovine α-lactalbumin. The phase diagram in free calcium and temperature coordinates for α-lactalbumin is shown in Figure 12.

The energies of charge–charge interaction in apo- and calci-α-lactalbumin were calculated using a Tanford–Kirkwood algorithm with either a solvent accessibility correction or using the finite difference Poisson–Boltzmann method (Permyakov et al., 2005a). These analyses revealed two major regions of bovine α-lactalbumin that possess highly unfavorable electrostatic potentials: (1) the calcium binding loop and its neighboring residues and (2) the N-terminal

TABLE 13.1. Apparent Dissociation Constants of Metal Ions for Bovine α-Lactalbumin

	Dissociation Constants (M)		
Cation	37°C	20–25°C	Reference
Ca^{2+}	5×10^{-8}	3.3×10^{-9}	Permyakov et al., 1985
Mn^{2+}	—	3.3×10^{-5}	Murakami et al., 1982
Mn^{2+}	—	1.4×10^{-6}	Desmet et al., 1991
Mg^{2+}	4.7×10^{-3}; 2.2×10^{-2}	5×10^{-4}; 5×10^{-3}	Permyakov et al., 1985
Cd^{2+}	—	2.5×10^{-6}	Murakami et al., 1982
Sr^{2+}	—	3.7×10^{-7}	Desmet et al., 1991
Ba^{2+}	—	1.1×10^{-5}	Desmet et al., 1991
Pb^{2+}	5×10^{-7}	—	Veprintsev et al., 1996
Na^{+}	2.7×10^{-2}	10^{-2}	Permyakov et al., 1985
K^{+}	0.167	0.125	Permyakov et al., 1985

region. Several individual mutants were prepared to neutralize specific individual surface acidic amino acids at both the N-terminus and in the calcium binding loop. These mutants were characterized by intrinsic fluorescence, differential scanning microcalorimetry, and circular dichroism. The structural and thermodynamic data agree in every case with the theoretical predictions, confirming that the N-terminal region is very sensitive to changes in charge. For example, D14N with enzymatically removed N-terminal Met destabilizes the protein and decreases its calcium affinity. On the other hand, desMet E1M and desMet D37N mutations increase the thermal stability and calcium affinity. The desMet E1Q has a marked increase in protein stability; whereas, desMet E7Q and desMet E11L display a slight increase in calcium affinity and thermal stability. We do not understand what selective advantage, if any, is derived from the unfavorable energy contributed by Glu1 and the energetically favorable consequences of neutralizing this residue.

Hakansson et al. (1995, 1999) and Svensson et al. (1999) have found that some multimeric forms of α-lactalbumin in complexes with oleic acid cause apoptosis of transformed embryonic and lymphoid cells. Apoptosis is a conserved mechanism of cell death involved in embryogenesis, normal cell turnover, and various pathological disorders. A family of proteases, the caspases (cysteine-containing aspartate-specific proteases), appears to play a central role in initiation and execution of the apoptotic process.

The multimeric α-lactalbumin forms were isolated from the casein fraction of milk (Svensson et al., 1999). It was found that the apoptosis-inducing fraction of α-lactalbumin contains oligomers of α-lactalbumin that undergone a conformational change toward a disordered state. Oligomerization appears to conserve α-lactalbumin in a state with disordered properties in physiological conditions. Multimeric α-lactalbumin was shown to bind to the cell surface, enter the cytoplasm, and accumulate in the cell nuclei (Svensson et al., 1999). This is consistent with the finding that apo and zinc forms of α-lactalbumin in disordered conformation interact with model phospholipid membranes more strongly than does calci-α-lactalbumin (Cawthern et al., 1996). Addition of the multimeric α-lactalbumin with lipids to isolated liver mitochondria induces a loss of the mitochondrial membrane potential, mitochondrial swelling, and the release of cytochrome *c* (Köhler et al., 2001).

Similar results were obtained with HAMLET (human α-lactalbumin made lethal to tumor cells), which is native human α-lactalbumin converted in vitro to the apoptosis-inducing conformer of the protein in a complex with oleic acid (Svensson et al., 2000; Svanborg et al., 2003; Gustafsson et al., 2005). Svensson et al. (2003) identified unsaturated C18 fatty acids in the *cis* conformation as the cofactors that bind to apo-α-lactalbumin and form HAMLET. The complexes, formed on an ion exchange column, are stable in a disordered conformation and possess novel biological activity. The protein, fatty acid interaction, is specific, since saturated C18 fatty acids, or unsaturated C18:1*trans* conformers are unable to form complexes with apo-α-lactalbumin, as are fatty acids with shorter or longer carbon chains. Unsaturated *cis* fatty acids other than C18:1:9*cis* are able to form stable complexes, but they are not active in the apoptosis assay. These results demonstrate that stereo-specific, lipid–protein interactions can stabilize partially unfolded conformations and form molecular complexes with novel biological activity. It should be noted that the binding of fatty acids to α-lactalbumin was first revealed in the work of Cawthern et al. (1997). They studied interactions of α-lactalbumin with stearic acid and its spin-labeled analog, doxylstearic acid. At least two classes of fatty acid binding sites with dissociation constants 10^{-6} to 10^{-4} M were found in α-lactalbumin.

A similar complex was formed by titration of human α-lactalbumin by oleic acid at pH 8.3 up to the critical micelle concentration of oleic acid (Knyazeva et al., 2008). The complex formation depends strongly on [Ca], ionic strength, and temperature. The spectrofluorimetrically estimated number of oleic acid molecules irreversibly bound per α-lactalbumin molecule (after dialysis of the OA-loaded preparation against water followed by lyophilization) depends on temperature: 2.9 at 17°C (native apo-α-lactalbumin; with the resulting complex referred to as the LA-OA-17 state) and 9 at 45°C (thermally unfolded apo-α-lactalbumin; LA-OA-45). Intrinsic Trp fluorescence measurements revealed substantially decreased thermal stability of Ca^{2+}-free forms of HAMLET, LA-OA-45, and OA-saturated protein. The irreversibly bound OA does not affect the calcium affinity of the protein. The far-UV CD spectra of apo-α-lactalbumin show that all oleic acid–bound forms of α-lactalbumin have an elevated content of α-helices. The various complexes possess similar cytotoxic activities against human epidermoid larynx carcinoma cells. Overall, the LA-OA-45 complex possesses physicochemical, structural, and cytotoxic properties closely resembling those of HAMLET. The fact that the HAMLET like complex can be formed in aqueous solution makes the process of its preparation more transparent and controllable, opening up opportunities for the formation of active complexes with specific properties.

Although near- and far-UV circular dichroism and fluorescence spectroscopy are not able to discriminate between HAMLET and apo-α-lactalbumin, H/D exchange experiments show clearly that they correspond to two distinct conformational states, with HAMLET incorporating a greater number of deuterium atoms than the apo and holo forms (Casbarra et al., 2004). Complementary proteolysis experiments revealed that HAMLET and apo-protein are both accessible to proteases in the β-domain but showed substantial differences in accessibility to proteases at specific sites. It was concluded that the conformational changes associated with the release of calcium from α-lactalbumin are not sufficient to induce the HAMLET conformation. Metal depletion might represent the first event to produce a partial unfolding in the β-domain of α-lactalbumin, but more unfolding is needed to generate the active conformation, HAMLET, very likely allowing the protein to bind the C18:1 fatty acid moiety. It was suggested that the aliphatic chain of oleic acid intercalates with the β-sheet, displaces the central β-strand, and exposes Tyr50 to protease action. This hypothesis was further supported by the greater accessibility to proteases exhibited by Asp37 in HAMLET as compared with the apo-protein. The side chain of this residue makes contact with the peptide bonds of Thr 38 and Gln 39. In the HAMLET complex, the carboxyl group of the fatty acid might replace the β-carboxyl group of Asp37, thus exposing this residue to proteolytic enzymes, and possibly making stabilizing interactions with the same residues or with either the ε-NH_2 or the α-NH_2 of Lys1. Nuclear Overhauser enhancement spectroscopy shows that the two ends of the fatty acid are in close proximity and close to the double bond, indicating that the oleic acid is bound to HAMLET in a compact conformation (Fast et al., 2005). The data show further that HAMLET is partly unfolded under physiological conditions.

The stability of HAMLET toward thermal and urea denaturation was measured by Fast et al. (2005) using circular dichroism and fluorescence spectroscopy as well as differential scanning calorimetry. Under all conditions examined, HAMLET appears to have the same or lower stability than intact α-lactalbumin. The largest difference is seen for thermal denaturation of the apo forms, whose transition midpoint is 15°C lower for apo-HAMLET than for intact apo-α-lactalbumin. The difference becomes progressively smaller as the calcium concentration increases. Denaturation of HAMLET is irreversible. It was suggested that HAMLET is a kinetic trap: it has lower stability than monomeric α-lactalbumin; its denaturation is irreversible and HAMLET is lost after denaturation; formation of HAMLET requires a specific conversion protocol.

Svensson et al. (2003) asked whether unfolding of α-lactalbumin is sufficient to induce cell death. The bovine α-lactalbumin mutant D87A is unable to bind calcium. It is inactive in the apoptosis assay but can readily be converted to a HAMLET-like complex in the presence of oleic acid. BAMLET (bovine α-lactalbumin made lethal to tumor cells) and D87A BAMLET complexes are both able to kill tumor cells. It was concluded that partial unfolding of α-lactalbumin is necessary but not sufficient to trigger cell death, and that the activity of HAMLET is defined by both the protein and the lipid cofactor. Furthermore, a functional calcium binding site is not required for conversion of α-lactalbumin to the active complex or to cause cell death. This suggests that the lipid cofactor stabilizes the altered fold without interferring with the calcium binding site.

It was found that HAMLET limits the progression of human glioblastomas in a xenograft model and removes skin papillomas in patients (Svensson et al., 2000; Svanborg et al., 2003; Gustafsson et al., 2004). Treatment with topical α-lactalbumin/oleic acid has a beneficial and lasting effect on skin papillomas. Intratumoral administration of HAMLET prolongs survival in a human glioblastoma (GBM) xenograft model, by selective induction of tumor cell apoptosis (Fischer et al., 2004). HAMLET causes apoptosis in vivo in the tumor but not in adjacent intact brain tissue or in nontransformed human astrocytes. No toxic side effects were observed. The broad antitumor activity includes >40 different lymphomas and carcinomas, and apoptosis is independent of p53 or bcl-2.

To explore the molecular mechanisms underlying interaction of the specific complexes of human α-lactalbumin with oleic acid, HAMLET and LA-OA-17, with the cell membrane, their interactions with small unilamellar dipalmitoylphosphatidylcholine (DPPC) vesicles and electroexcitable plasma membrane of internodal native and perfused cells of the green alga *Chara corallina* have been studied (Zherelova et al., 2009). The fractionation (Sephadex G-200) of mixtures of the complexes with the vesicles shows that oleic acid binding increases the affinity of α-lactalbumin to DPPC vesicles. Calcium association decreases protein affinity to the vesicles; the effect is less pronounced for LA-OA-17. Voltage clamp studies show that LA-OA-17, HAMLET, and their constituents produce different modifying effects on the plasmalemmal ionic channels of *C. corallina* cells. The irreversible binding of the complexes to the plasmalemma is accompanied by changes in the activation–inactivation kinetics of developing integral transmembrane currents, suppression of the Ca^{2+} current and calcium-activated Cl^- current, and increase in the nonspecific K^+ leakage currents. The latter reflects the development of nonselective permeability of the plasma membrane. The HAMLET-induced effects on the plasmalemmal currents are less pronounced and potentiated by LA-OA-17. The control experiments with oleic acid and intact α-lactalbumin show their qualitatively different and much less pronounced effects on transmembrane ionic currents.

In tumor cells HAMLET enters the cytoplasm, translocates to the perinuclear area, and enters the nucleus, where it accumulates. In the cytoplasm, HAMLET targets ribosomes and activates caspases. Histones were identified as targets of HAMLET among nuclear constituents (Düringer et al., 2003). HAMLET was found to bind histone H3 strongly and to a lesser extent, histones H4 and H2B. In vivo in tumor cells, HAMLET colocalizes with histones and perturbs the chromatin structure; HAMLET associates with chromatin in an insoluble nuclear fraction resistant to salt extraction. In vitro, HAMLET binds strongly to histones and impairs their deposition on DNA. It was concluded that HAMLET interacts with histones and chromatin in tumor cell nuclei and suggested that this interaction locks the cells into the death pathway by disrupting chromatin organization irreversibly.

Monomeric α-lactalbumin in the absence of fatty acids is also able to bind efficiently to the primary target of HAMLET, histone HIII, regardless of calcium content (Permyakov et al., 2004). Thus, the modification of α-lactalbumin by oleic acid is not required for binding to histones. It was suggested that interaction of negatively charged α-lactalbumin with the basic histone stabilizes apo-α-lactalbumin and destabilizes calci-α-lactalbumin, due to compensation for excess negative charge in the calcium binding loop of α-lactalbumin by positively charged residues of the histone. Spectrofluorimetric curves of titration of α-lactalbumin by histone H3 are well approximated by a scheme of cooperative binding of four α-lactalbumin molecules per molecule of histone, with an equilibrium dissociation constant of 1.0 μM. Such a stoichiometry of binding implies that the binding process is not site specific with respect to histone and is probably driven by electrostatic interactions. Coincubation of positively charged polyamino acids (poly-Lys and poly-Arg used as models of histones) with α-lactalbumin results in effects similar to those caused by histone HIII, confirming the electrostatic nature of the α-lactalbumin–histone interaction. In all cases that were studied, the binding was accompanied by aggregation. α-Lactalbumin can be used as a basis for the design of antitumor agents, acting through disorganization of chromatin structure due its interaction with histones.

Association of α-lactalbumin with histones or with poly-Lys(Arg) essentially changes its properties. Complex formation depends strongly on ionic strength, confirming its electrostatic nature (Permyakov et al., 2005b). Experiments with the polyamino acids of various molecular masses demonstrated a direct proportionality between the number of α-lactalbumin molecules bound per poly-Lys(Arg) and the surface area of the polyamino acid random coil. The binding of the polyamino acids to calci-α-lactalbumin decreases its thermal stability to the level of its free apo form and decreases calcium affinity by four orders of magnitude. The conformational state of α-lactalbumin in a complex with poly-Lys(Arg), termed α-lactalbumin modified by polyamino acid (LAMPA), differs from all other α-lactalbumin states characterized to date, representing an apolike (molten globulelike) state with substantially decreased affinity for calcium ion. The requirement for efficient conversion of α-lactalbumin to the LAMPA state is a poly-Lys(Arg) chain consisting of several tens of amino acid residues.

Svensson et al. (2000) suggested that the conditions under which α-lactalbumin is lethal to tumors resemble those in the stomach of the nursing child and that the binding of fatty acids to α-lactalbumin (Cawthern et al., 1997) could be important for lowering the incidence of cancer in breast-fed children by purging of tumor cells from the gut of the neonate. Low pH releases calcium from α-lactalbumin and activates lipases that hydrolyze free fatty acids from triglycerides of milk.

α-lactalbumin can alter its biological function depending on its conformation. Human milk has a high concentration of whey protein (70% of total protein). Of this, α-lactalbumin accounts for 41% of the whey and 28% of the total protein. Only 3% of the protein in bovine milk is α-lactalbumin. α-Lactalbumin passes from the mammary gland through the gastrointestinal tract of a baby. In the mammary gland α-lactalbumin functions as a specifier protein in lactose synthesis. The water content of milk is controlled through lactose, and this is very important for successful breast feeding. Stinnakre et al. (1999) created α-lactalbumin-deficient mice by disrupting the gene by homologous recombination in embryonic stem cells. Homozygous mutant mice were viable and fertile, but females could not feed their offspring. They produced highly viscous milk that pups appeared to be unable to remove from the mammary gland. This milk was rich in fat and protein and was devoid of α-lactalbumin and lactose.

The low pH in the stomach of a breast-fed child results in release of calcium from α-lactalbumin and precipitation of casein. The acid lipase hydrolyses triglycerides, and fatty acids are released. These events seem to create conditions for formation of the MAL state of α-lactalbumin. Breast feeding was proposed to protect both mother and child against cancer. MAL and HAMLET may have protective effects and may be naturally occurring surveillance molecules that purge unwanted cells from local tissues, thereby driving the intestinal mucosa toward maturity (Svensson et al., 2000). By inducing apoptosis, HAMLET and MAL may reduce the pool of potentially malignant cells that could serve as nuclei for future tumor development and explain the reduced frequency of cancer in breast-fed individuals.

Svensson et al. (2000) believe that there are obvious parallels between α-lactalbumin and the prion system. First, both proteins have multiple activities that depend on their folding state. Second, both proteins require a cofactor for the functional transition. Third, like the prion protein, α-lactalbumin changes its conformation to a disordered state but stays in this intermediate form rather than altering its secondary structure to a β-sheet-rich state.

13.2. CELL MATRIX PROTEINS

The extracellular matrix is a network built up from a variety of proteins and proteoglycans. Its interaction with cells is mediated by cell adhesion molecules. Most extracellular matrix proteins are mosaic multidomain proteins (reviewed by Maurer and Hohenester, 1997). Many of the interactions of these proteins involve calcium. The extracellular calcium binding domains of known structure can be divided into two groups: calcium can either be bound to a single domain or can mediate interactions between independently folded domains. Whether calcium serves only a structural function or whether it is involved in information transfer is not established; if some sort of signal transduction, there should be a meaningful change in calcium concentration.

13.2.1. BM-40 (Osteonectin or SPARC, Secreted Protein Acidic and Rich in Cysteine)

BM-40 is an EF-hand protein (Chapter 5); however, it functions in the extracellular environment as do other secreted glycoproteins. Nature seems unwilling to honor our categories. BM-40 might contribute to the organization of matrix in connective tissue as well as basement membranes. It is expressed abundantly in basement membranes and in capsules that surround a variety of organs and tissues. It is involved in the regulation of bone mineralization, tissue remodeling, and cell growth (Lane and Sage, 1994; reviewed by Bradshaw and Sage, 2001). BM-40 was first described as a major constituent of bovine and human bone (reviewed by Phan et al., 2007). However, it is also expressed in a variety of tissues during embryogenesis and repair. It is found only in specific organs of mouse embryo; whereas, in adult mice, BM-40 is limited to tissues undergoing normal remodeling, such as the gut or bone as well as injured tissues. BM-40 is present at a moderate level in human steroidogenic cells, chrondrocytes, placental trophoblast, vascular smooth muscle, and endothelial cells. It is strongly expressed in fibroblast, endothelial cells, epithelial cells, and macrophages in injured tissues. Furthermore, the increased level of BM-40 is also associated with increased capacity for invasion in vitro of prostate cancer, breast cancer, gastric cancer, glioblastoma, and malignant melanoma.

The main activities of BM-40 include modulating cells, extracellular matrix interactions, delaying cell cycle progression, inhibiting proliferation and angiogenesis, and regulating the expression of a number of growth factor and extracellular matrix proteins. These functions underlie the process of wound repair, which consists of inflammation, cell migration, proliferation, and angiogenesis.

BM-40 and peptides derived from proteolysis of BM-40 promote changes in cell shape, disrupt cell adhesion, inhibit cell cycle, regulate extracellular matrix, and modulate cell proliferation and migration (reviewed by Phan et al., 2007). They also influence cellular response through interaction with growth factors such as platelet-derived growth factor, vascular endothelial growth factor, basic fibroblast growth factor, and transforming growth factor.

The N-terminal domain, BM-40, binds several Ca^{2+} ions with low affinity [K_d(Ca) ~5 to 10 mM] (Maurer et al., 1992). It contains 15 Glu's and presumably adopts a defined tertiary structure only in the presence of a high concentration of calcium. Glu3 and Glu4 can act as amine acceptor sites in transglutaminase-catalyzed cross-linking modification (Hohenadl et al. 1995). Transglutaminases catalyzes a calcium-dependent transfer reaction between the γ-carboxamide of Gln's and various primary amines. BM-40 is a predominant glutaminyl substrate (amine acceptor) in the chondrocyte matrix and is coexpressed with transglutaminase c in maturing cartilage (Aeschlimann et al., 1993).

The first domain is followed by a domain homologous to follistatin and a C-terminal extracellular calcium binding domain (EC domain). The follistatinlike domain (FS, residues 53 to 137) is very elongated and consists of an N-terminal β-hairpin and a small core of mixed α/β structure; the EC domain is globular and almost entirely α-helical (Hohenester et al., 1996). The interface between the two domains is small and predominantly polar. The EC lobe of BM-40 has a compact, highly α-helical structure stabilized by two disulfide bonds (Figure 13.2) (Hohenester et al., 1996). This lobe (residues 138 to 286) contains a canonical pair of calcium binding EF-hands. The pair of the EF-hand calcium binding sites is elaborated by three additional helices that contain the collagen binding site of

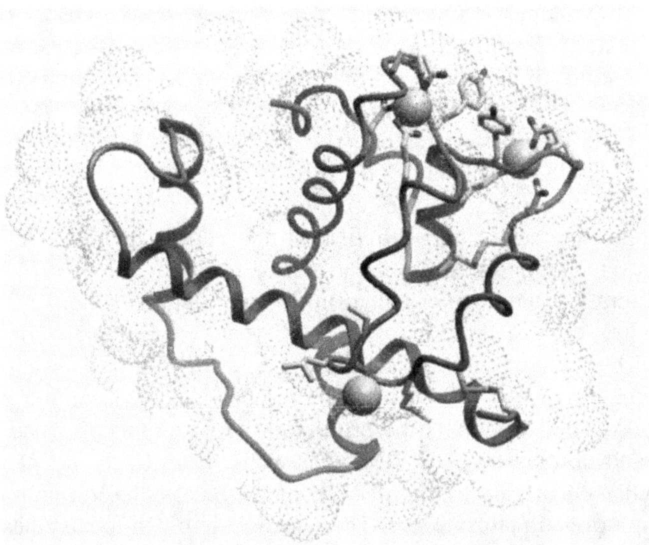

Figure 13.2. Structure of fragment (residues 136 to 286) of BM-40 (osteonectin) (PDB file 1SPA). The two EF-hands to the right are related by an approximate, local twofold axis. (From Hohenester et al., 1996.)

BM-40. One EF-hand of BM-40 is unusual: a single residue inserted into the EF-hand assumes a *cis*-peptide bond conformation. Another unusual feature of the BM-40 EC domain is the presence of an amphiphilic N-terminal α-helix that lies across the cleft formed by the helices of the EF-hand pair, just as do target peptides in calmodulin. An endogenous protease cleavage site is located at a Leu–Leu bond in position 197/198 in the α-helical region (Mann et al., 1987; Mayer et al., 1991).

Calcium binding to BM-40 is accompanied by a large increase in α-helix content and concomitant formation of binding sites for extracellular ligands such as collagen IV (Maurer et al., 1995). Since the affinity of the EF-hands for calcium ions is high, the EC domain should always be saturated with calcium in the extracellular space.

The follistatinlike domain of BM-40 contains two copper binding sites (Vernon and Sage, 1989). Copper binding to the sequence Lys119, Gly120, His121, Lys122 stimulates cell proliferation and angiogenesis (Funk and Sage, 1993). Glycosylation of BM-40 at Asn99 of the follistatinlike domain is a result of a posttranslational modification that alters the function of the protein. A difference in glycosylation between bone and platelet BM-40 has been claimed to account for their structural and functional heterogeneity (Kelm and Mann, 1991). Bone BM-40 has a high mannose-type oligosaccharide; whereas, platelet BM-40 has a complex oligosaccharide. Each form of BM-40 exhibits a different collagen binding specificity (Kelm and Mann, 1991).

Binding to cytokines is one of the major characteristic of BM-40. It binds the PDGF (platelet derived growth factor) dimers AB and BB, but not Å. This specific interaction of BM-40 with the B-chain prevents binding of PDGF to its receptors on fibroblasts (Raines et al., 1992). The affinity of BM-40 for this important growth factor could regulate the availability of PDGF dimers and thus affect the biological activity of PDGF.

BM-40 binds to types I, V, and VIII collagens (reviewed by Yan and Sage, 1999). Specific interaction between BM-40 and certain of the collagens could result in the remodeling of extracellular matrix.

Antiproliferation and counter adhesion are two major functions of BM-40 as defined in vitro; each of these involves a different signaling pathway (Motamed and Sage, 1998; reviewed by Yan and Sage, 1999). BM-40 is a potent cell cycle inhibitor that arrests cells in mid-G_1, independent of discernible changes in cell shape. BM-40 also inhibits the proliferation of endothelial, smooth muscle, mesangial, and fibroblastic cells. BM-40 is well known for its counter-adhesive function, achieved in part by the dissolution of focal adhesion complexes and reorganization of actin stress fibers. The antispreading and focal adhesion disassembly in bovine endothelial cells is mediated by BM-40 or a C-terminal peptide containing the calcium binding EF-hand and is believed to involve the tyrosine phosphorylation of focal adhesion-associated proteins (e.g., paxillin) (Motamed and Sage, 1998).

BM-40 has been shown to bind to the surface of endothelial cells. Scatchard analysis indicates 4.2×10^7 sites/cell and a K_i value of 2.4 nM. Iodinated proteins from plasma membranes were affinity chromatographed on a C-terminal BM-40 peptide (residues 254 to 273); several proteins with apparent masses ranging from 153 to 100 kDa (unreduced) or from 153 to 122 kDa (reduced) were eluted with the soluble peptide (Yost and Sage, 1993). These proteins represent candidates for a BM-40 receptor(s) that mediates the biological activity of this protein on endothelial cells. Characterization of cell surface receptors for BM-40 is critical to an understanding of the specific pathways mediated by this protein.

Four proteins have been described that contain the FS and EC domains, including the EF-hand. These proteins are the rat brain protein SC1 (I. G. Johnston et al., 1990) and its human homolog, hevin, isolated from high endothelial venules (Girard and Springer, 1995), the quail retina protein QR1 (Guermah et al., 1991), the TGF-β-induced protein TSC-36/glioma-secreted follistatin related protein FRP (Shibanuma et al., 1993), and the human testicular proteoglycan, testican (Alliel et al. 1993). Among these four, SC1 has the highest similarity to BM-40, with an identity between mouse BM-40 and mouse SC1 of 70% (Soderling et al., 1997).

BM-40 null mice display early cataractogenesis, a phenotype with 100% penetrance (Yan and Sage, 1999). Transmission electron microscopy of lens epithelial cells in BM-40 null mice shows an intrusion of cellular processes into the basement membrane of the lens capsule, whereas wild-type lens epithelial cells exhibit a precise border at the cell–matrix interface (Norose et al., 2000). Bradshaw and Sage (2001) have proposed that this phenotype reflects aberrant cell behavior or differentiation resulting from altered composition or structure of the basement membrane formed in the absence of BM-40.

BM-40 and its homolog, hevin (Sullivan and Sage, 2004) have deadhesive effects on cultured cells and have been characterized as antiproliferative factors in some cellular contexts (reviewed by Framson and Sage, 2004). Both proteins are produced at high levels in many types of cancers, especially by cells associated with tumor stroma and vasculature. BM-40 and hevin take part in the regulation of tumor cell growth, differentiation, and metastasis. BM-40 functions not only to modulate cell–cell and cell–matrix interactions, but its deadhesive and growth inhibitory properties in nontransformed cells have led to studies to assess its role in cancer (reviewed by Tai and Tang, 2008). It was found that in certain types of cancers, such as melanomas and gliomas, BM-40 is associated with a highly aggressive tumor phenotype, whereas in others, mainly ovarian neuroblastomas and colorectal cancers, BM-40 may function as a tumor suppressor. Recent studies have also demonstrated a role for BM-40 in sensitizing therapy-resistant cancers. Different studies, including expression array analysis, identified BM-40 as a marker of poor prognosis in different cancer types (reviewed by Podhajcer et al., 2008). Further evidence demonstrated that high BM-40 levels are often associated with the most aggressive and highly metastatic tumors.

13.2.2. Epidermal Growth Factor (EGF)–Like Domains

EGF-like domains are one of the most widely distributed modules in extracellular proteins; they have six conserved Cys's organized in a characteristic pattern (1–3, 2–4, and 5–6) and two small β-sheets. EGF-like domains are found in variety of proteins associated with biological functions such as blood coagulation and fibrinolysis, activation of complement, cell adhesion and signaling, neurite outgrowth, formation of neuromuscular junctions, and involvement in basement membranes and connective tissue microfibrils (reviewed by Maurer and Hohenester, 1997). A subset of EGF-like domains contains a bipartite consensus sequence with a D/N-X-D/N-E/Q motif preceding the first Cys and a X-D/N-X_n-Y/F-X motif between the third and fourth Cys's. This consensus was found to correlate with a calcium binding site in EGF-like domains (Handford et al., 1991).

The major constituents of extracellular microfibrils are fibrillin-1 and fibrillin-2, which are large, modular proteins composed of 54 domains, of which 43 are homologs of the calcium binding EGF-like domain. Mutations in fibrillin-1 and fibrillin-2 cause Marfan syndrome and the related disorders Shprintzen–Goldberg syndrome, *ectopia lentis*, and arachnodactyly. Downing et al. (1996) determined the solution structures of the EGF-modules 32 and 33 of fibrillin-1. The two domains are in a linear, rodlike arrangement and interact hydrophobically through a Tyr, contributed by the N-terminal domain, and an Ile and a Gly from the C-terminal domain. A Ca^{2+} ion is located in the domain interface and is coordinated by six oxygens provided by the C-terminal domain. A second, lower-affinity calcium binding site is located in the N-terminal EGF-like domain.

13.2.3. Cadherins

Cadherins (calcium-dependent adhesion molecules) are cell adhesion proteins involved in establishment and maintenance of intercellular connections and in controlling cell polarity and morphogenesis (reviewed by Ivanov et al., 2001). They are transmembrane proteins with large extracellular segments consisting of five repeated domains, each about 110 residues long, with their adhesive properties in the N-terminal domain. Cadherin-mediated cell adhesion depends on the presence of extracellular calcium (Takeichi, 1995).

Cadherins form a superfamily with at least six subfamilies, which can be distinguished on the basis of protein domain composition, genomic structure, and phylogenetic analysis of the protein sequences (reviewed by Foty and Steinberg, 2004). These subfamilies are: classic or type I cadherins, atypical or type II cadherins, desmocollins, desmogleins, protocadherins, and others. Classic cadherins all contain five cadherin ectodomains (ECs) of approximately 110 residues each. ECs 1 to 4 are all similar in sequence, while EC5 is less well conserved. Classic cadherins mediate homophilic adhesion between cells in a Ca^{2+}-dependent manner. All cadherins except T-cadherin (heart cadherin) contain a catenin binding domain.

Cadherins and cadherin-related molecules are defined by having an ectodomain in which at least two consecutive calcium binding ECs are present (reviewed by Hulpiau and van Roy, 2009). There are usually five or six domains. However, members of the Fat cadherin subfamily (*Drosophila fat* gene) have an extraordinarily large extracellular region, comprising 34 repeated domains, making them the largest cadherins (reviewed by Tanoue and Takeichi, 2005). Fat cadherin regulates cell proliferation and planar cell polarity. Additional modules in the ectodomains point to adaptive evolution. Despite the occurrence of several conserved motifs in subsets of cytoplasmic domains, these domains are even more diverse than ectodomains and probably evolved separately from the ectodomains (reviewed by Hulpiau and van Roy, 2009). They proposed a major branch in the cadherin family; these two groups were divided into eight subfamilies and into a cadherin-related major group, further subdivided into four subgroups and 11 subfamilies.

Epithelial cells express classical type I, E-cadherin, N-cadherin, and P-cadherin. The intracellular domains of classic cadherins interact with β-catenin, γ-catenin (also called plakoglobin), and p120ctn to assemble the cytoplasmic cell adhesion complex (CCC), which is essential for the formation of extracellular cell–cell adhesion (reviewed by Chan, 2006). β-Catenin and γ-catenin bind directly to α-catenin, which links the CCC to the actin cytoskeleton. The cadherins are responsible for homotypic cell–cell adhesion.

In humans there are more than 80 members of the cadherin superfamily. Sequencing the genome of *C. elegans* and *Drosophila* revealed the existence of 14 and 16 different genes, respectively.

The crystal structure of domains 1 and 2 of cadherin shows that both domains are folded independently and consist entirely of β-sheets (Figure 13.3) (Nagar

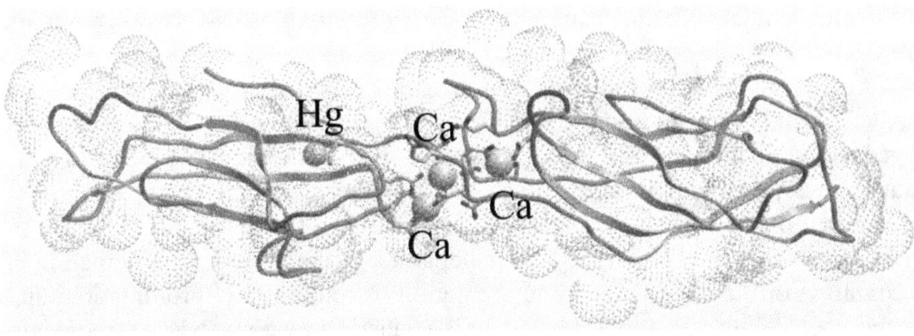

Figure 13.3. Structure of E-cadherin domains 1 and 2 in a complex with calcium (PDB file 1EDH). (From Nagar et al., 1996.)

et al., 1996). The structure of the ten residue interdomain linker is stabilized by three Ca^{2+} ions bound to a cluster of acidic residues contributed by both domains. The extracellular region of cadherins forms a rigid rodlike structure only in the presence of calcium.

In crystals, cadherin domains 1 and 2 form dimers in which both molecules are aligned in parallel. The dimer interface is formed by the calcium binding linker segments, suggesting that calcium is not only important for stabilizing the relative orientation of domains within one polypeptide chain but also plays a role in the quaternary structure of cadherin.

Classic cadherins are a family of calcium-dependent, homophilic cell–cell adhesion molecules (reviewed by S. C. Suzuki and Takeichi, 2008). They confer strong adhesiveness to animal cells when they are anchored to the actin cytoskeleton via their cytoplasmic binding partners, catenins. The cadherin/catenin adhesion system plays a key role in the morphogenesis and function of the vertebrate and invertebrate nervous systems. In early vertebrate development, cadherins are involved in multiple events of brain morphogenesis, including the formation and maintenance of the neuroepithelium–neurite extension, and migration of neuronal cells. In the invertebrate nervous system, cell–cell interaction mediated by classic cadherins plays an important role in wiring among neurons. For synaptogenesis, the cadherin–catenin system not only stabilizes cell–cell contacts at excitatory synapses but also assembles synaptic molecules at synaptic sites. Furthermore, this system is involved in synaptic plasticity. Additionally, recent findings have shown that N-cadherin may participate in synaptic physiology by regulating calcium influx through voltage-activated calcium currents (reviewed by Brusés, 2006). The diverse activities of N-cadherin stem from its ability to operate as both an adhesion molecule that links cytoskeletons across cell membranes and a ligand-activated homophilic receptor capable of initiating intracellular signaling. An important mechanism of cadherin signaling is the regulation of small Rho guanosine triphosphatase activity that affects cytoskeleton dynamics and calcium influx. Because both the regulation of cadherin adhesive activity and cadherin-mediated signaling are affected by the binding of molecules to the intracellular domain, changes in the composition of the N-cadherin complex are central to the regulation of cadherin-mediated functions.

T-cadherin (T-cad, H-cadherin, or cadherin-13) is an atypical member of the cadherin family, lacking transmembrane and cytosolic domains and possessing a glycosylphosphatidylinositol moiety that anchors T-cadherin to the plasma membrane (reviewed by Philippova et al., 2009). The structural characteristics of T-cadherin predict that it is unlikely to function as a true adhesion molecule in vivo. It was suggested that it may act rather as a signaling receptor participating in recognition of the environment and regulation of cell motility, proliferation, and phenotype. Cellular expression levels of T-cadherin in various tissues frequently correlate, negatively or positively, with the proliferative potential of the cells. Studies on the loss and gain of function demonstrate the ability of T-cadherin to modulate cell motility and growth.

The formation and maintenance of cadherin-mediated adhesions is a multistep process, and mechanisms have evolved to regulate each step (reviewed by Lilien et al., 2002). This suggests that switching the functional state plays an important role in development. One reason for the loss of cadherin-mediated adhesion, the "turn-off" of cadherin function, is the exit or "drop-out" of cells from neural and epithelial layers and their conversion to a motile phenotype. It was suggested that epithelial mesenchymal conversions may be initiated by signaling pathways that result in the loss of cadherin function. Tyrosine phosphorylation of β-catenin is one such mechanism. Enhanced phosphorylation of Tyr's on β-catenin is almost invariably associated with loss of the cadherin–actin connection concomitant with loss of adhesive function. Several tyrosine kinases and phosphatases have been shown to have the potential to alter the phosphorylation state of β-catenin and thus the function of cadherins. It was shown that nonreceptor tyrosine phosphatase PTP1B is crucial for maintenance of N-cadherin-mediated adhesions in embryonic neural retina cells.

Underexpression of the E-cadherin is found in gastric and other cancers and correlates with infiltrative and metastatic ability (Takeichi, 1993; reviewed by Chen et al., 2004). The loss of E-cadherin-mediated cell–cell adhesion appears to be a prerequisite for tumor cell invasion and metastasis formation (Birchmeier and Behrens, 1994). Reestablishing the functional cadherin complex (e.g., by forced expression of E-cadherin) results in a reversion from an invasive, mesenchymal to a benign, epithelial phenotype in cultured tumor cells. It is now being recognized increasingly that there is also a possible role of E-cadherin in modulating intracellular signaling and thus in promoting tumor growth. Cadherins control the balance between suppression and promotion of invasion by cancer cells (reviewed by Derycke and Bracke, 2004). E-cadherin functions as an invasion suppressor and is down-regulated in most carcinomas, whereas N-cadherin, as an invasion promoter, is frequently up-regulated. Expression of N-cadherin in epithelial cells induces changes in morphology to a fibroblastic phenotype, rendering the cells more motile and invasive. However, in some cancers, such as osteosarcoma, N-cadherin may behave as a tumor suppressor. N-cadherin can have multiple functions: promoting adhesion or induction of migration, dependent on the cellular context.

13.2.4. Pentraxins

Pentraxins are a family of highly conserved, multimeric pattern recognition proteins (reviewed by Deban et al., 2009). Pentraxins are divided into two groups: short and long. C-reactive protein, the first pattern recognition receptor identified, and serum amyloid P component (SAP) are short pentraxins produced in the liver. Long pentraxins, including the prototype PTX3, are expressed in a variety of cells and tissues, most notably dendritic cells and macrophages. Through interaction with several ligands, including selected pathogens and apoptotic cells, pentraxins play a role in complement activation, pathogen recognition, and apoptotic cell clearance. In addition, PTX3 is involved in the deposition of extracellular matrix and in female fertility.

SAP is the major DNA and chromatin binding glycoprotein of plasma (reviewed by Maurer and Hohenester, 1997). The interaction is strictly calcium dependent. The physiological role of SAP is not yet defined. SAP is localized in the glomerular basement membrane and in the peripheral, microfibrillar mantle of elastic fibers. SAP exhibits calcium-dependent binding to several ligands in vitro, including oligosaccharides with terminal *N*-acetylgalactosamine, 6-phosphate-mannose, glucuronic acid, and galactose.

In the crystal structure (Figure 13.4), five identical subunits of SAP are arranged in a flat ring with a substantial hole in the center (Emsley et al., 1994).

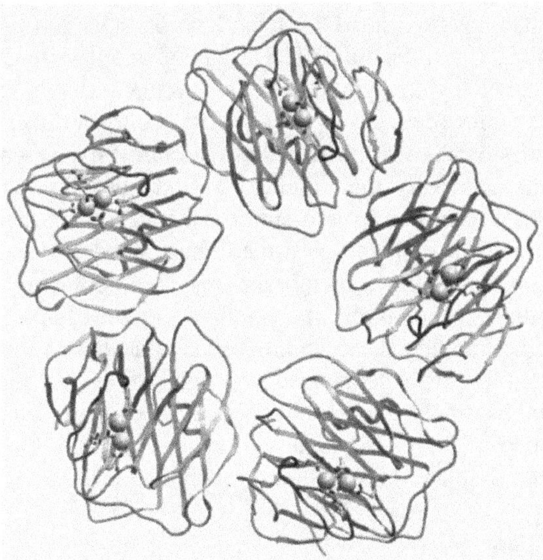

Figure 13.4. Structure of serum amyloid P component complexed with calcium (PDB file 1SAC). (From Emsley et al., 1994.)

The subunits are composed almost exclusively of β-sheets in a lectinlike fold. One face of the disk-shaped molecule contains the sites of interaction with ligands. An acidic functional group of large polyionic ligands bridge two Ca^{2+} ions; this accounts for the strict calcium dependence of all interactions.

13.3. BLOOD-CLOTTING PROTEINS

Most of the proteins in the blood-clotting cascade contain posttranslationally modified Glu's: 3-amino-1,1,3-propanetricarboxylic acid, commonly called 4-carboglutamic or γ-carboxyglutamic acid (Gla):

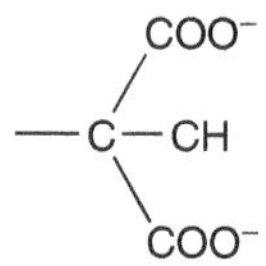

In the conversion of Glu to Gla, an extra carboxylate group is added to the γ-carbon of specific Glu's in the N-terminal Gla domain by a vitamin K dependent carboxylase. The carboxylase binds to a propeptide that is cleaved off before secretion of the proteins, but the signal that limits the carboxylation to Glu's within the first 40 residues has not been identified.

These side chains bind calcium mainly by bidentate chelation (Zell et al., 1985). The resulting Gla's have significantly higher affinity for Ca^{2+} ions than do the Glu's, and this ability to chelate calcium under physiological conditions is essential for biological activity.

Prothrombin, factor VII, factor IX, factor X, protein C, and protein Z contain Gla (van Eldik et al., 1982). Factors VII, IX, and X, and protein C share the same domain architecture. They are synthesized and circulate as zymogens, which can be activated to enzymes (denoted by an "a"; e.g., factor VIIa) by limited proteolysis. This conversion does not affect their calcium binding properties. These proteins are characterized by their ability to bind phospholipid in a calcium-dependent manner. These proteins are either substrates or, in their active forms, catalysts of reactions taking place on a membrane surface, usually in a macromolecular complex containing a membrane-associated cofactor. The accumulation of Gla-containing proteins on phospholipid surfaces requires a functional, calcium-loaded Gla domain. The calcium-loaded Gla domain mediates localization of the vitamin K–dependent plasma proteins to membrane surfaces; this is essential for the assembly into functional ternary enzyme–cofactor–substrate complexes. Deletion of the Gla domain or impaired calcium binding to the Gla domain results in a dramatic loss of biological function.

13.3.1. Prothrombin

Prothrombin, the best studied blood γ-carboxyglutamic acid containing protein, is 582 residues long; it has ten Gla's and three complex carbohydrate chains.

Prothrombin has two calcium binding sites with a K_d(Ca) value of 0.2 mM and several weaker sites (Nelsestuen, 1976; Bloom and Mann, 1978).

The bicarboxylate side chain of Gla can be considered as a substituted malonate. The K_d(Ca) value for malonate is 30 mM, which is more than 100 times higher than that found for the strong sites of prothrombin. To explain the higher affinity of prothrombin, it is assumed that the Ca^{2+} ions are bound not by one but by two or three carboxyglutamates (Martin, 1984).

Activation of prothrombin to thrombin requires removal of the N-terminal peptide consisting of 156 residues, known as fragment-1, which contains all ten Gla's and two of the carbohydrate chains of the intact prothrombin (Magnusson et al., 1975). Removal of most of the sugar residues from prothrombin does not alter its calcium affinity. Prothrombin binds to phospholipid dispersions in the presence of calcium (Nelsestuen, 1976); the activation of prothrombin to thrombin is accelerated in the calcium–protein–phospholipid complex. The phospholipid binding site in prothrombin is located in its fragment-1 region (Gitel et al., 1973). Most of the Gla-containing domain in fragment-1, which is homologous to the N-terminal regions of the other six blood proteins, cannot be identified in the electron density map of crystals of the apo structure (Figure 13.5) (Soriano-Garcia et al., 1989). In prothrombin crystals grown in the presence of calcium, the Gla domain possesses a well defined, folded structure. It consists of three turns of α-helix and five short β-strands organized into two β-sheets, whose β-strands are

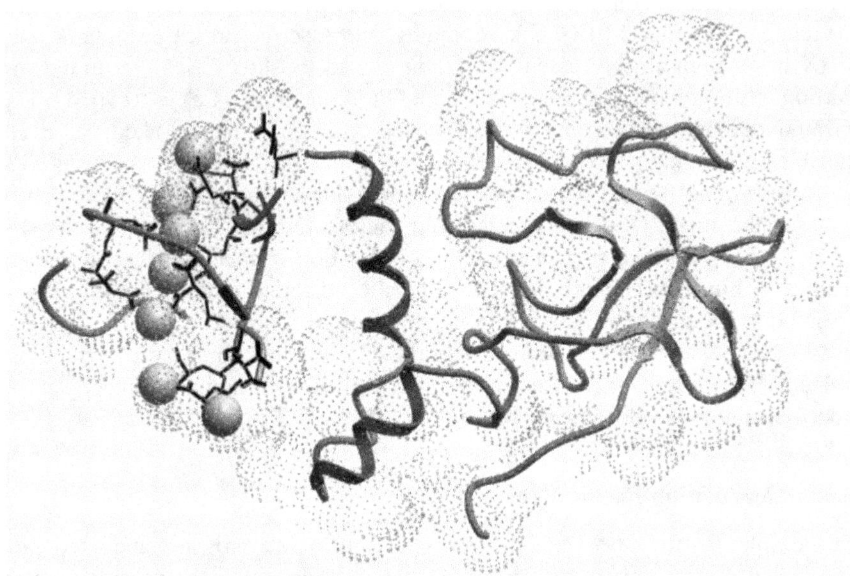

Figure 13.5. Structure of the Gla domain of prothrombin fragment 1 (PDB file 2 PF2). Seven Ca^{2+} ions are coordinated by the pairs of carboxyl groups of nine Gla's. The closest approach of two Ca^{2+} ions is 3.5 Å. (From Soriano-Garcia et al., 1992.)

antiparallel. A disulfide bond, Cys18–Cys23, in a short conserved loop in the Gla domain is located near the cluster of conserved aromatic residues. The Gla domain has a disklike shape, and all Gla residues are accessible to water. Moreover, paired Gla residues (7 and 8, 20 and 21, 26 and 27) are localized on the same side of the Gla domain, creating an electronegative surface, which may be responsible for the binding of phospholipids. The binding of calcium to this region anchors the protein to specific positions at the membrane surface; so that it can initiate the blood clotting process. Calcium binding results in the exposure of hydrophobic residues residing in the interior of the apo form and a concomitant internalization of charged Gla's, which are exposed in the apo form. Three hydrophobic residues, Leu6, Phe9, and Val10, close to the N-terminus were proposed to be involved in phospholipid binding (Sunnerhagen et al., 1995). The same calcium induced conformational transition was seen and a corresponding phospholipid interacting patch (Phe4, Leu5, and Val 8) was inferred when comparing the structures of the apo form of the Gla domain from factor X and the calci form of the Gla domain from prothrombin (Sunnerhagen et al., 1995). Although magnesium binds to the Gla domain in a manner similar to that of calcium, it does not make these proteins interact with membranes. This is presumably due to the inability of the amino teminal part of the magnesium loaded Gla domain to attain the conformation needed to properly present Phe4, Leu5, and Val 8.

In addition to the weak Gla-containing calcium binding sites, some blood coagulation proteins have stronger calcium binding sites. Huang et al. (1991) found such a site in the first epidermal growth factor (EGF) like domain of blood coagulation factor IX. It consists of three different regions: a Gla-containing domain, two tandemly connected EGF like domains, and a serine protease domain. Factor IX has a high-affinity calcium binding site in the first EGF like domain in addition to the calcium binding sites in the Gla-domain [K_d(Ca) ~100 μM]. The EGF like domain has two antiparallel β-sheets and two domains linked by Trp28. The calcium binding site involves residues on both strands of the β-sheets as well as the N-terminal region of the peptide. The solution structures of the calcium forms of the first EGF-like domain from factors IX and X show that Asp46, Gln49, and Asp63 take part in calcium coordination in factor X. The structural changes within the EGF-like domain induced by calcium binding are minor and restricted to the calcium ligands and neighboring residues. The calcium-induced interaction between the Gla domain and the EGF domains in factor IX requires 5 M urea for disruption; the compact structures of both domains are preserved under these nominal denaturing conditions (Medved et al., 1994).

13.3.2. Coagulation Factor VIIa

The complex formed between the procoagulant serine protease activated factor VII and the membrane protein tissue factor, exposed on the vascular lumen upon injury, triggers the initiation of blood clotting. Coagulation factor VIIa consists of 406 residues divided between a light chain (residues 1 to 152) and a heavy chain (153 to 406) (Furie and Furie, 1988). The light chain contains a domain rich in Gla residues (Gla domain), followed by two EGF like domains. The heavy chain has a serine protease fold, which includes the catalytic triad. The function of the serine protease domain is to convert the coagulation factors IX and X from zymogens to active enzymes. The Gla- and EGF like domains are required for the protein–protein and protein–membrane interactions that are responsible for formation of a complex between factor VII and the tissue factor on the cell membrane. Factor VII coordinates calcium in three distinct regions with different affinities; this calcium binding is needed for high-affinity binding of factor VII to the tissue factor. Using ^{1}H NMR spectroscopy, Muranyi et al. (1998) determined the solution structure of a synthetic N-terminal EGF-like domain of human coagulation factor VII (residues 45 to 85) in the absence of calcium. The peptide consists of two β-sheets, the first β-sheet comprising residues Ser60–Asp63 and Tyr68–Phe71, and a smaller one involving residues Phe76–Glu77 and Thr83–His84. The structural changes in the EGF domain upon calcium binding are minor and are concentrated near the calcium binding site.

13.4. OSTEOCALCIN

Although bone has long been recognized as a target for hormones influencing calcium and phosphorus homeostasis and bone structure, the skeleton itself produces at least two hormones, fibroblast growth factor 23 and osteocalcin (reviewed by Fukumoto and Martin, 2009). Fibroblast growth factor 23 is produced by osteocytes in bone and acts on the kidney to inhibit 1α-hydroxylation of vitamin D and promote phosphorus excretion. Mouse genetics studies revealed that the osteoblast product, osteocalcin, acts on pancreatic β-cells to enhance insulin production and on peripheral tissues to increase glucose utilization as a result of increased insulin sensitivity and to reduce visceral fat.

Osteocalcin is the most abundant noncollagenous protein in bone (reviewed by Hoang et al., 2003). In humans the osteocalcin gene is located on chromosome 1 and is regulated at the transcriptional level by 1,25-dihydroxyvitamin D_3. Osteocalcin is synthesized as 11-kDa pre-pro-osteocalcin of 98 amino acid residues. The molecule consists of three parts, a 23-residue signal peptide that is cleaved during translation, a 26-residue propeptide that targets the protein for γ-carboxylation and the 49-residue mature protein. Osteocalcin is one of the three known vitamin K–dependent proteins produced by osteoblasts. Vitamin K_1 or phylloquinone is an essential cofactor for posttranslational γ-carboxylation of osteocalcin. Porcine osteocalcin contains three Gla's at positions 17, 21, and 24.

Mature osteocalcin is largely unstructured in the absence of calcium and undergoes a transition to a folded state at physiological concentrations of calcium. The crystal structure of osteocalcin (Figure 13.6) (Hoang et al., 2003) shows three Ca^{2+} ions and a disulfide bridge between Cys23 and Cys29. Osteocalcin forms a tight globular structure comprised of three α-helices and a short extended β-strand. The three α-helices form a tightly packed core involving conserved hydrophobic residues Leu16, Leu32, Phe38, Ala41, Tyr42, Phe45, and Tyr46.

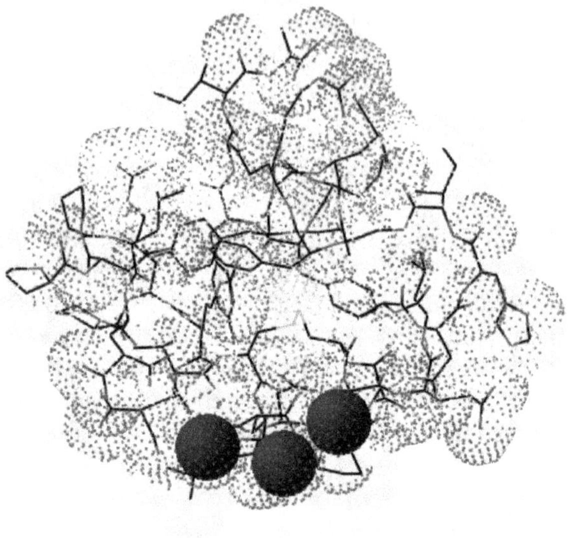

Figure 13.6. Structure of osteocalcin (PDB file 1Q8H). The compact structure is comprised of three α-helices and a short extended β-strand. Three Ca^{2+} ions are coordinated by the pairs of carboxyl groups of the side chains of Gla's. (From Hoang et al., 2003.)

All three Gla's implicated in hydroxyapatite binding are located on the same surface of helix 1 and, together with the conserved residue Asp30 from helix 2, coordinate five Ca^{2+} ions, which are sandwiched between two symmetry related osteocalcins and show both monodentate and bidentate modes of chelation with extensive bridging. In the osteocalcin crystal structure, the Ca^{2+} ions coordinated by Gla's have a periodic order reminiscent of a crystalline lattice. The coordination network of Ca–O atoms at the osteocalcin–hydroxyapatite interface closely mimics that in the hydroxyapatite crystal lattice.

Osteocalcin influences bone mineralization through its ability to bind to hydroxyapatite. Serum osteoclacin accounts for 10 to 40% of osteocalcin synthesized; it is not incorporated into the bone matrix (reviewed by Lee et al., 2000). In postmenopausal women, serum osteocalcin levels correlate with both the bone formation rate and the kinetically determined the calcium accretion rate but not with bone resorption. In osteomalacia, serum osteocalcin correlates with parameters of osteoid formation and is significantly elevated. Circulating osteocalcin is increased in hyperparathyroidism and decreased in hypoparathyroidism.

Since about 30% of the de novo synthesized osteocalcin is released to the bloodstream, it is widely used as a marker for bone formation (reviewed by Garnero, 2008). Circulating immunoreactive osteocalcin consists of several fractions, which may differ from each other with respect to size and calcium binding properties. Markers of bone turnover can be used to predict the rate of bone loss in postmenopausal women and can also be used to assess the risk of fractures.

13.5. CALCIUM BINDING LECTINS

Lectins have no known enzymic activity but exhibit numerous biological activities that are related to their ability to bind carbohydrates in the presence of divalent cations. They have specific binding sites for carbohydrates and thereby interact with specific cells, cell fractions, or glycoproteins. They are widely distributed; proposed functions include promotion of symbiosis, involvement in cell recognition, and organization of supramolecular structures (Barondes, 1981).

13.5.1. Concanavalin A

ConA is extracted from the jack bean, *Canavalia ensiformis*. It is the best characterized plant lectin. Carbohydrate binding serves as the initial step in a wide variety of biological functions, ranging from fertilization to viral infection. Although ConA, the legume lectins, pea lectin, and lentil lectin are very similar in structure, they have different binding affinities for saccharide derivatives of mannopyranoside vs. glucopyranoside configurations.

Each 25- to 30-kDa monomer, of 237 residues, can be visualized as an ellipsoid dome with a narrow, flat base (Figure 13.7) as seen in the crystal structure (Becker

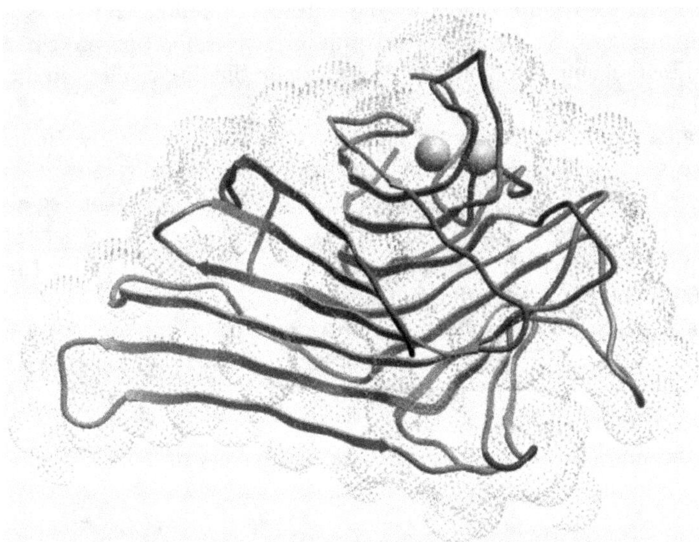

Figure 13.7. Structure of concanavalin A (PDB file 1DQ1). The two Ca^{2+} ions are 4.0 Å apart, near the apex of the ellipsoidal dome. Both must be bound for carbohydrate binding. (From Bouckaert et al., 2000.)

et al., 1975; Reeke et al., 1978). Each monomer has a carbohydrate binding site and two divalent cation binding sites. The two metal sites are closely spaced near the apex of the dome; they must both be occupied for carbohydrate binding. In solution, below pH 5.8 these plant lectins form dimers; above pH 6.9, tetramers.

In the presence of divalent cations, ConA preferentially binds α-D-mannopyranosyl, α-D-glucopyranosyl, and α-D-*N*-acetylglucosaminyl residues. The saccharide binding site is located within 10 to 14 Å of the two divalent cation binding sites. The residues that may participate in saccharide binding (14–16, 97–99, 168–169, 207–208, 224–228, and 235–237) are in close proximity in the three dimensional structure. The interaction force between a polymer tethered ConA and a similarly tethered mannose carbohydrate was measured as 47 ± 9 pN at a bond loading rate of ~ 10 nNs^{-1} (Ratto et al., 2004).

Most of the residues involved in metal binding in ConA are located in the N-terminus. The divalent cations are 4.5 Å apart and are bound in two adjoining octahedral sites that have a common edge. The ligands involved in the binding of divalent cations at the first site, called S1, are the side chains of Glu8, Asp10, Asp19, His24, and two water molecules. One of the water molecules appears to be hydrogen bonded to the carbonyl oxygen of Val32 and the side-chain hydroxyl of Ser34. The first site, S1, binds a Mn^{2+} ion (Liener, 1976) while the second site, S2, binds either calcium or magnesium (Shoham et al., 1973; Becker et al., 1975). Site S2 utilizes the side chains of Asp10, Asn14, and Asp19, the carbonyl oxygen of Tyr12, and two water molecules. The water molecules might be hydrogen bonded to the side-chain carboxyl of Asp208 and carbonyl of Arg228. Both of the side-chain oxygens of Asp10 are involved in metal binding, and Asp10 and Asp19 are shared by the two metal binding sites. It is assumed that the manganese, S1, site is mostly preformed in the apo-protein and that conformational changes in the protein occur upon binding calcium in the S2 site (Reeke et al., 1978).

There are other cation binding plant lectins that are functionally similar to and have some sequence similarity with ConA but differ in their subunit structures (Foriers et al., 1978, 1981; Richardson et al., 1978; Cunningham et al., 1979; Hemperly et al., 1979; Gebauer et al., 1981). These lectins are composed of two types of polypeptide chains termed α (molecular mass 6000) and β (molecular mass 15,000) and have a subunit structure of $(\alpha\beta)_2$.

Lei and Chang (2009) found that ConA from jack bean seeds has a potent antihepatoma effect. ConA binds to the mannose moiety on the cell membrane glycoprotein and is internalized preferentially to the mitochondria; autophagy is then triggered, which leads to cell death. ConA, as a T-cell mitogen, subsequently activates the immune response in the liver and results in the eradication of the tumor in a murine in situ hepatoma model.

13.5.2. Mannose Binding Proteins

Mannose binding proteins are members of the C-type lectin family, which are usually found in the serum and liver of mammals (reviewed by Zelensky and

Gready, 2005). The family of proteins that contain C-type lectinlike domains (CTLDs) is a large group of extracellular metazoan proteins with diverse functions. The CTLD structure has a characteristic double loop ("loop-in-a-loop") stabilized by two highly conserved disulfide bridges located at the bases of the loops, as well as a set of conserved hydrophobic and polar interactions. The second loop, called the *long loop region*, is structurally and evolutionarily flexible and is involved in calcium-dependent carbohydrate binding and interaction with other ligands.

Mannose binding proteins, pulmonary surfactant proteins, and macrophage mannose receptor bind to carbohydrate-rich cell surfaces of pathogens as a first step in an antibody-independent immune response. Selectins direct the movement of leukocytes to sites of inflammation, and hepatic lectins selectively remove serum glycoproteins by receptor-mediated endocytosis. The mannose binding proteins bind to various monosaccharides containing vicinal, equatorial hydroxyl groups such as those found at the 3 and 4 positions of mannose, *N*-acetylglucosamine, and fucose.

Carbohydrate recognition domains of all C-type lectins possess a similar core structure (Weis et al., 1992; Drickamer and Taylor, 1993; Kolatkar and Weis, 1996; Ng et al., 1996; Ng and Weis, 1997). The common arrangement of structural domains in C-type lectins includes a Cys-rich domain at the N-terminus followed by a collagenous domain, an oligomerization domain, and a C-terminal carbohydrate recognition domain. Individual monomers assemble into homotrimers, which further associate into larger oligomers (Figure 13.8). All C-type lectins

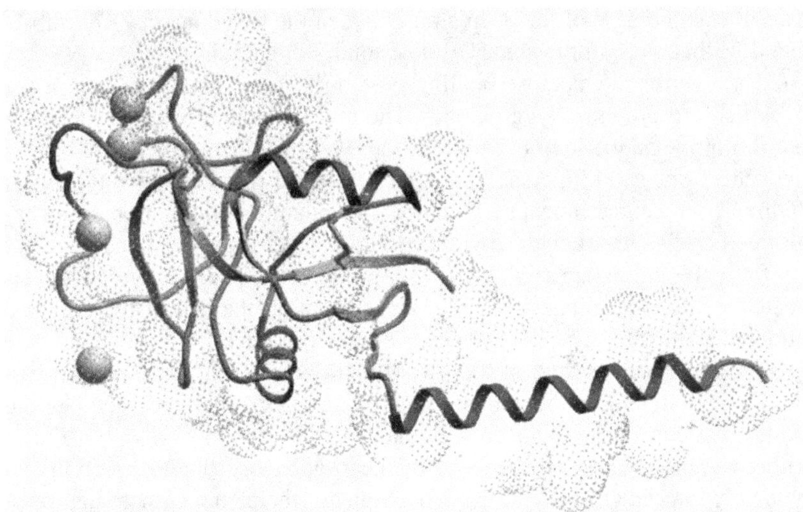

Figure 13.8. Structure of mannose binding protein with four Ca^{2+} ions (PDB file 1HUP). Individual monomers assemble into homotrimers, which further associate into larger oligomers. (From Sheriff et al., 1994.)

that bind carbohydrates share a calcium binding site (site 2 in mannose binding protein) at which carbohydrates interact directly with the bound calcium as well as with amino acids that serve as calcium ligands. Mannose binding proteins have three calcium binding sites per monomer. The amino acid residues that form the binding site are highly conserved, especially a tripeptide sequence of two calcium ligands flanking a *cis*-Pro. If the Pro is in the *trans* configuration, it is impossible to place the flanking amino acid residues in an orientation for calcium binding. The binding of calcium to mannose binding protein results in a conformational change that involves the *trans-cis* isomerization of a peptide bond (Ng and Weis, 1997).

13.5.3. D-Galactose Binding Protein

Sugar transport across the plasma membrane is one of the most important transport processes. D-Galactose binding protein is one of several bacterial sugar binding proteins (molecular masses 25 to 45 kDa) that are essential components of high affinity active transport systems for a large variety of carbohydrates, amino acids, and ions. They serve as initial receptors for the behavioral response of bacterial chemotaxis. There are significant similarities in both the tertiary structure and ligand binding properties of the L-arabinose-, sulfate-, D-galactose-, Leu/Ile/Val, and Leu-specific binding proteins. All five proteins are ellipsoidal in shape (axial ratios about 2 : 1) and are composed of two similar, yet distinct globular domains. The two domains in all of the structures are connected by three separate peptide segments. Although these interdomain connecting segments are widely separated in the sequence, they are spatially close together. These segments serve as the base of the deep cleft between the two domains. The substrate is bound in this cleft and becomes almost totally engulfed.

The Ca^{2+} ion in D-galactose binding protein is coordinated by seven ligands, all of which are protein oxygen atoms. The entire calcium binding site is composed of a nine-residue loop (134 to 142) and a dipeptide segment (204 to 205). Ligands are from Asp134, Asn136, Asp138, Gln140 (peptide carbonyl oxygen), and Gln142, as well as bidentate Glu205. The calcium coordination geometry is an almost perfect pentagonal bipyramid whose apices are occupied by oxygen atoms from the side chains of Asp134 and Gln 142. The calcium binding site is located in the C-terminal domain of the protein, at the end of the ellipsoidal, two domain structure. The nine residue loop is preceded by a reverse turn and followed by the first β-strand of the C-terminal domain. This reverse turn is preceded by the first helix in the domain. Leu204 and Glu205 are near the beginning of the third strand of the C-terminal domain.

Although the calcium binding site in the D-galactose binding protein is not immediately preceded and followed by helices, there are both structural and sequence similarities between this site and that of the canonical of EF-hand. In particular, the nine residue loop of the extracellular galactose binding protein adopts a conformation very similar to the first nine of the 12 residue EF-loop. Moreover, Glu205, providing bidentate ligands to the metal, occupies a position

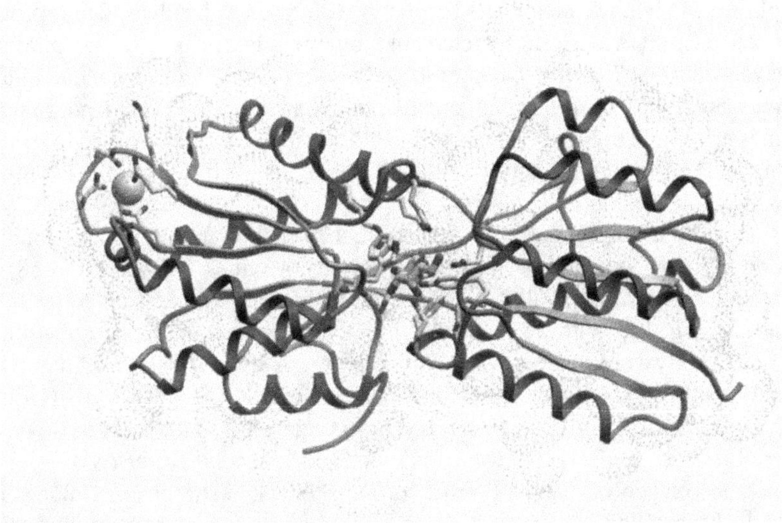

Figure 13.9. Structure of galactose binding protein with bound galactose and a Ca^{2+} ion (PDB file 1GCA). (From Zou et al., 1993.)

equivalent to the (near) invariant Glu at coordinating position 12 (−Z) of the EF-loop. Additionally, the backbone atoms of residue 204 partly superimpose on those of the residue at position 11 of the EF-loop (Figure 13.9) (Vyas et al., 1987; Quiocho et al., 1987).

13.6. CALCIUM BINDING HYDROLYTIC ENZYMES

Many enzymes that catalyze the hydrolysis of ester, phosphodiester, and peptide bonds bind calcium (reviewed by van Eldik et al., 1982). Crystallographic structures are available for phospholipase A_2, staphylococcal nuclease, thermolysin, trypsin, chymotrypsin, calpain, lipase, and others. The binding of calcium to these enzymes might serve three functions. First, calcium can stabilize an intermediate in the active site, as in phospholipase A_2 and staphylococcal nuclease. Second, calcium can stabilize the enzyme at high temperatures, as in thermolysin. Third, calcium can take part in the activation of a zymogen, as in trypsin, phospholipase A_2, and calpain.

13.6.1. Phospholipase A_2

Phospholipase A_2 catalyzes the specific hydrolysis of the fatty acid ester bonds at the C2 position of 1,2-diacyl-*sn*-phosphoglycerides. The family includes 15 groups comprising four main types, including the secreted phospholipase A_2 (sPLA2), cytosolic phospholipase A_2 (cPLA2), calcium-independent (iPLA2), and platelet-activating factor (PAF) acetylhydrolase/oxidized lipid lipoprotein-associated (Lp)PLA2. Both secreted and intracellular forms of phospholipase A_2 have been described in Chapter 12. Phospholipases A_2's from mammalian pancreas as well as from bee and snake venoms are secreted. These homologous enzymes are single polypeptides of 118 to 129 residues; those involved in calcium binding and in forming the active site are highly conserved (Dijkstra et al., 1981b, 1984).

^1H and ^{15}N NMR spectra (Dekker et al., 1991) show that porcine pancreatic phospholipase A_2 in solution has three α-helices, residues 1 to 13, 40 to 58, and 90 to 109; and an antiparallel β-sheets for residues 74 to 85 and for residues 25 and 26 together with 115 and 116. Calcium is an obligatory cofactor for interfacial catalysis by secreted phospholipase A_2. A Ca^{2+} ion is coordinated near the active site by six ligands that form an octahedron (Figure 13.10) (Dijkstra et al., 1981). The ligands are a carboxylate oxygen of Asp49, carbonyl oxygens of Tyr28, Glu30, and Glu32, water oxygen, and a second water oxygen or the carboxylate oxygen of Asp49 (Verheij et al., 1980; Dijkstra et al., 1981b). This relatively high-affinity calcium binding site is part of the catalytic site of the enzyme. The binding of calcium affects His48, which plays an essential role in catalysis but is not directly involved in calcium binding. The active site of the

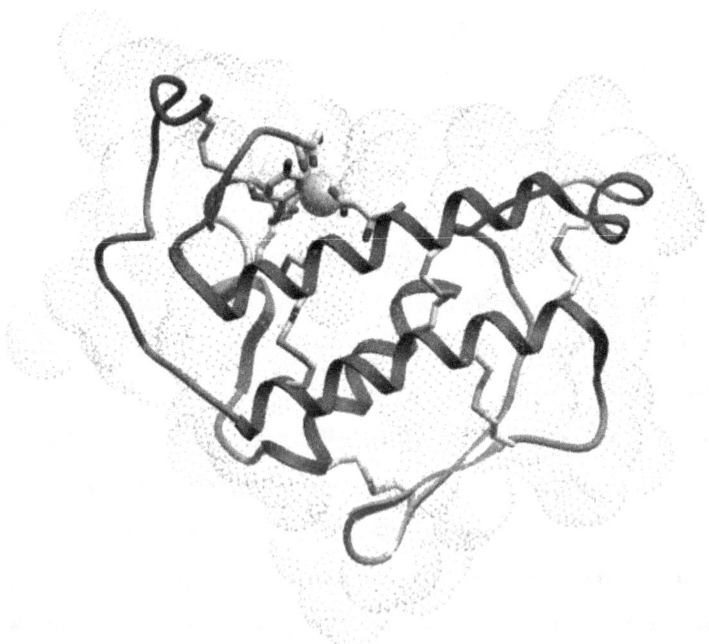

Figure 13.10. Structure of pancreatic phospholipase A2 (PDB file 1G4I) note the six disulfide bonds. (From Steiner et al., 2001.)

enzyme is composed of Phe5, Ile9, Phe22, Ala102, Ala103, Phe106, Cys29, and Cys45. The mechanism for the hydrolysis of phospholipids by phospholipase A_2 is similar to that for the hydrolysis of peptide bonds by serine proteases, in which calcium stabilizes the tetrahedral intermediate.

A low affinity calcium binding site in phospholipase A_2 is postulated to be involved in binding of micelles and is located near the N-terminus. Phospholipase A_2 has a ring of amino acids surrounding the entrance to the active site; this ring has been proposed to be involved in binding micelles. This region of the enzyme includes the N-terminal tripeptide and Tyr69, as well as Asp6, Leu19, Leu20, Asp23, Asp24, Leu31, and Thr70. The interaction of porcine pancreatic phospholipase A_2 with dodecylphosphocholine micelles changes conformation of the N-terminal part of the molecule, the end of the surface loop at residues Tyr69 and Thr70, the active site residue His48, and also the calcium binding loop (residues 28 to 32) (Peters et al., 1992).

Atherosclerosis has been regarded as a lipid driven disease; however, it is evident that it also involves the simultaneous and combined effect of inflammatory and immunological pathways. The secreted phospholipase, sPLA2, and the lipoprotein-associated phospholipase A_2, Lp-PLA2, have been associated with atherogenesis and its complications (reviewed by Garcia-Garcia and Serruys, 2009). These two enzymes produce biologically active metabolites that are involved in several phases of the atherosclerosis process. The increased levels of these two phospholipases in animals have been related to an increase in complex coronary lesions and an increase in major cardiovascular clinical events, respectively. Therefore, inhibition of these enzymes has become the focus of research in the past decade. Novel pharmacological inhibitors of those enzymes, such as darapladib and varespladib, emerge as promising therapeutical options for treating patients with coronary artery disease.

13.6.2. Staphylococcal Nuclease

Staphylococcal nuclease is an extracellular phosphodiesterase of *Staphylococcus aureus*. It is 149 residues long (molecular mass 16.8 kDa). It binds one Ca^{2+} ion (Figure 13.11) (Cotton et al., 1979) and requires millimolar concentrations of calcium for activity. The calcium site is octahedral and consists of oxygen-containing side chains of amino acids and peptide carbonyl groups of Asp19, Asp21, Asp40, Thr41, and Glu43. Its calcium coordination is similar to that in the high-affinity calcium binding site in phospholipase A_2. Calcium also stabilizes an intermediate during the hydrolysis of the phosphodiester bond by a mechanism, similar to that proposed for phospholipase A_2. The active sites of both enzymes are near the calcium site.

The proposed mechanism of action of staphylococcal nuclease involves nucleophilic attack on phosphorus by a water molecule, which is bound to Glu43, in line with the 5′-$CH_2O(H)$ leaving group (Cotton et al., 1979). The carboxylate of Glu43 promotes this attack by acting as a general base for the abstraction of a proton from the attacking water molecule. Nucleophilic attack is facilitated

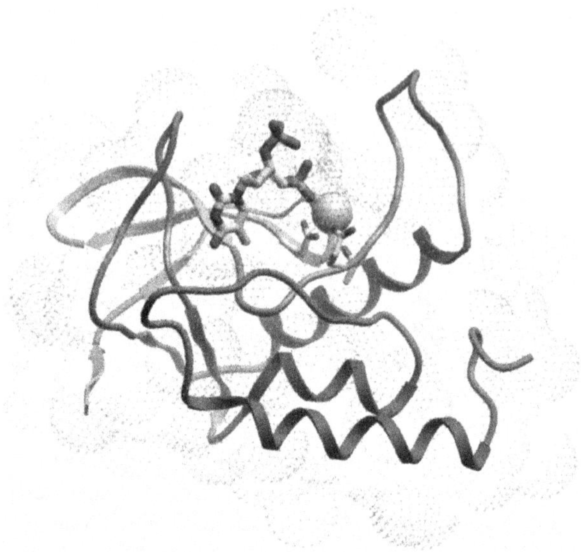

Figure 13.11. Structure of *staphylococcal* nuclease with bound pdTp inhibitor and a Ca^{2+} ion (PDB file 1SNC). (From Loll and Lattman, 1989.)

further by polarization of the phosphodiester by an ionic interaction between the bound Ca^{2+} ion and a phosphate oxygen atom and by four hydrogen bonds to phosphate oxygen atoms from guanidinium ions of Arg35 and Arg87. These interactions may also catalyze the reaction by lowering the energy of a trigonal bipyramidal transition state. The hydrolysis of nucleic acid substrate proceeds by cleavage of the 5′-P–O bond to yield a free 5′-hydroxyl group and a terminal, 3′-phosphate monoester group.

13.6.3. Thermolysin

Thermolysinlike proteases are a group of metalloendopeptidases which contain one catalytic Zn^{2+} ion and two to four Ca^{2+} ions that are important for stability. The thermolysin family is characterized by the amino acid sequence HEXXH, which includes two of the three zinc ligands and the catalytic Glu.

The N-terminal domain of thermolysin consists mainly of β-strands; whereas, the C-terminal domain is mainly α-helical (Figure 13.12) (Holmes and Matthews, 1982). Ca^{2+} ions 1 and 2 (Ca1,2) are found in the C-terminal domain in the double calcium binding site, close to the active site zinc. Ca3 is located at the surface in the N-terminal domain, and Ca4 is bound by a surface ω-loop (McPhalen et al., 1991) in the C-terminal domain. Removal of calcium by chelators results in a partially unfolded, flexible molecule and in rapid autolytic degradation.

Calcium binding isotherms were determined for thermolysin in the pH range 5.6 to 10.5 and in the temperature range 5 to 45°C (Buchanan et al., 1986). At

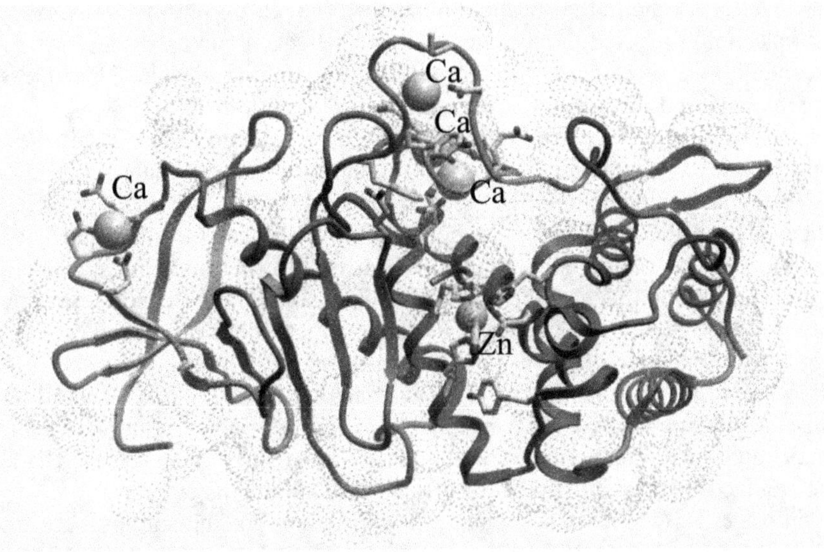

Figure 13.12. Structure of thermolysin (PDB file 1GXW). One Zn^{2+} and four Ca^{2+} ions are bound to the fully active form. (From Gaucher et al., 2002.)

least two of the four sites bind calcium with positive cooperativity and independently of the other two sites. Thermodynamic parameters obtained from a van't Hoff analysis (Chapter 5) indicate that calcium binding to both cooperative and independent sites is an entropy-driven process. At pH 7.0, $\Delta H_1 = 90.4$ kJ mol^{-1}; $\Delta H_2 = 97.5$ kJ mol^{-1}; $\Delta S_1 = 456$ J K^{-1} mol^{-1}; $\Delta S_2 = 262$ J K^{-1} mol^{-1}. Analysis of the pH dependence of the calcium binding constants indicates that the binding of four protons at the cooperative site and one to two protons at the independent sites modulates the calcium affinity.

Calcium sites 3 and 4 are more important than the double calcium site for thermal stability of thermolysinlike proteases. On the other hand, thermolysin is inactive without the double calcium site being occupied (Veltman et al., 1998). Removal of calcium causes a sharp decrease in thermostability and decreases the values of the activation enthalpy (ΔH^*) and entropy (ΔS^*) for heat inactivation.

13.6.4. Lipases

Lipases are versalite enzymes that have been isolated from a variety of eukaryotic and prokaryotic organisms. They hydrolyze the ester bonds in long chain triacylglycerols. More than 20 lipase structures have been determined. Although the overall sequence similarity of lipases is low and molecular masses vary from 20 to 60 kDa, all lipases share a comparable tertiary structure, which is known as the α/β *hydrolase fold* (Winkler et al., 1990). The region of highest conservation is the active site, which contains a classical Ser–His–Asp catalytic triad. The catalytic reaction involves a nucleophilic attack by the active site Ser on the cleavable ester bond in the substrate molecule. The reaction results in the release of an alcohol group and the formation of an acyl-enzyme intermediate, which is susceptible to a second nucleophilic attack by a water molecule. This releases the fatty acid and regenerates active enzyme.

For several lipases this reaction is enhanced by calcium (Noble et al., 1993). For example, *P. glutae* lipase consists of three domains (Figure 13.13). The largest domain contains a six stranded β-sheet flanked on one face by three α-helices and on the other by two α-helices. The β-sheet is mostly parallel, with exception of one of the edge strands, which is formed from a parallel and an antiparallel section. Each of the other domains is formed by secondary structural elements contiguous in sequence. The first of these domains (residues 118 to 165) consists of three α-helices, and the other one (214 to 271) consists of four α-helices and a β-hairpin. The catalytic residues are Ser87, Asp263, and His285. Three of the calcium ligands (carboxylate of Asp287 and carbonyl oxygens of Gln291 and Val295) are only a few residues downstream of the catalytic His285. It is assumed that calcium activates lipase by stabilizing its structure.

Simons et al. (1999) found a high-affinity calcium binding site in lipase from *Staphylococcus hyicus* with a K_d(Ca) value of 55 μM. Asp354 and Asp357 are calcium ligands in this lipase. Although the D357K, D357N, and D357A mutants do not bind calcium at room temperature, they are nearly as active as the wild-type lipase in the presence of calcium. The residual activity of the apo-enzyme compared to the activity of the calci-enzyme varies from 65% at 10°C to nearly zero at 40°C. Binding of calcium is important for the structural stabilization of *staphylococcal* lipases.

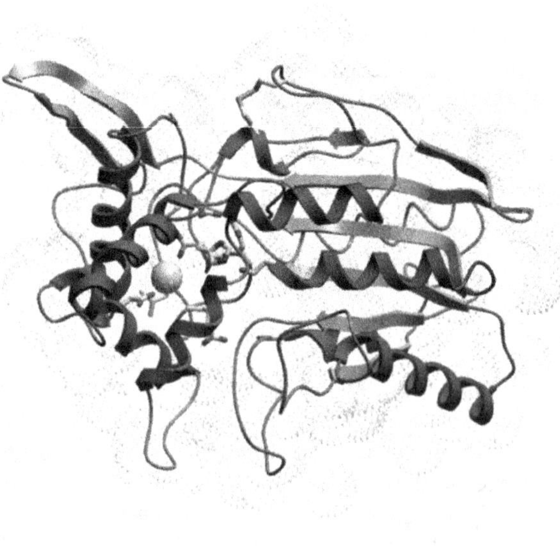

Figure 13.13. Ribbon model of *Pseudomonas glumae* lipase (PDB file 1TAH). (From Noble et al., 1993.)

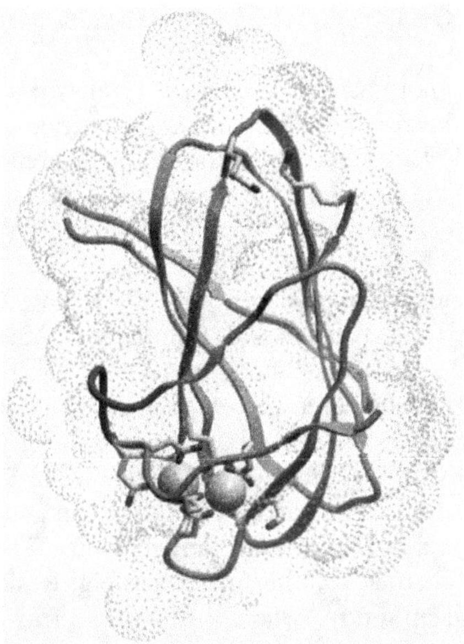

Figure 13.14. Crystal structure of *Clostridium histolyticum colg* collagenase, collagen binding domain 3B complexed with calcium (PDB file 2O8O). (From Wilson et al., 2003.)

13.6.5. Collagenase

Collagenase from the anaerobic spore-forming bacterium *Histotoxic clostridia* is responsible for the extensive tissue destruction of gas gangrene. The crystal structure (Figure 13.14) of the collagen binding domain of *Clostridium histolyticum* class I collagenase was determined in the absence and in the presence of calcium (Wilson et al., 2003). This domain binds two Ca^{2+} ions per molecule. The cations bind between two loops and have limited solvent access. Oxygen atoms of Glu899, Glu901, Asp927, and Asp930 side chains, main-chain carbonyl of Ser922, and one water molecule form a coordination, best described as a square antiprism. The water, Glu901, and Glu899 (bidentate) form one face, and Ser922, Asp927, and Asp930 (bidentate) form the other face. The peptide bond between residues Glu901 and Asn902 adopts a *cis*-peptide conformation that is stabilized by calcium chelation. This dual-ion structure is different from those in all other calcium binding proteins. The apparent K_d(Ca) of the domain is about 4 μM.

13.6.6. Other Hydrolytic Enzymes

Structural data are also available for several other hydrolytic enzymes. All of them possess rather low affinity for calcium, and their structures are different from the structures of other families of calcium binding proteins.

13.7. MISCELLANEOUS

Several extracellular polysaccharide-degrading enzymes from *Clostridium*, *Ruminococcus*, and *Bacteroides* have one or two regions, usually near the C-terminus, that resemble EF-hands. In contrast to the canonical helix–loop–helix structure, the first helix is generally absent in these proteins; hence homology is difficult to establish or structure to predict. D-galactose-binding protein was discussed in Section 13.5.3.

A similar calcium binding site is seen in Slt35, a proteolytic fragment of the membrane-bound lytic transglycosylase B from *E. coli* (Van Asselt and Dijkstra, 1999; Van Asselt et al., 1999). The structure reveals an ellipsoid molecule with three domains called α, β, and core; the core domain is sandwiched between the α and β domains. It resembles lysozyme, but it contains a single helix–loop–helix metal ion binding site that is similar to the EF-hand. The Slt35 loop consists of 15 residues instead of the 12 in a canonical EF-hand and coordinates a Ca^{2+} ion with only six ligands in a distorted octahedral geometry. The binding of Ca^{2+} to this site increases the thermal stability of the protein. Residues from the α and core domains form a deep groove where the substrate fragment GlcNAc can be bound.

Some gram-negative bacteria secrete proteins with hemolysin type calcium binding domain organized in a parallel β-roll structure (Baumann et al., 1993). An example of this structure is the structure of the alkaline protease of *Pseudomonas aeruginosa*, a zinc metalloprotease, which has been determined to a resolution of 1.64 Å. The molecule is elongated with overall dimensions of 90 × 35 × 25 Å; it has two distinct structural domains. The N-terminal domain is the proteolytic domain: it has an overall tertiary fold and active site zinc ligation similar to that of astacin, a metalloprotease isolated from a European freshwater crayfish. The C-terminal domain consists of a 21-strand β-sandwich. Within this domain there is a parallel β-roll structure in which successive β-strands are wound in a right-handed spiral and in which Ca^{2+} ions are bound within the turns between strands by a repeated GGXGXD sequence motif, a motif that is found in a diverse group of proteins secreted by gram-negative bacteria (X is any residue but preferably a small, hydrophilic residue). The proteins of this family have variable numbers of tandem repeats of the consensus sequence: from seven in *P. aeruginosa* alkaline protease to 40 in *B. pertussis* cyclolysin precursor. Calcium binding to *E. coli* hemolysin is absolutely required for cytotoxicity. The calcium binding regions have been suggested to be used as a receptor for binding of hemolysin to erythrocytes.

Phospholipase A is found in the outer membrane of *E. coli* (Snijder and Dijkstra, 2000). It is involved in transport across the bacterial outer membrane and has been implicated in bacterial virulence. It is calcium dependent and its activity is strictly regulated by reversible dimerization. In the monomeric form of this protein, a Ca^{2+} ion is bound between loops L3 and L4, with Asp149 and Asp184 as ligands. After dimerization, a second calcium binding site is formed at the dimer interface in the active site.

The acid shock protein, Asp24, may play a role during *Brucella abortus* infection in its host (Lin and Ficht, 1995). Expression of Asp24 is optimal at pH values below 4.0 and within the first 3 hours following a shift from pH 7.3 to 3.8. This corresponds directly with a period of optimal bacterial survival at a reduced pH and suggests an active role for this protein in resistance to such environments.

14

INTERACTIONS OF CALCIUM BINDING PROTEINS WITH OTHER METAL IONS

Magnesium and zinc, at physiological concentrations, sometimes compete for calcium binding sites. Given the precedents of known high affinity, high specificity calcium binding sites (e.g., in parvalbumin), one can safely assume that nature could have evolved high-specificity calcium binding sites in other proteins. These observations pose two challenges to the researcher: (1) to determine whether the competing cation is binding to the calcium binding site or to a different site, and (2) to determine whether the binding of the non-calcium cation exerts some sort of modulatory function on the calcium binding protein. These possible physiological functions, cytosolic or extracellular, might occur under normal or pathologic conditions.

14.1. MAGNESIUM

Magnesium is the second most abundant intracellular metal cation, exceeded only by potassium. Magnesium is essentially an intracellular ion: approximately 98% of nonskeletal magnesium is located inside the cell. The concentration of intracellular free Mg^{2+} ion is 0.3 to 0.6 mM (reviewed by Takaya et al., 2000). The concentration gradient of magnesium across the plasma membrane of most cell types is small. Magnesium homeostasis depends on the balance between intestinal absorption and renal excretion. Within cells, magnesium is involved in the synthesis of DNA and RNA as well as maintenance of their conformations. Magnesium forms complexes with phospholipids and affects membrane fluidity

Calcium Binding Proteins, By Eugene A. Permyakov and Robert H. Kretsinger

and permeability. Intracellular magnesium is also involved in the secretion of hormones, including insulin and prolactin.

Many of the actions of magnesium are due to its role as a cofactor for a wide range of enzymes. Magnesium activates nearly all of the enzymes involved in the metabolism of phosphorylated compounds, as well as many enzymes in the glycolytic and tricarboxylic acid pathways. The most common physiological role of magnesium is to bind ATP or other nucleotide triphosphates in the catalytic pocket of an enzyme. The function of magnesium binding to the phosphoryl groups of ATP in many cases appears to be activation of the phosphate ester to hydrolysis (reviewed by Maguire and Cowan, 2002). The relatively slow exchange rate of water in the hydration shell of the Mg^{2+} ion may also play a role, making it somewhat harder to lose a water molecule and therefore somewhat easier to allow formation of a structure with water in a particular orientation.

Magnesium competes for most calcium binding sites. Although calcium binding sites are subdivided into calcium, magnesium (they bind both Ca^{2+} and Mg^{2+} ions), and calcium specific (they do not bind Mg^{2+} ions) sites, this classification is conditional. As a rule, while the calcium–magnesium sites bind Ca^{2+} ions very strongly (dissociation constants 10^{-9} to 10^{-8} M), they bind magnesium ions four to five orders of magnitude more weakly. The calcium-specific sites bind calcium much less tightly (dissociation constants 10^{-7} to 10^{-6} M) and if they still bind magnesium four to five orders of magnitude more weakly, it would be rather difficult to measure such high dissociation constants. For this reason such binding sites are often called "calcium specific," even though the ratio of calcium to magnesium affinity differs little.

The binding of magnesium to calcium binding sites often causes conformational changes, similar to those induced by calcium binding. However, these magnesium induced changes are less pronounced and the magnesium loaded state of most calcium binding proteins can differ significantly from the calcium loaded state. For example, the thermal unfolding of pike α-parvalbumin is described by a model of two successive two-state transitions; this suggests the presence of two thermodynamic domains (S. E. Permyakov et al., 2009a). The first, low-temperature transition has midtemperature 90°C, while the second, high temperature transition occurs at 120°C. The intermediate state binds calcium and is characterized by a largely preserved secondary structure and increased solvent exposure of hydrophobic groups. Surprisingly, magnesium- and sodium-loaded forms of pike α-parvalbumin have a single two-state transition. This indicates that at least one of the calcium binding domains is essential for stabilization of the intermediate, and that calcium and magnesium loaded states of pike parvalbumin have different structures (Chapter 10).

Upon the binding of two Ca^{2+} ions, recoverin extrudes a myristoyl group (Chapter 11) and associates with lipid bilayers. Sr^{2+}, Ba^{2+}, Cd^{2+}, and Mn^{2+} ions cause similar large conformational changes and extrusion of the myristoyl group. In contrast, no such effect is induced by magnesium; even though recoverin binds two Mg^{2+} ions, it does not induce a large conformational change (Ozawa et al., 2000). Infrared spectroscopic analysis confirmed that Ca^{2+} ions, as well

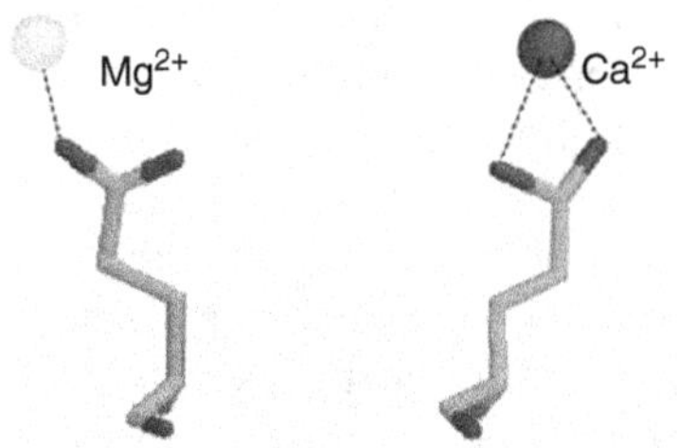

Figure 14.1. Bidentate coordination of a Ca^{2+} ion and monodentate coordination of a Mg^{2+} ion in parvalbumin. All three side-chain, dihedral angles—$\chi_1.\chi_2$, and χ_3—of Glu 12 (at the −Z position) change.

as Sr^{2+}, Ba^{2+}, Cd^{2+}, and Mn^{2+}, are coordinated to the carboxylate of Glu in a bidentate manner; whereas, the coordination of magnesium is monodentate with six-coordinate geometry as in parvalbumin (Chapter 10) (Figure 14.1). These data reveal that metal ions that are seven coordinate in the EF-hands induce a large conformational change in recoverin so that it extrudes the myristoyl group. Metal ions, such as Mg^{2+}, with six-coordinate geometry, do not cause the extrusion of the myristoyl group in recoverin.

Many calcium binding proteins of the cytosol are saturated by magnesium in resting cells when the concentration of free Ca^{2+} ions is low. For example, the two high affinity sites of parvalbumin seem to be filled with magnesium in resting skeletal muscle cells (reviewed by Permyakov et al., 2007). An increase in concentration of free calcium in response to muscle excitation results in an exchange of magnesium by calcium in parvalbumin. For cod parvalbumin the process takes about 1 s at 10°C; this allows Ca^{2+} ions to bind to troponin first and quickly switch on the muscle contraction (Permyakov et al., 1987a). Magnesium can play an important role in functioning of the calcium sensor proteins such as calmodulin. Magnesium is ineffective in replacing calcium as the stimulus for activation of calcium dependent enzymes; this is consistent with the inability of magnesium to cause exposure of either of the two hydrophobic cores of calmodulin. Nevertheless, magnesium increases the thermodynamic stability of calmodulin; under resting, non-stimulated conditions, magnesi-calmodulin is more stable than is apo-calmodulin (Martin et al., 2000). Apo-calmodulin binds skeletal muscle myosin light chain kinase and neuromodulin with dissociation constant 70 to 90 nM at low ionic strength. These affinities are virtually unchanged by addition of 5 mM magnesium, in marked contrast to the strong enhancement of binding induced by calcium. Under conditions of stimulation and increased $[Ca^{2+}]_{cyt}$, magnesium directs the initial target binding preferentially to the C-lobe of calmodulin, due to the relative affinities for calcium and magnesium of the two lobes (Martin et al., 2000). The higher affinity of magnesium for the N-lobe than the C-lobe increases the relative weakness in the calcium-dependent target affinity of the N-lobe. Numerical simulations show that this factor significantly enhances the specificity of the calcium-dependent target affinity for the C-lobe of calmodulin in a partially calcium-saturated intermediate. Magnesium thus amplifies the intrinsic differences of the two lobes in a target-specific manner.

Calcium influx in cells may be modulated by both intracellular and extracellular magnesium, and it has been proposed that magnesium can serve as a naturally occurring calcium antagonist (reviewed by Takaya et al., 2000). External calcium effectively prevents membrane permeability to magnesium (Nakayama and Tomita, 1991). Increased intracellular calcium can displace magnesium bound to common intracellular binding sites; this leads to an increase in intracellular free Mg^{2+} ion (Freudenrich et al., 1992). In cardiac myocytes $[Mg^{2+}]_{cyt}$ modulates calcium channel activity (Hartzell and White, 1989). An increase in $[Mg^{2+}]_{cyt}$ causes a marked increase in the rate and extent of voltage-dependent inactivation of calcium channels in cardiac myocytes.

14.2. SODIUM AND POTASSIUM

Sodium and potassium are necessary for organisms to maintain the balance of their fluid systems. They are also required for nerve and muscle functioning. The amount of sodium that a person consumes each day varies from person to person and from culture to culture; some people get as little as 2 g per day, some as much as 20 g per day.

The interaction between a Na^+ or K^+ ion and ligand is based solely on electrostatics and has no covalent bond character (reviewed by Page and Di Cera, 2006). Ligand exchange rates (k_{ex}) of Me^+ are very high and allow rapid association and dissociation. The charge density of any Me^+ is low. However, Me^+ coordination can play an important role in rate enhancement or allosteric regulation of an enzyme catalyzed reaction.

Li^+ and Na^+ ions are small and bind three or four water molecules (Van Geet, 1972) with reasonable affinity; this results in a larger apparent size in aqueous solution. K^+ favors six to eight water molecules coordinated more weakly (Mochiduki et al., 2001). Overall, sodium balance in all organisms, including humans, is driven largely through transport of water molecules rather than through regulated movement of sodium and stems from electronic differences of Me^+ ions. The primary hydration shells of Me^+ ions orient second-shell water molecules through electrostatic interactions and hydrogen bonding.

Alkali metal ions, Na^+ and K^+, are the most frequently occurring metal ions in biological systems. Na^+ ions are most abundant outside cells, while K^+ ions are present in higher concentration inside cells (Chapter 3). Concentration gradients of these ions across cell membranes provide energy sources for action potentials, for ATP synthesis, and for transport of various substances and other ions across membranes.

Na^+ and K^+, ions bind to calcium binding sites in proteins but with low affinities. Such interactions were found, for example, in parvalbumin (Permyakov et al., 1983, 1985; Henzl et al., 2000, 2003d) and in α-lactalbumin (E. A. Permyakov et al., 1985). Usually, proteins have about one order of magnitude higher affinity for sodium than for potassium. The equilibrium dissociation constants for sodium and potassium correspond to the physiological intracellular concentrations of these metal cations. This means that in cells, sodium and potassium can compete with calcium and magnesium for the same binding sites.

14.3. ZINC

Unlike other first row transition metals (Sc, Ti, V, Cr, Mn, Fe, Co, Ni, and Cu), the Zn^{2+} ion has a filled *d* orbital (d^{10}) and therefore does not participate in reduction–oxidation (redox) reactions but rather functions as a Lewis acid to accept a pair of electrons. This lack of redox activity makes Zn^{2+} a stable ion in a biological medium. Therefore, zinc is an ideal metal cofactor for reactions that require a redox stable ion to function as a Lewis acid catalyst, such as proteolysis and the hydration of carbon dioxide. Zinc serves as a cofactor for more than 300 various enzymes; these include representatives of all six classes of enzymes (reviewed by Auld, 2001).

Three primary types of zinc binding sites have been revealed from examination of the structures of zinc proteins: *structural, catalytic*, and *cocatalytic* (reviewed by Auld, 2001). The most common amino acids that provide ligands to these sites are His, Glu, Asp, and Cys. Structural zinc sites have four protein ligands, usually Cys, and no bound water molecules: for example, zinc finger domains (Chapter 9). Catalytic sites generally form complexes with water and any three of His, Glu, Asp, and Cys, with His being preferred; water is always present. Cocatalytic sites contain two or three Zn^{2+} ions in close proximity with two of the metals bridged by a side chain of His, Asp, or Glu and sometimes by water. Asp and His are preferred; Cys is not found in cocatalytic sites.

S100s usually have several zinc binding sites distinct from the calcium binding sites (Baudier and Gerard, 1983; Mani and Kay, 1983; Leung et al., 1986). The binding of zinc to calci-S100A results in a decrease in helical content. While calcium binding increases solvent accessibility of the single Trp (red shift emission maximum from 337 to 340 nm) and of one or two Tyr's and Phe's, the binding of zinc causes minor conformation changes affecting only Tyr's. The interaction of the α-subunit of S100A with trifluoroperazine depends on zinc (Ogoma et al., 1992).

S100A3, a rather unusual member of the S100s, has four Cys's and binds four Zn^{2+} ions per monomer with $[Zn^{2+}]_{0.5}$ about 11 μM and a Hill coefficient of 1.4 at physiological ionic strength (Föhr et al., 1995).

Brodersen et al. (1998, 1999) determined the crystal structures of the human S100A7 (psoriasin) dimer in the calci form with and without bound zinc (Chapter 9). They found that the two zinc binding sites are formed at the interface of the two monomers and that the Zn^{2+} ion is coordinated by His17, Asp24, His86, and His90, in tetrahedral configuration; two ligands belong to one monomer and two ligands belong to the other. The removal of zinc results in reorganization of the neighboring, empty distorted EF-hand loop, making it more similar to the calcium-loaded EF-hand. Zinc binding causes a compression of the protein structure near the zinc binding sites and substrate binding cleft and essential changes in the protein hydrophobic core (i.e., the binding of zinc induces not only local structural changes near the zinc binding site, but changes the conformation of the entire protein).

α-Lactalbumin has several zinc binding sites distinct from its calcium binding site (Murakami and Berliner, 1983). The binding of zinc to calci-α-lactalbumin induces the formation of an apolike conformation. According to the EPR data of Murakami and Berliner (1983), the dissociation constant of zinc for calci-α-lactalbumin is 5.6×10^{-6} M. Permyakov et al. (1991b) showed that the intrinsic protein fluorescence of α-lactalbumin at room temperatures is not sensitive to the binding of zinc to the strong sites; however, zinc binding shifts the thermal denaturation transition to lower temperatures. The results obtained by fluorescence and calorimetry methods showed that the zinc binding sites of α-lactalbumin are filled sequentially. The binding of zinc to the strong site(s) with dissociation constant 2×10^{-6} M, as evaluated by bis-ANS fluorescence, causes conformational changes. This zinc binding does not affect the environment of Trp's but does increase rates of hydrolysis by trypsin and chymotrypsin, decreases protein affinity to bis-ANS, and increases protein affinity for UDP-Gal. Infrared spectroscopy (Prestrelski et al., 1991) indicates that the binding of zinc to α-lactalbumin does not induce any major changes in main chain conformation.

One of the zinc binding sites in bovine α-lactalbumin is located in the "cleft" region (active site in the homologous lysozyme) (Ren et al., 1993). In the crystal structure of human α-lactalbumin a Zn^{2+} ion is also coordinated by Glu49 and Glu116 of a neighboring protein. Ren et al., (1993) suggested that this is the strong zinc binding site. The distance between this strong zinc binding site and the calcium binding site, measured by fluorescence energy transfer between Co(II) as energy acceptor and Tb(III) as energy donor, is 14 to 18 Å (Permyakov and Berliner, 1994). This agrees well with the 17.5 Å found in the crystal structure. However, further studies of α-lactalbumin mutants showed that the strong zinc binding site is located in the N-terminal part of the molecule. Substitution of Glu1 by Met results in disappearance of strong zinc binding in α-lactalbumin (Permyakov et al., 2000b). The substitution of Glu49 by Ala does not affect zinc binding. The putative zinc binding site includes Glu1, Glu7, Glu11, and Asp37. The distance between the strong calcium binding site and this putative N-terminal zinc binding site is about 14 Å; this is in a good agreement with the data on fluorescence energy transfer (E. A. Permyakov and Berliner, 1994). Some of the weak zinc binding sites in α-lactalbumin contain His's (E. A. Permyakov et al., 1988b; Veprintsev et al., 1996). These data show that α-lactalbumin has one relatively strong zinc binding site and several weak zinc binding sites.

The binding of zinc to α-lactalbumin can play a functional role. α-Lactalbumin is a component of the lactose synthase system, which consists of galactosyltransferase and α-lactalbumin. β-4-Galactosyltransferase catalyzes the transfer of the galactosyl group from UDP-galactose to N-acetylglucosamine on glycoproteins. In the presence of α-lactalbumin, galactosyltransferase catalyzes the coupling of galactose to glucose to yield lactose. Zinc binding to α-lactalbumin affects lactose synthase (Permyakov et al., 1993). Zinc induces a decrease in both apparent Michaelis constant K_m(app) and V_{max} for manganese. This results in an apparent increase followed by a decrease, in lactose synthase activity at manganese concentrations below saturation of the first manganese binding site of galactosyltransferase. Increasing zinc also decreases K_m(app) and V_{max} for both glucose and UDP-galactose in the lactose synthase reaction with either Ca^{2+}- or apo-α-lactalbumin, further suggesting interactions between zinc-α-lactalbumin and the lactose synthase complex, presumably mediated by a zinc-induced conformational change upon binding to α-lactalbumin.

Zn^{2+} and Cu^{2+} ions also bind to parvalbumins (Permyakov et al., 1988a, 1992). Calci-parvalbumin from cod binds one Cu^{2+} ion per molecule; this is reflected in changes in the quenching properties of the environment of the single Trp. Zinc titration of cupric calci-parvalbumin reverses the quenching effects almost completely; Cu^{2+} and Zn^{2+} ions compete for the same binding site. The effective dissociation constant of copper for cod parvalbumin is 2×10^{-6} M. The dissociation constant of zinc, evaluated from the copper–zinc competition experiments, is 5×10^{-5} M.

The copper dissociation constant for apo-parvalbumin is 10^{-6} M; it is more weakly bound than is calcium. Copper binding to parvalbumin decreases accessibility of the Cys's to 5,5′-dithiobis(2-nitrobenzoic acid) and increases its affinity for the fluorescent probe bis-ANS. Experiments on fluorescence energy transfer from Tb^{3+} bound to one of the EF-loops to cobalt, located in the zinc binding site, indicated an interion distance of about 10 Å.

There is also competition between copper and zinc for a binding site(s) in calmodulin (Permyakov et al., 1988a). Effective dissociation constants of copper and zinc for calmodulin are 10^{-4} and 10^{-3} M, respectively. Warren et al. (2007) determined the 1.3-Å crystal structure of the zinc bound N-terminal lobe (Figure 14.2). The first Zn^{2+} ion is tetrahedrally coordinated by Asp22 and Asp24 of EF-loop 1, which coordinates the Ca^{2+} ion in calci-calmodulin, and by two cacodylate molecules from the crystallization buffer. Also there is a highly ordered water molecule in site 1; its protons are involved in hydrogen bonds to Asp20 and Thr26, which are involved in calcium coordination in the first EF-loop. The second Zn^{2+} ion is also tetrahedrally coordinated by Asp64 (-X of EF-loop 2) and Glu67 (−Z) as well as by Glu7 and Glu11 of a symmetry-related molecule. Thus, three of five residues of the second EF-loop normally involved in calcium binding are not liganded to the second Zn^{2+} ion. The structure of the N-loop with Zn^{2+} ions bound to both EF-loops resembles the structure of apo-calmodulin. Structural comparisons of apo-calmodulin with calci-calmodulin and with cross-linked calci-calmodulin suggest that the zinc-bound N-loop reveals an intermediate step in the binding of the Ca^{2+} ion in the first EF-hand.

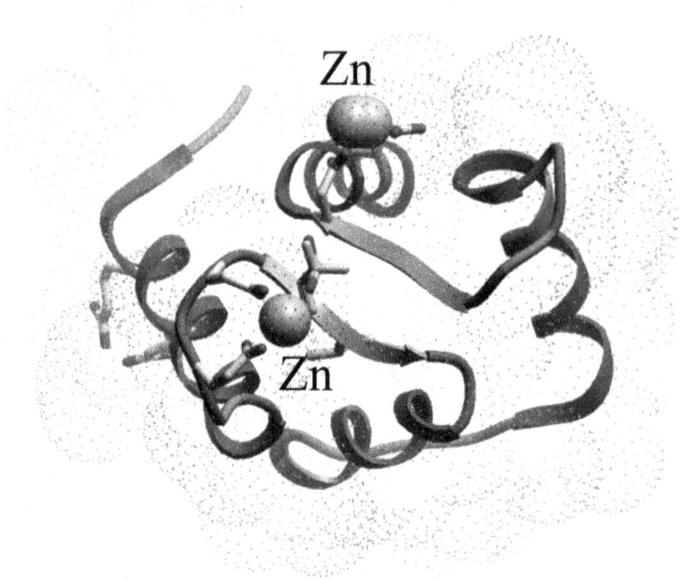

Figure 14.2. Structure of the N-terminal lobe of calmodulin with two Zn^{2+} ions bound (PDB file 2PQ3). The Zn^{2+} ion in EF-hand 1 is near the Ca^{2+} site but is not coordinated by Glu 12 (−Z). The Zn^{2+} ion in EF-hand 2 shows high anisotropy and is modeled as a double site, which is about 4 Å from the Ca^{2+} site; it is coordinated by Glu 12 (−Z) and by Asp 9 (−X), but not by the other residues involved in calcium coordination. (From Warren et al., 2007.)

14.4. TRACE METALS

Heavy metals such as mercury, lead, and cadmium have long been recognized as toxic. These cations can interact with calcium and zinc binding sites in calcium binding proteins; this influences their functions.

14.4.1. Mercury

Although Hg^{2+} ions bind to numerous nucleophilic groups on molecules, they have a greater predilection to bind to reduced sulfur atoms, especially those on endogenous thiol containing molecules, such as glutathione, cysteine, homocysteine, *N*-acetylcysteine, metallothionein, and other proteins with free thiolates; these include calcium binding proteins. The dissociation constant for mercury bonding to thiolate anions is of the order of 10^{-15} to 10^{-20} M. In contrast, the affinity of mercury for oxygen or nitrogen ligands (e.g., carbonyl or amino groups) is much weaker. For example, mercury binds to the strong zinc binding sites in α-lactalbumin with a dissociation constant of 10^{-4} M and also to some weaker sites (Veprintsev et al., 1996).

14.4.2. Lead

Lead is one of the metals that have the most damaging effects on human health. It can enter the human body through uptake of food, water, and air. Lead and calcium compete for calcium channels and calcium pumps of the plasma membrane. Lead interacts with many calcium-modulated proteins, such as calmodulin, protein kinase C, and Ca^{2+}-dependent K^+ channels (reviewed by Simons, 1993). The effects of lead on neurotransmission may be relevant to lead-induced human neuropathy and encephalopathy.

The interaction of lead with proteins is the fundamental mechanism by which lead exerts toxicity (reviewed by Goering, 1993). It interacts with enzyme functional groups and with high-affinity metal binding sites of various proteins, such as metallothioneins. Pb^{2+} ions bind to α-lactalbumin (Veprintsev et al., 1996). Pb^{2+} ions bind to the strong calcium binding site ($K_d \approx 5 \times 10^{-7}$ M), and also to the strong zinc binding site ($K_d = 10^{-5}$ M) as well as to weak secondary sites (probably containing His) with apparent dissociation constant 10^{-4} M; this is accompanied by strong protein aggregation. Lead exposure in vivo is able to displace and mimic the action of calcium on calmodulin; this may constitute one mechanism of lead neurotoxicity (Sandhir and Gill, 1994). Many factors influence lead protein interactions; these include binding affinity and specificity, protein folding, rates of protein synthesis and degradation, as well as intracellular localization.

Low concentrations of Pb^{2+} ions disrupt neurotransmission by causing aberrant augmentation of spontaneous transmitter release and suppression of evoked release (Suszkiw, 2004). These effects result from high-affinity interactions of lead with voltage-gated calcium channels as well as with calcium binding proteins, which regulate the synaptic vesicle mobilization, docking, and exocytosis. Augmentation of spontaneous release may involve stimulation of vesicle mobilization consequent to lead activation of CaMKII-dependent phosphorylation of synapsin I and/or stimulation of asynchronous exocytosis via direct lead activation of the putative exocytotic calcium sensor, synaptotagmin I. Synergistic stimulation of phospholipase C and DAG/Pb^{2+}-dependent activation of protein kinase C may enhance the secretagogue effects of lead by increasing the metal sensitivity of exocytosis and/or modulating calcium channel activity. In contrast to intracellularly mediated actions of lead, resulting in augmentation of spontaneous release, the inhibition of evoked transmitter release by lead is largely attributable to extracellular block of the voltage-gated calcium channels (Suszkiw, 2004).

14.4.3. Cadmium

Diets low in iron result in increased absorption of cadmium and lead, suggesting common molecular mechanisms of cadmium and lead transport. Cd^{2+} ions are taken up through calcium channels of the plasma membrane and cadmium is accumulated intracellularly due to its binding to cytoplasmic and nuclear proteins.

The intestinal transporter for nonheme iron, divalent metal transporter 1 (DMT1), mediates the transport of lead and calcium as well (reviewed by Bressler et al., 2004). In animals, cadmium induces calcium release from internal stores and activates calcium-signaling pathways (Waisberg et al., 2003). Cadmium in plants causes both the perturbation of the intracellular calcium level and interferes with calcium signaling by substituting calcium in calmodulin regulation (reviewed by Ghelis et al., 2000; Perfus-Barbeoch et al., 2002; Deckert, 2005).

14.4.4. Strontium

Strontium and calcium metabolism are remarkably similar (reviewed by Cohen-Solal, 2002; Nielsen, 2004). They are both absorbed in the gastrointestinal tract, concentrated in the bone, and excreted primarily in the urine. They seem to share a common transport system in renal tubules. Contraction of rat portal vein induced by noradrenaline is lost when calcium is replaced with strontium, which can enter through Ca^{2+} channels and be released from the sarcoplasmatic reticulum. Administered strontium is almost exclusively deposited in bone; it is incorporated by exchange with calcium. Strontium, as well as sodium and lead, can be substituted in the calcium positions of apatite. Radioactive strontium (^{85}Sr) has been used as a tracer for calcium in kinetic studies. ^{85}Sr and ^{44}Ca both have strong bone binding properties. In contrast to Ba and Ra, which are excreted primarily in the feces, Ca and Sr are largely excreted by the kidneys after intravenous injection.

Strontium binds to calcium binding sites in proteins. For example, strontium binds to the EF-hand calcium binding sites in calmodulin (Lepsík and Field, 2007). Bovine α-lactalbumin and bovine serum albumin also interact with strontium (Sandier et al., 1999).

Strontium ranelate and distrontium 5-[bis(2-oxido-2-oxoethyl)amino]-4-cyano-3-(2-oxido-2-oxoethyl)thiophene-2-carboxylate decrease bone resorption and increase bone formation in animals, resulting in increased bone mass (reviewed by Marie, 2005). The rather unique action of strontium ranelate as an osteoblast agonist (Chapter 9) could be explained by activation of a Ca-sensing receptor (Marie et al., 2001). Such receptors have been identified in different cells of the body including osteoblasts. It seems that strontium is an agonist of the calcium-sensing receptor in bone, although it binds with somewhat lower affinity than does calcium. Strontium ranelate has been approved for treatment of osteoporosis in Europe.

14.4.5. Lanthanides

The ionic radii of trivalent lanthanides range from 0.848 Å (Lu) to 1.034 Å (Ce), the values being relatively higher than those in other elements with the same oxidation number. The ionic radii of cerium, praseodymium, neodymium, and gadolinium are similar to that of calcium (0.99 Å). This similarity of ionic radius and higher charge density of lanthanide ions determine their high affinities for the calcium binding sites and physiologic effects of soluble lanthanide salts.

Lanthanides bind to the calcium binding sites in proteins and easily displace calcium from these sites. Pidcock and Moore (2001) compared crystal structures of calcium and lanthanide containing proteins and coordination complexes. They compared structural features of the metal binding sites, including coordination numbers of the metal ions, the identity of their ligands, the denticity of carboxylate ligands, and the types of secondary structures from which the ligands are derived. The average coordination number from protein ligands of a lanthanide ion in an intrinsic calcium binding site is 7.2, but for the adventitious sites it is only 4.4. A detailed comparison of the well defined Ca^{2+} and lanthanide ion binding sites suggests that a reduction of hydrogen bonding associated with the ligating residues of the binding sites containing lanthanide ions may be a response to the additional positive charge of the lanthanide ion.

The paramagnetism of lanthanide ions offers good opportunities for fast determinations of the coordination structures of protein–lanthanide complexes by nuclear magnetic resonance spectroscopy. The combination of pseudocontact shifts induced by a specifically bound lanthanide and prior knowledge of the tertiary structure of the protein can be used to achieve (1) rapid assignments of NMR spectra, (2) structure determinations of protein–protein complexes, and (3) identification of the binding mode of low-molecular-weight compounds in complexes with proteins (reviewed by Pintacuda et al., 2007). Dynamic measurements by NMR, also with the aid of calcium substitution with paramagnetic lanthanides, give information on the intrinsic amplitude of the conformational degrees of freedom sampled by the various EF-hand proteins, as well as information on the time scales of the motions (reviewed by Capozzi et al., 2006). NMR of lanthanide derivatives is especially valuable in capturing long-time-scale motions.

15

INTERACTIONS OF CALCIUM BINDING PROTEINS WITH OTHER PROTEINS AND WITH MEMBRANES

Most calcium binding proteins interact with other proteins in vivo. Many of these interactions have been studied in vitro. Two sorts of model systems have complemented these studies: the interactions of peptides representing the regions of contact with the calcium binding protein, and the interactions of parvalbumins, which do not naturally bind with these targets, with target peptides.

Many calcium binding proteins interact with membranes, either as integral membrane proteins or as extrinsic membrane proteins. Several studies address the questions of which membrane components, protein or lipid, are involved in these interactions.

One can anticipate that ever more studies will be directed toward these interactions in vivo. This will necessitate the development of specific antibodies, labels such as those used for fluorescent resonance energy transfer (FRET) spectoscopy, and mechanisms for precise control and measurement of calcium concentrations in microdomains of the cytosol.

15.1. INTERACTIONS WITH OTHER PROTEINS

15.1.1. Calmodulin and Targets

Calmodulin forms tight, calcium-dependent complexes with natural and synthetic peptides ($K_d < 10^{-7}$ M); these mimic the interactions of calmodulin with its natural targets involved in information transduction. Often, these peptides form

Calcium Binding Proteins, By Eugene A. Permyakov and Robert H. Kretsinger

basic amphiphilic α-helices with hydrophobic and positively charged residues on opposite faces; these are considered to be the main structural feature governing their calmodulin binding properties (Malencik and Anderson, 1983a,b; Cox, 1988). Exceptions will be described.

Melittin is a 26-residue acidic peptide from bee venom. Its binding to calmodulin, K_d 10^{-9} to 10^{-8} M, occurs only in the presence of calcium (Maulet and Cox, 1983) and was studied extensively as a model. The linker helix, which connects N- and C-terminal lobes, takes part in the binding of melittin (Caday and Steiner, 1986). The fragment 78–148, comprising a part of the linker helix, binds both melittin and mastoparan, $K_d = 2 \times 10^{-7}$ M, in a calcium-dependent manner. Fragment 1–77 has a much lower affinity for mastoparan ($K_d = 2 \times 10^{-5}$ M) (Sanyal et al., 1988). The overall structure of the calci-calmodulin/melittin complex is not a dumbbell but rather, a compact globule (Kataoka et al., 1989, 1991). The radius of gyration of calmodulin decreases from 20.9 Å to 18.0 Å, and the largest dimension decreases from 60 Å to 47.5 Å upon the binding of melittin. Calmodulin and troponin C interact with melittin and other peptides more strongly in the presence of calcium; in contrast, parvalbumin, calbindin D_{28K}, and α-lactalbumin form tight complexes with melittin only in the absence of divalent cations. This phenomenon might have a physiological significance; some calcium binding proteins (e.g., recoverin) activate their target enzymes only in the absence of calcium.

The calcium-dependent interactions of calmodulin with mastoparans, a group of three tetradecapeptides, and with melittin were studied most intensively (Malencik and Anderson, 1983b, 1984, 1986, 1988; Maulet and Cox, 1983; McDowell et al., 1985; Caday and Steiner, 1986; Linse et al., 1986; Muchmore et al., 1986). The dissociation constant of calci-calmodulin for mastoparan is 10^{-9} M (Malencik and Anderson, 1983a,b). The radius of gyration of calmodulin in solution decreases by 3.1 ± 0.3 Å upon binding of four equivalents of calcium and one equivalent of mastoparan as determined by small-angle x-ray scattering (Matsushima et al., 1989). The two lobes of calmodulin in the complex are 8.0 to 9.5 Å closer to each other than in the free apo-protein.

The radius of gyration of a complex of calmodulin with a peptide based on the calmodulin binding domain of myosin light chain kinase (Heidorn et al., 1989) and of complexes of calmodulin with two peptides corresponding to parts of the calmodulin binding domain of the plasma membrane Ca^{2+} pump (Kataoka et al., 1991) is up to 20% smaller than the radius of apo-calmodulin.

Most peptides bound to calmodulin adopt a helical conformation: for example, the calmodulin binding peptide of smooth muscle myosin light chain kinase (Roth et al., 1991; Ikura et al., 1992; Meador et al., 1993). Nonhelical backbone conformations are seen in the middle and at the C-terminal end of the bound peptide. In these complexes, the extended structure of calmodulin collapses to a more compact globule, with the two lobes facing each other in *cis* orientation and the linker helix disrupted by a flexible loop. The interaction is not just hydrophobic but involves the formation of hydrogen bonds between the basic amino terminal half of the peptide and Glu residues in the C-terminus of calmodulin.

Isothermal titration calorimetry (Wintrode and Privalov, 1997) showed that the interaction of calmodulin with a peptide corresponding to the calmodulin binding site of rabbit smooth muscle myosin light chain kinase proceeds with negative changes in enthalpy, heat capacity, and entropy at room temperatures. This indicates that it is an enthalpically driven process that is entropically unfavorable. It seems that the hydrophobic effect, an entropic effect, which favors the removal of nonpolar protein groups from water, is not a major force in calmodulin–peptide recognition. Although a large number of nonpolar side chains are buried upon binding, these stabilize the complex primarily by forming tightly packed van der Waals interactions with one another.

Peptides from calmodulin binding domains of myosin light chain kinase reduce the calcium dissociation rate constants for calmodulin's N-terminal sites from $405 \pm 75\ s^{-1}$ to $1.8–2.9\ s^{-1}$ and for the C-terminal sites of calmodulin from $2.4 \pm 0.2\ s^{-1}$ to $0.1–0.4\ s^{-1}$ at 10°C (Johnson et al., 1996). Since calcium dissociates 5 to 30 times faster from the N-terminal part in these complexes and both lobes are required for activation, calcium dissociation from the N-terminal lobe would control target protein inactivation. There are substantial differences between the processes of calcium dissociation from the complexes of calmodulin with myosin light chain kinase and nitric oxide synthase or with the respective peptides that are attributable to the distinct roles played by the calmodulin lobes in these complexes (Persechini et al., 1996).

Interaction of the 20-residue peptide derived from the calmodulin binding domain of the smooth muscle myosin light chain kinase with apo-calmodulin has been studied by a combination of isothermal titration calorimetry and differential scanning calorimetry. The dissociation constant for the apo-calmodulin/peptide complex is 10^{-6} M and the presence of salt weakens the affinity of the peptide for apo-calmodulin (Tsvetkov et al., 1999).

Small angle x-ray scattering and neutron scattering were used to obtain a model of calmodulin complexed with an enzymatically active, truncation mutant of skeletal muscle myosin light chain kinase (Krueger et al., 1997, 1998). Calmodulin undergoes an unhindered conformational collapse upon binding the enzyme and activates the enzyme by inducing a significant movement of the kinase's calmodulin binding and autoinhibitory sequences away from the surface of the catalytic core. The addition of substrates, a nonhydrolyzable analog of ATP (AMPPNP), and a peptide substrate for myosin light chain kinase, induces an overall compaction of the complex. Separation of the centers of mass of the calmodulin and kinase components is shortened by approximately 12 Å, thus bringing calmodulin closer to the catalytic site compared to a complex without substrates.

Determinants for calmodulin binding in the isolated peptide sequence and calmodulin binding to the same sequences within the enzyme are different (Krueger et al., 1998). For example, calmodulin binds to the myosin light chain kinase peptide I even in the presence of EGTA; whereas, binding of calmodulin to the enzyme requires calcium. The binding of calmodulin to the enzyme begins at substoichiometric concentrations of $\leq$2 mol of calcium per mole of calmodulin; however, the final compact structure of the calmodulin–enzyme complex requires saturating calcium.

It was long assumed that calmodulin interacts with different peptides in a similar way; calmodulin would engulf the helical or nonhelical peptides with its two lobes and form a compact ellipsoidal complex. This is not always the case (reviewed by Vetter and Leclerc, 2003). For example, when myristoylated alanine rich C kinase substrate (MARCKS) is complexed with calci-calmodulin, little change is observed in the circular dichroism spectra, suggesting that MARCKS binds to calmodulin in a nonhelical conformation (Matsubara et al., 1998). The crystal structure (Yamauchi et al., 2003) shows that the MARCKS peptide in this complex has an elongated structure with a short, one turn helix surrounded by two loops (Figure 15.1). Two hydrophobic residues, Phe157 and Leu159, that anchor the MARCKS peptide to the N- and C-lobes of calmodulin are separated by only one residue (not by 10, 14, or 16 residues as in "classical" complexes). Phe157, which interacts with many hydrophobic residues located in the hydrophobic pocket of calmodulin C-lobe, is buried deep in the pocket. In contrast, Leu159 interacts solely with a hydrophobic surface of the N-lobe.

The structure of a complex of calmodulin with the C20W peptide, corresponding to the N-terminal part of the calmodulin binding domain of the calcium pump, does not form a compact ellipsoidal structure but remains in

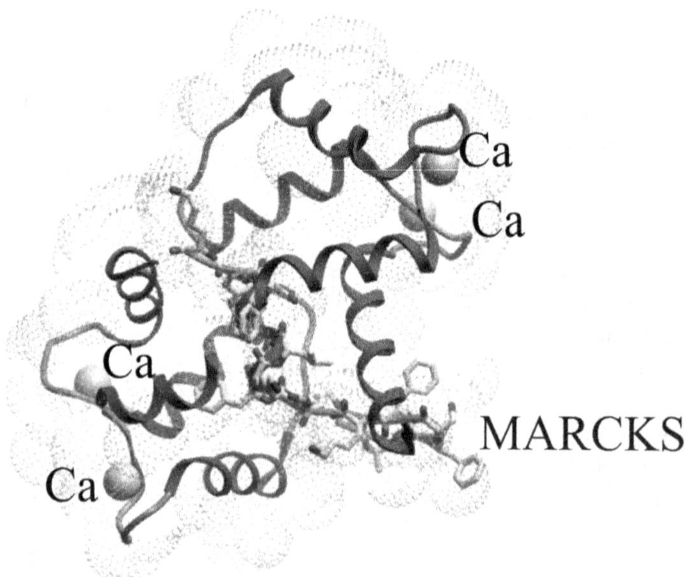

Figure 15.1. Structure of tetra-calci-calmodulin complexed with the MARCKS peptide (PDB file 1IWQ). (From Yamauchi et al., 2003.)

an elongated conformation, and the central, connecting region between Arg74 and Glu84 remains flexible (Elshorst et al., 1999). The two globular lobes do not contact each other (Figure 15.2). The C20W peptide is bound only to the C-terminal domain. The bound peptide is in α-helical conformation and is anchored to the peptide binding channel of the C-terminal domain, due to hydrophobic interactions with Trp1, Leu5, and Ile8. All the Met residues (109, 124, 144, and 145) of the C-terminal domain of calmodulin interact with the peptide. The hydrophilic part of the peptide is exposed to solvent.

In the complex of calmodulin with the 96 residue peptide corresponding to the luminal C-terminal part of the K^+-channel, calmodulin remains in an elongated conformation and forms a dimeric complex; two fragments of the K^+-channel are connected by two calmodulin molecules (Schumacher et al., 2001). The fragments of the K^+-channel consist of a long α-helix and of a shorter α-helix that lies antiparallel to the long helix. Within the complex the fragment of the K^+-channel is almost perpendicular to the long axis of the elongated calmodulin molecule.

The binding of calcium to calmodulin in its complexes with peptides and proteins causes cooperative interactions not only between the two calcium binding sites in each lobe but also between the two lobes. It was found, for example, that the binding of a 23-residue synthetic peptide, encompassing the calmodulin binding domain of rat cerebellar nitric oxide synthase, to calmodulin increases the affinity of calmodulin for metal ions and induces interlobe cooperativity in metal ion binding (Zhang et al., 1995).

Several recognition motifs for calmodulin interaction have been found (reviewed by Rhoads and Friedberg, 1997); these can be grouped into two major motifs, 1–8–14 and 1–5–10, based on the position of conserved hydrophobic

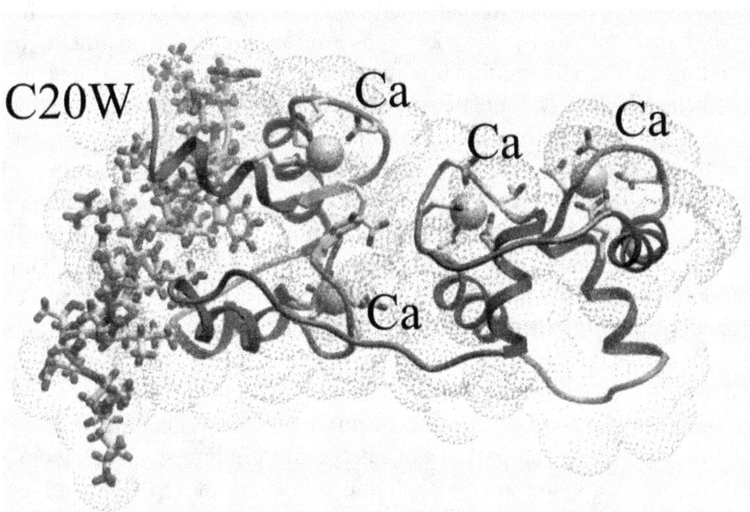

Figure 15.2. Structure of tetra-calci-calmodulin complexed with the C20W peptide (PDB file 1CFF). (From Elshorst et al., 1999.)

residues. Calmodulin binding domains of the target proteins usually have a hydrophobic sequence containing basic amino acids, mostly Arg; the basic and hydrophobic residues of the domain are normally spaced so as to produce an amphiphilic helix. An aromatic residue, frequently Trp, is almost always present in the N-terminal part of the domain (Vorherr et al., 1990, 1993).

Calmodulin recognizes positively charged amphiphilic α-helical peptides independent of their precise amino acid sequences. O'Neil and DeGrado (1990) found two features of calmodulin that facilitate its interaction with the α-helical peptides. The first feature is that 46% of the accessible surface area of the hydrophobic patches on the C- and N-lobes of calmodulin is contributed by Met side chains, including residues 36, 71, 72, 76, 106, 124, 144, and 145. This is an extraordinarily high Met content, since it is a relatively rare amino acid. Unlike other hydrophobic residues, the side chain of Met is unbranched, providing considerable conformational flexibility. Thus, a hydrophobic site containing a high proportion of Met residues might be able to alter its surface conformation while maintaining its backbone conformation relatively unchanged, allowing it to adjust to the fine topographical details of different bound peptides. The second feature that might explain the ability of calmodulin to bind so many different peptides is the flexibility of the central segment connecting the two lobes. One more important element in binding of specific peptides to calmodulin is an aromatic residue, frequently Trp, in the calmodulin binding region of the target. It has been suggested that the nonpolar aromatic residue acts as an anchor into one of the hydrophobic patches of calmodulin (Graeter et al., 1997; Gomes et al., 2000; Weljie and Vogel, 2000).

Interactions of calmodulin with its target proteins and peptides are usually calcium dependent, but a few calcium independent calmodulin–target protein interactions also have been found. For example, Yuan et al. (1998) found a calcium independent interaction between calmodulin and a synthetic peptide corresponding to the calmodulin binding domain of murine macrophage nitric oxide synthase (iNOS). The interaction of the iNOS peptide with apo-calmodulin changes its conformation from a β-sheet to a type II β-turn. In contrast, when the iNOS peptide is bound to calci-calcium, it acquires an α-helical conformation.

In some cases in which a peptide binds to both calci- and to apo-calmodulin, the binding is tighter in the presence of calcium. Nevertheless, the binding of peptides to apo-calmodulin can be strong enough to be physiologically significant.

Calmodulin interacts with over 40 target proteins (Chapter 9). Many calmodulin-regulated proteins contain a short calmodulin binding peptide of about 20 residues (Rhoads and Friedberg, 1997). The crystal structure of calmodulin kinase I shows that the calmodulin binding domain interacts with and obstructs the putative substrate binding sites of the inactive enzyme (Goldberg et al., 1996). Furthermore, the N-terminal part of the calmodulin binding sequence loops away from the enzyme, exposing the hydrophobic side chain of Trp303 to solvent and providing potential access for calci-calmodulin. It is supposed that this conserved hydrophobic residue interacts exclusively with the Met rich hydrophobic pocket in the C-terminal domain of calci-calmodulin.

Calmodulin activated enzymes often have a regulatory domain containing a calmodulin binding site and a catalytic domain. In the absence of calmodulin, the regulatory domain inhibits catalysis, but the binding of calmodulin removes this internal inhibition. According to James et al. (1995), calmodulin binding, autoinhibitory domains are two-faced; they regulate enzymes through a fine balance between internal binding, causing inhibition, and external calmodulin binding, producing activation. Calmodulin-activated enzymes should all have cationic, amphipatic regions, which are similar in these characteristics to melittin or mastoparan and at least one site that binds cationic, amphipathic regions of peptide sequence.

A label selection technique demonstrated that Lys75 of the linker helix is involved in interaction with myosin light chain kinase and also indicates that additional contacts near calcium binding site I occurs (Jackson et al., 1987). Photoaffinity labeling studies of the interaction of calmodulin with the plasma membrane calcium pump (Vorherr et al., 1992) showed that only the C-terminal lobe of calmodulin has high affinity for a 20-residue peptide corresponding to the calmodulin binding domain of the ATPase (K_d in the nanomolar range). These data are corroborated by the results of Bzdega and Kosk-Kosicka (1992), who studied interactions of mutant calmodulins with erythrocyte Ca^{2+}-ATPase. They found that there exist dramatic differences in the functional significance of the replacement of Glu at position 12 (−Z) in each of the four Ca^{2+} binding domains for activation of the Ca^{2+}-ATPase. The two Glu residues located in the C-terminal lobe of calmodulin are essential for activation of the Ca^{2+}-ATPase at physiological concentrations of calcium.

The participation of the linker helix of calmodulin in the interaction with target proteins was checked experimentally using mutant calmodulin (Putkey et al., 1988). For example, Thr79 was substituted by Pro; or the peptide Pro–Ser–Thr–Asp was inserted between Asp78 and Pro79. Intact and mutant calmodulins activate calcineurin and Ca^{2+}-ATPase equally well. The mutant proteins also activate cyclic nucleotide phosphodiesterase and myosin light chain kinase but at two and nine fold higher concentrations, respectively, than required for the native protein. Deletions of two, three, five, or eight residues in the central helix reduce the efficiency of phosphodiesterase activation (Van Berkum et al., 1990). Experiments with synthetic peptides and target proteins showed that the length of the helix, not its content, is important for the activation of the enzymes. These data indicate the existence in calmodulin of multiple sites of interaction with other proteins.

Special features of calmodulin are important in the activation of the catalytic subunit of phosphorylase kinase (Farrar et al., 1993). Calmodulin is an integral subunit, called δ, of phosphorylase kinase hexadecamer, and the activity of the isolated catalytic γ-subunit of the kinase is stimulated by calmodulin. The use of genetically engineered calmodulins, in which acidic residues in each of the α-helices of the calcium binding domains were changed to basic lysine residues, showed that activation of the γ-subunit is especially sensitive to substitutions in the second and third α-helices of calmodulin (residues 47–49 and 82–84).

Farrar et al. (1993) hypothesized that calmodulin interacts with an extended region or multiple regions of the γ-subunit and argued the importance of electrostatics in these interactions.

The affinity of calci-calmodulin for target peptides is reduced with increasing concentration of magnesium (Ohki et al., 1997). The calmodulin binding affinity of the myosin light chain kinase peptide in the presence of 50 mM Mg^{2+} is approximately 40-fold lower than that seen in the absence of magnesium.

Due to its central role as a calcium sensor in all types of cells, calmodulin is assumed to be involved in many diseases, including Parkinson's, Alzheimer's, and rheumatoid arthritis (reviewed by Schaub and Heizmann, 2008). However, no disease related mutations in any of the three calmodulin genes have been reported so far. Instead, defects in some of the many reaction partners of calmodulin might be responsible for the disease symptoms.

A particularly prominent regulator of cardiac function is the calmodulin-dependent protein kinase II (CaMKII) relaying the calcium signal to cardiac targets by phosphorylation (Zhang et al., 2005; reviewed by Schaub and Heizmann, 2008). Calci-calmodulin activates CaMKII and progressively phosphorylates its subunits (autophosphorylation of 6 to 12 subunits in the oligomeric CaMKII) in dependence on frequency and intensity of the calcium transients. Autophosphorylation increases the affinity of calci-calmodulin for CaMKII about 1000-fold up to $K_d \sim 10^{-10}$ M, keeping the enzyme on sustained activity far beyond the duration of a single calcium transient. CaMKII activates several targets: the ryanodine binding receptor, SERCA2A, phospholamban, and the L-type Ca^{2+}-channels. Activation of L-type Ca^{2+} channels produces a sustained increase in intracellular $[Ca^{2+}]$. Transgenic overexpression of CaMKII in mice causes myocardial dilation, contractile dysfunction, and abnormal intracellular calcium homeostasis that mimics changes in remodeled myocardium in heart failure patients and in animal models of structural heart disease (Zhang et al., 2005). CaMKII is a critical downstream element in the β-adrenergic signaling pathway; it may determine clinically relevant heart disease phenotypes. CaMKII inhibitory therapy could present a new approach for treating structural heart disease and myocardial dysfunction (reviewed by Schaub and Heizmann, 2008). This case also illustrates the necessity to focus therapeutic strategies on individual calmodulin targets rather than modulating calmodulin activity itself since it serves as a ubiquitous carrier of the intracellular calcium signal.

Khachaturian (1989) first proposed the "calcium hypothesis," which invoked calcium "dyshomeostasis" as the underlying cause of Alzheimer's disease. The hypothesis invokes the disruption of calcium signaling as the underlying cause of neuronal dysfunction and ultimately, apoptosis. As a primary calcium signal transducer, calmodulin responds to cytosolic calcium fluxes by binding to and regulating the activity of target calmodulin binding proteins. A search for calmodulin binding domains reveals that many of the proteins intimately linked to Alzheimer's disease may bind calmodulin; this opens new avenues for research on this devastating disease (O'Day and Myre, 2004). Presenilin (PSN-1 or

-2 homodimer), nicastrin (Nic), presenilin enhancer protein 2 (PEN-2), and presenilin-stabilizing factor APH-1 (APH-1 A or B) are a quartet of proteins that are considered to comprise the enzyme γ-secretase, which is the major source of the amyloid β-protein (Aβ) that is generated in Alzheimer's disease patients by the proteolysis of β-amyloid precursor protein. All of these subunit proteins possess one or more putative calcium dependent calmodulin binding domains. Presenilin-stabilizing factor APH-1a contains two calcium dependent calmodulin binding domains of the 1–5–10 motif (positions of conserved hydrophobic residues), while APH-1 b possesses a single 1–16 motif. Presenilin PSN-1 (1–5–10; 1–5–8–14) has two calcium dependent calmodulin binding domains of different motifs. On the other hand, PEN-2 (1–5–10; 1–8–14) and PSN-2 (1–5–10 and 1–16) each show a single calmodulin binding domains that fits two motifs. The final γ-secretase quartet member, nicastrin, possesses a single 1–16 motif calmodulin binding domain.

15.1.2. Troponin C and Targets

Like calmodulin, troponin C binds basic amphiphilic peptides (Malencik and Anderson, 1984; Cachia et al., 1986; Steiner and Norris, 1987; Iio, 1993). At physiological salt concentrations, melittin binds to calci-troponin C from skeletal muscle due to both hydrophobic and electrostatic interactions; it binds to apo-troponin C by electrostatic interactions (Iio, 1993). Fluorescence energy transfer data show that melittin is enfolded by the N- and C-lobes of troponin C, and the melittin helix is almost perpendicular to a line connecting Met25 and Cys98 in troponin C (Sano et al., 1998). The dissociation constant of the complex of melittin with the calci-troponin C is 2×10^{-10} M (Steiner and Norris, 1987). The binding of melittin to troponin C causes its contraction in a similar manner to that of calmodulin when it binds peptides known to form amphiphilic helices. In contrast, mastoparan binding to troponin C does not result in a contracted structure. The binding of synthetic peptides, corresponding to the inhibitory regions of skeletal (residues 104 to 115) and cardiac (residues 133 to 138) troponin I to skeletal and cardiac troponin C, alter calcium binding at the low-affinity sites of troponin C; however, the magnitude and direction of the pCa shifts depend on whether the peptides are bound to skeletal or cardiac troponin C (Van Eyk et al., 1991).

Troponin C interacts with both troponin I and troponin T (Evans and Levine, 1979, 1980; Barskaya and Gusev, 1982) (Chapter 9). There are three regions of troponin C—residues 51 to 62, 84 to 100, and 127 to 138—that interact with troponin I. Studies of interactions of troponin C with troponin I and troponin T in binary and tertiary complexes by intrinsic fluorescence and microcalorimetry (Morozova et al., 1991) indicate that isolated troponins I and T have poorly ordered structures. In contrast to other components, isolated troponin C is highly ordered. Binary complexes, containing troponin C, possess an intermediate structure in which the components affect each other but are unable to form a united highly cooperative ordered structure. In the heterotrimer, troponins C, I, and T interact with each other and form a highly cooperative calcium-dependent complex. The binding of calcium to troponin C changes the structure of the entire heterotrimer; this is reflected in changes of its thermal stability. The denaturation transition in the calci-troponin complex passes through several stages. The apo form of the troponin complex is less stable than the calci form, but this difference is not as pronounced as for isolated troponin C. The apo complex is a highly cooperative system that denaturates as a single domain, with a single transition.

Troponin I is organized into structural and regulatory regions, which interact in an antiparallel fashion with the corresponding structural and regulatory regions of troponin C (Farah et al., 1994; T. Kobayashi et al., 1994, 1996). The N-terminal region of troponin I interacts with the C-terminal domain of troponin C in the presence of calcium or magnesium. The inhibitory region plus the C-terminal region of troponin I (residues 103 to 182) interacts with the N-terminal domain of troponin C in a calcium-dependent manner (Krudy et al., 1994; Pearlstone et al., 1997). Krudy et al. (1994) proposed that the N-terminal domain of troponin I is anchored strongly to the C-terminal domain of troponin C in both the absence and presence of calcium, while the inhibitory and C-terminal regions of troponin I switch between actin-tropomyosin in the absence of calcium to binding sites in both N- and C-terminal domains of troponin C in the presence of calcium.

The inhibitory peptide of troponin I interacts with a F154W mutant of chicken recombinant troponin C. The binding interface involves residues 96 to 115 of troponin I; this interface probably differs for the apo, di-calci, and di-magnesi states (Chandra et al., 1994). Data on kinetics and energetics of the binding of three troponin I peptides (96–131, 96–139, and 96–148) to skeletal chicken troponin C indicate that residues 97 to 136 of troponin I are involved in the binding to the two domains of calci-troponin C (Kay et al., 1999). Two peptides of troponin I (96–116) and (104–116), interact with spectral probe mutants F29W and F105W of intact troponin C and of the isolated N (residues 1 to 90) and C (88 to 162) lobes of troponin C (Pearlstone and Smillie, 1995). The binding of 96–116 to the C-lobe of intact troponin C increases its calcium affinity without significantly affecting that of the N-lobe; only at high molar ratios of 96 to 116: troponin C is the calcium affinity of the N-lobe increased. Photo cross-linking of troponin I to troponin C thiol mutants indicate as well that in the troponin C–troponin I complex, both globular domains of troponin C, as well as the linker region between them, make contact with the inhibitory region of troponin I (Kobayashi et al. 1994, 1996).

The binding of calcium to the triggering sites in troponin C results in the opening of the N-terminal hydrophobic pocket bound by the B, C, and D helices (Gagne et al., 1995). This conformational change is believed to create an additional binding site for troponin I. Experiments on single Cys mutants with the use of photo cross-linking showed that the troponin I segment containing Met121 is in close contact with the N-terminal hydrophobic patch of troponin C (Luo et al., 1999). In thin filaments the segment containing this residue moves slightly away from the hydrophobic patch in the absence of calcium, possibly triggering the translocation of the actin-binding regions of troponin I toward actin.

15.1.3. Myosin Light Chains and Targets

Myosin light chains from both smooth and skeletal muscle also interact with melittin in a calcium-dependent manner with $K_d = 5$ to 12×10^{-9} M (Malencik and Anderson, 1988). The IQ motif (23 to 25 residues) with a consensus sequence, IQXXXRGXXXR (where X is any amino acid), is repeated tandemly (two to six times) in the heavy chains of many myosins. This motif confers on these proteins the capacity to bind myosin light chains or calmodulin (Houdusse and Cohen, 1995). The crystal structure showed that the lobes of the myosin light chains adopt three distinct conformations when bound to the heavy chain. Two of these conformations correspond to the *open* and *closed* forms that a lobe assumes when divalent cations are bound or absent, respectively. The third state is an unusual *semiopen* conformation in which no metal is bound. It was suggested that lobe conformation depends not only on divalent cation binding but also on the precise pattern of residues in the target peptide.

Myosin light chains are tightly bound to the heavy chains of myosin. The light chains are elongated and their C-terminii are near the hinge region that connects the myosin head with the helical tail. Its N-terminus is near the actin binding site (Kendrick-Jones and Jakes, 1977). In some types of muscle the binding of calcium to the regulatory myosin light chain induces changes in its conformation; this is transferred to the myosin heads and results in activation of the actomyosin interaction.

15.1.4. S100 and Targets

S100P can interact with melittin in both calcium-dependent and calcium-independent manners (Gribenko et al., 2002). Apo-S100P binds four melittin molecules per dimer with $K_d = 0.2$ mM. Calci-S100P binds two molecules of melittin per dimer with higher affinity, $K_d = 5$ μM. Melittin in complex with S100P adopts a helical conformation. S100B(ββ), binds the p53 peptide (residues 367 to 388) in a calcium dependent manner. The presence of the peptide increases the affinity for calcium, but not its affinity for zinc (Rustandi et al., 1998, 2000). The p53 peptide binds to a region that includes residues in the "hinge" (residues 44 to 47), in the C-terminal loop (residues 83 to 88), and in helix 3.

Up to 40 target proteins are reported to interact with S100 proteins (Chapter 9). For example, S100a and S100b interact with caldesmon (Polyakov et al., 1998). The list of calcium-dependent S100A1 target proteins includes twitchin kinase, brain-specific aldolase C, transcription factor MyoD, and others. Apo-S100a and S100b also bind target proteins (e.g., aldolase A). S100A1 utilizes distinct mechanisms for interaction with calcium-independent and calcium-dependent target proteins (Landar et al., 1998). It has two distinct target protein binding regions. The first is a calcium dependent hydrophobic region, which includes a cluster of three aromatic residues (F88–F89–W90). The second region is a calcium-independent binding site for target proteins, the specific residues involved in these interactions vary from one target protein to another.

Matsumura et al. (1998) found in crystal structures that upon calcium binding to S100B, dramatic changes occur in the terminal helices of the protein; these reveal a large hydrophobic surface, not present in the apo form. They suggested that S100B binds its target through hydrophobic interactions and possibly a Cys84 mediated disulfide bond. One of the calcium dependent target proteins of S100B is baculovirus p53 protein, which is a tumor suppressor (Delphin et al., 1999). The K_d value of calci-S100B for the C-terminal region of p53 (319–393) is 24 nM.

The putative target binding site in S100s includes helix 4 from each monomer paired in an antiparallel orientation in the dimer structure with helix 3 from each monomer subunit forming the sides of a helical bundle (Drohat et al., 1996; Kilby et al., 1996). The most important residues that are involved in putative target binding are those at the C-terminus of helix 4. Calcium binding to S100s results in structural changes in the linker, helix 3, and the second calcium binding loop region; these changes should give new characteristics to the potential binding cleft (Groves et al., 1998).

15.1.5. Parvalbumin and Targets

Calci- or magnesi-parvalbumin, pike pI 4.2, does not interact with melittin; however, apo-parvalbumin forms a 1:1 complex with melittin (Permyakov et al., 1989a). Similar effects were also found for pike pI 5.0 and its fragment 38–108 (parvalbumin without helices A and B). The formation of the complex transfers the single Trp residue of melittin from an aqueous to a more hydrophobic environment. The dissociation constant of the complex of melittin with pike parvalbumin pI 4.2 at 18°C is 10^{-6} M. Pike parvalbumin pI 5.0 has lower affinity for melittin; at 18°C, $K_d = 5.6 \times 10^{-6}$ M. The affinity of the fragment 38–108 for melittin is lower, $K_d = 1.7 \times 10^{-5}$ M.

15.1.6. Recoverin and Targets

Recoverin inhibits rhodopsin kinase (GRK1) at high calcium concentrations in the dark state of the photoreceptor cell (reviewed by Komolov et al., 2009). Decreasing free calcium concentration terminates inhibition and facilitates phosphorylation of illuminated rhodopsin (Rh*). When recoverin forms a complex with GRK1, it does not interfere with the phosphorylation of a C-terminal peptide of rhodopsin (S338–A348) by GRK1. Furthermore, while GRK1 competes with transducin on interaction with rhodopsin and thereby suppresses GTPase activity of transducin, recoverin in the complex with GRK1 does not influence this competition. Recoverin activates guanylyl cyclase below free calcium concentrations of 200 nM (Lambrecht and Koch, 1991). Recoverin is phosphorylated by an endogenous kinase and Mg-ATP, also at the decreased calcium concentration. At the same time, recoverin binds rhodopsin kinase, photoreceptor retinoid binding protein, and tubulin only in the presence of calcium (Chen et al., 1995).

Calci-recoverin binds to the N-terminus of GRK1 but does not bind to the other constructs (Komolov et al., 2009). The overall main chain structure of recoverin in the complex with the functional N-terminal fragment of GRK1, RK25, is similar to structures of calci-recoverin in the absence of GRK1. The first eight residues of recoverin are exposed to solvent, enabling the N-terminal myristoyl group to interact with target membranes, and calcium is bound at the second and third EF-hands (Ames et al., 2006). In the complex with recoverin, GRK1 forms an amphipathic α-helix that is located in the first 16 residues (Ames et al., 2006; Higgins et al., 2006). The recoverin–RK25 complex exhibits calcium-induced binding to rhodopsin immobilized on a concanavalin A resin. It was suggested that calci-recoverin is bound between rhodopsin and rhodopsin kinase in a ternary complex on rod outer-segment disk membranes, thereby blocking rhodopsin kinase interaction with rhodopsin at high calcium levels (Ames et al., 2006). Komolov et al. (2009) concluded that inhibition of GRK1 by recoverin is not the result of their direct competition for the same docking site on Rh*, although the interaction sites of GRK1/Rh* and GRK1/recoverin partially overlap. The N-terminus of GRK1 is recognized by Rh* leading to a conformational change that moves the C-terminus of Rh* into the catalytic kinase groove. Calci-recoverin interacting with the N-terminus of GRK1 prevents this conformational change and thus blocks Rh* phosphorylation by GRK1.

15.1.7. Other EF-Hand Proteins and Targets

Cytosolic, calcium dependent cysteine proteases, calpains, are ubiquitously distributed, along with their endogeneous inhibitor, calpastatin (120 kDa). The calcium dependent proteolytic system, which consists of calpain and calpastatin, is activated by an increase in the intracellular concentration of calcium and then kept under control by the inhibitory effects of calpastatin. It contains four tandem, inhibitory domains, each having three conserved regions—A, B, and C—and an N-terminal region, L. Each of the conserved regions has been shown to bind preferentially to separate domains of calpain. Regions A and C bind to the EF-hand domains, DIV and DVI, of calpain, while B binds to the protease domain, DII (Takano et al., 1995). Calpastatin has little secondary structure under physiological conditions, but the conserved sequences of regions A and C indicate their ability to form amphipathic α-helices.

Calcineurin, the major soluble calmodulin binding protein in brain extracts, calci-calmodulin regulated protein phosphatase, is a heterodimer consisting of two subunits (reviewed by Klee et al., 1988). Its small subunit, calcineurin B, has four EF-hands. The regulation of calcineurin activity by micromolar calcium is mediated by two different calcium binding proteins, calcineurin B and calmodulin, the former being held in a tight complex with calcineurin A.

Wills et al. (1994) studied the interaction of a smooth muscle, calcium binding protein, caltropin, with calponin, which is a component of smooth muscle that also binds to actin, tropomyosin, and calmodulin. Caltropin is a dimer under native conditions and interacts with calponin in a calcium dependent fashion in the ratio of 2 mol of dimer per mole of calponin.

Apo-calbindin D_{28k} binds melittin; the complex can be dissociated reversibly by addition of calcium (La Bella et al., 1996). The molar calbindin/melittin ratio in this complex is 1 : 4, suggesting that calbindin interacts with tetrameric form of melittin.

15.1.8. Annexin and Targets

Members of the annexin family form complexes with several EF-hand proteins, especially those of the S100 subfamily (reviewed by Gerke and Moss, 2002). S100A6, S100A10, and S100A11 were shown to bind specifically to annexins A11, A2, and A1, respectively. The best characterized of these annexin–S100 complexes is the annexin A2, S100A10 (p11) heterotetramer. It was clearly established that complex formation is highly specific, occurs in vivo, can be regulated by posttranslational modifications in the annexin, and modulates properties displayed by the isolated subunits (Waisman, 1995; reviewed by Gerke and Moss, 1997).

S100A10 is unique among S100s in having lost its ability to bind calcium. The binding region in annexin II is located in the first 14 amino acids in the N-tail, while in S100A10 a hydrophobic sequence near the C-terminus is essential for high-affinity binding to annexin. In the crystal structure of a complex of S100A10 with the annexin II N-tail, S100A10 is a tight, noncovalent dimer, and each N-tail interacts with both S100A10 monomers by hydrophobic contacts (Rety et al., 1999). The crystal structures of S100A10 in complex with an annexin II peptide (Rety et al., 1999) and that of S100A11 complexed with a peptide from annexin I (Rety et al., 2000) are quite similar; however, there is no significant sequence similarity between these two annexin peptides. Further, S100A11 binds calcium; whereas, S100A10 does not. In both structures, two annexin peptides per S100 dimer bind to symmetrically positioned sites formed by the C-terminal end of helix 4 and the linker loop 2 of one monomer and the N-terminal end of helix 1 of the second monomer. S100B binds the regulatory peptide, p53, at a different site, formed by helices 3 and 4 of the same monomer; the p53 peptide is nearly perpendicular to the annexin peptides (Rustandi et al., 2000). Moreover, it remains to be established whether this interaction occurs in vivo (Mailliard et al., 1996).

The third annexin–S100 interaction described to date is that between annexin A11 and S100A6 (Tokumitsu et al., 1992; reviewed by Gerke and Moss, 2002). Although the S100 binding site in annexin A11 is also located in the N-terminus, the mode of complex formation is likely to be different. Annexin A11 contains a long N-terminal tail of almost 200 residues rich in Gly, Tyr, and Pro and is probably inherently disordered. However, sequences within the N-tail of annexin A11 do not resemble the amphipathic helices found in annexins A1 and A2 and thus are unlikely to fit in a similar manner in a binding pocket formed by the S100 dimer. The physiological consequences of the annexin A11–S100A6 interaction remain to be established.

It was found that sorcin, the penta-EF-hand protein (Chapter 11), binds in a calcium-dependent manner to the GYP-rich N-terminal domain of annexin A7 (Brownawell and Creutz, 1997); the N-terminal domain is required for the interaction. Complex formation can recruit sorcin to the membrane of chromaffin granules, which are a prime site of annexin A7 localization. Moreover, the binding of sorcin inhibits the chromaffin granule aggregation mediated by annexin A7.

Several annexins have been described as part of the cytoskeleton, in particular as F-actin binding proteins. It has been suggested that at least some members of the family could participate in regulating membrane–cytoskeleton interactions (reviewed by Gerke and Moss, 2002). Annexin A1 binds to F-actin and also interacts with profilin, a G-actin binding protein and regulator of actin polymerization. Complex formation between annexin A1 and profilin modifies the profilin effect on actin polymerization. Annexin A2 is another F-actin binding annexin that also has a calcium dependent filament bundling activity. This bundling activity is particularly pronounced in the case of the heterotetrameric annexin A2–S100A10 complex (reviewed by Gerke and Moss, 1997). The F-actin binding site has been mapped to the C-terminus of annexin A2, underscoring the specificity of the interaction (Filipenko and Waisman, 1991).

Annexin A5 has been observed to relocate to the cortical membrane cytoskeleton after activation of platelets (reviewed by Gerke and Moss, 2002). This relocation appears to involve both binding to the plasma membrane and to a specific actin isoform, γ-actin, and is paralleled by an association with the platelet membrane of cytosolic phospholipase A2, suggesting an interaction between this phospholipase and annexin A5 (Tzima et al., 2000). Annexin A6, another actin-binding annexin, has been implicated in mediating in a calcium-dependent manner membrane–cytoskeleton contacts in smooth muscle cells (Babiychuk et al., 1999).

15.1.9. C2-Domain Proteins and Targets

The C2 domain can also drive protein–protein interactions. Synaptotagmin C2 domains have been demonstrated to bind, either constitutively or in a calcium-regulated fashion, to other intracellular proteins. The second C2 domains of most synaptotagmin isoforms bind in a calcium independent manner with high affinity (dissociation constant of 0.1 to 1.0 nM) to clathrin–AP2 (Zhang et al., 1994; Li et al., 1995) and to form a complex involved in coated pit assembly. Synaptotagmin also interacts with syntaxin (Chapman et al., 1995; Li et al., 1995), a plasma membrane protein involved in the exocytosis of synaptic vesicles. The first C2 domains of isoforms I, II, III, V, and VII also bind in a calcium-dependent manner to syntaxin (Li et al., 1995; Kee and Scheller, 1996; Sugita et al., 1996). Point mutations in the calcium binding site of the first C2 domain that eliminate calcium-dependent phospholipid binding also destroy syntaxin binding; this shows that the same residues may be important for both interactions (Li et al., 1995). The second C2 domain of synaptotagmin associates with itself in a calcium-dependent manner (Sugita et al., 1996). Finally, the full-length protein has been reported to bind to intracellular receptors for protein kinase C (Mochly-Rosen et al., 1992).

Several proteins have been identified as regulators of cytosolic phospholipase $A_2\alpha$ ($cPLA_2\alpha$) in either a positive or negative way (reviewed by Hirabayashi et al., 2004). First, vimentin, which is the major component protein of the intermediate filaments and enriched in the perinuclear region, binds to the C2 domain of $cPLA_2\alpha$ in the presence of calcium (Nakatani et al., 2000). Overexpression of vimentin in cells lacking the protein increases the arachidonic acid release twofold; whereas, overexpression of the isolated amino-terminal domain of vimentin in fibroblasts partially inhibits arachidonic acid release, suggesting that the C2 domain–vimentin complex may stabilize the binding of $cPLA_2\alpha$ to the membrane in the perinuclear region. Second, $cPLA_2\alpha$-interacting protein (PLIP) interacts with the N-terminal region of $cPLA_2\alpha$ (Sheridan et al., 2001). In renal mesangial cells, $cPLA_2\alpha$ colocalizes with PLIP in the nucleus. Overexpression of PLIP potentiates $cPLA_2\alpha$-mediated prostaglandin E2 production and serum deprivation–associated apoptosis. Third, p11/calpactin light chain, a member of the S100 protein family, interacts directly with the C-terminal region of $cPLA_2\alpha$ and inhibits its activity in vitro (Wu et al., 1997). The p11–cPLA2a complex is formed in a human bronchial epithelial cell line, and reduction of p11 expression in the cells by antisense RNA increases $cPLA_2\alpha$ activity and arachidonic acid release. Fourth, some annexins can function as negative regulators of $cPLA_2\alpha$ in cellular signal transduction. The C-terminal region of annexin 1 binds to the C2 domain of $cPLA_2\alpha$ and inhibits phospholipase activity by specific interaction (Kim et al., 2001b), whereas annexin 5 inhibits $cPLA_2\alpha$ activity mainly by substrate depletion (Mira et al., 1997). Annexins 2 and 3 do not have any inhibitory effects. Caspases-3 and -8, the general executors of apoptosis, cleave $cPLA_2\alpha$ at Asp522 and inactivate its enzymatic activity (Wissing et al., 1997).

15.1.10. Gla-Containing Proteins and Targets

Gla-containing blood proteins (Chapter 11) interact with each other in the blood coagulation cascade. The interactions of the proteins are summarized in a Y-shaped scheme, with distinct "intrinsic" and "extrinsic" pathways initiated by factor XII (FXII) and FVIIa/tissue factor (TF), respectively (reviewed by Monroe and Hoffman, 2006). The coagulation complexes require phospholipid and calcium for their activity. It is widely acknowledged that the coagulation reaction occurs on specific cell surfaces in vivo rather than on phospholipid vesicles.

Gla-containing blood proteins also interact with other proteins. For example, Sekiya et al. (1995b) found that IX/X-bp, an anticoagulant protein from snake venom, binds to the Gla domains of coagulation factors IX and X in a calcium-dependent fashion. The binding requires about 1.0 mM calcium at pH 7.5.

Matrix Gla protein (MGP) regulates calcification in cartilage and arteries. MGP binds vitronectin but not collagen, fibromodulin, heparin, osteocalcin, chondroitin sulfate, laminin, ovalbumin, or albumin (Nishimoto and Nishimoto, 2005). This binding is saturable and consistent with a single class of binding sites. A vitronectin binding site was identified within a 17 residue peptide, 61–77, near the C-terminus that corresponds to the C-terminus of MGP. MGP and the 61–77 peptide of vitronectin also bind to fibronectin. MGP and vitronectin are colocalized in embryonic tissues; this suggests that their interactions may modify cell–matrix interactions.

15.1.11. α-Lactalbumin and Targets

α-Lactalbumin forms a complex with galactosyltransferase, the catalytic domain of the enzyme component of lactose synthase. The N-terminal region of the catalytic domain and possibly part of the Pro rich "stem" region, are altered by the association with α-lactalbumin (Yadav and Brew, 1991). The crystal structures of galactosyltransferase bound with various substrates were solved at 2.0 Å resolution (Ramakrishnan et al., 2001; Ramakrishnan and Qasba, 2001; Boeggeman and Qasba, 2002). These structures reveal that upon substrate binding to galactosyltransferase, a large conformational change occurs in the region of residues 345 to 365. This moves His347 in such a way that it can coordinate a metal ion and creates both sugar and α-lactalbumin binding sites. At the sugar binding site a hydrophobic N-acetyl group binding pocket is formed by Arg359, Phe360, and Ile363. Binding of galactose (Glc) to galactosyltransferase causes a reorientation of the Arg359 side chain; this blocks the hydrophobic pocket and maximizes the interactions with Glc. The role of α-lactalbumin is to hold Glc by hydrogen bonding with its O-1 hydroxyl group in the acceptor binding site on galactosyltransferase, while the binding pocket for the N-acetyl group adjusts to maximize the interactions with Glc. The effects of calcium binding on the structure of the α-lactalbumin/galactosyltransferase complex remain unknown. The catalytic domain of bovine β-1,4-galactosyltransferase has two metal binding sites, each with a distinct binding affinity. Site I binds manganese with high affinity and does not bind calcium; whereas, site II binds a variety of metal ions, including Ca^{2+} (Chapter 12). In the primary metal binding site of galactosyltransferase, the Mn^{2+} ion is coordinated to five ligands, two supplied by the phosphates of the sugar nucleotide and the other three by Asp254, His347, and Met344. The residue Asp254 in the D(252)VD(254) sequence in region I is the only residue that is coordinated to the Mn^{2+} ion. Region II forms a loop and contains the E(317)DDD(320) sequence in which residues Asp318 and Asp319 are directly involved in GlcNAc binding.

Using fluorescent and circular dichroism spectroscopy as well as affinity chromatography, Permyakov et al. (1991a) found that melittin binds to the apo and acidic states of human α-lactalbumin, while the presence of calcium makes the interaction weaker. The dissociation constant for the complex of melittin with apo-α-lactalbumin determined from spectropolarimetric titration data is 5×10^{-8} M. The binding of melittin by α-lactalbumin decreases its affinity for calcium by three orders of magnitude. The interaction of apo-α-lactalbumin with melittin causes some changes in the environment of its aromatic amino acid residues and drastically changes the conformation of melittin, increasing its α-helical content but leaving its single Trp residue accessible to solvent. The binding of melittin by α-lactalbumin in its acidic state does not increase its α-helical content.

15.1.12. Other Calcium Binding Proteins and Targets

Calcium binding plays an important role in interactions between extracellular matrix proteins. The C-type lectin subdomain (CLD) of aggrecan binds the fibronectin type III domains ($^{4-5}$FnIII) of tenasin (Diao and Tajkhorshid, 2008). The crystal structure of this complex was determined by Lundell et al. (2004). Aggrecan is responsible for generating a large osmotic swelling pressure in tissues such as cartilage; this is essential to the mechanical properties of these tissues.

Tenascins are expressed in the central nervous system and connective tissues during development as well as during wound healing. They bind to a variety of other components of the extracellular matrix (e.g., fibronectins and lecticans), and to cell surface receptors such as integrin and cell adhesion molecules of the immunoglobulin family. Tenascins are involved in a wide range of physiological processes, such as embryonic development and tissue remodeling. The binding between CLD and $^{4-5}$FnIII is essential in cross-linking aggrecan, hyaluronan, and tenascin to form extended protein networks in tissues such as cartilage. The interaction between aggrecan and tenascin is highly calcium dependent (Aspberg et al., 1995, 1997). In the crystal structure of the complex of the CLD and $^{3-5}$FnIII, three Ca^{2+} ions have been resolved, termed Ca1, Ca2, and Ca3. None of these Ca^{2+} ions in the complex of the CLD and $^{4-5}$FnIII bridges the two proteins directly. However, one of the Ca^{2+} ions (Ca2) is found to play the role of maintaining the structure of the L4 loop at the CLD binding surface, thus facilitating a high-affinity binding between the CLD and $^{4-5}$FnIII. Removal of this Ca^{2+} ion causes a drastic structural change in the L4 loop, which presumably hinders the binding of the CLD to the $^{4-5}$FnIII. Ca1 and Ca3 have no significant effect on the structure of the CLD binding surface and thus are not expected to affect the binding.

15.2. INTERACTIONS WITH MEMBRANES

Membranes are essential components of all cells. To understand the effects of calcium on membranes, one must characterize the calcium binding of membrane-associated proteins and, especially, the bridging of Ca^{2+} ions between these proteins and phospholipid headgroups of the lipid bilayer.

15.2.1. EF-Hand Proteins and Membranes

Calmodulin interacts with many membrane associated proteins. For example, it modulates the interactions of membrane associated guanylate kinase homologs

(MAGUKs) and other scaffolding proteins of the membrane cytoskeleton by binding to the Src homology 3 (SH3) domain, the guanylate kinase (GK) domain, or the connecting HOOK region of MAGUKs (Paarmenn et al., 2008). The K_d value for MAGUK–calmodulin interaction is 50 to 180 nM. Calci-calmodulin may serve as a general regulator to affect the interactions of MAGUKs and various components of the cytoskeleton.

Naoi et al. (1989) studied the conformation and properties of calmodulin in liposomes of dipalmitoylphosphatidylcholine. Since the transition temperature of the liposomes containing calmodulin was the same as that of the liposome without the protein and the enthalpy increased with the increment of calmodulin content, they concluded that calmodulin might be a peripheral proteins. Calci-calmodulin plays an important role in the membrane fusion processes; it interacts with soluble protein receptors of *N*-ethylmaleimid-sensitive factor (SNARE) and vesicle-associated membrane protein 2 (VAMP2). Calmodulin takes part in exo- and endocytosis, which involves syntaxin 13 and antigene 1.

Recoverin has a unique calcium–myristoyl switch, which acts as a calcium sensor involved in the recovery of retinal rods from photoactivation. Calcium binding to recoverin induces the extrusion of its N-terminal myristoyl tail to the solvent; this enables it to bind disk membranes of rod outer segments and to inactivate peripheral protein rhodopsin kinase (Ames et al., 1995a,b, 1997). Light induces lowering of intracellular calcium; this causes a conformation change in recoverin and sequestration of the myristoyl group in its hydrophobic cleft. As a result, recoverin loses its affinity for membranes and moves to the cytosol; this allows rhodopsin kinase to phosphorylate light activated rhodopsin.

The interaction of recoverin with membranes has been studied by several methods. For example, Lange and Koch (1997) studied the dynamics of its calcium dependent membrane association by surface plasmon resonance spectroscopy (Chapter 6). The binding of recoverin to artificial liposomes or rod outer segment membranes is strictly dependent on calcium and the myristoyl group. The association and dissociation of recoverin to membranes is fast and biphasic with time constants of 0.1 s^{-1} and 0.01 s^{-1}. Atomic force spectroscopy was used to determine the extent of membrane binding of recoverin in the absence and presence of calcium (Desmeules et al., 2002). An adhesion force between recoverin and dipalmitoylphosphatidylcholine bilayers in the presence of calcium is 48 pN; this is consistent with extraction forces of lipids with membranes. No binding was observed in the absence of calcium.

Calcineurin binding to phospholipid monolayers is myristoyl independent and is mediated by anionic phospholipids and/or diacylglycerol (Kennedy et al., 1997). Although the effect of calcium on calcineurin lipid binding is minor, calmodulin alters the binding of calcineurin to the lipid membrane in a calcium-dependent manner.

Calcium plays very important roles in the processes of fusion of secretory granules and synaptic vesicles with plasmatic membrane and also in the processes of intracellular membrane fusion (reviewed by Hay, 2007). All these calcium-dependent processes occur with participation of calcium binding proteins. For example, synaptotagmins are suggested to confer calcium dependence to the regulated exocytosis. Nevertheless, these proteins do not participate in early secretory events and hardly take part in calcium regulation of the transport of various substances.

15.2.2. Annexins and Membranes

Annexins (Chapter 12) are a large family of proteins that share the common property of binding to membranes in a calcium-dependent manner (Klee, 1988; Meers et al., 1991; Bandorowicz-Pikula, 1997; Chen et al., 1993; Evans and Nelsestuen, 1994; Raynal and Pollard, 1994; Moss, 1997; Moss and Morgan, 2004). These proteins associate with membranes in the presence of calcium and nonionic detergents and are released by addition of the calcium chelator EGTA. Interactions with specific phospholipids increase the rather low affinity for calcium of these proteins (dissociation constant 10^{-5} to 10^{-4} M) by one to two orders of magnitude. In the presence of micromolar calcium, these proteins specifically bind acidic phospholipids. The interaction between the proteins and membranes is supposedly maintained by calcium bridges between the proteins and acidic phospholipids.

Annexins are inferred to be involved in many processes. For example, the most extensively studied function of annexins II and V is to mediate calcium-dependent exocytosis. Annexin V leaves large unilamellar phospholipid vesicle membranes intact when added in the presence of 1 mM calcium, and it causes loss of vesicle integrity in the absence of calcium (Goossens et al., 1995). Phospholipid-dependent inhibition of blood coagulation has been suggested as an extracellular function of annexin V (Reutelingsperger et al., 1985).

All annexins are monomeric, with the exception of annexin II, which is usually associated with p11, a member of the S100 subfamily. They are built of four (eight in the case of annexin VI) repeating domains, each comprising about 75 amino acids (Chapter 10). Each domain folds into a bundle of four antiparallel helices, each topped off with fifth perpendicular helix. In the presence of anionic phospholipids such as phosphatidylinositol, the affinity of annexins for calcium is greatly enhanced; therefore, annexins may bind to the plasma membrane at physiological concentrations of intracellular calcium. The binding of annexins to membranes in vitro sometimes causes their aggregation.

The roles of individual domains of annexins in calcium-dependent membrane binding and aggregation are different despite their similarities in sequence and structure (Nelson and Creutz, 1995). Mutations in individual domains of annexin IV have preferential effects on the binding or aggregating activities of the protein. For example, mutation of the first or fourth domains has a greater effect on membrane binding than aggregation; conversely, mutation of the second domain has a more dramatic effect on membrane aggregation. Mutagenesis studies of all six calcium binding sites in annexin I (Bitto and Cho, 1998) showed the importance of domain II in vesicle binding and the involvement of domains III and IV in vesicle aggregation.

The annexin induced aggregation of lipid vesicles can be regulated by other proteins. The aggregation of lipid vesicles caused by annexin II is inhibited by phosphorylation of annexin II by protein kinase C (S. A. Johnston et al., 1992). Protein kinase C catalyzes the incorporation of about 2 mol of phosphate per mole of annexin II. This phosphorylation causes a dramatic decrease in the rate and extent of lipid vesicle aggregation without significantly affecting calcium-dependent lipid binding by the phosphorylated protein. This suggests that protein kinase C may play an important role in regulation of the function of annexin II. Annexin II is also phosphorylated by pp60 (Hubaishy et al., 1995). These results suggest that Tyr phosphorylation of annexin II is a negative modulator and that dephosphorylation of annexin II might be necessary for its activation. Both salt composition and concentration modulate annexin II–dependent chromaffin granule aggregation as well as binding to the membranes of these secretory granules (Jones et al., 1994). Annexin V inhibits phosphorylation of annexin II by endogenous protein kinase C and phosphorylation of myelin basic protein by protein kinase Cα (Dubois et al., 1998). A similar effect was also found for annexin I (Wang and Creutz, 1994).

De la Fuente and Parra (1995) investigated the mechanism of vesicle aggregation performing experiments in which annexin I bound to phosphatidylserine liposomes was allowed to aggregate additional liposomes. They showed that annexin I might bind two membranes in a bivalent fashion. Phosphatidylserine liposomes containing annexin I bind phosphatidylcholine, demonstrating that membrane-bound annexin I binds phosphatidylcholine, in contrast to soluble annexin I. Coaggregation with phosphatidylserine liposomes was stimulated by both millimolar calcium and magnesium, whereas the rate of coaggregation with phosphatidylcholine was not stimulated by calcium above 50 μM. The N-terminal domain of annexin I was especially important for interactions with other cellular proteins (Mailliard et al., 1996). It was detected by calcium-dependent binding of S100C to a fusion protein containing residues 2 to 47 of human annexin I.

The interaction of annexins with membranes requires calcium. The association of annexin II with phospholipid membranes is of high affinity and involves 11 to 12 equivalents of calcium per heterotetramer of two annexin II and two p11 subunits (Evans and Nelsestuen, 1994). This heterotetramer requires less calcium to induce membrane binding than does annexin II alone. Dimerization of the membrane binding unit may be needed for annexins to function at intracellular calcium levels. In the case of annexin V, the binding to membranes requires 10 to 100 μM calcium. With membranes containing phosphatidylcholine, protein–membrane association is rapid and is complete within seconds after the reagents are mixed (Bazzi and Nelsestuen, 1991, 1992). At low calcium levels, the kinetics of the binding reaction are sensitive to calcium concentration. However, in the presence of saturating levels of calcium and at relatively low protein/vesicle ratios, the binding reactions are rapid and the rate constants are comparable to the collisional limit: 1.4×10^{10} M^{-1} s^{-1} for large unilamellar vesicles and 2.7×10^{10} M^{-1} s^{-1} for small unilamellar vesicles (Lu et al., 1995). Protein dissociation is so slow that the complexes can be regarded as permanent. Calcium chelation causes rapid dissociation of protein–membrane complexes (first-order rate constants ranging from 10 to 50 s^{-1}).

In order to interact with membranes, annexins need not only Ca^{2+} ions, but also lipids. For example, the ability of lung annexin I to significantly penetrate monolayers of some major lipid components of lung and several synthetic lipids is calcium dependent and occurs only in the presence of dipalmitoylphosphatidylglycerol or palmitoyloleyoylphosphatidylglycerol (Koppenol et al., 1998). A lipid is inserted into the calcium binding loop of annexin I, and the conserved Lys128 of domain II of annexin I is involved in the specific binding of annexins to anionic phospholipids (Trave et al., 1994). About five Ca^{2+} ions are bound per annexin V monomer, and apparent calcium affinity is greatly increased in the presence of acidic phospholipids (Schlaepfer et al., 1987). Calcium chelators rapidly reverse the binding to phospholipids.

The interaction of annexins with membranes causes changes in the protein structure. As indicated by Fourier transform infrared spectroscopy, the change in conformation of annexin V upon binding to lipid monolayer membranes involves the formation of new β-strands in which interstrand hydrogen bonds orient parallel to the membrane surface (Silvestro and Axelsen, 1999). The role of the N-terminus of annexin V in governing annexin membrane interaction properties is very important (Hoekstra et al., 1993). On phospholipid monolayers, annexin V trimers assemble to form a hexagonal lattice with a unit vector length of about 180 Å (Vogel et al., 1994). Annexin V binding to phosphatidylserine liposomes is consistent with a binding surface area of about 60 phospholipid molecules per annexin. Mutational and crystallographic analysis showed that the side chains of highly conserved interfacial residues Trp185 and Thr or Ser at positions 72, 144, 228, and 303 in rat annexin V contribute to phospholipid binding and participate directly in intermolecular contacts with phospholipid membrane components (Campos et al., 1998).

A membrane inserted form of annexin XII with dramatically perturbed structure was found at pH 4 (Langen et al., 1998). Upon insertion, a continuous transmembrane α-helix is reversibly formed from the helix–loop–helix domain in the solution structure. Such helices might come together to form an aqueous pore that mediates the ion channel activity reported for some annexins. Formation of the calcium–anionic phospholipid–annexin V ternary complex significantly protects the protein from thermal denaturation as compared with annexin V alone, Ca^{2+}/annexin V, or lipid/annexin V mixtures (Wu et al., 1999). In the bulk phase, annexin V undergoes an irreversible thermal denaturation at 45 to 50°C, as shown by infrared spectroscopy. The interaction with the phospholipid monolayer does not practically influence the position of the thermal transition but changes its extent.

The interaction of annexins with membranes changes the properties of the membranes. The binding of annexins might decrease the lateral mobility of membrane phospholipids without inducing phase separation (Meers et al., 1991). EPR studies of fluidity of the hydrophobic interior of phospholipid vesicles (phosphatidylcholine or phosphatidylethanolamine) after calcium-dependent binding of human annexin V showed a membrane rigidification because of the "crystallizing" tendency of the vesicle bound annexin V (Megli et al., 1998). Fluorescence and NMR studies on phosphatidylcholine/phosphatidylserine showed that even at complete coverage of the lipid bilayers by annexin V, there is no influence on the lipid molecular packing and the acyl chain flexibility of both phospholipids (Saurel et al., 1998). The presence of the protein induces the appearance of slow motions (milliseconds to microseconds) of lipid molecules. At the same time, annexin V reduces the lateral diffusion rate of phosphatidylcholine by 40% and nearly blocks the diffusion of phosphatidylserine. Annexin IV also decreases the rate of lateral lipid diffusion and changes the fluid-phase structure of the lipid bilayer when it binds to negatively charged membranes in the presence of calcium (Gilmanshin et al., 1994). Association of annexin VI with membranes induces extensive clustering of acidic phospholipids as detected by self quenching of fluorescent labeled acidic phospholipids (Bazzi and Nelsestuen, 1991). Moreover, Kaetzel et al. (1994) found an inhibitory effect of annexin IV on calcium dependent Cl^- conductance. It is a mechanism by which calcium binding proteins can modulate membrane channel activity.

In some cases the binding of calcium results in self-association of annexins. For example, it was found by Zaks and Creutz (1991) that annexin IV (p32), annexin VI (p67), and annexin VII (synexin) self-associate in a calcium-dependent manner in both solution and in chromaffin granule membranes and phospholipid vesicles. The calcium level required for self-association of the annexins in membranes is about ten fold lower than that required when in solution (pCa 5.0 to 4.5). The authors suggested that the annexin self-association might be necessary for membrane contact at low calcium levels, but not at high levels, where the protein might bind two membranes as a monomer.

15.2.3. C2-Domain Proteins and Membranes

Protein kinases C (α, β, and γ) and related proteins (Allen and Katz, 1991; Bazzi and Nelsestuen, 1991; Newton, 1995) require calcium for membrane and phorbol ester binding as well as kinase activity. The data of M. C. Maurer et al. (1992) obtained by proton NMR in the presence of diheptanoylphosphatidylcholine/1, 2-dioleoyl-*sn*-glycerol are consistent with a single class of metal binding sites on protein kinase Cα.

Other proteins containing C2 calcium binding domains also interact with membranes. The most studied is synaptotagmin interaction with artificial liposomes containing phosphatidylserine (Brose et al., 1992; Davletov and Südhof, 1993; Chapman and Jahn, 1994). This interaction is induced by micromolar calcium and requires high concentrations of phosphatidylserine in liposomes (>25%). The C2A domain of synaptotagmin is responsible for phosphatidylserine binding. Synaptotagmin also interacts with phosphoinositides. Only highly phosphorylated members of this lipid family bind synaptotagmin with high affinity. The specificity of the interaction with different phosphoinositides is strictly dependent on the calcium concentration.

Neuronal communication is known to be mediated by calcium triggered fusion of synaptic vesicles filled with neurotransmitter with the presynaptic plasma membrane. Upon depolarization, calcium influx through voltage-activated channels triggers fusion of synaptic vesicles, releasing transmitters into the synaptic cleft, where they bind and activate postsynaptic receptors. The fusion process is strictly regulated by calcium. Synaptotagmin I functions as a calcium sensor that regulates exocytosis, whereas soluble *N*-ethylmaleimide-sensitive factor attachment protein (SNAP) receptor (SNARE) proteins in the vesicle and target membrane complexes that directly facilitate bilayer fusion (reviewed by Chicka et al., 2008). SNARE proteins are believed to form the core of a membrane fusion complex. The tandem C2 domains of synaptotagmin mediate calcium-independent and calcium-promoted interactions with the t-SNAREs syntaxin 1A and SNAP-25, providing a bridge between the putative calcium sensor for synaptic vesicle exocytosis and the core of the synaptic vesicle fusion apparatus. It was found that synaptotagmin interacts with SNARE proteins in the target membrane to halt SNARE complex assembly at a step after donor vesicles attach, or dock, to target membranes (Chicka et al., 2008). This results in fusion complexes that when subsequently triggered by calcium, drive rapid, highly efficient lipid mixing. Calcium independent interactions with SNAREs also predispose synaptotagmin to penetrate the target membrane selectively in response to calcium. Calci-synaptotagmin inserts into the target membrane to accelerate SNARE-facilitated fusion.

15.2.4. Gla-Containing Proteins and Membranes

Some vitamin K–dependent proteins (e.g., factor X and prothrombin) interact with membrane surfaces in a calcium-dependent manner, and these interactions play a vital role in blood coagulation. Blood clotting factor X and prothrombin binding to large unilamellar phospholipid vesicles is primarily entropy driven, whereas the binding to small unilamellar vesicles is primarily enthalpy driven (Plager and Nelsestuen, 1994). The exothermic process for calcium dependent factor X or prothrombin binding to small unilamellar vesicles may result from protein induced changes in the phospholipid packing/calcium interaction, possibly related to changes in how calcium is bound to the phospholipid. The binding to phosphatidylserine large unilamellar vesicles results in an enthalpy loss in the main endotherm of prothrombin denaturation ($T_m \approx 57$ to 58°C) and a comparable enthalpy gain in the minor endotherm ($T_m \approx 59^\circ$C) accompanying an upward shift in peak temperature ($T_m \approx 73^\circ$C) (Lentz et al., 1994). Kalafatis et al. (1994) found two regions of factor V_a light chains interacting with the lipid bilayer and having different requirements for the interaction. The binding site, located in the A3 domain, interacts with vesicles containing neutral phospholipid and is probably hydrophobic, whereas the binding site located on the C1–C2 domains interacts with membranes composed of anionic and neutral phospholipid and displays ionic binding characteristics.

15.2.5. α-Lactalbumin and Membranes

Some of the proteins with the highest affinity for calcium can also interact with membranes. α-Lactalbumin is able to interact with vesicles from dimyristoylphosphatidylcholine (Herreman et al., 1981; Hanssens et al., 1983, 1985; Amelloot et al., 1984); however, these studies were carried out without taking into consideration the effects of metal cation binding. Permyakov et al., (1988b) studied the interaction of α-lactalbumin and parvalbumin with dipalmitoylphosphatidylcholine vesicles by microcalorimetry, intrinsic fluorescence, and gel chromatography. According to the electron microscopy data, the diameter of the liposomes used in these works ranged from several hundred to several thousand angstroms. The narrow calorimetric peak with maximum at 41°C for pure liposomes from dipalmitoylphosphatidylcholine corresponds to the phase transition from the gel to the liquid crystalline state. Interactions of liposomes with various substances cause changes in this transition. In order to increase the sensitivity of microcalorimetry to changes in the position and shape of the thermogram for liposomes in the presence of proteins, Permyakov et al. (1988b) measured the difference between the heat sorption curves for liposomes in the presence and in absence of protein. The thermal denaturation of the proteins was measured by microcalorimetry and intrinsic fluorescence. The data obtained showed the existence of interactions between parvalbumin or α-lactalbumin and dipalmitoylphosphatidylcholine vesicles. The interactions change the physical properties of the proteins and the liposomes, increasing or decreasing their thermostability. These interactions are modulated by calcium and by magnesium.

Chromatography on Sephadex G-75 in another regime, in which the column is equilibrated with protein solution of a known concentration and the liposomes are applied to it (the Hummel Dreyer method), demonstrates that the total quantity of parvalbumin bound to the liposomes is high: roughly about 30 molecules per liposome for the calci-protein, about 50 molecules per liposome for the magnesi-protein, and about 70 molecules per liposome for the apo-protein. The dissociation constant of the complexes of liposomes with the protein evaluated from the Scatchard plots is 10^{-5} M (Permyakov et al., 1989a).

Cawthern et al. (1996) and Grishchenko et al. (1996) found that the thermal transitions in α-lactalbumin in isolated dimyristoylphosphatidylcholine (DMPC), dipalmitoylphosphatidyl-choline (DPPC), or egg lecithin liposomes depend on the metal state of the protein. The intrinsic fluorescence of DMPC- or DPPC-α-lactalbumin is sensitive to two thermal transitions: the phase transition from the gel to the liquid crystalline state in the lipid vesicles and the unfolding of the protein. Quenching experiments demonstrated that Trp accessibility increases upon the protein–liposome association. Above the protein transition (70°C), Trp's of α-lactalbumin interact significantly with the apolar phase of the phospholipids. At pH 2, where the protein inserts rapidly into the bilayer, the isolated DMPC/α-lactalbumin complex shows a distinct fluorescence thermal transition between 40 and 60°C, consistent with a partially inserted form that possesses some degree of tertiary structure and unfolds cooperatively. These results suggest a model in which a limited change of conformation occurs upon association with the membrane at neutral pH and physiological temperatures, with a concomitant increase in the exposure of tryptophan to external quenchers.

These data are compatible with those of Hanssens et al. (1983), who believe that at neutral pH α-lactalbumin is electrostatically adsorbed on the external surface of dimyristoylphosphatidylcholine vesicles, while at pH 4 the protein is able to penetrate the vesicles and to interact with the hydrophobic tails of fatty acids of the bilayer (Amelloot et al., 1984). Kim and Kim (1989) showed that a segment of α-lactalbumin, which penetrates into phosphatidylserine/phosphatidylethanolamine vesicle bilayers under acidic condition, exists in the membrane as an α-helix and that only one surface of this α-helix is exposed to the hydrophobic interior of the bilayer.

Banuelos and Muga (1995) studied the effect of membrane binding on the structure of several conformers of α-lactalbumin by infrared spectroscopy, circular dichroism, and fluorescence spectroscopy. They found that the membrane-associated conformations (when membranes are negatively charged) are characterized by high helical content, higher than any of those found in solution; a lack of stable tertiary structure; and the disappearance of their thermotropic transition. Their results indicate that the affinity of α-lactalbumin for negatively charged vesicles depends strongly on its conformational properties in solution (Banuelos and Muga, 1996). They revealed that native, calcium bound, ordered conformations associate with lipid bilayers through electrostatic forces. Substantial protein destabilization facilitates membrane binding by α-lactalbumin, and the association process is favored for a set of conformers having significant secondary structure, but lacking nativelike, stable tertiary structure. It was revealed that apo-α-lactalbumin has a higher affinity for dimiristoyl (and dipalmitoyl) phosphatidylcholine vesicles than the calcium loaded form at pH 7.4 and 26 to 28°C (Berliner and Koga, 1987). This is not a very good choice of temperature for the measurements, since 26 to 28°C is almost the middle of the denaturation thermal transition for apo-α-lactalbumin. This complicates interpretation of the data.

Electron spin resonance spectroscopy was used to study interactions of α-lactalbumin with dioleoyl and dimyristoyl phosphatidylglycerol bilayers (Montich and Marsh, 1995). It was found that the association of α-lactalbumin with dimyristoylphosphatidylglycerol bilayers increases chain mobility at temperatures in the lipid gel phase, restricts the chain mobility at temperatures corresponding to the lipid fluid phase, and abolishes the cooperative lipid chain melting transition. The binding of α-lactalbumin to dioleoyl phosphatidylglycerol bilayers at pH 4.0 causes a motional restriction throughout the full length of the lipid acyl chain. The protein might traverse the lipid bilayer.

15.2.6. Other Calcium Binding Proteins and Membranes

Calcium plays very important roles in the processes of fusion of secretory granules and synaptic vesicles with plasmatic membrane and also in the processes of intracellular membrane fusion (reviewed by Hay, 2007). All these calcium-dependent processes occur with participation of calcium binding proteins. For example, synaptotagmins are suggested to confer calcium dependence to the regulated exocytosis. Nevertheless, these proteins do not participate in early secretory events and hardly take part in calcium regulation of the transport of various substances. In contrast, calci-calmodulin plays an important role in the membrane fusion processes; it interacts with soluble protein receptors of *N*-ethylmaleimid-sensitive factor (SNARE), vesicle associated membrane protein 2 (VAMP2), taking part in exo- and endocytosis, syntaxin 13, and antigene 1. One more calcium sensitive protein taking part in the endosome fusion is the Hrs protein, which interacts directly with SNARE via a SNARE-like superhelix. Hrs is a large multifunctional peripheral membrane protein; its addition to the endosome fusion system inhibits this process. Hrs is supposed to prevent formation of the SNARE complex until a calcium signal from the inner part of endosome eliminates the inhibition; this permits formation of the SNARE complex. Another calcium sensitive protein taking part in the endosome fusion is the Hrs protein, which interacts directly with SNARE via a SNARE like superhelix. Hrs is a large multifunctional peripheral membrane protein; its addition to the endosome fusion system inhibits this process. Hrs is supposed to prevent formation of the SNARE complex until a calcium signal from the inner part of endosome eliminates the inhibition; this permits formation of the SNARE complex.

Chicken gizzard calponin is sedimented with phosphatidylserine and phosphatidylinositol vesicles but not with phosphatidylcholine vesicles (Fujii et al., 1995). The phospholipid binding site in this protein is located within the N-terminal 22-kDa fragment, in which the binding of actin, calmodulin, S100, and tropomyosin also occur. The amount of calponin bound to the vesicles decreases with increasing ionic strength or calcium concentration.

Actin interacts with distearoylphosphatidylcholine liposomes (Gicquaud, 1993). Differential scanning calorimetry shows that when actin is in contact with lipids, it undergoes a major conformational change, which results in the complete disappearance of its phase transition. In contrast, the phase transition of the membrane lipids is only weakly affected by the presence of actin. Electron microscopy revealed the formation of paracrystalline arrays of actin filaments at the surface of the liposomes.

16

GENETIC ENGINEERING OF CALCIUM BINDING PROTEINS

Genetic engineering has a range of meanings; as used in the context of calcium binding proteins, it refers to the synthesis in a prokaryote or eukaryotic cell of a protein(s) having amino acid sequence different, or mutated, from the wild type protein of reference amino acid sequence. This protein is subsequently purified and characterized; often, the wild type is also expressed as a control. One can also express the mutant protein in a cell of interest to see how it functions in vivo. Several problems include confirmation of the correct amino acid sequence, correct folding of the expressed protein into its native conformation, removal (if desired) of the N-terminal Met or fMet, and removal of undesired posttranslational modifications.

One can insert reporter groups, notably Trp for its spectral properties or Cys for its unique reactivity. Calcium binding sites can be modified and analyzed, one residue at a time, by point mutations and/or indels. Mutations elsewhere than in calcium binding sites yield information about the involvement, structural and thermodynamic, of the entire protein in calcium binding kinetics and affinity as well as interactions with other proteins. Chimeric proteins can be generated by splicing together parts of genes encoding different proteins. All of these techniques can be directed toward engineering calcium binding sites, as well as target sites, in proteins that previously had none.

Calcium Binding Proteins, By Eugene A. Permyakov and Robert H. Kretsinger

16.1. PROBLEMS

The user or critic should be aware of several problems in genetic engineering. The correct folding of a mutant or even wild type protein expressed in cell culture should be confirmed. This is especially important for proteins with several disulfide bonds; they must be formed correctly to give a native (like) structure. Folding procedures for expressed proteins are sometimes rather complicated. One should confirm by various methods the correctness of the polypeptide chain folding. This is relatively easy to do for the wild type recombinant protein, which should be expressed as a control, comparing its properties with the properties of the native protein. For mutant proteins it can be difficult to distinguish between effects of the mutation and incorrect folding. In some cases, researchers make mutations in proteins in order to induce only local changes in their structure, but one of the results of these mutations can be incorrect, unanticipated folding. For example, mutant α-lactalbumins and lyzozymes (Chapter 13) can have seriously distorted conformations due to incorrect formation of their four disulfide bonds. In this case one should check the correctness of the disulfide bonds' formation. Even for proteins without disulfide bonds, the correct global folding of mutants can be a serious problem; mutations can change the folding pathway of proteins.

Proteins expressed in both prokaryotes and eukaryotes have *N*-formyl methionine attached to the N-terminus of the synthesized protein. Within most cells *N*-deformylases and aminopeptidases remove the N-fMet (reviewed by Meinnel et al., 1993). Generally, the presence or absence of the N-fMet is ignored, even though such protein modification may have significant consequences. The effects of N-Met on the properties of α-lactalbumin were described by Veprintsev et al. (1999). Wild-type recombinant α-lactalbumin with N-Met is partially unfolded. Moreover, in the absence of calcium, the recombinant wild type protein has practically no thermal transition as monitored by differential scanning calorimetry; whereas, the native apo-protein melts with a half transition temperature around 30°C. In the presence of calcium, the recombinant wild type protein with fMet is 10 K less thermostable than the native protein. The calcium binding constant for the recombinant wild type α-lactalbumin is more then one order of magnitude lower than that for the wild-type native protein, extracted from tissue. The properties of the expressed protein with the N-Met enzymatically removed by aminopeptidase are close to those of the wild-type native protein. Fluorescence, circular dichroism, and differential scanning calorimetry show that expressed wild-type α-lactalbumin in the absence of calcium is in a molten globule state. Similar results were obtained for recombinant ribosomal protein S6 (Uversky et al., 1999). These results underscore the importance of either removing this N-Met with aminopeptidase or assuring that the Met does not significantly affect any properties of the protein.

16.2. INSERTION OF REPORTER GROUPS

One of the most sensitive and widely used methods for studying conformational changes in proteins is intrinsic fluorescence. However, some EF-hand proteins have no Trp's, the major reporter group in both fluorescent and circular dichroic spectroscopy. Moreover, most parvalbumins have no Tyr's. The insertion of Trp's, whose fluorescence and circular dichroism parameters are usually sensitive to the conformation of the protein, allows one to obtain more information about the protein.

Kilhoffer et al. (1988) made several F/W mutants of calmodulin, at residues 26 (calcium binding loop 1), 62 (loop 2), 99 (loop 3), and 135 (loop 4) to study calcium binding (Kilhoffer et al., 1992). Wu and Reid (1997a,b) made the F92W mutant, which is sensitive to the binding of calcium to the C-lobe of calmodulin.

A F29W mutation was made in the C-terminus of helix A of chicken troponin C; its fluorescence changes with the binding of calcium but not of magnesium (Pearlstone et al., 1992a,b; Trigo-Gonzalez et al., 1992). In contrast, fluorescence from W105 (F105W) in EF-hand 3 is sensitive to the binding of both calcium and magnesium (Trigo-Gonzalez et al., 1992). Similarly, the fluorescence of a F22W mutant is sensitive only to the binding of calcium to the low affinity sites, whereas fluorescence of Trp's at positions 52 (between B and C helices), 78, or 90 (linker helix) responds to the binding of calcium to both high and low affinity calcium binding sites (She et al., 1998a; Moncrieffe et al., 1999). This allows discrimination of binding of cations to the low and the high affinity sites.

The fluorescence of Trp's substituted into positions 29 and 105 was used to study properties of recombinant N-terminal (residues 1 to 90) and C-terminal (residues 88 to 162) lobes of troponin C (M. X. Li et al., 1994) as well as their interactions with fragments of troponin I (Pearlstone and Smillie, 1995; Pearlstone et al., 1997). To determine the residues responsible for the high calcium affinity in trout cardiac troponin C, trout and mammalian cardiac troponin mutants were characterized by monitoring fluorescence of F27W that is sensitive to calcium binding to site 2 (Gillis et al., 2000; 2003).

She et al. (1998a,b) substituted Trp into position 22 or 90 and Cys into position 52 or 101 of troponin C. They then attached the fluorescent label 5-(iodacetamidethyl)aminonaphthalene-1-sulfonic acid to the Cys. Fluorescence energy transfer from Trp to the fluorescent label allowed determination of the distances between residues 22 and 52, and 90 and 101 for various protein states.

Calcium binding to chicken recombinant skeletal muscle troponin C and its mutants containing Trp (F29W), 5-hydroxytryptophan (F29HTrp), or 7-azatryptophan (F29ZTrp) at position 29 was measured by flow dialysis and by fluorescence (Fortes de Valencia et al., 2003). Analysis of the results allowed the determination of the effects of each amino acid on the calcium binding properties of the N-terminal regulatory domain of troponin C. Compared with intact troponin C, the calcium affinities of the N-terminal sites were (1) increased sixfold in F29W, (2) increased threefold in F29ZTrp, and (3) decreased slightly in F29HTrp.

In addition to Trp, Tyr is also used as a reporter group. For example, Pearlstone et al. (1992) made a M48Y mutant (C-terminus of the B helix) in chicken troponin C and used the mutant protein for a circular dichroic study of the binding of calcium to the low affinity sites.

Pauls et al. (1993, 1994) made F102W rat parvalbumin; it is very sensitive to conformational changes (Permyakov et al., 1980). The fluorescence of the inserted Trp was used for studies of cation binding by parvalbumins with substitutions in their calcium binding sites (Pauls et al., 1994, 1996). The cation binding properties of F102W are similar to those of wild-type parvalbumin. A similar F102W mutant was made for oncomodulin (Durussel et al., 1996).

So called *atomic mutations* involve incorporation of chemically modified Trp's. For example, the single Trp187 of annexin was substituted by a Trp in which hydrogens in position 4, 5, or 6 of the indole ring were substituted by fluorine atoms (Minks et al., 1999). Such chemical modifications slightly increase the lengths of the covalent bonds and invert polarities of various parts of the indole ring. These minimal changes in local geometry do not influence the secondary and tertiary structures of the mutants; nevertheless, they differ from the wild type by their stability, cooperativity of folding, biological activity, and fluorescent properties. The authors explained these rather global effects induced by minimal local changes by relatively large changes in polar interactions of the indole ring or by changes in van der Waals radii, or by a combination of these two effects.

The insertion of fluorescent residues or the attachment of fluorescent labels may perturb the structure and/or reactivity of the targeted protein. It is often difficult to distinguish possible perturbations from the effect to be monitored by fluorescence. It is certainly reassuring to get similar results from different reporters or from different techniques.

16.3. MUTATIONS IN CALCIUM BINDING SITES

Calmodulin and troponin C have four calcium binding sites per molecule; this complicates the interpretation of results. They also play significant roles, hence justifying their investigations by genetic engineering. Maune et al. (1992) made several mutants of calmodulins, substituting the Glu [position 12 (−X) of each of the four calcium binding loops] with Gln to eliminate the negative charge or with Lys to change the negative to positive charge. Addition of excess calcium to the calmodulins with E/Q or E/K mutations in the N-lobe causes a more pronounced increase in helical content than the addition of excess calcium to the calmodulins with the mutations in the C-lobe. Maune et al. concluded that sites 3 and 4 of native calmodulin are filled by calcium first. Evenas et al. (1997, 1998) also studied properties of calmodulin with substitutions of the bidentate Glu (12, −X): E104Q and E140Q. Both mutatations decrease calcium affinities of sites 3 and 4 by two and four orders of magnitude, respectively.

Waltersson et al. (1993) made Asp/Asn mutations at position 3 (+Y) of EF-hands 3 and 4 of calmodulin. These D48N and D95N mutations have opposite effects. D95N decreases calcium affinity and positive cooperativity of calcium binding in the C-lobe. In contrast, D48N causes a slight increase in calcium affinity of site 2 and an increase in the cooperativity of calcium binding in the N-lobe.

Reid and Hodges (1980) postulated that the calcium affinity of the EF-hand depends on the number of acidic ligands and their locations in the loop, high affinity being reached when pairs of anionic ligands are located at the opposite vertexes of the calcium binding octahedron. To test this hypothesis Wu and Reid (1997a,b) made five mutant calmodulins with substitutions in site 3 designed so that that they contained three or four acidic residues with acidic pairs along the X and/or Z axes (Chapter 9). They showed that calcium affinity of site 3 increased with the increase in number of the chelating residues from zero to two, when the number of the acidic chelating pairs increased from three to four, and when the location of the pairs changed from X to Z axes. Black et al. (2000) made a series of mutants to probe the relationship between calcium affinity and the number of paired, negatively charged residues in EF-loop 1. When the number of acid pairs on the X-axis, the Y-axis, and the Z-axis was increased from zero to one, then to two, there is a progressive increase in calcium affinity of EF-hand 1. Addition of a sixth acidic chelating residue, by replacing the residue at position −Y (7), which coordinates calcium with its peptide carbonyl oxygen, with Asp, to both sites 1 and 2 reduces the calcium affinity of the N-lobe. Increases in calcium affinity are caused by increases in the calcium association rates for the Y- and Z-axis acid pairs, while the X-axis acid pair causes a reduction in the calcium dissociation rate.

Tikunova et al. (2001) made seven calmodulin mutants to examine the role of acidic residues in magnesium binding in the first EF-hand. A mutant with acidic residues in all six of the chelating positions has magnesium affinity similar to that of F19W calmodulin, which has four acidic residues in chelating positions, two of which form a Z-axis acid pair. Only calmodulin mutants that have Z-acid pairs were able to bind magnesium with physiologically relevant affinities. Removal of the Z-acid pair from the first EF-hand produced a dramatic 58-fold decrease in its magnesium affinity and a 1.8-fold increase in the rate of magnesium dissociation. Addition of an X- or Y-acid pair could not restore the high magnesium affinity lost with removal of the Z-acid pair. Therefore, the Z-acid pair in the first EF-hand supports high magnesium affinity primarily by increasing its rate of association. In contrast, the D133E mutant in EF-hand 4, which creates a Glu–Glu pair along the Z axis, decreases the calcium affinity of this site by more then three orders of magnitude and decreases the calcium affinity of site 3 (Wu and Reid, 1997a,b). This effect may be explained by difference in size between the side chains of Asp and Glu.

Browne et al. (1997) demonstrated the importance of the Ile or Val in positions 8 of each of the calcium binding loops in calmodulin for maintenance of its structure. These highly conservative residues take part of the hydrophobic cores of both lobes of calmodulin and connect the two calcium binding sites by means of a short antiparallel β-sheet. In the apo state, substitutions of Ile or of Val with Gly destabilizes the protein. The binding of calcium restores the native α-helical structure in the proteins, with mutations in sites 1 to 3 but not in the site 4. Gly can assume a broader range of ϕ,ψ dihedral value since it lacks a side chain. This permits Gly to form reverse turns of the polypeptide chain, but its substitution for other amino acids is generally destabilizing.

NMR spectroscopy was used to study calcium binding to mutant troponin C in which Glu41 in site 1 was substituted by Ala (Li et al., 1997; Gagne et al., 1997). This mutation decreases calcium affinity of sites 1 and 2 by two and one order of magnitude, respectively. Mutations in the high- and low-affinity sites of troponin C were used to reveal roles of these two types of binding sites in the regulation of muscle contraction (Szczesna et al., 1996). The authors inactivated binding sites 1 and 2, site 3, site 4, or sites 3 and 4 by means of substitution of the Asp at position X with Ala. They found that although the mutant with inactivated sites 1 and 2 binds to troponin-depleted muscle fibers, it could not restore their calcium dependent tension. All the proteins with mutations in the C-lobe bind to the muscle fibers but could not restore their calcium-dependent tension. The authors corroborated the conclusion that calcium-dependent activation of contraction of skeletal muscles is regulated by the low-affinity sites 1 and 2 in troponin C.

To investigate the roles of sites 1 and 2 invariant Glu residues 41 and 77 in the functional properties and calcium-induced structural opening of skeletal muscle troponin C regulatory domain, Pearlstone et al. (2000) replaced them with Ala in intact F29W troponin C and in wild-type and F29W N-domains (residues 1 to 90). Introduction of E41A/F29W and E77A/F29W mutants into skinned muscle fibers depleted of troponin C showed that calcium-induced tension is greatly reduced compared with the F29W control. The major structural change in the N-domain of troponin C, including the closed-to-open transition, is triggered by calcium binding at site 2.

Feng and Stemmer (2001) made four E/K mutants of calcineurin B, one in the 12 (−Z) position in each of the four EF-hands. The E/K mutants in EF-hands 1, 2, or 3 reduce calcium-dependent phosphatase activity in the absence of calmodulin; the site 2 mutation is most effective. The E/K mutant in binding site 4 does not change enzyme activity or sensitivity to calcium in either the absence or presence of calmodulin. The decrease in calcium sensitivity for the enzyme with site 2 of calcineurin B inactivated is apparently due to a decrease in the affinity of that mutant for calmodulin at low calcium concentrations.

Chicken (gizzard) smooth muscle myosin regulatory light chain (RLC) is a good model to study the properties of the EF-hand sites because it has only one site that binds both calcium and magnesium and retains its metal binding properties even when the RLC is dissociated from the other myosin chains. The protein is divided into two lobes, linked by a helix. The N-lobe contains the EF-hand that binds calcium. Blumenschein and Reinach (2000) made all possible combinations of three single mutations: D5S (coordination position Z), D9E (−X), and D12E (−Z). The single mutants D5S and D9E and the double mutant D5S/D9E have a low affinity for calcium. All the mutants containing mutation D12E are of low affinity and have higher affinities than does the wild-type RLC, even when containing mutations D5S or D9E. All of the mutants studied have a lower affinity for magnesium than that of the wild-type protein. Since high-affinity sites have the same coordinating positions as low-affinity mutants, the authors confirmed that the key for specificity is in the tertiary structure, not in any particular position.

Permyakov et al. (2000a) studied interactions of calcium with recoverin and mutants with substitutions in its two calcium binding sites by means of intrinsic fluorescence. Glu/Gln mutations in one (second or third) or two (second and third) calcium binding sites changes the calcium affinity of recoverin differently. E121Q and E85Q/E123Q result in a total loss calcium affinity; whereas, E85Q causes only a moderate decrease in calcium affinity. They suggested that the binding of calcium to recoverin occurs sequentially; filling of the second site becomes possible only after filling of the third site.

Matsuda et al. (1998) made E85M and E121M mutants of S-modulin (EF-2 and EF-3 calcium binding sites). E121M neither binds calcium nor inhibits phosphorylation of rhodopsin. In contrast, E85M binds one equivalent of calcium and has the same membrane affinity as that of wild-type S-modulin; however, it lost the ability to inhibit rhodopsin phosphorylation. The authors suggested that the binding of calcium to the EF-3 site is required for the EF-2 site to bind calcium and to induce exposure of the myristoyl group and that the binding of calcium to EF-2 is important for the interaction with rhodopsin kinase.

A 33-residue synthetic peptide mimic of the EF-hand cation binding motif was used to try to increase magnesium selectivity of the calcium binding site (Reid and Procyshyn, 1995). Decreasing the size of the cation binding cavity in the Z-axis acid-paired motifs through replacement of chelating residues in the +Z or −X metal ion coordinating positions in the loop region by glutamic acid was successful in decreasing the calcium ion affinity. The same changes did not create or enhance magnesium binding in the 33-residue model cation binding motif. The −X Glu in a 33-residue model peptide, ParvCD, comprising the CD site of carp parvalbumin 4.25 was replaced with Asp (ParvCD-XD) and the effect on calcium-dependent dimerization and calcium affinity was assessed (Franchini and Reid, 1999). The peptide ParvCD has a 105-fold lower calcium affinity than does the same site in the native protein. Both the ParvCD and ParvCD-XD model peptides fail to bind magnesium. The authors believe that the low calcium affinity and failure of the model ParvCD site to bind magnesium may be due to higher enthalpic costs of chelation by the −X Glu. Replacement of the −X Glu with an Asp resulted in a twofold increase in the calcium affinity of both the monomer and dimer forms and a twofold increase in the calcium-dependent dimerization of the peptide. In rat oncomodulin a −X Glu/Asp replacement increases calcium affinity (Hapak et al., 1989). The different effect of a −X Glu/Asp substitution in the different sites suggests site specific factors dictating the thermodynamic contribution of the −X glutamate to calcium affinity.

Pauls et al. (1994) made D51A/E62V/F102W and D90A/E101V/F102W mutants of rat parvalbumin. Both mutants bind one Ca^{2+} ion with dissociation constants 9.1×10^{-8} M and 3.1×10^{-7} M, respectively; native parvalbumin has two binding sites with $K_d(\mathrm{Ca}) = 3.7 \times 10^{-8}$ M. The mutations in the CD site result in its total inactivation but did not influence calcium affinity of the EF site; the mutations in the EF site caused its total inactivation and reduced the calcium affinity of the CD site.

Rhyner et al. (1996) made nine mutants—D51A, G62V, D51A/G62V; D90A, E101V, D90A/E101V; and D51A/D90A, E62V/E101V, D51A/G62V/D90A/E101 V—of human α-parvalbumin. The three mutants with an inactivated CD site bind one Ca^{2+} ion with $K_d(\mathrm{Ca}) = 4.3 \times 10^{-8}$ M and one Mg^{2+} ion with $K_d(\mathrm{Mg}) = 2.2 \times 10^{-4}$ M, very similar to the affinities of wild type parvalbumin. The mutants with an inactivated EF site bind one Ca^{2+} ion with $K_d(\mathrm{Ca}) = 1.3 \times 10^{-7}$, 2.2×10^{-7}, and 2.8×10^{-7} M for D/91A, E102V, and D/91A/E102V respectively. The three mutants with both sites inactivated bind neither calcium nor magnesium.

Cates et al. (1999) determined the crystal structures of parvalbumin mutants designed to explore the role of the last coordinating residue of the calcium binding loop. E101D has 100 fold lower calcium affinity and 10 fold higher magnesium affinity. E101D does not affect the magnesium coordination geometry of the binding loop, but it does pull the F helix 1.1 Å toward the loop. E101D cannot make the sevenfold coordination preferred by calcium, presumably because of limits imposed by the tertiary structure. The authors concluded that the last coordinating residue (−Z) and the calcium binding loop delimit the allowable geometries for the coordinating sphere.

Henzl et al. (1996) found that the insertion of a fifth carboxylate ligand (S55D and G98D) into the CD and EF sites of oncomodulin increases its affinity for both calcium and magnesium by almost one order of magnitude. Both mutations destabilize the apo form and stabilize the calci form of oncomodulin as seen in scanning calorimetry. Henzl et al. (1998) then change the sequence of the CD loop (low affinity) of oncomodulin into the sequence of its EF-loop (high affinity), and vice versa. They concluded from these experiments that the calcium affinity of a site depends not only on its amino acid sequence, but also on such factors as the binding loop flexibility and electrostatic potentials. It was found that although both mutations increase calcium affinity significantly, they utilize distinct energetic strategies. S55D improves calcium binding enthalpy; while the mutation G98A improves the binding entropy (Henzl et al., 2002). They also found different peptide backbone dynamics. G98D has an average order parameter (0.87), slightly greater than that of the wild-type oncomodulin; the S55D also has an average order parameter (0.82), slightly lower than that of the wild type. Furthermore, whereas just two backbone N–H bonds in G98D show internal motion on the 20- to 200-ps time scale, fully 52 of the 93 residues analyzed in S55D show this behavior. These findings suggest that the increased electrostatic repulsion attendant to introduction of an additional carboxylate into the CD site impedes backbone vibrational motion throughout the molecule.

Parvalbumin is the major fish allergen (Chapter 8). Dot blot assays and immunoblot inhibition experiments performed with sera from 21 fish allergic patients showed that mutant parvalbumin with substitutions in both EF and CD calcium binding sites had a 95% reduced IgE reactivity and represented the derivative with the least allergenic activity (Swoboda et al., 2007). This was confirmed by in vitro basophil histamine release assays and in vivo skin prick testing. The potential applicability for immunotherapy of the mutant parvalbumin was demonstrated by the fact that mouse IgG Abs could be raised by immunization with the mutated molecule, which cross reacts with parvalbumins from various species of fish and inhibits the binding of IgE from fish allergic patients to the wild type allergen. Using the hypoallergenic carp parvalbumin mutant, it may be possible to treat fish allergy by immunotherapy.

Malmendal et al. (1998) mutate in the calcium binding loop of the N-terminal noncanonical EF-hand of calbindin D_{9k} (A14D/A15Δ/P20Δ/N21G/P43M), trying to convert it to the consensus loop of the EF-hand. Using NMR methods, they showed that the resulting EF-hand is a rigid structure allowing binding of calcium with minimal reorganization and that the structural and dynamic properties of the entire EF-hand, not amino acid sequence of its calcium binding loop, dictate flexibility of the loop in this system.

In order to study contributions of electrostatic interactions to the total calcium binding energy, one can choose proteins with a single calcium binding site; this makes the interpretation of experimental data much easier. Substitution of Asp and Glu with Ala can result in formation of incorrectly folded mutant proteins (Pauls et al., 1994; Szczesna et al., 1996; Anderson et al., 1997). For example, some Asp/Ala mutations in α-lactalbumin (e.g., D87A) result in an incorrectly folded mutant, characterized by total absence of both calcium binding and cooperative thermal transitions (Anderson et al., 1997). In contrast, D87N α-lactalbumin binds calcium and has a cooperative thermal transition (Permyakov et al., 1997). The N-terminal Met of D87N was removed enzymatically to avoid its destabilizing action (Veprintsev et al., 1999). The calcium dissociation constant of this mutant is 5×10^{-6} M; this is two orders of magnitude higher than the value for the native protein. This difference in the dissociation constants corresponds to a difference of 11 kJ mol^{-1} in free energy of calcium binding. Taking into account that free energy change upon calcium binding to the native protein is about 54 kJ mol^{-1}, one can conclude that each ligand in the binding site gives approximately the same contribution to the change in free energy.

The calcium binding loop in galactose binding protein from *E. coli* connects an α-helix and a β-strand. Drake et al. (1997) made a set of mutant proteins with Gly substituted side chains of the coordinating residues. They found that Gly in loop positions 1, 3, 5, or 8, which are occupied by calcium coordinating or by hydrophobic side chains, destabilizes the protein. In contrast, Gly in positions 2, 4, 6, and 7, which do not take part directly in the calcium coordination, does not influence the calcium affinity of the protein. The authors concluded that an increase in flexibility of the calcium binding loop, as a result of the Gly substitutions, does not cause any significant change in affinity and selectivity of the site.

Substitution of the calcium binding amino acid residues in the calcium binding sites of proteins can be a powerful tool for studies of details of the structure and metal binding mechanism of these sites.

16.4. MUTATIONS ELSEWHERE THAN IN CALCIUM BINDING SITES

Calcium binding proteins have been mutated throughout to evaluate the effects on calcium binding, conformation, energetics, and interactions with targets.

Selective [^{15}N]Ile labeling of apo-calmodulin provides NMR probes in eight positions, allowing monitoring of the equilibrium thermal unfolding of the individual domains (Biekofsky et al., 2002). The approach was applied to apo-I63G and apo-V136G calmodulin. These mutations involve residues in position 8 of the calcium binding loops of calcium binding sites 1 and 2, respectively. These highly conserved residues contribute to the hydrophobic cores of the N- and C-lobes, respectively, and are involved in the functional link of the two calcium binding sites within each lobe via a short β-sheet. Both of these mutations cause significant destabilization of apo-calmodulin. These mutant proteins provide interesting models for studying the unfolding of a folded region (C- or N-folds) of calmodulin in the presence of the other, destabilized mutant fold. The results allow one to estimate the free energy of stabilization of a given folding unit due to its interaction with a neighboring unfolded portion of the calmodulin structure.

Genetic engineering can be used to test hypotheses about the function of a calcium binding protein. Tan et al. (1996) made Q41C/K75C and I85C/L112C mutants of calmodulin. They assumed that the Cys's in these positions would easily form disulfide bonds and block the calcium-induced transition in conformation. As predicted, the mutants bound calcium poorly and lost their ability to activate phosphodiesterase and calcineurin.

Vanscyoc and Shea (2001) expressed a fragment, residues 1 to 75 of calmodulin, and calmodulin mutated between calcium binding sites 1 and 2 in the N-lobe. These mutated calmodulins bind calcium as well as or better than does the wild-type fragment (1 to 75). There is a strong correlation between an increase in calcium affinity, an increase in Stokes radius, and a decrease in the thermal stability of the apo state, indicating that increased calcium affinity correlates with increased disorder. The mutations that have the greatest effect are farthest from the 12-residue calcium binding loops, in the linker between helices B and C (i.e., in the intralobe hinge between the two EF-hands). This indicates the importance of residues outside those ligands participating in the acid pairs that are well known to affect the calcium affinities of EF-hand proteins. The fluorescence was used to monitor the calcium induced changes in structure.

Mutations in various parts of calcium binding proteins allow one to see rather subtle details of their structures and dynamics. For example, mutations in the N- and D-helices of the N-lobe of troponin C were used to reveal interactions between the two lobes (Smith et al., 1999). In addition, it was shown that the troponin C mutant lacking residues 1 to 7 of the N-helix (des1 to 7) has a normal conformation and fulfills its regulatory function. Mutants R11C in the N-terminal helix and E76C in helix D (these residues form a hydrogen bond in the native protein) disturb the protein's functioning, decrease the stability of the N-lobe in the absence of bivalent cations, and increase the calcium affinity of the C-lobe.

Ramakrishnan and Hitchcock-DeGregori (1996) made three mutant troponins C: D89A, a protein in which the linker helix was substituted by an artificial helix with sequence ^{87}AEAALKAAMEA^{97}, and another troponin C with ^{87}AE*D*ALKAAMEA^{97}. The first two mutants are defective in activation of the actomyosin ATPase, whereas the third has higher activity. The authors suggested an important role of Asp89 in calcium-dependent signal transduction.

Tikunova et al. (2002) and Tikunova and Davis (2004) substituted 27 individual Phe's, Ile's, Leu's, Val's, and Met's in troponin C with Gln to examine the role of hydrophobic residues in calcium binding and energy exchange with the N-domain of a fluorescent troponin C, F29W. The effective calcium affinities of these troponin C F29W mutants vary $\sim$ 2340 fold, while calcium association and dissociation rates vary less than 70-fold and more than 45-fold, respectively. No correlation was found between the calcium binding properties of these F29W mutants of troponin C and the solvent accessibility of the hydrophobic amino acids in the apo state, in the calci state, or in the difference between the two states. The authors suggested that their results demonstrate that a single hydrophobic residue which does not ligate calcium directly can play an important role in controlling calcium binding within an EF-hand system. To sensitize the N-lobe of cardiac troponin C F27W to calcium, they individually substituted the hydrophobic residues F20, V44, M45, L48, and M81 with polar Gln (the regulatory N-lobe of cardiac troponin C is capable of binding only one Ca^{2+} ion, that with its second EF-hand. The first EF-hand of cardiac troponin C is unable to bind calcium because of a single residue insertion (Val28) and two chelating residue substitutions Asp29 $\rightarrow$ Leu and Asp31 $\rightarrow$ Ala). Tikunova and Davis tried to increase the calcium affinity of the N-domain of cardiac troponin C by facilitating the movement of helices B and C (BC unit) away from helices N, A, and D (NAD unit); this would be a large conformational transition from a closed to an open state. As anticipated, these selected hydrophobic residue substitutions increased the calcium affinity of the regulatory domain of cardiac troponin C F27W- two to 15-fold. Surprisingly, the increased calcium affinity caused by the hydrophobic residue substitutions was largely due to faster calcium association rates (2.6- to 8.7-fold faster) rather than to slower calcium dissociation rates (1.2- to 2.9-fold slower). The regulatory N-lobe of cardiac troponin C F27W and its mutants were also able to bind magnesium competitively and with physiologically relevant affinities (1.2 to 2.7 mM).

Genetic engineering can reveal functional differences between troponin C and calmodulin. Two principal differences in their amino acid sequences are a short N-terminal helix and the triplet ^{91}KGK^{93} in troponin C, both of which are absent in calmodulin. Gulati et al. (1995) made a R11A mutant in troponin C (this residue connects the N-terminal helix with the rest of troponin C). It has 50% calmodulin activity as judged by the phosphodiesterase activation test. The removal of ^{91}KGK^{93} increases the calmodulin activity up to 80%. Both mutants fulfilled 100% of the function of troponin C.

The binding of calcium to recoverin causes conformation changes, resulting in transfer of the N-terminal myristoyl group from the hydrophobic pocket to the protein surface to the water environment (Chapter 11). A. N. Baldwin and Ames (1998) made mutant recoverins W31K and I52A/Y53A to decrease the hydrophobicity of the myristoyl binding pocket to make inclusion of the myrystoyl group less energetically favorable; this could result in an increase in calcium affinity. As expected, myristoylated forms of W31K and I52A/Y53A have slightly higher calcium affinity and decreased binding cooperativity compared with wild-type recoverin.

Julenius et al. (1998) and Kragelund et al. (1998) made 11 mutant calbindin D_{9k}'s. They found a good correlation between the difference in unfolding free energy and the surface area change induced by mutation. They also found that substitutions of Ala and Gly with more bulky residues in the hydrophobic core increases the dissociation rate of calcium, and at a constant association rate, decreases affinity. In contrast, substitutions of residues far from the calcium binding site or near the surface of the protein influence calcium affinity only weakly. F66W causes a sixfold decrease in the calcium dissociation rate and a 25-fold increase in calcium affinity.

The conserved Arg75–Glu81 hydrogen bond in parvalbumin is buried in a hydrophobic interface near the AB domain. In the crystal structure of the rat parvalbumin mutant R75A, the main chain of the AB loop is displaced by 7.8 Å; however, the residues that flank residue 75Ala are not displaced (Hoh et al., 2009). The thermal stability of calci-R75A is lower ($\Delta T_m = -11.4$ K) than that of the wild-type protein, and apo-R75A is unfolded at room temperature. Both calcium and magnesium affinities are lower for R75A (Ca^{2+}: $K_1 = 2.4 \times 10^{-8}$ M, $K_2 = 1.6 \times 10^{-7}$ M; Mg^{2+}: $K_1 = 3.4 \times 10^{-5}$ M, $K_2 = 3.2 \times 10^{-4}$ M). Comparison of the crystallographic B factors showed an increase in the flexibility of the AB loop, consistent with this region being more exposed to solvent in the mutant. The mutant structure therefore demonstrates the role of this hydrogen bond in attaching the AB domain to the protein core. Normal-mode analysis indicates an altered orientation of the AB domain with regard to the CD–EF lobe.

Uchiyama et al. (1995) made five mutant α-lactalbumins—T29I, A30I, A30T, T33I, and A30I/T33I—all in helix B. They found that the stability of the molten globule state against denaturation by urea increases as a result of mutations that increase the hydrophobicity of the inner aspect of helix B. Substitution of all hydrophobic residues in the helical domain of α-lactalbumin with Leu's resulted in a protein that forms a molten globule with conservation of many structural features characteristic for the native protein (Wu and Kim, 1997).

Horii et al. (2001) made two mutant α-lactalbumins: T29V and T29I. These mutants are 1.9 and 3.2 kcal mol^{-1} more stable than the wild type to unfolding induced by guanidine hydrochloride. The overall structures of the mutants as seen in crystals are almost identical to that of the wild-type protein, despite the disruption of the hydrogen bond between the side chain hydroxyl group of Thr29 and the main chain carbonyl group of Glu25. Stabilization of the mutants was caused primarily by loss of solvation in the denatured state.

Greene et al. (1999) made 23 mutant α-lactalbumins; only three of them, in flexible regions adjacent to the region of interaction with galactosyl transferase, increase the stability of α-lactalbumins. Two of them are substitutions of Leu110, and the third is K114N. Both substitutions of Leu110 change functional activity. K114N increases its thermal stability by 10°C and decreases its activity.

Saeki et al. (2004) made 16 mutant goat α-lactalbumins and studied their equilibrium unfolding transitions and kinetic refolding reactions. The results suggest that the transition state structure is localized around the calcium binding site and at the interface between the C-helix and the β-domain.

Amino acid substitutions at the surface, as well as in the hydrophobic core, can stabilize calcium binding proteins. One example is E1M in α-lactalbumin (Veprintsev et al., 1999). Both apo-E1M and calci-E1M undergo a thermal transition 10 K higher. E1M also has a 10-fold higher affinity for calcium than that of the wild type. These effects probably reflect the elimination of the carboxlyate that pulls the N-terminus to the solvent.

16.5. GENERATION OF CHIMERIC PROTEINS

In contrast to mutant proteins having one or a few point mutations or short indels, one can create chimeric proteins in which (partial) domains from different proteins are spliced together.

Nakashima et al. (1996) made several chimeras combining yeast and chicken calmodulins. The C-lobe of chicken calmodulin was substituted with the C-lobe of yeast calmodulin. The chimeras were designed so that each calcium binding site of the yeast calmodulin was replaced in series from the C-terminus. Resulting mutant proteins showed calcium binding properties inherent to the original calcium binding site. Cooperative calcium binding and a suitable rearrangement of the two EF-hand sites in each half of calmodulin were shown to be important for high affinity interaction with calmodulin dependent cyclic nucleotide phosphodiesterase. Residues 88 to 128 and 129 to 148 of chicken calmodulin are required for effective activation of myosin light chain kinase (skMLCK and smMLCK) and phosphodiesterase, respectively. The differences in the structural requirements indicate different manners of interaction. While phosphodiesterase was activated to similar levels by different chimeras, the maximum activity (V_{max}) achieved by chicken calmodulins was not achieved by any chimeric calmodulins with MLCKs. Residues in chicken calmodulin sequences 1 to 50 and 88 to 129, in addition to calcium binding to the fourth site, are important for high values of V_{max} of skMLCK. On the other hand, Met51 and residues in chicken calmodulin sequence 88 to 129 are essential for the high V_{max} value of smMLCK.

Wang et al. (1998a) showed that substitution of the entire third EF-hand in troponin C with the corresponding domain of calmodulin results in creation of a mutant protein with calcium and magnesium affinities 10 times greater than those of the native troponin C. Substitution of loop 3 in troponin C with loop 3 of calmodulin gives a chimera with calcium affinity 10 times higher than that for native calmodulin. The authors suggested that the high calcium and magnesium affinity of the third EF-hand in troponin C results from the two (X and Z) acid pairs in loop 3, coupled with the higher hydrophobicity of helices E and F compared to those of the third EF-hand of calmodulin.

Franz et al. (1998) made a chimera of S100A2 in which the amino terminal calcium binding site 1 was replaced by the canonical EF-loop of α-parvalbumin (N mutant) and a second chimera in which both calcium binding loops of S100A2 were replaced by the canonical EF-loop of α-parvalbumin (NC mutant). Size exclusion chromatography and circular dichroism spectroscopy showed that irrespective of the state of cation binding, wild-type S100A2 and both mutants are dimers and are rich in α-helical structure. Wild type S100A2 binds four Ca^{2+} ions per dimer with pronounced positive cooperativity. Both mutants also bind four Ca^{2+} ions but with a higher affinity than does wild-type S100A2 and with negative cooperativity. The binding of the first two Ca^{2+} ions to the N mutant occurred with 100 fold higher affinity than in wild type S100A2 and a twofold increase for the last two Ca^{2+} ions. A further two to three fold increase in affinity was observed for respective binding steps of the NC mutant. Wild type S100A2 and both mutants bind four Zn^{2+} atoms per dimer with similar affinity, as determined by the Hummel–Dryer method.

Durussel et al. (1996) exchanged the entire CD sites of parvalbumin (high affinity) and oncomodulin (low affinity). The parvalbumin with the oncomodulin CD site binds only one Mg^{2+} ion per molecule (instead of two for the wild type parvalbumin), while the chimeric oncomodulin binds two Mg^{2+} ions per molecule (instead of one). The authors concluded that the high affinity depends upon the amino acid sequence of the CD site.

Pardon et al. (1995) substituted residues 76 to 102 in human lysozyme with residues 72 to 97 of bovine α-lactalbumin. In α-lactalbumin these residues comprise the calcium binding loop and the central helix C. This chimeric protein binds calcium with high affinity, $K_d(\text{Ca}) = 4 \times 10^{-9}$ M, and conserved 10% of the muramidase activity of lysozyme.

Masaki et al. (2000) made a chimeric α-lactalbumin composed of the α-domain of the human protein and β-domain of the bovine α-lactalbumin. This chimeric protein is in the molten globule state at neutral pH, in the absence of calcium and neutral pH. Its stability is identical to the stability of recombinant human α-lactalbumin and higher than that of the recombinant bovine protein. They concluded that the stability of α-lactalbumin is defined by the stability of the α-domain.

Mizuguchi et al. (1999) made a chimeric α-lactalbumin by replacing the flexible loop (residues 105 to 110) in human α-lactalbumin with helix D (residues 109 to 114) of equine lysozyme. The stability of the molten globule of the chimeric protein to the action of guanidine hydrochloride is identical to that of equine lysozyme and higher than that of human α-lactalbumin, despite the fact that the chimeric protein contains only six residues of lysozyme.

16.6. CREATION OF CALCIUM BINDING SITES IN PROTEINS

Cox et al. (1999) substituted the 10-residue sequence of the AB loop of several parvalbumins, which do not bind calcium, with a 12-residue canonical calcium binding loop. None of the AB sites inserted bind calcium.

Permyakov et al. (2000a) made a G160D/K161E/K162N/K166Q mutant in the fourth EF-hand of recoverin, which does bind calcium as the wild type. In addition to the wild-type EF-hands 2 and 3; this mutant EF-hand 4 does bind calcium. The G160D/K161E/K162N/K166Q mutant has higher calcium capacity, lower thermostability, and changed secondary and tertiary structure compared with wild-type recoverin.

Leontiev et al. (1993) mutated the loop in phage T4 lysozyme to the canonical calcium binding loop (G51D/N53D). The mutant protein has 60% of its initial activity and binds calcium but with a rather low binding constant (10^3 to 10^4 M^{-1}). The stability of this mutant T4 lysozyme to urea is lower than that of the wild type protein, both in the presence of calcium.

Yang et al. (2002) showed that all the canonical and pseudo-EF-hand calcium binding sites can be described using a set of geometric descriptions based on ideal pentagonal bipyramid geometry. In addition, two different natural magnesium binding sites in calbindin D_{9k} and parvalbumin can also be described using an octahedral coordinate system. They designed de novo calcium-binding sites into the scaffold of non-calcium binding proteins CD2 and Rop. Domain 1 of CD2 is an all β-sheet protein with an IgG fold, while Rop has a four-helix bundle. Both proteins contain many amino acids that may serve as calcium binding ligands. Glu was used as the anchor residue to design calcium binding sites in CD2 and Rop. A Ca^{2+} ion is (conceptually) placed on the same plane as two of the side-chain oxygen atoms and the Cδ atom with a distance of 2.4 Å between the calcium and oxygen atoms. The O–Ca–O angle and Ca–O–Cδ–Cγ dihedral angle are set to 53.8° and 93.5°, respectively. This anchor (Glu–Ca) has the same values for the Glu side chain with 24 configurations in the rotamer library. For non-EF-hand calcium binding sites, an anchor residue was generated by attaching a Ca^{2+} ion to an Asp with an O–Ca distance of 2.4 Å, using a Ca–Cδ–Cγ angle of 139°, and a Cγ–Oδ1–Oδ 2–Ca dihedral angle of 168°. For magnesium binding sites, an Asp with bound Mg^{2+} ion was used as the anchor. The Mg^{2+} ion is placed 2.1 Å away from the side-chain oxygen atom Asp with an Mg–Oδ–Cγ angle of 141° and an Mg–Oδ–Cγ–Cβ dihedral angle of 62.5°. After attaching the anchor residue to the backbone of the protein, the calcium binding geometry or positions of other ligands are then defined relative to the anchor. The configurations generated from this procedure are optimized using the Polak–Ribiere nonlinear conjugate gradient algorithm. All of the designed calcium binding sites in these proteins contain at least three or more mutations to the original structure with the calcium binding ligand residues. Four of the designed calcium binding sites in CD2 bind calcium strongly.

Jones et al. (2008) have successfully introduced a novel calcium binding site into the noncalcium binding cell adhesion protein CD2 using criteria tailored specifically to the structural and functional properties of the protein environment and charged adhesion surface. First, they introduced the bidentate ligand Asp, since it has a strong β-sheet propensity, which is important for stabilizing the native conformation. Second, to modulate the adhesion, the site was engineered adjacent to the CD48 (physiological binding partner) binding site. Third, residues known to be essential for CD48 binding were not mutated. Fourth, solvent accessibility, side-chain steric conflicts with existing atoms, disruption of hydrogen bonding, conserved folding residues, and hydrophobic interactions were also considered. The proposed calcium binding site of this mutant has a coordination number of six, with D79 as its bidentate ligand. The proposed calcium binding pocket is formed by the side-chain carboxyl oxygen atoms of two charge mutations, T79D and A92E, and two wild-type amino acids, N77 and N90. Residues D79 and N77 are from the F β-strand, while E92 and N90 are from the G β-strand of the protein. This designed site with ligand residues exclusively from the β-sheets selectively binds calcium and lanthanides over other mono- and divalent cations. Calcium and lanthanide binding specifically alters the local environment of the designed calcium binding site; however, the designed protein undergoes a significantly smaller conformation change compared with those observed in naturally occurring calcium binding sites that are composed of at least part of a flexible loop and helical regions. The CD2–CD48 binding affinity increases approximately threefold after this modification, suggesting that the cell adhesion of CD2 can be modulated by altering the local electrostatic environment.

Maniccia et al. (2006) investigated the effect of local charge on calcium affinity of designed CD2 proteins. Whereas mutation and calcium binding do not alter the native structure of the protein, calcium binding specifically induced changes around the designed binding site. The designed protein with a −5 charge at the binding site has a 14-, 20-, and 12-fold increase in affinity for calcium, terbium, and lanthanum, respectively, compared to the designed protein with a −3 charge; this indicates that greater local charges gives stronger calcium and lanthanide binding. The localized charged residues significantly decrease the thermal stability of the designed protein. Wild type CD2 has a T_m of 61°C, which is similar to the designed protein with a −3 charge. The designed protein with a −5 charge has a T_m value of 41°C. This decrease in melting temperature is partially restored by Ca^{2+} binding. The effect on protein stability is modulated by the environment and the locations of the charged mutations.

Ye et al. (2003) inserted the calcium binding loop 3, with Gly linkers, from calmodulin into the scaffold protein CD2.D1 at three locations to study site-specific calcium binding properties of EF-hand motifs. They followed three criteria in their work. First, the host protein must retain its native structure in the presence and absence of calcium so that the large change in conformational entropy that arises from a conformational change in the protein is reduced. Second, the calcium binding loop should be able to maintain its native calcium binding and structural properties in a foreign host protein. Third, the influence of the host protein environment on the calcium binding properties of the grafted EF-loop needs to be minimized, if not completely decoupled. Three different locations in CD2.D1 were chosen for insertion of the loop based on several considerations. First, CD2.D1, a noncalcium binding cell adhesion molecule, has been shown to be an excellent host protein with a strong ability to maintain its native structure upon large changes of electrostatic interactions. The three insertion positions, 22–23, 52–53, and 83–84, are well solvated and seem to be tolerant to mutations. In addition, these loops in the host protein have significantly different relative electrostatic potentials predicted using AMMP (Another Molecular Mechanics Program). While position 52–53 of the CD loop has a positive potential of 47 kcal mol^{-1}, position 22–23 of loop strands BC and position 83–84 in loop FG have negative potentials of about −42 and −21 kcal mol^{-1}, respectively. Furthermore, these loops connect with different β-strands and at opposite sides or ends of the protein structure, representing different protein environments with different hydrogen bonding and hydrophobic interactions. To provide the flexibility required for calcium binding and native geometry, three Gly residues were used as attachments at either side of the EF-loop. The host protein forms a 1 : 1 calcium, or lanthanum, protein complex and retains its native structure. Tyr-sensitized terbium energy transfer showed that lanthanum and calcium compete for the same site. The EF-loop grafted into three in different environments has similar lanthanum binding affinities, suggesting that it is largely solvated and functions independent of the host protein.

The creation of artificial calcium binding sites is very promising since it could result in development of new technologies of creation of metal binding proteins with properties appropriate to various applications in research and in technology: for example, selective sensors or filters for water purification.

16.7. STUDIES OF PROTEIN–PROTEIN AND PROTEIN–MEMBRANE INTERACTIONS

Vogel and Zhang (1995) made a set of mutant calmodulins with substitutions of Met's by Leu's. After confirming that the mutations do not produce any essential distortions of the protein secondary and tertiary structure, they studied their activation of enzymes. M36L does not activate phosphodiesterase, but activates myosin light chain kinase 100% and activates calcineurin 75%. In contrast, M124L activates phosphodiesterase and calcineurin 100% and activates myosin light chain kinase 50%. The authors suggested that although all three enzymes bind to the same common site of calmodulin, there are essential differences in the details of their binding. Meyer et al. (1996) made the F92A mutant and found that it cannot activate phosphodiesterase and activates calcineurin 60%.

Yin et al. (1999) made several Gln/Met mutants at multiple positions in calmodulin. The substitution of multiple Gln's in either lobe results in a decrease in the affinity of calmodulin for PM-Ca-ATPase. However, despite multiple substitutions of Gln for Met's at positions 36, 51, 71, and 72 in the N-lobe or for Met's at positions 124, 144, and 145 in the C-lobe, these mutants are able to fully activate PM-Ca-ATPase. The role of individual Met's in modulating the affinity between the carboxyl terminus and PM-Ca-ATPase was further investigated by the substitution of individual Met's with Gln. The results showed differences in the interactions between individual Met mutants in calmodulin and different target enzymes, and suggest that hydrophobic interactions between Met's in calmodulin and the binding site on PM-Ca-ATPase are not necessary for enzyme activation. These results indicate that the initial associations between calmodulin and the calmodulin binding sequence in PM-Ca-ATPase are guided by nonspecific hydrophobic interactions involving both lobes of calmodulin.

In most well-characterized cases, both lobes of calmodulin interact with a target protein. However, physiologically separable roles for the two domains were demonstrated by mutants of *Paramecium tetraurelia* calmodulin. To determine whether these mutants can associate with canonical targets in a calcium-dependent manner, their abilities to bind melittin were assessed using analytical gel permeation chromatography, analytical ultracentrifugation, and fluorescence spectroscopy (Sorensen et al., 2001). The Stokes radius of wild-type *P. tetraurelia* calmodulin and 11 of the mutants decreased dramatically upon binding melittin in the presence of calcium. Melittin binds to wild type *P. tetraurelia* calmodulin and its mutants in a calcium independent manner. However, there are domain-specific perturbations. Mutations in the N-domain of *P. tetraurelia* calmodulin do not affect the spectrum of Trp19 of melittin under apo or calci conditions, whereas most of the mutations in the C-lobe do. These data are consistent with a calcium dependent model of sequential target association, whereby melittin (1) binds to the C-lobe of *P. tetraurelia* calmodulin in the absence of calcium, (2) remains associated with the C-lobe upon calcium binding to sites 3 and 4, and (3) subsequently binds to the N-lobe upon calcium binding to sites 1 and 2, causing collapse of the tertiary structure of calmodulin.

Computational algorithms have been used to optimize interactions of calmodulin with its targets. For example, Yosef et al. (2009) used this computational approach to modify calmodulin binding specificity for two targets: calmodulin-dependent protein kinase II (CaMKII) and calcineurin (CaN). Starting from the crystal structure of calmodulin in complex with the calmodulin binding domain of CaMKII, they optimized calmodulin interactions with CaMKII by introducing mutations into the calmodulin sequence. The optimization was performed with a protein design program, ORBIT, using a modified energy function that emphasize intermolecular interactions in the sequence selection procedure. Several calmodulin mutants were constructed and tested for binding to CaMKII and CaN peptides using surface plasmon resonance. Most of these calmodulin mutants have increased affinity for the CaMKII peptide and substantial decreases in affinity for the CaN peptide compared to that of wild-type calmodulin. The best calmodulin design exhibited about a 900-fold increase in binding specificity toward the CaMKII peptide.

Kobayashi et al. (1999) made five mutations—R44A, R81A, E53A/E54A, E60A/E61A, and E85A/D86A—in the N-lobe of rabbit fast skeletal muscle troponin C. These residues surround a hydrophobic pocket in the N-lobe which, when calcium binds, is exposed and interacts with the inhibitory region of troponin I. Studies of these interactions of these mutants with troponin I and with troponin T and studies of the inhibition of actomyosin S1 ATPase activity showed an elaborate network of interactions formed by troponin I, troponin T, and the N-lobe of troponin C, all of which are important in the calcium-dependent regulation of muscle contraction.

Landar et al. (1997) made point mutations of Cys68 and of Cys84 in S100. They showed that these mutations do not influence the dimerization of S100; however, they are very important for interactions with some of its target proteins. Landar et al. (1998) made both deletion and point mutants of S100A1—lacking the carboxyl terminal nine residues, Δ85 to 93, or containing Ala substitutions at Phe88 (F88A), Phe 89 (F89A), or Trp90 (W90A), or F88A/F89A, or F88A/F89A/W90A. Fluorescence studies showed that the Δ85–93 S100A1 has reduced calcium dependent interaction with the dodecyl CapZ peptide, TRTK, while W90A binds TRTK with a K_d of 5.55 μM. These results demonstrate that a calcium dependent target binding site and a phenyl-Sepharose binding region are indistinguishable. In contrast to the calcium dependent binding of TRTK, activation of the calcium independent target protein aldolase A by these S100A1 mutants is indistinguishable from binding by native S100A1. These results demonstrate that these nine carboxyl terminal residues are not required for S100A1 modulation of the calcium independent target, aldolase A. Altogether, these results indicate that S100A1 utilizes distinct mechanisms for interaction with calcium independent and calcium dependent target proteins.

Peflin and ALG-2 belong to the penta-EF-hand subfamily (Chapter 11). ALG-2 forms either a homo- or a heterodimer with peflin, as do other penta-EF-hand proteins. Kitaura et al. (2002) found that the fifth EF-hand (EF-5) regions of both peflein and ALG-2 are essential for dimerization and for stability. Exogenously expressed EF-5-deletion (ΔEF-5) mutants of ALG-2 and peflin are unstable and are not detected in HEK293 cells by Western blotting. In a pulse chase experiment, the ΔEF-5 mutants are rapidly degraded, but they are stabilized by treatment with a proteasome inhibitor, MG132. In MG132-treated cells, ΔEF-5 mutants are recovered in the insoluble fractions. These results indicate that the absence of this fifth EF-hand results in rapid degradation by the proteasome.

Interactions that stabilize the native state of a protein were studied by measuring the affinity between subdomain fragments with and without site specific residue substitutions (Berggård et al., 2001). A calbindin D_{9k} variant with a single CNBr cleavage site at position 43 between its two EF-hands was used as a starting point for the study. Eleven substitutions involving residues at the hydrophobic interface between the two EF-hands were introduced. The native calbindin D_{9k} and its mutants were cleaved with CNBr to produce wild-type and mutated single EF-hands: EF1 (residues 1 to 43) and EF2 (44 to 75). For the 11 mutants of the core, a strong correlation was found between the affinity of EF1 for EF2 and the stability toward denaturation of the corresponding intact protein. The correlation observed implies that the factors governing the stability of the intact protein also contribute to the affinity of the bimolecular EF1–EF2 complex.

The calcium binding sites of both annexin IV and surfactant protein A (SP-A) were investigated using mutants (Sohma et al., 1999). SP-A binds in a calcium-dependent manner to an annexin IV truncation mutant consisting of the N-tail and the first three of four domains, T_{N-1-3}. SP-A also binds to T_{3-4}, but this interaction is not calcium dependent. SP-A binds weakly to the other truncation mutants (T_{N-1-2}, T_{2-3}, and T_{2-3-4}). Each domain of annexin IV has conserved acidic residues (Glu70, Asp142, Glu226, and Glu301) that putatively ligate calcium. Annexin IV mutants in which one, two, or three residues of the four Asp/Glu's were substituted by Ala in helices D and E showed that calcium binding in the third domain is more important than in the other calcium binding sites for SP-A binding. SP-A is a member of the animal lectin group and is homologous with mannose binding protein A. The mutations R197D or R197N in SP-A eliminate binding to annexin IV, whereas E195Q has no effect. These results indicate that calcium binding to domain 3 of annexin IV is required for binding by SP-A and that Arg197 of SP-A is important in this binding.

Annexins I, II, IV, and VII, promote membrane aggregation. To identify the residues involved, Bitto and Cho (1999) generated truncated mutants of human annexin I lacking various parts of the N-terminal tail. The in vitro vesicle binding and aggregation activities of these mutants indicate that residues Lys26 and Lys29 of the N-tail and the C-terminal domain are involved in membrane aggregation.

Annexin II binds and aggregates membranes in a calcium dependent manner. Protein kinase C phosphorylates Ser's 11 and 25, thereby modifying its oligomeric structure and its ability to aggregate membranes.

Ayala-Sanmartin et al. (2000) made mutants in which Ser11 and Ser25 of annexin II were replaced by Ala and/or Glu. S11E or S25E mutants, in either monomeric or tetrameric form, have the same affinities for membranes. However, S11E and S25E affect the aggregation of the two forms. The aggregation efficiency of the tetramer is decreased but not the calcium affinity; whereas, this affinity is affected in the monomer. These effects are stronger in the S11E mutant and they are cumulative in the double mutant. This indicates a different conformation of the N-tail in the mutants (and in the phosphorylated protein), an inference that is supported by proteolysis experiments.

REFERENCES

Ababou, A., and Desjarlais, J. R. Solvation energetics and conformational change in EF-hand proteins. Protein Sci. (2001) 10, 301–312.

Abe, Y., Yamaguchi, H., and Yazawa, M. Ca^{2+}-dependent changes in helical structure in calmodulin. Proc. Jpn. Acad. (1988) 64, Ser. 8, 6, 163–166.

Acharya, K. R., Stuart, D. I., Walker, N. P. C., Lewis, M., and Phillips, D. C. Refined structure of baboon α-lactalbumin at 1.7 Å resolution: comparison with C-type lysozyme. J. Mol. Biol. (1989) 208, 1, 99–127.

Acharya, K. R., Stuart, D. I., Phillips, D. C., and Scheraga, H. A. A critical evaluation of the predicted and x-ray structures of α-lactalbumin. J. Protein Chem. (1990) 9, 5, 549–563.

Acharya, K. R., Ren, J., Stuart, D. I., Phillips, D. C., and Fenna, R. E. Crystal structure of human α-lactalbumin at 1.7 Å resolution. J. Mol. Biol. (1991) 221, 3, 571–581.

Acharya, K. R., Stuart, D. I., Phillips, D. C., McKenzie, H. A., and Teahan, C. G. Models of the three-dimensional structures of echidna, horse, and pigeon lysozymes: calcium-binding lysozymes and their relationship with α-lactalbumins. J. Protein Chem. (1994) 13, 6, 569–584.

Addy, C., Ohara, M., Kawai, F., Kidera, A., Ikeguchi, M., Fuchigami, S., Osawa, M., Shimada, I., Park, S. Y., Tame, J. R., and Heddle, J. G. Nickel binding to NikA: an additional binding site reconciles spectroscopy, calorimetry and crystallography. Acta Crystallogr. D (2007) 63, Pt. 2, 221–229.

Adman, E. T., and Jensen, L. H. Structural features of azurin at 2.7 Angstroms resolution. Isr. J. Chem. (1981) 21, 8–18.

Aeschlimann, D., Wetterwald, A., Fleisch, H., and Paulsson, M. Expression of tissue transglutaminase in skeletal tissues correlates with events of terminal differentiation of chondrocytes. J. Cell. Biol. (1993) 120, 1461–1470.

Calcium Binding Proteins, By Eugene A. Permyakov and Robert H. Kretsinger

Afonso, C. L., Tulman, E. R., Lu, Z., Oma, E., Katish, G. F., and Rock, D. L. The genome of *Melanoplus sanguinipes* entomopoxvirus. J. Virol. (1999) 73, 533–552.

Ahmed, F. R., Przybylska, M., Rose, D. R., Birnbaum, G. I., Pippy, M. E., and MacManus, J. P. Structure of oncomodulin refined at 1.85 Å resolution: an example of extensive molecular aggregation via Ca^{2+}. J. Mol. Biol. (1990) 216, 1, 127–140.

Airaksinen, M. S., Eilers, J., Garaschuk, O., Thoenen, H., Konnerth, A., and Meyer, M. Ataxia and altered dendritic calcium signaling in mice carrying a targeted null mutation of the calbindin D_{28k} gene. Proc. Natl. Acad. Sci. USA (1997) 94, 1488–1493.

Aisen, P., and Listowsky, I. Iron transport and storage proteins. Annu. Rev. Biochem. (1980) 49, 357–393.

Aitio, H., Annila, A., Heikkinen, S., Thulin, E., Drakenberg, T., and Kilpelainen, I. NMR assignments, secondary structure, and global fold of calerythrin, an EF-hand calcium-binding protein from *Saccharopolyspora erythraea*. Protein Sci. (1999) 8, 2580–2588.

Aitio, H., Laakso, T., Pihlajamaa, T., Torkkeli, M., Kilpelainen, I., Drakenberg, T., Serimaa, R., and Annila, A. Characterization of apo and partially saturated states of calerythrin, an EF-hand protein from *S. erythraea*: a molten globule when deprived of Ca^{2+}. Protein Sci. (2001) 10, 74–82.

Ajioka, J., and Swindle, J. The calmodulin-ubiquitin associated genes of *Trypanosoma cruzi*: their identification and transcription. Mol. Biochem. Parasitol. (1993) 57, 1, 127–136.

Akke, M., Skelton, N. J., Kordel, J., Palmer, A. G., and Chazin, W. J. Effects of ion binding on the backbone dynamics of calbindin D_{9k} determined by ^{15}N NMR relaxation. Biochemistry (1993) 32, 37, 9832–9844.

Alagem, N., Dvir, M., and Reuveny, E. Mechanism of Ba^{2+} block of a mouse inwardly rectifying K^+ channel: differential contribution by two discrete residues. J. Physiol. (2001) 534, 2, 381–393.

Aleksenko, S. S., Hartinger, C. G., Semenova, O., Meelich, K., Timerbaev, A. R., and Keppler, B. K. Characterization of interactions between human serum albumin and tumor-inhibiting amino alcohol platinum(II) complexes using capillary electrophoresis. J. Chromatogr. A. (2007) 1155, 2, 218–221.

Alexandrescu, A. T., Broadhurst, R. W., Wormald, C., Chyan, C. L., Baum, J., and Dobson, C. M. ^1H-NMR assignments and local environments of aromatic residues in bovine, human and guinea pig variants of α-lactalbumin. Eur. J. Biochem. (1992) 210, 3, 699–709.

Alford, R. H. Metal cation requirements for phytohemagglutinin-induced transformation of human periferal blood lymphocytes. J. Immunol. (1970) 104, 698–703.

Allen, B. G., and Katz, S. Isolation and characterization of the calcium- and phospholipid-dependent protein kinase (protein kinase C) subtypes from bovine heart. Biochemistry, (1991) 30, 117, 4334-4343.

Alliel, P. M., Perin, J. P., Jolles, P., and Bonnet, F. J. Testican, a multidomain testicular proteoglycan resembling modulators of cell social behaviour. Eur. J. Biochem. (1993) 214, 347–350.

Altschul, S. F., Gish, W., Miller, W., Myers, E. W., and Lipman, D. J. Basic local alignment search tool. J. Mol. Biol. (1990) 215, 3, 403–410.

Alvarez, J., and Montero, M. Measuring [Ca^{2+}] in the endoplasmic reticulum with aequorin. Cell Calcium (2002) 32, 5–6, 251–260.

Amelloot, M., Hendickx, H., Herreman, W., Pottel, H., Van Cauvelaert, F., and Van der Meer, W. Effect of orientational order on the decay of the fluorescence anisotropy in membrane suspensions: experimental verification on unilamellar vesicles and lipid α-lactalbumin complexes. Biophys. J. (1984) 46, 4, 525–540.

Amend, J. P., Edwards, K. J., and Lyons, T. W. Sulfur Biogeochemistry: Past and Present. Geological Society of America, Boulder, CO, 2004.

Ames, J. B., Ishima, R., Tanaka, T., Gordon, J. I., Stryer, L., and Ikura, M. Molecular mechanics of calcium–myristoyl switches. Nature (1997) 389, 6647, 198–202.

Ames, J. B., Tanaka, T., Stryer, L., and Ikura, M. Secondary structure of myristoylated recoverin determined by three-dimensional heteronuclear NMR: implications for the calcium–myristoyl switch. Biochemistry (1994) 33, 35, 10743–10753.

Ames, J. B., Porumb, T., Tanaka, T., Ikura, M., and Stryer, L. J. Amino-terminal myristoylation induces cooperative calcium-binding to recoverin. Biol. Chem. (1995a) 270, 4526–4533.

Ames, J. B., Tanaka, T., Ikura, M., and Stryer, L. Nuclear magnetic resonance for Ca^{2+}-induced extrusion of the myristoyl group of recoverin. J. Biol. Chem. (1995b) 270, 52, 30909–30913.

Ames, J. B., Levay, K., Wingard, J. N., Lusin, J. D., and Slepak, V. Z. Structural basis for calcium-induced inhibition of rhodopsin kinase by recoverin. J. Biol. Chem. (2006) 281, 48, 37237–37245.

Amicosante, M., and Fontenot, A. P. T cell recognition in chronic beryllium disease. Clin. Immunol. (2006) 121, 2, 134–143.

Amos, L. A. Molecular motors: not quite like clockwork. Cell. Mol. Life Sci. (2008) 65, 4, 509–515.

Amos, W. B. Reversible mechanochemical cycle in the contaction of *Vorticella*. Nature (1971) 229, 127–128.

Anagli, J., Hofmann, F., Quadroni, M., Vorherr, T., and Carafoli, E. The calmodulin-binding domain of the inducible (macrophage) nitric oxide synthase. Eur. J. Biochem. (1995) 233, 701–708.

Anderson, P. J., Brooks, C. L., and Berliner, L. J. Functional identification of calcium binding residues in bovine α-lactalbumin. Biochemistry (1997) 36, 39, 11648–11654.

Andersson, C. E., and Mowbray, S. L. Activation of ribokinase by monovalent cations. J. Mol. Biol. (2002) 315, 3, 409–419.

Andrade, J., Zhao, H., Titus, B., Timm Pearce, S., and Barroso, M. The EF-hand Ca^{2+}-binding protein p22 plays a role in microtubule and endoplasmic reticulum organization and dynamics with distinct Ca^{2+}-binding requirements. Mol. Biol. Cell. (2004) 15, 2, 481–496.

André, B., Noegel, A. A., and Schleicher, M. *Dictyostelium discoideum* contains a family of calmodulin-related EF-hand proteins that are developmentally regulated. FEBS Lett. (1996) 382, 198–202.

Andre, I., and Linse, S. Measurements of Ca^{2+}-binding constants of proteins and presentation of the CaLigator software. Anal. Biochem. (2002) 305, 195–205.

André, J. P., and Mäcke, H. R. NMR spectroscopy of group 13 metal ions: biologically relevant aspects. J. Inorg. Biochem. (2003) 97, 4, 315–323.

Andrews, S. C., Robinson, A. K., and Rodriguez-Quiñones, F. Bacterial iron homeostasis. FEMS Microbiol. Rev. (2003) 27, 215–237.

Anglister, J., Grzesiek, S., Wang, A. C., Ren, H., Klee, C. B., and Bax, A. ^{1}H, ^{13}C, ^{15}N nuclear magnetic resonance backbone assignments and secondary structure of human calcineurin B. Biochemistry (1994) 33, 12, 3540–3547.

Aoba, T., Shimazu, Y., Taya, Y., Soeno, Y., Sato, K., and Miake, Y. Fluoride and apatite formation in vivo and in vitro. J. Electron Microsc. (Tokyo) (2003) 52, 6, 615–625.

Aoki, S. T., Settembre, E. C., Trask, S. D., Greenberg, H. B., Harrison, S. C., and Dormitzer, P. R. Structure of rotavirus outer-layer protein VP7 bound with a neutralizing Fab. Science (2009) 324, 5933, 1444–1447.

Aoyagi, M., Arvai, A. S., Tainer, J. A., and Getzoff, E. D. Structural basis for endothelial nitric oxide synthase binding to calmodulin. EMBO J. (2003) 22, 4, 766–775.

Armstrong, C. M., and Taylor, S. R. Interaction of barium ions with potassium channels in squid giant axons. Biophys. J. (1980) 30, 473–488.

Aron, H. Eine einfache Methode zur Bestimmung des Calciums in organischen Substanzen. Biochem. Z. (1907) 4, 268–270.

Arredondo, M., and Nuñez, M. T. Iron and copper metabolism. Mol. Aspects Med. (2005) 26, 313–327.

Arthus, M., and Pagès, C. Nouvelle théorie chimique de sang. Arch. Physio. Normale (1890) 5, 2, 739–746.

Ash, D. E. Structure and function of arginases. J. Nutr. (2004) 134, 2760S–2764S.

Aspberg, A., Binkert, C., and Ruoslahti, E. The versican C-type lectin domain recognizes the adhesion protein tenascin-R. Proc. Natl. Acad. Sci. USA (1995) 92, 10590–10594.

Aspberg, A., Miura, R., Bourdoulous, S., Shimonaka, M., Heinegard, D., Schachner, M., Ruoslahti, E., and Yamaguchi, Y. The C-type lectin domains of lecticans, a family of aggregating chondroitin sulfate proteoglycans, bind tenascin-R by protein–protein interactions independent of carbohydrate moiety. Proc. Natl. Acad. Sci. USA (1997) 94, 10116–10121.

Atack, J. R. Lithium, phosphatidylinositol signaling, and bipolar disorder. In: Manji, H. K., Bowden, C. L., Belmaker, R. H., eds., Bipolar Medications: Mechanism of Action. American Psychiatric Press, Washington, DC, 2000, pp. 1–30.

Atiyeh, B. S., Costagliola, M., Hayek, S. N., and Dibo, S. A. Effect of silver on burn wound infection control and healing: review of the literature. Burns (2007) 33, 2, 139–148.

Atkinson, M. J., Spanner, M. T., Rosemann, M., Linzner, U., Müller, W. A., and Gössner, W. Intracellular sequestration of ^{223}Ra by the iron-storage protein ferritin. Radiat. Res. (2005) 164, 2, 230–233.

Auld, D. S. Zinc coordination sphere in biochemical zinc sites. Biometals (2001) 14, 271–313.

Autieri, M. V. cDNA cloning of human allograft inflammatory factor-1: tissue distribution, cytokine induction, and mRNA expression in injured rat carotid arteries. Biochem. Biophys. Res. Commun. (1996) 228, 29–37.

Avery, S. V. Caesium accumulation by microorganisms: uptake mechanisms, cation competition, compartmentalization and toxicity. J. Ind. Microbiol. (1995) 14, 2, 76–84.

Avila-Sakar, A. J., Creutz, C. E., and Kretsinger, R. H. Crystal structure of bovine annexin VI in a calcium-bound state. Biochim. Biophys. Acta (1998) 1387, 103–116.

Avila-Sakar, A. J., Kretsinger, R. H., and Creutz, C. E. Membrane-bound 3D structures reveal the intrinsic flexibility of annexin VI. J. Struct. Biol. (2000) 130, 1, 54–62.

Ayala-Sanmartin, J., Gouache, P., and Henry, J.-P. N-Terminal domain of annexin 2 regulates Ca^{2+}-dependent membrane aggregation by the core domain: a site directed mutagenesis study. Biochemistry (2000) 39, 15190–15198.

Babiychuk, E. B., Palstra, R. J., Schaller, J., Kampfer, U., and Draeger, A. Annexin VI participates in the formation of a reversible, membrane cytoskeleton complex in smooth muscle cells. J. Biol. Chem. (1999) 274, 35191–35195.

Babu, Y. S., Sack, J. S., Greenhough, T. J., Bugg, C. E., Means, A. R., and Cook, W. J. Three-dimensional structure of calmodulin. Nature (1985) 315, 6014, 37–40.

Bai, J., and Chapman, E. R. The C2 domains of synaptotagmin: partners in exocytosis. Trends Biochem. Sci. (2004) 29, 3, 143–151.

Bailey, K. Myosin and adenosinetriphosphatase. Biochem. J. (1942) 36, 121–139.

Baker, H. M., Anderson, B. F., and Baker, E. N. Dealing with iron: common structural principles in proteins that transport iron and heme. Proc. Natl. Acad. Sci. USA (2003) 100, 7, 3579–3583.

Baker, P. F., Blaustein, M. P., Hodgkin, A. L., and Steinhardt, R. A. The influence of calcium on sodium efflux in squid axons. J. Physiol. (1969) 200, 431–458.

Baksh, S., Spamer, C., Heilmann, C., and Michalak, M. Identification of the Zn^{2+} binding region in calreticulin. FEBS Lett. (1995) 376, 1–2, 53–57.

Baldellon, C., Alattia, J. R., Strub, M. P., Berchtold, M. W., Cave, A., and Padilla, A. ^{13}N NMR relaxation studies of calcium-loaded parvalbumin show tight dynamics compared to those of other EF-hand proteins. Biochemistry (1998) 37, 28, 9964–9975.

Baldrian, P. Fungal laccases: occurrence and properties. FEMS Microbiol. Rev. (2006) 30, 215–242.

Baldwin, A. N., and Ames, J. B. Core mutations that promote the calcium-induced allosteric transition of bovine recoverin. Biochemistry (1998) 37, 17408–17419.

Baldwin, G. S., Curtain, C. C., and Sawyer, W. H. Selective, high-affinity binding of ferric ions by glycine-extended gastrin(17). Biochemistry (2001) 40, 36, 10741–10746.

Baldwin, R. L. Energetics of protein folding. J. Mol. Biol. (2007) 371, 2, 283–301.

Balk, S. D. Calcium as a regulator of the proliferation of a normal, but not of transformed chicken fibroblasts in a plasma-containing medium. Proc. Natl. Acad. Sci. USA (1971) 68, 271–275.

Balla, T. Regulation of Ca^{2+} entry by inositol lipids in mammalian cells by multiple mechanisms. Cell Calcium (2009) 45, 6, 527–534.

Ballangrud, A. M., Yang, W. H., Palm, S., Enmon, R., Borchardt, P. E., Pellegrini, V. A., McDevitt, M. R., Scheinberg, D. A., and Sgouros, G. Alpha-particle emitting atomic generator (actinium-225)-labeled trastuzumab (herceptin) targeting of breast cancer spheroids: efficacy versus HER2/neu expression. Clin. Cancer Res. (2004) 10, 13, 4489–4497.

Ballatori, N. Transport of toxic metals by molecular mimicry. Environ. Health Perspect. (2002) 110, Suppl. 5, 689–694.

Bangham, A. D., and Papahadjopoulos, D. Biophysical properties of phospholipids: I. Interaction of phosphatidylserine monolayers with metal ions. Biochim. Biophys. Acta (1966) 126, 181–184.

Banci, L., Bertini, I., Ciofi-Baffoni, S., Finney, L. A., Outten, C. E., and O'Halloran, T. V. A new zinc–protein coordination site in intracellular metal trafficking: solution structure of the Apo and Zn(II) forms of ZntA(46–118). J. Mol. Biol. (2002) 323, 5, 883–897.

Bandyopadhyay, J., Lee, J., and Bandyopadhyay, A. Regulation of calcineurin, a calcium/calmodulin-dependent protein phosphatase, in *C. elegans*. Mol. Cells (2004) 18, 1, 10–16.

Banuelos, S., and Muga, A. Binding of molten globule-like conformations to lipid bilayers: structure of native and partially folded α-lactalbumin bound to model membranes. J. Biol. Chem. (1995) 270, 50, 29910–29915.

Banuelos, S., and Muga, A. Structural requirements for the association of native and partially folded conformations of α-lactalbumin with model membranes. Biochemistry (1996) 35, 13, 3892–3898.

Barbato, G., Ikura, M., Kay, L. E., Pastor, R. W., and Bax, A. Backbone dynamics of calmodulin studied by N^{15} relaxation using inverse detected two-dimensional NMR spectroscopy: the central helix is flexible. Biochemistry (1992) 31, 5269–5278.

Barceloux, D. G. Chromium. J. Toxicol. Clin. Toxicol. (1999a) 37, 2, 173–194.

Barceloux, D. G. Vanadium. J. Toxicol. Clin. Toxicol. (1999b) 37, 2, 265–278.

Barkay, T., Miller, S. M., and Summers, A. O. Bacterial mercury resistance from atoms to ecosystems. FEMS Microbiol. Rev. (2003) 27, 2–3, 355–84.

Barondeau, D. P., and Getzoff, E. D. Structural insights into protein–metal ion partnerships. Curr. Opin. Struct. Biol. (2004) 14, 765–774.

Barondes, S. H. Lectins: their multiple endogenous cellular functions. Annu. Rev. Biochem. (1981) 50, 207–231.

Barroso, M. B., Bernd, K. K., DeWitt, N. D., Chang, A., Mills, K., and Stzul, E. S. A novel Ca^{2+}-binding protein, p22, is required for constitutive membrane traffic. J. Biol. Chem. (1996) 271, 10183–10187.

Barskaya, N. V., and Gusev, N. B. Biological activities of beef cardiac troponin C C-terminal peptide (residues 84–161). Biochem. J. (1982) 207, 2, 185–192.

Barski, J. J., Morl, K., and Meyer, M. Conditional inactivation of the calbindinD-28k (*Calb1*) gene by Cre/loxP-mediated recombination. Genesis (2002) 32, 165–168.

Barth, A. Infrared spectroscopy. In: Uversky V. N., and Permyakov E. A., eds, Methods in Protein Structure and Stability Analysis. Vibrational Spectroscopy. Nova Sci. Publishers, New York, 2007a, 69–151.

Barth, A. Infrared spectroscopy of proteins. Biochim. Biophys. Acta (2007b) 1767, 1073–1101.

Barth, A., and Zscherp, C. Substrate binding and enzyme function investigated by infrared spectroscopy. FEBS Lett. (2000) 477, 3, 151–156.

Barth, A., and Zscherp, C. What vibrations tell us about proteins. Q. Rev. Biophys. (2002) 35, 4, 369–430.

Bartling, D., Bülter, H., and Weiler, E. W. *Arabidopsis thaliana* cDNA encoding a novel member of the EF-hand superfamily of calcium-binding proteins. Plant Physiol. (1993) 102, 1059–1060.

Bastianelli, E. Distribution of calcium-binding proteins in the cerebellum. Cerebellum (2003) 2, 4, 242–262.

Batistic, O., and Kudla, J. Plant calcineurin B-like proteins and their interacting protein kinases. Biochim. Biophys. Acta (2009) 1793, 6, 985–992.

Baudier, J., Glasser, N., Gerard, D. Ions binding to S100 proteins: I. Calcium- and zinc-binding properties of bovine brain S100 αα, S100a (αβ), and S100b (ββ) protein: Zn^{2+} regulates Ca^{2+} binding on S100b protein. J. Biol. Chem. (1986) 261, 18, 8192–8203.

Baudier, J., and Gerard, D. Ions binding to S100 proteins: structural changes induced by calcium and zinc on S100a and S100b proteins. Biochemistry (1983) 22, 14, 3360–3369.

Baudier, J., Haglid, K., Haiech, J., and Gerard, D. Zinc ion binding to human brain calcium binding proteins, calmodulin and S100b protein. Biochem. Biophys. Res. Commun. (1983) 114, 3, 1138–1146.

Bauer, M. C., Nilsson, H., Thulin, E., Frohm, B., Malm, J., and Linse, S. Zn^{2+} binding to human calbindin D_{28k} and the role of histidine residues. Protein Sci. (2008) 17, 4, 760–767.

Baumann, U., Wu, S., Flaherty, K. M., and McKay, D. B. Three-dimensional structure of the alkaline protease of *Pseudomonas aeruginosa*: a two-domain protein with a calcium binding parallel beta roll motif. EMBO J. (1993) 12, 9, 3357–3364.

Bazzi, M. D., and Nelsestuen, G. L. Extensive segregation of acidic phospholipids in membranes induced by protein kinase C and related proteins. Biochemistry (1991) 30, 32, 7961–7969.

Bazzi, M. D., and Nelsestuen, G. L. Interaction of annexin VI with membranes: highly restricted dissipation of clustered phospholipids in membranes containing phosphatidylethanolamine. Biochemistry (1992) 31, 42, 10406–10413.

Beale, J. E., Jeebhay, M. F., and Lopata, A. L. Characterisation of purified parvalbumin from five fish species and nucleotide sequencing of this major allergen from Pacific pilchard, *Sardinops sagax*. Mol. Immunol. (2009) 46, 15, 2985–2993.

Beard, N. A., Laver, D. R., and Dulhunty, A. F. Calsequestrin and the calcium release channel of skeletal and cardiac muscle. Prog. Biophys. Mol. Biol. (2004) 85, 1, 33–69.

Beard, N. A., Wei, L., and Dulhunty, A. F. Ca^{2+} signaling in striated muscle: the elusive roles of triadin, junctin, and calsequestrin. Eur. Biophys. J. (2009) 39, 27–36.

Bearer, E. L., DeGiorgis, J. A., Jaffer, H., Medeiros, N. A., and Reese, T. S. An axoplasmic myosin with a calmodulin-like light chain. Proc. Natl. Acad. Sci. USA (1996) 93, 6064–6068.

Becker, G. L., Fiskum, G., and Lehninger, A. L. Regulation of free Ca^{2+} by liver mitochondria and endoplasmic reticulum. J. Biol. Chem. (1980) 255, 19, 9009–9012.

Becker, J. W., Reeke, G. N., Wang, J. L., Cunningham, B. A., and Edelman, G. M. Covalent and three-dimensional structure of concanavalin A: III. Structure of the monomer and its interactions with metals and saccharides. J. Biol. Chem. (1975) 250, 4, 1513–1524.

Bedard, K., Szabo, E., Michalak, M., and Opas, M. Cellular functions of endoplasmic reticulum chaperones calreticulin, calnexin, and ERp57. Int. Rev. Cytol. (2005) 245, 91–121.

Beinert, H., Holm, R. H., and Munck, E. Iron–sulfur clusters: nature's modular, multipurpose structures. Science (1997) 277, 5326, 653–659.

Belchier, J. An account of the bones of animals being changed to a red colour by aliment only. Philos. Trans. (1736) 39, 287–288.

Belhadri, A. Presence of centrin in the human parasite *Giardia:* a further indication of its ubiquity in eukaryotes. Biochem. Biophys. Res. Commun. (1995) 214, 597–601.

Belle, C., Rammal, W., and Pierre, J. L. Sulfur ligation in copper enzymes and models. J. Inorg. Biochem. (2005) 99, 1929–1936.

Bellido, T., Huening, M., Raval-Pandya, M., Manolagas, S. C., and Christakos, S. Calbindin-D_{28k} is expressed in osteoblastic cells and suppresses their apoptosis by inhibiting caspase-3 activity. J. Biol. Chem. (2000) 275, 26328–26332.

Bendall, J. R. Effect of the "marsh factor" on the shortening of muscle fibre models in the presence of adenosine triphosphate. Nature (1952) 170, 1058–1060.

Bento, I., Martins, L. O., Lopes, G. G., Carrondo, M. A., and Lindley, P. F. Dioxygen reduction by multi-copper oxidases: a structural perspective. Dalton Trans. (2005) 7, 21, 3507–3513.

Bento, I., Peixoto, C., Zaitsev, V. N., and Lindley, P. F. Ceruloplasmin revisited: structural and functional roles of various metal cation-binding sites. Acta Crystallogr. D (2007) 63, 240–247.

Berchtold, M. W., Brinkmeier, H., and Muntener, M. Calcium ion in skeletal muscle: its structural role for muscle function, plasticity, and disease. Physiol. Rev. (2000) 80, 3, 1215–1265.

Berger, J., Rachlin, A. I., Scott, W. E., Sternback, L. H., and Goldberg, J. The isolation of three new crystalline antibiotics from streptomyces. J. Am. Chem. Soc. (1951) 73, 5295–5298.

Berggård, T., Silow, M., Thulin, E., and Linse, S. Ca^{2+}- and H^{+}-dependent conformational changes of calbindin D_{28k}. Biochemistry (2000a) 39, 23, 6864–6873.

Berggård, T., Julenius, K., Ogard, A., Drakenberg, T., and Linse, S. Fragment complementation studies of protein stabilization by hydrophobic core residues. Biochemistry (2001) 40, 1257–1264.

Berggård, T., Szczepankiewicz, O., Thulin, E., and Linse, S. Myo-inositol monophosphatase is an activated target of calbindin D_{28k}. J. Biol. Chem. (2002) 277, 41954–41959.

Berggård, T., Thulin, E., Akerfeldt, K. S., and Linse, S. Fragment complementation of calbindin D_{28k}. Protein Sci. (2000b) 9, 11, 2094–2108.

Bergmann, U., Glatzel, P., Robblee, J. H., et al. High-resolution x-ray spectroscopy of rare events: a different look at local structure and chemistry. J. Synchrotron Radiat. (2001) 8, Pt. 2, 199–203.

Berliner, L. J., and Koga, K. α-Lactalbumin binding to membranes: evidence for partially buried protein. Biochemistry (1987) 26, 11, 3006–3009.

Bernstein, L. R. Mechanisms of therapeutic activity for gallium. Pharmacol. Rev. (1998) 50, 4, 665–682.

Berridge, M. J. Temporal aspects of calcium signalling. Adv. Second Messenger Phosphoprotein Res. (1990) 24, 108–114.

Berridge, M. J. The versatility and complexity of calcium signalling. Novartis Found. Symp. (2001) 239, 52–64.

Berridge, M. J. The endoplasmic reticulum: a multifunctional signaling organelle. Cell Calcium (2002) 32, 5–6, 235–249.

Berridge, M. J. Calcium microdomains: organization and function. Cell Calcium (2006) 40, 5–6, 405–412.

Berridge, M. J. Inositol trisphosphate and calcium oscillations. Biochem. Soc. Symp. (2007) 74, 1–7.

Berridge, M. J. Smooth muscle cell calcium activation mechanisms. J. Physiol. (2008) 586, Pt. 21, 5047–5061.

Berridge, M. J. Inositol trisphosphate and calcium signalling mechanisms. Biochim. Biophys. Acta (2009) 1793, 6, 933–940.

Berridge, M., Lipp, P., and Bootman, M. Calcium signalling. Curr. Biol. (1999) 9, 5, R157–R159.

Bers, D. M. Measurement of calcium transport in heart using modern approaches. New Horiz. (1996) 4, 1, 36–44.

Bers, D. M., Patton, C. W., and Nuccitelli, R. A practical guide to the preparation of Ca^{2+} buffers. Methods Cell Biol. (1994) 40, 3–29.

Bershitsky, S. Y., Ferenczi, M. A., Koubassova, N. A., and Tsaturyan, A. K. Insight into the actin–myosin motor from x-ray diffraction on muscle. Front. Biosci. (2009) 14, 3188–3213.

Bertini, I., Gelis, I., Katsaros, N., Luchinat, C., and Provenzani, A. Tuning the affinity for lanthanides of calcium binding proteins. Biochemistry (2003) 42, 8011–8021.

Beyersmann, D., and Haase, H. Functions of zinc in signaling, proliferation and differentiation of mammalian cells. Biometals (2001) 14, 331–341.

Beyersmann, D., and Hechtenberg, S. Cadmium, gene regulation, and cellular signalling in mammalian cells. Toxicol. Appl. Pharmacol. (1997) 144, 2, 247–261.

Bezanilla, F. Voltage-gated ion channels. IEEE Trans. Nanobiosci. (2005) 4, 34–48.

Bezdekova, A., and Budesinsky, B. W. Spektralphotometrische Untersuchung der Reaktion der Calciumionen mit Kalcion IREA und mit Calcichrom. Collect. Czech. Commun. (1965) 30, 811–817.

Biekofsky, R. R., Martin, S. R., McCormick, J. E., Masino, L., Fefeu, S., Bayley, P. M., and Feeney, J. Thermal stability of calmodulin and mutants studied by $^{1}H^{15}N$ HSQC NMR measurements of selectively labeled [^{15}N]Ile proteins. Biochemistry (2002) 41, 6850–6859.

Biggins, S., and Rose, M. D. Direct interaction between yeast spindle pole body components. KAR1P is required for CDC31P localization to the spindle pole body. J. Cell Biol. (1994) 1235, 843–852.

Bihler, I., Charles, P., and Sawh, P. C. Effects of strontium on calcium-dependent hexose transport in muscle. Can. J. Physiol. Pharmacol. (1986) 64, 176–179.

Birchmeier, W., and Behrens, J. Cadherin expression in carcinomas: role in the formation of cell junctions and the prevention of invasiveness. Biochim. Biophys. Acta (1994) 1198, 11–26.

Birks, R. I., and MacIntosh, F. C. Acetylcholine metabolism at nerve-endings. Br. Med. Bull. (1957) 13, 157–161.

Bitto, E., and Cho, W. Roles of individual domains of annexin I in its vesicle binding and vesicle aggregation: a comprehensive mutagenesis study. Biochemistry (1998) 37, 28, 10231–10237.

Bitto, E., and Cho, W. Structural determinant of the vesicle aggregation activity of annexin I. Biochemistry (1999) 38, 14094–14100.

Bitto, E., Bingman, C. A., Bittova, L., Frederick, R. O., Fox, B. G., and Phillips, G. N. X-ray structure of *Danio rerio* secretagogin: a hexa-EF-had calcium sensor. Proteins (2009) 76, 477–483.

Black, D. J., Tikunova, S. B., Johnson, J. D., and Davis, J. P. Acid pairs increase the N-terminal Ca^{2+} affinity of CaM by increasing the rate of Ca^{2+} association. Biochemistry (2000) 39, 13831–13837.

Blackmore, P. F., Hughes, B. P., Shuman, E. A., and Exton, J. H. α-Adrenergic activation of phosphorylase in liver cells involves mobilization of intracellular calcium without influx of extracellular calcium. J. Biol. Chem. (1982) 257, 1, 190–197.

Blaine, S. A., Wick, M., Dessev, C., and Nemenoff, R. A. Induction of $cPLA_2$ in lung epithelial cells and non-small cell lung cancer is mediated by Sp1 and c-Jun. J. Biol. Chem. (2001) 276, 46, 42737–42743.

Blamey, C. J., Ceccarelli, C., Naik, U. P., and Bahnson, B. J. The crystal structure of calcium- and integrin-binding protein 1: insights into redox regulated functions. Protein Sci. (2005) 14, 5, 1214–1221.

Blanchard, H., Grochulski, P., Li, Y., Arthur, S. C., Davies, P. L., Elce, J. S., and Cygler, M. Structure of calpain Ca^{2+}-binding domain reveals a novel EF-hand and Ca^{2+}-induced conformational changes. Nat. Struct. Biol. (1997) 4, 532–538.

Blancuzzi, Y., Padilla, A., Parello, J., and Cave, A. Symmetrical rearrangement of the cation-binding sites of parvalbumin upon Ca^{2+}/Mg^{2+} exchange: a study by ^1H ^2D NMR. Biochemistry (1993) 32, 1302–1309.

Blatow, M., Caputi, A., Burnashev, N., Monyer, H., and Rozov, A. Ca^{2+} buffer saturation underlies paired pulse facilitation in calbindin-D_{28k}-containing terminals. Neuron (2003) 38, 79–88.

Blaustein, M. P., and Hodgkin, A. L. The effect of cyanide on the efflux of calcium from squid axons. J. Physiol. (1969) 200, 497–527.

Bloom, J. W., and Mann, K. G. Metal ion induced conformations of prothrombin and prothrombin fragment 1. Biochemistry (1978) 17, 21, 4430–4438.

Blum, J., Handmaker, H., and Rinne, N. A. Technetium labeled small peptide radiopharmaceuticals in the identification of lung cancer. Curr. Pharm. Des. (2002) 8, 20, 1827–1836.

Blumenschein, T. M. A., and Reinach, F. C. Analysis of affinity and specificity in an EF-hand site using double mutant cycles. Biochemistry (2000) 39, 3603–3610.

Boeggeman, E., and Qasba, P. K. Studies on the metal binding sites in the catalytic domain of β 1,4-galactosyltransferase. Glycobiology (2002) 12, 7, 395–407.

Boersma, H. H., Kietselaer, B. L. J. H., Stolk, L. M. L., Bennaghmouch, A., Hofstra, L., Narula, J., Heidendal, G. A. K., and Reutelingsperger, C. P. M. Past, present, and future of annexin A5: from protein discovery to clinical applications. J. Nucl. Med. (2005) 46, 2035–2050.

Bogert, L. J., and Hastings, A. B. The chemical salts of bone. J. Biol. Chem. (1931) 94, 473–481.

Bolliger, A. The volumetric micro-determination of calcium with methylene blue following its precipitation as calcium picrolonate. Aust. J. Exp. Biol. Med. Sci. (1935) 13, 75–78.

Bollmann, J. H., Sakmann, B., and Borst, J. G. Calcium sensitivity of glutamate release in a calyx-type terminal. Science (2000) 289, 953–957.

Bonnevier, J., Malmqvist, U., Sonntag, D., Schroeter, M., Nilsson, H., Pfitzer, G., and Arner, A. Sustained norepinephrine contraction in the rat portal vein is lost when Ca^{2+} is replaced with Sr^{2+}. Am. J. Physiol. Cell Physiol. (2002) 282, 4, C845–C852.

Bonucci, E. Calcification and silicification: a comparative survey of the early stages of biomineralization. J. Bone Miner. Metab. (2009) 27, 3, 255–264.

Bose, R. N. Biomolecular targets for platinum antitumor drugs. Mini Rev. Med. Chem. (2002) 2, 2, 103–111.

Bottrill, M., Kwok, L., and Long, N. J. Lanthanides in magnetic resonance imaging. Chem. Soc. Rev. (2006) 35, 6, 557–571.

Bozler, E. Relaxation in extracted muscle fibers. J. Gen. Physiol. (1954) 38, 149–159.

Brabec, V. DNA modifications by antitumor platinum and ruthenium compounds: their recognition and repair. Prog. Nucleic Acid Res. Mol. Biol. (2002) 71, 1–68.

Brachet, J., Baltus, E., de Schutter-Pays, A., Hanocq-Quertier, J., Hubert, E., and Steinert, G. Induction of maturation (meiosis) in *Xenopus laevis* oocytes by three organomercurials. Proc. Natl. Acad. Sci. USA (1975) 72, 1574–1578.

Bradley, P., Misura, K. M., and Baker, D. Toward high-resolution de novo structure prediction for small proteins. Science (2005) 309, 1868–1871.

Bradshaw, A. D., and Sage, E. H. SPARC, a matricellular protein that functions in cellular differentiation and tissue response to injury. J. Clin. Invest. (2001) 107, 9, 1049–1054.

Brandt, B. B., and Harney, J. An overview of interstitial brachytherapy and hyperthermia. Oncol. Nurs. Forum (1989) 16, 6, 833–841.

Bremel, R. D. Myosin linked calcium regulation in vertebrate smooth muscle. Nature (1974) 252, 5482, 405–407.

Bressler, J. P., Olivi, L., Cheong, J. H., Kim, Y., and Bannona, D. Divalent metal transporter 1 in lead and cadmium transport. Ann. N.Y. Acad. Sci. (2004) 1012, 142–152.

Briggs, K. K., Silvers, A. J., Johansen, K. M., and Johansen, J. Calsensin: a novel calcium-binding protein expressed in a subset of peripheral leech neurons fasciculationg in a single axon tract. J. Cell Biol. (1995) 129, 1355–1362.

Brini, M., Pinton, P., Pozzan, T., and Rizzuto, R. Targeted recombinant aequorins: tools for monitoring [Ca^{2+}] in the various compartments of a living cell. Microsc. Res. Tech. (1999) 46, 380–389.

Brink, F. The role of calcium ions in neural processes. Pharmacol. Rev. (1954) 6, 243–298.

Brito, R. M. M., Putkey, J. A., Strynadka, N. C., James, M. N. G., and Rosenvear, P. R. Comparative NMR studies on troponin C and a mutant inacpable of binding calcium at site II. Biochemistry (1991) 30, 42, 10236–10245.

Brittain, T. Root effect hemoglobins. J. Inorg. Biochem. (2005) 99, 120–129.

Brodersen, D. E., Etzerodt, M., Madsen, P., Celis, J. E., Thogersen, H. C., and Nyborg, J. EF-hands at atomic resolution: the structure of human psoriasin (S100A7) solved by MAD phasing. Structure (1998) 6, 477–489.

Brodersen, D. E., Nyborg, J., and Kjeldgaard, M. Zinc-binding site of an S100 protein revealed. Two crystal structures of Ca^{2+}-bound human psoriasin (S100A7) in the Zn^{2+}-loaded and Zn^{2+}-free states. Biochemistry (1999) 38, 6, 1695–1704.

Bronner, F. Mechanisms of intestinal calcium absorption. J. Cell. Biochem. (2003) 88, 2, 387–393.

Bronner, F., Botnick, F., and Freund, T. Calcium transport in *Bacillus megaterium*. Isr. J. Med. Sci. (1971) 7, 1224–1229.

Brose, N., Petrenko, A. G., Sudhof, T. C., and Jahn, R. Synaptotagmin: a calcium sensor on the synaptic vesicle surface. Science (1992) 256, 5059, 1021–1025.

Brostrom, C. O., and Wolff, D. I. Properties and function of calmodulin. Biochem. Pharmacol. (1981) 30, 12, 1395–1405.

Brown, H. M., Hagiwara, S., Koike, H., and Meech, R. W. Membrane properties of a barnacle photoreceptor examined by the voltage clamp technique. J. Physiol. (1970) 208, 385–413.

Brown, K., Tegoni, M., Prudencio, M., Pereira, A. S., Besson, S., Moura, J. G., Moura, I., and Cambillau, C. A novel type of catalytic copper cluster in nitrous oxide reductase. Nat. Struct. Biol. (2000) 7, 191–195.

Brown, L. J., MacDonald, M. J., Lehn, D. A., and Moran, S. M. Sequence of rat mitochondrial glycerol-3-phosphate dehydrogenase cDNA: evidence for EF-hand calcium-binding domains. J. Biol. Chem. (1994) 269, 14363–14366.

Brownawell, A., and Creutz, C. Calcium-dependent binding of sorcin to the N-terminal domain of synexin (annexin VII). J. Biol. Chem. (1997) 272, 22182–22190.

Browne, J. P., Strom, M., Martin, S. R., and Bayley, P. M. The role of β-sheet interactions in domain stability, folding, and target recognition reactions of calmodulin. Biochemistry (1997) 36, 9550–9561.

Brunger, A. T. Structure and function of SNARE and SNARE-interacting proteins. Q. Rev. Biophys. (2005) 38, 1, 1–47.

Bruns, C. M., Nowalk, A. J., Arvai, A. S., McTigue, M. A., Vaughan, K. G., Mietzner, T. A., and McRee, D. E. Structure of *Haemophilus influenzae* Fe^{3+}-binding protein reveals convergent evolution within a superfamily. Nat. Struct. Biol. (1997) 4, 11, 919–924.

Brusés, J. L. *N*-Cadherin signaling in synapse formation and neuronal physiology. Mol. Neurobiol. (2006) 33, 3, 237–252.

Buchanan, J. D., Corbett, R. J., and Roche, R. S. The thermodynamics of calcium binding to thermolysin. Biophys. Chem. (1986) 23, 3–4, 183–199.

Buckingham, J. C., John, C. D., Solito, E., Tierney, T., Flower, R. J., Christian, H., and Morris, J. Annexin 1, glucocorticoids, and the neuroendocrine-immune interface. Ann. N.Y. Acad. Sci. (2006) 1088, 396–409.

Budesinsky, B. W. Photometric determination of calcium with antipyrylazo III. Anal. Chim. Acta (1974) 71, 343–347.

Bugajska-Schretter, A., Elfman, L., Fuchs, T., et al. Parvalbumin, a crossreactive fish allergen, contains IgE binding epitopes sensitive to periodate treatment and Ca^{2+} depletion. J. Allergy Clin. Immunol. (1998) 101, 67–74.

Bulleid, A. An experimental study of *Leptothrix buccalis*. Br. Dent. J. (1925) 46, 289–300.

Bunney, T. D., and Katan, M. Phospholipase C epsilon: linking second messengers and small GTPases. Trends Cell Biol. (2006) 16, 640–648.

Burdette, S. C., and Lippard, S. J. Meeting of the minds: metalloneurochemistry Proc. Natl. Acad. Sci. USA (2003) 100, 7, 3605–3610.

Burgoyne, R. D., and Weiss, J. L. The neuronal calcium sensor family of Ca^{2+}-binding proteins. Biochem. J. (2001) 353, 1–12.

Burley, S. K., and Bonanno, J. B. Structuring the universe of proteins. Annu. Rev. Genom. Hum. Genet. (2002) 3, 243–262.

Burstein, E. A., Permyakov, E. A., Yashin, V. A., Burkhanov, S. A., and Finazzi-Agro, A. The fine structure of luminescence spectra of azurin. Biochim. Biophys. Acta (1977) 491, 1, 155–159.

Busenlehner, L. S., Apuy, J. L., and Giedroc, D. P. Characterization of a metalloregulatory bismuth(III) site in *Staphylococcus aureus* pI258 CadC repressor. J. Biol. Inorg. Chem. (2002) 7, 4–5, 551–559.

Butler, A. Vanadium haloperoxidases. Curr. Opin. Chem. Biol. (1998) 2, 2, 279–285.

Buxbaum, J. D., Choi, E.-K., Luo, Y., Lilliehook, C., Crowley, A. C., Merriam, O. E., and Wasco, W. Calsenilin: a calcium-binding protein that interacts with the presenilins and regulates the levels of a presenilin fragment. Nat. Med. (1998) 4, 1177–1181.

Bzdega, T., and Kosk-Kosicka, D. Regulation of the erythrocyte Ca^{2+}-ATPase by mutant calmodulins with Glu Ala substitutions in the Ca^{2+}-binding domains. J. Biol. Chem. (1992) 267, 7, 4394–4397.

Cachia, P. J., Van Eyk, J., Ingraham, R. H., McCubbin, W. D., Kay, C. M., and Hodges, R. S. Calmodulin and troponin C: a comparative study of the interaction of mastoparan and troponin I inhibitory peptide 104–115. Biochemistry (1986) 25, 12, 3553–3562.

Caday, C. G., and Steiner, R. F. The interaction of calmodulin with melittin. Biochem. Biophys. Res. Commun. (1986) 135, 2, 419–425.

Caires, A. C. Recent advances involving palladium(II) complexes for the cancer therapy. Anticancer Agents Med. Chem. (2007) 7, 5, 484–491.

Cairo, G., and Pietrangelo, A. Iron regulatory proteins in pathobiology. Biochem. J. (2000) 352, 241–250.

Cairo, G., Recalcati, S., Pietrangelo, A., and Minotti, G. The iron regulatory proteins: targets and modulators of free radical reactions and oxidative damage. Free Radic. Biol. Med. (2002) 32, 12, 1237–1243.

Cama, E., Pethe, S., Boucher, J.-L., Han, S., Emig, F. A., Ash, D. E., Viola, R. E., Mansuy, D., and Christianson, D. W. Inhibitor coordination interactions in the binuclear manganese cluster of arginase. Biochemistry (2004) 43, 8987–8999.

Cambell, A. K. Chemiluminescence: Principles and Applications in Biology and Medicine. Horwood/VCH, Chichester, UK, 1988.

Camors, E., Monceau, V., and Charlemagne, D. Annexins and Ca^{2+} handling in the heart. Cardiovasc. Res. (2005) 65, 4, 793–802.

Campos, B., Mo, Y. D., Mealy, T. R., Li, C. W., Swairjo, M. A., Balch, C., Head, J. F., Retzinger, G., Dedman, J. R., and Seaton, B. A. Mutational and crystallographic analyses of interfacial residues in annexin V suggest direct interactions with phospholipid membrane components. Biochemistry (1998) 37, 22, 8004–8010.

Caohuy, H., and Pollard, H. B. Activation of annexin 7 by protein kinase C in vitro and in vivo. J. Biol. Chem. (2001) 276, 16, 12813–12821.

Capozzi, F., Casadei, F., and Luchinat, C. EF-hand protein dynamics and evolution of calcium signal transduction: an NMR view. J. Biol. Inorg. Chem. (2006) 11, 8, 949–962.

Carafoli, E. Calcium signaling: A tale for all seasons. Proc. Natl. Acad. Sci. USA (2002) 99, 3, 1115–1122.

Carafoli, E., Santella, L., Branca, D., and Brini, M. Generation, control, and processing of cellular calcium signals. Crit. Rev. Biochem. Mol. Biol. (2001) 36, 107–260.

Caramelo, J. J., and Parodi, A. J. Getting in and out from calnexin/calreticulin cycles. J. Biol. Chem. (2008) 283, 16, 10221–10225.

Carr, C. M., and Munson, M. Tag team action at the synapse. EMBO Rep. (2007) 8, 9, 834–838.

Carrión, A. M., Link, W. A., Ledo, F., Mellstrom, B., and Naranjo, J. R. DREAM is a Ca^{2+}-regulated transcriptional repressor. Nature (1999) 398, 80–84.

Carrondo, M. A. Ferritins, iron uptake and storage from the bacterioferritin viewpoint. EMBO J. (2003) 22, 9, 1959–1968.

Carruthers, C., and Suntzeff, V. The role of calcium in carcinogenesis. Science (1944) 99, 245–247.

Caruso, F., and Rossi, M. Antitumor titanium compounds. Mini Rev. Med. Chem. (2004) 4, 1, 49–60.

Casbarra, A., Birolo, L., Infusini, G., Dal Piaz, F., Svensson, M., Pucci, P., Svanborg, C., and Marino, G. Conformational analysis of HAMLET, the folding variant of human α-lactalbumin associated with apoptosis. Protein Sci. (2004) 13, 5, 1322–1330.

Cates, M. S., Berry, M. B., Ho, E. L., Li, Q., Potter, J. D., and Phillips, G. N. Metal-ion affinity and specificity in EF-hand proteins: coordination geometry and domain plasticity in parvalbumin. Structure (1999) 7, 1269–1278.

Cates, M. S., Teodoro, M. L., and Phillips, G. N., Jr. Molecular mechanisms of calcium and magnesium binding to parvalbumin. Biophys. J. (2002) 82, 1133–1146.

Cawthern, K. M., Permyakov, E. A., and Berliner, L. J. Membrane-bound states of α-lactalbumin: implications for the protein stability and conformation. Protein Sci. (1996) 5, 7, 1394–1405.

Cawthern, K. M., Narayan, M., Chaudhuri, D., Permyakov, E. A., and Berliner, L. J. Interactions of α-lactalbumin with fatty acids and spin label analogs. J. Biol. Chem. (1997) 272, 49, 30812–30816.

Celio, M. R. (Ed.) Guidebook to the Calcium-Binding Proteins. Sambrook and Tooze with Oxford University Press, Oxford, UK, 1996.

Cervantes, C., Campos-García, J., Devars, S., Gutiérrez-Corona, F., Loza-Tavera, H., Torres-Guzmán, J. C., and Moreno-Sánchez, R. Interactions of chromium with microorganisms and plants. FEMS Microbiol. Rev. (2001) 25, 3, 335–347.

Chabre, M. Aluminofluoride and beryllofluoride complexes: a new phosphate analogs in enzymology. Trends Biochem. Sci. (1990) 15, 1, 6–10.

Chamnongpol, S., and Groisman, E. A. Mg^{2+} homeostasis and avoidance of metal toxicity. Mol. Microbiol. (2002) 44, 2, 561–571.

Chamnongpol, S., Cromie, M., and Groisman, E. A., Mg^{2+} Sensing by the Mg^{2+} Sensor PhoQ of *Salmonella enterica*. J. Mol. Biol. (2003) 325, 795–807.

Chan, A. O. E-cadherin in gastric cancer. World J. Gastroenterol. (2006) 12, 2, 199–203.

Chandra, B. R., Yogavel, M., and Sharma, A. Structural analysis of ABC-family periplasmic zinc binding protein provides new insights into mechanism of ligand uptake and release. J. Mol. Biol. (2007) 367, 4, 970–982.

Chandra, M., McCubbin, W. D., Oikawa, K., Kay, C. M., and Smillie, L. B. Ca^{2+}, Mg^{2+}, and troponin I inhibitory peptide binding to a Phe-154 to Trp mutant of chicken skeletal muscle troponin C. Biochemistry (1994) 33, 10, 2961–2969.

Chapman, E. R., and Jahn, R. Calcium-dependent interaction of the cytoplasmic region of synaptotagmin with membranes: autonomous function of a single C_2-homologous domain. J. Biol. Chem. (1994) 269, 8, 5735–5741.

Chapman, E. R., Hanson, P. I., An, S., and Jahn, R. Ca^{2+} regulates the interaction between synaptotagmin and syntaxin 1. J. Biol. Chem. (1995) 270, 23667–23671.

Chattopadhyaya, R., Meador, W. E., Means, A. R., and Quiocho, F. A. Calmodulin structure refined at 1.7 Å resolution. J. Mol. Biol. (1992) 228, 4, 1177–1192.

Chen, C. K., Inglese, J., Lefkowitz, R. J., and Hurley, J. B. Ca^{2+}-dependent interaction of recoverin with rhodopsin kinase. J. Biol. Chem. (1995) 270, 30, 18060–18066.

Chen, J. M., Sheldon, A., and Pincus, M. R. Structure–function correlations of calcium binding and calcium channel activities based on 3-dimensional models of human annexins I, II, III, V and VII. J. Biomol. Struct. Dyn. (1993) 10, 6, 1067–1089.

Chen, J. Z., Settembre, E. C., Aoki, S. T., Zhang, X., Bellamy, A. R., Dormitzer, P. R., Harrison, S. C., and Grigorieff, N. Molecular interactions in rotavirus assembly and uncoating seen by high-resolution cryo-EM. Proc. Natl. Acad. Sci. USA (2009) 106, 26, 10644–10648.

Chen, P., and Solomon, E. I. O_2 activation by binuclear Cu sites: noncoupled versus exchange coupled reaction mechanisms. Proc. Natl. Acad. Sci. USA (2004) 101, 36, 13105–13110.

Chen, Q., Lipkina, G., Song, Q., and Kramer, R. H. Promoter methylation regulates cadherin switching in squamous cell carcinoma. Biochem. Biophys. Res. Commun. (2004) 315, 4, 850–856.

Cheng, H. Y., Pitcher, G. M., Laviolette, S. R., et al. DREAM is a critical transcriptional repressor for pain modulation. Cell (2002) 108, 1, 31–43.

Cherrier, M. V., Cavazza, C., Bochot, C., Lemaire, D., and Fontecilla-Camps, J. C. Structural characterization of a putative endogenous metal chelator in the periplasmic nickel transporter NikA. Biochemistry (2008) 47, 38, 9937–9943.

Chiasson, D., Ekengren, S. K., Martin, G. B., Dobney, S. L., and Snedden, W. A. Calmodulin-like proteins from *Arabidopsis* and tomato are involved in host defense against *Pseudomonas syringae* pv. *tomato*. Plant Mol. Biol. (2005) 58, 6, 887–897.

Chicka, M. C., Hui, E., Liu, H., and Chapman, E. R. Synaptotagmin arrests the SNARE complex before triggering fast, efficient membrane fusion in response to Ca^{2+}. Nat. Struct. Mol. Biol. (2008) 15, 8, 827–835.

Chikh, Z., Ha-Duong, N. T., Miquel, G., and El Hage Chahine, J. M. Gallium uptake by transferrin and interaction with receptor 1. J. Biol. Inorg. Chem. (2007) 12, 1, 90–100.

Chin, D., and Means, A. R. Calmodulin: a prototypical calcium sensor. Trends Cell Biol. (2000) 10, 8, 322–328.

Choi, K. C., and Jeung, E. B. Molecular mechanism of regulation of the calcium-binding protein calbindin-D_{9k}, and its physiological role(s) in mammals: a review of current research. J. Cell. Mol. Med. (2008) 12, 2, 409–420.

Choi, K. C., Leung, P. C., and Jeung, E. B. Biology and physiology of calbindin-D9k in female reproductive tissues: involvement of steroids and endocrine disruptors. Reprod. Biol. Endocrinol. (2005) 3, 66.

Chourey, K., Thompson, M. R., Morrell-Falvey, J., Verberkmoes, N. C., Brown, S. D., Shah, M., Zhou, J., Doktycz, M., Hettich, R. L., and Thompson, D. K. Global molecular and morphological effects of 24-hour chromium(VI) exposure on *Shewanella oneidensis* MR-1. Appl. Environ. Microbiol. (2006) 72, 9, 6331–6344.

Christakos, S., Dhawan, P., Peng, X., Obukhov, A. G., Nowycky, M. C., Benn, B. S., Zhong, Y., Liu, Y., and Shen, Q. New insights into the function and regulation of vitamin D target proteins. J. Steroid Biochem. Mol. Biol. (2007) 103, 405–410.

Christianson, D. W. Structural biology of zinc. Adv. Protein Chem. (1991) 42, 281–355.

Christianson, D. W., and Cox, J. D. Catalysis by metal-activated hydroxide in zinc and manganese metalloenzymes. Annu. Rev. Biochem. (1999) 68, 33–57.

Christova, P., Cox, J. A., and Craescu, C. T. Ion-induced conformational and stability changes in *Nereis* sarcoplasmic calcium binding protein: evidence that the APO state is a molten globule. Proteins: Struct. Funct. Genet. (2000) 40, 177–184.

Chrysina, E. D., Brew, K., and Acharya, R. Crystal structures of apo- and holo-bovine α-lactalbumin at 2.2-Å resolution reveal an effect of calcium on inter-lobe interactions. J. Biol. Chem. (2000) 275, 47, 37021–37029.

Chuang, D. M. Neuroprotective and neurotrophic actions of the mood stabilizer lithium: Can it be used to treat neurodegenerative diseases? Crit. Rev. Neurobiol. (2004) 16, 1–2, 83–90.

Cizewski Culotta, V., Mei Yanga, and O'Halloran, T. V. Activation of superoxide dismutases: putting the metal to the pedal. Biochim. Biophys. Acta. (2006) 1763, 7, 747–758.

Clark, E. P., and Collip, J. B. A study of the Tisdall method for the determination of blood serum calcium with a suggested modification. J. Biol. Chem. (1925) 63, 461–464.

Clark, G. B., Sessions, A., Eastburn, D. J., and Roux, S. J. Differential expression of members of the annexin multigene family in *Arabidopsis*. Plant Physiol. (2001) 126, 1072–1084.

Clarkson, J., and Campbell, I. D. Studies of protein–ligand interactions by NMR. Biochem. Soc. Trans. (2003) 31, 1006–1009.

Claude, A. Particulate components of cytoplasm. Cold Spring Harbor Symp. Quant. Biol. (1941) 9, 263–271.

Clausen, J., Edeling, C.-J., and Fogh, J. ^{67}Ga binding to human serum proteins and tumor components. Cancer Res. (1974) 34, 1931–1937.

Clemens, S. Molecular mechanisms of plant metal tolerance and homeostasis. Planta (2001) 212, 475–486.

Clement, S., Dumont, J. E., and Schurmans, S. Loss of calciphosine gene expression in mouse and other rodents. Biochem. Biophys. Res. Commun. (1997) 232, 407–413.

Cliff, M. J., and Ladbury, J. E. A survey of the year 2002 literature on applications of isothermal titration calorimetry. J. Mol. Recognit. (2003) 16, 6, 383–391.

Clipstone, N. A., and Crabtree, G. R. Identification of calcineurin as a key signalling enzyme in T-lymphocyte activation. Nature (1992) 357, 695–697.

Close, R. A., and West, T. S. A new selective metallochromic reagent for the determination and chelatometric determination of calcium. Talanta (1960) 5, 221–230.

Cockcroft, S. The latest phospholipase C, PLCη, is implicated in neuronal function. Trends Biochem Sci. (2006) 31, 1, 4–7.

Codd, R., Irwin, J. A., and Lay, P. A. Sialoglycoprotein and carbohydrate complexes in chromium toxicity. Curr. Opin. Chem. Biol. (2003) 7, 2, 213–219.

Cohen-Solal, M. Strontium overload and toxicity: impact on renal osteodystrophy. Nephrol. Dial. Transplant. (2002) 17, Suppl. 2, 30–34.

Collin, T., Chat, M., Lucas, M. G., Moreno, H., Racay, P., Schwaller, B., Marty, A., and Llano, I. Developmental changes in parvalbumin regulate presynaptic Ca^{2+} signaling. J. Neurosci. (2005) 25, 96–107.

Collins, J. H., Potter, J. D., Horn, M. J., Wilshire, G., and Jackman, N. The amino acid sequence of rabbit skeletal muscle troponin C: gene replication and homology with calcium binding proteins from carp and hake muscle. FEBS Lett. (1973) 36, 268–272.

Collip, J. B. The extraction of a parathyroid hormone which will prevent or control parathyroid tetany and which regulates the level of blood calcium. J. Biol. Chem. (1925) 63, 395–438.

Coman, D. R. Decreased mutual adhesiveness: a property of cells from squamous cell carcinoma. Cancer Res. (1944) 4, 625–629.

Coman, D. R. Cellular adhesiveness in relation to the invasiveness of cancer: electron microscopy of liver perfused with a chelating agent. Cancer Res. (1954) 14, 519–521.

Cook, W. J., Ealik, S. E., Babu, Y. S., Cox, J. A., and Vijay Kumar, S. 3-Dimensional structure of a sarcoplasmic calcium-binding protein from *Nereis diversicolor*. J. Biol. Chem. (1991) 266, 652–656.

Cook, W. J., Jeffrey, L. C., Cox, J. A., and Vijay Kumar, S. Structure of a sarcoplasmic calcium-binding protein from *Amphioxus* refined at 2.4 Angstrom resolution. J. Mol. Biol. (1993) 229, 461–471.

Cook, W. J., Walter, L. J., and Walter, M. R. Drug binding by calmodulin: crystal structure of a calmodulin–trifluoperazine complex. Biochemistry (1994) 33, 51, 15259–15265.

Cooper, C. R., Graves, B., Pruitt, F., et al. Novel surface expression of reticulocalbin 1 on bone endothelial cells and human prostate cancer cells is regulated by TNF-alpha. J. Cell. Biochem. (2008) 104, 2298–2309.

Cooper, E. A., and Knutson, K. Fourier transform infrared spectroscopy investigations of protein structure. Pharm. Biotechnol. (1995) 7, 101–143.

Copp, D. H. Parathyroids, calcitonin, and control of plasma calcium. Rec. Prog. Horm. Res. (1964) 20, 59–88.

Corbalan-Garcia, S., Rodriguez-Alfaro, J. A., and Gomez-Fernandez, J. C. Determination of the calcium-binding sites of the C2 domain of protein kinase Calpha that are critical for its translocation to the plasma membrane. Biochem. J. (1999) 337, 513–521.

Corbin, J. A., Evans, J. H., Landgraf, K. E., and Falke, J. J. Mechanism of specific membrane targeting by C2 domains: localized pools of target lipids enhance Ca^{2+} affinity. Biochemistry (2007) 46, 14, 4322–4336.

Corson, D. C., Williams, T. C., Kay, L. E., and Sykes, B. D. Proton NMR spectroscopic studies of calcium-binding proteins: 1. Stepwise proteolysis of the carboxyl-terminal α-helix of a helix–loop–helix metal-binding domain. Biochemistry (1986) 25, 7, 1817–1826.

Cotton, F. A., Hazen, E. E., Jr., and Legg, M. J. Staphylococcal nuclease: proposed mechanism of action based on structure of enzyme-thymidine 3′,5′-bis phosphate-calcium ion complex of 1.5 resolution. Proc. Natl. Acad. Sci. USA (1979) 76, 6, 2551–2555.

Coussens, L., Parker, P. J., Rhee, L., Yang-Feng, T. L., Chen, E., Waterfield, M. D., Francke, U., and Ullrich, A. Multiple, distinct forms of bovine and human protein kinase C suggest diversity in cellular signaling pathways. Science (1986) 233, 859–866.

Cox, J. A. Interactive properties of calmodulin. Biochem. J. (1988) 249, 2, 621–629.

Cox, J. A., and Stein, E. A. Characterization of the new sarcoplasmic calcium binding protein with magnesium-induced cooperativity in the binding of calcium. Biochemistry (1981) 20, 5430–5436.

Cox, J. A., Wingle, D. R., and Stein, E. A. Calcium, magnesium and conformation of parvalbumin during muscular activity. Biochimie (1979) 61, 7, 601–605.

Cox, J. A., Alard, P., and Schaad, O. Comparative molecular modeling of *Amphioxus* calcium vector protein with calmodulin and troponin-C. Protein Eng. (1990) 4, 23–32.

Cox, J. A., Durussel, I., Scott, D. J., and Berchtold, M. W. Remodeling of the AB site of rat parvalbumin and oncomodulin into a canonical EF-hand. Eur. J. Biochem. (1999) 264, 790–799.

Cozens, B., and Reithmeier, R. A. F. Size and shape of rabbit skeletal muscle calsequestrin. J. Biol. Chem. (1984) 259, 10, 6248–6252.

Cribbs, L. L. T-type Ca^{2+} channels in vascular smooth muscle: multiple functions. Cell Calcium (2006) 40, 2, 221–230.

Crichton, R. R., and Charloteaux-Wauters, M. Iron transport and storage. Eur. J. Biochem. (1987) 164, 485–506.

Crobu, S., Marchetti, M., and Sanna, G. Studies of interaction between Rh(I) and human serum albumin in a "nanostructured biocatalyst" active in the hydroformylation reaction. J. Inorg. Biochem. (2006) 100, 9, 1514–1520.

Crumpton, M. J., and Dedman, J. R. Protein terminology tangle. Nature (1990) 345, 212.

Cunningham, B. A., Hemperly, J. J., Hopp, T. P., and Edelman, G. M. Flavin versus concanavalin A: circularly permuted amino acid sequences. Proc. Natl. Acad. Sci. USA (1979) 76, 7, 3218–3222.

Curran, H. R., Brunsetter, B. C., and Myers, A. T. Spectrochemical analysis of vegetative cells and spores of bacteria. J. Bacteriol. (1943) 45, 485–494.

Cutler, C. S., Smith, C. J., Ehrhardt, G. J., Tyler, T. T., Jurisson, S. S., and Deutsch, E. Current and potential therapeutic uses of lanthanide radioisotopes. Cancer Biother. Radiopharm. (2000) 15, 6, 531–545.

D'Acquisto, F., Perretti, M., and Flower, R. J. Annexin-A1: a pivotal regulator of the innate and adaptive immune systems. Br. J. Pharmacol. (2008) 155, 2, 152–169.

Dai, H., Shen, N., Arac, D., and Rizo J. A quaternary SNARE-synaptotagmin-Ca^{2+}-phospholipid complex in neurotransmitter release. J. Mol. Biol. (2007) 367, 848–863.

Dai, S., Schwendtmayer, C., Johansson, K., Ramaswamy, S., Schurmann, P., and Eklund, H. How does light regulate chloroplast enzymes? Structure–function studies of the ferredoxin/thioredoxin system. Q. Rev. Biophys. (2000) 33, 1, 67–108.

Dalby, A. R. A comparative proteomic analysis of the simple amino acid repeat distributions in *Plasmodia* reveals lineage specific amino acid selection. PLoS One (2009) 4, 7, e6231.

Dalgarno, D. C., Levine, B. A., Williams, R. J., Fullmer, C. S., and Wasserman, R. H. Proton-NMR studies of the solution conformations of vitamin-D-induced bovine intestinal calcium-binding protein. Eur. J. Biochem. (1983) 137, 3, 523–529.

Dameron, C. T., and Harrison, M. D. Mechanisms for protection against copper toxicity. Am. J. Clin. Nutr. (1998) 67, 1091S–1097S.

Darwish, H. M., and DeLuca, H. F. Identification of a 1,25-dihydroxyvitamin D_3-response element in the 5′-flanking region of the rat calbindin D-9k gene. Proc. Natl. Acad. Sci. USA (1992) 89, 603–607.

Davidson, A. L., Dassa, E., Orelle, C., and Chen, J. Structure, function, and evolution of bacterial ATP-binding cassette systems. Microbiol. Mol. Biol. Rev. (2008) 72, 2, 317–364.

Davis, C. M., and Vincent, J. B. Chromium in carbohydrate and lipid metabolism. J. Biol. Inorg. Chem. (1997) 2, 675–679.

Davis, C. M., Sumrall, K. H., and Vincent, J. B. The biologically active form of chromium may activate a membrane phosphotyrosine phosphatase (PTP). Biochemistry (1996) 35, 12963–12969.

Davis, C. M., Royer, A. C., and Vincent, J. B. Synthetic multinuclear chromium assembly activates insulin receptor kinase activity: functional model for low-molecular-weight chromium-binding substance. Inorg. Chem. (1997) 36, 5316–5320.

Davletov, B. A., and Südhof, T. C. A single C2 domain from synaptotagmin I is sufficient for high affinity Ca^{2+}/phospholipid binding. J. Biol. Chem. (1993) 268, 35, 26386–26390.

Dayan, A. D., and Paine, A. J. Mechanisms of chromium toxicity, carcinogenicity and allergenicity: review of the literature from 1985 to 2000. Hum. Exp. Toxicol. (2001) 20, 9, 439–451.

D'Costa, A. M., and Denning, M. F. A caspase-resistant mutant of PKC-δ protects keratinocytes from UV-induced apoptosis. Cell Death Differ. (2005) 12, 224–232.

de Arruda, M. V., Waston, S., Lin, C.-S., Leavitt, J., and Matsudaira, P. Fimbrin is a homolog of the cytoplasmic phosphoprotein plastin and has domains homologous with calmodulin and actin gelation proteins. J. Cell Biol. (1990) 111, 1069–1079.

Deban, L., Bottazzi, B., Garlanda, C., de la Torre, Y. M., and Mantovani, A. Pentraxins: multifunctional proteins at the interface of innate immunity and inflammation. Biofactors (2009) 35, 2, 138–145.

de Bont, J. M., den Boer, M. L., Kros, J. M., Passier, M. M., Reddingius, R. E., Smitt, P. A., Luider, T. M., and Pieters, R. Identification of novel biomarkers in pediatric primitive neuroectodermal tumors and ependymomas by proteome-wide analysis. J. Neuropathol. Exp. Neurol. (2007) 66, 6, 505–516.

Deckert, J. Cadmium toxicity in plants: Is there any analogy to its carcinogenic effect in mammalian cells? Biometals (2005) 18, 475–481.

Declercq, J. P., Tinant, B., Parello, J., Etienne, G., and Huber, R. Crystal structure determination and refinement of pike 4.10 parvalbumin (minor component from *Esox lucius*). J. Mol. Biol. (1988) 202, 2, 349–353.

Declercq, J. P., Tinant, B., Parello, J., and Rambaud, J. Ionic interactions of parvalbumin: crystal structure of pike 4.10 parvalbumin in four different ionic environments. J. Mol. Biol. (1991) 220, 1017–1039.

Declercq, J. P., Evrard, C., Lamzin, V., and Parello, J. Crystal structure of the EF-hand parvalbumin at atomic resolution (0.91 Å) and at low temperature (100 K): evidence for conformational multistates within the hydrophobic core. Protein Sci. (1999) 8, 2194–2204.

Dejgaard, S., Nicolay, J., Taheri, M., Thomas, D. Y., and Bergeron, J. J. The ER glycoprotein quality control system. Curr. Issues Mol. Biol. (2004) 6, 1, 29–42.

Deka N., Wong, E., Matera, A. G., Kraft, R., Leinwand, L. A., and Schmid, C. W. Repetitive nucleotide-sequence insertions into a novel calmodulin-related gene and its processed pseudogene. Gene (1988) 71, 123–124.

Dekker, N., Peters, A. R., Slotboom, A. J., Boelens, R., Kaptein, R., and de Haas, G. Porcine pancreatic phospholipase A_2: sequence-specific ^1H and ^{15}N NMR assignements and secondary structure. Biochemistry (1991) 30, 12, 3135–3147.

De la Fuente, M., and Parra, A. V. Vesicle aggregation by annexin I: role of a secondary membrane binding site. Biochemistry (1995) 34, 33, 10393–10399.

Dell'Angelica, E. C., Schleicher, C. H., and Santome, J. A. Primary structure and binding properties of calgranulin C, a novel S100-like calcium-binding protein from pig granulocytes. J. Biol. Chem. (1994) 269, 28929–28936.

De Long, R. P., Coman, D. R., and Zeidman, I. The significance of low calcium and high potassium content in neoplastic tissue. Cancer (1950) 3, 718–721.

Delphin, C., Ronjat, M., Deloulme, J. C., Garin, G., Debussche, L., Higashimoto, Y., Sakaguchi, K., and Baudier, J. Calcium-dependent interaction of S100B with the C-terminal domain of the tumor suppressor p53. J. Biol. Chem. (1999) 274, 15, 10539–10544.

DeLuca, H. F., and Engstrom, G. W. Calcium uptake by rat kidney mitochondria. Proc. Natl. Acad. Sci. USA (1961) 47, 1744–1750.

Demaurex, N., and Frieden, M. Measurements of the free luminal ER Ca^{2+} concentration with targeted "cameleon" fluorescent proteins. Cell Calcium (2003) 34, 2, 109–119.

Den, Z., Pelias, M. Z., Ditta, G. M., Keats, B. J., Deininger, P. L., and Batzer, M. A. Assembly of a high-resolution map of the Acadian Usher syndrome region and localization of the nuclear EF-hand acidic gene. Biochim. Biophys. Acta (1998) 1407, 84–91.

Denisov, V. P., and Halle, B. Direct observation of calcium-coordinated water in calbindin D_{9k} by nuclear magnetic relaxation dispersion. J. Am. Chem. Soc. (1995) 117, 32, 8456–8465.

Denning, M. F. Epidermal keratinocytes: regulation of multiple cell phenotypes by multiple protein kinase C isoforms. Int. J. Biochem. Cell Biol. (2004) 36, 1141–1146.

Denton, R. M. Regulation of mitochondrial dehydrogenases by calcium ions. Biochim. Biophys. Acta (2009) 1787, 11, 1309–1316.

De Rienzo, F., Gabdoulline, R. R., Wade, R. C., Sola, M., and Menziani, M. C. Computational approaches to structural and functional analysis of plastocyanin and other blue copper proteins. Cell. Mol. Life Sci. (2004) 61, 1123–1142.

Derycke, L. D. M., and Bracke, M. E. *N*-Cadherin in the spotlight of cell–cell adhesion, differentiation, embryogenesis, invasion and signaling. Int. J. Dev. Biol. (2004) 48, 463–476.

Desmeules, P., Grandbois, M., Bondarenko, V. A., Yamazaki, A., and Salesse, C. Measurement of membrane binding between recoverin, a calcium–myristoyl switch protein, and lipid bilayers by AFM-based force spectroscopy. Biophys. J. (2002) 82, 6, 3343–3350.

Desoize, B. Metals and metal compounds in cancer treatment. Anticancer Res. (2004) 24, 3a, 1529–1544.

Dharamsi, A., Tessarolo, D., Coukell, B., and Pun, J. CBP1 associates with the *Dictyostelium* cytoskeleton and is important for normal cell aggregation under certain developmental conditions. Exp. Cell Res. (2000) 258, 298–309.

Diao, J., and Tajkhorshid, E. Indirect role of Ca^{2+} in the assembly of extracellular matrix proteins. Biophys. J. (2008) 95, 1, 120–127.

Di Cera, E. A structural perspective on enzymes activated by monovalent cations. J. Biol. Chem. (2006) 281, 3, 1305–1308.

Digonnet, C., Aldon, D., Leduc, N., Dumas, C., and Rougier, M. First evidence of a calcium transient in flowering plants at fertilization. Development (1997) 124, 2867–2874.

Dijkstra, B. W., Drenth, J., and Kalk, K. H. Active site and catalytic mechanism of phosphodiesterase A_2. Nature (1981a) 289, 5798, 604–606.

Dijkstra, B. W., Kalk, K. H., Hol, W. G. J., and Drenth, J. Structure of bovine pancreatic phospholipase A_2 at 1.7 Å resolution. J. Mol. Biol. (1981b) 147, 1, 97–123.

Dijkstra, B. W., Kalk, K. H., Drenth, J., et al. Role of the N-terminus in the interaction of pancreatic phospholipase A_2 with aggregated substrates: properties and crystal structure of transaminated phospholipase A_2. Biochemistry (1984) 23, 12, 2759–2766.

Dionisio, N., Garcia-Mediavilla, M. V., Sanchez-Campos, S., Majano, P. L., Benedicto, I., Rosado, J. A., Salido, G. M., and Gonzalez-Gallego, J. Hepatitis C virus NS5A and core proteins induce oxidative stress-mediated calcium signalling alterations in hepatocytes. J. Hepatol. (2009) 50, 5, 872–882.

Dipolo, R., Requena, J., Brinley, F. J. Jr, Mullins, L. J., Scarpa, A., and Tiffert, T. Ionized calcium concentrations in squid axons. J. Gen. Physiol. (1976) 67, 4, 433–467.

DiSaia, P. J. Pregnancy and delivery of a patient with Starr–Edwards mitral valve prosthesis: report of a case. Obstet. Gynecol. (1966) 28, 469–472.

Dizhoor, A. M., Ray, S., Kumar, S., Niemi, G., Spencer, M., Brolley, D., Walsh, K. A., Philipov, P. P., Hurley, J. B., and Stryer, L. Recoverin: a calcium sensitive activator of retinal rod guanylate cyclase. Science (1991) 251, 915–918.

Domingo, J. L. Aluminum and other metals in Alzheimer's disease: a review of potential therapy with chelating agents. J. Alzheimers Dis. (2006) 10, 2–3, 331–341.

Dominguez, D. C. Calcium signalling in bacteria. Mol. Microbiol. (2004) 54, 2, 291–297.

Donato, R. S100: a multigenic family of calcium-modulated proteins of the EF-hand type with intracellular and extracellular functional roles. Int. J. Biochem. Cell Biol. (2001) 33, 637–668.

Donato, R., Sorci, G., Riuzzi, F., Arcuri, C., Bianchi, R., Brozzi, F., Tubaro, C., and Giambanco, I. S100B's double life: intracellular regulator and extracellular signal. Biochim. Biophys. Acta (2009) 1793, 6, 1008–1022.

Dong, H., Li, X., Lou, Z., Xu, X., Su, D., Zhou, X., Zhou, W., Bartlam, M., and Rao, Z. Crystal-structure and biochemical characterization of recombinant human calcyphosine delineates a novel EF-hand-containing protein family. J. Mol. Biol. (2008) 383, 3, 455–464.

Dong, K. C., and Berger, J. M. Structural basis for gate-DNA recognition and bending by type IIA topoisomerases. Nature (2007) 450, 1201–1205.

Dong, Q., Patel, M., Scott, K. F., Graham, G. G., Russell, P. J., and Sved, P. Oncogenic action of phospholipase A_2 in prostate cancer. Cancer Lett. (2006) 240, 1, 9–16.

Doniach, S. Changes in biomolecular conformation seen by small angle x-ray scattering. Chem. Rev. (2001) 101, 6, 1763–1778.

Douglas, W. W. A possible mechanism of neurosecretion: release of vasopressin by depolarization and its dependence on calcium. Nature (1963) 197, 81–82.

Douglas, W. W., and Rubin, R. P. The role of calcium in the secretory response of the adrenal medulla to acetylcholine. J. Physiol. (1961) 159, 40–57.

Downing, A. K., Knott, V., Werner, J. M., Cardy, C. M., Campbell, I. D., and Handford, P. A. Solution structure of a pair of Ca^{2+}-binding epidermal growth factor-like domains: implications for the Marfan syndrome on other genetic disorders. Cell (1996) 85, 597–605.

Doyle, D. A., Cabral, J. M., Pfuetzner, R. A., Kuo, A., Gulbis, J. M., Cohen, S. L., Chait, B. T., and MacKinnon, R. The structure of the potassium channel: molecular basis of K^+ conduction and selectivity. Science (1998) 280, 5360, 69–77.

Drake, S. K., Zimmer, M. A., Miller, C. L., and Falke, J. J. Optimizing the metal binding parameters of an EF-hand-like calcium chelation loop: coordinating side chains play a more important tuning role than chelation loop flexibility. Biochemistry (1997) 36, 9917–9926.

Drakenberg, T., Hofman, T., and Chazin, W. J. ^{1}H NMR studies of porcine calbindin 9kD in solution: sequential resonance assignment, secondary structure, and global fold. Biochemistry (1989) 28, 14, 5946–5954.

Drickamer, K., and Taylor, M. E. V Biology of animal lectins. Annu. Rev. Cell Biol. (1993) 9, 237–264.

Drohat, A. C., Amburgey, J. C., Abildgaard, F., Starich, M. R., Baldisseri, D., and Weber, D. J. Solution structure of rat apo-S100B (ββ) as determined by NMR spectroscopy. Biochemistry (1996) 35, 11577–11588.

Drum, C. L., Yan, S.-Z., Bard, J., Shen, Y.-Q., Lu, D., Soelaiman, S., Grabarek, Z., Bohm, A., and Tang, W.-J. Structural basis for the activation of anthrax adenylyl cyclase exotoxin by calmodulin. Nature (2002) 415, 396–402.

Dubois, T., Mira, J. P., Feliers, D., Solito, E., Russo-Marie, F., and Oudinet, J. P. Annexin V inhibits protein kinase C activity via mechanism of phospholipid sequestration. Biochem. J. (1998) 330, 1277–1282.

Dubreuil, G. Mitochondries des ostéoclastes et des cellules de bizzozero. C. R. Soc. Biol. (1910) 69, 71–73.

Dudev, T., and Lim, C. Principles governing Mg, Ca, and Zn binding and selectivity in proteins. Chem. Rev. (2003) 103, 773–787.

Dudev, T., and Lim, C. Metal binding affinity and selectivity in metalloproteins: insights from computational studies. Annu. Rev. Biophys. (2008) 37, 97–116.

Duffy, B., Schwietert, C., France, A., Mann, N., Culbertson, K., Harmon, B., and McCue, J. P. Transition metals as protease inhibitors. Biol. Trace Elem. Res. (1998) 64, 1–3, 197–213.

du Hamel, M. Sur une racine qui a la faculté de teindre en rouge les os des animaux vivant. Hist. Acad. R. Sci. Mem. Math. Phys. (1739) 52, 1–13.

Dundas, J., Ouyang, Z., Tseng, J., Binkowski, A., Turpaz, Y., and Liang, J. CASTp: computed atlas of surface topography of proteins with structural and topographical mapping of functionally annotated residues. Nucleic Acids Res. (2006) 34, W116–W118.

Dunham, E. T., and Glynn, I. M. Adenosinetriphosphatase activity and the active movements of alkali metal ions. J. Physiol. (1961) 156, 274–293.

Düringer, C., Hamiche, A., Gustafsson, L., Kimura, H., and Svanborg, C. HAMLET interacts with histones and chromatin in tumor cell nuclei. J. Biol. Chem. (2003) 278, 43, 42131–42135.

Durussel, I., Pauls, T. L., Cox, J. A., and Berchtold, M. W. Chimeras of parvalbumin and oncomodulin involving exchange of the complete CD site show that the Ca^{2+}/Mg^{2+} specificity is an intrinsic property of the site. Eur. J. Biochem. (1996) 242, 256–263.

Dyson, H. J., and Wright, P. E. Unfolded proteins and protein folding studied by NMR. Chem. Rev. (2004) 104, 8, 3607–3622.

Eaton, D. C., and Brodwick, M. S. Effects of barium on the potassium conductance of squid axon. J. Gen. Physiol. (1980) 75, 727–750.

Eaton, W. A., Henry, E. R., Hofrichter, J., and Mozzarelli A. Is cooperative oxygen binding by hemoglobin really understood? Nat. Struct. Biol. (1999) 6, 4, 351–358.

Ebashi, S. Calcium binding activity of vesicular relaxing factor. J. Biochem. (1961) 50, 236–244.

Ebashi, S. Third component participating in the superprecipitation of "natural actomyosin." Nature (1963) 200, 1010–1010.

Ebashi, S., Ebashi, F., and Kodama, A. Troponin as the Ca^{++}-receptive protein in the contractile system. J. Biochem. (1967) 62, 137–138.

Eberspach, J., Strasburger, W., Glatter, U., Gerday, C., and Wollmer, A. Interaction of parvalbumin of pike II with calcium and terbium ion. Biochim. Biophys. Acta. (1988) 952, 1, 67–76.

Ebinu, J. O., Bottorff, D. A., Chan, E. Y. W., Stang, S. L., Dunn, R. J., and Stone, J. C. RasGRP, a Ras guanyl nucleotide-releasing protein with calcium- and diacylglycerol-binding motifs. Science (1998) 280, 1082–1086.

Ebrey, T., and Koutalos, Y. Vertebrate photoreceptors. Prog. Retin. Eye Res. (2001) 20, 1, 49–94.

Eckert, R. L., Broome, A. M., Ruse, M., Robinson, N., Ryan, D., and Lee, K. S100 proteins in the epidermis. J. Invest. Dermatol. (2004) 123, 1, 23–33.

Edwards, A. Large-scale structural biology of the human proteome. Annu. Rev. Biochem. (2009) 78, 541–568.

Einspahr, H., and Bugg, C. E. Crystal structure studies of calcium complexes and implications for biological systems. In: Sigel, A., and Sigel, H., eds., Metal Ions in Biological Systems, Vol. 17 (1984) New York: Marcel Dekker, Inc., Basel, 51–97.

Eis, P. S. Fluorescence studies of zinc finger peptides and proteins. Methods Enzymol. (1997) 278, 330–343.

Eisler, R. Mammalian sensitivity to elemental gold (Au degrees). Biol. Trace Elem. Res. (2004) 100, 1, 1–18.

Eitinger, T., Suhr, J., Moore, L., and Smith, J. A. Secondary transporters for nickel and cobalt ions: theme and variations. Biometals (2005) 18, 4, 399–405.

El-Aziza, N. A., Khatera, A. E. M., and Al-Sewaidan, H. A. Natural radioactivity contents in tobacco. Int. Congr. Ser. (2005) 1276, 407–408.

Elkins, K. M., Gatzeva-Topalova, P. Z., and Nelson, D. J. Molecular dynamic study of Ca^{2+} binding loop variants of parvalbumin with modifications at the "gateway" position. Protein Eng. (2001) 14, 2, 115–126.

Ellgaard, L., Riek, R., Herrmann, T., Guntert, P., Braun, D., Helenius, A., and Wutrich, K. NMR structure of the calreticulin P-domain. Proc. Natl. Acad. Sci. USA (2001) 98, 6, 3133–3138.

Elsayed, S., and Aas, K. Characterization of a major allergen (cod). Observations on effect of denaturation on the allergenic activity. J. Allergy (1971) 47, 283–291.

Elsayed, S., and Bennich, H. The primary structure of allergen M from cod. Scand J. Immunol. (1974) 3, 683–686.

Elsayed, S., and Bennich, H. The primary structure of allergen M from cod. Scand. J. Immunol. (1975) 4, 203–208.

Elshorst, B., Hennig, M., Forsterling, H., et al. NMR solution structure of a complex of calmodulin with a binding peptide of the Ca^{2+} pump. Biochemistry (1999) 38, 12320–12332.

Emberley, E. D., Murphy, L. C., and Watson, P. H. S100A7 and the progression of breast cancer. Breast Cancer Res. (2004) 6, 4, 153–159.

Emsley, J., White, H. E., O'Hara, B. P., Oliva, G., Srinivasan, N., Tickle, I. J., Blundell, T. I., Pepys, M. B., and Wood, S. P. Structure of pentameric human serum amyloid component. Nature (1994) 367, 338–345.

Engelhardt, W. N., and Ljubimowa, M. N. Myosine and adenosinetriphosphatase. Nature (1939) 144, 668–669.

Engman, D. M., Krause, K.-H., Blumin, J. H., Kim, K. S., Kirchhoff, L. V., and Donelson, J. E. A novel flagellar Ca^{2+}-binding protein in trypanosomes. J. Biol. Chem. (1989) 264, 18627–18631.

Ennever, J. Intracellular calcification by oral filamentous microorganisms. J. Periodontol. (1960) 31, 304–307.

Ennever, J., and Creamer, A. Microbiologic calcification: bone mineral and bacteria. Calcif. Tissue Res. (1967) 1, 87–93.

Espósito, B. P., Faljoni-Alário, A., de Menezes, J. F., de Brito, H. F., and Najjar, R. A circular dichroism and fluorescence quenching study of the interactions between rhodium(II) complexes and human serum albumin. J. Inorg. Biochem. (1999) 75, 1, 55–61.

Esposito, L., Vitagliano, L., and Mazzarella, L. Recent advances in atomic resolution protein crystallography. Protein Pept. Lett. (2002) 9, 2, 95–106.

Essen, L. O., Perisic, O., Cheung, R., Katan, M., and Williams, R. L. Crystal structure of a mammalian phosphoinositide-specific phospholipase C delta. Nature (1996) 380, 6575, 595–602.

Essen, L. O., Perisic, O., Lynch, D. E., Katan, M., and Williams, R. L. A ternary metal binding site in the C_2 domain of phosphoinositide-specific phospholipase C-δ 1. Biochemistry (1997) 36, 10, 2753–2762.

Evans, J. S., and Levine, B. A. Protein–protein interaction sites of the troponin complex. Biochem. Soc. Trans. (1979) 7, 4, 701–702.

Evans, J. S., Levine, B. A., Leavis, D. C., Gergely, J., Grabarek, Z., and Drabikowski, W. Proton magnetic studies on proteolytic fragments of troponin C: structural homology with the native molecule. Biochim. Biophys. Acta (1980) 623, 1, 10–21.

Evans, T. C., and Nelsestuen, G. L. Calcium and membrane-binding properties of monomeric and multimeric annexin II. Biochemistry (1994) 33, 45, 13231–13238.

Evenas, J., Thulin, E., Malmendal, A., Forsen, S., and Carlstrom, G. NMR studies of the E140Q mutant of the carboxy-terminal domain of calmodulin reveal global conformational exchange in the Ca^{2+}-saturated state. Biochemistry (1997) 36, 3448–3457.

Evenas, J., Malmendal, A., Thulin, E., Carlstrom, G., and Forsen, S. Ca^{2+} binding and conformational changes in a calmodulin domain. Biochemistry (1998) 37, 13744–13754.

Ewbank, J. J., and Creighton, T. E. Pathway of disulfide-coupled unfolding and refolding of bovine α-lactalbumin. Biochemistry (1993a) 32, 14, 3677–3693.

Ewbank, J. J., and Creighton, T. E. Structural characterization of the disulfide folding intermediates of bovine α-lactalbumin. Biochemistry (1993b) 32, 14, 3694–3707.

Eyles, S. J., and Kaltashov, I. A. Methods to study protein dynamics and folding by mass spectrometry. Methods (2004) 34, 1, 88–99.

Fahie, K., Pitts, R., Elkins, K., and Nelson, D. J. Molecular dynamic study of the Ca^{2+} binding loop variants of silver hake parvalbumin with aspartic acid at the "gateway" position. J. Biomol. Struct. Dyn. (2002) 5, 821–837.

Fährmann, M. Targeting protein kinase C (PKC) in physiology and cancer of the gastric cell system. Curr. Med. Chem. (2008) 15, 12, 1175–1191.

Falke, J. J., Drake, S. K., Hazard, A. L., and Peersen, O. B. Molecular tuning of ion binding to calcium signaling proteins. Q. Rev. Biophys. (1994) 27, 219–290.

Fallon, J. L., and Quiocho, F. A. A closed compact structure of native Ca^{2+}-calmodulin. Structure (2003) 11, 1303–1307.

Fan, X., White, I. M., Shopova, S. I., Zhu, H., Suter, J. D., and Sun, Y. Sensitive optical biosensors for unlabeled targets: a review. Anal. Chim. Acta (2008) 620, 1–2, 8–26.

Fantus, I. G., Deragon, G., Lai, R., and Tang, S., Modulation of insulin action by vanadate: evidence of a role for phosphotyrosine phosphatase activity to alter cellular signaling. Mol. Cell. Biochem. (1995) 153, 103–112.

Farah, C. S., Miyamoto, C. A., Ramos, C. H. I., da Silva, A. C., Quaggio, R. B., Fujimori, K., Smillie, L. B., and Reinach, F. C. Structural and regulatory functions of the N- and C-terminal regions of skeletal muscle troponin I. J. Biol. Chem. (1994) 269, 7, 5230–5240.

Farrar, Y. J. K., Lukas, T. J., Craig, T. A., Watterson, D. M., and Carlson, G. M. Features of calmodulin that are important in the activation of the catalytic subunit of phosphorylase kinase. J. Biol. Chem. (1993) 268, 6, 4120–4125.

Fast, J., Mossberg, A. K., Nilsson, H., Svanberg, C., Akke, M., and Linse, S. Compact oleic acid in HAMLET. FEBS Lett. (2005) 579, 27, 6095–6100.

Fatt, P., and Katz, B. Spontaneous subthreshold activity at motor nerve endings. J. Physiol. (1952) 117, 109–128.

Fazioli, F., Minichiello, L., Matoskova, B., Wong, W. T., and Di Fiore, P. P. EPS15, a novel tyrosine kinase substrate, exhibits transforming activity. Mol. Cell. Biol. (1993) 13, 5814–5828.

Fedrizzi, L., Lim, D., Carafoli, E., and Brini, M. Interplay of the Ca^{2+}-binding protein DREAM with presenilin in neuronal Ca^{2+} signaling. J. Biol. Chem. (2008) 283, 41, 27494–27503.

Feenstra, T. P. La Signification du calcium pour les courants rétiniens. Arch. Neerland. Physiol. (1926) 11, 139–140.

Feng, B., and Stemmer, P. M. Ca^{2+} binding site 2 in calcineurin-B modulates calmodulin-dependent calcineurin phosphatase activity. Biochemistry (2001) 40, 8808–8814.

Ferguson, A. D., Hofmann, E., Coulton, J. W., Diederichs, K., and Welte, W. Siderophore-mediated iron transport: crystal structure of FhuA with bound lipopolysaccharide. Science (1998) 282, 5397, 2215–2220.

Ferguson, A. D., Chakraborty, R., Smith, B. S., Esser, L., van der Helm, D., and Deisenhofer, J. Structural basis of gating by the outer membrane transporter FecA. Science (2002) 295, 5560, 1715–1719.

Ferreira, F., Bucchini, D., Martin, M. E., Levi, S., Arosio, P., Grandchamp, B., and Beaumont, C. J. Biol. Chem. (2000) 275, 3021–3024.

Filatov, V. L., Katrukha, A. G., Bulargina, T. V., and Gusev, N. B. Troponin: structure, properties, and mechanism of functioning. Biochemistry (Mosc.) (1999) 64, 9, 969–985.

Filipenko, N. R., and Waisman, D. M. The C-terminus of annexin II mediates binding to F-actin. J. Biol. Chem. (2001) 276, 5310–5315.

Findlay, W. A., and Sykes, B. D. ^{1}H-NMR resonance assignments, secondary structure, and global fold of the TR1C fragment of turkey skeletal troponin C in the calcium-free state. Biochemistry (1993) 32, 13, 3461–3467.

Findlay, W. A., Sonnichsen, F. D., and Sykes, B. D. Solution structure of the TR1C fragment of skeletal muscle troponin-C. J. Biol. Chem. (1994) 269, 9, 6773–6778.

Finer, J. T., Simmons, R. M., and Spudich, J. A. Single myosin molecule mechanics: piconewton forces and nanometre steps. Nature (1994) 368, 6467, 113–119.

Finney, L. A., and O'Halloran, T. V. Transition metal speciation in the cell: insights from the chemistry of metal ion receptors. Science (2003) 300, 931–936.

Flaherty, K. M., Zozulya, S., Stryer, L., and McKay, D. B. Three-dimensional structure of recoverin, a calcium sensor in vision. Cell (1993) 75, 4, 709–716.

Föhr, U. G., Heizmann, C. W., Engelkamp, D., Schafer, B. W., and Cox, J. A. Purification and cation binding properties of the recombinant human S100 calcium-binding protein A3, an EF hand motif protein with high affinity for zinc. J. Biol. Chem. (1995) 270, 36, 21056–21061.

Fontán-Lozano, A., Romero-Granados, R., del-Pozo-Martín, Y., Suárez-Pereira, I., Delgado-García, J. M., Penninger, J. M., and Carrión, A. M. Lack of DREAM protein enhances learning and memory and slows brain aging. Curr. Biol. (2009) 19, 1, 54–60.

Foriers, A., De Neve, R., Kanarek, L., and Strosberg, A. D. Common ancestor for concanavalin A and lentil lectin? Proc. Natl. Acad. Sci. USA (1978) 75, 3, 1136–1139.

Foriers, A., Lebrun, E., Van Rapenbusch, R., De Neve, R., and Strosberg, A. D. Structure of the lentil (*Lens culinaris*) lectin. Amino acid sequence determination and prediction of the secondary structure. J. Biol. Chem. (1981) 256, 11, 5550–5560.

Forsén, S., Linse, S., Thulin, E., Lindegaro, B., Martin, S. R., Bayley, P. M., Brodin, P., and Grundstrom, T. Kinetics of calcium binding to calbindin mutants. Eur. J. Biochem. (1988) 177, 1, 47–52.

Fortes de Valencia, F., Paulucci, A. A., Quaggio, R. B., Rasera da Silva, A. C., Farah, C. S., and de Castro Reinach, F. Parallel measurement of Ca^{2+} binding and fluorescence emission upon Ca^{2+} titration of recombinant skeletal muscle troponin C: measurement of sequential calcium binding to the regulatory sites. J. Biol. Chem. (2003) 278, 11007–11014.

Foster, G. B., Baghdiantz, A., Kumar, M. A., Slack, E., Soliman, H. A., and MacIntyre, I. Thyroid origin of calcitonin. Nature (1964) 202, 1303–1305.

Foster, M. Text-book of Physiology, 4th ed., Macmillan and Co., New York, 1883.

Foty, R. A., and Steinberg, M. S. Cadherin-mediated cell–cell adhesion and issue segregation in relation to malignancy. Int. J. Dev. Biol. (2004) 48, 397–409.

Fowler, B. A. Roles of lead-binding proteins in mediating lead bioavailability. Environ. Health Perspect. (1998) 106, Suppl. 6, 1585–1587.

Fowler, B. A., and DuVal, G. E. Effects of lead on the kidney: roles of high affinity lead-binding proteins. Environ. Health Perspect. (1991) 91, 77–80.

Framson, P. E., and Sage, E. H. SPARC and tumor growth: where the seed meets the soil? J. Cell. Biochem. (2004) 92, 4, 679–690.

Franchini, P. L., and Reid, R. E. Investigating site-specific effects of the −X glutamate in a parvalbumin CD site model peptide. Arch. Biochem. Biophys. (1999) 372, 80–88.

Franz, C., Durussel, I., Cox, J. A., Schäfer, B. W., and Heizmann, C. W. Binding of Ca^{2+} and Zn^{2+} to human nuclear S100A2 and mutant proteins. J. Biol. Chem. (1998) 273, 18826–18834.

Franzini-Armstrong, C., and Porter, K. R. Sarcolemmal invaginations constituting the T system in fish muscle fibers. J. Cell Biol. (1964) 22, 675–696.

Frederickson, C. J., and Bush, A. J. Synaptically released zinc: physiological functions and pathological effects. Biometals (2001) 14, 353–366.

Freudenrich, C. C., Muphy, E., Liu, S., and Lieberman, M. Magnesium homeostasis in cardiac cells. Mol. Cell. Biochem. (1992) 114, 97–103.

Friedberg, F. Centrin isoforms in mammals: relation to calmodulin. Mol. Biol. Rep. (2006) 33, 4, 243–252.

Friedrich, P. The intriguing Ca^{2+} requirement of calpain activation. Biochem. Biophys. Res. Commun. (2004) 323, 4, 1131–1133.

Fritz, G., Heizmann, C. W., and Kroneck, P. M. Probing the structure of the human Ca^{2+}- and Zn^{2+}-binding protein S100A3: spectroscopic investigations of its transition metal ion complexes, and three-dimensional structural model. Biochim. Biophys. Acta. (1998) 1448, 264–276.

Fritz, G., Mittl, P. R., Vasak, M., Grutter, M. G., and Heizmann, C. W. The crystal structure of metal-free human EF-hand protein S100A3 at 1.7-Å resolution. J. Biol. Chem. (2002) 277, 36, 33092–33098.

Fujii, H., and Zhu, J. K. An autophosphorylation site of the protein kinase SOS2 is important for salt tolerance in *Arabidopsis*. Mol. Plant (2009) 2, 1, 183–190.

Fujii, T., Yamana, K., Ogoma, Y., and Kondo, Y. Interaction of calponin with phospholipids. J. Biochem. (1995) 117, 5, 999–1003.

Fujisawa, T., Ueki, T., and Iida, S. Structural change of the troponin C molecule upon Ca^{2+}-binding measured in solution by x-ray scattering technique. J. Biochem. (1989) 105, 2, 377–383.

Fujisawa, T., Ueki, T., and Iida, S. Structural change of troponin C molecule and its domains upon Ca^{2+} binding in the presence of Mg^{2+} ions measured by a solution x-ray scattering technique. J. Biochem. (1990) 107, 3, 343–351.

Fukuda, H., Irie, K., Nakahara, A., Ohigashi, H., and Wender, P. A. Solid-phase synthesis, mass spectrometric analysis of the zinc-folding, and phorbol ester-binding studies of the 116-mer peptide containing the tandem cysteine-rich C1 domains of protein kinase C gamma. Bioorg. Med. Chem. (1999) 7, 6, 1213–1221.

Fukumoto, S., and Martin, T. J. Bone as an endocrine organ. Trends Endocrinol. Metab. (2009) 20, 5, 230–236.

Fulkerson, J. F., Jr., and Mobley, H. L. Membrane topology of the NixA nickel transporter of *Helicobacter pylori*: two nickel transport-specific motifs within transmembrane helices II and III. J. Bacteriol. (2000) 182, 6, 1722–1730.

Fulpius, B., and Baumann, F. Effects of sodium, potassium and calcium ions on slow and spike potentials in single photoreceptor cells. J. Gen. Physiol. (1969) 53, 541–561.

Funk, S. E., and Sage, E. H. Differential effects of SPARC and cationic SPARC peptides on DNA synthesis by endothelial cells and fibroblasts. J. Cell. Physiol. (1993) 154, 53–63.

Furie, B., and Furie, B. C. The molecular basis of blood coagulation. Cell (1988) 53, 4, 505–518.

Gagne, S. M., Tsuda, S., Li, M. X., Smillie, L. B., and Sykes, B. D. Structures of the troponin C regulatory domains in the apo and calcium-saturated states. Nat. Struct. Biol. (1995) 2, 9, 784–789.

Gagne, S. M., Li, M. X., and Sykes, B. D. Mechanism of direct coupling between binding and induced structural change in regulatory calcium binding proteins. Biochemistry (1997) 36, 4386–4392.

Gaither, L. A., and Eide, D. J. Eukaryotic zinc transporters and their regulation. Biometals (2001) 14, 251–270.

Gallagher, W. H., and Lauffer, M. A. Calcium ion binding by tobacco mosaic virus, J. Mol. Biol. (1983) 170, 905–919.

Gallegos, L. L., and Newton, A. C. Spatiotemporal dynamics of lipid signaling: protein kinase C as a paradigm. IUBMB Life (2008) 60, 12, 782–789.

Galstoff, P. S. Regeneration after dissociation, (an experimental study on sponges): I. Behavior of dissociated cells of microciona prolifera under normal and altered conditions. J. Exp. Zool. (1925) 42, 183–221.

Gangola, P., and Rosen, B. P. Maintenance of intracellular calcium in *Escherichia coli*. J. Biol. Chem. (1987) 262, 26, 12570–12574.

Garcia, R. A., Pantazatos, D., and Villarreal, F. J. Hydrogen/deuterium exchange mass spectrometry for investigating protein–ligand interactions. Assay Drug Dev. Technol. (2004) 2, 1, 81–91.

Garcia-Garcia, H. M., and Serruys, P. W. Phospholipase A_2 inhibitors. Curr. Opin. Lipidol. (2009) 20, 4, 327–332.

Garnero, P. Biomarkers for osteoporosis management: utility in diagnosis, fracture risk prediction and therapy monitoring. Mol. Diagn. Ther. (2008) 12, 3, 157–170.

Garone, L., Theibert, J. L., Miegel, A., Maeda, Y., Murphy, C., and Collins, J. H. Lobster troponin-C amino acid sequences of 3 isoforms. Arch. Biochem. Biophys. (1991) 291, 89–91.

Garrett, S. C., Varney, K. M., Weber, D. J., and Bresnick, A. R. S100A4, a mediator of metastasis. J. Biol. Chem. (2006) 281, 2, 677–680.

Gazaryan, I. G., Krasinskaya, I. P., Kristal, B. S., and Brown, A. M. Zinc irreversibly damages major enzymes of energy production and antioxidant defense prior to mitochondrial permeability transition. J. Biol. Chem. (2007) 17, 282, 33, 24373–24380.

Ge, R., Sun, X., Gu, Q., Watt, R. M., Tanner, J. A., Wong, B. C., Xia, H. H., Huang, J. D., He, Q. Y., and Sun, H. A proteomic approach for the identification of bismuth-binding proteins in *Helicobacter pylori*. J. Biol. Inorg. Chem. (2007) 12, 6, 831–842.

Gebauer, G., Schiltz, E., and Rudiger, H. The amino acid sequence of the subunit of the mitogenic lectin from *Vicia sativa*. Eur. J. Biochem. (1981) 113, 2, 319–325.

Gellatly, K. S., and Lefebvre, D. D. Identification of a cDNA clone coding for a novel calcium-binding protein from potato-tuber. Plant Physiol. (1993) 101, 1405–1406.

Gerber, G. B., and Léonard, A. Mutagenicity, carcinogenicity and teratogenicity of germanium compounds. Mutat. Res. (1997) 387, 3, 141–146.

Gerber, J., and Lill, R. Biogenesis of iron–sulfur proteins in eukaryotes: components, mechanism and pathology. Mitochondrion (2002) 2, 71–86.

Gerke, V., and Moss, S. E., Annexins and membrane dynamics. Biochim. Biophys. Acta (1997) 1357, 129–154.

Gerke, V., and Moss, S. E., Annexins: from structure to function. Physiol. Rev. (2002) 82, 331–371.

Gerke, V., Creutz, C. E., and Moss, S. E. Annexins: linking Ca^{2+} signaling to membrane dynamics. Nat. Rev. Mol. Cell. Biol. (2005) 6, 449–461.

Ghelis, T., Dellis, O., Jeannette, E., Bardat, F., Miginiac, E., and Sott, B. Abscisic acid plasmalemma perception triggers a calcium influx essential for RAB18 gene expression in *Arabidopsis thaliana* suspension cells. FEBS Lett. (2000) 483, 67–70.

Ghosh, M., Tucker, D. E., Burchett, S. A., and Leslie, C. C. Properties of the group IV phospholipase A_2 family. Prog. Lipid Res. (2006) 45, 6, 487–510.

Ghosh, S., Sharma, A., and Talukder, G. Zirconium. An abnormal trace element in biology. Biol. Trace Elem. Res. (1992) 35, 3, 247–271.

Gicquaud, C. Actin conformation is drastically altered by direct interaction with membrane lipids: a differential scanning calorimetry study. Biochemistry (1993) 32, 44, 11873–11877.

Gifford, J. L., Walsh, M. P., and Vogel, H. J. Structures and metal-ion-binding properties of the Ca^{2+}-binding helix–loop–helix EF-hand motives. Biochem. J. (2007) 405, 199–221.

Gilbert, D. L., and Fenn, W. O. Calcium equilibrium in muscle. J. Gen. Physiol. (1957) 40, 393–408.

Gillis, J. M., Thomason, D., Lefèvre, J., and Kretsinger, R. H. Parvalbumins and muscle relaxation: a computer simulation study. J. Muscle Res. Cell. Motil. (1982) 3, 4, 377–398.

Gillis, J. M., Thomason, D., LeFevre, J., and Kretsinger, R. H. Formation of calcium parvalbumin complex during contraction: a source of "unexplained heat"? In: Pollack, G. H., and Sugi, H., eds., Contractile Mechanisms in Muscle. Plenum Press, New York, 1984, pp. 573–579.

Gillis, T. E., Marshall, C. R., Xue, X.-H., Borgford, T. J., and Tibbits, G. F. Ca^{2+} binding to cardiac troponin C: effects of temperature and pH on mammalian and salmonid isoforms. Am. J. Physiol. Regul. Integr. Comp. Physiol. (2000) 279, R1707–R1715.

Gillis, T. E., Moyes, C. D., and Tibbits, G. F. Sequence mutations in teleost cardiac troponin C that are permissive of high Ca^{2+} affinity of site II. Am. J. Physiol. Cell Physiol. (2003) 284, C1176–C1184.

Gilmanshin, R., Creutz, C. E., and Tamm, L. K. Annexin IV reduces the rate of lateral lipid diffusion and changes the fluid phase structure of the lipid bilayer when it binds to negatively charged membranes in the presence of calcium. Biochemistry (1994) 33, 27, 8225–8232.

Girard, J. P., and Springer, T. A. Cloning from purified high endothelial venule cells of hevin, a close relative of the antiadhesive extracellular matrix protein SPARC. Immunity (1995) 2, 113–123.

Gitel, S. N., Owen, W. G., Esmon, C. T., and Jackson, C. M. A polypeptide region of bovine prothrombin specific for binding to phospholipids. Proc. Natl. Acad. Sci. USA (1973) 70, 5, 1344–1348.

Gley, M. E. Sur les fonctions de la gland thyroïde chez le lapin et chez le chien. C. R. Sceances Soc. Biol. (1891) 13, 843–847.

Glusker, J. P., Katz, A. K., and Bock, C. W. Metal ions in biological systems. Rigaku J. (1999) 16, 2, 8–16.

Goering, P. L. Lead–protein interactions as a basis for lead toxicity. Neurotoxicology (1993) 14, 2–3, 45–60.

Goldberg, J., Nairn, A. C., and Kuriyan, J. Structural basis for the autoinhibition of calcium/calmodulin-dependent protein kinase I. Cell (1996) 84, 6, 875–887.

Goldsmith, D., Kothawala, P., Chalian, A., Bernal, M., Robbins, S., and Covic, A. Systematic review of the evidence underlying the association between mineral metabolism disturbances and risk of fracture and need for parathyroidectomy in CKD. Am. J. Kidney Dis. (2009) 53, 6, 1002–1013.

Goldstein, D., and Stark-Mayer, C. New specific test for calcium. Anal. Chim. Acta (1958) 19, 437–439.

Gomes, A. V., Barnes, J. A., and Vogel, H. J. Spectroscopic characterization of the interaction between calmodulin-dependent protein kinase I and calmodulin. Arch. Biochem. Biophys. (2000) 379, 1, 28–36.

Gomme, P. T., McCann, K. B., and Bertolini, J. Transferrin: structure, function and potential therapeutic actions. Drug Discov. Today (2005) 10, 4, 267–273.

Gonzales, F., and Karnovsky, M. Electron microscopy of osteoclasts in healing fractures of rat bone. J. Biophys. Biochem. Cytol. (1961) 9, 299–316.

Goodman, J., Neil, J. J., and Ackerman, J. J. Biomedical applications of ^{133}Cs NMR. NMR Biomed. (2005) 18, 2, 125–134.

Goossens, E. L. J., Reutelingsperger, C. P. M., Jongsma, F. H. M., Kraayenhof, R., and Hermens, W. T. Annexin V perturbs or stabilises phospholipid membranes in a calcium-dependent manner. FEBS Lett. (1995) 359, 1, 155–158.

Gordon, T., and Bowser, D. Beryllium: genotoxicity and carcinogenicity. Mutat. Res. (2003) 533, 1–2, 99–105.

Gorski, J. P. Acidic phosphoproteins from bone matrix: a structural rationalization of their role in biomineralization. Calcif. Tissue Int. (1992) 50, 5, 391–396.

Gosselin-Rey, C., Bernard, N., and Gerday, C. Conformation and immunochemistry of parvalbumin III from pike white muscle: modification of the arginine with 1,2-cyclohexandione. Biochim. Biophys. Acta (1973) 303, 90–104.

Gouaux, E., and Mackinnon, R. Principles of selective ion transport in channels and pumps. Science (2005) 310, 5753, 1461–1465.

Goulding, N. J., Jefferiss, C. M., Pan, L., Rigby, W. F., and Guyre, P. M. Specific binding of lipocortin-1 (annexin I) to monocytes and neutrophils is decreased in rheumatoid arthritis. Arthritis Rheum. (1992) 35, 1395–1397.

Grabarek, Z. Structural basis for diversity of the EF-hand calcium binding proteins. J. Mol. Biol. (2006) 359, 509–525.

Graether, S. P., Heinonen, T. Y., Raharjo, W. H., Jin, J. P., and Mak, A. S. Tryptophan residues in caldesmon are major determinants for calmodulin binding. Biochemistry (1997) 36, 2, 364–369.

Gramolini, A. O., Trivieri, M. G., Oudit, G. Y., Kislinger, T., Li, W., Patel, M. M., Emili, A., Kranias, E. G., Backx, P. H., and Maclennan, D. H. Cardiac-specific overexpression of sarcolipin in phospholamban null mice impairs myocyte function that is restored by phosphorylation, Proc. Natl. Acad. Sci. USA (2006) 103, 2446–2451.

Grass, G., Fricke, B., and Nies, D. H. Control of expression of a periplasmic nickel efflux pump by periplasmic nickel concentrations. Biometals (2005) 18, 4, 437–448.

Gray, H. B., and Winkler, J. R. Electron tunneling through proteins. Q. Rev. Biophys. (2003) 36, 3, 341–372.

Green, J. R. On certain points connected with the coagulation of the blood. J. Physiol. (1887) 8, 354–371.

Greenberg, D. M., and Gunther, L. On the determination of diffusable and non-diffusable serum calcium. J. Biol. Chem. (1930) 85, 491–508.

Greene, L. H., Grobler, J. A., Malinovskii, V. A., Tian, J., Acharya, K. R., and Brew, K. Stability, activity and flexibility in α-lactalbumin. Protein Eng. (1999) 12, 581–587.

Greenfield, N. J. Analysis of circular dichroism data. Methods Enzymol. (2004) 383, 282–317.

Greenwald, I. The effect of the administration of calcium salts and sodium phosphate upon the calcification and phosphorus metabolism of thyroparathyroidectomized dogs, with a consideration of the nature of the calcium compounds of blood and their relation to the pathogenesis of tetany. J. Biol. Chem. (1926) 67, 1–28.

Grewal, T., and Enrich, C. Annexins: modulators of EGF receptor signalling and trafficking. Cell Signal. (2009) 21, 347–358.

Gribenko, A. V., Guzman-Casado, M., Lopez, M. M., and Makhatadze, G. I. Conformational and thermodynamic properties of peptide binding to the human S100P protein. Protein Sci. (2002) 11, 6, 1367–1375.

Griffith, J. P., Kim, J. L., Kim, E. E., Sintchak, M. D., Thomson, J. A., Fitzgibbon, M. J., Fleming, M. A., Caron, P. R., Hsiao, K., and Navia, M. A. X-ray structure of calcineurin inhibited by the immunophilin-immunosuppressant FKBP12–FK506 complex. Cell (1995) 82, 3, 507–522.

Griko, Y. V., and Remeta, D. P. Energetics of solvent and ligand-induced conformational changes in α-lactalbumin. Protein Sci. (1999) 8, 3, 554–561.

Grishchenko, V. M., Kalinichenko, L. P., Deikus, G. Y., Veprintsev, D. B., Cawthern, K., Berliner, L. J., and Permyakov E. A. Interactions of α-lactalbumin with lipid vesicles studied by tryptophan fluorescence. Biochem. Mol. Biol. Int. (1996) 38, 3, 453–466.

Grobler, J. A., and Hurley, J. H. Catalysis by phospholipase C δ(1) requires that Ca^{2+} bind to the catalytic domain, but not the C2 domain. Biochemistry (1998) 37, 5020–5028.

Groenendyk, J., and Michalak, M. Endoplasmic reticulum quality control and apoptosis. Acta Biochim. Pol. (2005) 52, 2, 381–395.

Gross, M. D., Gosnell, M., Tsarbopoulos, A., and Hunziker, W. A functional and degenerate pair of EF hands contains the very high affinity calcium-binding site of calbindin–D28K. J. Biol. Chem. (1993) 268, 28, 20917–20922.

Groves, P., Finn, B. E., Kuznicki, J., and Forsen, S. A model for target protein binding to calcium-activated S100 dimers. FEBS Lett. (1998) 421, 3, 175–179.

Guerini, D. Calcineurin: not just a simple protein phosphatase. Biochim. Biophys. Res. Commun. (1997) 235, 271–275.

Guermah, M., Crisanti, P., Laugier, D., Dezelee, P., Bidou, L., Pessac, B., and Calothy, G. Transcription of a quail gene expressed in embryonic retinal cells is shut off sharply at hatching. Proc. Natl. Acad. Sci. USA (1991) 88, 4503–4507.

Guhaniyogi, J., Robinson, V. L., and Stock, A. M. Crystal structures of beryllium fluoride-free and beryllium fluoride-bound CheY in complex with the conserved C-terminal peptide of CheZ reveal dual binding modes specific to CheY conformation. J. Mol. Biol. (2006) 359, 3, 624–645.

Guha Thakurta, P., Choudhury, D., Dasgupta, R., and Dattagupta, J. K. Structure of diferric hen serum transferrin at 2.8 Å resolution. Acta Crystallogr. D (2003) 59, 1773–1781.

Guinto, E. R., Caccia, S., Rose, T., Fütterer, K., Waksman, G., and Di Cera, E. Unexpected crucial role of residue 225 in serine proteases. Proc. Natl. Acad. Sci. USA. (1999) 96, 5, 1852–1857.

Gulati, J., Babu, A., Su, H., and Zhang, Y. F. Identification of the regions conferring calmodulin-like properties to troponin C. J. Biol. Chem. (1993) 268, 16, 11685–11690.

Gulati, J., Akella, A. B., Su, H., Mehler, E. L., and Weinstein, H. Functional role of arginine-11 in the N-terminal helix of skeletal troponin C: combined mutagenesis and molecular dynamics investigation. Biochemistry (1995) 34, 22, 7348–7355.

Gulec, S. A., and Fong, Y. Yttrium 90 microsphere selective internal radiation treatment of hepatic colorectal metastases. Arch. Surg. (2007) 142, 7, 675–682.

Gupta, A., Matsui, K., Lo, J. F., and Silver, S. Molecular basis for resistance to silver cations in *Salmonella*. Nat. Med. (1999) 5, 183–188.

Gustafsson, L., Hallgren, O., Mossberg, A. K., Pettersson, J., Fischer, W., Aronsson, A., and Svanborg, C. HAMLET kills tumor cells by apoptosis: structure, cellular mechanisms, and therapy. J. Nutr. (2005) 135, 5, 1299–1303.

Hackney, D. D. The kinetic cycles of myosin, kinesin, and dynein. Annu. Rev. Physiol. (1996) 58, 731–750.

Hagins, W. A. The visual process: excitatory mechanisms in the primary receptor cells. Annu. Rev. Biophys. Bioeng. (1972) 1, 131–158.

Hagiwara, S., and Nakajimas, S. Effects of the intracellular calcium ion concentration upon the excitability of the muscle fiber membrane of a barnacle. J. Gen. Physiol. (1966) 49, 807–818.

Haiduc, I., and Silvestru, C. Rhodium, iridium, copper and gold antitumor organometallic compounds (review). In Vivo (1989) 3, 4, 285–293.

Haiech, J., Derancourt, J., Pechére, J. F., and Demaille, J. G. Magnesium and calcium binding to parvalbumins: evidence for differences between parvalbumins and an explanation of their relaxing function. Biochemistry (1979) 18, 13, 2752–2758.

Haiech, J., Claude, B., Klee, C. B., and Demaille, J. G. Effects of cation on affinity of calmodulin for calcium: ordered binding of calcium allows the specific activation of calmodulin stimulated enzymes. Biochemistry (1981) 20, 13, 3890–3897.

Hainfeld, J. F. Uranium-loaded apoferritin with antibodies attached: molecular design for uranium neutron-capture therapy. Proc. Natl. Acad. Sci. USA (1992) 89, 22, 11064–11068.

Hakansson, A., Zhivotovsky, B., Orrenius, S., Sabharwal, H., and Svanborg, C. Apoptosis induced by a human milk protein. Proc. Natl. Acad. Sci. USA (1995) 92, 17, 8064–8068.

Hakansson, A., Andreasson, J., Zhivotovsky, B., Karpman, D., Orrenius, S., and Svanborg, C. Multimeric α-lactalbumin from human milk induces apoptosis through a direct effect on cell nuclei. Exp. Cell Res. (1999) 246, 2, 451–460.

Hakansson, A., Svensson, M., Mossberg, A. K., Sabharwal, H., Linse, S., Lazou, I., Lonnerdal, B., and Svanborg, C. A folding variant of α-lactalbumin with bactericidal activity against *Streptococcus pneumoniae*. Mol. Microbiol. (2000) 35, 3, 589–600.

Hales, S. Vegetable Statics. 1769 (first edition, 1727).

Hall, D. R., Bond, C. S., Leonard, G. A., Watt, C. I., Berry, A., and Hunter, W. N. Structure of tagatose-1,6-bisphosphate aldolase. Insight into chiral discrimination, mechanism, and specificity of class II aldolases. J. Biol. Chem. (2002a) 277, 24, 22018–22024.

Hall, D. R., Hadden, J. M., Leonard, G. A., Bailey, S., Neu, M., Winn, M., and Lindley, P. F. The crystal and molecular structures of diferric porcine and rabbit serum transferrins at resolutions of 2.15 and 2.60 Å, respectively. Acta Crystallogr. D Biol. Crystallogr. (2002b) 58, Pt. 1, 70–80.

Hall, D. R., Leonard, G. A., Reed, C. D., Watt, C. I., Berry, A., and Hunter, W. N. The crystal structure of *Escherichia coli* class II fructose-1,6-bisphosphate aldolase in complex with phosphoglycolohydroxamate reveals details of mechanism and specificity. J. Mol. Biol. (1999) 287, 2, 383–394.

Hall, J. L. Cellular mechanisms for heavy metal detoxification and tolerance. J. Exp. Bot. (2002c) 53, 1–11.

Hamada, T., Asanuma, M., Ueki, T., et al. Solution structure of vanabin2, a vanadium(IV)-binding protein from the vanadium-rich ascidian *Ascidia sydneiensis samea*. J. Am. Chem. Soc. (2005) 127, 4216–4222.

Hamburger, H. J. The influence of small amounts of calcium on the motion of phagocytes. Proc. Konink. Akad. (1910) 13, 66–79.

Hamer, D. H. Metallothionein. Annu. Rev. Biochem. (1986) 55, 913–951.

Hammarsten, O. Om det Chemiska Förlåppet vid Caseinets Coagulation med Löpe. Upsala Läkarefor. Förhand. (1873/74) 9, 363–399 and 451–486.

Han, S., Tang, R., Anderson, L. K., Woerner, T. E., and Pei, Z. M. A cell surface receptor mediates extracellular Ca^{2+} sensing in guard cells. Nature (2003) 425, 196–200.

Handford, P., Mayhew, M., Baron, M., Winship, P. R., Campbell, I. D., and Brownlee, G. G. Key residues involved in Ca^{2+}-binding motifs in EGF-like domains. Nature (1991) 351, 164–167.

Hanssens, I., Herreman, W., Van Ceunebroeck, J. C., Dangreau, H., Gielens, C., Preaux, G., and Van Cauwelaert, F. Interaction of α-lactalbumin with dimyristoylphosphatidylcholine vesicles. III. Influence of the lipid-to-protein molar ratio on the complex formation. Biochim. Biophys. Acta (1983) 728, 3, 293–304.

Hanssens, I., Van Ceunebroeck, J. C., Pottel, H., Preaux, G., and Van Cauwelaert, F. Influence of the protein conformation on the interaction between α-lactalbumin and dimiristoylphosphatidylcholine vesicles. Biochim. Biophys. Acta (1985) 817, 1, 154–164.

Hapak, R. C., Lammers, P. J., Palmisano, W. A., Birnbaum, E. R., and Henzl, M. T. Site-specific substitution of glutamate for aspartate at position 59 of rat oncomodulin. J. Biol. Chem. (1989) 264, 18751–18760.

Hardin, P. E., Angerer, L. M., Hardin, S. H., Angerer, R. C., and Klein, W. H. *SPEC2* genes of *Strongylocentrotus purpuratus* structure and differential expression in embryonic aboral ectoderm cells. J. Mol. Biol. (1988) 202, 417–431.

Harmon, A. C. Calcium-regulated protein kinases of plants. Gravit. Space Biol. Bull. (2003) 16, 2, 83–90.

Harris, D. C. Quantitative Chemical Analysis. W.H. Freeman, New York, 1999.

Harris, W. R. Thermodynamics of gallium complexation by human lactoferrin. Biochemistry (1986) 25, 803–808.

Harris, W. R. Binding and transport of nanoferrous metals by serum transferrin. Struct. Bonding (1998) 92, 121–162.

Harris, W. R., and Pecoraro, V. L. Thermodynamic binding constants for gallium transferrin. Biochemistry (1983) 22, 292–299.

Harris, W. R., Carrano, C. J., Pecoraro, V. L., and Raymond, K. N. Siderophilin metal coordination: 1. Complexation of Th by transferrin: structure–function implications. J. Am. Chem. Soc. (1981) 103, 2231–2237.

Harris, W. R., Berthon, G., Day, J. P., Exley, C., Flaten, T. P., Forbes, W. F., Kiss, T., Orvig, C., and Zatta, P. F. Speciation of aluminum in biological systems. J. Toxicol. Environ. Health (1996) 48, 6, 543–568.

Harrison, A., and King, S. M. The molecular anatomy of dynein. Essays Biochem. (2000) 35, 75–87.

Harrison, J., Leggett, R., Lloyd, D., Phipps, A., and Scott, B. Polonium-210 as a poison. J. Radiol. Prot. (2007) 27, 1, 17–40.

Harrison, M. D., Jones, C. E., and Dameron, C. T. Copper chaperones: function, structure and copper-binding properties. J. Biol. Inorg. Chem. (1999) 4, 145–153.

Hartzell, H. C., and White, R. E. Effects of magnesium on activation of the voltage-gated calcium current in cardiac myocytes. J. Gen. Physiol. (1989) 94, 745–767.

Harvey, A. M., and MacIntosh, F. C. Calcium and synaptic transmission in a sympathetic anglion. J. Physiol. (1940) 408–416.

Hashimoto, Y., and Soderling, T. R. Regulation of calcineurin by phosphorylation: identification of the regulatory site phosphorylated by Ca^{2+}/calmodulin-dependent protein kinase II and protein kinase C. J. Biol. Chem. (1989) 264, 16524–16529.

Hasselbach, W., and Makinose, M. Über den Mechanismus des Calcium Transportes durch die Membranen des sarkoplasmatischen Reticulums. Biochem. Z. (1963) 339, 94–111.

Hasselbach, W., and Weber, H. H. Der Einfluss des MB-Faktors auf die Kontraktion des Fasermodells. Biochim. Biophys. Acta (1953) 11, 160–161.

Hauschka, P. V., and Gallop, P. M. Purification and calcium-binding properties of osteocalcin, the γ-carboxyglutamate-containing protein of bone. In: Wasserman, R. H. et al., eds., Calcium Binding Proteins and Calcium Function. Elsevier, Amsterdam, 1977, 338–347.

Hauschka, P. V., Lian, J. B., and Gallop, P. M. Direct identification of the calcium-binding amino acid, γ-carboxyglutamate, in mineralized tissue. Proc. Natl. Acad. Sci. USA (1975) 72, 3925–3929.

Hay, J. C. Calcium: a fundamental regulator of intracellular membrane fusion. EMBO Rep. (2007) 8, 3, 236–240.

Hayes, M. J., and Moss, S. E. Annexins and disease. Biochim. Biophys. Res. Commun. (2004) 322, 1166–1170.

Head, J. F. Squidulin. In: Celio, M. R., Pauls, T., and Schwaller, B., eds., Guidebook to the Calcium-Binding Proteins. Oxford University Press, Oxford, UK, 1996, pp. 84–85.

Head, J. F., Inouye, S., Teranishi, K., and Shimomura, O. The crystal structure of the photoprotein aequorin at 2.3 Å resolution. Nature (2000) 405, 6784, 372–376.

Heck, A. J., and Van Den Heuvel, R. H. Investigation of intact protein complexes by mass spectrometry. Mass Spectrom. Rev. (2004) 23, 5, 368–389.

Heddle, J., Scott, D. J., Unzai, S., Park, S. Y., and Tame, J. R. Crystal structures of the liganded and unliganded nickel-binding protein NikA from *Escherichia coli*. J. Biol. Chem. (2003) 278, 50, 50322–50329.

Heffron, J. K., and Moerland, T. S. Parvalbumin characterization from the euryhaline stingray *Dasyatis sabina*. Comp. Biochem. Physiol. A (2008) 150, 3, 339–346.

Heidelberger, R., Heinemann, C., Neher, E., and Matthews, G. Calcium dependence of the rate of exocytosis in a synaptic terminal. Nature (Lond.) (1994) 371, 513–515.

Heidorn, D. B., Seeger, P. A., Rokop, S. E., Blumenthal, D. K., Means, A. R., Crespi, H., and Trewhella, J. Changes in the structure of calmodulin induced by a peptide based on the calmodulin-binding domain of myosin light chain kinase. Biochemistry (1989) 29, 16, 6757–6764.

Heilbrunn, L. V. Studies in artificial parthenogenesis: II. Physical changes in the egg of arbacia. Biol. Bull. (1915) 29, 149–203.

Heilbrunn, L. V. An experimental study of cell division I. The physical conditions which determine the appearance of the spindle in sea urchin eggs. J. Exp. Zool. (1920) 30, 211–237.

Heilbrunn, L. V. The colloid chemistry of protoplasm: I. general considerations; II. The electrical charges of protoplasm. Am. J. Physiol. (1923) 64, 481–498.

Heilbrunn, L. V. The viscosity of protoplasm. Q. Rev. Biol. (1927) 2, 230–248.

Heilbrunn, L. V. The Colloid Chemistry of Protoplasm. Verlag Gebrüder Borntraeger, Berlin, 1928.

Heilbrunn, L. V. The action of calcium on muscle protoplasm. Physiol. Zool. (1940) 13, 88–94.

Heilbrunn, L. V. An Outline of General Physiology, 3rd ed. W.B. Saunders, Philadelphia, 1952.

Heilbrunn, L. V., and Daugherty, K. The action of ultraviolet rays of amoeba protoplasm. Protoplasma (1933) 18, 596–619.

Heilbrunn, L. V., and Young, R. A. The action of ultraviolet rays on arbacia egg protoplasm. Physiol. Zool. (1930) 3, 330–341.

Heilmann, C., and Spamer, C. In: Celio, M. R., Pauls, T., Schwaller, B., eds., Guidebook to the Calcium-Binding Proteins. Oxford University Press, Oxford, UK, 1996.

Heinemann, U., Illing, G., and Oschkinat, H. High-throughput three-dimensional protein structure determination. Curr. Opin. Biotechnol. (2001) 12, 4, 348–354.

Heizmann, C. W. Calcium signaling in the brain. Acta Neurobiol. Exp. (Wars.) (1993) 53, 1, 15–23.

Heizmann, C. W., and Cox, J. A. New perspectives on S100 proteins: a multi-functional Ca^{2+}-, Zn^{2+}- and Cu^{2+}-binding protein family. Biometals (1998) 11, 4, 383–397.

Heizmann, C. W., and Hunziker, W. Intracellular calcium-binding proteins: more sites than insights. Trends Biochem. Sci. (1991) 16, 98–103.

Heizmann, C. W., Ackermann, G. E., and Galichet, A. Pathologies involving the S100 proteins and RAGE. Subcell. Biochem. (2007) 45, 93–138.

Helenius, A., Trombetta, E. S., Hebert, D. N., and Simons, J. F. Calnexin, calreticulin and the folding of glycoproteins. Cell Biol. (1997) 7, 193–200.

Hemperley, J. J., Hopp, T. P., Becker, J. W., and Cunningham, B. A. The chemical characterization of flavin, a lectin isolated from *Vicia laba*. J. Biol. Chem. (1979) 254, 14, 6803–6810.

Henkel, G., and Krebs, B. Metallothioneins: zinc, cadmium, mercury, and copper thiolates and selenolates mimicking protein active site features: structural aspects and biological implications. Chem. Rev. (2004) 104, 2, 801–824.

Henquin, J. C. Regulation of insulin secretion: a matter of phase control and amplitude modulation. Diabetologia (2009) 52, 5, 739–751.

Henzl, M. T., and Tanner, J. J. Solution structure of Ca^{2+}-free rat β-parvalbumin (oncomodulin). Protein Sci. (2007) 16, 1914–1926.

Henzl, M. T., and Tanner, J. J. Solution structure of Ca^{2+}-free rat α-parvalbumin. Protein Sci. (2008) 17, 3, 431–438.

Henzl, M. T., Hapak, R. C., and Goodpasture, E. A. Introduction of a fifth carboxylate ligand heightens the affinity of the oncomodulin CD and EF sites for Ca^{2+}. Biochemistry (1996) 35, 5856–5869.

Henzl, M. T., Hapak, R. C., and Likos, J. J. Interconversion of the ligand arrays in the CD and EF sites of oncomodulin: Influence on Ca^{2+}-binding affinity. Biochemistry (1998) 37, 9101–9111.

Henzl, M. T., Agah, S., and Larson, J. D. Association of the AB and CD-EF domains from rat α- and β-parvalbumins. Biochemistry (2004) 43, 34, 10906–10917.

Henzl, M. T., Larson, J. D., and Agah, S. Estimation of parvalbumin Ca^{2+}- and Mg^{2+}-bindng constants by global least-squares analysis of isothermal titration calorimetry data. Anal. Biochem. (2003) 319, 216–233.

Henzl, M. T., Larson, J. D., and Agah, S. Influence of monovalent cations on rat α- and β-parvalbumin stabilities. Biochemistry (2000) 39, 19, 5859–5867.

Henzl, M. T., Davis, M. E., and Tan, A. Divalent ion binding properties of the timothy grass allergen, Phl p 7. Biochemistry (2008) 47, 30, 7846–7856.

Henzl, M. T., Wycoff, W. G., Larson, J. D., and Likos, J. J. ^{15}N nuclear magnetic resonance relaxation studies on rat β-parvalbumin and the pantacarboxylate variants, S55D and G98D. Protein Sci. (2002) 11, 158–173.

Herbaud, M. L., Guiseppi, A., Denizot, F., Haiech, J., and Kilhoffer, M. C. Calcium signalling in *Bacillus subtilis*. Biochim. Biophys. Acta (1998) 1448, 2, 212–226.

Hernandez, O. M., Jones, M., Guzman, G., and Szczesna-Cordary, D. Myosin essential light chain in health and disease. Am. J. Physiol. Heart Circ. Physiol. (2007) 292, H1643–H1654.

Herreman, W., Van Tornout, P., Van Cauwelaert, F., and Hanssens, I. Interaction of α-lactalbumin with dimiristoylphosphatidyl-choline vesicles: II. A fluorescence polarization study. Biochim. Biophys. Acta (1981) 640, 2, 419–429.

Herrero-Lancina, M., and West, T. S. Sensitive and selective spectrophotometric reaction of determination of trace amounts of calcium. Anal. Chem. (1963) 35, 2131–2135.

Hertadi, R., and Ikai, A. Unfolding mechanics of holo- and apocalmodulin studied by the atomic force microscope. Protein Sci. (2002) 11, 6, 1532–1538.

Herzberg, O., and James, M. N. G. Structure of the calcium regulatory muscle protein troponin C at 2.8 Å resolution. Nature (1985) 313, 6004, 653–659.

Herzberg, O., Moult, J., and James, M. H. G. A model for the Ca^{2+}-induced conformational transition of troponin C. J. Biol. Chem. (1986) 261, 2638–2644.

Hess, A. F., and Weinstock, M. Antirachitic properties imparted to inert fluids and to green vegetables by ultraviolet irradiation. J. Biol. Chem. (1924) 62, 301–313.

Heuft-Dorenbosch, L. L. J., de Vet, H. C. W, and van der Linden, S. Yttrium radiosynoviorthesis in the treatment of knee arthritis in rheumatoid arthritis: a systematic review. Ann. Rheum. Dis. (2000) 59, 583–586.

Higgins, M. K., Oprian, D. D., and Schertler, G. F. X. Recoverin binds exclusively to an amphipathic peptide at the N terminus of rhodopsin kinase, inhibiting rhodopsin phosphorylation without affecting catalytic activity of the kinase. J. Biol. Chem. (2006) 281, 19426–19432.

Hill, A. V. The abrupt transition from rest to activity in muscle. Proc. R. Soc. B (1949) 136, 399–420.

Hille, R. Molybdenum and tungsten in biology. Trends Biochem. Sci. (2002) 27, 7, 360–367.

Himeno, S., Yanagiya, T., Enomoto, S., Kondo, Y., and Imura, N. Cellular cadmium uptake mediated by the transport system for manganese. Tohoku J. Exp. Med. (2002) 196, 1, 43–50.

Hirabayashi, T., Murayama, T., and Shimizu, T. Regulatory mechanism and physiological role of cytosolic phospholipase A_2. Biol. Pharm. Bull. (2004) 27, 8, 1168–1173.

Hirano, S., and Suzuki, K. T., Exposure metabolism and toxicity for rare earths and related compounds. Environ. Health. Perspect. (1996) 104, 85–95.

Hirokawa, Y., Oda, N., Cox, N. J., et al. Genetic variation in the gene encoding calpain-10 is associated with type 2 diabetes mellitus. Nat. Genet. (2000) 26, 163–175.

Hirose, M. The structural mechanism for iron uptake and release by transferrins. Biosci. Biotechnol. Biochem. (2000) 64, 7, 1328–1336.

Hitchcock, S. E. Study of the structure of troponin C by measuring the relative reactivities of lysines with acetic anhydride. J. Mol. Biol. (1981) 147, 1, 153–173.

Hoang, Q. Q., Sicheri, F., Howard, A. J., and Yang, D. C. Bone recognition mechanism of porcine osteocalcin from crystal structure. Nature (2003) 425, 977–980.

Hohenester, E., Mauer, P., Hohenadl, D., Timpl, R., Jansonius, J. N., and Engel, J. Structure of a novel extracellular Ca^{2+}-binding module in BM-40. Nat. Struct. Biol. (1996) 3, 67–73.

Hodgkin, A. L., and Katz, B. The effect of calcium on the axoplasm of giant nerve fibers. J. Exp. Biol. (1949) 26, 292–294.

Hodgkin, A. L., and Keynes, R. D. Movements of labelled calcium in squid giant axons. J. Physiol. (1957) 138, 253–281.

Hoeflich, K. P., and Ikura, M. Calmodulin in action: diversity in target recognition and activation mechanisms. Cell (2002) 108, 6, 739–742.

Hoekstra, D., Buist-Arkema, R., Klappe, K., and Reutelingsperger, C. P. M. Interaction of annexins with membranes: the N-terminus as a governing parameter as revealed with a chimeric annexin. Biochemistry (1993) 32, 51, 14194–14202.

Hof, P. R., Glezer, I. I., Flagg, R. A., Rubin, M. B., Nimchinsky, E. A., and Vogt Weisenhorn, D. M. Cellular distribution of the calcium-binding proteins parvalbumin, calbindin, and calretinin in the neurocortex of mammals: phylogenic and developmental patterns. J. Chem. Neuroanat. (1999) 16, 2, 77–116.

Hofer, A. M. Another dimension to calcium signaling: a look at extracellular calcium. J. Cell Sci. (2005) 118, 855–862.

Hofer, A. M., and Brown, E. M. Extracellular calcium sensing and signalling. Nat. Rev. Mol. Cell. Biol. (2003) 4, 530–538.

Hofmannn, T., Eng, S., Lilja, H., Drakenberg, T., Vogel, H. J., and Forsen, S. Site–site interactions in EF-hand calcium-binding proteins: laser-exited europium luminescence studies of 9kDa calbindin, the pig intestinal calcium-binding protein. Eur. J. Biochem. (1988) 172, 2, 307–313.

Hogeboom, G. H., Schneider, W. C., and Pallade, G. E. Cytochemical studies of mammalian tissues: I. Isolation of intact mitochondria from rat liver; some biochemical properties of mitochondria and submicroscopic particulate matter. J. Biol. Chem. (1948) 172, 619–639.

Hogstrand, C., Kille, P., Nicholson, R. I., and Taylor, K. M. Zinc transporters and cancer: a potential role for ZIP7 as a hub for tyrosine kinase activation. Trends Mol. Med. (2009) 15, 3, 101–111.

Hoh, F., Cavé, A., Strub, M. P., Banères, J. L., and Padilla, A. Removing the invariant salt bridge of parvalbumin increases flexibility in the AB-loop structure. Acta Crystallogr. D (2009) 65, Pt. 8, 733–743.

Hohenadl, C., Mann, K., Mayer, U., Timpl, R., Paulsson, M., and Aeschlimann, D. Two adjacent N-terminal glutamines of BM-40 (osteonectin, SPARC) act as amine acceptor sites in transglutaminasec-catalyzed modification. J. Biol. Chem. (1995) 270, 23415–23420.

Hohenester, E., Maurer, P., Hohenadl, C., Timpl, R., Jansonius, J. N., and Engel, J. Structure of a novel extracellular Ca^{2+}-binding module in BM-40. Nat. Struct. Biol. (1996) 3, 67–73.

Holmes, M. A., and Matthews, B. W. Structure of thermolysin refined at 1.6 Å resolution. J. Mol. Biol. (1982) 160, 4, 623–639.

Holroyde, M. J., Potter, J. D., and Solaro, R. J., The calcium binding properties of phosphorylated and unphosphorylated cardiac and skeletal myosins, J. Biol. Chem. (1979) 254, 6478–6482.

Homouz, D., Sanabria, H., Waxham, M. N., and Cheung, M. S. Modulation of calmodulin plasticity by the effect of macromolecular crowding. J. Mol. Biol. (2009) 391, 933–943.

Honoré, B. The rapidly expanding CREC protein family: members, localization, function, and role in disease. Bioessays (2009) 31, 3, 262–277.

Hoofnagle, A. N., Resing, K. A., and Ahn, N. G. Protein analysis by hydrogen exchange mass spectrometry. Annu. Rev. Biophys. Biomol. Struct. (2003) 32, 1–25.

Horii, K., Saito, M., Yoda, T., Tsumoto, K., Matsushima, M., Kuwajima, K., and Kumagai, I. Contribution of Thr29 to the thermodynamic stability of goat α-lactalbumin as determined by experimental and theoretical approaches. Proteins (2001) 45, 16–29.

Houben, R., Soute, B. A., Knapen, M. H., and Vermeer, C. Strategies for developing human osteocalcin standards: a critical evaluation. Scand. J. Clin. Lab. Invest. Suppl. (1997) 227, 100–104.

Houdusse, A., and Cohen, C. Target sequence recognition by the calmodulin superfamily: implications from light chain binding to the regulatory domain of scallop myosin. Proc. Natl. Acad. Sci. USA (1995) 92, 10644–10647.

Houdusse, A., Love, M. L., Dominguez, R., Grabarek, Z., and Cohen, C. Structures of four Ca^{2+}-bound troponin C at 2.0 Å resolution: further insights into the Ca^{2+}-switch in the calmodulin superfamily. Structure (1997) 5, 12, 1695–1711.

Howard, N., Abell, C., Blakemore, W., Chessari, G., Congreve, M., Howard, S., Jhoti, H., Murray, C. W., Seavers, L. C. A., and Van Montfort, R. L. M. Application of fragment screening and fragment linking to the discovery of novel thrombin inhibitors. J. Med. Chem. (2006) 49, 1346–1356.

Hu, H., and Chazin, W. J. Unique features in the C-terminal domain provide caltractin with target specificity. J. Mol. Biol. (2003) 330, 3, 473–484.

Hu, H., Sheehan, J. H., and Chazin, W. J. The mode of action of centrin. Binding of Ca^{2+} and a peptide fragment of Kar1p to the C-terminal domain. J. Biol. Chem. (2004) 279, 49, 50895–50903.

Hu, Y., Rech, S., Gunsalus, R. P., and Rees, D. C. Crystal structure of the molybdate binding protein ModA. Nat. Struct. Biol. (1997) 4, 9, 703–707.

Hu, Y., Faham, S., Roy, R., Adams, M. W., and Rees, D. C. Formaldehyde ferredoxin oxidoreductase from *Pyrococcus furiosus*: the 1.85 Å resolution crystal structure and its mechanistic implications. J. Mol. Biol. (1999) 286, 899–914.

Huang, L. H., Cheng, H., Pardi, A., Tam, J. P., and Sweeney, W. V. Sequence-specific 1H NMR assignments, secondary structure, and location of the calcium binding site in the first epidermal growth factor like domain of blood coagulation factor IX. Biochemistry (1991) 30, 30, 7402–7409.

Hubaishy, I., Jones, P. G., Bjorge, J., Bellagamba, C., Fitzpatrick, S., Fujita, D. J., and Waisman, D. M. Modulation of annexin II tetramer by tyrosine phosphorylation. Biochemistry (1995) 34, 44, 14527–14534.

Huber, M., Siegenthaler, G., Mirancea, N., Marenholz, I., Nizetic, D., Breitkreutz, D. Mischke, D., and Hohl, D. Isolation and characterization of human repetin, a member of the fused gene family of the epidermal differentiation complex. J. Investig. Dermatol. (2005) 124, 998–1007.

Huber, R., Berendes, R., Burger, A., Schneider, M., Karshikov, A., Leucke, A., Romisch, J., and Paques, E. Crystal and molecular structure of human annexin V after refinement. J. Mol. Biol. (1992) 223, 683–704.

Huffman, D. L., and O'Halloran, T. V. Function, structure, and mechanism of intracellular copper trafficking proteins. Annu. Rev. Biochem. (2001) 70, 677–701.

Huggins, C. The composition of bone and the function of the bone cell. Physiol. Rev. (1937) 17, 119–143.

Hughes, R. E., Brzovic, P. S., Klevit, R. E., and Hurley, J. B. Calcium-dependent solvation of the myristoyl group of recoverin. Biochemistry (1995) 34, 36, 11410–11416.

Hulpiau, P., and van Roy, F. Molecular evolution of the cadherin superfamily. Int. J. Biochem. Cell Biol. (2009) 41, 2, 349–369.

Hume, E. M., and Smith, H. H. XLX. The effect of air, which has been exposed to the radiations of the mercury–vapour quartz lamp, in promoting the growth of rats, fed on a diet deficient in fat-soluble vitamins. Biochem. J. (1923) 17, 364–372.

Hunter, G. K., Kyle, C. L., and Goldberg, H. A. Modulation of crystal formation by bone phosphoproteins: structural specificity of the osteopontin-mediated inhibition of hydroxyapatite formation. Biochem. J. (1994) 300, 723–728.

Hunter, J. The Works of John Hunter, F.R.S. with Notes, Vol. 4. Palmer, J. F. (Ed.) Longman, London, 1837, pp. 315–318.

Huxley, A. F., and Niedergerke, R. Structural changes in muscle during contraction: interference microscopy of living muscle fibres. Nature (1954) 173, 971–973.

Huxley, H. E., and Hanson, J. Changes in the cross-striations of muscle during contraction and stretch and their structural interpretation. Nature (1954) 173, 973–976.

Hwang, W., and Lang, M. J. Mechanical design of translocating motor proteins. Cell Biochem. Biophys. (2009) 54, 1–3, 11–22.

Ignoul, S., and Eggermont, J. CBS domains: structure, function, and pathology in human proteins. Am. J. Physiol. Cell Physiol. (2005) 289, 6, C1369–C1378.

Iio, T. Melittin-binding of troponin C. J. Biochem. (1993) 114, 773–778.

Ikeguchi, M., Sugai, S., Fujino, M., Sugawara, T., and Kuwajima, K. Contribution of the 6–120 disulfide bond of α-lactalbumin to the stabilities of its native and molten globule states. Biochemistry (1992) 31, 50, 12695–12700.

Ikura, M. Calcium binding and conformational response in EF-hand proteins. Trends Biochem. Sci. (1996) 21, 1, 14–17.

Ikura, M., Clore, G. M., Gronenborn, A. M., Zhu, G., Klee, C. B., and Bax, A. Solution structure of a calmodulin-target peptide complex by multidimensional NMR. Science (1992) 256, 5057, 632–638.

Ikura, M., Osawa, M., and Ames, J. B. The role of calcium-binding proteins in the control of transcription: structure to function. Bioessays (2002) 24, 7, 625–636.

Ilari, A., Johnson, K. A., Nastopoulos, V., Verzili, D., Zamparelli, C., Colotti, G., Tsernoglou, D., and Chiancone, E. The crystal structure of the sorcin calcium binding domain provides a model of Ca^{2+}-dependent processes in the full-length protein. J. Mol. Biol. (2002) 317, 3, 447–458.

Imlay, J. A. Iron–sulphur clusters and the problem with oxygen. Mol. Microbiol. (2006) 59, 4, 1073–1082.

Inesi, G., Prasad, A. M., and Pilankatta, R. The Ca^{2+} ATPase of cardiac sarcoplasmic reticulum: physiological role and relevance to diseases. Biochem. Biophys. Res. Commun. (2008) 369, 1, 182–187.

Ingraham, R. H., and Hodges, R. S. Effects of Ca^{2+} and subunit interaction on surface accessibility of cystein residues in cardiac troponin C. Biochemistry (1988) 27, 16, 5891–5898.

Ishikawa, K., Nakagawa, A., Tanaka, I., Suzuki, M., and Nishihira, J. The structure of human MRP8, a member of the S100 calcium-binding protein family, by MAD phasing at 1.9 Å resolution. Acta Crystallogr. D (2000) 56, 559–566.

Ishitani, M., Liu, J., Halfter U, Kim, C.-S., Shi, W., and Zhu, J.-K. SOS3 function in plant salt tolerance requires N-myristoylation and calcium binding. Plant Cell (2000) 12, 1667–1677.

Itabashi, T., Mikamib, K., and Asaia, H. Characterization of the *spasmin 1* gene in *Zoothamnium arbuscula* strain Kawagoe (protozoa, ciliophora) and its relation to other spasmins and centrins. Res. Microbiol. (2003) 154, 361–367.

Itou, H., Yao, M., Fujita, I., Watanabe, N., Suzuki, M., Nishihira, J., and Tanaka, I. The crystal structure of human MRP14 (S100A9), a Ca^{2+}-dependent regulator protein in inflammatory process. J. Mol. Biol. (2002) 316, 265–276.

Ivanov, D. B., Philippova, M. P., and Tkachuk, V. A. Structure and functions of classical cadherins. Biochemistry (Mosc.) (2001) 66, 10, 1174–1186.

Ivshina, I. B., Peshkur, T. A., and Korobov, V. P. Efficient uptake of cesium ions by *Rhodococcus* cells. Microbiology (2002) 71, 3, 357–361.

Iwata, S., Ostermeier, C., Ludwig, B., and Michel, H., Structure at 2.8 Å resolution of cytochrome *c* oxidase from *Paracoccus denitrificans*. Nature (1995) 376, 660–669.

Iznaga-Escobar, N. ^{188}Re-direct labeling of monoclonal antibodies for radioimmunotherapy of solid tumors: biodistribution, normal organ dosimetry, and toxicology. Nucl. Med. Biol. (1998) 25, 5, 441–447.

Jabri, E., Carr, M. B., Hausinger, R. P., and Karplus, P. A. The crystal structure of urease from *Klebsiella aerogenes*. Science (1995) 268, 5213, 998–1004.

Jackson, A. E., Carraway, K. L., Payne, M. E., Means, A., Puett, D., and Brew, K. Association of calmodulin and smooth muscle myosin light chain kinase: application of a label selection technique with trace acetylated calmodulin. Proteins: Struct. Funct. Genet. (1987) 2, 202–209.

Jackson, K. W., and Mahmood, T. M. Atomic absorption, atomic emission, and flame emission spectrometry. Anal. Chem. (1994) 66, 12, 252R–279R.

Jackson, M., and Mantsch, H. H. The use and misuse of FTIR spectroscopy in the determination of protein structure. Crit. Rev. Biochem. Mol. Biol. (1995) 30, 2, 95–120.

Jakobek, J. L., Smith-Becker, J. A., and Lindgren, P. B. A bean cDNA expressed during a hypersensitive reaction encodes a putative calcium-binding protein. Mol. Plant Microbe Interact. (1999) 12, 712–719.

Jakubovics, N. S., and Jenkinson, H. F. Out of the Iron Age: new insights into the critical role of manganese homeostasis in bacteria. Microbiology (2001) 147, 1709–1718.

James, P., Vorherr, T., and Carafoli, E. Calmodulin-binding domains: just twofaced or multifaceted? Trends Biochem. Sci. (1995) 20, 38–42.

Jardetzky, O. The study of molecular switching mechanism. Life Sci. (1978) 22, 1245–1252.

Jarret, H. W., and Madhavan, R. Calmodulin-binding proteins also have a calmodulin-like binding site within their structure: the flip-flop model. J. Biol. Chem. (1991) 266, 1, 362–371.

Jeffery, E. H., Abreo, K., Burgess, E., Cannata, J., and Greger, J. L. Systemic aluminum toxicity: effects on bone, hematopoietic tissue, and kidney. J. Toxicol. Environ. Health (1996) 48, 6, 649–665.

Jelesarov, I., and Bosshard, H. R. Isothermal titration calorimetry and differential scanning calorimetry as complementary tools to investigate the energetics of biomolecular recognition. J. Mol. Recognit. (1999) 12, 1, 3–18.

Jelikić-Stankov, M., Uskokovi ć-Marković, S., Holclajtner-Antunović, I., Todorović, M., and Djurdjević, P. Compounds of Mo, V and W in biochemistry and their biomedical activity. J. Trace Elem. Med. Biol. (2007) 21, 1, 8–16.

Jeong, J. M., and Chung, J. K. Therapy with ^{188}Re-labeled radiopharmaceuticals: an overview of promising results from initial clinical trials. Cancer Biother. Radiopharm. (2003) 18, 5, 707–717.

Jia, J., Han, Q., Borregaard, N., Lollike, K., and Cygler, M. Crystal structure of human grancalcin, a member of the penta-EF-hand protein family. J. Mol. Biol. (2000) 300, 1271–1281.

Jia, J., Tarabykina, S., Hansen, C., Berchtold, M., and Cygler, M. Structure of apoptosis-linked protein ALG-2: insites into Ca^{2+}-induced changes in penta-EF-hand proteins. Structure (2001) 9, 267–275.

Jiang, Y., and MacKinnon, R. The barium site in a potassium channel by x-ray crystallography. J. Gen. Physiol. (2000) 115, 269–272.

Jockusch, H., Friedrich, G., and Zippel, M. Serum parvalbumin, an indicator of muscle disease in murine dystrophy and myotonia. Muscle Nerve (1990) 13, 6, 551–555.

Johansson, C., Ullner, M., and Drakenberg, T. The solution structures of mutant calbindin D9k's, as determined by NMR show that the calcium binding site can adopt different folds. Biochemistry (1993) 32, 33, 8429–8438.

Johnson, D. C., Dean, D. R., Smith, A. D., and Johnson, M. K. Structure, function, and formation of biological iron–sulfur clusters. Annu. Rev. Biochem. (2005) 74, 247–281.

Johnson, J. D., and Mills, J. S. Calmodulin. Med. Res. Rev. (1986) 6, 3, 431–363.

Johnson, J. D., and Potter, J. D. Detection of two classes of Ca^{2+}- binding sites in troponin C with circular dichroism and tyrosine fluorescence. J. Biol. Chem. (1978) 253, 11, 3775–3777.

Johnson, J. D., Snyder, C., Walsh, M., and Flynn, M. Effects of myosin light chain kinase and peptides on Ca^{2+} exchange with the N- and C-terminal Ca^{2+} binding sites of calmodulin. J. Biol. Chem. (1996) 271, 2, 761–767.

Johnson, S. M., Herrin, J., Liu, J. S., and Paul, I. C. Crystal and molecular structure of the barium salt of an antibiotic containing a high proportion of oxygen. J. Am. Chem. Soc. (1970) 92, 4428–4435.

Johnson, W. C. Jr., Palczewski, K., Gorczyca, W. A., Riazance-Lawrence, J. H., Witkowska, D., and Polans, A. S. Calcium binding to recoverin: implications for secondary structure and membrane association. Biochim. Biophys. Acta (1997) 1342, 2, 164–174.

Johnston, I. G., Paladino, T., Gurd, J. W., and Brown, I. R. Molecular cloning of SC1: a putative brain extracellular matrix glycoprotein showing partial similarity to osteonectin/BM40/SPARC. Neuron (1990) 4, 165–176.

Johnston, S. A., Hubaishy, I., and Waisman, D. M. Phosphorylation of annexin II tetramer by protein kinase C inhibits aggregation of lipid vesicles by the protein. J. Biol. Chem. (1992) 267, 36, 25976–25981.

Jones, H. E., Holland, I. B., Baker, H. L., and Campbell, A. K. Slow changes in cytosolic free Ca^{2+} in *Escherichia coli* highlight two putative influx mechanisms in response to changes in extracellular calcium. Cell Calcium (1999) 25, 3, 265–274.

Jones, L. M., Yang, W., Maniccia, A. W., Harrison, A., van der Merwe, P. A., and Yang, J. J. Rational design of a novel calcium-binding site adjacent to the ligand-binding site on CD2 increases its CD48 affinity. Protein Sci. (2008) 17, 3, 439–449.

Jones, P. J., Fitzpatrick, S., and Waisman, D. M. Salt dependency of chromaffin granule aggregation by annexin II tetramer. Biochemistry (1994) 33, 46, 13751–13760.

Jorgensen, P. L., and Pedersen, P. A. Structure-function relationships of Na^+, K^+, ATP, or Mg^{2+} binding and energy transduction in Na,K-ATPase. Biochim. Biophys. Acta (2001) 1505, 57–74.

Joseph, N. Interaction of amino acids and salts: III. The determination of the activities of calcium, barium, and strontium chloride in amino acid solutions by means of electrodes of the third kind. J. Biol. Chem. (1939) 130, 203–217.

Jost, M., Thiel, C., Weber, K., and Gerke, V. Mapping of three unique Ca^{2+}-binding sites in human annexin II. Eur. J. Biochem. (1992) 207, 923–930.

Juers, D. H., Jacobson, R. H., Wigley, D., Zhang, X. J., Huber, R. E., Tronrud, D. E., and Matthews, B. W. High resolution refinement of β-galactosidase in a new crystal form reveals multiple metal-binding sites and provides a structural basis for α-complementation. Protein Sci. (2000) 9, 9, 1685–1699.

Julenius, K., Thulin, E., Linse, S., and Finn, B. E. Hydrophobic core substitutions in calbindin D_{9k}: effects on stability and structure. Biochemistry (1998) 37, 25, 8915–8925.

Julenius, K., Robblee, J., Thulin, E., Finn, B. E., Fairman, R., and Linse, S. Coupling of ligand binding and dimerization of helix-loop-helix peptides: spectroscopic and sedimentation analyses of calbindin D_{9k} EF-hands. Proteins (2002) 47, 3, 323–333.

Jung, C. Insight into protein structure and protein–ligand recognition by Fourier transform infrared spectroscopy. J. Mol. Recognit. (2000) 13, 6, 325–351.

Jurado, L. A., Chockalingam, P. S., and Jarrett, H. W. Apocalmodulin. Physiol. Rev. (1999) 79, 3, 661–682.

Kaetzel, M. A., Chan, H. C., Dubinsky, W. P., Dedman, J. R., and Nelson, D. J. A role for annexin IV in epithelial cell function: inhibition of calcium-activated chloride conductance. J. Biol. Chem. (1994) 269, 7, 5297–5302.

Kakalis, L. T., Kennedy, M., Sikkink, R., Rusnak, F., and Armitage, I. M. Characterization of the calcium-binding sites of calcineurin B. FEBS Lett. (1995) 362, 1, 55–58.

Kalamajski, S., Aspberg, A., Lindblom, K., Heinegård, D., and Oldberg, A. Asporin competes with decorin for collagen binding, binds calcium and promotes osteoblast collagen mineralization. Biochem. J. (2009) 423, 1, 53–59.

Kalifatis, M., Rand, M. D., and Mann, K. G. Factor Va-membrane interaction is mediated by two regions located on the light chain of the cofactor. Biochemistry (1994) 33, 2, 486–493.

Kamada, T. Intracellular calcium and ciliary reversal in *Paramecium*. Proc. Imp. Acad. Tokyo (1938) 14, 260–262.

Kanayama, Y., Enomoto, S., Irie, T., and Amano, R. Axonal transport of rubidium and thallium in the olfactory nerve of mice. Nucl. Med. Biol. (2005) 32, 5, 505–512.

Kanda, T., Nose, Y., Wuchiyama, J., Uyama, T., Moriyama, Y., and Michibata, H. Identification of a vanadium-associated protein from the vanadium-rich ascidian, *Ascidia sydneiensis samea*. Zool. Sci. (1997) 14, 37–42.

Kaneko, N., Ago, H., Matsuda, R., Inagaki, E., and Miyano, M. Crystal structure of annexin V with its ligand K-201 as a calcium channel activity inhibitor. J. Mol. Biol. (1997) 274, 16–20.

Kantardjieff, K. A., Rupp, B. Matthews coefficient probabilities: Improved estimates for unit cell contents of proteins, DNA, and protein-nucleic acid complex crystals. Protein Sci. (2003) 12, 9, 1865–1871.

Kanuru, M., Samuel, J. J., Balivada, L. M., and Aradhyam, G. K. Ion-binding properties of Calnuc, Ca^{2+} versus Mg^{2+}: Calnuc adopts additional and unusual Ca^{2+}-binding sites upon interaction with G-protein. FEBS J. (2009) 276, 9, 2529–2546.

Kanyo, Z. F., Scolnick, L. R., Ash, D. E., and Christianson, D. W. Structure of a unique binuclear manganese cluster in arginase. Nature (1996) 383, 554–557.

Kasturi, R., Vasulka, C., Johnson, J. D. Ca^{2+}, caldesmon, and myosin light chain kinase exchange with calmodulin. J. Biol. Chem. (1993) 268, 11, 7958–7964.

Kataoka, M., Head, J. F., Seaton, B. A., and Engelman, D. M. Melittin binding causes a large calcium-dependent conformational change in calmodulin. Proc. Natl. Acad. Sci. USA (1989) 86, 18 6944–6948.

Kataoka, M., Head, J. F., Vorherr, T., Krebs, J., and 165arafoli, E. Small-angle x-ray scattering study of calmodulin bound to two peptides corresponding to parts of calmodulin-binding domain of plasma membrane Ca^{2+} pump. Biochemistry (1991) 30, 25, 6247–6251.

Kataoka, M., Mihars, K., and Tokunaga, F. Recoverin alters its surface properties depending on both calcium-binding and N-terminal myristoylation. J. Biochem. (1993) 114, 535–540.

Kato, M., Tanaka, N., Ishii, H., Yin, C., and Sakuma, Y. Ca^{2+} channels and Ca^{2+}-activated K^+ channels in adult rat gonadotrophin-releasing hormone neurones. J. Neuroendocrinol. (2009) 21, 4, 312–315.

Katz, A. K., Glusker, J. P., Beebe, S. A., and Bock, C. W. Calcium ion coordination: a comparison with that of beryllium, magnesium, and zinc. J. Am. Chem. Soc. (1996) 118, 24, 5752–5763.

Kawasaki, H., Nakayama, S., and Kretsinger, R. H. Classification and evolution of EF-hand proteins. Biometals (1998) 11, 277–295.

Kay, R. T. M., Tripet, B. P., Pearlstone, J. R., Smillie, L. B., and Sykes, B. D. Defining the region of troponin I that binds to troponin C. Biochemistry (1999) 38, 17, 5478–5489.

Kee, Y., and Scheller, R. H. Localization of synaptotagmin-binding domains on syntaxin. J. Neurosci. (1996) 16, 1975–1981.

Keilin, D. Cytochrome and respiratory enzymes. Proc. R. Soc. B (1929) 104, 206–252.

Kelly, S. M., and Price, N. C. The use of circular dichroism in the investigation of protein structure and function. Curr. Protein Pept. Sci. (2000) 1, 4, 349–384.

Kelm, R. J., Jr., and Mann, K. G. The collagen binding specificity of bone and platelet osteonectin is related to differences in glycosylation. J. Biol. Chem. (1991) 266, 9632–9639.

Kendrick-Jones, J., and Jakes, R. Regulatory light chains in myosin. In: Casteels, R. C., Godfraind, T., and Rüegg, J. C., eds., Excitation–Contraction Coupling in Smooth Muscle. North-Holland, New York, 1977, pp. 343–367.

Kennedy, M. T., Brockman, H., and Rusnak, F. Contributions of myristoylation to calcineurin structure/function. J. Biol. Chem. (1996) 271, 26517–26521.

Kennedy, M. T., Brockman, H., and Rusnak, F. Determinants of calcineurin binding to model membranes. Biochemistry (1997) 36, 44, 13579–13585.

Kerber, I. J., Warr, O. S., and Richardson, C. Pregnancy in a patient with a prosthetic mitral valve: associated with a fetal anomaly attributed to warfarin sodium. J. Am. Med. Assoc. (1968) 203, 223–225.

Kerrick, W. G. L., Hoar, P. E., and Cassidy, P. S. Calcium-activated tension: the role of myosin light chain phosphorylation. Fed. Proc. (1980) 39, 5, 1558–1563.

Khachaturian, Z. S. Calcium, membranes, aging, and Alzheimer's disease: introduction and overview. Ann. N. Y. Acad. Sci. (1989) 568, 1–4.

Kheifets, V., and Mochly-Rosen, D. Insight into intra- and inter-molecular interactions of PKC: design of specific modulators of kinase function. Pharmacol. Res. (2007) 55, 6, 467–476.

Kielhorn, J., Melber, C., Keller, D., and Mangelsdorf, I. Palladium: a review of exposure and effects to human health. Int. J. Hyg. Environ. Health. (2002) 205, 6, 417–432.

Kielly, W. W., and Meyerhof, O. Studies on the adenosinetriphosphatase of muscle: II. A new magnesium-activated adenosinetriphosphatase. J. Biol. Chem. (1948) 176, 591–601.

Kilby, P. M., Van Eldik, L. J., and Roberts, G. C. The solution structure of the bovine S100B protein dimer in the calcium-free state. Structure (1996) 4, 9, 1041–1052.

Kilhoffer, M. C., Haiech, J., and Demaille, J. G. Ion binding to calmodulin: a comparison with other intracellular calcium binding proteins. Mol. Cell. Biochem. (1983) 51, 1, 33–54.

Kilhoffer, M. C., Roberts, D. M., Adibi, A. O., Watterson, D. M., and Haiech, J. Investigation of the mechanism of calcium binding to calmodulin: use of an isofunctional mutant with a tryptophan introduced by site-directed mutagenesis. J. Biol. Chem. (1988) 263, 17023–17029.

Kilhoffer, M. C., Kubina, M., Travers, F., and Haiech, J. Use of engineered proteins with internal tryptophan reporter groups and pertubation techniques to probe the mechanism of ligand–protein interactions: investigation of the mechanism of calcium binding to calmodulin. Biochemistry (1992) 31, 8098–8106.

Kim, J., and Kim, H. Interaction of α-lactalbumin with phospholipid vesicles as studied by photoactivated hydrophobic labeling. Biochim. Biophys. Acta (1989) 983, 1–8.

Kim, S., Cullis, D. N., Feig, L. A., and Baleja, J. D. Solution structure of the Reps1 EH domain and characterization of its binding to NPF target sequences. Biochemistry (2001a) 40, 6776–6785.

Kim, S., Ko, J., Kim, J. H., Choi, E. C., and Na, D. S. Differential effects of annexins I, II, III, and V on cytosolic phospholipase A_2 activity: specific interaction model. FEBS Lett. (2001b) 489, 2–3, 243–248.

King, M. M., and Huang, C. Y. The calmodulin-dependent activation and deactivation of the phosphoprotein phosphatase, calcineurin, and the effect of nucleotides, pyrophosphate, and divalent metal ions. J. Biol. Chem. (1984) 259, 8847–8856.

King, W. A., Stone, D. B., Timmins, P. A., Narayan, T., Von Brasch, A. A. M., Mendelson, R. A., and Curmi, P. M. G. Solution structure of the chicken skeletal muscle troponin complex via small-angle and x-ray scattering. J. Mol. Biol. (2005) 345, 797–815.

Kini, R. M. Structure–function relationships and mechanism of anticoagulant phospholipase A_2 enzymes from snake venoms. Toxicon (2005) 45, 8, 1147–1161.

Kirberger, M., Wang, X., Deng, H., Yang, W., Chen, G., and Yang, J. J. Statistical analysis of structural characteristics of protein Ca^{2+}-binding sites. J. Biol. Inorg. Chem. (2008) 13, 7, 1169–1181.

Kissinger, C. R., Parge, H. E., Knighton, D. R., et al. Crystal structures of human calcineurin and the human FKBP12–FK506–calcineurin complex. Nature (1995) 378, 6557, 641–644.

Kitaura, Y., Watanabe, M., Satoh, H., Kawai, T., Hitomi, K., and Maki, M. Peflin, a novel member of the five-EF-hand-protein family, is similar to the apoptosis-linked gene 1 (ALG-2) protein but possesses nonapeptide repeats in the N-terminal hydrophobic region. Biochem. Biophys. Res. Commun. (1999) 263, 68–75.

Kitaura, Y., Satoh, H., Takahashi, H., Shibata, H., and Maki, M. Both ALG-2 and peflin, penta-EF-hand (PEF) proteins, are stabilized by dimerization through their fifth EF-hand regions. Arch. Biochem. Biophys. (2002) 399, 12–18.

Klaassen, C. D., Liu, J., and Choudhuri, S. Metallothionein: an intracellular protein to protect against cadmium toxicity. Annu. Rev. Pharmacol. Toxicol. (1999) 39, 267–294.

Klee, C. B. Ca^{2+}-dependent phospholipid- (and membrane-) binding proteins. Biochemistry (1988) 27, 18, 6645–6653.

Klee, C. B., and Cohen, P. The calmodulin-regulated protein phosphatase. In: Cohen, P., and Klee, C. B. (Eds.), Molecular Aspects of Cellular Regulation: Calmodulin, Vol. 5. Elsevier, Amsterdam, 1988, pp. 225–248.

Klee, C. B., Crouch, T. H., and Krinks, M. H. Subunit structure and catalytic properties of bovine brain Ca^{2+}-dependent cyclic nucleotide phosphodiesterase. Biochemistry (1979) 18, 722–729.

Klee, C. B., Draetta, G. F., and Hubbard, M. J. Calcineurin. Adv. Enzymol. Relat. Areas Mol. Biol. (1988) 61, 149–200.

Klee, C. B., Ren, H., Wang, X. Regulation of the calmodulin-stimulated protein phosphatase, calcineurin. J. Biol. Chem. (1998) 273, 22, 13367–13370.

Klein, P. S., and Melton, D. A. A molecular mechanism for the effect of lithium on development. Proc. Natl. Acad. Sci. USA (1996) 93, 8455–8459.

Kleindienst, A., Hesse, F., Bullock, M. R., and Buchfelder, M. The neurotrophic protein S100B: value as a marker of brain damage and possible therapeutic implications. Prog. Brain Res. (2007) 161, 317–325.

Kletzin, A., Mukund, S., Kelley-Crouse, T. L., Chan, M. K., Rees, D. C., and Adams, M. W. W. Molecular characterization of the genes encoding the tungsten-containing aldehyde ferredoxin oxidoreductase from *Pyrococcus furiosus* and formaldehyde ferredoxin oxidoreductase from *Thermococcus litoralis*. J. Bacteriol. (1995) 177, 16, 4817–4819.

Knollmann, B. C. New roles of calsequestrin and triadin in cardiac muscle. J. Physiol. (2009) 587, Pt. 13, 3081–3087.

Knyazeva, E. L., Grishchenko, V. M., Fadeev, R. S., Akatov, V. S., Permyakov, S. E., and Permyakov, E. A. Who Is Mr. HAMLET? Interaction of human α-lactalbumin with monomeric oleic acid. Biochemistry (2008) 47, 49, 13127–13137.

Kobayashi, C., and Takada, S. Protein grabs a ligand by extending anchor residues: molecular simulation for Ca^{2+} binding to calmodulin loop. Biophys. J. (2006) 90, 3043–3051.

Kobayashi, M., and Shimizu, S. Cobalt proteins. Eur. J. Biochem. (1999) 261, 1–9.

Kobayashi, M., Takamatsu, K., Saitoh, S., Miura, M., and Noguchi, T. Molecular cloning of hippocalcin, a novel calcium-binding protein of the recoverin family exclusively expressed in hippocampus. Biochem. Biophys. Res. Commun. (1992) 189, 1, 511–517.

Kobayashi, R., Hidaka, H., and Tashima, Y. Purification, characterization, and partial sequence analysis of 32-kDa calcimedin from chicken gizzard. Arch. Biochem. Biophys. (1990) 277, 203–210.

Kobayashi, T., Tao, T., Gergely, J., and Collins, J. H. Structure of the troponin complex: Implications of photocross-linking of troponin I to troponin C thiol mutants. J. Biol. Chem. (1994) 269, 8, 5725–5729.

Kobayashi, T., Grabarek, Z., Gergely, J., and Collins, J. H. Extensive interactions between troponins C and I: zero-length cross-linking of troponin I and acetylated troponin C. Biochemistry (1995) 34, 34, 10946–10952.

Kobayashi, T., Leavis, P. C., and Collins, J. H. Interaction of a troponin I inhibitory peptide with both domains of troponin C. Biochim. Biophys. Acta (1996) 1294, 25–30.

Kobayashi, T., Zhao, X., Wade, R., and Collins, J. H. Involvement of conserved, acidic residues in the N-terminal domain of troponin C in calcium-dependent regulation. Biochemistry (1999) 38, 5386–5391.

Koh, J., Suh, S. W., Gwag, B. J., He, Y. Y., Hsu, C. Y., and Choi, D. W. The role of zinc in selective neuronal death after transient global cerebral ischemia. Science (1996) 272, 1013–1016.

Köhler, C., Gogvadze, V., Håkansson, A., Svanborg, C., Orrenius, S., and Zhivotovsky, B. A folding variant of human α-lactalbumin induces mitochondrial permeability transition in isolated mitochondria. Eur. J. Biochem. (2001) 268, 1, 186–191.

Köhler, C., Hakansson, A., Svanborg, C., Orrenius, S., and Zhivotovsky, B. Protease activation in apoptosis induced by MAL. Exp. Cell Res. (1999) 249, 2, 260–268.

Kohout, S. C., Corbalán-García, S., Gómez-Fernández, J. C., and Falke, J. J. C2 domain of protein kinase Cα: elucidation of the membrane docking surface by site-directed fluorescence and spin labeling. Biochemistry (2003) 42, 5, 1254–1265.

Kohse, R. P., and Heilmeyer, J. The effect of Mg^{2+} on the Ca^{2+}- binding properties and Ca^{2+}-induced tyrosine fluorescence changes of calmodulin isolated from rabbit skeletal muscle. Eur. J. Biochem. (1981) 117, 3, 507–513.

Kojetin, D. J., Venters, R. A., Kordys, D. R., Thompson, R. J., Kumar, R., and Cavanagh, J. Structure, binding interface and hydrophobic transitions of Ca^{2+}-loaded calbindin D_{28k}. Nat. Struct. Mol. Biol. (2006) 13, 7, 641–647.

Kolatkar, A. R., and Weis, W. I. Structural basis of galactose recognition by C-type animal lectins. J. Biol. Chem. (1996) 271, 12, 6679–6685.

Komolov, K. E., Senin, I. I., Kovaleva, N. A., Christoph, M. P., Churumova, V. A., Grigoriev, I. I., Akhtar, M., Philippov, P. P., and Koch, K. W. Mechanism of rhodopsin kinase regulation by recoverin. J. Neurochem. (2009) 110, 1, 72–79.

Koneshan, S., Rasaiah, J. C., Lynden-Bell, R. M., and Lee, S. H. Solvent structure, dynamics, and ion mobility in aqueous solutions at 25°C. J. Phys. Chem. B (1998) 102, 4193–4204.

Koppenol, S., Tsao, F. H. C., Yu, H., and Zografi, G. The interaction of lung annexin I with phospholipid monolayers at the air/water interface. Biochim. Biophys. Acta (1998) 1369, 221–232.

Kördel, J., Pearlman, D. A., and Chazin, W. J. Protein solution structure calculations in solution: solvated molecular dynamics refinement of calbindin D_{9k}. J. Biomol. NMR (1997) 10, 3, 231–243.

Korfias, S., Stranjalis, G., Papadimitriou, A., Psachoulia, C., Daskalakis, G., Antsaklis, A., and Sakas, D. E. Serum S-100B protein as a biochemical marker of brain injury: a review of current concepts. Curr. Med. Chem. (2006) 13, 30, 3719–3731.

Kortvely, E., and Gulya, K. Calmodulin, and various ways to regulate its activity. Life Sci. (2004) 74, 9, 1065–1070.

Kostova, I. Gold coordination complexes as anticancer agents. Anticancer Agents Med. Chem. (2006) 6, 1, 19–32.

Kouchi, Z., Shikano, T., Nakamura, Y., Shirakawa, H., Fukami, K., and Miyazaki, S. The role of EF-hand domains and C2 domain in regulation of enzymatic activity of phospholipase C ζ. J. Biol. Chem. (2005) 280, 21015–21021.

Kragelund, B. B., Jonsson, M., Bifulco, G., Chazin, W. J., Nilsson, H., Finn, B. E., and Linse, S. Hydrophobic core substitutions in calbindin D_{9k}: effects on Ca^{2+} binding and dissociation. Biochemistry (1998) 37, 25, 8926–8937.

Kramer, B., and Tisdall, F. F. A simple technique for the determination of calcium and magnesium in small amounts of serum. J. Biol. Chem. (1921) 47, 475–481.

Krause, K. H., and Michalak, M. Calreticulin. Cell (1997) 88, 4, 439–443.

Krause, K. H., Milos, M., Luan-Rilliet, Y., Lew, D. P., and Cox, J. A. Thermodynamics of cation binding to rabbit skeletal muscle calsequestrin: evidence for distinct Ca^{2+}- and Mg^{2+}-binding sites. J. Biol. Chem. (1991) 266, 15, 9453–9459.

Krebs, J., and Carafoli, E. Influence of Ca^{2+} and trifluoroperazine on the structure of calmodulin: a ^{1}H-nuclear magnetic resonance study. Eur. J. Biochem. (1982) 124, 3, 619–627.

Krem, M. M., and Di Cera, E. Molecular markers of serine protease evolution. EMBO J. (2001) 20, 3036–3045.

Kretsinger, R. H. Gene triplication deduced from the tertiary structure of a muscle calcium binding protein. Nat. New Biol. (1972) 240, 85–88.

Kretsinger, R. H. Hypothesis: calcium modulated proteins contain EF hands. In: Carafoli, E. (Ed.), Calcium Transport in Contraction and Secretion. North-Holland, New York, 1975, pp. 469–478.

Kretsinger, R. H. Calcium-binding proteins. Annu. Rev. Biochem. (1976) 45, 239–266.

Kretsinger, R. H. Evolution of the informational role of calcium in eukaryotes. In: Wasserman, R. Corradino, R., Carafoli, E., Kretsinger, R. H., MacLennan, D., and Siegel, F., eds., Calcium Binding Proteins and Calcium Function. Elsevier North-Holland, New York, 1977, pp. 63–72.

Kretsinger, R. H. Why cells must export calcium. In: Bronner, F., ed., Intracellular Calcium Regulation. Alan R. Liss, New York, 1990, pp. 439–457.

Kretsinger, R. H., and Nelson, D. J. Calcium in biological Systems. Coord. Chem. Rev. (1976) 18, 29–124.

Kretsinger, R. H., and Nockolds, C. E. Carp muscle calcium binding protein: II. Structural determination and general description. J. Biol. Chem. (1973) 248, 9, 3313–3326.

Kretsinger, R. H., Rudnick, S. E., and Weissman, L. J. Crystal structure of calmodulin. J. Inorg. Biochem. (1986) 28, 2–3, 289–302.

Krewulak, K. D., and Vogel, H. J. Structural biology of bacterial iron uptake. Biochim. Biophys. Acta (2008) 1778, 9, 1781–1804.

Krezel, A., and Maret, W. Zinc-buffering capacity of a eukaryotic cell at physiological pZn. J. Biol. Inorg. Chem. (2006) 11, 8, 1049–1062.

Krieg, P., Schuppler, M., Koester, R., Mincheva, A., Lichter, P., and Marks, F. Repetin (Rptn), a new member of the "fused gene" subgroup within the S100 gene family encoding a murine epidermal differentiation protein. Genomics (1997) 43, 339–348.

Krnjević, K., and Lisiewicz, A. Injections of calcium ions into spinal motoneurons. J. Physiol. (1972) 225, 363–390.

Krudy, G. A., Kleerekoper, Q., Guo, X., Howarth, J. W., Solaro, R. J., and Rosevear, P. R. NMR studies delineating spatial relationships within the cardiac troponin I–troponin C complex. J. Biol. Chem. (1994) 269, 38, 23731–23735.

Krueger, J. K., Olah, G. A., Rokop, S. E., Zhi, G., Stull, J. T., and Trewhella, J. Structures of calmodulin and a functional myosin light chain kinase in the activated complex: a neutron scattering study. Biochemistry (1997) 36, 20, 6017–6023.

Krueger, J. K., Bishop, N. A., Blumenthal, D. K., Zhi, G., Beckingham, K., Stull, J. T., and Trewhella, J. Calmodulin binding to myosin light chain kinase begins at substoichiometric Ca^{2+} concentrations: a small-angle scattering study of binding and conformational transitions. Biochemistry (1998) 37, 51, 17810–17817.

Kubal, G., Mason, A. B., Patel, S. U., Sadler, P. J., and Woodworth, R. C. Oxalate and Ga^{3+}-induced structural changes in human serum transferrin and its recombinant N-lobe: ^1H NMR detection of preferential C-lobe Ga^{3+} binding. Biochemistry (1993) 32, 3387–3395.

Kuboniwa, H., Tjandra, N., Grzesiek, S., Ren, H., Klee, C. B., and Bax, A. Solution structure of calcium-free calmodulin. Nat. Struct. Biol. (1995) 2, 768–776.

Kumar, S., Harrylock, M., Walsh, K. A., Cormier, M. J., and Charbonneau, H. Amino-acid sequence of the Ca^{2+}-triggered luciferin binding-protein of *Renilla reniformis*. FEBS Lett. (1990) 268, 287–290.

Kumar, T. K., and Yu, C. Monitoring protein folding at atomic resolution. Acc. Chem. Res. (2004) 37, 12, 929–936.

Kupriyanov, V. V., and Gruwel, M. L. Rubidium-87 magnetic resonance spectroscopy and imaging for analysis of mammalian K^+ transport. NMR Biomed. (2005) 18, 2, 111–124.

Kuwajima, K., Harushima, Y., and Sugai, S. Influence of Ca^{2+} binding on the structure and stability of bovine α-lactalbumin studied by circular dichroism and nuclear magnetic resonance spectra. Int. J. Pept. Protein Res. (1986) 27, 1, 18–27.

Kuwajima, K., Ikeguchi, M., Sugawara, T., Hiraoka, Y., and Sugai, S. Kinetics of disulfide bond reduction in α-lactalbumin by dithiothreitol and molecular basis of superreactivity of the Cys6–Cys120 disulfide bond. Biochemistry (1990) 29, 36, 8240–8249.

Kwon, H., Goodwin, E. B., Nyitray, L., Berliner, E., O'Neall-Hennessey, E., Melandri, F. D., and Szent-Gyorgyi, A. G. Isolation of the regulatory domain of scallop myosin: role of the essential light chain in calcium binding. Proc. Natl. Acad. Sci. USA (1990) 87, 4771–4775.

La Bella, V., Ho, B. K., Appel, S. H., and Smith, R. G. Calbindin D_{28k} forms a Ca^{2+}-dissociable complex with melittin in vitro. Biochem. Mol. Biol. Int. (1996) 38, 6, 1199–1210.

Lackowicz, J. Principles of Fluorescence Spectroscopy, Kluwer Academic/Plenum Publishers, New York, 1999.

Lafitte, D., Heck, A. J., Hill, T. J., Jumel, K., Harding, S. E., and Derrick, P. J. Evidence of noncovalent dimerization of calmodulin. Eur. J. Biochem. (1999) 261, 1, 337–344.

La Greca, N., Hibbs, A. R., Riffkin, C., Roley, M., and Tilley, L. Identification of an endoplasmic reticulum-resident calcium-binding protein with multiple EF-hand motifs in asexual stages of *Plasmodium falciparum*. Mol. Biochem. Parasitol. (1997) 89, 283–293.

Lahat, H., Pras, E., Olender, T., et al. A missense mutation in a highly conserved region of CASQ2 is associated with autosomal recessive catecholamine-induced polymorphic ventricular tachycardia in Bedouin families from Israel. Am. J. Hum. Genet. (2001) 69, 1378–1384.

Laidlaw, P. P., and Payne, W. W. A method for the estimation of small quantities of calcium. Biochem. J. (1922) 16, 494–498.

Laine, E., Yoneda, J. D., Blondel, A., and Malliavin, T. E. The conformational plasticity of calmodulin upon calcium complexation gives a model of its interaction with the oedema factor of *Bacillus anthracis*. Proteins (2008) 71, 4, 1813–1829.

Laity, J. H., Lee, B. M., and Wright, P. E. Zinc finger proteins: new insights into structural and functional diversity. Curr. Opin. Struct. Biol. (2001) 11, 39–46.

Laizé, V., Viegas, C. S., Price, P. A., and Cancela, M. L. Identification of an osteocalcin isoform in fish with a large acidic prodomain. J. Biol. Chem. (2006) 281, 22, 15037–15043.

Lambers, T. T., Mahieu, F., Oancea, E., et al. Calbindin-D_{28K} dynamically controls TRPV5-mediated Ca^{2+} transport. EMBO J. (2006) 25, 2978–2988.

Lambrecht, H. G., and Koch, K. W. Phosphorylation of recoverin, the calcium-sensitive activator of photoreceptor guanylyl cyclase. FEBS Lett. (1991) 294, 3, 207–209.

Landar, A., Hall, T. L., Cornwall, E. H., Correia, J. J., Drohat, A. C., Weber, D. J., and Zimmer, D. B. The role of cysteine residues in S100B dimerization and regulation of target protein activity. Biochim. Biophys. Acta (1997) 1343, 117–129.

Landar, A., Rustandi, R. R., Weber, D. J., and Zimmer, D. B. S100A1 utilizes different mechanisms for interacting with calcium-dependent and calcium-independent target proteins. Biochemistry (1998) 37, 50, 17429–17438.

Lander, E. S., Linton, L. M., Birren, B., et al. Initial sequencing and analysis of the human genome. Nature (2001) 409, 860–921.

Landro, J. A., Gerlt, J. A., Kozarich, J. W., Koo, C. W., Shah, V. J., Kenyon, G. L., and Neidhart, D. J. The role of lysine 166 in the mechanism of mandelate racemase from *Pseudomonas putida*: mechanistic and crystallographic evidence for stereospecific alkylation by (*R*)-alpha-phenylglycidate. Biochemistry (1994) 33, 635–643.

Lane, D. A., Philippou, H., and Huntington, J. A. Directing thrombin. Blood (2005) 106, 2605–2612.

Lane, T. F., and Sage, H. The biology of SPARC, a protein that modulates cell–matrix interactions. FASEB J. (1994) 8, 163–173.

Lange, C., and Koch, K. W. Calcium-dependent binding of recoverin to membranes monitoted by surface plasmon resonance spectroscopy in real time. Biochemistry (1997) 36, 40, 12019–12026.

Langen, R., Isas, J. M., Hubbell, W. L., and Haigler, H. T. A transmembrane form of annexin XII detected by site-directed spin labeling. Proc. Natl. Acad. Sci. USA (1998) 95, 24, 14060–14065.

Laroche, A., Lemieux, G., and Pallotta, D. The nucleotide-sequence of a developmentally regulated cDNA from *Physarum polycephalum*. Nucleic Acids Res. (1989) 17, 10502–10502.

Larsen, K. S., and Auld, D. S. Carboxypeptidase A: mechanism of zinc inhibition. Biochemistry (1989) 28, 25, 9620–9625.

Laufer, E. M., Reutelingsperger, C. P., Narula, J., and Hofstra, L. Annexin A5: an imaging biomarker of cardiovascular risk. Basic Res. Cardiol. (2008) 103, 2, 95–104.

Leavitt, S., and Freire, E. Direct measurement of protein binding energetics by isothermal titration calorimetry. Curr. Opin. Struct. Biol. (2001) 11, 5, 560–566.

Le Brun, N. E., Wilson, M. T., Andrews, S. C., Guest, J. R., Harrison, P. M., Thomson, A. J., and Moore, G. R. Kinetic and structural characterization of an intermediate in the biomineralization of bacterioferritin. FEBS Lett. (1993) 333, 197–202.

Le Brun, N. E., Andrews, S. C., Guest, J. R., Harrison, P. M., Moore, G. R., and Thomson, A. J. Identifcation of the ferroxidase centre of *Escherichia coli* bacterioferritin. Biochem. J. (1995) 312, 385–392.

Lecocq, R., Lamy, F., and Dumont, J. E. Pattern of protein phosphorylation in intact stimulated cells: thyrotropin and dog thyroid. Eur. J. Biochem. (1979) 102, 1, 147–152.

Lecocq, R., Lamy, F., and Dumont, J. E. Use of two-dimensional gel electrophoresis and autoradiography as a tool in cell biology: the example of the thyroid and the liver. Electrophoresis (1990) 11, 3, 200–212.

Lecomte, C., Guillot, B., Muzet, N., Pichon-Pesme, V., and Jelsch, C. Ultra-high-resolution x-ray structure of proteins. Cell. Mol. Life Sci. (2004) 61, 7–8, 774–782.

Lee, A. J., Hodges, S., and Eastell, R. Measurement of osteocalcin. Ann. Clin. Biochem. (2000) 37, Pt. 4, 432–446.

Lee, C. H., Jeong, S. Y., Kim, B. J., Choi, C. H., Kima, J. S., Koo, B. M., Seoka, Y. J., Yim, H. S., and Kang, S. O. *Dictyostelium* CBP3 associates with actin cytoskeleton and is related to slug migration. Biochim. Biophys. Acta (2005) 1743, 281–290.

Lee, M. G.-S., Chen, J., Ho, A. W. M., D'Alesandro, P. A., and Van der Ploeg, L. H. T. A putative flagellar Ca^{2+}-binding protein of the flagellum of trypanosomatid protozoan parasites. Nucleic Acids Res. (1990) 18, 4252–4252.

Lee, S. C., Kim, I. G., Marekov, L. N., O'Keefe, E. J., Parry, D., and Steinert, P. M. The structure of human trychoyalin: potential multiple roles as a functional EF-hand-like calcium binding protein, a cornifield cell envelope precursor, and an intermediate filament-associated (crosslinking) protein. J. Biol. Chem. (1993) 268, 12164–12176.

Lee, S. H., Rosenmund, C., Schwaller, B., and Neher, E. Differences in Ca^{2+} buffering properties between excitatory and inhibitory hippocampal neurons from the rat. J. Physiol. (2000a) 525, 405–418.

Lee, S. H., Schwaller, B., and Neher, E. Kinetics of Ca^{2+} binding to parvalbumin in bovine chromaffin cells: implications for [Ca^{2+}] transients of neuronal dendrites. J. Physiol. (2000b) 525, 419–432.

LeGeros, R. Z. Calcium Phosphates in Oral Biology and Medicine. S. Karger, Basel, Switzerland, 1991, pp. 1–94.

Leggett, R. W., and Pellmar, T. C. The biokinetics of uranium migrating from embedded DU fragments. J. Environ. Radioact. (2003) 64, 2-3, 205–225.

Lehninger, A. L. Mitochondria and calcium ion transport. Biochem. J. (1970) 119, 129–138.

Lehninger, A. L., and Kennedy, E. P. Intracellular structures and the fatty acid oxidase system of rat liver. J. Biol. Chem. (1948) 172, 847–848.

Lei, H. Y., and Chang, C. P. Lectin of concanavalin A as an anti-hepatoma therapeutic agent. J. Biomed. Sci. (2009) 16, 10.

LeMaster, D. M., Minnich, M., Parsons, P. J., Anderson, J. S., and Hernández, G. Tetrathiolate coordination of germanium(IV) in a protein active site. J. Inorg. Biochem. (2006) 100, 8, 1410–1412.

Lemnius, L. De Miraculis Occultis Naturae. 1581. 1658 translation, published by J. Straeter, book IV, chapter 9.

Lentz, B. R., Zhou, C. M., and Wu, J. R. Phosphatidylserine-containing membranes alter the thermal stability of prothrombin's catalytic domain: a differential scanning calorimetric study. Biochemistry (1994) 33, 18, 5460–5468.

Lenz, S. E., Henschel, Y., Zopf, D., Voss, B., and Gundelfinger, E. D. VILIP, a cognate protein of the retinal calcium binding proteins visinin and recoverin, is expressed in the developing chicken brain. Brain Res. Mol. Brain Res. (1992) 15, 1–2, 133–140.

Leontiev, V. V., Uversky, V. N., Permyakov, E. A., and Murzin, A. G. Introduction of Ca^{2+}-binding amino-acid sequence into the T4 lysozyme. Biochim. Biophys. Acta (1993) 1162, 84–88.

Lepsík, M., and Field, M. J. Binding of calcium and other metal ions to the EF-hand loops of calmodulin studied by quantum chemical calculations and molecular dynamics simulations. J. Phys. Chem. (2007) 111, 33, 10012–10022.

Leung, I. K. M., Mani, R. S., and Kay, C. M. Isolation, characterication and metal-ion-binding properties of the α-subunit from S-100a protein. Biochem. J. (1986) 237, 3, 757–764.

Levay, P. F., and Viljoen, M. Lactoferrin: A general review. Haematologica (1995) 80, 252–267.

Levine, R. M., and Weiss, B. Binding of trifluoroperazine to the calcium-dependent activator of cyclic nucleotide phospho-diesterase. Mol. Pharmacol. (1977) 13, 2, 690–697.

Levine, R. M., and Weiss, B. Specificity of the binding of trifluoperazine to the calcium-dependent activator of phosphodiesterase and to a series of other calcium-binding proteins. Biochim. Biophys. Acta (1978) 540, 2, 197–204.

Levinskas, G. J., and Neuman, W. F. The solubility of bone mineral: I. Solubility studies of synthetic hydroxylapatite. J. Phys. Chem. (1955) 89, 164–168.

Levy, R. The role of cytosolic phospholipase $A_2 - \alpha$ in regulation of phagocytic functions. Biochim. Biophys. Acta (2006) 1761, 11, 1323–1334.

Levy, Y. Y., Lai, E. Y., Remillard, S. P., Heintzelman, M. B., and Fulton, C. Centrin is a conserved protein that forms diverse associations with centrioles and MTOCs in *Naegleria* and other organisms. Cell Motil. Cytoskelet. (1996) 33, 4, 298–323.

Li, C., Ullrich, B., Zhang, J. Z., Anderson, R. G. W., Brose, N., and Siidhof, T. C. Ca^{2+}-dependent and -independent activities of neural and non-neural synaptotagmins. Nature (1995) 375, 594–599.

Li, M. X., Chandra, M., Pearlstone, J. R., Racher, K. I., Trigo-Gonzalez, G., Borgford, T., Kay, C. M., and Smillie, L. B. Properties of isolated recombinant N and C domains of chicken troponin C. Biochemistry (1994) 33, 917–925.

Li, M. X., Gagne, S. M., Spyracopoulos, L., Kloks, C. P. A. M., Audette, G., Chandra, M., Solaro, R. J., Smillie, L. B., and Sykes, B. D. NMR studies of Ca^{2+} binding to the regulatory domains of cardiac and E41A skeletal muscle troponin C reveal the importance of site I to energetics of the induced structural changes. Biochemistry (1997) 36, 41, 12519–12525.

Li, Y., and Zamble, D. B. Nickel homeostasis and nickel regulation: an overview. Chem. Rev. (2009) 109, 4617–4643.

Liener, I. E. Phytohemagglutinins (Phytolectins). Annu. Rev. Plant Physiol. (1976) 27, 291–319.

Lilien, J., Balsamo, J., Arregui, C., and Xu, G. Turn-off, drop-out: functional state switching of cadherins. Dev. Dyn. (2002) 224, 1, 18–29.

Lim, L. H., and Pervaiz, S. Annexin 1: the new face of an old molecule. FASEB J. (2007) 21, 4, 968–975.

Lin, C. S., Shen, W. Y., Chen, Z. P., Tu, Y. H., and Matsudaira, P. Identification of I-plastin, a human fimbrin isoform expressed in intestine and kidney. Mol. Cell Biol. (1994) 14, 2457–2467.

Lin, G., Chattopadhyay, D., Maki, M., Wang, K. K. W., Carson, M., and Jin, L. Crystal structure of calcium bound domain VI of calpain at 1.9 Å resolution and its role in enzyme assembly, regulation, and inhibitor binding. Nat. Struct. Biol. (1997) 4, 539–546.

Lin, J., and Ficht, T. A. Protein synthesis in *Brucella abortus* induced during macrophage infection. Infect. Immun. (1995) 63, 4, 1409–1414.

Lin, P., Le-Niculescu, H., Hofmeister, R., McCaffery, J. M., Jin, M., Hennemann, H., McQuistan, T., De Vries, L., and Farquhar, M. The mammalian calcium-binding protein, nucleobindin (CALNUC), is a Golgi resident protein. J. Cell Biol. (1998) 141, 1515–1527.

Lin, X., and Barber, D. L. A calcineurin homologous protein inhibits GTPase-stimulated Na–H exchange. Proc. Natl. Acad. Sci. USA (1996) 93, 12631–12636.

Linberg, R., Conover, C. D., Shum, K. L., and Shorr, R. G. Hemoglobin based oxygen carriers: How much methemoglobin is too much? Artif. Cells Blood Substit. Immobil. Biotechnol. (1998) 26, 2, 133–148.

Linder, M. C., Wooten, L., Cerveza, P., Cotton, S., Shulze, R., and Lomeli, N. Copper transport. Am. J. Clin. Nutr. (1998) 67, Suppl., 965S–971S.

Ling, V., and Zielinski, R. E. Isolation of an *Arabidopsis* sequence encoding a 22-kDa calcium-binding protein (CaBP-22) related to calmodulin. Plant Mol. Biol. (1993) 22, 207–214.

Linse, S., Drakenberg, T., and Forsén, S. Mastoparan binding induces a structural change affecting both the N-terminal and C-terminal domains of calmodulin. A ^{113}Cd-NMR study. FEBS Lett. (1986) 199, 1, 28–32.

Linse, S., Helmersson, A., and Forsén, S. Calcium binding to calmodulin and its globular domains. J. Biol. Chem. (1991) 266, 13, 8050–8054.

Linse, S., Thulin, E., Gifford, L. K., Radzewsky, D., Hagan, J., Wilk, R. R., and Åkerfeldt, K. S. Domain organization of calbindin D-28k as determined from the association of six synthetic EF-hand fragments. Protein Sci. (1997) 6, 2385–2396.

Lisman, J. E., and Brown, J. E. The effects of intracellular iontrophoretic injection of calcium and sodium ions on the light response of *Limulus* ventral photoreceptors. J. Gen. Physiol. (1972) 59, 701–719.

Liu, J., and Zhu, J. K. A calcium sensor homolog required for plant salt tolerance. Science (1998) 280, 1943–1945.

Liu, J., Farmer, J. D., Jr., Lane, W. S., Friedman, J., Weissman, I., and Schreiber, S. L. Calcineurin is a common target of cyclophilin–cyclosporin A and FKBP–FK506 complexes. Cell (1991) 66, 807–815.

Liu, J., Ishitani, M., Halfter, U., Halfter, U., Kim, C.-S., and Zhu, J.-K. The *Arabidopsis thaliana* SOS2 gene encodes a protein kinase that is required for salt tolerance. Proc. Natl. Acad. Sci. USA (2000) 97, 3703–3734.

Liu, X., and Farley, J. M. Acetylcholine-induced chloride current oscillations in swine tracheal smooth muscle cells. J. Pharmacol. Exp. Ther. (1996a) 276, 178–186.

Liu, X., and Farley, J. M. Acetylcholine-induced Ca^{2+}-dependent chloride current oscillations are mediated by inositol 1,4,5-trisphosphate in tracheal myocytes. J. Pharmacol. Exp. Ther. (1996b) 277, 796–804.

Liu, X., and Theil, E. C. Ferritin as an iron concentrator and chelator target. Ann. N.Y. Acad. Sci. (2005) 1054, 136–140.

Liu, Z.-J., Vysotski, E. S., Chen, C.-J., Rose, J. P., Lee, J., and Wang, B.-C. Structure of the Ca^{2+}-regulated photoprotein obelin at 1.7 Å resolution determined directly from its sulfur substructure. Protein Sci. (2000) 9, 2085–2093.

Llorens, I., Den Auwer, C., Moisy, P., Ansoborlo, E., Vidaud, C., and Funke, H. Neptunium uptake by serum transferrin. FEBS J. (2005) 272, 7, 1739–1744.

Lo, K. W., Zhang, Q., Li, M., and Zhang, M. Apoptosis-linked gene product ALG-2 is a new member of the calpain small subunit subfamily of Ca^{2+}-binding proteins. Biochemistry (1999) 38, 23, 7498–7508.

Locke, F. S. Die Wirkung der Physiologischen Kochsalzlösung auf quergestreifte Muskel. Pflüger's Arch. (1893) 54, 501–524.

Locke, F. S. Notiz über des Einfluss Physiologischen Kochsalzlösung auf die Electrische Erregbarkeit von Muskel und Nerv. Zbl. Physiol. (1894) 8, 166–167.

Lockhart, J. C., and Gray, H. Molecular graphics in the study of the calcium-binding sites of carp parvalbumin and other proteins. J. Inclusion Phenomena (1987) 5, 1, 113–116.

Lockless, S. W., Zhou, M., and Mackinnon, R. Structural and thermodynamic properties of selective ion binding in a K^+ channel. PLoS Biol. (2007) 5, e121–e121.

Loeb, J. On an apparently new form of muscular irritability (contact irritability) produced by solutions of salts (preferably sodium salts) whose anions are liable to form insoluble calcium compounds. Am. J. Physiol. (1901) 5, 362–373.

Logan, K. J., Ng, P. K., Turner, C. J., Schmidt, R. P., Terner, U. K., Scott, J. R., Lentle, B. C., and Noujaim, A. A. Comparative pharmacokinetics of ^{67} Ga and ^{59}Fe in humans. Int. J. Nucl. Med. Biol. (1981) 8, 271–276.

Lokuta, A. J., Meyers, M. B., Sander, P. R., Fishman, G. I., and Valdivia, H. H. Modulation of cardiac ryanodine receptors by sorcin. J. Biol. Chem. (1997) 272, 25333–25338.

Long, S. B., Campbell, E. B., and Mackinnon, R. Crystal structure of a mammalian voltage dependent Shaker family K^+ channel. Science (2005) 309, 897–903.

Lorand, L. Factor XIII and the clotting of fibrinogen: from basic research to medicine. J. Thromb. Haemost. (2005) 3, 7, 1337–1348.

Lovley, D. R., and Phillips, E. J. P. Reduction of uranium by *Desulfovibrio desulfuricans*. Appl. Environ. Microbiol. (1992) 58, 850–856.

Lovley, D. R., Widman, P. K., Woodward, J. C., and Phillips, E. J. Reduction of uranium by cytochrome c_3 of *Desulfovibrio vulgaris*. Appl. Environ. Microbiol. (1993) 59, 11, 3572–3576.

Lowenstam, H. A. Minerals formed by organisms. Science (1981) 211, 1126–1131.

Lowenstein, W. R. Permeability of membrane junctions. Ann. N.Y. Acad. Sci. (1966) 137, 441–472.

Lowther, W. T., Orville, A. M., Madden, D. T., Lim, S., Rich, D. H., and Matthews, B. W. *Escherichia coli* methionine aminopeptidase: implications of crystallographic analyses of the native, mutant, and inhibited enzymes for the mechanism of catalysis. Biochemistry (1999) 38, 7678–7688.

Lu, Y., Bazzi, M. D., and Nelsestuen, G. L. Kinetics of annexin VI, calcium, and phospholipid association and dissociation. Biochemistry (1995) 34, 34, 10777–10785.

Luan, S. The CBL-CIPK network in plant calcium signaling. Trends Plant Sci. (2009) 14, 1, 37–42.

Lucosz, W. Principles and sensitivities of integrated-optical and SPR sensors for direct affinity sensing and immunoassay. Biosens. Bioelectron. (1991) 6, 215–233.

Lucosz, W., Integrated-optical and SPR sensors for direct affinity sensing: II. Anisotropy of adsorbed adlayers. Biosens. Bioelectron. (1997) 12, 175–190.

Lu Deng, Vysotski, E. S., Markova, S. V., Zhi-Jie Liu, Lee, J., Rose, J., and Wang, B. C. All three Ca^{2+}-bindig loops of photoproteins bind calcium ions: the crystal structures of calcium-loaded apo-aequorin and apo-obelin. Protein Sci. (2005) 14, 663–675.

Lukianova, O. A., and Davis, S. S. A role for iron–sulfur clusters in DNA repair. Curr. Opin. Chem. Biol. (2005) 9, 145–151.

Lundberg, S., Buevich, A. V., Sethson, I., Edlund, U., and Backman, L. Calcium-binding mechanism of human nonerythroid α-spectrin EF-structures. Biochemistry (1997) 36, 23, 7199–7208.

Lundell, A., Olin, A. I., Mörgelin, M., al-Karadaghi, S., Aspberg, A., and Logan, D. T. Structural basis for interactions between tenascins and lectican C-type lectin domains: evidence for a crosslinking role for tenascins. Structure (2004) 12, 1495–1506.

Luo, Y., Leszyk, J., Qian, Y., Gergely, J., and Tao, T. Residues 48 and 82 at the N-terminal hydrophobic pocket of rabbit skeletal muscle troponin-C photo-cross-link to Met121 of troponin-I. Biochemistry (1999) 38, 20, 6678–6688.

Lutz, W., Frank, E. M., Craig, T. A., Thompson, R., Venters, R. A., Kojetin, D., Cavanagh, J., and Kumar, R. Calbindin D_{28K} interacts with Ran-binding protein M: identification of interacting domains by NMR spectroscopy. Biochem. Biophys. Res. Commun. (2003) 303, 1186–1192.

L'vov, N. P., Nosikov, A. N., and Antipov, A. N. Tungsten-containing enzymes. Biochemistry (Mosc.) (2002) 67, 2, 196–200.

Lyman, H. A rapid method for determining calcium in blood and milk. J. Biol. Chem. (1917) 29, 169–178.

Lynch, J., and Michalak, M. Calreticulin is an upstream regulator of calcineurin. Biochem. Biophys. Res. Commun. (2003) 311, 4, 1173–1179.

Ma, Y., Griesmeier, U., Susani, M., et al. Comparison of natural and recombinant forms of the major fish allergen parvalbumin from cod and carp. Mol. Nutr. Food Res. (2008) 52, Suppl. 2, S196–S207.

Ma, Z., Jacobsen, F. E., and Giedroc, D. P. Coordination chemistry of bacterial metal transport and sensing. Chem. Rev. (2009) 109, 4644–4681.

Macario, A. J., and Conway de Macario, E. Molecular chaperones: multiple functions, pathologies, and potential applications. Front. Biosci. (2007) 12, 2588–2600.

MacCallum, A. B. The inorganic composition of the blood in vertebrates and invertebrates. Proc. R. Soc. B (1910) 82, 602–624.

MacCallum, S. G., and Voegtlin, C. On the relationship of tetany to the parathyroid glands and to calcium metabolism. J. Exp. Med. (1909) 11, 118–151.

MacDonald, M. J., and Brown, L. J. Calcium activation of mitochondrial glycerol phosphate dehydrogenase restudied. Arch. Biochem. Biophys. (1996) 326, 1, 79–84.

Maciejewski, J. J., Vacchiano, E. J., McCutcheon, S. M., and Buhse, H. E., Jr. Cloning and expression of a cDNA encoding a *Vorticella convallaria* spasmin: an EF-hand calcium-binding protein. J. Eukaryot. Microbiol. (1999) 46, 2, 165–173.

Mackall, J., and Klee, C. B. Calcium-induced sensitization of the central helix of calmodulin to proteolysis. Biochemistry (1991) 30, 29, 7242–7247.

MacPherson, I. S., and Murphy, M. E. P. Type-2 copper-containing enzymes. Cell. Mol. Life Sci. (2007) 64, 22, 2887–2899.

Mäcke, H. R., Hofmann, M., and Haberkorn, U. ^{68}Ga-labeled peptides in tumor imaging. J. Nucl. Med. (2005) 46, 172S–178S.

Magnusson, S., Petersen, T. E., Sottrup-Jensen, L., and Clays, H. Complete primary structure of prothrombin: isolation, structure and reactivity of carboxylated glutamic acid residues and regulation of prothrombin activation by thrombin. In: Reich, E., et al., eds., Proteases and Biological Control, Vol. 2. Cold Spring Harbor Laboratories Press, Cold Spring Harbor, NY, 1975, pp. 123–149.

Maguire, M. E. The structure of CorA: a Mg^{2+}-selective channel. Curr. Opin. Struct. Biol. (2006) 16, 432–438.

Maguire, M. E., and Cowan, J. A. Magnesium chemistry and biochemistry. Biometals (2002) 15, 203–210.

Magyar, J. S., Weng, T. C., Stern, C. M., et al. Reexamination of lead(II) coordination preferences in sulfur-rich sites: implications for a critical mechanism of lead poisoning. J. Am. Chem. Soc. (2005) 127, 26, 9495–9505.

Mailliard, W. S., Haigler, H. T., and Schlaepfer, D. D. Calcium-dependent binding of S100C to the N-terminal domain of annexin I. J. Biol. Chem. (1996) 271, 2, 719–725.

Makarovsky, I., Markel, G., Hoffman, A., Schein, O., Finkelstien, A., Brosh-Nissimov, T., Tashma, Z., Dushnitsky, T., and Eisenkraft, A. Osmium tetroxide: a new kind of weapon. Isr. Med. Assoc. J. (2007) 9, 10, 750–752.

Maki, M., Kitaura, Y., Satoh, H., Ohkouchi, S., and Shibata, H. Structures, functions and molecular evolution of the penta-EF-hand Ca^{2+}-binding proteins. Biochim. Biophys. Acta (2002) 1600, 1–2, 51–60.

Malencik, D. A., and Anderson, S. R. Binding of hormones and neuropeptides by calmodulin. Biochemistry (1983a) 22, 8, 1995–2001.

Malencik, D. A., and Anderson, S. R. High affinity binding of the mastoparans by calmodulin. Biochem. Biophys. Res. Commun. (1983b) 114, 1, 50–56.

Malencik, D. A., and Anderson, S. R. Peptide binding by calmodulin and its proteolytic fragments and by troponin C. Biochemistry (1984) 23, 11, 2420–2428.

Malencik, D. A., and Anderson, S. R. Demonstration of a fluorimetrically distinguishable intermediate in calcium binding by calmodulin–mastoparan complexes. Biochem. Biophys. Res. Commun. (1986) 135, 3, 1050–1057.

Malencik, D. A., and Anderson, S. R. Association of melittin with the isolated myosin light chains. Biochemistry (1988) 27, 6, 1941–1949.

Malencik, D. A., Anderson, S. R., Shatalin, Y., and Schimerlik, M. I. Rapid kinetic studies on calcium interactions with native and fluorescently labelled calmodulin. Biochem. Biophys. Res. Commun. (1981) 101, 2, 390–395.

Mäler, L., Blankenship, J., Rance, M., and Chazin, W. J. Site-site communication in the EF-hand Ca^{2+}-binding protein calbindin D_{9K}. Nat. Struct. Biol. (2000) 7, 3, 245–250.

Mäler, L., Sastry, M., and Chazin, W. J. A structural basis for S100 protein specificity derived from comparative analysis of apo and Ca^{2+}-calcyclin. J. Mol. Biol. (2002) 317, 2, 279–290.

Malmendal, A., Carlstrom, G., Hambraeus, C., Drakenberg, T., Forsen, S., and Akke, M. Sequence and context dependence of EF-hand loop dynamics: an ^{15}N relaxation study of a calcium-binding site mutant of calbindin D_{9k}. Biochemistry (1998) 37, 8, 2586–2595.

Malmendal, A., Evenäs, J., Thulin, E., Gippert, G. P., Drakenberg, T., and Forsèn, S. When size is important: Accommodation of magnesium in a calcium binding regulatory domain. J. Biol. Chem. (1998) 273, 44, 28994–29001.

Mandal, R., Kalke, R., and Li, X. F. Interaction of oxaliplatin, cisplatin, and carboplatin with hemoglobin and the resulting release of a heme group. Chem. Res. Toxicol. (2004) 17, 10, 1391–1397.

Mani, R. S., and Kay, C. M. Circular dichroism studies on the zinc-induced conformational changes in S-100a and S-100b proteins. FEBS Lett. (1983) 163, 2, 282–286.

Maniccia, A. W., Yang, W., Li, S. Y., Johnson, J. A., and Yang, J. J. Using protein design to dissect the effect of charged residues on metal binding and protein stability. Biochemistry (2006) 45, 18, 5848–5856.

Mann, K., Deutzmann, R., Paulsson, M., and Timpl, R. Solubilization of protein BM-40 from a basement membrane tumor with chelating agents and evidence for its identity with osteonectin and SPARC. FASEB J. (1987) 218, 167–172.

Marcon, G., Messori, L., Orioli, P., Cinellu, M. A., and Minghetti, G. Reactions of gold(III) complexes with serum albumin. Eur. J. Biochem. (2003) 270, 23, 4655–4661.

Maret, W., and Li, Y. Coordination dynamics of zinc in proteins. Chem. Rev. (2009) 109, 4682–4707.

Margulis, L. Symbiotic theory of the origin of eukaryotic organelles; criteria for proof. Symp. Soc. Exp. Biol. (1975) 29, 21–38.

Margulis, L. Evolution of mitosis and the late appearance of metozoa, metaphyta, and fungi. In: Ponnamperuma, C. (Ed.), Chemical Evolution of the Early Pre-Cambrian. Academic Press, New York, 1977, pp. 187–189.

Marie, P. J. Strontium ranelate: a novel mode of action optimizing bone formation and resorption. Osteoporos. Int. (2005) 16, S7–S10.

Marie, P. J., Ammann, P., Boivin, P., and Rey, C. Mechanisms of action and therapeutic potential of strontium in bone. Calcif. Tissue Int. (2001) 69, 121–129.

Marilley, D., and Schwaller, B. Accociation between the calcium-binding protein calretinin and cytoskeletal components in the human colon adenocarcinoma cell line WiDr. Exp. Cell Res. (2000) 259, 12–22.

Marin, F., Luquet, G., Marie, B., and Medakovic, D. Molluscan shell proteins: primary structure, origin, and evolution. Curr. Top. Dev. Biol. (2008) 80, 209–276.

Markoff, A., and Gerke, V. Expression and functions of annexins in the kidney. Am. J. Physiol. Renal Physiol. (2005) 289, 5, F949–F956.

Marriot, W. M., and Howland, J. Phosphate retention as a factor in the production of acidosis in nephritis. Arch. Intern. Med. (1916) 18, 708–711.

Marsh, B. B. A factor modifying muscle fiber synaeresis. Nature (1951) 167, 1065–1066.

Marsh, B. B. The effects of adenosine triphosphate on the fibre volume of a muscle homogenate. Biochim. Biophys. Acta (1952) 9, 247–260.

Marston, S., Lehman, W., Woody, C., Pritchard, K., and Smith, C. Caldesmon and Ca^{2+} regulation in smooth muscles. In: Gerday, C., Bolis, L., and Giles, R., eds., Calcium and Calcium Binding Proteins. Springer-Verlag, Berlin, 1988, pp. 69–81.

Martell, A., and Smith, J. Critical Stability Constants, Vol. 3. Plenum Press, New York, 1977.

Martin, B. Bioinorganic chemistry of calcium. In: Metal Ions in Biological Systems, New York, 1984, pp. 2–28.

Martin, C. J. A review of factors affecting patient doses for barium enemas and meals. Br. J. Radiol. (2004) 77, 864–868.

Martin, R. B. Practical hardness scales for metal ion complexes. Inorg. Chim. Acta (2002) 339, 27–33.

Martin, S. R., Andersson Teleman, A., Bayley, P. M., and Drakenberg, T. F. Kinetics of calcium dissociation from calmodulin and its tryptic fragments: a stopped flow fluorescence study using Quin 2 reveals a two-domain structure. Eur. J. Biochem. (1985) 151, 3, 543–550.

Martin, S. R., Masino, L., and Bayley, P. M. Enhancement by Mg^{2+} of domain specificity in Ca^{2+}-dependent interactions of calmodulin with target sequences. Protein Sci. (2000) 9, 12, 2477–2488.

Masaki, K., Masuda, R., Takase, K., Kawano, K., and Nitta, K. Stability of the molten globule state of a domain-exchanged chimeric protein between human and bovine α-lactalbumins. Protein Eng. (2000) 13, 1–4.

Mast, S. O., and Nadler, J. E. Reversal of ciliary action in *Paramecium caudatum*. J. Morphol. (1926) 43, 105–117.

Matei, E., Miron, S., Blouquit, Y., Duchambon, P., Durussel, I., Cox, J. A, and Craescu, C. T. C-terminal half of human centrin 2 behaves like a regulatory EF-hand domain. Biochemistry (2003) 42, 6, 1439–1450.

Mathias, R. S., and Brown, E. M. Divalent cations modulate PTH-dependent 3′,5′-cyclic adenosine monophosphate production in renal proximal tubular cells. Endocrinology (1991) 128, 3005–3012.

Matsubara, M., Yamauchi, E., Hayashi, N., and Taniguchi, H. MAECKS, a major protein kinase C substrate, assumes non-helical conformations both in solution and in complex with Ca^{2+}-calmodulin. FEBS Lett. (1998) 421, 203–207.

Matsuda, S., Hisatomi, O., Ishino, T., Kobayashi, Y., and Tokunaga, F. The role of calcium-binding sites in S-modulin function. J. Biol. Chem. (1998) 273, 20223–20227.

Matsumoto, F., Makino, K., Maeda, K., Patzelt, H., Maeda, Y., and Fujiwara, S. Conformational changes of troponin C within the thin filaments detected by neutron scattering. J. Mol. Biol. (2004) 342, 1209–1221.

Matsumura, H., Shiba, T., Inoue, T., Harada, S., and Kai, Y. A novel mode of target recognition suggested by the 2.0 Å structure of holo S100B from bovine brain. Structure (1998) 6, 2, 233–241.

Matsushima, N., Tzumi, V., Matsuo, T., Yoshido, H., Ueki, T., and Miyaka, Y. Binding of both Ca^{2+} and mastoparan to calmodulin induces a large change in the tertiary structure. J. Biochem. (1989) 105, 6, 883–887.

Matsuura, I., Ishihara, K., Nakai, Y., Yazawa, M., Toda, H., and Yagi, K. A site-directed mutagenesis study of yeast calmodulin. J. Biochem. (1991) 109, 1, 190–197.

Matsuura, I., Kimura, E., Tai, K., and Yazawa, M. Mutagenesis of the fourth calcium-binding domain of yeast calmodulin. J. Biol. Chem. (1993) 268, 18, 13267–13273.

Matthew, W. D., Tsavaler, L., and Reichardt, L. F. Identification of a synaptic vesicle-specific membrane protein with a wide distribution in neuronal and neurosecretory tissue. J. Cell Biol. (1981) 91, 1, 257–269.

Matthews, J. M., and Sunde, M. Zinc fingers: folds for many occasions. Life (2002) 54, 351–355.

Maulet, Y., and Cox, J. A. Structural changes in melittin and calmodulin upon complex formation and their modulation by calcium. Biochemistry (1983) 22, 24, 5680–5686.

Maune, J. F., Beckingham, K., Martin, S. R., and Bayley, P. M. Circular dichroism studies on calcium binding to two series of Ca^{2+} binding site mutants of *Drosophila melanogaster* calmodulin. Biochemistry (1992) 31, 7779–7786.

Maurer, M. C., Sando, J. J., and Grisham, C. M. High-affinity Ca^{2+}- and substrate-binding sites on protein kinase C as determined by nuclear magnetic resonance spectroscopy. Biochemistry (1992) 31, 33, 7714–7721.

Maurer, P., and Hohenester, E. Structural and functional aspects of calcium binding in extracellular matrix proteins. Matrix Biol. (1997) 15, 8–9, 569–580.

Maurer, P., Mayer, U., Bruch, M., Jenoe, P., Mann, K., Landwehr, R., and Engel, J. High and low affinity Ca^{2+} binding and stability of the multidomain extracellular glycoprotein BM40/SPARC/osteonectin. Eur. J. Biochem. (1992) 205, 233–240.

Maurer, P., Hohenadl, C., Hohenester, E., Goehring, W., Timpl, R., and Engel, J. The C-terminal portion of BM-40/SPARC/osteonectin is an autonomously folded crystallazable domain that binds Ca^{2+} and collagen. J. Mol. Biol. (1995) 253, 347–357.

Maximov, E. E., Zapevalova, N. P., and Mitin, Y. V. On the calcium binding ability of the evolutionary ancestor of calcium binding proteins. FEBS Lett. (1978) 88, 1, 80–82.

Mayer, U., Aumailley, M., Mann, K., Timpl, R., and Engel, J. Calcium-dependent binding of basement membrane protein BM-40 (osteonectin, SPARC) to basement membrane collagen type IV. Eur. J. Biochem. (1991) 198, 141–150.

Mazia, D. The release of calcium in arbacia eggs on fertilization. J. Cell Comp. Physiol. (1937) 10, 291–304.

Mazia, D., and Clark, J. M. Free calcium in the action of stimulating agents on *Elodea* cells. Biol. Bull. (1936) 71, 306–323.

McCall, K. A., Huang, C., and Fierke, C. A. Function and mechanism of zinc metalloenzymes. J. Nutr. (2000) 130, 1437S–1446S.

McCollum, E. V., Simmonds, N., Parsons, H. T., Shipley, P. G., and Park, E. A. Studies on experimental rickets: I. The production of rachitis and similar diseases in the rat by deficient diets. J. Biol. Chem. (1920) 45, 333–342.

McConnell, D. A structural investigation of the isomorphism of the apatite group. Am. Min. (1938) 23, 1–19.

McDowell, L., Sanyal, G., and Prendergast, F. G. Probable role of amphiphilicity in the binding of mastoparan to calmodulin. Biochemistry (1985) 24, 12, 2979–2984.

McDowell, H., Gregory, T. M., and Brown, W. E. Solubility of $Ca_5(PO_4)_3OH$ in the system $Ca(OH)_2$–H_3PO_4–H_2O at 5, 15, 25 and 37°C. J. Res. Natl Bur. Stand. (USA) (1977) 81A, 273–281.

McLean, F. C. The parathyroid hormone and bone. Clin. Orthopaed. (1957) 9, 46–60.

McLean, F. C., and Budy, A. M. Connective and supporting tissues: bone. Annu. Rev. Physiol. (1959) 21, 69–90.

McLean, F. C., and Hastings, A. B. A biological method for the estimation of calcium ion concentration. J. Biol. Chem. (1934) 107, 337–350.

McLean, F. C., and Hastings, A. B. Clinical estimation and significance of calcium-ion concentrations in the blood. Am. J. Med. Sci. (1935) 189, 601–613.

McPhalen, C. A., Strynadka, N. C., and James, M. N. Calcium-binding sites in proteins: a structural perspective. Adv. Protein Chem. (1991) 42, 77–144.

Meador, W. E., Means, A. R., and Quiocho, F. A. Modulation of calmodulin plasticity in molecular recognition on the basis of x-ray structures. Science (1993) 262, 1718–1721.

Medkova, M., and Cho, W. Mutagenesis of the C2 domain of protein kinase C-alpha: differential roles of Ca^{2+} ligands and membrane binding residues. J. Biol. Chem. (1998) 273, 28, 17544–17552.

Medved, L. V., Vysotchin, A., and Ingham, K. C. Ca^{2+}-dependent interactions between Gla and EGF domains in human coagulation factor IX. Biochemistry (1994) 33, 2, 478–485.

Medvedeva, M. V., Djemuchadze, D. R., Watterson, D. M., Marston, S. B., and Gusev, N. B. Replacement of Lys-75 of calmodulin affects its interaction with smooth muscle caldesmon. Biochim. Biophys. Acta (2001) 1544, 143–150.

Medvedkin, V. N., Mitin, Y. V., and Permyakov, E. A. Influence of Arg-74 on formation of structure of the C-terminal domain of pike parvalbumin III. Bioorg. Khim. (Mosc.) (1987) 13, 2, 177–182.

Meech, R. W. Intracellular calcium injection causes increased potassium conductance in *Aplysia* nerve cells. Comp. Biochem. Physiol. (1972) 42A, 493–499.

Meers, P., Daleke, D., Hong, K., and Papahadjopoulos, D. Interactions of annexins with membrane phospholipids. Biochemistry (1991) 30, 11, 2903–2908.

Megli, F. M., Selvaggi, M., Liemann, S., Quagliariello, E., and Huber, R. The calcium-dependent binding of annexin V to phospholipid vesicles influences the bilayer inner fluidity gradient. Biochemistry (1998) 37, 29, 10540–10546.

Mehmel, M. Über die Struktur des Apatits. I. Z. Kristal. (1930) 75, 323–331.

Meier, M., Menne, J., and Haller, H. Targeting the protein kinase C family in the diabetic kidney: lessons from analysis of mutant mice. Diabetologia (2009) 52, 5, 765–775.

Meinnel, T., Mechulam, Y., and Blanquet, S. Methionine as translation start signal: a review of the enzymes of the pathway in *Escherichia coli*. Biochimie (1993) 75, 12, 1061–1075.

Meléndez, E. Titanium complexes in cancer treatment. Crit. Rev. Oncol. Hematol. (2002) 42, 3, 309–315.

Mellanby, E. Experimental rickets. Spec. Rep. Ser. Med. Coun. (1921) 61, 1–78.

Mendel, R. R., and Bittner, F. Cell biology of molybdenum. Biochim. Biophys. Acta (2006) 1763, 621–635.

Meng, W., Sawasdikosol, S., Burakoff, S., and Eck, M. J. Structure of the amino-terminal domain of Cbl complexed to its binding site on ZAP-70 kinase. Nature (1999) 398, 84–90.

Meng, Y. L., Liu, Z., and Rosen, B. P. As(III) and Sb(III) uptake by GlpF and efflux by ArsB in *Escherichia coli*. J. Biol. Chem. (2004) 279, 18, 18334–18341.

Méplan, C., Verhaegh, G., Richard, M. J., and Hainaut, P. Metal ions as regulators of the conformation and function of the tumour suppressor protein p53: implications for carcinogenesis. Proc. Nutr. Soc. (1999) 58, 3, 565–571.

Mercogliano, C. P., and DeRosier, D. J. Gold nanocluster formation using metallothionein: mass spectrometry and electron microscopy. J. Mol. Biol. (2006) 355, 2, 211–23.

Mercogliano, C. P., and DeRosier, D. J. Concatenated metallothionein as a clonable gold label for electron microscopy. J. Struct. Biol. (2007) 160, 1, 70–82.

Messerschmidt, A. Multi-copper Oxidases. World Scientific, Singapore, 1997.

Messerschmidt, A., Prade, L., and Wever, R. Implications for the catalytic mechanism of the vanadium-containing enzyme chloroperoxidase from the fungus *Curvularia inaequalis* by x-ray structures of the native and peroxide form. Biol. Chem. (1997) 378, 309–315.

Meyer, D. F., Mabuchi, Y., and Grabarek, Z. The role of Phe-92 in the Ca^{2+}-induced conformational transition in the C-terminal domain of calmodulin. J. Biol. Chem. (1996) 271, 11284–11290.

Meyer, J. Ferredoxins of the third kind. FEBS Lett. (2001) 509, 1, 1–5.

Meyers, M. B., Pickel, V. M., Sheu, S. S., Sharma, V. K., Scotto, K. W., and Fishman, G. I. Association of sorcin with the cardiac ryanodine receptor. J. Biol. Chem. (1995a) 270, 44, 26411–26418.

Meyers, M. B., Zamparelli, C., Verzili, D., Dicker, A. P., Blanck, T. J., and Chiancone, E. Calcium-dependent translocation of sorcin to membranes: functional relevance in contractile tissue. FEBS Lett. (1995b) 357, 230–234.

Michael, S. F., Kilfoil, V. J., Schmidt, M. H., Amann, B. T., and Berg, J. M. Metal binding and folding properties of a minimalist Cys_2His_2 zinc finger peptide. Proc. Natl. Acad. Sci. USA (1992) 89, 11, 4796–4800.

Michalak, M., Guo, L., Robertson, M., Lozak, M., and Opas, M. Calreticulin in the heart. Mol. Cell. Biochem. (2004) 263, 1–2, 137–142.

Michalak, M., Groenendyk, J., Szabo, E., Gold, L. I., and Opas, M. Calreticulin, a multi-process calcium-buffering chaperone of the endoplasmic reticulum. Biochem. J. (2009) 417, 3, 651–666.

Michetti, F., and Gazzolo, D. S100B protein in biological fluids: a tool for perinatal medicine. Clin. Chem. (2002) 48, 12, 2097–2104.

Michibata, H., Yamaguchi, N., Uyama, T., and Ueki, T. Molecular biological approaches to the accumulation and reduction of vanadium by ascidians. Coord. Chem. Rev. (2003) 237, 41–51.

Michiels, J., Xi, C., Verhaert, J., and Vanderleyden, J. The functions of Ca^{2+} in bacteria: A role for EF-hand proteins? Trends Microbiol. (2002) 10, 87–93.

Miguel, R. N., Shi, J., and Mizuguchi, K. Protein fold recognition and comparative modeling using HOMSTRAD, JOY, and FUGUE. In: Tsigelny, I. F. (Ed.), Protein Structure Prediction: Bioinformatic Approach. International University Line Biotechnology Series, 2002, pp. 143–165.

Milacic, V., Fregona, D., and Dou, Q. P. Gold complexes as prospective metal-based anticancer drugs. Histol. Histopathol. (2008) 23, 1, 101–108.

Millecchia, R., and Mauro, A. The ventral photoreceptor cells of *Limulus*: III. A voltage-clamp study. J. Gen. Physiol. (1969) 54, 331–351.

Miller, J. E., Di Bilio, A. J., Wehbi, W. A., Green, M. T., Museth, A. K., Richards, J. R., Winkler, J. R., and Gray, H. B. Electron tunneling in rhenium-modified *Pseudomonas aeruginosa* azurins. Biochim. Biophys. Acta (2004) 1655, 1–3, 59–63.

Milović, N. M., Dutca, L. M., and Kostić, N. M. Combined use of platinum(II) complexes and palladium(II) complexes for selective cleavage of peptides and proteins. Inorg. Chem. (2003) 42, 13, 4036–4045.

Minks, C., Huber, R., Moroder, L., and Budisa, N. Atomic mutations at the single tryptophan residue of human recombinant annexin V: effects on structure, stability, and activity. Biochemistry (1999) 38, 10649–10659.

Miquel, G., Nekaa, T., Kahn, P. H., Hémadi, M., and El Hage Chahine, J. M. Mechanism of formation of the complex between transferrin and bismuth, and interaction with transferrin receptor 1. Biochemistry (2004) 43, 46, 14722–14731.

Mira, J. P., Dubois, T., Oudinet, J. P., Lukowski, S., Russo-Marie, F., and Geny, B. Inhibition of cytosolic phospholipase A_2 by annexin V in differentiated permeabilized HL-60 cells: evidence of crucial importance of domain I type II Ca^{2+}-binding site in the mechanism of inhibition. J. Biol. Chem. (1997) 272, 16, 10474–10482.

Mitchell, R. D., Simmerman, H. K. B., and Jones, L. R. Ca^{2+} binding effects on protein conformation and protein interactions of canine cardiac calsequestrin. J. Biol. Chem. (1988) 263, 3, 1376–1381.

Miura, K., Titani, K., Kurosawa, Y., and Kanai, Y. Molecular-cloning of nucleobindin, a novel DNA-binding protein that contains both a signal peptide and a leucine zipper structure. Biochem. Biophys. Res. Commun. (1992) 187, 375–380.

Miura, K., Hirai, M., Kanai, Y., and Kurosawa, Y. Organization of the human gene for nucleobindin (NUC) and its chromosomal assignment to 19q13.2-q13.4. Genomics (1996) 34, 181–186.

Miwa, N., and Kawamura, S. Frog p26olf, a molecule with two S100-like regions in a single peptide. Microsc. Res. Tech. (2003) 60, 6, 593–599.

Miwa, N., Kobayashi, M., Takamaztsu, K., and Kawamura, S. Purification and molecular cloning of a novel calcium-binding protein, p26olf, in the frog olfactory epithelium. Biochem. Biophys. Res. Commun. (1998a) 251, 3, 860–867.

Miwa, N., Shinmyo, Y., and Kawamura, S. Calcium-binding by p26olf, an S100-like protein in the frog olfactory epithelium. Eur. J. Biochem. (2001) 268, 23, 6029–6036.

Miwa, N., Shinmyo, Y., and Kawamura, S. Cloning and characterization of *Xenopus* dicalcin, a novel S100-like calcium-binding protein in *Xenopus* eggs. DNA Seq. (2007) 18, 5, 400–404.

Miwa, N., Uebi, T., and Kawamura, S. Characterization of p26olf, a novel calcium-binding protein in the frog olfactory epithelium. Biochem. Biophys. Res. Commun. (1998a) 25, 860–867.

Miwa, N., Uebi, T., and Kawamura, S. Characterization of p26olf, a novel calcium-binding protein in the frog olfactory epithelium. Biochem. Biophys. Res. Commun. (1998b) 25, 860–867.

Miwa, N., Uebi, T., and Kawamura, S. S100-annexin complexes: biology of conditional association. FEBS J. (2008) 275, 20, 4945–4955.

Miyawaki, A., Llopis, J., Heim, R., McCaffery, J. M., Adams, J. A., Ikura, M., and Tsien, R. Y. Fluorescent indicators for Ca^{2+} based on green fluorescent proteins and calmodulin. Nature (1997) 388, 6645, 882–887.

Miyazaki, S., and Ito, M. Calcium signals for egg activation in mammals. J. Pharmacol. Sci. (2006) 100, 5, 545–552.

Mizuguchi, M., Masaki, K., and Nitta, K. The molten globule state of a chimera of human α-lactalbumin and equine lysozyme. J. Mol. Biol. (1999) 292, 1137–1148.

Mobley, H. L., Island, M. D., and Hausinger, R. P. Molecular biology of microbial ureases. Microbiol. Rev. (1995) 59, 451–480.

Mochiduki, K., Sugawara, K., Kameda, Y., Usuki, T., and Uemura, O. Hydration structure of the potassium ion determined by combined analysis of neutron and x-ray diffraction data. J. Phys. Soc. Jpn. (2001) 70, Suppl. A, 371–373.

Mochly-Rosen, D., Miller, K. G., Scheller, R. H., Khaner, H., Lopez, H., and Smith, B. L. p65 fragments, homologous to the C2 region of protein kinase C, bind to the intracellular receptors for protein kinase C. Biochemistry (1992) 31, 8120–8124.

Moews, P. C., and Kretsinger, R. H. Refinement of the structure of carp muscle calcium binding parvalbumin by model building and difference Fourier analysis. J. Mol. Biol. (1975) 91, 2, 201–228.

Mohs, A. M., and Lu, Z. R. Gadolinium(III)-based blood-pool contrast agents for magnetic resonance imaging: status and clinical potential. Expert Opin. Drug Deliv. (2007) 4, 2, 149–164.

Moldoveanu, T., Hosfield, C. M., Lim, D., Elce, J. S., Jia, Z., and Davies, P. L. A Ca^{2+} switch aligns the active site of calpain. Cell (2002) 108, 5, 649–660.

Møller, J. V., Nissen, P., Sørensen, T. L., and le Maire, M. Transport mechanism of the sarcoplasmic reticulum Ca^{2+}-ATPase pump. Curr. Opin. Struct. Biol. (2005) 15, 4, 387–393.

Moncrieffe, M. C., Venyaminov, S. Y., Miller, T. E., Guzman, G., Potter, J. D., and Prendergast, F. G. Optical spectroscopic characterization of single tryptophan mutants of chicken skeletal troponin C: evidence for interdomain interaction. Biochemistry (1999) 38, 11973–11983.

Moncrieffe, M. C., Juranic, N., Kemple, M. D., Potter, J. D., Macura, S., and Prendergast, F. G. Structure–fluorescence correlations in a single tryptophan mutant of carp parvalbumin: solution structure, backbone and side-chain dynamics. J. Mol. Biol. (2000) 297, 1, 147–163.

Monroe, D. M., and Hoffman, M. What does it take to make the perfect clot? Arterioscler. Thromb. Vasc. Biol. (2006) 26, 41–48.

Montich, G. G., and Marsh, D. Interaction of α-lactalbumin with phosphatidylglycerol: influence of protein binding on the lipid phase transition and lipid acyl chain mobility. Biochemistry (1995) 34, 40, 13139–13145.

Moore, C. P., Rodney, G., Zhang, J. Z., Santacruz-Toloza, L., Strasburg, G., and Hamilton, S. L. Apocalmodulin and Ca^{2+} calmodulin bind to the same region on the skeletal muscle Ca^{2+} release channel. Biochemistry (1999) 38, 26, 8532–8537.

Moore, S., and Pressman, B. C. Mechanism of action of valinomycin on mitochondria. Biochem. Biophys. Res. Commun. (1964) 15, 562–567.

Mooren, F. C., and Kinne, R. K. Cellular calcium in health and disease. Biochim. Biophys. Acta (1998) 1406, 2, 127–151.

Morand, E. F., Jefferiss, C. M., Dixey, J., Mitra, D., and Goulding, N. J. Impaired glucocorticoid induction of mononuclear leukocyte lipocortin-1 in rheumatoid arthritis. Arthritis Rheum. (1994) 37, 207–211.

Morgan, D. W., Welton, A. F., Heick, A. E., and Christakos, S. Specific in vitro activation of Ca,Mg-ATPase by vitamin D–dependent rat renal calcium binding protein (calbindin D_{28K}). Biochem. Biophys. Res. Commun. (1986) 138, 547–553.

Moroz, O. E., Antson, A. A., Murshudov, G. N., Maitland, N. J., Dodson, G. G., and Wilson, K. S. The three-dimensional structure of human S100A12. Acta Crystallogr. D (2001) 57, 20–29.

Morozova, L. A., Shnyrov, V. L., Gusev, N. B., and Permyakov, E. A. Thermal transitions in cardiac troponin and its subunits. In: Heilmeyer, L. M. G., Jr. (Ed.), Cellular Regulation by Protein Phosphorylation. NATO ASI Series, Vol. H56. Springer-Verlag, Berlin, 1991, pp. 67–71.

Morris, B. W., Gray, T. A., and MacNeil, S. Glucose-dependent uptake of chromium in human and rat insulin-sensitive tissues. Clin. Chem. (1993) 84, 477–482.

Mortimer, J. C., Laohavisit, A., MacPherson, N., Webb, A., Brownlee, C., Battey, N. H., and Davies, J. M. Annexins: multifunctional components of growth and adaptation. J. Exp. Bot. (2008) 59, 3, 533–544.

Moss, S. E. Annexins. Trends Cell Biol. (1997) 7, 87–89.

Moss, S. E., and Morgan, R. O. The annexins. Genome Biol. (2004) 5, 4, 219.

Motamed, K., and Sage, E. H. SPARC inhibits endothelial cell adhesion but not proliferation through a tyrosine phosphorylation-dependent pathway. J. Cell. Biochem. (1998) 70, 543–552.

Moulis, J. M., Sieker, L. C., Wilson, K. S., and Dauter, Z. Crystal structure of the 2[4Fe–4S] ferredoxin from *Chromatium vinosum*: evolutionary and mechanistic inferences for [3/4Fe–4S] ferredoxins. Protein Sci. (1996) 5, 9, 1765–1775.

Moult, J., Fidelis, K., Kryshtafovych, A., Rost, B., Hubband, T., and Tramontano, A. Critical assessment of methods of protein structure prediction (CASP): round VII. Proteins (2007) 69, Suppl. 8, 3–7.

Moussu, G. Sur la fonction parathyroídienne. C. R. Soc. Biol. (1898) 5, 867–869.

Muchmore, D. C., Malencik, D. A., and Anderson, S. R. ^{1}H NMR studies of mastoparan binding by calmodulin. Biochem. Biophys. Res. Commun. (1986) 137, 3, 1069–1078.

Mueller, M., Martens, L., and Apweiler, R. Annotating the human proteome: beyond establishing a parts list. Biochim. Biophys. Acta (2007) 1774, 2, 175–191.

Mularoni, L., Guigó, R., and Albà, M. M. Mutation patterns of amino acid tandem repeats in the human proteome. Genome Biol. (2006) 7, 4, R33.

Muñoz, V. (Ed.) Protein Folding, Misfolding and Aggregation: Classical Themes and Novel Approaches. RSC Biomolecular Sciences Series. Royal Society of Chemistry, London, 2008.

Muñoz-Páez, A., Pappalardo, R. R., and Sanchez Marcos, E. Determination of the second hydration shell of Cr^{3+} and Zn^{2+} in aqueous solutions by extended x-ray absorption fine structure. J. Am. Chem. Soc. (1995) 117, 47, 11710–11720.

Murakami, K., and Berliner, L. J. A distinct zinc binding site in the α-lactalbumin regulates calcium binding: Is there a physiological role for this control? Biochemistry (1983) 23, 14, 3370–3374.

Muralikrishna Adibhatla, R., and Hatcher, J. F. Phospholipase A_2, reactive oxygen species, and lipid peroxidation in cerebral ischemia. Free Radic. Biol. Med. (2006) 40, 3, 376–387.

Muranyi, A., Finn, B. E., Gippert, G. P., Forsen, S., Stenflo, J., and Drakenberg, T. Solution structure of the N-terminal EGF-like domain from human factor VII. Biochemistry (1998) 37, 10605–10615.

Murthy, R., Nunez, R., Szklaruk, J., et al. Yttrium-90 microsphere therapy for hepatic malignancy: devices, indications, technical considerations, and potential complications. RadioGraphics (2005) 25, S41–S55.

Musat, G. V., Neumann, J. M., Smith, J. C., and Sanson, A. Structure of human annexin I: comparison of homology modelling and crystallographic experiment. Biochimie (1997) 79, 691–703.

Mutti, A., and Corradi, M. Recent developments in human biomonitoring: non-invasive assessment of target tissue dose and effects of pneumotoxic metals. Med. Lav. (2006) 97, 2, 199–206.

Nagai, T., Makinose, M. and Hasselbach, W. Der physiologische Erschlaffungs-Faktor und die Muskelgrana. Biochem. Biophys. Acta (1960) 43, 223–238.

Nagaoka, M. H., and Maitani, T. Binding affinity of aluminium to human serum transferrin and effects of carbohydrate chain modification as studied by HPLC/high-resolution ICP-MS: speciation of aluminium in human serum. J. Inorg. Biochem. (2005) 99, 9, 1887–1894.

Nagar, B., Overduin, M., Ikura, M., and Rini, J. M. Structural basis of calcium-induced E-cadherin rigidification and dimerization. Nature (1996) 380, 6572, 360–364.

Naik, U. P., Patel, P. M., and Parise, L. V. Identification of a novel calcium-binding protein that interacts with the integrin alpha(IIb) cytoplasmic domain. J. Biol. Chem. (1997) 272, 4651–4654.

Naitoh, Y. Ionic control of the reversal response of cilia in *Paramecium caudatum*: a calcium hypothesis. J. Gen. Physiol. (1968) 51, 85–103.

Nakagawa, T., Fujiu, K., Cole, E. S., and Numata, O. Involvement of a 25 kDa tetrahymena Ca^{2+}-binding protein in pronuclear exchange. Cell Struct. Funct. (2008) 33, 151–162.

Nakamura, K., and Go, N. Function and molecular evolution of multicopper blue proteins. Cell. Mol. Life Sci. (2005) 62, 2050–2066.

Nakamura, K., and Watanabe, S. Myosin-like protein and actin-like protein from *Eschericia coli* K12 C600. J. Biochem. (1978) 83, 1459–1470.

Nakanishi, M., and Rosenberg, D. W. Roles of $cPLA_2\alpha$ and arachidonic acid in cancer. Biochim. Biophys. Acta. (2006) 1761, 11, 1335–1343.

Nakashima, K., Maekawa, H., and Yazawa, M. Chimeras of yeast and chicken calmodulin demonstrate differences in activation mechanisms of target enzymes. Biochemistry (1996) 35, 17, 5602–5610.

Nakashima, K., Ishida, H., Ohki, S., Hikichi, K., and Yazawa, M. Calcium binding induces interaction between the N- and C-terminal domains of yeast calmodulin and modulates its overall conformation. Biochemistry (1999) 38, 1, 98–104.

Nakatani, Y., Tanioka, T., Sunaga, S., Murakami, M., and Kudo, I. Identification of a cellular protein that functionally interacts with the C2 domain of cytosolic phospholipase $A_2\alpha$. J. Biol. Chem. (2000) 275, 2, 1161–1168.

Nakayama, S., and Tomita, T. Regulation of intracellular free magnesium concentration in the taenia of guinea-pig caecum. J. Physiol. (Lond.) (1991) 435, 559–572.

Nakayama, S., Moncrief, N. D., and Kretsinger, R. H. Evolution of EF-hand calcium-modulated proteins: II. Domains of several subfamilies have diverse evolutionary histories. J. Mol. Evol. (1992) 34, 5, 416–448.

Nalefski, E. A., and Falke, J. J. The C2 domain calcium-binding motif: structural and functional diversity. Protein Sci. (1996) 5, 2375–2390.

Naoi, M., Odori, S., Ogoma, Y., Fujii, T., Fujii, H., Hachimori, A., Kondo, Y., and Hayakawa, T. Conformation and the properties of calmodulin and polypeptides in liposomes. Rep. Prog. Polym. Phys. Jpn. (1989) 32, 667–670.

Nar, H., Messerschmidt, A., Huber, R., van de Kamp, M., and Canters, G. W. Crystal structure analysis of oxidized *Pseudomonas aeruginosa* azurin at pH 5.5 and pH 9.0. A pH-induced conformational transition involves a peptide bond flip. J. Mol. Biol. (1991) 221, 3, 765–772.

Nar, H., Messerschmidt, A., Huber, R., van de Kamp, M., and Canters, G. W. Crystal structure of *Pseudomonas aeruginosa* apo-azurin at 1.85 Å resolution. FEBS Lett. (1992) 306, 2–3, 119–124.

Nara, M., and Tanokura, M. Infrared spectroscopic study of the metal coordination structures of calcium binding proteins. Biochem. Biophys. Res. Commun. (2008) 369, 225–239.

Nara, M., Tasumi, M., Tanokura, M., Hiraoki, T., Yazawa, M., and Tsutsumi, A. Infrared studies of interaction between metal ions and Ca^{2+}-binding proteins: marker bands for identifying the types of coordination of the side-chain COO^- groups to metal ions in pike parvalbumin ($pI = 4.10$). FEBS Lett. (1994) 349, 1, 84–88.

Narasimhan, J., Antholine, W. E., and Chitambar, C. R. Effect of gallium on the tyrosyl radical of the iron-dependent M2 subunit of ribonucleotide reductase. Biochem. Pharmacol. (1992) 44, 2403–2408.

Náray-Szabó, St., Taylor, W. H., and Jackson, W. W. VIII. The structure of cyanite. Z. Kristal. (1929) 71, 117–130.

Narayan, V. A., Kriwacki, R. W., and Caradonna, J. P. Structures of zinc finger domains from transcription factor Sp1: insights into sequence-specific protein–DNA recognition. J. Biol. Chem. (1997) 272, 7801–7809.

Navarro, C., Wu, L. F., and Mandrand-Berthelot, M. A. The nik operon of *Escherichia coli* encodes a periplasmic binding-protein-dependent transport system for nickel. Mol. Microbiol. (1993) 9, 6, 1181–1191.

Naylor, C. E., Jepson, M., Crane, D. T., Titball, R. W., Miller, J., Basak, A. K., and Bolgiano, B. Characterisation of the calcium-binding C-terminal domain of *Clostridium perfringens* α-toxin. J. Mol. Biol. (1999) 294, 3, 757–770.

Nef, S., Fiumelli, H., de Castro, E., Raes, M. B., and Nef, P. Identification of neuronal calcium sensor (NCS-1) possibly involved in the regulation of receptor phosphorylation. J. Recept. Signal Transduct. Res. (1995) 15, 1-4, 365–378.

Nelsestuen, G. L. Role of γ-carboxyglutamic acid: an unusual protein transition required for the calcium-dependent binding of prothrombin to phospholipid. J. Biol. Chem. (1976) 251, 18, 5648–5656.

Nelsestuen, G. L. Calcium function in blood coagulation. In: Metal Ions in Biological Systems, Vol. 17. New York, 1984, pp. 353–380.

Nelson, D. J., Opella, S. J., and Jardetsky, O. ^{13}C nuclear magnetic resonance study of molecular motions and conformational transitions in muscle calcium binding proteins. Biochemistry (1976) 15, 25, 5552–5560.

Nelson, M. R., and Creutz, C. E. Combinatorial mutagenesis of the four domains of annexin IV: effects on chromaffin granule binding and aggregating activities. Biochemistry (1995) 34, 9, 3121–3132.

Neuman, W. F., and Neuman, M. W. The nature of the mineral phase of bone. Chem. Rev. (1953) 53, 1–45.

Newman, R. A., Van Scyoc, W. S., Sorensen, B. R., Jaren, O. R., and Shea, M. A. Interdomain cooperativity of calmodulin bound to melittin preferentially increases calcium affinity of sites I and II. Proteins (2008) 71, 4, 1792–1812.

Newton, A. C. Protein kinase C: structure, function, and regulation. J. Biol. Chem. (1995) 270, 48, 28495–28498.

Newton, A. C. Regulation of the ABC kinases by phosphorylation: protein kinase C as a paradigm. Biochem. J. (2003) 370, 361–371.

Ng, K. K., and Weis, W. I. Structure of a selectin-like mutant of mannose-binding protein complexed with sialylated and sulfated Lewis(x) oligosaccharides. Biochemistry (1997) 36, 5, 979–988.

Ng, K. K., Drickamer, K., and Weis, W. I. Structural analysis of monosaccharide recognition by rat liver mannose-binding protein. J. Biol. Chem. (1996) 271, 2, 663–674.

Nicolaysen, R. Studies upon the mode of action of vitamin D: III. The influence of vitamin D on the absorption of calcium and phosphorous in the rat. Biochem. J. (1937) 31, 122–129.

Niedergerke, R. Movements of Ca in frog heart ventricles at rest and during contractures. J. Physiol. (1963) 167, 515–550.

Nielsen, S. P. The biological role of strontium. Bone (2004) 35, 583–588.

Nies, D. H. Efflux-mediated heavy metal resistance in prokaryotes. FEMS Microbiol. Rev. (2003) 27, 2–3, 313–339.

Nishad Fathima, N., Balaraman, M., Raghava Rao, J., and Unni Nair, B. Effect of zirconium(IV) complexes on the thermal and enzymatic stability of type I collagen. J. Inorg. Biochem. (2003) 95, 1, 47–54.

Nishimoto, S. K., and Nishimoto, M. Matrix Gla protein C-terminal region binds to vitronectin: co-localization suggests binding occurs during tissue development. Matrix Biol. (2005) 24, 5, 353–361.

Nishizuka, Y. Intracellular signaling by hydrolysis of phospholipids and activation of PKC. Science (1992) 258, 607–614.

Noble, M. E., Cleasby, A., Johnson, L. N., Egmond, M. R., and Frenken, L. G. The crystal structure of triacylglycerol lipase from *Pseudomonas glumae* reveals a partially redundant catalytic aspartate. FEBS Lett. (1993) 331, 1–2, 123–128.

Nockolds, C. E., Kretsinger, R. H., Coffee, C. J., and Bradshaw, R. A. Structure of a calcium-binding carp myogen. Proc. Natl. Acad. Sci. USA (1972) 69, 3, 581–584.

Nomikos, M., Blayney, L. M., Larman, M. G., Campbell, K., Rossbach, A., Saunders, C. M., Swann, K., and Lai, F. A. Role of phospholipase C-ζ domains in Ca^{2+}-dependent phosphatidylinositol 4,5-bisphosphate hydrolysis and cytoplasmic Ca^{2+} oscillations. J. Biol. Chem. (2005) 280, 35, 31011–31018.

Norose, K., Lo, W.-K., Clark, J. I., Sage, E. H., and Howe, C. C. Lenses of SPARC-null mice exhibit an abnormal cell surface-basement membrane interface. Exp. Eye Res. (2000) 71, 295–307.

Norris, V., Chen, M., Goldberg, M., Voskuil, J., McGurk, G., and Holland, I. B. Calcium in bacteria: A solution to which problem? Mol. Microbiol. (1991) 5, 4, 775–778.

Norris, V., Grant, S., Freestone, P., Canvin, J., Sheikh, F. N., Toth, I., Trinei, M., Modha, K., and Norman, R. I. Calcium signalling in bacteria. J. Bacteriol. (1996) 178, 13, 3677–3682.

Nugent, J. H., Rich, A. M., and Evans, M. C. Photosynthetic water oxidation: towards a mechanism. Biochim. Biophys. Acta (2001) 1503, 1–2, 138–146.

Numata, O., Hanyu, K., Takeda, T., and Watanabe, Y. Tetrahymena calcium-binding proteins, TCBP-25 and TCBP-23. Methods Cell Biol. (2000) 62, 455–465.

Oberholtzer, J. C., Buettger, C., Summers, M. C., and Matschinsky, F. M. The 28-kDa calbindin-D is a major calcium-binding protein in the basilar papilla of the chick. Proc. Natl. Acad. Sci. USA (1988) 85, 10, 3387–3390.

Oberwinkler, J., and Stavenga, D. G. Calcium transients in the rhabdomeres of dark- and light-adapted fly photoreceptor cells. J. Neurosci. (2000) 20, 5, 1701–1709.

Ochoa, W. F., Corbalán-Garcia, S., Eritja, R., Rodríguez-Alfaro, J. A., Gómez-Fernández, J. C., Fita, I., and Verdaguer, N. Additional binding sites for anionic phospholipids and calcium ions in the crystal structures of complexes of the C_2 domain of protein kinase Cα. J. Mol. Biol. (2002) 320, 2, 277–291.

O'Day, D. H., and Myre, M. A. Calmodulin-binding domains in Alzheimer's disease proteins: extending the calcium hypothesis. Biochem. Biophys. Res. Commun. (2004) 320 1051–1054.

O'Doherty, J., Youmans, S. J., Armstrong, W. M., Stark, R. J. Calcium regulation during stimulus-secretion coupling: continuous measurement of intracellular calcium activities. Science (1980) 209, 4455, 510–513.

Ogata, H., Mizoguchi, Y., Mizuno, N., Miki, K., Adachi, S., Yasuoka, N., Yagi, T., Yamauchi, O., Hirota, S., and Higuchi, Y. Structural studies of the carbon monoxide complex of [NiFe]hydrogenase from *Desulfovibrio vulgaris*. Miyazaki F: Suggestion for the initial activation site for dihydrogen. J. Am. Chem. Soc. (2002) 124, 11628–11635.

Ogawa, H., and Toyoshima, C. Homology modeling of the cation binding sites of Na^+ K^+-ATPase. Proc. Natl. Acad. Sci. USA (2002) 99, 15977–15982.

Ogoma, Y., Kobayashi, H., Fujii, T., Kondo, Y., Hachimori, A., Shimizu, T., and Hatano, M. Binding study of metal ions to S100 protein: ^{43}Ca, ^{25} Mg, ^{67}Zn and ^{39}K n.m.r. Int. J. Biol. Macromol. (1992) 14, 279–286.

Ohashi, W., Inouye S., Yamazaki, T., and Hirota, H. NMR analysis of the Mg^{2+}-binding properties of aequorin, a Ca^{2+}-binding photoprotein. J. Biochem. (2005) 138, 613–620.

Ohki, S., Ikura, M., and Zhang, M. Identification of Mg^{2+}-binding sites and the role of Mg^{2+} on target recognition by calmodulin. Biochemistry (1997) 36, 14, 4309–4316.

Okazaki, K., Watanabe, M., Ando, Y., Hagiwara, M., Terasawa, M., and Hidaka, H. Full sequence of neurocalcin, a novel calcium-binding protein abundant in central nervous system. Biochem. Biophys. Res. Commun. (1992) 185, 1, 147–153.

O'Keefe, S. J., Tamura, J., Kincaid, R. L., Tocci, M. J., and O'Neill, E. A. FK-506- and CsA-sensitive activation of the interleukin-2 promoter by calcineurin. Nature (1992) 357, 692–694.

Olah, G. A., and Trewhella, J. A model structure of the muscle protein complex $4Ca^{2+}$. troponin C.troponin I derived from small-angle scattering data: implications for regulation. Biochemistry (1994) 33, 43, 12800–12806.

Olah, G. A., Rokop, S. E., Wang, C. L. A., Blechner, S. L., and Trewhella, J. Troponin I encompasses an extended troponin C in the Ca^{2+}-bound complex: a small-angle x-ray and neutron scattering study. Biochemistry (1994) 33, 27, 8233–8239.

Omelon, S., Georgiou, J., Henneman, Z. J., Wise, L. M., Sukhu, B., Hunt, T., Wynnyckyj, C., Holmyard, D., Bielecki, R., and Grynpas, M. D. Control of vertebrate skeletal mineralization by polyphosphates. PLoS One (2009) 4, 5, e5634.

O'Neil, K. F., and DeGrado, W. F. How calmodulin binds its targets: sequence independent recognition of amphiphilic α-helices. Trends Biochem. Sci. (1990) 15, 2, 59–64.

Onck, L. A., and Smith, R. J. Calmodulin and calcium mediated regulation in prokaryotes. J. Gen. Microbiol. (1992) 138, 6, 1039–1049.

Opella, S. J., and Marassi, F. M. Structure determination of membrane proteins by NMR spectroscopy. Chem. Rev. (2004) 104, 8, 3587–3606.

Opella, S. J., Nelson, D. J., and Jardetsky, O. Carbon magnetic resonance study of the conformational changes in carp muscle calcium-binding parvalbumin. J. Am. Chem. Soc. (1974) 96, 22, 7157–7158.

Ordal, G. W. Calcium ion regulates chemotactic behaviour in bacteria. Nature (1977) 270, 66–67.

Osawa, M., Swindells, M. B., Tanakawa, J., Tanaka, T., Mase, T., Furuya, T., and Ikura, M. Solution structure of calmodulin-W-7 complex: the basis of diversity in molecular recognition. J. Mol. Biol. (1998) 276, 165–176.

Osawa, M., Tong, K. I., Lilliehook, C., Wasco, W., Buxbaum, J. D., Cheng, H.-Y. M., Penninger, J. M., Ikura, M., and Ames, J. B. Calcium-regulated DNA binding and oligomerization of the neuronal calcium-sensing protein, calsenin/DREAM/KChIP3. J. Biol. Chem. (2001) 276, 44, 41005–41013.

Osterhout, W. J. V. On the penetration of inorganic salts into living protoplasm. Z. Phys. Chem. (1910) 70, 408–413.

Ostrovsky, A. V., Kalinichenko, L. P., Emelyanenko, V. I., Klimanov, A. V. and Permyakov, E. A. Environment of tryptophan residues in various conformational states of α-lactalbumin studied by time-resolved and steady-state fluorescence spectroscopy. Biophys. Chem. (1988) 30, 2, 105–112.

Otterbein, L. R., Kordowska, J., Witte-Hoffman, C., Wang, C. L., and Dominguez, R. Crystal structures of S100A6 in the Ca^{2+}-free and Ca^{2+}-bound states: the calcium sensor mechanism of S100 proteins revealed at atomic resolution. Structure (2002) 10, 557–567.

Ou, W. J., Bergeron, J. J. M., Li, Y., Kang, C. Y., and Thomas, D. Y. Conformational changes induced in the endoplasmic reticulum luminal domain of calnexin by Mg-ATP and Ca^{2+}. J. Biol. Chem. (1995) 270, 30, 18051–18059.

Outten, C. E., and O'Halloran, T. V. Femtomolar sensitivity of metalloregulatory proteins controlling zinc homeostasis. Science (2001) 292, 2488–2492.

Outten, C. E., Tobin, D. A., Penner-Hahn, J. E., and O'Halloran, T. V. Characterization of the metal receptor sites in *Escherichia coli* Zur, an ultrasensitive zinc(II) metalloregulatory protein. Biochemistry (2001) 40, 35, 10417–10423.

Ozawa, T., Fukuda, M., Nara, M., Nakamura, A., Komine, Y., Kohama, K., and Umezawa, Y. How can Ca^{2+} selectively activate recoverin in the presence of Mg^{2+}? Surface plasmon resonance and FT-IR spectroscopic studies. Biochemistry (2000) 39, 47, 14495–14503.

Paarmann, I., Lye, M. F., Lavie, A., and Konrad, M. Structural requirements for calmodulin binding to membrane-associated guanylate kinase homologs. Protein Sci. (2008) 17, 11, 1946–1954.

Pabo, C. O., Peisach, E., and Grant, R. A. Design and selection of novel Cys_2His_2 zinc finger proteins. Annu. Rev. Biochem. (2001) 70, 313–340.

Page, J. S., Masselon, C. D., and Smith, R. D. FTICR mass spectrometry for qualitative and quantitative bioanalyses. Curr. Opin. Biotechnol. (2004) 15, 3.

Page, M. J., and Di Cera, E. Role of Na^+ and K^+ in enzyme function. Physiol. Rev. (2006) 86, 1049–1092.

Pałasz, A., and Czekaj, P. Toxicological and cytophysiological aspects of lanthanides action. Acta Biochim. Pol. (2000) 47, 4, 1107–1114.

Palczewska, M., Groves, P., Batta, G., Heise, B., and Kuznicki, J. Calretinin and calbindin D28k have different domain organizations. Protein Sci. (2003) 12, 180–184.

Palmgren, M. G., and Axelsen, K. B. Evolution of P-type ATPases. Biochim. Biophys. Acta (1998) 1365, 37–45.

Palmiter, R. D., and Findley, S. D. Cloning and functional characterization of a mammalian zinc transporter that confers resistance to zinc. EMBO J. (1995) 14, 639–649.

Pannequin, J., Kovac, S., Tantiongco, J. P., Norton, R. S., Shulkes, A., Barnham, K. J., and Baldwin, G. S. A novel effect of bismuth ions: selective inhibition of the biological activity of glycine-extended gastrin. J. Biol. Chem. (2004) 279, 4, 2453–2460.

Paoletti, P., Vergnano, A. M., Barbour, B., and Casado, M. Zinc at glutamatergic synapses. Neuroscience (2009) 158, 1, 126–136.

Paoli, M., Liddington, R., Tame, J., Wilkinson, A., and Dodson, G. Crystal structure of T state haemoglobin with oxygen bound at all four haems. J. Mol. Biol. (1996) 256, 775–792.

Papillon, M. F. Recherches expérimentales sur les modifications de la composition immédiate des os. C. R. Acad. Sci. (1870) 71, 372–374.

Papp-Wallace, K. M., and Maguire, M. E. Manganese transport and the role of manganese in virulence. Annu. Rev. Microbiol. (2006) 60, 187–209.

Pardon, E., Haezebrouck, P., De Baetselier, A., Hooke, S. D., Fancourt, K. T., Desmet, J., Dobson, C. M., Van Dael, H., and Joniau, M. A Ca^{2+}-binding chimera of human lysozyme and bovine α-lactalbumin that can form a molten globule. J. Biol. Chem. (1995) 270, 10514–10524.

Park, H., Wu, S., Dunker, A. K., and Kang, C. Polymerization of calsequestrin: implications for Ca^{2+} regulation. J. Biol. Chem. (2003) 278, 18, 16176–16182.

Park, H. Y., Kim, S. A., Korlach, J., Rhoades, E., Kwok, L. W., Zipfel, W. R., Waxham, M. N., Webb, W. W., and Pollack, L. Conformational changes of calmodulin upon Ca^{2+} binding studied with a microfluidic mixer. Proc. Natl. Acad. Sci. USA (2008) 105, 2, 542–547.

Park, J. H., Pulvermüller, A., Scheerer, P., Rausch, S., Giessl, A., Höhne, W., Wolfrum, U., Hofmann, K. P., Ernst, O. P., Choe, H. W., and Krauss, N. Insights into functional aspects of centrins from the structure of N-terminally extended mouse centrin 1. Vision Res. (2006) 46, 27, 4568–4674.

Pastrana-Rios, B., Ocaña, W., Rios, M., Vargas, G. L., Ysa, G., Poynter, G., Tapia, J., and Salisbury, J. L. Centrin: its secondary structure in the presence and absence of cations. Biochemistry (2002) 41, 22, 6911–6919.

Patil, S., Takezawa, D., and Poovaiah, B. W. Chimeric plant calcium-calmodulin-dependent protein-kinase gene with a neural visinin-like calcium-binding domain. Proc. Natl. Acad. Sci. USA (1995) 92, 4897–4901.

Pattanayek, R., Elrod, M., and Stubbs, G. Characterization of a putative calcium-binding site in tobacco mosaic virus. Proteins (1992) 12, 128–132.

Patzer, S. I., and Hantke, K. The zinc-responsive regulator Zur and its control of the znu gene cluster encoding the ZnuABC zinc uptake system in *Escherichia coli*. J. Biol. Chem. (2000) 275, 32, 24321–24332.

Pauls, T. L., Durussel, I., Cox, J. A., Clark, I. D., Szabo, A. G., Gagne, S. M., Sykes, B. D., and Berchtold, M. W. Metal binding properties of recombinant rat parvalbumin wild-type and F102W mutant. J. Biol. Chem. (1993) 268, 20897–20903.

Pauls, T. L., Durussel, I., Berchtold, M. W., and Cox, J. A. Inactivation of individual Ca^{2+}-binding sites in the paired EF-hand sites of parvalbumin reveals asymmetrical metal-binding properties. Biochemistry (1994) 33, 10393–10400.

Pauls, T. L., Durussel, I., Clark, I. D., Szabo, A. G., Berchtold, M. W., and Cox, J. A. Site-specific replacement of amino acid residues in the CD site of rat parvalbumin changes the metal specificity of this Ca^{2+}/Mg^{2+}-mixed site toward a Ca^{2+}-specific site. Eur. J. Biochem. (1996) 242, 249–255.

Payandeh, J., and Pai, E. F. A structural basis for Mg^{2+} homeostasis and the CorA translocation cycle. EMBO J. (2006) 25, 3762–3773.

Pearlstone, J. R., and Smillie, L. B. Evidence for two-site binding of troponin I inhibitory peptides to the N and C domains of troponin C. Biochemistry (1995) 34, 21, 6932–6940.

Pearlstone, J. R., Borgford, T., Chandra, M., Oikawa, K., Kay, C. M., Herzberg, O., Moult, J., Herklotz, A., Reinach, F. C., and Smillie, L. B. Construction and characterization of a spectral probe mutant of troponin C: application to analyses of mutants with increased Ca^{2+} affinity. Biochemistry (1992a) 31, 6545–6553.

Pearlstone, J. R., McCubbin, W. D., Kay, C. M., Sykes, B. D., and Smillie, L. B. Spectroscopic analysis of a methionine-48 to tyrosine mutant of chicken troponin C. Biochemistry (1992b) 31, 9703–9708.

Pearlstone, J. R., Sykes, B. D., and Smillie, L. B. Interactions of structural C and regulatory N domains of troponin C with repeated sequence motifs in troponin I. Biochemistry (1997) 36, 24, 7601–7606.

Pearlstone, J. R., Chandra, M., Sorenson, M. M., and Smillie, L. B. Biological function and site II Ca^{2+}-induced opening of the regulatory domain of skeletal troponin C are impaired by invariant site I or II Glu mutations. J. Biol. Chem. (2000) 275, 35106–35115.

Pearson, M. A., Michel, L. O., Hausinger, R. P., and Karplus, P. A. Structures of Cys319 variants and acetohydroxamate-inhibited *Klebsiella aerogenes* urease. Biochemistry (1997) 36, 8164–8172.

Pechére, J. F., Capony, J. P., and Demaille, J. Evolutionary aspects of the structure of muscular parvalbumins. Syst. Zool. (1973) 22, 2, 533–548.

Peck, E. J., Jr., and Ray, W. J., Jr. Metal complexes of phosphoglucomutase in vivo. J. Biol. Chem. (1971) 246, 1160–1167.

Pellegrini, A., Thomas, U., Bramaz, N., Hunziker, P., and von Fellenberg, R. Isolation and identification of three bactericidal domains in the bovine α-lactalbumin molecule. Biochim. Biophys. Acta (1999) 1426, 3, 439–448.

Pelton, J. T., and McLean, L. R. Spectroscopic methods for analysis of protein secondary structure. Anal. Biochem. (2000) 277, 2, 167–176.

Perez, J. F., and Sanderson, M. J. The contraction of smooth muscle cells of intrapulmonary arterioles is determined by the frequency of Ca^{2+} oscillations induced by 5-HT and KCl. J. Gen. Physiol. (2005) 125, 555–567.

Perfus-Barbeoch, L., Leonhardt, N., Vavaddeur, A., and Forestier, C. Heavy metal toxicity: cadmium permeates through calcium channels and disturbs the plant water status. Plant J. (2002) 32, 539–548.

Perisic, O., Fong, S., Lynch, D. E., Bycroft, M., and Williams, R. L. Crystal structure of a calcium–phospholipid binding domain from cytosolic phospholipase A2. J. Biol. Chem. (1998) 273, 3, 1596–1604.

Perkins, S. J., Ashton, A. W., Boehm, M. K., and Chamberlain, D. Molecular structures from low angle x-ray and neutron scattering studies. Int. J. Biol. Macromol. (1998) 22, 1, 1–16.

Permyakov, E. A. Parvalbumin and Related Calcium Binding Proteins. Nauka, Moscow, 1985.

Permyakov, E. A. Calcium Binding Proteins. Nauka, Moscow, 1993a.

Permyakov, E. A. Luminescent Spectroscopy of Proteins. CRC Press, Boca Raton, FL, 1993b.

Permyakov, E. A. Method of Intrinsic Protein Luminescence. Nauka, Moscow, 2003a.

Permyakov, E. A. Method of Intrinsic Protein Luminescence. Moscow, Nauka, 2003b.

Permyakov, E. A. α-Lactalbumin. Nova Science Publishers, New York, 2005.

Permyakov, E. A. Parvalbumin. Nova Science Publishers, New York, 2006.

Permyakov, E. A. Metalloproteomics. Wiley, Hoboken, NJ, 2009.

Permyakov, E. A., and Berliner, L. J. Co^{2+} binding to α-lactalbumin. J. Protein Chem. (1994) 13, 3, 277–281.

Permyakov, E. A., Burstein, E. A., Sawada, Y., and Yamazaki, Y. Luminescence of phenylalamine residues in superoxide dismutase from green pea. Biochim. Biophys. Acta (1977) 491, 149–154.

Permyakov, E. A., and Kretsinger, R. H. Cell signaling, beyond cytosolic calcium in eukaryotes. J. Inorg. Biochem. (2009) 103, 77–86.

Permyakov, E. A., Yarmolenko, V. V., Emelyanenko, V. I., Burstein, E. A., Closset, J., and Gerday, C. Fluorescence studies of the calcium binding to whiting (*Gadus merlangus*) parvalbumin. Eur. J. Biochem. (1980) 109, 307–315.

Permyakov, E. A., Yarmolenko, V. V., Kalinichenko, L. P., Morozova, L. A. and Burstein, E. A. Calcium binding to α-lactalbumin: structural rearrangement and association constant evaluation by means of intrinsic protein fluorescence changes. Biochem. Biophys. Res. Commun. (1981) 100, 1, 191–197.

Permyakov, E. A., Medvedkin, V. N., Kalinichenko, L. P., and Burstein, E. A. Comparative study of physiochemical properties of two pike parvalbumins by means of their intrinsic tyrosyl and phenylalanyl fluorescence. Arch. Biochem. Biophys. (1983) 227, 9–20.

Permyakov, E. A., Morozova, L. A., and Burstein, E. A. Cation binding effects on the pH, thermal and urea denaturation transitions in α-lactalbumin. Biophys. Chem. (1985) 21, 1, 21–31.

Permyakov, E. A., Ostrovsky, A. V., and Kalinichenko, L. P. Stopped flow kinetic studies of Ca^{2+} and Mg^{2+} dissociation in cod parvalbumin and bovine α-lactalbumin. Biophys. Chem. (1987a) 28, 2, 225–233.

Permyakov, E. A., Ostrovsky, A. V., Kalinichenko, L. P., and Deikus, G. Y. Kinetics of dissociation of complexes of parvalbumin with calcium and magnesium ions. Mol. Biol. (Mosc.) (1987b) 2, 4, 1017–1022.

Permyakov, E. A., Kalinichenko, L. P., Morozova, L. A., Derezhov, V. Y., Bagelova, J., and Antalik, M. Interactions of copper and zinc cations with calcium binding proteins. Mol. Biol. (Mosc.) (1988a), 22, 4, 984–991.

Permyakov, E. A., Kreimer, D. I., Kalinichenko, L. P., and Shnyrov, V. L. Interaction of calcium binding proteins, parvalbumin and α-lactalbumin with dipalmitoylphosphatidylcholine vesicles. Gen. Physiol. Biophys. (1988b) 7, 1, 95–107.

Permyakov, E. A., Kreimer, D. I., Kalinichenko, L. P., Orlova, A. A., and Shnyrov, V. L. Interaction of parvalbumins with model phospholipid vesicles. Cell Calcium (1989a) 10, 2, 71–79.

Permyakov, E. A., Medvedkin, V. N., Kornejchuk, G. A., Kostzhevskaya, E. G., and Murzin, A. G. Calcium inhibited binding of melittin to parvalbumin: A new role for parvalbumin? Mol. Biol. (Mosc.) (1989b) 23, 3, 693–698.

Permyakov, E. A., Grishchenko, V. M., Kalinichenko, L. P., Orlov, N. Y., Kuwajima, K., and Sugai, S. Calcium-regulated interactions of human α-lactalbumin with bee venom melittin. Biophys. Chem. (1991a) 39, 1, 111–117.

Permyakov, E. A., Medvedkin, V. N., Mitin, Y. V., and Kretsinger, R. H. Noncovalent complex between domain AB and domains CD*EF of parvalbumin. Biochim. Biophys. Acta (1991b) 1076, 67–70.

Permyakov, E. A., Kalinichenko, L. P., Derezhkov, V. Y., Antalik, M., Meinholtz, D. C., and Berliner, L. J. Interaction of cupric ion with parvalbumin. Biophys. Chem. (1992), 42, 1, 189–194.

Permyakov, E. A., Reyzer, I. L., and Berliner, L. J. Effects of Zn(II) on galactosyltransferase activity. J. Protein Chem. (1993) 12, 5, 633–638.

Permyakov, E. A., Permyakov, S. E., Deikus, G. Y., Morozova-Roche, L. A., Grishchenko, V. M., Kalinichenko, L. P., and Uversky, V. N. Ultraviolet illumination-induced reduction of α-lactalbumin disulfide bridges. Proteins: Struct. Funct. Genet. (2003) 51, 498–503.

Permyakov, S. E., and Permyakov, E. A. The use of the free metal–temperature "phase diagrams" for studies of single site metal binding proteins. Protein J. (2006) 26, 1, 1–12.

Permyakov, S. E., Cherskaya, A. M., Senin, I. I., Zargarov, A. A., Shulga-Moskoy, S. V., Alexeev, A. M., Zinchenko, L. V., Lipkin, V. M., Philippov, P. P., Uversky, V. N., and Permyakov, E. A. Effects of mutations in the calcium-binding sites of recoverin on its calcium affinity: evidence for successive filling of the calcium binding sites. Protein Eng. (2000a) 13, 11, 783–790.

Permyakov, S. E., Veprintsev, D. B., Brooks, C. L., Permyakov, E. A., and Berliner, L. J. Zinc binding in bovine α-lactalbumin: sequence homology may not be a predictor of subtle functional features. Proteins (2000b) 40, 1, 106–111.

Permyakov, S. E., Uversky, V. N., Veprintsev, D. B., Cherskaya, A. M., Brooks, C. L., Permyakov, E. A., and Berliner, L. J. Mutating aspartate in the calcium-binding site of α-lactalbumin: effects on the protein stability and cation binding. Protein Eng. (2001) 14, 785–789.

Permyakov, S. E., Pershikova, I. V., Khokhlova, T. I., Uversky, V. N., and Permyakov, E. A. No need to be HAMLET or BAMLET to interact with histones: binding of monomeric α-lactalbumin to histones and basic poly-amino acids. Biochemistry (2004) 43, 19, 5575–5582.

Permyakov, S. E., Makhatadze, G. I., Owenius, R., Uversky, V. N., Brooks, C. L., Permyakov, E. A., and Berliner, L. J. How to improve nature: study of the electrostatic properties of the surface of α-lactalbumin. Protein Eng. Des. Sel. (2005a) 18, 9, 425–433.

Permyakov, S. E., Pershikova, I. V., Zhadan, A. P., Goers, J., Bakunts, A. G., Uversky, V. N., Berliner, L. J., and Permyakov, E. A. Conversion of human α-lactalbumin to an apo-like state in the complexes with basic poly-amino acids: toward understanding of the molecular mechanism of antitumor action of HAMLET. J. Proteome Res. (2005b) 4, 2, 564–569.

Permyakov, S. E., Khokhlova, T. I., Nazipova, A. A., Zhadan, A. P., Morozova-Roche, L. A., and Permyakov, E. A. Calcium-binding and temperature induced transitions in equine lysozyme: new insights from the pCa–temperature "phase diagrams." Proteins: Struct. Funct. Bioinf. (2006) 65, 4, 984–998.

Permyakov, S. E., Nazipova, A. A., Denesyuk, A. I., Bakunts, A. G., Zinchenko, D. V., Lipkin, V. M., Uversky, V. N., and Permyakov, E. A. Recoverin as a redox-sensitive protein. J. Proteome Res. (2007) 6, 1855–1863.

Permyakov, S. E., Bakunts, A. G., Denesyuk, A. I., Knyazeva, E. L., Uversky, V. N., and Permyakov, E. A. Apo-parvalbumin as an intrinsically disordered protein. Proteins (2008) 72, 822–836.

Permyakov, S. E., Bakunts, A. G., Permyakova, M. E., Denesyuk, A. I., Uversky, V. N., and Permyakov, E. A. Metal-induced changes of interdomain cooperativity in parvalbumin. Cell Calcium (2009a) 46, 163–175.

Permyakov, S. E., Karnoup, A. S., Bakunts, A. G., and Permyakov, E. A. Sequence microheterogeneity of parvalbumin pI 5.0 of pike: a mass spectrometric study. Biochim. Biophys. Acta (2009b) 1794, 129–136.

Perry, S. V., and Grey, T. C. A study of the effects of substrate concentration and certain relaxing factors on the magnesium-activated myofibrillar adenosine triphosphatase. Biochem. J. (1956) 64, 184–192.

Persechini, A., Kretsinger, R. H., and Davis, T. N. Calmodulin with deletions in the central helix functionally replace the native protein in yeast cells. Proc. Natl. Acad. Sci. USA (1991) 88, 449–452.

Persechini, A., White, H. D., and Gansz, K. J. Different mechanisms for Ca^{2+} dissociation from complexes of calmodulin with nitric oxide synthase or myosin light chain kinase. J. Biol. Chem. (1996) 271, 1, 62–67.

Persson, M., Tars, K., and Liljas, L. The capsid of the small RNA phage PRR1 is stabilized by metal ions. J. Mol. Biol. (2008) 383, 4, 914–922.

Peter, A. L., and Viraraghavan, T. Thallium: a review of public health and environmental concerns. Environ. Int. (2005) 31, 4, 493–501.

Peters, A. R., Dekker, N., Van den Berg, L., Boelens, R., Kaptein, R., Slotboom, A. J., and De Haas, G. H. Conformational changes in phospholipase A2 upon binding to micellar interfaces in the absence and presence of competitive inhibitors: a ^{1}H and ^{15}N NMR study. Biochemistry (1992) 31, 41, 10024–10030.

Petraglio, G., Bartolini, M., Branduardi, D., Andrisano, V., Recanatini, M., Gervasio, F. L., Cavalli, A., and Parrinello, M. The role of Li^+, Na^+, and K^+ in the ligand binding inside the human acetylcholinesterase gorge. Proteins (2008) 70, 3, 779–785.

Petsko, G. A. Chemistry and biology. Proc. Natl. Acad. Sci. USA (2000) 97, 2, 538–540.

Pettifor, J. M., and Benson, R. Congenital malformations associated with the administration of oral anticoagulants during pregnancy. J. Pediatr. (1975) 86, 459–462.

Phan, E., Ahluwalia, A., and Tarnawski, A. S. Role of SPARC: matricellular protein in pathophysiology and tissue injury healing—implications for gastritis and gastric ulcers. Med. Sci. Monit. (2007) 13, 2, 25–30.

Phiel, C. J., and Klein, P. S. Molecular targets of lithium action. Annu. Rev. Pharmacol. Toxicol. (2001) 41, 789–813.

Philippov, P., and Koch, K. (Eds.) Neuronal Calcium Sensor Proteins. Nova Science Publishers, New York, 2006.

Philippova, M., Joshi, M. B., Kyriakakis, E., Pfaff, D., Erne, P., and Resink, T. J. A guide and guard: the many faces of T-cadherin. Cell Signal. (2009) 21, 7, 1035–1044.

Pickford, A. R., and Campbell, I. D. NMR studies of modular protein structures and their interactions. Chem. Rev. (2004) 104, 8, 3557–3566.

Pidcock, E., and Moore, G. R. Structural characteristics of protein binding sites for calcium and lanthanide ions. J. Biol. Inorg. Chem. (2001) 6, 5–6, 479–489.

Pienaar, R. N. Coccolith production in hymenomonas carterae. Protoplasm (1971) 73, 217–224.

Piermarini, P. M., and Evans, D. H. Osmoregulation of the Atlantic stingray (*Dasyatis sabina*) from the freshwater Lake Jesup of the St. Johns River, Florida. Physiol. Zool. (1998) 71, 5, 553–560.

Piliarik, M., Vaisocherová, H., and Homola, J. Surface plasmon resonance biosensing. Methods Mol. Biol. (2009) 503, 65–88.

Pineda, A. O., Carrell, C. J., Bush, L. A., Prasad, S., Caccia, S., Zhi-Wei Chen, Mathews, F. S., and Di Cera, E. Molecular dissection of Na^+ binding to thrombin. J. Biol. Chem. (2004) 279, 30, 31842–31853.

Pintacuda, G., John, M., Su, X. C., and Otting, G. NMR structure determination of protein–ligand complexes by lanthanide labeling. Acc. Chem. Res. (2007) 40, 3, 206–212.

Pinto, A. P. A., Campana, P. T., Beltramini, L. M., Silber, A. M., and Araújo, A. P. U. Structural characterization of a recombinant flagellar calcium-binding protein from *Trypanosoma cruzi*. Biochim. Biophys. Acta (2003) 1652, 107–114.

Piontek, K., Antorini, M., and Choinowski, T. Crystal structure of a laccase from the fungus *Trametes versicolor* at 1.90-Å resolution containing a full complement of coppers. J. Biol. Chem. (2002) 277, 37663–37673.

Plager, D. A., and Nelsestuen, G. L. Direct enthalpy measurements of factor X and prothrombin association with small and large unilamellar vesicles. Biochemistry (1994) 33, 22, 7005–7013.

Plum, G. E, and Breslauer, K. J. Calorimetry of proteins and nucleic acids. Curr. Opin. Struct. Biol. (1995) 5, 5, 682–690.

Pochet, R., Blachier, F., Malaisse, W., Parmentier, M., Pasteels, B., Pohl, V., Résibois, A., Rogers, J., and Roman, A. Calbindin-D28 in mammalian brain, retina, and endocrine pancreas: immunohistochemical comparison with calretinin. Adv. Exp. Med. Biol. (1989) 255, 435–443.

Podhajcer, O. L., Benedetti, L. G., Girotti, M. R., Prada, F., Salvatierra, E., and Llera, A. S. The role of the matricellular protein SPARC in the dynamic interaction between the tumor and the host. Cancer Metast. Rev. (2008) 27, 4, 691–705.

Politno, M., and King, M. M. Calcineurin-phospholipid interactions: identification of the phospholipid-binding subunit and analyses of a two-stage binding process. J. Biol. Chem. (1990) 265, 7619–7622.

Pollack, H. Microsurgical studies in cell physiology: VI. Calcium ions in living protoplasm. J. Gen. Physiol. (1928) 11, 539–545.

Pollmann, K., Raff, J., Merroun, M., Fahmy, K., and Selenska-Pobell, S. Metal binding by bacteria from uranium mining waste piles and its technological applications. Biotechnol. Adv. (2006) 24, 1, 58–68.

Polpong, P., and Bovornkitti, S. Indoor radon. J. Med. Assoc. Thail. (1998) 81, 1, 47–57.

Polyakov, A. A., Huber, P. A., Marston, S. B., and Gusev, N. B. Interaction of isoforms of S100 protein with smooth muscle caldesmon. FEBS Lett. (1998) 422, 2, 235–239.

Pongs, O., Lindemeier, J., Zhu, X. R., et al. Frequenin: a novel calcium-binding protein that modulates synaptic efficacy in the *Drosophila* nervous system. Neuron (1993) 11, 1, 15–28.

Poole, C. P., Jr., and Farach, H. A. Handbook of Electron Spin Resonance. American Institute of Physics Press, New York, 1994.

Porter, K. R., and Palade, G. E. Studies on the endoplasmic reticulum: III. Its form and distribution in striated muscle cells. J. Biophys. Biochem. Cytol. (1957) 3, 269–300.

Portzehl, H., Cadwell, D. C., and Rüegg, J. C. The dependence of contraction and relaxation of muscle fibres from the crab *Maia squinado* on the internal concentration of free calcium ions. Biochim. Biophys. Acta (1964) 79, 581–591.

Porumb, T. Determination of calcium-binding constants by flow dialysis. Anal. Biochem. (1994) 220, 227–237.

Potts, B., Smith, J., Akke, M., Macke, T., Okazaki, K., and Hidaka, H. The structure of calcyclin reveals a novel homodimeric fold for S100 Ca^{2+}-binding proteins. Nat. Struct. Biol. (1995) 2, 790–796.

Prakash, Y. S., Kannan, M. S., and Sieck, G. C. Regulation of intracellular calcium oscillations in porcine tracheal smooth muscle cells. Am. J. Physiol. (1997) 272, C966–C975.

Prasad, S., Cantwell, A. M., Bush, L. A., Shih, P., Hong Xu, and Di Cera, E. Residue Asp189 controls both substrate binding and the monovalent cation specificity of thrombin. J. Biol. Chem. (2004) 279, 11, 10103–10108.

Prasher, D. C., McCann, R. O., Longiaru, M., and Cormier, M. J. Sequence comparisons of complementary DNAs encoding aequorin isotypes. Biochemistry (1987) 26, 1326–1332.

Prêcheur, B., Cox, J. A., Petrova, T., Mispelter, J., and Craescu, C. T. *Nereis* sarcoplasmic Ca^{2+}-binding protein has a highly unstructured apo state which is switched to the native state upon binding of the first Ca^{2+} ion. FEBS Lett. (1996) 395, 89–94.

Presland, R. B., Bassuk, J. A., Kimball, J. R., and Dale, B. A. Characterization of two distinct calcium binding sites in the amino terminus of human profillarggrin. J. Invest. Dermatol. (1995) 104, 218–222.

Prestrelski, S. J., Byler, D. M., and Thompson, M. P. Effect of metal ion binding on the secondary structure of bovine α-lactalbumin as examined by infrared spectroscopy. Biochemistry (1991) 30, 36, 8797–8804.

Príbram, R. Eine neue Methode zur Bestimmung des Kalkes and der Phosphorsäure in Blutserum. Ber. Veh. Sachs. Akad. Wiss. Math.-Phys. Kl. (1871) 23, 279–284.

Price, P. A., Poser, J. W., and Raman, N. Primary structure of the γ-carboxyglutamic acid-containing protein from bovine bone. Proc. Natl. Acad. Sci. USA (1976) 73, 3374–3375.

Pulvermüller, A., Giessl, A., Heck, M., Wottrich, R., Schmitt, A., Ernst, O. P., Choe, H. W., Hofmann, K. P., and Wolfrum U. Calcium-dependent assembly of centrin-G-protein complex in photoreceptor cells. Mol. Cell. Biol. (2002) 22, 7, 2194–2203.

Putkey, J. A., Ono, T., Vanberkum, M. F. A., and Means, A. R. Functional significance of the central helix in calmodulin. J. Biol. Chem. (1988) 263, 23, 11242–11249.

Putkey, J. A., Liu, W., and Sweeney, H. L. Function of the N-terminal calcium-binding sites in cardiac/slow troponin C assessed in fast skeletal muscle fibers. J. Biol. Chem. (1991) 266, 23, 14881–14884.

Putney, J. W., and Bird, G. S. Cytoplasmic calcium oscillations and store-operated calcium influx. J. Physiol. (2008) 586, 13, 3055–3059.

Quinton, R. Perméabilité de la paroi extérieure de l'invertibré marin, non seulemnet à l'eau encore aus sels. C. R. Acad. Sci. (1900) 131, 905–952.

Quiocho, F. A., Vyas, N. K., Sack, J. S., and Vyas, M. N. Atomic protein structures reveal basic features of binding of sugars and ionic substrates, and calcium cation. Cold Spring Harbor Symp. Quant. Biol. (1987) 52, 453–463.

Radtke, F., Heuchel, R., Georgiev, O., Hergersberg, M., Gariglio, M., Dembic, Z., and Schaffner, W. Cloned transcription factor MTF-1 activates the mouse metallothionein I promoter. EMBO J. (1993) 12, 4, 1355–1362.

Raines, E. W., Lane, T. F., Iruela-Arispe, M. L., Ross, R., and Sage, E. H. The extracellular glycoprotein SPARC interacts with platelet-derived growth factor (PDGF)-AB and -BB and inhibits the binding of PDGF to its receptors. Proc. Natl. Acad. Sci. USA (1992) 89, 1281–1285.

Rakowski, R. F., and Sagar, S. Found: Na^+ and K^+ binding sites of the sodium pump. News Physiol. Sci. (2003) 18, 164–168.

Rall, T. W., and Sutherland, E. W. Formation of a cyclic adenine ribonucleotide by tissue paricles. J. Biol. Chem. (1958) 232, 1065–1076.

Ramachandiran, S., Takezawa, D., Wang, W., and Poovaiah, B. W. Functional domains of plant chimeric calcium/calmodulin-dependent protein kinase: regulation by autoinhibitory and visinin-like domains. J. Biochem. (1997) 121, 5, 984–990.

Ramachandran, G. N., Ramakrishnan, C., and Sasisekharan, V. Stereochemistry of polypeptide chain configurations. J. Mol. Biol. (1963) 7, 95–99.

Ramakrishnan, B., and Qasba, P. K. Crystal structure of lactose synthase reveals a large conformational change in its catalytic component, the β 1,4-galactosyltransferase-I. J. Mol. Biol. (2001) 310, 1, 205–218.

Ramakrishnan, B., Shah, P. S., and Qasba, P. K. α-Lactalbumin (LA) stimulates milk β-1,4-galactosyltransferase I (β 4Gal-T1) to transfer glucose from UDP-glucose to N-acetylglucosamine: crystal structure of β 4Gal-T1× LA complex with UDP-Glc. J. Biol. Chem. (2001) 276, 40, 37665–37671.

Ramakrishnan, S., and Hitchcock-DeGregori, S. E. Structural and functional significance of aspartic acid 89 of the troponin C central helix in Ca^{2+} signaling. Biochemistry (1996) 35, 15515–15521.

Ramana, V. V., and Sastry, K. S. Chromium toxicity in *Neurospora crassa*. J. Inorg. Biochem. (1994) 56, 87–95.

Rasmussen, H. Parathyroid hormone: nature and mechanism of action. Am. J. Med. (1961) 30, 112–128.

Rasmussen, H. Cell communication, calcium ion, and cyclic adenosine monophosphate. Science (1970) 170, 404–412.

Rasmussen, H., and Weisman, D. M. The messenger function of calcium in endocrine systems. Biochem. Act. Horm. (1981) VIII, 1–115.

Ratto, T. V., Langry, K. C., Rudd, R. E., Balhorn, R. L., Allen, M. J., and McElfresh, M. W. Force spectroscopy of the double-tethered concanavalin-A mannose bond. Biophys. J. (2004) 86, 2430–2437.

Raymackers, J. M., Gailly, P., Schoor, M. C., Pette, D., Schwaller, B., Hunziker, W., Celio, M. R. and Gillis, J. M. Tetanus relaxation of fast skeletal muscles of the mouse made parvalbumin deficient by gene inactivation. J. Physiol. (2000) 527, 355–364.

Raymond, K. N., Kappel, M. J., Pecoraro, V. L., Harris, W. R., Carrano, C. J., Weitl, F. L., and Durbin, P. W. Specific sequestering agents for actinide ions. In: Perspective Proceedings of the Actinides Conference. Pergamon Press, Oxford, UK, 1982, pp. 491–507.

Raynal, P., and Pollard, H. B. Annexins: the problem of assessing the biological role for a gene family of multifunctional calcium- and phospholipid-binding proteins. Biochim. Biophys. Acta (1994) 1197, 63–93.

Rebecchi, M. J., and Pentyala, S. N. Structure, function, and control of phosphoinositide-specific phospholipase C. Physiol. Rev. (2000) 80, 1291–1335.

Reddy, G. R., Devi, B. C., and Chetty, C. S. Developmental lead neurotoxicity: alterations in brain cholinergic system. Neurotoxicology (2007) 28, 2, 402–407.

Redfield, C. Using nuclear magnetic resonance spectroscopy to study molten globule states of proteins. Methods (2004) 34, 1, 121–132.

Reed, P. W., and Hardy, H. A. A23187: a divalent cation ionophore. J. Biol. Chem. (1972) 247, 6970–6977.

Reeke, G. N., Becker, J. W., and Edelman, G. M. Changes in the three-dimensional structure of concanavalin A upon demetallization. Proc. Natl. Acad. Sci. USA (1978) 75, 5, 2286–2290.

Rees, D. C., Lewis, M., and Lipscomb, W. N. Refined crystal structure of carboxypeptidase A at 1.54 Å resolution. J. Mol. Biol. (1983) 168, 367–387.

Rees, D. C., Johnson, E., and Lewinson, O. ABC transporters: the power to change. Nat. Rev. Mol. Cell. Biol. (2009) 10, 3, 218–227.

Reese, J. T., and Pearson, W. R. Empirical determination of effective gap penalties for sequence comparison. Bioinformatics (2002) 18, 1500–1507.

Reid, R. E., and Hodges, R. S. Co-operativity and calcium/magnesium binding to troponin C and muscle calcium binding parvalbumin: a hypothesis. J. Theor. Biol. (1980) 84, 401–444.

Reid, R. E., and Procyshyn, R. M. Engineering magnesium selectivity in the helix–loop–helix calcium-binding motif. Arch. Biochem. Biophys. (1995) 323, 115–119.

Reisner, P. D., Christakos, S., and Vanaman, T. C. In Vitro enzyme activation with calbindin-D_{28k}, the vitamin D-dependent 28kDa calcium binding protein. FEBS Lett. (1992) 297, 127–131.

Reitz, F. B., and Pollack, G. H. Labview virtual instruments for calcium buffer calculations. Comput. Methods Prog. Biomed. (2003) 70, 1, 61–69.

Ren, J., Stuart, D. I., and Acharya, K. R. α-Lactalbumin possesses a distinct zinc binding site. J. Biol. Chem. (1993) 268, 26, 19292–19298.

Rescher, U., and Gerke, V. Annexins: unique membrane binding proteins with diverse functions. J. Cell Sci. (2004) 117, Pt. 13, 2631–2639.

Rester, U., Moser, M., Huber, R., and Bode, W. L-Isoaspartate 115 of porcine β-trypsin promotes crystallization of its complex with bdellastasin. Acta Crystallogr. D (2000) 56, 581–588.

Rety, S., Sopkova, J., Renouard, M., Osterloh, D., Gerke, V., and Tabaries, S. The crystal structure of the complex of p11 with the annexin II N-terminal peptide. Nat. Struct. Biol. (1999) 6, 89–95.

Rety, S., Osterloh, D., Arie, J. P., Tabaries, S., Seeman, J., and Russo-Marie, F. Structural basis of the Ca^{2+}-dependent association between S100C (S100A11) and its target, the N-terminal part of annexin I. Struct. Fold. Des. (2000) 8, 175–184.

Reutelingsperger, C. P. M., Hornstra, G., and Hemker, U. C. Isolation and partial purification of a novel anticoagulant from arteries of human umbilical cord. Eur. J. Biochem. (1985) 151, 625–629.

Reuter, H., and Seitz, N. The dependence of calcium efflux from cardiac muscle on temperature and external ion composition. J. Physiol. (1968) 195, 451–470.

Revett, S. P., King, G., Shabanowitz, J., Hunt, D. F., Hartman, K. L., Laue, T. M., and Nelson, D. J. Characterization of a helix–loop–helix (EF-hand) motif of silver hake parvalbumin isoform B. Protein Sci. (1997) 6, 11, 2397–2408.

Reyland, M. E. Protein kinase C isoforms: multi-functional regulators of cell life and death. Front. Biosci. (2009) 14, 2386–2399.

Rhoads, A. R., and Friedberg, F. Sequence motifs for calmodulin recognition. FASEB J. (1997) 11, 331–340.

Rhyner, J. A., Durussel, I., Cox, J. A., Ilg, E. C., Schäfer, B. W., and Heizmann, C. W. Human recombinant α-parvalbumin and nine mutants with individually inactivated calcium- and magnesium-binding sites: biochemical and immunological properties. Biochim. Biophys. Acta (1996) 1313, 3, 179–186.

Ribeiroa, A. C. F., Barrosa, M. C. F., Telesa, A. S. N., Valentea, A. J. M., Loboa, V. M. M., Sobrala, A. J. F. N., and Esteso, M. A. Diffusion coefficients and electrical conductivities for calcium chloride aqueous solutions at 298.15 K and 310.15 K. Electrochim. Acta (2008) 54, 2, 192–196.

Richard, I., Broux, O., Allamand, V., et al. Mutations in the proteolytic enzyme calpain 3 cause limb-girdle muscular dystrophy type 2A. Cell (1995) 81, 27–40.

Richards, T. W., McCaffrey, C., and Bisbee, H. Die Okklusion von Magnesiumoxalat durch Calciuoxalat und die Löslichkeit von Calciumoxalat. Z. Anorg. Chem. (1901) 28, 71–89.

Richardson, C., Behnke, W. D., Freisheim, J. H., and Blumenthal, K. M. The complete amino acid sequence of the α-subunit of pea lectin, *Pisum sativum*. Biochim. Biophys. Acta (1978) 537, 2, 310–319.

Richardson, R. C., King, N. M., Harrington, D. J., Sun, H., Royer, W. E., and Nelson, D. J. X-ray crystal structure and molecular dynamics simulations of silver hake parvalbumin (isoform B). Protein Sci. (2000) 9, 73–82.

Righetti, P. G., Campostrini, N., Pascali, J., Hamdan, M., and Astner, H. Quantitative proteomics: a review of different methodologies. Eur. J. Mass Spectrom. (2004) 10, 3, 335–348.

Ringer, S. Regarding the action of hydrate of soda, hydrate of ammonia, and hydrate of potash on the ventricle of the frog's heart. J. Physiol. (1881) 3, 195–202.

Ringer, S. A further contribution regarding the influence of the different constituents of the blood on the contraction of the heart. J. Physiol. (1882) 4, 29–42.

Ringer, S., and Buxton, D. W. Concerning the action of calcium, potassium, and sodium salts upon the eel's heart and upon the skeletal muscles of the frog. J. Physiol. (1887) 8, 15–19.

Rintala-Dempsey, A. C., Rezvanpour, A., Shaw, G. S. S100–annexin complexes: structural insights. FEBS J. (2008) 275, 20, 4956–4966.

Rivas, M., Mellström, B., Torres, B., Cali, G., Ferrara, A. M., Terracciano, D., Zannini, M., Morreale de Escobar, G., and Naranjo, J. R. The DREAM protein is associated with thyroid enlargement and nodular development. Mol. Endocrinol. (2009) 23, 6, 862–870.

Rizo, J., and Rosenmund, C. Synaptic vesicle fusion. Nat. Struct. Mol. Biol. (2008) 15, 7, 665–674.

Rizo, J., and Südhof, T. C. C_2-domains, structure and function of a universal Ca^{2+}-binding domain. J. Biol. Chem. (1998) 273, 26, 15879–15882.

Rizo, J., Ubach, J., and García, J. Ca^{2+}-binding mode of the C_2A-domain of synaptotagmin. In: Vogel, H. J., ed., Methods in Molecular Biology, Vol. 172, Calcium-Binding Protein Protocols, Vol. 1, Reviews and Case Studies. Humana Press, Totowa, NJ, 2002, pp. 305–316.

Rizzuto, R., Pinton, P., Ferrari, D., Chami, M., Szabadkai, G., Magalhães, P. J., Di Virgilio, F., and Pozzan, T. Calcium and apoptosis: facts and hypotheses. Oncogene (2003) 22, 53, 8619–8627.

Robblee, J. H., Cinco, R. M., and Yachandra, V. K. X-ray spectroscopy-based structure of the Mn cluster and mechanism of photosynthetic oxygen evolution. Biochim. Biophys. Acta (2001) 1503, 1–2, 7–23.

Robison, R. The possible significance of hexosephorphoric esters in ossification. Biochem. J. (1923) 17, 286–293.

Rodriguez-Navarro, A., Quintero, F. J., and Garciadeblas, B. Na^+-ATPases and Na^+/H^+ antiporters in fungi. Biochim. Biophys. Acta (1994) 1187, 203–205.

Roe, M. W., Lemasters, J. J., and Herman, B. Assessment of fura-2 for measurements of cytosolic free calcium. Cell Calcium (1990) 11, 2–3, 63–73.

Rogers, M. S., and Dooley, D. M. Copper-tyrosyl radical enzymes. Curr. Opin. Chem. Biol. (2003) 7, 189–196.

Rogstam, A., Linse, S., Lindqvist, A., Wagner, L., and Berggard, T. Binding of calcium ions and SNAP-25 to the hexa EF-hand protein secretagogin. Biochem. J. (2007) 401, 353–363.

Rohacs, T., Thyagarajan, B., and Lukacs V. Phospholipase C mediated modulation of TRPV1 channels. Mol. Neurobiol. (2008) 37, 2–3, 153–163.

Rona, P., and Kleinmann, H. Eine Methode zur nephelometrischen Bestimmung Kleinster Calciummengen. Biochem. Z. (1923) 137, 157–183.

Rona, P., and Takahashi, D. Über das Verhalten des Calciums im Serum und über der Gehalt der Blutkörperchen an Calcium. Biochem. Z. (1911) 31, 336–344.

Rooney, J. P. The role of thiols, dithiols, nutritional factors and interacting ligands in the toxicology of mercury. Toxicology (2007) 234, 3, 145–156.

Rose, G. D., Fleming, P. J., Banavar, J. R., and Maritan, A. A backbone-based theory of protein folding. Proc. Natl. Acad. Sci. USA (2006) 103, 45, 16623–16633.

Rosen, B. P. Transport and detoxification systems for transition metals, heavy metals and metalloids in eukaryotic and prokaryotic microbes. Comp. Biochem. Physiol. A (2002), 133, 689–693.

Rosengarth, A., Gerke, V., and Luecke, H. X-ray structure of full-length annexin 1 and implications for membrane aggregation. J. Mol. Biol. (2001a) 306, 489–498.

Rosengarth, A., Rosgen, J., Hinz, H. J., and Gerke, V. Folding energetic of ligand binding proteins: II. Cooperative binding of Ca^{2+} to annexin I. J. Mol. Biol. (2001b) 306, 825–835.

Rosenzweig, A. C. Copper delivery by metallochaperone proteins. Acc. Chem. Res. (2001) 34, 119–128.

Rossi, C. S., and Lehninger, A. L. Stoichiometry of respiratory stimulation, accumulation of Ca^{++} and phosphate, and oxidative phosphorylation in rat liver mitochondria. J. Biol. Chem. (1964) 239, 3971–3980.

Roth, J. A. Homeostatic and toxic mechanisms regulating manganese uptake, retention, and elimination. Biol. Res. (2006) 39, 45–57.

Roth, S. M., Schneider, D. M., Strobel, L. A., VanBerkum, M. F., Means, A. R., and Wand, A. J. Structure of the smooth muscle myosin light-chain kinase calmodulin-binding domain peptide bound to calmodulin. Biochemistry (1991) 30, 42, 10078–10084.

Routledge, L. M. Calcium-binding proteins in the vorticellid spasmoneme. J. Cell Biol. (1978) 77, 2, 358–370.

Roux, B. Ion conduction and selectivity in K^+ channels. Annu. Rev. Biophys. Biomol. Struct. (2005) 34, 153–171.

Ruan, X., Bhattacharjee, H., and Rosen, B. P. Cys-113 and Cys-422 form a high affinity metalloid binding site in the ArsA ATPase. J. Biol. Chem. (2006) 281, 15, 9925–9934.

Rüdel, H. Case study: bioavailability of tin and tin compounds. Ecotoxicol. Environ. Saf. (2003) 56, 1, 180–189.

Rusnak, F., and Mertz, P. Calcineurin: form and function. Physiol. Rev. (2000) 80, 4, 1483–1521.

Rustandi, R. R., Drohat, A. C., Baldisseri, D. M., Wilder, P. T., and Weber, D. J. The Ca^{2+}-dependent interaction of S100B(ββ) with a peptide derived from p53. Biochemistry (1998) 37, 7, 1951–1960.

Rustandi, R. R., Baldisseri, D. M., and Weber, D. J. Structure of the negative regulatory domain of p53 bound to S100B(ββ). Nat. Struct. Biol. (2000) 7, 7, 570–574.

Sabatini, B. L., and Regehr, W. G. Timing of neurotransmission at fast synapses in the mammalian brain. Nature (Lond.) (1996) 384, 170–172.

Sabbatani, L. Azióne tòssica et anticoàgulante del sali. Arch. Fisil. (1904) 1, 536–538.

Saeki, K., Arai, M., Yoda, T., Nakao, M., and Kuwajima, K. Localized nature of the transition-state structure in goat α-lactalbumin folding. J. Mol. Biol. (2004) 341, 589–604.

Saijo, Y., Hata, S., Sheen, J., and Izui, K. cDNA cloning and prokaryotic expression of maize calcium-dependent protein kinases. Biochim. Biophys. Acta (1997) 1350, 109–114.

Sakane, F., Imai, S., Kai, M., Yasuda, S., and Kanoh, H. Diacylglycerol kinases: Why so many of them? Biochim. Biophys. Acta (2007) 1771, 793–806.

Sakane, H., Muñoz-Páez, A., Díaz-Moreno, S., Martínez, J. M., Pappalardo, R. R., and Sánchez Marcos, E. Second hydration shell single scattering versus first hydration shell multiple scattering in $M(H_2O)_6^{3+}$ EXAFS Spectra. J. Am. Chem. Soc. (1998) 120, 40, 10397–10401.

Salama, I., Malone, P. S., Mihaimeed, F., and Jones, J. L. A review of the S100 proteins in cancer. Eur. J. Surg. Oncol. (2008) 34, 4, 357–364.

Salvato, M., Sulston, J., Albertson, D., and Brenner, S. A novel calmodulin-like gene from the nematode *Caenorrhabditis elegans*. J. Mol. Biol. (1986) 190, 281–290.

Sánchez-Barrena, M. J., Fujii, H., Angulo, I., Martínez-Ripoll, M., Zhu, J. K., and Albert, A. The structure of the C-terminal domain of the protein kinase AtSOS2 bound to the calcium sensor AtSOS3. Mol. Cell. (2007) 26, 3, 427–435.

Sanchez-Ruiz, J. M. Differential scanning calorimetry of proteins. Subcell. Biochem. (1995) 24, 133–176.

Sanderson, M. J., Delmotte, P., Bai, Y., and Perez-Zogbhi, J. F. Regulation of airway smooth muscle cell contractility by Ca^{2+} signaling and sensitivity. Proc. Am. Thorac. Soc. (2008) 5, 1, 23–31.

Sanderson, P. H., Marshall, F., and Wilson, R. E. Calcium and phosphorus homeostasis in the parathyroidectomized dog; evaluation by means of ethylenediamine tetraacetate and calcium tolerance tests. J. Clin. Invest. (1960) 39, 662–670.

Sandhir, R., and Gill, K. D. Effect of lead on the biological activity of calmodulin in rat brain. Exp. Mol. Pathol. (1994) 61, 1, 69–75.

Sandier, A., Amiel, C., Sebille, B., Rouchaud, J. C., and Fedoroff, M. A study of strontium binding to albumins, by a chromatographic method involving atomic emission spectrometric detection. Int. J. Biol. Macromol. (1999) 24, 1, 43–48.

Sandow, A. Excitation-contraction coupling in muscular responses. Yale J. Biol. Med. (1952) 25, 176–201.

Sandström, J. Om en ny Körtel hos menniskan och Åtsilliga döggdjur. Upsala Läkareför. Förh. (1880) 15, 441–471.

Sano, H., Takahashi, S., and Iio, T. Fluorescence energy transfer of complexes of skeletal muscle troponin C and melittin derivatives. J. Biochem. (1998) 124, 602–608.

Santamaria-Kisiel, L., Rintala-Dempsey, A. C., and Shaw, G. S. Calcium-dependent and -independent interactions of the S100 protein family. Biochem. J. (2006) 396, 2, 201–214.

Santana, L. F., and Navedo, M. F. Molecular and biophysical mechanisms of Ca^{2+} sparklets in smooth muscle. J. Mol. Cell. Cardiol. (2009) 47, 4, 436–444.

Sanyal, G., Richard, L. M., Carraway, R. L., and Puett, D. Binding of amphiphilic peptides to a carboxy-terminal tryptic fragment of calmodulin. Biochemistry (1988) 27, 17, 6229–6236.

Sárdy, M., Ruzicka, T., and Kuhn, A. Topical calcineurin inhibitors in cutaneous lupus erythematosus. Arch. Dermatol. Res. (2009) 301, 1, 93–98.

Sastry, M., Ketchem, R. R., Crescenzi, O., Weber, C., Lubienski, M. J., Hidaka, H., and Chazin, W. J. The three-dimensional structure of Ca^{2+}-bound calcyclin: implications for Ca^{2+}-signal transduction by S100 proteins. Structure (1998) 6, 223–231.

Sato, T., Nakajima, H., Fujio, K., and Mori, Y. Enhancement of prostaglandin E2 production by epidermal growth factor requires the coordinate activation of cytosolic phospholipase A_2 and cyclooxygenase 2 in human squamous carcinoma A431 cells. Prostaglandins (1997) 53, 5, 355–369.

Satyshur, K. A., Rao, S. T., Pyzalska, D., Drendel, W., Greaser, M., and Sundaralingam, M. Refined structure of chicken skeletal muscle troponin C in the two-calcium state at 2-Å resolution. J. Biol. Chem. (1988) 263, 4, 1628–1647.

Saunders, C. M., Larman, M. G., Parrington, J., et al. PLCζ: a sperm-specific trigger of Ca^{2+} oscillations in eggs and embryo development. Development (2002) 129, 3533–3544.

Saurel, O., Cezanne, L., Milon, A., Tocanne, J. F., and Demange, P. Influence of annexin V on the structure and dynamics of phosphatidylcholine/phosphstidylserine bilayers: a fluorescence and NMR study. Biochemistry (1998) 37, 5, 1403–1410.

Sauter, A., Staudenmann, W., Hughes, G. J., and Heizmann, C. W. A novel EF-hand Ca^{2+}-binding protein from abdominal muscle of crustaceans with similarity to calcyphosine from dog thyroidea. Eur. J. Biochem. (1995) 227, 97–101.

Savory, J., Herman, M. M., and Ghribi, O. Mechanisms of aluminum-induced neurodegeneration in animals: implications for Alzheimer's disease. J. Alzheimer's Dis. (2006) 10, 2–3, 135–144.

Schäfer, B. W., Wicki, R., Engelkamp, D., Mattei, M. G., and Heizmann, C. W. Isolation of a YAC clone covering a cluster of nine S100 genes on human chromosome 1q21: rationale for a new nomenclature of the S100 calcium-binding protein family. Genomics (1995) 25, 638–643.

Schäfer, B. W., Fritschy, J. M., Murmann, P., Troxler, H., Durussel, I., Heizmann, C. W., and Cox, J. A. Brain S100A5 is a novel calcium-, zinc-, and copper ion-binding protein of the EF-hand superfamily. J. Biol. Chem. (2000) 275, 39, 30623–30630.

Schatzmann, H. J. ATP-dependent Ca^{++}-extrusion from human red cells. Experientia (1966) 22, 364–365.

Schaub, J. R., Lu, X., Doneske, B., Shin, Y. K., and McNew, J. A. Hemifusion arrest by complexin is relieved by Ca^{2+}-synaptotagmin 1. Nat. Struct. Mol. Biol. (2006) 13, 748–750.

Schaub, M. C., and Heizmann, C. W. Calcium, troponin, calmodulin, S100 proteins: from myocardial basics to new therapeutic strategies. Biochim. Biophys. Res. Commun. (2008) 369, 1, 247–264.

Schaub, M. C., Hefti, M. A., Zuellig, R. A., and Morano, I. Modulation of contractility in human cardiac hypertrophy by myosin essential light chain isoforms. Cardiovasc. Res. (1998) 37, 381–404.

Schauer, K., Rodionov, D. A., and de Reuse, H. New substrates for TonB-dependent transport: Do we only see the "tip of the iceberg"? Trends Biochem. Sci. (2008) 33, 7, 330–338.

Scheiner-Bobis, G. The sodium pump. Its molecular properties and mechanics of ion transport. Eur. J. Biochem. (2002) 269, 10, 2424–2433.

Schiavo, G., Osborne, S. L., and Sgouros, J. G. Synaptotagmins: More isoforms than functions? Biochem. Biophys. Res. Commun. (1998) 248, 1, 1–8.

Schlaepfer, D. D., Mehlman, T., Burgess, W. H., and Haigler, H. T. Structural and functional characterization of endonexin II, a calcium- and phospholipid-binding protein. Proc. Natl. Acad. Sci. USA (1987) 84, 17, 6078–6082.

Schmidt, H., Schwaller, B., and Eilers, J. Calbindin D_{28k} targets myo-inositol monophosphatase in spines and dendrites of cerebellar Purkinje neurons. Proc. Natl. Acad. Sci. USA (2005) 102, 5850–5855.

Schneeberger, P. R., and Heizmann, C. W. Parvalbumin in rat kidney: purification and localization. FEBS Lett. (1986) 201, 51–56.

Scholey, J. M., Taylor, K. A., and Kendrick-Jones, J. The role of myosin light chains in regulating actin–myosin interaction. Biochimie (1981) 63, 4, 255–271.

Schou, M. Lithium treatment at 52. J. Affect. Disord. (2001) 67, 21–32.

Schrag, J. D., Bergeron, J. J., Li, Y., Borisova, S., Hahn, M., Thomas, D. Y., and Cygler, M. The structure of calnexin, an ER chaperone involved in quality control of protein folding. Mol. Cell (2001) 8, 633–644.

Schultz, J., Milpetz, F., Bork, P., and Ponting, C. P. SMART, a simple modular architecture research tool: identification of signaling domains. Proc. Natl. Acad. Sci. USA. (1998) 95, 11, 5857–5864.

Schumacher, M. A., Rivard, A. F., Bachinger, H. P., and Adelman, J. P. Structure of the gating domain of a Ca^{2+-} activated K^+ channel complexed with Ca^{2+}/calmodulin. Nature (2001) 410, 1120–1124.

Schumann, F., Hoffmeister, H., Bader, R., Schmidt, M., Witzgall, R., and Kalbitzer, H. R. Ca^{2+}-dependent conformational changes in a C-terminal cytosolic domain of polycystin-2. J. Biol. Chem. (2009) 284, 36, 24372–24383.

Schutt, C. Hands on the calcium switch. Nature (1985) 15, 6014, 15.

Schwab, F. J., Appel, H., Neu, M., and Thies, W. G. Influence of protein dynamics on the metal-sites of ovotransferrin. Eur. Biophys. J. (1992) 21, 2, 147–154.

Schwaller, B. Calretinin. In: Celio, M. R., Pauls, T., Schwaller, B., eds., Guidebook to the Calcium-Binding Proteins. Oxford University Press, Oxford, UK, 1996, pp. 26–28.

Schwaller, B. Ca^{2+}-buffers. In: Bradshaw, R. A., and Dennis, E. A., eds., Handbook of Cell Signaling, Vol. 2. Academic Press, Oxford, UK, 2003, pp. 67–71.

Schwaller, B. The continuing disappearance of "pure" Ca^{2+} buffers. Cell. Mol. Life Sci. (2008) 66, 2, 275–300.

Schwaller, B., Durussel, I., Jermann, D., Herrmann, B., and Cox, J. A. Comparison of the Ca^{2+}-binding properties of human recombinant calretinin-22k and calretinin. J. Biol. Chem. (1997) 272, 47, 29663–29671.

Schwaller, B., Meyer, M., and Schiffmann, S. "New" functions for "old" proteins: the role of the calcium-binding proteins calbindin D-28k, calretinin and parvalbumin, in cerebellar physiology—studies with knockout mice. Cerebellum (2002) 1, 4, 241–258.

Schwarzenbach, G., and Gysling, H. Metallindikatoren: I. Murexid als Indikator auf Calcium- und andere Metall-Ionen. Komplexbildung und Lichtabsorption. Helv. Chim. Acta (1949) 32, 1314–1325.

Schweimer, K., Petersen, A., Suck, R., Becker, W. M., Rösch, P., and Matecko, I. Solution structure of Phl p 3, a major allergen from timothy grass pollen. Biol. Chem. (2008) 389, 7, 919–923.

Schwenk, C. F., Loeffler, H. H., and Rode, B. M. Dynamics of the solvation process of Ca^{2+} in water. Chem. Phys. Lett. (2001) 349, 1–2, 99–103.

Scott, B. L., and Pease, D. C. Electron microscopy of the epiphyseal apparatus. Anat. Rec. (1956) 126, 465–495.

Scott, G. H. Distribution of mineral ash in striated muscle. Proc. Soc. Exp. Biol. Med. (1932) 29, 349–351.

Sekiya, F., Yamashita, T., Atoda, H., Komiyama, Y., and Morita, T. Regulation of the tertiary structure and function of coagulation factor IX by magnesium(II) ions. J. Biol. Chem. (1995a) 270, 24, 14325–14331.

Sekiya, F., Yamashita, T., and Morita, T. Role of calcium(II) ions in the recognition of coagulation factors IX and X by IX/X-bp, an anticoagulant from snake venom. Biochemistry (1995b) 34, 31, 10043–10047.

Semplicini, A., Sartori, M., Ceolotto, G., and Calò, L. A. The Li^+/Na^+ exchange in hypertension. Front. Biosci. (2003) 8, d912–d929.

Senin, I. I., Churumova, V. A., Philippov, P. P., and Koch, K. W. Membrane binding of the neuronal calcium sensor recoverin: modulatory role of the charged carboxy-terminus. BMC Biochem. (2007) 8, 24.

Senolt, L., Grigorian, M., Lukanidin, E., Michel, B. A., Gay, R. E., Gay, S., Pavelka, K., and Neidhart, M. S100A4 (Mts1): is there any relation to the pathogenesis of rheumatoid arthritis? Autoimmun. Rev. (2006) 5, 2, 129–131.

Serohijos, A. W., Tsygankov, D., Liu, S., Elston, T. C., and Dokholyan, N. V. Multiscale approaches for studying energy transduction in dynein. Phys. Chem. Chem. Phys. (2009) 11, 24, 4840–4850.

Shao, X., Li, C., Fernandez, I., Zhang, X., Südhof, T. C., and Rizo, J. Synaptotagmin–syntaxin interaction: the C_2 domain as a Ca^{2+}-dependent electrostatic switch. J. Neuron (1997) 18, 133–142.

Shao, X., Fernandez, I., Südhof, T. C., and Rizo, J. Solution structures of the Ca^{2+}-free and Ca^{2+}-bound C_2A domain of synaptotagmin I: Does Ca^{2+} induce a conformational change? Biochemistry (1998) 37, 46, 16106–16115.

Sharma, R., Rensing, C., Rosen, B. P., and Mitra, B. The ATP hydrolytic activity of purified ZntA, a Pb(II)/Cd(II)/Zn(II)-translocating ATPase from *Escherichia coli*. J. Biol. Chem. (2000) 275, 3873–3878.

She, M., Dong, W. J., Umeda, P. K., and Cheung, H. C. Tryptophan mutants of troponin C from skeletal muscle: an optical probe of the regulatory domain. Eur. J. Biochem. (1998a) 252, 600–607.

She, M., Xing, J., Dong, W. J., Umeda, P. K., and Cheung, H. C. Calcium binding to the regulatory domain of skeletal muscle troponin C induces a highly constrained open conformation. J. Mol. Biol. (1998b) 281, 445–452.

Sheehan, J. H., Bunick, C. G., Hu, H., Fagan, P. A., Meyn, S. M., and Chazin, W. J. Structure of the N-terminal calcium sensor domain of centrin reveals the biochemical basis for domain-specific function. J. Biol. Chem. (2006) 281, 5, 2876–2881.

Shelling, J. G., and Sykes, B. D. [1]H nuclear magnetic resonance study of the two calcium-binding sites of porcine intestinal calcium-binding protein. J. Biol. Chem. (1985) 260, 14, 8342–8347.

Sheridan, A. M., Force, T., Yoon, H. J., O'Leary, E., Choukroun, G., Taheri, M. R., and Bonventre, J. V. PLIP, a novel splice variant of Tip60, interacts with group IV cytosolic phospholipase A_2, induces apoptosis, and potentiates prostaglandin production. Mol. Cell. Biol. (2001) 21, 14, 4470–4481.

Shibanuma, M., Mashimo, J., Mita, A., Kuroki, T., and Nose, K. Cloning from a mouse osteoblastic cell line of a set of transforming-growth-factor-beta 1-regulated genes, one of which seems to encode a follistatin-related polypeptide. Eur. J. Biochem. (1993) 217, 13–19.

Shibata, N., Masuda, J., Tobimatsu, T., Toraya, T., Suto, K., Morimoto, Y., and Yasuoka, N. A new mode of B_{12} binding and the direct participation of a potassium ion in enzyme catalysis: x-ray structure of diol dehydratase. Structure (1999) 7, 8, 997–1008.

Shieh, R. C., Chang, J.-C., and Arreola, J. Interaction of Ba^{2+} with the pores of the cloned inward rectifier K^+ channels Kir2.1 expressed in *Xenopus* oocytes. Biophys. J. (1998) 75, 2313–2322.

Shimomura, O., Johnson, F. H., and Saiga, Y. Extraction, purification and properties of aequorin, a bioluminescent protein from the luminous hydromedusan, *Aequorea*. J. Cell. Comp. Physiol. (1962) 59, 223–239.

Shimoni-Livny, L., Glusker, J. P., and Bock, C. W. Lone pair functionality in divalent lead compounds. Inorg. Chem. (1998) 37, 1853–1867.

Shnyrov, V. L., Sanchez-Ruiz, J. M., Boiko, B. N., Zhadan, G. G., and Permyakov, E. A. Applications of scanning microcalorimetry in biophysics and biochemistry. Thermochim. Acta (1997) 302, 1–2, 165–180.

Shoham, M., Kalb, A. J., and Pecht, I. Specificity of metal ion interaction with concanavalin A. Biochemistry (1973) 12, 10, 1914–1917.

Shrivastava, R., Upreti, R. K., Seth, P. K., and Chaturvedi, U. C. Effects of chromium on the immune system. FEMS Immunol. Med. Microbiol. (2002) 34, 1, 1–7.

Sia, S. K., Li, M. X., Spyracopoulos, L., Garne, S. M., Liu, W., Putkey, J. A., and Sykes, B. D. Structure of cardiac muscle troponin C unexpectedly reveals a closed regulatory domain. J. Biol. Chem. (1997) 272, 29, 18216–18221.

Silver, S. Bacterial silver resistance: molecular biology and uses and misuses of silver compounds. FEMS Microbiol. Rev. (2003) 27, 2–3, 341–353.

Silver, S., Toth, K., and Schribner, H. Facilitated transport of calcium by cells and subcellular membranes of *Bacillus subtilis* and *Escherichia coli*. J. Bacteriol. (1975) 122, 880–885.

Silvestro, L., and Axelsen, P. H. Fourier transform infrared linked analysis of conformational changes in annexin V upon membrane binding. Biochemistry (1999) 38, 1, 113–121.

Silvestry-Rodriguez, N., Sicairos-Ruelas, E. E., Gerba, C. P., and Bright, K. R. Silver as a disinfectant. Rev. Environ. Contam. Toxicol. (2007) 191, 23–45.

Simon, L. S. Osteoporosis. Clin. Geriatr. Med. (2005) 21, 603–629.

Simons, J. W. F. A., Van Kampen, M. D., Ubarretxena-Belandia, I., Cox, R. C., Alves dos Santos, C. M., Egmond, M. R., and Verheij, H. M. Identification of a calcium binding site in *Staphylococcus hyicus* lipase: generation of calcium-independent variants. Biochemistry (1999) 38, 1, 2–10.

Simons, T. J. Lead–calcium interactions in cellular lead toxicity. Neurotoxicology (1993) 14, 2–3, 77–85.

Simons, T. J. B. Intracellular free zinc and zinc buffering in human red blood cells. J. Membr. Biol. (1991) 123, 63–71.

Simpson, A. W. M. Fluorescent measurement of $[Ca^{2+}]_c$. In: Lambert, D. G., ed., Methods in Molecular Biology, Vol. 114, Calcium Signaling Protocols. Humana Press, Totowa, NJ, 2002, pp. 3–47.

Sissi, C., and Palumbo, M. Effects of magnesium and related divalent metal ions in topoisomerase structure and function. Nucleic Acids Res. (2009) 37, 3, 702–711.

Sitailo, L. A., Tibudan, S. S., and Denning, M. F. Bax activation and induction of apoptosis in human keratinocytes by the protein kinase Cδ catalytic domain. J. Invest. Dermatol. (2004) 123, 434–443.

Sjöblom, B., Salmazo, A., and Djinović-Carugo, K. α-Actinin structure and regulation. Cell. Mol. Life Sci. (2008) 65, 17, 2688–2701.

Skelton, N. J., Kordel, J., and Chazin, W. J. Determination of the solution structure of apo calbindin D $_{9K}$ by NMR spectroscopy. J. Mol. Biol. (1995) 249, 441–462.

Skoog, D. A., Holler, E. J., and Nieman, T. A. Principles of Instrumental Analysis. Brooks/Cole Thomson Learning, Pacific Grove, CA, 1998.

Slater, E. C., and Cleland, K. W. The effect of calcium on the respiratory and phosphorylative activities of heart-muscle sarcosomes. Biochem. J. (1953) 55, 566–580.

Slepecky, R. A., and Foster, W. W. Alterations in metal content of spores of *Bacillus megaterium* and the effect on some spore properties. J. Bacteriol. (1959) 78, 117–123.

Slupsky, J. R., Ohnishi, M., Carpenter, M. R., and Reithmeier, R. A. Characterization of cardiac calsequestrin. Biochemistry (1987) 26, 20, 6539–6544.

Smaili, S., Hirata, H., Ureshino, R., et al. Calcium and cell death signaling in neurodegeneration and aging. An. Acad. Bras. Cienc. (2009) 81, 3, 467–475.

Smith, M. Vanadium biochemistry: the unknown role of vanadium-containing cells in ascidians (sea squirts). Experientia (1989) 45, 5, 452–457.

Smith, P. D., Davies, A., Crumpton, M. J., and Moss, S. E. Structure of the human annexin VI gene. Proc. Natl. Acad. Sci. USA (1994) 91, 2713–2717.

Smith, R. L., and Maguire, M. E. Microbial magnesium transport: unusual transporters searching for identity. Mol. Microbiol. (1998) 28, 217–226.

Smith, S. A. The cell-based model of coagulation. J. Vet. Emerg. Crit. Care (San Antonio) (2009) 19, 1, 3–10.

Smith, S. P., and Shaw, G. S. A novel calcium-sensitive switch revealed by the structure of human S100B in the calcium-bound form. Structure (1998) 6, 2, 211–222.

Snijder, H. J., and Dijkstra, B. W. Bacterial phospholipase A: structure and function of an integral membrane phospholipase. Biochim. Biophys. Acta (2000) 1488, 1–2, 91–101.

Soderling, J. A., Reed, M. J., Corsa, A., and Sage, E. H. Cloning and expression of murine SC1, a gene product homologous to SPARC. J. Histochem. Cytochem. (1997) 45, 823–835.

Soding, J. Protein homology detection by HMMM-HMMM comparison. Bioinformatics (2005) 21, 951–960.

Sohma, H., Creutz, C. E., Saitoh, M., Sano, H., Kuroki, Y., Voelker, D. R., and Akino, T. Characterization of the Ca^{2+}-dependent binding of annexin IV to surfactant protein A. Biochem. J. (1999) 341, 203–209.

Solodukhin, A. S., Kretsinger, R. H., and Sando, J. J. Initial three-dimensional reconstructions of protein kinase C delta from two-dimensional crystals on lipid monolayers. Cell Signal. (2007) 19, 10, 2035–2045.

Solomon, E. I., Sundaram, U. M., and Machonkin, T. E. Multicopper oxidases and oxygenases. Chem. Rev. (1996) 96, 2563–2606.

Somlyo, A. P., and Somlyo, A. V. Ca^{2+} sensitivity of smooth muscle and nonmuscle myosin II: modulated by G proteins, kinases, and myosin phosphatase. Physiol. Rev. (2003) 83, 1325–1358.

Sommerville, L. E., and Hartshorne, D. J. Intracellular calcium and smooth muscle contraction. Cell Calcium (1986) 7, 353–364.

Sorensen, B. R., and Shea, M. A. Interactions between domains of apo calmodulin alter calcium binding and stability. Biochemistry (1998) 37, 12, 4244–4253.

Sorensen, B. R., Eppel, J.-T., and Shea, M. A. *Paramecium* calmodulin mutants defective in ion channel regulation associate with melittin in the absence of calcium but require it for tertiary collapse. Biochemistry (2001) 40, 896–903.

Soriano-Garcia, M., Padmanabhan, K., de Vos, A. M., and Tulinsky, A. The Ca^{2+} ion and membrane binding structure of the Gla domain of Ca-prothrombin fragment 1. Biochemistry (1992) 31, 9, 2554–2566.

Soriano-Garcia, M., Park, C. H., Tulinsky, A., Ravichandran, K. G., and Skrzypczak-Jankun, E. Structure of Ca^{2+} prothrombin fragment 1 including the conformation of the Gla domain. Biochemistry (1989) 28, 17, 6805–6810.

Sorimachi, H., and Suzuki, K. The structure of calpain. J. Biochem. (2001) 129, 653–664.

Sorimachi, H., Ishiura, S., and Suzuki, K. Structure and physiological function of calpains. Biochem. J. (1997) 328, 721–732.

Sreerama, N., and Woody, R. W. Computation and analysis of protein circular dichroism spectra. Methods Enzymol. (2004) 383, 318–351.

Sri Krishna, S., Majumdar, I., and Grishin, N. V. Structural classification of zinc fngers. Nucleic Acids Res. (2003) 31, 2, 532–550.

Stebbings, J. H. Health risks from radium in workplaces: an unfinished story. Occup. Med. (2001) 16, 2, 259–270.

Steckoll, S. H., Goffer, Z., Nathan, H., and Haas, N. Red-stained human bones from Qumran. Isr. J. Med. Sci. (1971) 7, 1219–1223.

Steele, F. R., Washburn, T., Rieger, R., and O'Toussa, J. E. *Drosophila* retinal-degeneration-C (RDGC) encodes a novel serine threonine protein phosphatase. Cell (1992) 69, 669–676.

Steenbock, H., and Black, A. Fat-soluble vitamins. XVII. The induction of growth-promoting and calcifying properties in a cation by exposure to ultraviolet light. J. Biol. Chem. (1924) 61, 405–422.

Steinberg, M. S. Calcium complexing by embryonic cell surfaces: relation to intercellular adhesiveness. In: Brennan, M. S., and Simpson, W. L., eds., Biological Interaction in Normal and Neoplastic Growth. 1962, pp. 127–140.

Steiner, R. F., and Norris, L. The interaction of melittin with troponin C. Arch. Biochem. Biophys. (1987) 254, 1, 342–352.

Steiner, R. A., Rozeboom, H. J., de Vries, A., Kalk, K. H., Murshudov, G. N., Wilson. K. S., and Dijkstra, B. W. X-ray structure of bovine pancreatic phospholipase A2 at atomic resolution. Acta Crystallogr. D Biol. Crystallogr. (2001) 57, 516–526.

Steinhardt, R. A., and Epel, D. Activation of sea-urchin eggs by a calcium ionophore. Proc. Natl. Acad. Sci. USA (1974) 71, 1915–1919.

Stemmer, P. M., and Klee, C. B. Dual calcium ion regulation of calcineurin by calmodulin and calcineurin B. Biochemistry (1994) 33, 22, 6859–6866.

Sternberg, M. J. E., ed., Protein Structure Prediction. Oxford University Press, New York, 2002.

Stewart, A. J., Morgan, K., Farquharson, C., and Millar, R. P. Phospholipase C-eta enzymes as putative protein kinase C and Ca^{2+} signalling components in neuronal and neuroendocrine tissues. Neuroendocrinology (2007) 86, 243–248.

Stewart, L., Ireton, G. C., Parker, L. H., Madden, K. R., and Champoux, J. J. Biochemical and biophysical analyses of recombinant forms of human topoisomerase I. J. Biol. Chem. (1996) 271, 7593–7601.

Sticht, H., and Rosch, P. The structure of iron-sulfur proteins. Prog. Biophys. Mol. Biol. (1998) 70, 2, 95–136.

Stie, J., and Fox, D. Calcineurin regulation in fungi and beyond. Eukaryot. Cell (2008) 7, 2, 177–186.

Stinnakre, M. G., Soulier, S., Schibler, L., Lepourry, L., Mercier, J. C., and Vilotte, J. L. Position-independent and copy-number-related expression of a goat bacterial artificial chromosome α-lactalbumin gene in transgenic mice. Biochem. J. (1999) 339, Pt. 1, 33–36.

Stokes, D. L., and Green, N. M. Structure and function of the calcium pump. Annu. Rev. Biophys. Biomol. Struct. (2003) 32, 445–468.

Strange, R. W., Antonyuk, S. V., Hough, M. A., Doucette, P. A., Valentine, J. S., and Hasnain, S. S. Variable metallation of human superoxide dismutase: atomic resolution crystal structures of Cu–Zn, Zn–Zn and as-isolated wild-type enzymes. J. Mol. Biol. (2006) 356, 1152–1163.

Strobl, S., Fernandez-Catalan, C., Braun, M., et al. The crystal structure of calcium-free human m-calpain suggests an electrostatic switch mechanism for activation by calcium. Proc. Natl. Acad. Sci. USA (2000) 97, 2, 588–592.

Stryer, L. Visual excitation and recovery. J. Biol. Chem. (1991) 266, 17, 10711–10714.

Stuart, D. I., Archarya, K. R., Walker, N. P. C., Smith, S. G., Lewis, M., and Phillips, D. C. α-Lactalbumin possesses a novel calcium binding loop. Nature (1986) 324, 84–87.

Stull, J. T., Blumenthal, D. K., and Cooke, R. Regulation of contraction by myosin phosphorylation: a comparison between smooth and skeletal muscles. Biochem. Pharmacol. (1980) 29, 2537–2543.

Subramanian, L., and Polans, A. S. Cancer-related diseases of the eye: the role of calcium and calcium-binding proteins. Biochem. Biophys. Res. Commun. (2004) 322, 1153–1165.

Südhof, T. C. Synaptotagmins: Why so many? J. Biol. Chem. (2002) 277, 10, 7629–7632.

Sugita, S., Hata, Y., and Südhof, T. C. Distinct Ca^{2+}-dependent properties of the first and second C2-domains of synaptotagmin. J. Biol. Chem. (1996) 271, 1262–1265.

Suko, J., Pidlich, J., and Bertel, O. Calcium release from intact calmodulin and calmodulin fragment 78–148 measured by stopped flow fluorescence with 2-*p*-toluidinylnaphthalene sulfonate: effect of calmodulin fragments on cardiac sarcoplasmic reticulum. Eur. J. Biochem. (1985) 153, 3, 451–457.

Sullivan, D. S., Biggins, S., and Rose, M. D. The yeast centrin, cdc31p, and the interacting protein kinase, Kic1p, are required for cell integrity. J. Cell Biol. (1998) 143, 751–765.

Sullivan, M. M., and Sage, E. H. Hevin/SC1, a matricellular glycoprotein and potential tumor-suppressor of the SPARC/BM-40/østeonectin family. Int. J. Biochem. Cell Biol. (2004) 36, 6, 991–996.

Sun, G. Y., Xu, J., Jensen, M. D., and Simonyi, A. Phospholipase A_2 in the central nervous system: implications for neurodegenerative diseases. J. Lipid Res. (2004) 45, 2, 205–213.

Sun, H., Li, H., Harvey, I., and Sadler, P. J. Interactions of bismuth complexes with metallothionein(II). J. Biol. Chem. (1999) 274, 41, 29094–29101.

Sun, J. Y., and Wu, L. G. Fast kinetics of exocytosis revealed by simultaneous measurements of presynaptic capacitance and postsynaptic currents at a central synapse. Neuron (2001) 30, 171–182.

Sunnerhagen, M., Forsén, S., Hoffrén, A.-M., Drakenberg, T., Teleman, O., and Stenflo, J. Structure of the Ca^{2+}-free Gla domain sheds light on membrane binding of blood coagulation proteins. Nat. Struct. Biol. (1995) 2, 504–509.

Suszkiw, J. B. Presynaptic disruption of transmitter release by lead. Neurotoxicology (2004) 25, 4, 599–604.

Sutherland, E. W., Öye, I., and Butcher, R. W. The action of epinephrine and the role of the adenyl cyclase system in hormone action. Rec. Prog. Horm. Res. (1965) 21, 623–646.

Sutton, R. B., Davletov, B. A., Berghuis, A. M., Südhof, T. C., and Sprang, S. R. Structure of the first C2 domain of synaptotagmin I: a novel Ca^{2+}/phospholipid-binding fold. Cell (1995) 80, 6, 929–938.

Suzuki, H., Kawasaki, M., Inuzuka, T., Okumura, T., Kakiuchi, T., Shibata, H., Wakatsuki, S., and Maki, M. Structural basis for Ca^{2+}-dependent formation of ALG-2/Alix peptide complex: Ca^{2+}/EF3-driven arginine switch mechanism. Structure (2008) 16, 1562–1573.

Suzuki, K., Hata, S., Kawabata, Y., and Sorimachi, H. Structure, activation, and biology of calpain. Diabetes (2004) 53, Suppl. 1, S12–S18.

Suzuki, S. C., and Takeichi, M. Cadherins in neuronal morphogenesis and function. Dev. Growth Differ. (2008) 50, Suppl. 1, S119–S130.

Svanborg, C., Agerstam, H., Aronson, A., et al. HAMLET kills tumor cells by an apoptosis-like mechanism: cellular, molecular, and therapeutic aspects. Adv. Cancer Res. (2003) 88, 1–29.

Svensson, M., Sabharwal, H., Hakansson, A., Mossberg, A. K., Lipniunas, P., Leffler, H., Svanborg, C., and Linse, S. Molecular characterization of α-lactalbumin folding variants that induce apoptosis in tumor cells. J. Biol. Chem. (1999) 274, 10, 6388–6396.

Svensson, M., Hakansson, A., Mossberg, A. K., Linse, S., and Svanborg, C. Conversion of α-lactalbumin to a protein inducing apoptosis. Proc. Natl. Acad. Sci. USA (2000) 97, 8, 4221–4226.

Svensson, M., Mossberg, A. K., Pettersson, J., Linse, S., and Svanborg, C. Lipids as cofactors in protein folding: stereo-specific lipid–protein interactions are required to form HAMLET (human α-lactalbumin made lethal to tumor cells). Protein Sci. (2003) 12, 12, 2805–2814.

Svergun, D. I., and Koch, M. H. Advances in structure analysis using small-angle scattering in solution. Curr. Opin. Struct. Biol. (2002) 12, 654–660.

Svoboda, K., Schmidt, C. F., Schnapp, B. J., and Block, S. M. Direct observation of kinesin stepping by optical trapping interferometry. Nature (1993) 365, 6448, 721–727.

Swan, D. G., Hale, R. S., Dhillon, N., and Leadlay, P. F. A bacterial calcium-binding protein homologous to calmodulin. Nature (1987) 329, 6134, 84–85.

Swan, D. G., Cortes, J., Hale, R. S., and Leadlay, P. F. Cloning, characterization, and heterologous expression of the *Saccharopolysporta erythreaea* (*Streptomyces erythraeus*) gene encoding an EF-hand calcium-binding protein. J. Bacteriol. (1989) 171, 5614–5619.

Swann, K., and Yu, Y. The dynamics of calcium oscillations that activate mammalian eggs. Int. J. Dev. Biol. (2008) 52, 5–6, 585–594.

Swann, K., Saunders, C. M., Rogers, N. T., and Lai, F. A. PLCζ (zeta): a sperm protein that triggers Ca^{2+} oscillations and egg activation in mammals. Semin. Cell. Dev. Biol. (2006) 17, 264–273.

Sweadner, K. J., and Donnet, C. Structural similarities of Na,K-ATPase and SERCA, the Ca^{2+}-ATPase of the sarcoplasmic reticulum. Biochem. J. (2001) 356, 685–704.

Swoboda, I., Bugajska-Schretter, A., Linhart, B., et al. A recombinant hypoallergenic parvalbumin mutant for immunotherapy of IgE-mediated fish allergy. J. Immunol. (2007) 178, 10, 6290–6296.

Szczesna, D., Guzman, G., Miller, T., Zhao, J., Farokhi, K., Ellemberger, H., and Potter, J. D. The role of the four Ca^{2+} binding sites of troponin C in the regulation of skeletal muscle contraction. J. Biol. Chem. (1996) 271, 8381–8386.

Szebenyi, D. M., and Moffat, K. The refined structure of vitamin D–dependent calcium-binding protein from bovine intestine: molecular details, ion binding, and implications for the structure of other calcium-binding proteins. J. Biol. Chem. (1986) 261, 19, 8761–8777.

Szebenyi, D. M. E., Obendorf, S. K., and Moffat, K. Structure of vitamin D–dependent calcium binding protein from bovine intestine. Nature (1981) 294, 327–332.

Tada, M., Yamamoto, T., and Tonomura, Y. Molecuilar mechanism of active calcium transport by sarcoplasmic reticulum. Physiol. Rev. (1978) 58, 1, 1–79.

Tai, I. T., and Tang, M. J. SPARC in cancer biology: its role in cancer progression and potential for therapy. Drug Resist. Update (2008) 11, 6, 231–246.

Takamatsu, K., Saitoh, S., and Kobayashi, M. Hippocalcin. In: Celio, M. R., Pauls, T., Schwaller, B., eds., Guidebook to the Calcium-Binding Proteins. Oxford University Press, Oxford, UK, 1996, pp. 100–102.

Takano, E., Ma, H., Yang, H. Q., Maki, M., and Hatanaka, M. Preference of calcium-dependent interactions between calmodulin-like domains of calpain and calpastatin subdomains. FEBS Lett. (1995) 362, 1, 93–97.

Takashima, S. Phosphorylation of myosin regulatory light chain by myosin light chain kinase, and muscle contraction. Circ. J. (2009) 73, 2, 208–213.

Takaya, J., Higashino, H., and Kobayashi, Y. Can magnesium act as a second messenger? Current data on translocation induced by various biologically active substances. Magnesium Res. (2000) 13, 2, 139–146.

Takeichi, M. Cadherins in cancer: implications for invasion and metastasis. Curr. Opin. Cell Biol. (1993) 5, 806–811.

Takeichi, M. Morphogenetic roles of classic cadherins. Curr. Opin. Cell Biol. (1995) 7, 619–627.

Takemasa, T., Takagi, T., Kobayashi, T., Konishi, K., and Watanabe, Y. The 3rd calmodulin family protein in *Tetrahymena*: cloning of the cDNA for *Tetrahymena* calcium-binding protein of 23kDa (TCBP-23). J. Biol. Chem. (1990) 265, 2514–2517.

Takenaka, K., Fukami, K., Otsuki, M., Nakamura, Y., Kataoka, Y., Wada, M., Tsuji, K., Nishikawa, S., Yoshida, N., and Takenawa, T. Role of phospholipase C-L2, a novel phospholipase C-like protein that lacks lipase activity, in B-cell receptor signaling. Mol. Cell Biol. (2003) 23, 7329–7338.

Taketo, M. M., and Sonoshita, M. Phospolipase A_2 and apoptosis. Biochim. Biophys. Acta (2002) 1585, 2–3, 72–76.

Talib, J., Beck, J. L., and Ralph, S. F. A mass spectrometric investigation of the binding of gold antiarthritic agents and the metabolite $[Au(CN)_2]^-$ to human serum albumin. J. Biol. Inorg. Chem. (2006) 11, 5, 559–570.

Talukdar, T., Gorecka, K. M., de Carvalho-Niebel, F., Downie, J. A., Cullimore, J., and Pikula, S. Annexins—calcium- and membrane-binding proteins in the plant kingdom: potential role in nodulation and mycorrhization in *Medicago truncatula*. Acta Biochim. Pol. (2009) 56, 2, 199–210.

Tam, R., and Saier, M. H., Jr. Structural, functional, and evolutionary relationships among extracellular solute-binding receptors of bacteria. Microbiol. Rev. (1993) 57, 2, 320–346.

Tamagnini, P., Axelsson, R., Lindberg, P., Oxelfelt, F., Wünschiers, R., and Lindblad P. Hydrogenases and hydrogen metabolism of cyanobacteria. Microbiol. Mol. Biol. Rev. (2002) 66, 1, 1–20.

Tamm, L. K., and Tatulian, S. A. Infrared spectroscopy of proteins and peptides in lipid bilayers. Q. Rev. Biophys. (1997) 30, 4, 365–429.

Tan, R. Y., Mabuchi, Y., and Grabarek, Z. Blocking the Ca^{2+}-induced conformational transitions in calmodulin with disulfide bonds. J. Biol. Chem. (1996) 271, 7479–7483.

Tanaka, T., Ames, J. B., Harvey, T. S., Stryer, L., and Ikura, M. Sequestration of the membrane-targeting myristoyl group of recoverin in the calcium-free state. Nature (1995) 376, 6539, 444–447.

Tanaka, T., Ames, J. B., Kainosho, M., Stryer, L., and Ikura, M. Differential isotype labelling strategy for determining the structure of myristoylated recoverin by NMR spectroscopy. J. Biomol. NMR (1998) 11, 2, 135–152.

Tanaka, T., Miwa, N., Kawamura, S., Sohma, H., Nitta, K., and Matsushima, N. Molecular modeling of single polypeptide chain of calcium-binding protein p26olf from dimeric S100B(ββ). Protein Eng. (1999) 12, 5, 395–405.

Tang, J., Maximov, A., Shin, O. H., Dai, H., Rizo, J., and Südhof, T. C. A complexin/synaptotagmin 1 switch controls fast synaptic vesicle exocytosis. Cell (2006) 126, 1175–1187.

Tange, Y., and Niwa, O. Identification of the *ure1+* gene encoding urease in fission yeast. Curr. Genet. (1997) 32, 244–246.

Tanokura, M., Tawada, Y., Ono, A., and Ohtsuki, I. Chymotryptic subfragments of troponin T from rabbit skeletal muscle: interaction with tropomyosin, troponin I and troponin C. J. Biochem. (1983) 93, 2, 331–337.

Tanoue, T., and Takeichi, M. New insights into Fat cadherins. J. Cell Sci. (2005) 118, 11, 2348–2353.

Tatur, J., Hagen, W. R., and Matias, P. M. Crystal structure of the ferritin from the hyperthermophilic archaeal anaerobe *Pyrococcus furiosus*. J. Biol. Inorg. Chem. (2007) 12, 615–630.

Taves, D. R. Similarity of octacalcium phosphate and hydroxyapatite structures. Nature (1963) 200, 1312–1313.

Taylor, D. M. A bioinorganic chemsitry of actinides in blood. J. Alloys Comp. (1998) 271, 6–10.

Taylor, K. A., and Taylor, D. W. Projection image of smooth muscle α-actinin from 2-dimensional crystals formed on positively charged lipid layers. J. Mol. Biol. (1993) 230, 196–205.

Taylor, T. P., Ding, M., Ehler, D. S., Foreman, T. M., Kaszuba, J. P., and Sauer, N. N. Beryllium in the environment: a review. J. Environ. Sci. Health A (2003) 38, 2, 439–469.

Teahan, C. G, Totty, N. F., and Segal, A. W. Isolation and characterization of grancalcin, a novel 28 kDa EF-hand calcium-binding protein from human neutrophils. Biochem. J. (1992) 286, 549–554.

Teicher, B. A. Protein kinase C as a therapeutic target. Clin. Cancer Res. (2006) 12, 18, 5336–3545.

Tengvall, P., and Lundström, I. Physico-chemical considerations of titanium as a biomaterial. Clin. Mater. (1992) 9, 2, 115–134.

Theil, E. C. Ferritin. In Messerschmidt, A., Poulos, H. R. and Weighardt, K., eds., Handbook of Metalloproteins, Vol. 2. Wiley, Chichester, UK, 2001, pp. 771–781.

Theil, E. C. Ferritin: at the crossroads of iron and oxygen metabolism. J. Nutr. (2003) 133, 1549S–1553S.

Theil, E. C. Iron, ferritin, and nutrition. Annu. Rev. Nutr. (2004) 24, 327–343.

Theil, E. C., Matzapetakis, M., and Liu, X. Ferritins: iron/oxygen biominerals in protein nanocages. J. Biol. Inorg. Chem. (2006) 11, 803–810.

Then, G. M., Appel, H., Duffield, J., Taylor, D. M., and Thies, W. G. In Vivo and in vitro studies of hafnium-binding to rat serum transferrin. J. Inorg. Biochem. (1986) 27, 4, 255–270.

Thepaut, M., Strub, M. P., Cave, A., Baneres, J. L., Berchtold, M. W., Dumas, C., and Padilla, A. Structure of rat parvalbumin with deleted AB domain: implications for the evolution of EF hand calcium-binding proteins and possible physiological relevance. Proteins: Struct. Funct. Genet. (2001) 45, 117–128.

Théret, I., Baladi, S., Cox, J. A., Gallay, J., Sakamoto, H., and Craescu, C. T Solution structure and backbone dynamics of the defunct domain of calcium vector protein. Biochemistry (2001) 40, 13888–13897.

Thompson, J. R., Ryan, Z. C., Salisbury, J. L., and Kumar, R. The structure of the human centrin 2-*Xeroderma pigmentosum* group C protein complex. J. Biol. Chem. (2006) 281, 18746–18752.

Tikunova, S. B., and Davis, J. P. Designing calcium-sensitizing mutations in the regulatory domain of cardiac troponin C. J. Biol. Chem. (2004) 279, 34, 35341–35352.

Tikunova, S. B., Black, D. J., Johnson, J. D., and Davis, J. P. Modifying Mg^{2+} binding and exchange with the N-terminal of calmodulin. Biochemistry (2001) 40, 3348–3353.

Tikunova, S. B., Rall, J. A., and Davis, J. P. Effect of hydrophobic residue substitutions with glutamine on Ca^{2+} binding and exchange with the N-domain of troponin C. Biochemistry (2002) 41, 6697–6705.

Tipton, S. R. Calcium content of nerve. Am. J. Physiol. (1934) 109, 457–466.

Tjoelker, L. W., Sayfried, C. E., Eddy, R. L., Byers, M. G., Shows, T. B., Calderon, J., Schreiber, R. B., and Gray, P. W. Human, mouse, and rat calnexin cDNA cloning: identification of potential calcium binding motif and gene localization to human chromosome 5. Biochemistry (1994) 33, 11, 3229–3236.

Todd, B., Moore, D., Deivanayagam, C. C., Lin, G. D., Chattopadhyay, D., Maki, M., Wang, K. K., and Narayana, S. V. A structural model for the inhibition of calpain by calpastatin: crystal structures of the native domain VI of calpain and its complexes with calpastatin peptide and a small molecule inhibitor. J. Mol. Biol. (2003) 328, 1, 131–146.

Tokumitsu, H., Mizutani, A., Minami, H., Kobayashi, R., and Hidaka, H. A calcyclin-associated protein is a newly identified member of the Ca^{2+}/phospholipid-binding proteins, annexin family. J. Biol. Chem. (1992) 267, 8919–8924.

Tombola, F., Pathak, M. M., and Isacoff, E. Y. How does voltage open an ion channel? Annu. Rev. Cell. Dev. Biol. (2006) 22, 23–52.

Tomita, A. ^{47}Ca kinetic study of vitamin K_2. Clin. Endocrinol. Tokyo (1971) 19, 731–736.

Tossavainen, H., Permi, P., Annila, A., Kilpelainen, I., and Drakenberg, T. NMR solution structure of calerythrin, an EF-hand calcium-binding protein from *Saccharopolyspora erythraea*. Eur. J. Biochem. (2003) 270, 11, 2505–2512.

Toyoshima, C., Nakasako, M., Nomura, H., and Ogawa, H. Crystal structure of the calcium pump of sarcoplasmic reticulum at 2.6 Å resolution. Nature (2000) 405, 6787, 647–655.

Toyoshima, C., and Nomura, H. Structural changes in the calcium pump accompanying the dissociation of calcium. Nature (2002) 418, 605–611.

Traaseth, N. J., Ha, K. N., Verardi, R., Shi, L., Buffy, J. J., Masterson, L. R., and Veglia, G. Structural and dynamic basis of phospholamban and sarcolipin inhibition of Ca^{2+}-ATPase. Biochemistry (2008) 47, 1, 3–13.

Trave, G., Quignard, J. F., Lionne, C., Widada, J. S., and Liautard, J. P. Interdependence of phospholipid specificity and calcium binding in annexin I as shown by site-directed mutagenesis. Biochim. Biophys. Acta (1994) 1205, 215–222.

Trave, G., Lacombe, P.-J., Pfuhl, M., Saraste, M., and Pastore, A. Molecular mechanism of the calcium-induced conformational change in the spectrin EF-hands. EMBO J. (1995a) 14, 4922–4931.

Trave, G., Pastore, A., Hyvonen, M., and Saraste, M. The C-terminal domain of α-spectrin is structurally related to calmodulin. Eur. J. Biochem. (1995b) 227, 1–2, 35–42.

Trewhella, J., Liddle, W. K., Heidorn, D. B., and Strynadka, N. Calmodulin and troponin C structures studied by Fourier transform infrared spectroscopy: effects of Ca^{2+} and Mg^{2+} binding. Biochemistry (1989) 28, 3, 1294–1301.

Trifaro, J. M., Bader, M. F., and Doucet, J. P. Chromaffin cell cytoskeleton: its possible role in secretion. Can. J. Biochem. Cell. Biol. (1985) 63, 2, 661–679.

Trigo-Gonzalez, G., Racher, K., Burtnik, L., and Borgford, T. A comparative spectroscopic study of tryptophan probes engineered into high- and low-affinity domains of recombinant chicken troponin C. Biochemistry (1992) 31, 7009–7015.

Trojan, P., Krauss, N., Choe, H. W., Giessl, A., Pulvermüller, A., and Wolfrum, U. Centrins in retinal photoreceptor cells: regulators in the connecting cilium. Prog. Retin. Eye Res. (2008a) 27, 3, 237–259.

Trojan, P., Rausch, S., Giessl, A., Klemm, C., Krause, E., Pulvermüller, A., and Wolfrum, U. Light-dependent CK2-mediated phosphorylation of centrins regulates complex formation with visual G-protein. Biochim. Biophys. Acta (2008b) 1783, 6, 1248–1260.

Tsigelny, I. F. (Ed.) Protein Structure Prediction: Bioinformatic Approach. International University Line Biotechnology Series, 2002.

Tsuchiya, T., and Rosen, B. P. Characterization of an active transport system for calcium in inverted membrane vesicles of *Eschericia coli*. J. Biol. Chem. (1975) 250, 7687–7692.

Tsukihara, T., Aoyama, H., Yamashita, E., Tomizaki, T., Yamaguchi, H., Shinzawa-Itoh, K., Nakashima, R., Yaono, R., and Yoshikawa, S. The whole structure of the 13-subunit oxidized cytochrome c oxidase at 2.8 Å. Science (1996) 272, 1136–1144.

Tsvetkov, P. O., Protasevich, I. I., Gilli, R., Lafitte, D., Lobachov, V. M., Haiech, J., Briand, C., and Makarov, A. A. Apocalmodulin binds to the myosin light chain kinase calmodulin target site. J. Biol. Chem. (1999) 274, 26, 18161–18164.

Tucker, W. C., and Chapman, E. R. Role of synaptotagmin in Ca^{2+}-triggered exocytosis. Biochem. J. (2002) 366, Pt. 1, 1–13.

Tugarinov, V., Hwang, P. M., and Kay, L. E. Nuclear magnetic resonance spectroscopy of high-molecular-weight proteins. Annu. Rev. Biochem. (2004) 73, 107–146.

Tzima, E., Trotter, P. J., Hastings, A. D., Orchard, M. A., and Walker, J. H. Investigation of the relocation of cytosolic phospholipase A2 and annexin V in activated platelets. Thromb. Res. (2000) 97, 421–429.

Ubach, J., Zhang, X., Shao, X., Sudhof, T. C., and Rizo J. Ca^{2+} binding to synaptotagmin: how many Ca^{2+} ions bind to the tip of a C2-domain? EMBO J. (1998) 17, 3921–3930.

Uchiyama, H., Perez-Prat, E. M., Watanabe, K., Kumagai, I., and Kuwajima, K. Effects of amino acid substitutions in the hydrophobic core of α-lactalbumin on the stability of the molten globule state. Protein Eng. (1995) 8, 1153–1161.

Ueki, T., Adachi, T., Kawano, S., et al. Vanadium-binding proteins (vanabins) from a vanadium-rich ascidian *Ascidia sydneiensis samea*. Biochim. Biophys. Acta (2003a) 1626, 43–50.

Ueki, T., Sakamoto, Y., Yamaguchi, N., and Michibata, H. Bioaccumulation of copper ions by *Escherichia coli* expressing vanabin genes from the vanadium-rich ascidian *Ascidia sydneiensis samea*. Appl. Environ. Microbiol. (2003b) 69, 6442–6446.

Umekawa, H., Naka, M., Inagaki, M., Onishi H., Wakabayashi, T., and Hidaka, H. Conformational studies on myosin phosphorylated by protein kinase C. J. Biol. Chem. (1985) 260, 22, 9833–9837.

Uversky, V. N. Natively unfolded proteins: a point where biology waits for physics. Protein Sci. (2002) 11, 4, 739–756.

Uversky, V. N., Abdullaev, Z. K., Arseniev, A. S., Bocharov, E. V., Dolgikh, D. A., Latypov, R. F., Melnik, T. N., Vassilenko, K. S., and Kirpichnikov, M. P. Structure and stability of recombinant protein depend on the extra N-terminal methionine residue: S6 permutein from direct and fusion expression systems. Biochim. Biophys. Acta (1999) 1432, 324–332.

Valenta, R., Hayek, B., Seiberler, S., et al. Calcium-binding allergens: from plants to man. Int. Arch. Allergy Immunol. (1998) 117, 3, 160–166.

Vallabhajosula, S. R., Harwig, J. F., Siemsen, J. K., and Wolf, W. Radiogallium localization in tumors: blood binding and transport and the role of transferrin. J. Nucl. Med. (1980) 21, 650–656.

Vallee, B. L., and Falchuk, K. H. The biochemical basis of zinc physiology. Physiol. Rev. (1993) 73, 79–118.

Van Asselt, E. J., and Dijkstra, B. W. Binding of calcium in the EF-hand of *Escherichia coli* lytic transglycosylase Slt35 is important for stability. FEBS Lett. (1999) 458, 3, 429–435.

Van Asselt, E. J., Dijkstra, A. J., Kalk, K. H., Takacs, B., Keck, W., and Dijkstra BW. Crystal structure of *Escherichia coli lytic* transglycosylase Slt35 reveals a lysozyme-like catalytic domain with an EF-hand. Struct. Fold. Des. (1999) 7, 10, 1167–1180.

Vanbelle, C., Halgand, F., Cedervall, T., Thulin, E., Akerfeldt, K. S., Laprevote, O., and Linse, S. Deamidation and disulfide bridge formation in human calbindin D_{28k} with effects on calcium binding. Protein Sci. (2005) 14, 4, 968–979.

VanBerkum, M. F., George, S. E., and Means, A. R. Calmodulin activation of target enzymes. Consequences of deletions in the central helix. J. Biol. Chem. (1990) 265, 7, 3750–3756.

Vandonselaar, M., Hickie, R. A., Quail, J. W., and Delbaere, L. T. J. Trifluoperazine-induced conformational change in Ca^{2+}-calmodulin. Nat. Struct. Biol. (1994) 1, 11, 795–801.

van Eldik, L. J., Zendegui, J. G., Marshak, D. K., and Watterson, D. M. Calcium binding proteins and molecular basis of calcium action. Int. Rev. Cytol. N.Y. (1982) 77, 1–61.

Van Eyk, J. E., Kay, C. M., and Hodges, R. S. A comparative study of the interactions of synthetic peptides of the skeletal and cardiac troponin I inhibitory region with skeletal and cardiac troponin C. Biochemistry (1991) 30, 41, 9974–9981.

Van Geet, A. L. Hydration number of sodium ions determined by sodium magnetic resonance. J. Am. Chem. Soc. (1972) 94, 16, 5583–5587.

Vanhooren, A., Devreese, B., Vanhee, K., Van Beeumen, J., and Hanssens, I. Photoexcitation of tryptophan groups induces reduction of two disulfide bonds in goat α-lactalbumin. Biochemistry (2002) 41, 36, 11035–11043.

Van Rossum, G. S. A. T., Bijvelt, J. J. M., van den Bosch, H., Verkleij, A. J., and Boonstra, J. Cytosolic phospholipase A_2 and lipoxygenase are involved in cell cycle progression in neuroblastoma cells. Cell. Mol. Life Sci. (2002) 59, 181–188.

Vanscyoc, W. S., and Shea, M. A. Phenylalanine fluorescence studies of calcium binding to N-domain fragments of *Paramecium* calmodulin mutants show increased calcium affinity correlates with increased disorder. Protein Sci. (2001) 10, 1758–1768.

Van Slyke, L. L., and Bosworth, A. W. Condition of casein and salts in milk. J. Biol. Chem. (1915) 20, 135–152.

Vasington, F. D., and Murphy, J. V. Ca^{++} uptake by rat kidney mitochondria and its dependence on respiration and phosphorylation. J. Biol. Chem. (1962) 237, 2670–2677.

Vassiliev, I. R., Antonkine, M. L., and Golbeck, J. H. Iron–sulfur clusters in type I reaction centers. Biochim. Biophys. Acta (2001) 1507, 139–160.

Vassole, G., and Generali, F. Fonction parathyroïdienne et fonction thyroïdienne. Arch. Ital. Biol. (1900) 33, 154–156.

Vassylyev, D. G., Takeda, S., Wakatsuki, S., Maeda, K., and Maeda, Y. Crystal structure of troponin C in complex with troponin I fragment at 2.3-Å resolution. Proc. Natl. Acad. Sci. USA (1998) 95, 4847–4852.

Veenstra, T. D., Gross, M. D., Hunziker, W., and Kumar, R. Identification of metal-binding sites in rat brain calcium-binding protein. J. Biol. Chem. (1995) 270, 51, 30353–30358.

Veenstra, T. D., Johnson, K. L., Tomlinson, A. J., Naylor, S., and Kumar, R. Determination of calcium-binding sites in rat brain calbindin D_{28k} by electrospray ionization mass spectrometry. Biochemistry (1997) 36, 12, 3535–3542.

Veeraraghavan, S., Fagan, P. A., Hu, H., Lee, V., Harper, J. F., Huang, B., and Chazin, W. J. Structural independence of the two EF-hand domains of caltractin. J. Biol. Chem. (2002) 277, 32, 28564–28571.

Veltman, O. R., Vriend, G., Berendsen, H. J. C., Van den Burg, B., Venema, G., and Eijsink, V. G. H. A single calcium binding site is crucial for the calcium-dependent thermal stability of thermolysin-like proteases. Biochemistry (1998), 37, 15, 5312–5319.

Vendrell, J., Querol, E., and Avilés, F. X. Metallocarboxypeptidases and their protein inhibitors: structure, function and biomedical properties. Biochim. Biophys. Acta (2000) 1477, 284–298.

Venetucci, L. A., and Eisner, D. A. Calsequestrin mutations and sudden death: A case of too little sarcoplasmic reticulum calcium buffering? Circ. Res. (2008) 103, 223–225.

Venkitaramani, D. V., Fulton, D. B., Andreotti, A. H., Johansen, K. M., and Johansen, J. Solution structure and backbone dynamics of calsensin, an invertebrate neuronal calcium-binding protein. Protein Sci. (2005) 14, 7, 1894–1901.

Venkitaramani, D. V., Fulton, D. B., Andreotti, A. H., Johansen, K. M., and Johansen, J. Mapping the Ca^{2+}-dependent binding of an invertebrate homolog of protein phosphatase 4 regulatory subunit 2 to the small EF-hand protein, calsensin. Biochim. Biophys. Acta (2006) 1763, 3, 322–329.

Veprintsev, D. B., Permyakov, E. A., Kalinichenko, L. A., and Berliner, L. J. Pb^{2+} and Hg^{2+} binding to α-lactalbumin. Biochem. Mol. Biol. Int. (1996) 39, 6, 1255–1265.

Veprintsev, D. B., Narayan, M., Permyakov, S. E., Uversky, V. N., Brooks, C. L., Cherskaya, A. M., Permyakov, E. A., and Berliner, L. J. Fine tuning the N-terminus of a calcium binding protein: α-lactalbumin. Proteins (1999) 37, 65–72.

Verdino, P., Westritschnig, K., Valenta, R., and Keller, W. The cross-reactive calcium-binding pollen allergen, Phl p 7, reveals a novel dimer assembly. EMBO J. (2002) 21, 5007–5016.

Verdino, P., Barderas, R., Villalba, M., Westritschnig, K., Valenta, R., Rodriguez, R., and Keller, W. Three-dimensional structure of the cross-reactive pollen allergen Che a 3: visualizing cross-reactivity on the molecular surfaces of weed, grass, and tree pollen allergens. J. Immunol. (2008) 180, 4, 2313–2321.

Verheij, H. M., Volwerk, J. J., Jansen, E. H. J. M., et al. Methylation of histidine-48 in pancreatic phospholipase A2: role of histidine and calcium ion in the catalytic mechanism. Biochemistry (1980) 19, 4, 743–750.

Vernon, R., and Sage, E. H. The calcium-binding protein SPARC is secreted by Leydig and Sertoli cells of the adult mouse testis. Biol. Reprod. (1989) 40, 1329–1340.

Verzili, D., Zamparelli, C., Mattei, B., Noegel, A. A., and Chiancone, E. The sorcin-annexin VII calcium-dependent interaction requires the sorcin N-terminal domain. FEBS Lett. (2000) 471, 2–3, 197–200.

Vetter, S. W., and Leclerc, E. Novel aspects of calmodulin target recognition and activation. Eur. J. Biochem. (2003) 270, 404–414.

Viel, A. α-Actinin and spectrin structures: an unfolding family story. FEBS Lett. (1999) 460, 391–394.

Viel, A., and Branton, D. Interchain binding at the tail end of the *Drosophila* spectrin molecule. Proc. Natl. Acad. Sci. USA (1994) 91, 10839–10843.

Viel, A., and Branton, D. Spectrin: on the path from structure to function. Curr. Opin. Cell Biol. (1996) 8, 49–55.

Vignais, P. M., Billoud, B., and Meyer, J. Classification and phylogeny of hydrogenases. FEMS Microbiol. Rev. (2001) 25, 4, 455–501.

Viitanen, P. V., Lubben, T. H., Reed, J., Goloubinoff, P., O'Keefe, D. P., and Lorimer, G. H. Chaperonin-facilitated refolding of ribulosebisphosphate carboxylase and ATP hydrolysis by chaperonin 60 (groEL) are K^+ dependent. Biochemistry (1990) 29, 24, 5665–5671.

Vijay-Kumar, S., and Cook, W. J. Structure of a sarcoplasmic calcium-binding protein from *Nereis diversicolor refined at 2.0 Å resolution*. J. Mol. Biol. (1992) 224, 2, 413–426.

Vijay-Kumar, S., and Kumar, V. D. Neurocalcin. Role in neuronal sygnaling. In: Vogel, H. J. (Ed.), Methods in Molecular Biology, Vol. 172; Calcium-Binding Protein Protocols, Vol. 1, Reviews and Case Studies. Humana Press, Totowa, NJ, 2002, pp. 261–279.

Villarroya, M., Lopez, M. G., Cano-Abad, M. F., and Garcia, A. G. Measurement of Ca^{2+} entry using $^{45}Ca^{2+}$. Methods Mol. Biol. (1999) 114, 137–147.

Villeret, V., Huang, S., Frommt, H. J., and Lipscomb, W. N. Crystallographic evidence for the action of potassium, thallium, and lithium ions on fructose-1,6-bisphosphatase Proc. Natl. Acad. Sci. USA (1995) 92, 8916–8920.

Vincent, J. B. The biochemistry of chromium. J. Nutr. (2000a) 130, 4, 715–718.

Vincent, J. B. Elucidating a biological role for chromium at a molecular level. Acc. Chem. Res. (2000b) 33, 7, 503–510.

Vine, A. K. Recent advances in haemostasis and thrombosis. Retina (2009) 29, 1, 1–7.

Vines, H. W. C. The coagulation of the blood: I. The role of calcium. J. Physiol. (1921) 55, 86–99.

Vinogradova, M. V., Stone, D. B., Malanina, G. G., Karatzaferi, C., Cooke, R., Mendelson, R. A., and Fletterick, R. J. Ca^{2+}-regulated structural changes in troponin. Proc. Natl. Acad. Sci. USA (2005) 102, 5038–5043.

Virgolini, I., Traub, T., Novotny, C., et al. Experience with indium-111 and yttrium-90-labeled somatostatin analogs. Curr. Pharm. Des. (2002) 8, 20, 1781–1807.

Vito, P., Lacaná, E., and D'Adamio, L. Interfering with apoptosis: Ca^{2+}-binding protein ALG-2 and Alzheimer's disease gene *ALG-3*. Science (1996) 271, 521–525.

Voets, T., Nilius, B., Hoefs, S., van der Kemp, A. W., Droogmans, G., Bindels, R. J., and Hoenderop, J. G. TRPM6 forms the Mg^{2+} influx channel involved in intestinal and renal Mg^{2+} absorption. J. Biol. Chem. (2004) 279, 19–25.

Vogel, H. J. Calmodulin: a versalite calcium mediator protein. Biochem. Cell Biol. (1994) 72, 357–376.

Vogel, H. J., and Zhang, M. Protein engineering and NMR studies of calmodulin. Mol. Cell Biochem. (1995) 149–150, 3–15.

Voges, D., Berendes, R., Burger, A., Demange, P., Baumeister, W., and Huber, R. Three-dimensional structure of membrane-bound annexin V: a correlative electron microscopy–x-ray crystallography study. J. Mol. Biol. (1994) 238, 2, 199–213.

Volbeda, A., Fontecilla-Camps, J. C., and Frey M. Novel metal sites in protein structures. Curr. Opin. Struct. Biol. (1996) 6, 6, 804–812.

Vongvatcharanon, S., Vongvatcharanon, U., and Boonyoung, P. Immunohistochemical localization of parvalbumin calcium-binding protein in the heart tissues of various species. Acta Histochem. (2008) 110, 1, 26–33.

von Meysenburg, L., Pappenheimer, A. M., Zucker, T. F., and Murray, M. F. The diffusable calcium of the blood serum: I. A method for its determination. J. Biol. Chem. (1921) 47, 529–539.

Vorherr, T., James, P., Krebs, J., Enyedi, A., McCormick, A., Penniston, J. T., and Carafoli, E. Interaction of calmodulin with the calmodulin binding domain of the plasma membrane Ca^{2+} pump. Biochemistry (1990) 29, 2, 355–365.

Vorherr, T., Knopfel, L., Hofmann, F., Mollner, S., Pfeuffer, T., and Carafoli, E. Calmodulin binding domain of nitric oxide synthase and adenylyl cyclase. Biochemistry (1993) 32, 23, 6081–6088.

Vrettos, J. S., Limburg, J., and Brudvig, G. W. Mechanism of photosynthetic water oxidation: combining biophysical studies of photosystem II with inorganic model chemistry. Biochim. Biophys. Acta (2001) 1503, 1–2, 229–245.

Vyas, N. K., Vyas, M. N., and Quiocho, F. A. A novel calcium binding site in the galactose-binding protein of bacterial transport and chemotaxis. Nature (1987) 327, 635–638.

Vysotski, E. S., and Lee, J. Ca^{2+}-regulated photoproteins: structural insight into the bioluminescence mechanism. Acc. Chem. Res. (2004) 37, 405.

Vysotski, E. S., Markova, S. V., and Frank, L. A. Calcium-regulated photoproteins of marine coelenterates. Mol. Biol. (Mosc.) (2006) 40, 3, 404–417.

Waalkes, M. P. Cadmium carcinogenesis. Mutat. Res. (2003) 533, 107–120.

Wahr, P. A., Michele, D. E., and Metzger, J. M. Parvalbumin gene transfer corrects diastolic dysfunction in diseased cardiac myocytes. Proc. Natl. Acad. Sci. USA. (1999) 96, 21, 11982–11985.

Waisberg, M., Joseph, P., Hale, B., and Beyersmann, D. Molecular and cellular mechanisms of cadmium carcinogenesis. Toxicology (2003) 192, 2–3, 95–117.

Wall, C. M., Grand, R. J. A., and Perry, S. V. Biological activities of the peptides obtained by digestion of troponin C and calmodulin with thrombin. Biochem. J. (1981) 195, 1, 307–316.

Wall, J. D., and Krumholz, L. R. Uranium reduction. Annu. Rev. Microbiol. (2006) 60, 149–166.

Wallander, M. L., Leibold, E. A., and Eisenstein, R. S. Molecular control of vertebrate iron homeostasis by iron regulatory proteins. Biochim. Biophys. Acta (2006) 1763, 668–689.

Wallis, C. J., Babitch, J. A., and Wenegieme, E. F. Divalent cation binding to erythrocyte spectrin. Biochemistry (1993) 32, 19, 5045–5050.

Walser, M. Renal excretion of alkaline earths. In: Comar, C. L., Bronner, F., eds., Mineral Metabolism: An Advanced Treatise. Academic Press, New York, 1969, pp. 235–320.

Walsh, C. Posttranslational Modification of Proteins: Expanding Nature's Inventory. Roberts and Company Publishers, Portland, OR, 2007.

Walsh, C., Barrow, S., Voronina, S., Chvanov, M., Petersen, O. H., and Tepikin, A. Modulation of calcium signalling by mitochondria. Biochim. Biophys. Acta (2009) 1787, 11, 1374–1382.

Waltersson, Y., Linse, S., Brodin, P., and Grundstrom, T. Mutational effects on the cooperativity of Ca^{2+} binding in calmodulin. Biochemistry (1993) 32, 7866–7871.

Wandersman, C., and Delepelaire, P. Bacterial iron sources: from siderophores to hemophores. Annu. Rev. Microbiol. (2004) 58, 611–647.

Wang, J., and Boisvert, D. C. Structural basis for GroEL-assisted protein folding from the crystal structure of $(GroEL\text{-}KMgATP)_{14}$ at 2.0 Å resolution. J. Mol. Biol. (2003) 327, 843–855.

Wang, J., and Chen, C. Biosorption of heavy metals by *Saccharomyces cerevisiae*: a review. Biotechnol. Adv. (2006) 24, 427–451.

Wang, K. C., Mutus, B., Sharma, R. K., Lam, H. Y., and Wang, J. H. On the mechanism of interaction between calmodulin and calmodulin-dependent proteins. Can. J. Biochem. Cell Biol. (1983) 61, 8, 911–920.

Wang, S., George, S. E., Davis, J. P., and Johnson, J. D. Structural determinants of Ca^{2+} exchange and affinity in the C terminal of cardiac troponin C. Biochemistry (1998a) 37, 14539–14544.

Wang, S., Trumble, W. R., Liao, H., Wesson, C. R., Dunker, A. K., and Kang, C. H. Crystal structure of calsequestrin from rabbit skeletal muscle sarcoplasmic reticulum. Nat. Struct. Biol. (1998b) 5, 6, 476–483.

Wang, W., and Creutz, C. E. Regulation of the chromaffin granule aggregating activity of annexin I by phosphorylation. Biochemistry (1992) 31, 41, 9934–9939.

Warkany, J. A warfarin embryopathy? Am. J. Dis. Child. (1975) 129, 287–288.

Warren, J. T., Guo, Q., and Tang, W. J. A 1.3-A structure of zinc-bound N-terminal domain of calmodulin elucidates potential early ion-binding step. J. Mol. Biol. (2007) 374, 2, 517–527.

Watabe, N. Crystallographic analysis of the coccolith of *Coccolithus huxleyi*. Calcif. Tissue Res. (1967) 1, 114–121.

Watkins, N. J., Knight, M. R., Trewavas, A. J., and Campbell, A. K. Free calcium transients in chemotactic and non-chemotactic strains of *Escherichia coli* determined by using recombinant aequorin. Biochem. J. (1995) 306, Pt. 3, 865–869.

Wawrzynow, A., Theibert, J. L., Murphy, C., Jona, I., Martonosi, A., and Collins, J. H. Sarcolipin, the "proteolipid" of skeletal muscle sarcoplasmic reticulum, is a unique, amphipathic, 31-residue peptide, Arch. Biochem. Biophys. (1992) 298, 620–623.

Weber, A. On the role of calcium in the activity of adenosine 5′-triphosphate hydrolysis by actomyosin. J. Biol. Chem. (1959) 234, 2764–2769.

Wegener, A. D., Simmerman, H. K., Lindemann, J. P., and Jones, L. R. Phospholamban phosphorylation in intact ventricles: phosphorylation of serine 16 and threonine 17 in response to β-adrenergic stimulation, J. Biol. Chem. (1989) 264, 11468–11474.

Weiner, S. Biomineralization: a structural perspective. J. Struct. Biol. (2008) 163, 3, 229–234.

Weis, W. I., Drickamer, K., and Hendrickson, W. A. Structure of a C-type mannose-binding protein complexed with an oligosaccharide. Nature (1992) 360, 6400, 127–134.

Weiss, T. S., Chamberlain, C. E., Takeda, T., Lin, P., Hahn, K. M., and Farquhar, M. G. Gαi3 binding to calnuc on Golgi membranes in living cells monitored by fluorescence resonance energy transfer of green fluorescent protein fusion proteins. Proc. Natl. Acad. Sci. USA (2001) 98, 14961–14966.

Weljie, A. M., and Vogel, H. J. Tryptophan fluorescence of calmodulin binding domain peptides interacting with calmodulin containing unnatural methionine analogues. Protein Eng. (2000) 13, 1, 59–66.

Wells, H. G. Calcification and ossification. Arch. Intern. Med. (1911) 7, 721–753.

Wendel, M., Sommarin, Y., Bergman, T., and Heinegard, D. Isolation, characterization, and primary structure of a calcium-binding 63-kDa bone protein. J. Biol. Chem. (1995) 270, 6125–6133.

Wenegieme, E. F., Babitch, J. A., and Naren, A. P. Cation binding to chicken gizzard α-actinin. Biochim. Biophys. Acta (1994) 1205, 308–316.

Werber, M. M. Metal binding to myosin and to myosin DTNB-light chain. Experientia (1978) 34, 5, 575–576.

Westerhoff, H. V., Wiechmann, A. H. C. A., Van Dam, K., and Hellingwerf, K. J. On the evaluation of data from flow-dialysis experiments. J. Biochem. Biophys. Methods (1989) 18, 53–64.

Wheate, N. J., Brodie, C. R., Collins, J. G., Kemp, S., and Aldrich-Wright, J. R. DNA intercalators in cancer therapy: organic and inorganic drugs and their spectroscopic tools of analysis. Mini Rev. Med. Chem. (2007) 7, 6, 627–648.

Whitaker, M. Calcium at fertilization and in early development. Physiol. Rev. (2006) 86, 1, 25–88.

Whitfield, J. F., and Dixon, R. H. Prevention of postirradiation mitotic delay in cultures of L mouse cells by calcium salts. Exp. Cell Res. (1962) 27, 154–157.

Whittam, R. Control of membrane permeability in red blood cells. Nature (1968) 219, 610–610.

Wiencek, J. M. New strategies for protein crystal growth. Annu. Rev. Biomed. Eng. (1999) 1, 505–534.

Wilbrant, W. Die Ionenpermeabilität der Erythrocyten in Nichtleiterlösungern. Pflügers Arch. Physiol. (1940) 243, 537–556.

Wilbrant, W., and Koller, H. Die Calciumirkung am Froschherzen als Funktion des Ionengleichgewichts zwischen Zellmembran und Umgebung. Helv. Physiol. Acta (1948) 6, 208–221.

Wilder, P. T., Baldisseri, D. M., Udan, R., Vallely, K. M., Weber, D. J. Location of the Zn^{2+}-binding site on S100B as determined by NMR spectroscopy and site-directed mutagenesis. Biochemistry (2003) 42, 46, 13410–13421.

Williams, D. B. Beyond lectins: the calnexin/calreticulin chaperone system of the endoplasmic reticulum. J. Cell Sci. (2006) 119, Pt. 4, 615–623.

Williams, R. J. P. Calcium. In: Vogel, H. J. (Ed.), Methods in Molecular Biology, Vol. 172; Calcium-Binding Protein Protocols, Vol. 1, Reviews and Case Studies. Humana Press, Totowa, NJ, 2002, pp. 21–49.

Williams, T. C., Corson, D. C., Oikawa, K., McCubbin, W. D., Kay, C. M., and Sykes, B. D. ^{1}H NMR spectroscopic studies of calcium-binding proteins: 3. Solution conformations of rat apo-α-parvalbumin and metal-bound rat α-parvalbumin. Biochemistry (1986) 25, 7, 1835–1846.

Wills, F. L., McCubbin, W. D., and Kay, C. M. Smooth muscle calponin–caltropin interaction: effects on biological activity and stability of calponin. Biochemistry (1994) 33, 18, 5562–5569.

Wilson, J. J., Matsushita, O., Okabe, A., and Sakon, J. A bacterial collagen-binding domain with novel calcium-binding motif controls domain orientation. EMBO J. (2003) 22, 8, 1743–1752.

Wilson, M. A., and Brunger, A. T. The 1.0Å crystal structure of Ca^{2+}-bound calmodulin: an analysis of disorder and implications for functionally relevant plasticity. J. Mol. Biol. (2000) 301, 5, 1237–1256.

Wimberly, B., Thulin, E., and Chazin, W. J. Characterization of the N-terminal half-saturated state of calbindin D_{9K}: NMR studies of the N56A mutant. Protein Sci. (1995) 4, 6, 1045–1055.

Winkler, F. K., D'Arcy, A., and Hunziker, W. Structure of human pancreatic lipase. Nature (1990) 343, 6260, 771–774.

Wintrode, P. L., and Privalov, P. L. Energetics of target peptide recognition by calmodulin: a calorimetric study. J. Mol. Biol. (1997) 266, 5, 1050–1062.

Wissing, D., Mouritzen, H., Egeblad, M., Poirier, G. G., and Jäättelä, M. Involvement of caspase-dependent activation of cytosolic phospholipase A_2 in tumor necrosis factor-induced apoptosis. Proc. Natl. Acad. Sci. USA (1997) 94, 10, 5073–5077.

Wittung-Stafshede, P. Role of cofactors in folding of the blue-copper protein azurin. Inorg. Chem. (2004) 43, 7926–7933.

Woehl, E. U., and Dunn, M. F. Monovalent metal ions play an essential role in catalysis and intersubunit communication in the tryptophan synthase bienzyme complex. Biochemistry (1995) 34, 29, 9466–9476.

Wolfrum, U., Giessl, A., and Pulvermüller, A. Centrins, a novel group of Ca^{2+}-binding proteins in vertebrate photoreceptor cells. Adv. Exp. Med. Biol. (2002) 514, 155–178.

Wong, S., Morales, T. H., Neigel, J. E., and Campbell, D. A. Genomic and transcriptional linkage of the genes for calmodulin, EF-hand 5-protein, and ubiquitin extension protein-52 in *Trypanosoma brucei*. Mol. Cell. Biol. (1993) 13, 207–216.

Woo, H. N., Chang, J. W., Choi, Y. H., Gwon, A. R., Jung, Y. K., and Jo, D. G. Characterization of subcellular localization and Ca^{2+} modulation of calsenilin/DREAM/KChIP3. Neuroreport (2008) 19, 12, 1193–1197.

Wood, T. C., and Pearson, W. R. Evolution of protein sequences and structures. J. Mol. Biol. (1999) 291, 977–995.

Woods, N. M., Cuthbertson, K. S., and Cobbold, P. H. Agonist-induced oscillations in cytoplasmic free calcium concentration in single rat hepatocytes. Cell Calcium (1987) 8, 1, 79–100.

Wriggers, W., Mehler, E., Pitici, F., Weinstein, H., and Schulten, K. Structure and dynamics of calmodulin in solution. Biophys. J. (1998) 74, 1622–1639.

Wu, F., Flach, C. R., Seaton, B. A., Mealy, T. R., and Mendelsohn, R. Stability of annexin V in ternary complexes with Ca^{2+} and anionic phospholipids: IR studies of monolayer and bulk phases. Biochemistry (1999) 38, 2, 792–799.

Wu, H. Y., Tomizawa, K., and Matsui, H. Calpain–calcineurin signaling in the pathogenesis of calcium-dependent disorder. Acta Med. Okayama (2007) 61, 3, 123–137.

Wu, L. C., and Kim, P. S. Hydrophobic sequence minimization of the α-lactalbumin molten globule. Proc. Natl. Acad. Sci. USA (1997) 94, 14314–14319.

Wu, S. T., Skoknick, J., and Zhang, Y. *Ab initio* modeling of small proteins by iterative TASSER simulations. BMC Biol. (2007) 5, 17.

Wu, T., Angus, C. W., Yao, X. L., Logun, C., and Shelhamer, J. H. P11, a unique member of the S100 family of calcium-binding proteins, interacts with and inhibits the activity of the 85-kDa cytosolic phospholipase A_2. J. Biol. Chem. (1997) 272, 27, 17145–17153.

Wu, X., and Reid, R. E. Conservative D133E mutation of calmodulin site IV drastically alters calcium binding and phosphodiesterase regulation. Biochemistry (1997a) 36, 3608–3616.

Wu, X., and Reid, R. E. Structure/calcium affinity relationships of site III of calmodulin: testing the acid pair hypothesis using calmodulin mutants. Biochemistry (1997b) 36, 8649–8656.

Wu, Y., Haghighat, N. G., and Ruben, L. The predominant calcimedins from *Trypanosoma brucei* comprise a family of flagellar EF-hand calcium-binding proteins. Biochem. J. (1992) 287, 187–193.

Wuchiyama, J., Nose, Y., Uyama, T., and Michibata, H. Preparation and localization of a monoclonal antibody against a vanadium-associated protein extracted from the blood cells of the vanadium-rich ascidian, *Ascidia sydneiensis samea*. Zool. Sci. (1997) 14, 409–414.

Wuthier, R. E. Calcification of vertebrate hard tissues. In: Sigel, H., ed., Metal Ions in Biological Systems, Vol. 17., Marcel Dekker, Inc., New York, 1984, pp. 411–472.

Wüthrich, K. NMR - this other method for protein and nucleic acid structure determination. Acta Crystallogr. D (1995) 51, 3, 249–270.

Xi, C., Schoeters, E., Vanderleyden, J., and Michiels, J. Symbiosis-specific expression of *Rhizobium etli* casA encoding a secreted calmodulin-related protein. Proc. Natl. Acad. Sci. USA (2000) 97, 20, 11114–11119.

Xiang, M. Q., Ge, T., Tomlinson, C. R., and Klein, W. H. Structure and promoter activity of the *LPS1* genes of *Lytechinus pictus*. Duplicted exons accout for LPS1 proteins with 8 calcium-binding domains. J. Biol. Chem. (1991) 266, 10524–10533.

Xie, X., Dwyer, M. D., Swenson, L., Parker, M. H., and Botfield, M. C. Crystal structure of calcium-free human sorcin: a member of the penta-EF-hand protein family. Protein Sci. (2001) 10, 12, 2419–2425.

Xiong, H., Feng, X., Gao, L., Xu, L., Pasek, D. A., Soek, J. H., and Meissner, G. Identification of a two EF-hand Ca^{2+} binding domain in lobster skeletal muscle ryanodine receptor/Ca^{2+} release channel. Biochemistry (1998) 37, 4804–4814.

Xu, Y., Feng, L., Jeffrey, P. D., Shi, Y., and Morel, F. M. Structure and metal exchange in the cadmium carbonic anhydrase of marine diatoms. Nature (2008) 452, 56–61.

Yadav, S. P., and Brew, K. Structure and function in galactosyltransferase: sequence locations of α-lactalbumin binding site, thiol groups, and disulfide bond. J. Biol. Chem. (1991) 266, 698–703.

Yamada, K., Sakane, F., Matsushima, N., and Kanoh, H. EF-hand motifs of alpha, beta and gamma isoforms of diacylglycerol kinase bind calcium with different affinities and conformational changes. Biochem. J. (1997) 321, 59–64.

Yamamoto, A., Wada, O., and Ono, T. Isolation of a biologically active low-molecular-mass chromium compound from rabbit liver. Eur. J. Biochem. (1987) 165, 627–631.

Yamamoto, A., Wada, O., and Manabe, S. Evidence that chromium is an essential factor for biological activity of low-molecular-weight chromium-binding substance. Biochem. Biophys. Res. Commun. (1989) 163, 189–193.

Yamashita, A., Singh, S. K., Kawate, T., Jin, Y., and Gouaux, E. Crystal structure of a bacterial homologue of Na^+/Cl−-dependent neurotransmitter transporters. Nature (2005) 437, 7056, 215–223.

Yamauchi, E., Nakatsu, T., Matsubara, M., Kato, H., and Taniguchi, H. Crystal structure of a MARCKS peptide containing the calmodulin-binding domain in complex with Ca^{2+}-calmodulin. Nat. Struct. Biol. (2003) 10, 3, 226–231.

Yamniuk, A. P., Anderson, K. L., Fraser, M. E., and Vogel, H. J. Auxiliary Ca^{2+} binding sites can influence the structure of CIB1. Protein Sci. (2009) 18, 5, 1128–1134.

Yan, Q., and Sage, E. H. SPARC, a matricellular glycoprotein with important biological functions. J. Histochem. Cytochem. (1999) 47, 12, 1495–1505.

Yan, X., Watson, J., Ho, P. S., and Deinzer, M. L. Mass spectrometric approaches using electrospray ionization charge states and hydrogen–deuterium exchange for determining protein structures and their conformational changes. Mol. Cell. Proteom. (2004) 3, 10.

Yang, A., Miron, S., Duchambon, P., Assairi, L., Blouquit, Y., and Craescu, C. T. The N-terminal domain of human centrin 2 has a closed structure, binds calcium with a very low affinity, and plays a role in the protein self-assembly. Biochemistry (2006) 45, 3, 880–889.

Yang, H. Q., Ma, H., Takano, E., Hatanaka, M., and Maki, M. Analysis of calcium-dependent interaction between amino-terminal conserved region of calpastatin functional domain and calmodulin-like domain of μ-calpain large subunit. J. Biol. Chem. (1994) 269, 29, 18977–18984.

Yang, W., Lee, H. W., Hellinga, H., and Yang, J. J. Structural analysis, identification, and design of calcium-binding sites in proteins. Proteins (2002) 47, 344–356.

Yano, J., and Yachandra, V. K. X-ray absorption spectroscopy. Photosynth. Res. (2009) in press.

Yao, R., Davidson, D. D., Lopez-Beltran, A., MacLennan, G. T., Montironi, R., and Cheng, L. The S100 proteins for screening and prognostic grading of bladder cancer. Histol. Histopathol. (2007) 22, 9, 1025–1032.

Yap, K. L., Ames, J. B., Swindells, M. B., and Ikura, M. Diversity of conformational states and changes within the EF-hand protein superfamily. Proteins (1999) 37, 3, 499–507.

Yates, J. R., 3rd. Mass spectral analysis in proteomics. Annu. Rev. Biophys. Biomol. Struct. (2004) 33, 297–316.

Ye, Y., Shealy, S., Lee, H. W., Torshin, I., Harrison, R., and Yang, J. J. A grafting approach to obtain site-specific metal-binding properties of EF-hand proteins. Protein Eng. (2003) 16, 6, 429–434.

Yin, D., Sun, H., Weaver, R. F., and Squier, T. C. Nonessential role for methionines in the productive association between calmodulin and the plasma membrane Ca-ATPase. Biochemistry (1999) 38, 13654–13660.

Yocum, C. F., and Pecoraro, V. L. Recent advances in the understanding of the biological chemistry of manganese. Curr. Opin. Chem. Biol. (1999) 3, 182–187.

Yokel, R. A. The toxicology of aluminum in the brain: a review. Neurotoxicology (2000) 21, 5, 813–828.

Yonezawa, T., Kurata, R., Kimura, M., and Inoko, H. PKC delta and epsilon in drug targeting and therapeutics. Recent Pat. DNA Gene Seq. (2009) 3, 2, 96–101.

Yosef, E., Politi, R., Choi, M. H., and Shifman, J. M. Computational design of calmodulin mutants with up to 900-fold increase in binding specificity. J. Mol. Biol. (2009) 385, 5, 1470–1480.

Yoshimatsu, S. I. A new rapid method for the determination of magnesium with one c.c. of blood, without preliminary elimination of calcium. (1929) 14, 29–35.

Yost, J. C., and Sage, E. H. Specific interaction of SPARC with endothelial cells is mediated through a carboxyl-terminal sequence containing a calcium-binding EF hand. J. Biol. Chem. (1993) 268, 25790–25796.

Young, H. S., and Stokes, D. L. The mechanics of calcium transport. J. Membr. Biol. (2004) 198, 2, 55–63.

Yu, L., Golbeck, J., Yao, J., and Rusnak, F. Spectroscopic and enzymatic characterization of the active site dinuclear metal center of calcineurin: implications for a mechanistic role. Biochemistry (1997) 36, 10727–10734.

Yu, W. H., and Fraser, P. E. S100β interaction with tau is promoted by zinc and inhibited by hyperphosphorylation in Alzheimer's disease. J. Neurosci. (2001) 21, 7, 2240–2246.

Yuan, T., Vogel, H. J., Sutherland, C., and Walsh, M. P. Characterization of the Ca^{2+} -dependent and -independent interactions between calmodulin and its binding domain of inducible nitric oxide synthase. FEBS Lett. (1998) 431, 210–214.

Zaks, W. J., and Creutz, C. E. Ca^{2+}-dependent annexin self-association on membrane surfaces. Biochemistry (1991) 30, 40, 9607–9615.

Zalups, R. K. Molecular interactions with mercury in the kidney. Pharmacol. Rev. (2000) 52, 1, 113–143.

Zamponi, G. W. Regulation of presynaptic calcium channels by synaptic proteins. J. Pharmacol. Sci. (2003) 92, 2, 79–83.

Zawia, N. H., Crumpton, T., Brydie, M., Reddy, G. R., and Razmiafshari, M. Disruption of the zinc finger domain: a common target that underlies many of the effects of lead. Neurotoxicology (2000) 21, 6, 1069–1080.

Zelensky, A. N., and Gready, J. E. The C-type lectin-like domain superfamily. FEBS J. (2005) 272, 24, 6179–6217.

Zell, A., Einspahr, H., and Bugg, C. E. Model for calcium binding to γ-carboxyglutamic acid residues of proteins: crystal structure of calcium α-ethylmalonate. Biochemistry (1985) 24, 2, 533–537.

Zhang, H., Wang, G., Ding, Y., Wang, Z., Barraclough, R., Rudland, P. S., Fernig, D. G., and Rao, Z. The crystal structure at 2 Å resolution of the Ca^{2+}-binding protein S100P. J. Mol. Biol. (2003) 325, 785–794.

Zhang, J. Z., Davletov, B. A., Siidhof, T. C., and Anderson, R. G. W. Synaptotagmin I is a high affinity receptor for clathrin AP-2: implications for membrane recycling. Cell (1994) 78, 751–760.

Zhang, L., Szeto, K. Y., Wong, W. B., Loh, T. T., Sadler, P. J., and Sun, H. Interactions of bismuth with human lactoferrin and recognition of the Bi(III)–lactoferrin complex by intestinal cells. Biochemistry (2001) 40, 44, 13281–13287.

Zhang, L., Mulrooney, S. B., Leung, A. F., Zeng, Y., Ko, B. B., Hausinger, R. P., and Sun, H. Inhibition of urease by bismuth(III): implications for the mechanism of action of bismuth drugs. Biometals (2006) 19, 5, 503–511.

Zhang, M., and Yuan, T. Molecular mechanisms of calmodulin's functional versatility. Biochem. Cell Biol. (1998) 76, 2–3, 313–323.

Zhang, M., Yuan, T., Aramini, J. M., and Vogel, H. J. Interaction of calmodulin with its binding domain of rat cerebellar nitric oxide synthase. J. Biol. Chem. (1995) 270, 36, 20901–20907.

Zhang, M., Gumerov, D. R., Kaltashov, I. A., and Mason, A. B. Indirect detection of protein–metal binding: interaction of serum transferrin with In^{3+} and Bi^{3+}. J. Am. Soc. Mass Spectrom. (2004) 15, 11, 1658–1664.

Zhang, R., Khoo, M. S. C., Wu, Y., et al. Calmodulin kinase II inhibition protects against structural heart disease, Nat. Med. (2005) 11, 409–417.

Zhang, S. L., Yu, Y., Roos, J., Kozak, J. A., Deerinck, T. J., Ellisman, M. H., Stauderman, K. A., and Cahalan, M. D. STIM1 is a Ca^{2+} sensor that activates CRAC channels and migrates from the Ca^{2+} store to the plasma membrane. Nature (2005) 437, 902–905.

Zhang, Y. Progress and challenges in protein structure prediction. Curr. Opin. Struct. Biol. (2008) 18, 342–348.

Zhang, Y., Dougherty, M., Downs, D. M., and Ealick, S. E. Crystal structure of an aminoimidazole riboside kinase from *Salmonella enterica*: implications for the evolution of the ribokinase superfamily. Structure (2004) 12, 10, 1809–1821.

Zhao, G., Bou-Abdallah, F., Yang, X., Arosio, P., and Chasteen, N. D. Is hydrogen peroxide produced during iron(II) oxidation in mammalian apoferritins? Biochemistry (2001) 40, 10832–10838.

Zhao, Y., Kappes, B., and Franklin, R. M. Gene structure and expression of an unusual protein-kinase from *Plasmodium falciparum* homologous at its carboxyl terminus with the EF hand calcium-binding proteins. J. Biol. Chem. (1993) 268, 4347–4354.

Zherelova, O. M., Kataev, A. A., Grishchenko, V. M., Knyazeva, E. L., Permyakov, S. E., and Permyakov, E. A. Interaction of antitumor α-lactalbumin–oleic acid complexes with artificial and natural membranes. J. Bioenerg. Biomembr. (2009) 41, 3, 229–237.

Zhou, H.-X., Rivas, G., and Minton, A. P. Macromolecelar crowding and confinement: biochemical, biophysical, and potential physiological consequences. Annu. Rev. Biophys. (2008) 37, 375–397.

Zhou, Y., Morais-Cabral, J. H., Kaufman, A., and MacKinnon, R. Chemistry of ion coordination and hydration revealed by a K^+ channel-Fab complex at 2.0 Å resolution. Nature (2001) 414, 6859, 43–48.

Zhou, Y., Frey, T. K., and Yang, J. J. Viral calciomics: interplays between Ca^{2+} and virus. Cell Calcium (2009) 46, 1, 1–17.

Zhu, M. M., Rempel, D. L., and Gross, M. L. Modeling data from titration, amide H/D exchange, and mass spectrometry to obtain protein–ligand binding constants. J. Am. Soc. Mass Spectrom. (2004) 15, 3, 388–397.

Zhu, Y. G., and Smolders, E. Plant uptake of radiocaesium: a review of mechanisms, regulation and application. J. Exp. Bot. (2000) 51, 351, 1635–1645.

Zou, J. Y., Flocco, M. M., and Mowbray, S. L. The 1.7 Å refined x-ray structure of the periplasmic glucose/galactose receptor from *Salmonella typhimurium*. J. Mol. Biol. (1993) 233, 739–752.

INDEX

Calcium Binding Proteins, By Eugene A. Permyakov and Robert H. Kretsinger

www.ingramcontent.com/pod-product-compliance
Lightning Source LLC
LaVergne TN
LVHW011540200926
848484LV00001B/284
* 9 7 8 0 4 7 0 5 2 5 8 4 5 *